# Summability, Fixed Point Theory and Generalized Integrals with Applications

This book presents contemporary mathematical concepts and techniques, including theories of summability, fixed point and non-absolute integration, and applications, providing an overview of recent developments in the foundations of the field as well as its applications. It discusses the recent results of double sequence spaces as the four-dimensional forward difference matrix in double sequence spaces, several new fixed points on Hadamard-type fractional integral and differential operator related to the qualitative properties of solutions like, existence and uniqueness, stability, continuous dependence, controllability, oscillations, etc. It also includes several new areas of nonabsolute integration theory that are introduced and their applications to other fields. This reference text is for researchers, academics and professionals in the field of pure and applied mathematics.

- Covers recent research breakthroughs in this field, offering new approaches and methods for both theoretical exploration and practical application
- Presents insights into functional analytic methods in summability, absolute and strong summability, direct theorems on summability, special and general summability methods, and their applications
- Highlights fixed-point theory's application to real-world problems and offers solutions to various complex challenges
- Introduces new areas of non-absolute integration theory, such as the Henstock-Kurzweil integral and generalised Riemann integral
- Discusses sequence spaces and functional analysis, including the exploration of double sequence spaces and the four-dimensional forward difference matrix, offering valuable contributions to ongoing research

# Summability, Fixed Point Theory and Generalized Integrals with Applications

Edited by
Hemanta Kalita, Talat Nazir,
Mehmet Gurdal, and Sanket Tikare

CRC Press
Taylor & Francis Group
Boca Raton London New York

CRC Press is an imprint of the
Taylor & Francis Group, an **informa** business
A CHAPMAN & HALL BOOK

Designed cover image: ShutterStock

First edition published 2025
by CRC Press
2385 NW Executive Center Drive, Suite 320, Boca Raton FL 33431

and by CRC Press
4 Park Square, Milton Park, Abingdon, Oxon, OX14 4RN

*CRC Press is an imprint of Taylor & Francis Group, LLC*



ISBN: 978-1-032-98002-7 (hbk)
ISBN: 978-1-032-98004-1 (pbk)
ISBN: 978-1-003-59657-8 (ebk)

DOI: 10.1201/9781003596578

Typeset in Nimbus font
by KnowledgeWorks Global Ltd.

Publisher's note: This book has been prepared from camera-ready copy provided by the authors.

# Contents

**Preface** .......... xi

**Editors** .......... xv

**Contributors** .......... xvii

**Chapter 1** On Summability, Fixed Point Theory and Generalized Integrals .... 1

*Mehmet Gurdal, Talat Nazir, Hemanta Kalita*

1.1 Introduction to Summability Theory .......... 1
1.1.1 Summability and Convergence .......... 3
1.1.2 Matrix Mappings and Duals .......... 4
1.2 Introduction to Fixed Point Theory .......... 5
1.2.1 Metric Space .......... 5
1.2.2 Banach Contraction Principle .......... 6
1.2.3 Ciric Generalized Contraction .......... 8
1.2.4 Ćirić Quasi-Contraction .......... 10
1.3 Non-Absolute Integrals and their Applications .......... 11
1.4 How We Replace Henstock-Kurzweil Integral with Lebesgue Measure? .......... 13
1.5 Application of Henstock-Kurzweil Integral and Non Absolute Integrable Function Spaces .......... 15

**Chapter 2** $q-$Bell Sequence Spaces and Associated Matrix Transformation .......... 21

*Koray Ibrahim Atabey, Muhammed Cinar, Mikail Et*

2.1 Introduction .......... 21
2.2 Main Result .......... 23
2.3 $\alpha-$, $\beta-$, $\gamma-$ Dual Space of $\ell_p(\widehat{B}_q)$ .......... 30
2.4 Matrix Transformations Related to the Space $\ell_p(\widehat{B}_q)$ .......... 32
2.5 The $\ell_p(\widehat{B}_q)$'s Certain Geometric Characteristics .......... 34

**Chapter 3** Almost Convergence via Bell Matrix ........ 39

*Muhammet Cihat Dagli, Murat Karakas*

3.1 Introduction ........ 39
3.2 A New Almost Convergent Sequence Space ........ 42
3.3 The Bell Core ........ 44

**Chapter 4** On Ideal Lacunary Statistical Convergence of Sequence in Gradual Normed Linear Spaces order $\alpha$ ........ 51

*Rabia Savas*

4.1 Introduction ........ 51
4.2 Definitions and Preliminaries ........ 54
4.3 Main Results ........ 56
4.4 Conclusion ........ 66

**Chapter 5** On the Four-Dimensional Forward Difference Matrix ........ 70

*Orhan Tug, Eberhard Malkowsky, Vladimir Rakocevic, Bipan Hazarika*

5.1 Introduction ........ 70
5.2 New Double Sequence Spaces ........ 75
5.3 Dual Spaces of the New Double Sequence Spaces ........ 80
5.4 Matrix Transformations ........ 87
5.5 Conclusion ........ 90

**Chapter 6** Absolute and Strong $\sigma-$Convergence defined by Modulus Function ........ 92

*Ekrem Savas*

6.1 Introduction ........ 92
6.2 Absolutely $\sigma-$Convergence with Indices ........ 97

**Chapter 7** Remarks on Ideal Variations of the Star-Hurewicz and Related Properties ........ 107

*Debraj Chandra, Nur Alam*

7.1 Introduction ........ 107
7.2 Definitions and Terminologies ........ 108
7.3 Main Results ........ 111
7.3.1 Groupability and Alster covers ........ 114
7.3.2 The cardinal $\mathfrak{b}(\mathscr{I})$ ........ 118
7.3.3 Games ........ 123

**Chapter 8** Controlled $v$-Generalized Metric-type Space and Some Fixed Point Theorems with Application ........ 128

*Kushal Roy, Mantu Saha, Vahid Parvaneh, Farhan Golkarmanesh, Choonkil Park*

8.1 Introduction and Preliminaries ........ 128
8.2 Introduction to Controlled $v$-Generalized Metric-type Space ........ 131
8.3 Fixed Point Results over Controlled $v$-Generalized Metric type Space ........ 135
8.4 An Application to Nonlinear Fredholm-type Integral Equations ........ 147

**Chapter 9** Fixed Points of Graphic Contractions in $b-$Metric Spaces with a Directed Graph ........ 151

*Talat Nazir, Zakaria Ali*

9.1 Introduction ........ 151
9.2 Graph Structure on $b-$Metric Space ........ 154
9.3 Fixed Point Results in $b-$Metric Space with Directed Graph ........ 156
9.4 Well-Posedness of Fixed Point Problem in $b-$Metric Space ........ 167
9.5 Conclusions ........ 171

**Chapter 10** Common Fixed Point Results via Simulation Functions and Solution of an Anti-Periodic Ordinary Differential Equation in Bipolar Metric Spaces ........ 173

*Pankaj Kumar, Manoj Kumar*

10.1 Introduction ........ 173
10.2 Main Results ........ 177
10.3 Consequences ........ 185
10.4 Solution for Nonlinear First-Order Caratheodory's type Ordinary Differential Equations ........ 186
10.5 Conclusion ........ 188

**Chapter 11** $wt$-Distance in Extended-Branciari $b$-Distance Metric Spaces and Application ........ 190

*Reena Jain, Hemant Kumar Nashine, Vahid Parvaneh*

11.1 Introduction and Auxiliary Facts ........ 190
11.2 Extension of $wt$-Distance and Simulation Function to Extended Branciari $b$-Metric Space ........ 192

11.3 Main Result ... 194
11.4 Results on Periodic Point ... 200
11.5 Application ... 202
11.6 Open Problem ... 203

**Chapter 12** Solution and Stability of New Type of Quadratic Multiplicative Functional Equation ... 205

*Jyotsana Jakhar, Shalu Sharma, Jagjeet Jakhar*

12.1 Introduction ... 205
12.2 General Solution for Quadratic Multiplicative Functional Equation ... 207
12.3 Stability of the Quadratic Multiplicative Functional Equation using Direct Method ... 208
12.4 Stability of the Quadratic Multiplicative Functional Equation using Fixed Point Method ... 212
12.5 Conclusion ... 216

**Chapter 13** Existence and Uniqueness of Solution of Nonlinear Fractional Dynamic Equation Involving Initial Condition on Time Scales ... 218

*Bikash Gogoi, Walickta Singkai, Satyabrata Gogoi*

13.1 Introduction ... 218
13.2 Preliminaries ... 219
13.3 Main Findings ... 221
13.4 Example ... 227
13.5 Conclusion ... 229

**Chapter 14** Remarks on Absolute and Non-absolute Integrability of Different Integrals for Multifunctions ... 231

*Cristina Stamate, Anca Croitoru*

14.1 Introduction ... 231
14.2 Preliminaries ... 232
14.3 A survey on real integrals for functions and multifunctions ... 237
14.4 Properties for integrals of $\mathbb{R}^n$-valued functions and multifunctions $(n \geq 1)$ ... 240
14.4.1 Some properties for the Sugeno integral ... 241
14.4.2 Some properties for the Choquet integral ... 244

14.4.3 Some properties for the Henstock-Kurzweil integral ........ 246
14.4.4 Some properties for the Brooks integral ........ 249
14.5 Conclusion ........ 251

**Chapter 15** Statistically Riemann Integrable Sequence of Functions on Time Scales ........ 254

*Hemen Bharali, Vikuozonuo Sekhose, Hemanta Kalita*

15.1 Literary Preface and Motivation ........ 254
15.2 Statistical Riemann Integrability of Sequence of Functions ........ 257
15.2.1 Riemann Δ-integrability of sequence of functions ........ 257
15.2.2 Riemann ∇-integrability of sequence of functions ........ 258
15.2.3 Statistically Riemann Δ-integrable sequence of functions ........ 258
15.2.4 Statistically Riemann ∇-integrable sequence of functions ........ 259
15.3 Uniform Statistical Riemann Integrability ........ 263
15.4 Conclusion ........ 266

**Chapter 16** On McShane-type Integrals for Functions with Values in Topological Vector Spaces ........ 268

*Hemanta Kalita*

16.1 Introduction and Preliminaries ........ 268
16.2 *ap*-McShane Integrals on Topological Vector Spaces ........ 270
16.3 Convergence Theorem on ap-McShane Integrals via Equi-integrability ........ 275
16.4 Conclusion ........ 278

**Chapter 17** On Existence and Uniqueness of Solutions of Volterra Integrals in Sense of McShane and Strong McShane ........ 280

*Hemanta Kalita, Talat Nazir*

17.1 Introduction and Preliminaries ........ 280
17.2 McShane and Strong McShane Integration of Vector-valued Functions ........ 283
17.2.1 McShane-Stieltjes integral ........ 291
17.3 Volterra-McShane Linear Integral Equations ........ 293

**Chapter 18** Weighted H-Orlicz Spaces and their Inclusion Properties .......... 298

*Hemanta Kalita, Anca Croitoru*

18.1 Introduction .......... 298
18.2 Preliminaries .......... 299
18.3 Weighted H-Orlicz Spaces .......... 300
18.4 Inclusion Relation between two Weighted H-Orlicz Spaces .......... 302
18.4.1 Inclusion relation between H-Orlicz spaces .......... 305
18.5 Conclusions .......... 307

**Index .......... 309**

# Preface

In practical mathematics and analysis, summability theory is typically employed. It is crucial to the engineering sciences, and scholars from all over the world have long researched different facets of the theory. Fixed point theory is an intriguing topic with many applications in several mathematical domains. It may be difficult to locate publications that present the argument in a cohesive manner due to this crosscutting nature. The majority of the time, fixed points appear when needed most. In 1956, Jaroslav Kurzweil worked on peculiar occurrences involving fast oscillating entries in ordinary differential equations that were not supported by theories or methods available at the time. His tool, which he created to explain the observed results, was somewhat reminiscent of the construction of the Perron integral employing minor and major functions. After a successful conclusion, the tool—the generalized Perron integral—became a stand-alone, self-contained entity. Defining the Lebesgue integral in Euclidean spaces from McShane's point of view has a clear pedagogical advantage, since the initial stages of development are both conceptually and technically simpler. The McShane integral evolves directly from the basic ideas about integration taught in elementary calculus.

This book is intended for graduate students and researchers with an interest in functional analysis in general and in summability theory, fixed point theory and generalized integrals in particular. The content is divided into two main parts. The first, which is more theoretical concept of summability, develops the main abstract theorems on the existence and uniqueness of fixed points of maps. The second part dedicates several recent trends of generalized integrals, their applications in integral equations, function spaces. Notable scholars in the fields of summability, fixed point theory, and integration theory, along with related applications, have authored these works. The book includes eighteen original research chapters that highlight the progress made in the fields of summability, fixed point theory, and integration theory.

In Chapter 1, we go over summability literature, fixed point theory, and generalized integrals. Discussions are held regarding duals, matrix mapping, summability, and convergence. Banach contraction principle, a fixed point theory, discusses a number of generalized contraction principles. Finally, we talk about using the Lebesgue measure in place of the Henstock-Kurzweil integral. There is also discussion of a number of Henstock-Kurzweil integral applications.

In Chapter 2, in this work, we discuss $q-$analogue of the $q-$Bell matrix $\widehat{B}_q = \widehat{B}_{nk}(q)$. Later, we discuss the sequence spaces $c_0(\widehat{B}_q)$, $c(\widehat{B}_q)$, $\ell_\infty(\widehat{B}_q)$, $\ell_p(\widehat{B}_q)$ $(1 \leq p < \infty)$ and study some topological properties and give some inclusion relations for these spaces. In addition, we build bases for the space $\ell_p(\widehat{B}_q)$, compute $\alpha$-, $\beta$-, $\gamma$-duals of the same space, characterize some matrix classes and give some geometric properties.

In Chapter 3, we present a new sequence space that is almost convergent and is obtained using the Bell matrix. We demonstrate that the space of all nearly convergent sequences is linearly isomorphic to this new sequence space. We also express this new space's $\beta-$dual. Additionally, we provide a number of inclusion theorems for the Bell core of a complex-valued sequence after defining it.

In Chapter 4, we introduce the ideas of ideal lacunary statistical convergence of sequences in gradual normed linear spaces of order $\alpha$. We will explore some basic characteristics of the recently proposed concept and its connection to ideal statistical convergence in gradual normed linear spaces in order to achieve this aim. There are also some inclusion results shown.

In Chapter 5, we investigate the spaces $\mathscr{M}_u(\Delta)$ and $\mathscr{C}_\vartheta(\Delta)$ of double sequences as domain of the four-dimensional forward difference matrix in the double sequence spaces $\mathscr{M}_u$ and $\mathscr{C}_\vartheta$, respectively, where $\vartheta \in \{bp, bp0, r, r0\}$. Then, we present some topological and algebraic properties of the new double sequence spaces $\mathscr{M}_u(\Delta)$ and $\mathscr{C}_\vartheta(\Delta)$. Moreover, we determine the $\alpha-$, $\beta(\vartheta)-$, and $\gamma-$duals of the new double sequence spaces $\mathscr{M}_u(\Delta)$ and $\mathscr{C}_\vartheta(\Delta)$ after proving some significant lemmas. Finally, we characterize some four-dimensional matrix classes $(\lambda(\Delta), \mu)$ and $(\mu, \lambda(\Delta))$, where $\lambda, \mu = \{\mathscr{M}_u, \mathscr{C}_\vartheta\}$. The main purpose of this article is to define some new double sequence spaces by calculating the four-dimensional forward difference matrix domain in the double sequence spaces $\mathscr{M}_u$ and $\mathscr{C}_\vartheta$, respectively, where $\vartheta \in \{bp, bp0, r, r0\}$.

In Chapter 6, we generalize the spaces $l^\sigma$ and $m^\sigma$ simultaneously in two directions: by using modulus function and index. We also prove some inclusion theorems.

In Chapter 7, we study variants of $\mathscr{I}$-Hurewicz property. We follow the approach of neighbourhood assignments to study the star-$\mathscr{I}$-Hurewicz and strongly star-$\mathscr{I}$-Hurewicz properties. We obtain certain results in the context of Alexandroff duplicate, groupability and Alster covers which have not been investigated in prior works. A number of observations concerning the cardinality $\mathfrak{b}(\mathscr{I})$ are presented. Few related games are also studied.

In Chapter 8, we present the notion of controlled $v$-generalized metric type space, which is a generalization of controlled metric-type spaces and $b_v(s)$-metric spaces. For different mappings of contractive type over such spaces, some fixed point theorems have been proved. Regarding our new metric type structure and stated theorems, several instances have been provided. Additionally, our fixed point theorem is used to solve non-linear integral equations.

In Chapter 9, fixed point results utilizing generalized graph set-valued contractions with domain of sets are discussed in $b$-metric spaces. The well-posedness of fixed point problem in $b$-metric space is also established.

In Chapter 10, we define $\alpha$-admissibility of one covariant map with respect to another covariant map and using this prove some common fixed point theorems. We also prove common fixed point theorems for contravariant, one covariant and another contravariant maps. Then, some consequences of our main results are mentioned. In the end, we provide an example to support our result and as an application discuss the solution of an anti-periodic ordinary differential equation.

In Chapter 11, in the extended Branciari $b$-distance space, we extend the concept of $wt$-distance and simulation function. As a result, we demonstrate several fundamental fixed point and periodic point theorems for newly defined simulation functions in the underlying space, which are then illustrated with examples. In conclusion, we apply the results to a differential equation with boundary values and offer an integral equation solution, which is the equivalent form of the differential equation.

In Chapter 12, we find the solution and stability of the following quadratic-multiplicative functional equation in eight variables. Our approach involves the utilization of the direct method and fixed point method to derive the desired results.

In Chapter 13, we discuss the existence and uniqueness of the solution of an implicit type fractional dynamic equation on a time scale using the Caputo nabla derivative (Caputo derivative in the nabla sense) based on the fixed point theorem. First, we discussed the existence of the solution by using Schauder's fixed point theorem, then uniqueness has been addressed by employing Banach fixed point theorem. Additionally, a practical example is provided to illustrate the application of the theoretical findings.

In Chapter 14, we present some properties concerning the absolute and nonabsolute integrability for some real and vector integrals for functions and multifunctions by respect to a submeasure or a multi-submeasure of the Sugeno, Choquet, Aumann, Brooks and Henstock-Kurzweil type.

In Chapter 15, we explore the idea of Riemann integrable sequence of functions which converges statistically on time scales and establish some interesting results pertaining to this. Efficient definitions of statistically Riemann $\Delta$-integrable and statistically Riemann $\nabla$-integrable sequence of functions are put forward. A relation between Riemann integrable sequence of functions and its corresponding statistical convergence is obtained. Notion of statistically Cauchy-Riemann sequence of functions is defined and its equivalence with statistically Riemann integrable sequence of functions is formulated.

In Chapter 16, we present an ap-McShane integrable function on vector spaces that are topological. This chapter discusses the fundamental findings of ap-McShane integrable functions. The equi-integrability of ap-McShane integrals is used to examine a number of convergence theorems of ap-McShane integrable functions. It is demonstrated that a convergence theorem based on the idea of equi-integrability is identical to the Vitali convergence theorem for ap-McShane integrable sequences of functions.

In Chapter 17, McShane and strong McShane integrals of a vector-valued function endowed with the Alexiewicz norm are discussed. Several properties of vector-valued McShane and strong McShane integrals including Fundamental theorem of calculus are enclosed in this manuscript. We extend this theory to study the existence and uniqueness of solutions of

$$x(t) - \int_a^b \mathscr{A}(t,s)x(s)d(s) = f(t),\ t \in [a,b]$$

where $x$ and $f$ are McShane integrable functions defined on $[a,b]$ with values in a Banach space.

In Chapter 18, we introduce the weak H-Orlicz spaces generated by convex functions. We present sufficient and necessary conditions for the inclusion relation between two weighted Orlicz spaces. In particular, we obtain some sufficient and necessary conditions for the inclusion properties of H-Orlicz spaces.

Each chapter of the current book has been carefully peer-reviewed by the team of experts. The proof-reading has been done by all the authors and the editors.

Hemanta Kalita, VIT Bhopal University, India
Talat Nazir, University of South Africa, South Africa
Mehmet Gurdal, Suleyman Demirel University, Turkey
Sanket Tikare, Ramniranjan Jhunjhunwala College, India

# Editors

**Dr. Hemanta Kalita** is a Doctorate from Gauhati University. He has an academic experience of 13 years and has worked at various levels up to Assistant Professor. He has 36 publications in peer-reviewed International journals with high impact factor and has 2 publications in various International Conferences held in India and abroad. Dr. Kalita has been an expert speaker at various International Conferences and has organized International Conference (ICNAA-2022), and International Conference (ICNACT-2024). Dr. Kalita was an Invited researcher at Usak University, Turkey, 2023. Dr. Kalita has published two issues in Journal of nonlinear and convex analysis as a guest editor in 2023 and 2024. He is working on Function spaces, integration theory, measure theory and generalized fractal calculus. Dr. Kalita has research collaborators in several countries of Europe, USA, Asian and Africa.

**Prof Dr. Talat Nazir** is a Professor of Mathematics in Department of Mathematical Sciences, University of South Africa, South Africa. Prof Nazir obtained a PhD Degree in Mathematics in the research area of Nonlinear Analysis in in 2012. Then, he did two Post Doctorates of Mathematics, one from the Netherlands in 2014 and other from Sweden in 2016. Prof Nazir is a famous researcher of Mathematics and have research collaboration with several international researchers worldwide. He published over hundred research articles, book chapters, and books in the peer-reviewed international journals with high impact factors. Prof Nazir is also working on some research projects of Sciences, Engineering and Technology. He is a member of several international mathematical societies. He is also the editorial board member of several reputed journal of mathematics. He has also delivered several contributed talks at various international conferences and research seminars.

**Dr. Mehmet Gurdal** has served Suleyman Demirel University in the capacity of Lecturer to Full Professor since 1997. He received his PhD degree from Suleyman Demirel University, Isparta, Turkey. His main research interests are in the fields of sequence spaces, summability theory, mathematical analysis, Berezin symbol and operator theory. Within the framework of international France-Turkey cooperation, he completed a research stay at the University of Claude Bernard Lyon 1 in Lyon and carried out an international project. He has lectured nationally and internationally. He has participated as a reviewer for the American Mathematical Society. He has published one book, seven book chapters and more than 160 research papers in several reputed journals. He is a regular reviewer of many scientific journals and member of the editorial board of various scientific journals, international scientific bodies and organizing committees. He became a Governing Board-Advisory Board member of the scientific and technological research council of Turkey (TUBITAK). He was deemed worthy of many awards, especially the Turkish Academy of Sciences Outstanding Young Scientist award (GEBIP-2015). He has directed bachelor, master, and doctoral theses and has directed various research stays. https://scholar.google.com.tr/citations?user=U5y5eisAAAAJ &hl=tr

**Dr. Sanket Tikare** is an Assistant Professor in the Department of Mathematics, at Ramniranjan Jhunjhunwala College, Mumbai, India, which is an empowered autonomous college affiliated with the University of Mumbai, India. He did his M.Sc. and Ph.D. in Mathematics, both from Shivaji University, Kolhapur, India in the years 2008 and 2012 respectively. He is specialized in Applied Analysis. His research area of interest includes time scale theory, dynamic equations on time scales, differential equations, difference equations, fractional equations. He has published several research papers in international journals. He has collaborated with several eminent researchers from various institutes around the globe. He is member of a life member of the Indian Mathematical Society, a life member of the International Society of Difference Equations. Sanket Tikare is a regular reviewer of many highly-reputed international journals. Moreover, he has written reviews on several articles for Mathematical Reviews (AMS) and Zentralblatt Math (Germany). He has delivered invited talks at 20 international conferences. He has also delivered several contributed talks at various conferences in India and abroad.

# Contributors

**Nur Alam**
University of Gour Banga
Malda, India

**Zakaria Ali**
University of South Africa
Florida 0003, South Africa

**Hemen Bharali**
Assam Don Dosco University
Sonapur, Assam, India

**Debraj Chandra**
University of Gour Banga
Malda, India

**Muhammet Cihat Dagli**
Akdeniz University
Antalya, Turkey

**Muhammed Cinar**
Mus Alparslan University
Muş, Turkey

**Anca Croitoru**
University of Alexandru Ioan Cuza
Iasi, Romania

**Mikail Et**
Firat University
Elazig, Turkey

**Bikash Gogoi**
Sibsagar University, Sibsagar
Assam, India

**Satyabrata Gogoi**
National Institute of Technology
Arunachal Pradesh, Jote, India

**Farhan Golkarmanesh**
Islamic Azad University
Sanandaj, Iran

**Mehmet Gurdal**
Suleyman Demirel University
Isparta, Turkey

**Bipan Hazarika**
Gauhati University
Gauhati, India

**Atabey Koray Ibrahim**
Muş Nizamulmulk Girl Anatolia Imam Hatip High School
Muş, Turkey

**Jyotsana Jakhar**
Department of Mathematics
M.D. University
Rohtak, Haryana, India

**Jagjeet Jakhar**
Department of Mathematics
Central University of Haryana
Mahendragarh, India

**Reena Jain**
VIT Bhopal University
Bhopal-Indore Highway, Madhya Pradesh, India

**Hemanta Kalita**
VIT Bhopal University
Bhopal-Indore Highway, Madhya Pradesh, India

**Murat Karakas**
Bitlis Eren University
Bitlis, Turkey

**Manoj Kumar**
Maharishi Markandeshwar (Deemed to be University)
Mullana, Ambala, Haryana, India

**Pankaj Kumar**
Baba Mastnath University
Asthal Bohar, Rohtak, Haryana, India

**Eberhard Malkowsky**
State University Novi Pazar
Novi Pazar, Serbia

**Hemant Kumar Nashine**
VIT Bhopal University
Bhopal-Indore Highway, Madhya Pradesh, India

**Talat Nazir**
Department of Mathematical Sciences
University of South Africa
Florida 0003, South Africa

**Choonkil Park**
Hanyang University
Seoul, Korea

**Vahid Parvaneh**
Gilan-E-Gharb Branch
Islamic Azad University
Gilan-E-Gharb, Iran

**Vladimir Rakocevic**
University of Nis
Visegradska, Nis-Serbia

**Roy Kushal**
Supreme Knowledge Foundation Group of Institutions
Chandannagar, Hooghly, West Bengal, India

**Mantu Saha**
The University of Burdwan
Purba Bardhaman, West Bengal, India

**Ekrem Savas**
Usak University
Usak, Turkey

**Rabia Savas**
Istanbul Medeniyet University
Istanbul, Turkey

**Vikuozonuo Sekhose**
Assam Don Bosco University
Sonapur, Assam, India

**Shalu Sharma**
Department of Mathematics
Central University of Haryana
Mahendragarh, India

**Walickta Singkai**
Department of Basic and Applied Science
National Institute of Technology
Arunachal Pradesh, Jote, India

**Cristina Stamate**
Institute of Mathematics
Romanian Academy
Iasi, Romania

**Orhan Tug**
Tishk International University
Erbil, Iraq

# 1 On Summability, Fixed Point Theory and Generalized Integrals

*Mehmet Gurdal, Talat Nazir, Hemanta Kalita*

In this chapter, we discuss the literature of summability, fixed point theory and generalized integrals. Summability and convergence, matrix mapping and duals are discussed. In fixed point theory: Banach contraction principle and several generalized contraction principles are covered. Finally, we discuss how one can replace Henstock-Kurzweil integral with Lebesgue measure. Further, several applications of Henstock-Kurzweil integrals are discussed.

## 1.1 INTRODUCTION TO SUMMABILITY THEORY

The notion of *infinite series* arose with the advent of algebraic operations in the field of number systems. Gregory focused on infinite series throughout the seventeenth century and wrote several articles about Maclaurin series. The former class of infinite series was referred to by him as *convergent series* (see [4]). Euler advanced the theory of $q$-series and hyper-geometric series throughout the eighteenth century, contributing to our understanding of infinite and non-convergent series. The notion of infinite was first introduced to mathematical analysis by Gauss. But Cauchy deserves the praise for his precise definition of the sum of an infinite series. In 1821, Cauchy formalized his ideas on convergence and divergence of infinite series, explicitly defining the sum of an infinite series, known as the Cauchy sum, based on the concept of limit developed in his book *Analyse Algébrique*. Another significant figure in the early nineteenth century who contributed concepts on convergence and divergence was Abel [1], Mathematicians developed a number of different techniques for allocating sums to infinite series by the end of the nineteenth century. *Summability Methods* was the moniker given to these summarizing techniques. The summability techniques that bear the names of eminent mathematicians such as Abel, Borel, Cesáro, Euler, Hausdorff, Hölder, Lambert, Nörlund, Reisz, Riemann and Lebesgue are among the most well-known.

DOI: 10.1201/9781003596578-1

Since summability theory was first based on classical analysis, it was regarded as a subfield of classical analysis. Hardy's book [17] is considered the pinnacle of that particular period. Functional analysis methods were first applied in foundational research by Zeller and others (see [29]) and were further advanced by Willansky and others (see [28]). It is encouraging to see that functional analysis basic textbooks have included the topic (see [22]). The theory and applications of summability theory have advanced significantly in the last century, both in the fields of functional analysis and classical analysis. In response, summability theory has given rise to several intriguing spaces of summable sequences and double sequences as well as new classes of matrices.

The framework of the summability techniques was often designed to assign convergent series with the same sum that Cauchy assigned since his idea of the sum of a convergent series survived all mathematical scrutiny. A function from the set of sequences of partial sums of a series to a value is known as a summability technique. Therefore, summability, in its widest sense, is the theory of assigning limits, which is essential to topology, functional analysis, analysis and function theory. If a summability technique adds up all convergent series to its Cauchy's sum, it is considered regular. If two summability techniques provide the same sum to the same series, they are considered consistent (see [17]).

The fundamental idea behind all summability techniques is to convert an infinite series or sequence of partial sums into another series or sequence that may be used with Cauchy's approach. When applied to convergent series, the transformation selected is typically linear and ensures that Cauchy's sum is preserved. All summability methods are considered to be derived from two general basic processes, methods based on sequence-to-sequence transformation and methods based on sequence-to-function transformation.

The two broad processes described above provide the basis for a variety of summability approaches that are crucial to summability theory and practice. Because of these approaches' advantages and applicability in other relevant sectors, they have become more important. Below is a quick description of matrix summability, one of these specialized techniques.

Let $\sum u_n$ be an infinite series with sequence of partial sums $\{s_n\}$. Let $T = (t_{nk})$ be an infinite matrix with real or complex elements, then the matrix summability method transform $\{t_n\}$ of $\{s_n\}$ is given by

$$t_n = \sum_{k=0}^{\infty} t_{nk} s_k, \text{ for } n = 0, 1, 2, \ldots$$

If $t_{nk} = 0$, for $k > n$, then the method is called a triangular matrix method. If $\lim_{n\to\infty} a_n = s$, then the sequence $\{s_n\}$, or the series $\sum u_n$, is said to be summable $(t_{nk})$, or simply $T$-summable to $s$. A series $\sum u_n$ is said to be absolutely summable by a $T$-method or simply $|T|$-summable, if $\sum |a_n - a_{n-1}| < \infty$.

### 1.1.1 SUMMABILITY AND CONVERGENCE

Summability theory and the study of sequence convergence have been among the most significant and active areas of pure mathematics research for the past few decades. These areas of study have found widespread applications in computer science, topology, functional analysis, Fourier analysis, measure theory, applied mathematics and mathematical modelling, among other fields. Over the past twenty years, one of the primary areas of focus in summability theory has been the extension of the concepts of sequence convergence.

It is not always the case that the qualities of a large class of sequences that are not convergent are precisely captured by the conventional idea of convergence. Additionally, in numerous mathematical investigations, we encounter sequences that are not convergent, but almost all of its terms, in many respects, exhibit the properties of a convergent sequence. Thus, it often seems advantageous to consider a broader spectrum of sequences when discussing convergence. One method of incorporating more sequences into this consideration is to examine those sequences that are convergent when limited to some *big* set of natural numbers, defined as large in a certain common sense. If by a *big* set of natural numbers, one refers to a co-finite subset of all natural numbers, then the conventional notion of convergence applies. If by a *big* set of natural numbers, one refers to a subset of all natural numbers with natural (or asymptotic) density equal to zero, then the concept of statistical convergence emerges.

Let $|K|$ denote the cardinal number of a subset $K$ of the set $\mathbb{N}$ of natural numbers, i.e. $|K| := \mathrm{card}K$, and $K_n := \{k \leq n : k \in K\}$. Accordingly, the lower and upper asymptotic densities of the set $K$ are given by

$$\underline{\delta}(K) = \liminf_{n\to\infty} \frac{|K_n|}{n} \text{ and } \overline{\delta}(K) = \limsup_{n\to\infty} \frac{|K_n|}{n},$$

respectively. If the limit of the sequence $\frac{|K_n|}{n}$ exists, this limit is called the natural density of the set $K$ and is denoted by $\delta(K)$. That is, if $\delta(K) = \underline{\delta}(K) = \overline{\delta}(K)$ then the natural density of the set $K \subset \mathbb{N}$ is $\delta(K) = \lim_{n\to\infty} \frac{|K_n|}{n}$. Freedman and Sember [13] provided an axiomatic way to introduce the notion of density of sets of natural numbers. Research in this direction was first initiated by Steinhaus [26] at a conference at the University of Wroclaw and later by Fast [11] in 1951. However, it became one of the main topics of summability theory after the works of Salat [25] and, in particular, Fridy [14]. Since then, a great deal of research has been done on statistical convergence and its subsequent extensions.

To establish the connection between convergence in the ordinary sense and statistical convergence, let us give the mathematical idea between these two concepts: Let us assume that outside each neighbourhood $\varepsilon$ of a point $L$, there may be not only finitely many but infinitely many elements of the sequence. But the number of such elements will be very small compared to the number of all elements of the sequence. So we can say that almost all elements of the sequence are inside the $\varepsilon$-neighbourhood of $L$. From this, we know that the sequence $x$ is *almost convergent* to $L$. The concept of statistical convergence is one of the mathematically precise expressions of this idea. Here, the *small* number of elements outside the $\varepsilon$-neighbourhood of $L$ is expressed by the fact that the natural density of such elements is zero. After giving the connection between these two convergences, we can give the definition of statistical convergence: We say that the sequence $x = (x_k)$ is statistically convergent to the number $L$ if the density of the set $\{k \in \mathbb{N} : |x_k - L| \geq \varepsilon\}$ is zero for every $\varepsilon > 0$.

The concept of Banach limit arises from the application of the well-known Hahn-Banach extension theorem in sequence space theory. This concept further leads to the lovely idea of *almost convergence*. In other words, the whole set of bounded sequence spaces $\ell_\infty$ may be covered by the lim functional defined on the set of convergent sequence spaces $c$; this extended functional is referred to as the Banach limit [3]. This idea of the weak limit was utilized by Lorentz [21] to define a novel form of convergence known as almost convergence. A strong relation between the intersection of a special class of classical matrix methods and matrix summability and statistical convergence was given by Fridy and Miller [15] in 1991. In another direction, the idea of statistical convergence was extended to double and multiple sequences by Mursaleen and Edely [23] and Şahiner et al. [27] which generalizes the well-known idea of Pringshiem convergence of double sequences. Since double natural density is used to define statistical convergence of double sequences, the investigations of double sequences are much more complicated and challenging.

### 1.1.2 MATRIX MAPPINGS AND DUALS

The theory of sequence spaces is a fundamental subject in summability that has several applications, essentially in functional analysis. The classical theory concerns with the generalization of the concept of convergence for series and sequences. The purpose is to assign a limit for nonconvergent series and sequences by applying a transformation, defined as infinite special matrices. Since the most general linear mapping from a sequence space into another sequence space can be represented by an infinite matrix, it will be advantage to focus on matrices instead of a general linear mapping.

The set

$$X_T = \{z \in \omega : Tz \in X\}$$

is referred to as the domain of the infinite matrix $T$ in the space $X$.

In recent years, the approach of constructing a new sequence space by means of a matrix domain has been employed by many scholars. In general, the first aim is to identify its special duals and some matrix transformations and to characterize its compact operators. The multiplier space of $X$ and $Y$ is the set $S(X,Y)$ defined by

$$S(X,Y) = \{u \in \omega : zu \in Y \text{ for all } z \in X\}.$$

Besides continuous duals, there are some other types of duals, namely, $\alpha-, \beta-$ and $\gamma-$duals (cf. [4], [5], [28]), which are very useful in the study of sequence spaces and matrix transformations. Taking into consideration this notation, the $\alpha$-dual, $\beta$-dual and $\gamma$-dual (also called as *Köthe-Toeplitz dual, generalized Köthe-Toeplitz dual* and *Garling dual*, respectively) of a sequence space $X$ are defined by

$$X^{\alpha} = S(X, \ell_1), \quad X^{\beta} = S(X, cs) \text{ and } X^{\gamma} = S(X, bs).$$

Here, $cs$ and $bs$ correspond to the spaces of sequences with convergent and bounded series, respectively.

## 1.2 INTRODUCTION TO FIXED POINT THEORY

Fixed point theory is one of the important, traditional and famous theories in mathematics and has a broad number of applications. In this theory, contraction is one of the main tools to prove the existence and uniqueness of a fixed point. Banach's contraction principle [2] which gives an answer on the existence and uniqueness of a solution of an operator equation $Tx = x$, is the most widely used fixed point theorem in all of analysis. This principle is constructive in nature and is one of the most useful tools in the study of nonlinear equations. There are many generalizations of Banach's contraction mapping principle in the literature. These generalizations were made either by using the contractive condition or by imposing some additional conditions on an ambient space. There have been a number of generalizations of metric spaces such as $b$-metric spaces, quasi-metric spaces, semi-metric spaces, $G$-metric spaces, and cone metric spaces (see [7, 10, 20, 19]).

The development of fixed point theory is a cardinal branch of nonlinear analysis, and has given great effects on the advancement of nonlinear analysis. Nonlinear analysis as an independent branch of mathematics was developed in the 1950's by mathematician such as Browder as a combination of functional analysis. Nonlinear results are applicable to a wide range of areas. Several problems in physics, chemistry, biology and economics lead to nonlinear models. Nonlinear differential and integral equations, variational inequalities, and more general optimization problems are some of the important subjects in nonlinear analysis.

The study of fixed points of mappings satisfying certain contractive conditions has been at the centre of rigorous research activity. Over the last 50 years or so, the theory of fixed points has been revealed as a very powerful and important tool in the study of nonlinear phenomena. In particular, fixed point techniques have been applied in such diverse fields as biology, chemistry, economics, engineering, game theory and physics.

### 1.2.1 METRIC SPACE

Metric spaces are playing an increasing role in mathematics and the applied sciences. Over the past two decades, the development of fixed point theory in metric spaces has attracted considerable attention due to numerous applications in areas such as variational and linear inequalities, optimization and approximation theory. In 1906, Fréchet [12] gave the formal definition of the distance by introducing a function $d$ that assigns a nonnegative real number $d(x,y)$ (the distance between $x$ and $y$) to every $x, y$ of elements of a nonempty set $X$ in the following way.

**Definition 1.2.1.** *Let $X$ be a nonempty set. A function $d : X \times X \mapsto \mathbb{R}$ is called a metric on $X$ if:*

$(d_1)$ $0 \leq d(x,y)$ *for all* $x,y \in X$ *and* $d(x,y) = 0 \Leftrightarrow x = y$;
$(d_2)$ $d(x,y) = d(y,x)$ *for all* $x,y \in X$;
$(d_3)$ $d(x,y) \leq d(x,z) + d(z,y)$ *for all* $x,y,z \in X$.

If $d$ is a metric on $X$, then $(X,d)$ is called a metric spaces.

**Example 1.2.2.** *A function $d : \mathbb{R} \times \mathbb{R} \mapsto \mathbb{R}$ defined with $d(x,y) = |x-y|$, $x,y \in \mathbb{R}$, is a metric on $\mathbb{R}$, often called the usual metric on $\mathbb{R}$.*

**Example 1.2.3.** *If $C[0,1] = \{x : [0,1] \mapsto \mathbb{R} \mid x \text{ is a continuous function}\}$ and*

$$d(x,y) = \sup_{t \in [0,1]} |x(t) - y(t)|, \; x,y \in C[0,1],$$

*then $(C[0,1],d)$ is a metric space and such defined metric is well-known as supremum metric.*

In the setting of metric space, we introduce a concept of Cauchy and convergent sequence.

**Definition 1.2.4.** *If $\{x_n\}$ is a sequence in $X$ such that, for any $\varepsilon > 0$, there exists $n_0 \in \mathbb{N}$ such that $d(x_n, x_m) < \varepsilon$, $m,n \geq n_0$, then $\{x_n\}$ is a Cauchy sequence in $X$.*

**Definition 1.2.5.** *If every Cauchy sequence in a metric space $(X,d)$ is convergent in $X$, then $(X,d)$ is a complete metric space.*

### 1.2.2 BANACH CONTRACTION PRINCIPLE

The main focus of this section is on Banach fixed point theorem also known as the Banach contraction principle. It is well-established opinion that research in the area of metric fixed point theory started in 1922 [2] with this result.

**Definition 1.2.6.** *Let $X$ be a nonempty set and $f : X \to X$ be a function. Then $x_0 \in X$ is called a fixed point of $f$ if $f(x_0) = x_0$. We denote the set of all fixed points of $f$ by $Fix(f)$.*

**Definition 1.2.7.** *Let $(X,d)$ be a metric space. A mapping $f : X \to X$ is a* contraction *if there exists some $q \in [0,1)$ such that*

$$d(f(x), f(y)) \leq q \cdot d(x,y) \text{ for all } x,y \in X. \tag{1.1}$$

We shall also use the term $q$–contraction for contraction.

Observe that every contraction is a continuous mapping. The following theorem shows the existence and uniqueness of a fixed point of arbitrary contraction on a complete metric space. It is important to mention that there exists a continuous mapping without fixed point property.

**Example 1.2.8.** *A mapping $f:\mathbb{R}\mapsto\mathbb{R}$ defined with $f(x)=x+1$, $x\in\mathbb{R}$, is continuous, but $f(x)\neq x$ for each $x\in\mathbb{R}$.*

**Theorem 1.2.9.** ***(Banach contraction principle)*** *If $(X,d)$ is a complete metric space and $f:X\mapsto X$ a contraction, then a mapping $f$ has a unique fixed point in $X$.*

**Remark 1.2.10.** *There exist various approaches in proving the Banach fixed point theorem, but this proof gives a method of finding a fixed point for a contraction $f$. It is also known as Picard iteration method or fixed point iteration. It is based on idea of forming a sequence of successive iterations. Start with any $x_0\in X$ and define $x_n=fx_{n-1}$, $n\in\mathbb{N}$.*

**Corollary 1.2.11.** *Let $S$ be a closed subset of a complete metric space $(X,d)$ and $f: S\mapsto S$ a contraction. For an arbitrary $x_0\in S$, the iterative sequence $x_n=f(x_{n-1})$, $n\in\mathbb{N}$, converges to a fixed point of a mapping $f$.*

The following example will show that the closeness condition is necessary.

**Example 1.2.12.** *Define a usual metric $d$ on $\mathbb{R}$, $d(x,y)=|x-y|$, $x,y\in\mathbb{R}$, let $S=B(0,1)=\{x\in\mathbb{R}:|x|<1\}$, then a mapping*

$$f:S\mapsto S,\ \text{defined as}\ f(x)=\frac{x+1}{2},$$

*is a contraction without fixed point in $S$.*

**Corollary 1.2.13.** *Let $f:X\mapsto X$ be a $q-$contraction on a complete metric space $(X,d)$ and $z\in X$ the fixed point of $f$. Then:*

*(i) the sequence $\{f^n(x)\}$ converges to $z$ for any $x\in X$;*
*(ii) $d(x,z)\leq\frac{1}{1-q}d(x,fx)$;*
*(iii) $d(f^nx,z)\leq\frac{q^n}{1-q}d(x,fx)$, $n\in\mathbb{N}$;*
*(iv) $d(f^{n+1}x,z)\leq\frac{q}{1-q}d(f^nx,f^{n+1}x)$, $n\in\mathbb{N}$.*

Banach fixed point theorem has wide and diverse applications e.g. solving different kinds of equations, inclusions etc.

The following consequences show a relation between $f^n$ and $f$ in the case when $f^n$ is a contraction.

**Corollary 1.2.14.** *Bryant [6] If $(X,d)$ is a complete metric space, $f:X\mapsto X$ a mapping such that $f^n$ is a contraction for some $n\geq1$, then $f$ has a unique fixed point in $X$.*

**Proof:** By Banach fixed point theorem, there exists unique $z\in X$, $f^nz=z$. Since $f^n(fz)=f(f^n(z))=fz$, it follows $fz=z$. Every fixed point of $f$ is, at the same time, a fixed point of $f^n$, thus $z$ is unique fixed point of $f$.

As observed in [6], a mapping $f$ mentioned in this consequence needs not to be continuous unlike in Theorem 1.2.9.

**Example 1.2.15.** *(Bryant [6]) Define $f : [0,2] \mapsto [0,2]$ with $f(x) = 1, x \in [0,1)$, i $f(x) = 2$, $x \in [1,2]$. Then $f^2(x) = 2$, $x \in [0,2]$ and $f^2 : [0,2] \mapsto [0,2]$ is a contraction despite that $f$ isn't continuous mapping.*

Since the proof of Banach theorem is based on an iterative sequence for a point $x \in X$, next reasonable step in the research was to check local properties and modify this result.

**Theorem 1.2.16.** *Let $(X,d)$ be a complete metric space and for some $x_0 \in X$, $r > 0$, $B(x_0,r) = \{x \in X : d(x,x_0) < r\}$ is an open ball in $X$. Let $f : K(x_0,r) \mapsto X$ be a contraction, i.e.,*

$$d(fx,fy) \leq q \cdot d(x,y), x,y \in B(x_0,r), \tag{1.2}$$

*for some $q \in [0,1)$ and*

$$d(fx_0,x_0) < (1-q)r. \tag{1.3}$$

*Then a mapping $f$ has a unique fixed point in $B(x_0,r)$.*

### 1.2.3 CIRIC GENERALIZED CONTRACTION

Ćirić in [8] generalized well-known contractive condition and introduced a concept of a generalized contraction defined as follows.

**Definition 1.2.17.** *(Ćirić [8]) Let $(X,d)$ be a metric space. A mapping $f : X \mapsto X$ is $\lambda-$generalized contraction if for any $x,y \in X$ there exist some nonnegative numbers $q(x,y)$, $r(x,y)$, $s(x,y)$ and $t(x,y)$, such that*

$$\sup_{x,y \in X} \{q(x,y) + r(x,y) + s(x,y) + 2t(x,y)\} = \lambda < 1,$$

*and for any $x,y \in X$,*

$$\begin{aligned} d(fx,fy) \leq\ & q(x,y)d(x,y) + r(x,y)d(x,fx) + s(x,y)d(y,fy) \\ & + t(x,y)(d(x,fy) + d(y,fx)). \end{aligned} \tag{1.4}$$

Obviously, this condition is equivalent to the fact that there exists a constant $h \in [0,1)$, such that for any $x,y \in X$,

$$d(fx,fy) \leq h \max\left\{d(x,y), d(x,fx), d(y,fy), \frac{d(x,fy)+d(y,fx)}{2}\right\}. \tag{1.5}$$

This example shows that (1.4) indeed generalizes (1.1).

**Example 1.2.18.** *Let* $X = [0,2] \subseteq \mathbb{R}$ *and*

$$f(x) = \begin{cases} \frac{x}{9} & if\ 0 \leq x \leq 1, \\ \frac{x}{10} & if\ 1 < x \leq 2. \end{cases}$$

*Mapping* $f$ *does not satisfy (1.1) since, for* $x = \frac{999}{1000}$ *and* $y = \frac{1001}{1000}$,

$$d(fx, fy) = \frac{981}{90000} > 5 \cdot \frac{180}{90000} = 5d(x,y).$$

*But, (1.4) holds for* $q(x,y) \equiv \frac{1}{10}$, $r(x,y) = s(x,y) \equiv \frac{1}{4}$ *and* $t(x,y) \equiv \frac{1}{6}$, $x, y \in X$.

Let $(X,d)$ be a metric space, $f : X \mapsto X$ a mapping and $x \in X$ an arbitrary point. An $f$-orbit of element $x$ is a subset $O(x;f)$ of $X$ defined by

$$O(x;f) = \{f^n(x) \mid n \in \mathbb{N}_0\}.$$

In the case $f$ is determined, usual notation is $O(x)$. Furthermore, for any $n \in \mathbb{N}$, define a set

$$O(x,n) = \{x, fx, f^2x, \ldots, f^n x\}.$$

**Definition 1.2.19.** *Let* $(X,d)$ *be a metric space and* $f : X \mapsto X$ *a mapping. Then* $X$ *is a* $f$*-orbitally complete metric space if any Cauchy sequence in* $O(x;f)$, $x \in X$, *converges in* $X$.

Obviously, every complete metric space is $f$-orbitally complete, but the converse is not true. It is evident from the proof of Banach theorem that it is enough to assume that $(X,d)$ is $f$-orbitally complete instead of complete metric space. Same remark is applicable for $\lambda-$generalized contraction as proved in the following theorem.

**Theorem 1.2.20.** *(Ćirić [8]) If* $f : X \mapsto X$ *is* $\lambda-$*generalized contraction on* $f-$*orbitally complete metric space* $X$, *then, for any* $x \in X$, *an iterative sequence* $(f^n x)$ *converges to a unique fixed point* $z$ *of* $f$ *and*

$$d(f^n x, u) \leq \frac{\lambda^n}{1-\lambda} \cdot d(x, fx).$$

As in the case of the Banach theorem, we may consider the local properties of Theorem 1.2.20.

**Theorem 1.2.21.** *Let* $f : B \mapsto X$ *be a mapping of* $f-$*orbitally complete metric space* $(X,d)$ *where*

$$B = B(x_0, r) = \{x \in X : d(x_0, x) \leq r\},$$

*for some* $x_0 \in X$ *and* $r > 0$. *If* $f$ *is* $\lambda-$*generalized contraction on* $B$ *and*

$$d(x_0, fx_0) \leq (1-\lambda) \cdot r, \tag{1.6}$$

*then, the sequence* $(f^n x_0)$ *converges to a unique fixed point* $z$ *of* $f$ *in* $B$ *and*

$$d(f^n x_0, u) \leq \lambda^n \cdot r,$$

*for* $\lambda = \{q(x,y) + r(x,y) + 2t(x,y) : x, y \in B\}$.

**Proof:** Evidently, $x_n \in B, n \in \mathbb{N}$, due to (1.6) and mathematical induction. Analogously to the proof of Theorem 1.2.20, $(f^n x_0)$ is a Cauchy sequence in $B$ and it's limit is a fixed point of $f$. Inequality (1.4) guarantees uniqueness.

### 1.2.4 CIRIC QUASI-CONTRACTION

Ćirić [9] in 1971. used a concept of generalized contraction to replace a linear combination of distances observed in (1.4) with their maximum and defined a new class of contractive mappings called quasi-contractions.

**Definition 1.2.22.** *(Ćirić [9]) A mapping $f : X \mapsto X$ of a metric space $(X,d)$ is a quasi-contraction if exists some $\lambda$, $0 < \lambda < 1$, such that*

$$d(fx,fy) \leq \lambda \cdot \max\{d(x,y),d(x,fx),d(y,fy),d(x,fy),d(y,fx)\}, \tag{1.7}$$

*for all $x,y \in X$.*

Obviously if a mapping $f$ satisfies condition (1.1), then (1.7) also holds. An example presented by Ćirić shows that the reverse doesn't apply.

**Example 1.2.23.** *(Ćirić [9]) Let*

$$M_1 = \{\frac{m}{n} : m = 0, 3^k, n = 3k+1, k \in \mathbb{N}_0\}$$
$$M_2 = \{\frac{m}{n} : m = 3^k, n = 3k+2, k \in \mathbb{N}_0\}$$

*and $M = M_1 \cup M_2$ a metric space with the usual metric $d(x,y) = |x-y|$, $x,y \in M$. A mapping $f : X \mapsto X$, defined with*

$$f(x) = \begin{cases} \frac{3}{5}x, & if \ x \in M_1, \\ \frac{1}{8}x, & if \ x \in M_2. \end{cases}$$

*is a quasi-contraction for $\lambda = \frac{3}{5}$, but not contraction ($x = 1, y = \frac{1}{2}$).*

**Theorem 1.2.24.** *(Ćirić [9]) If $f : X \mapsto X$ is a quasi-contraction on a $f-$orbitally complete metric space $(X,d)$, then $f$ has a unique fixed point $z$ in $X$ and iterative sequence $(f^n x)$ converges to $z$ for any $x \in X$. Moreover,*

$$d(f^n x, z) \leq \frac{\lambda^n}{1-\lambda} d(x, fx).$$

Hardy and Rogers [18] improved some fixed point results by presenting the following theorem.

**Theorem 1.2.25.** *(Hardy and Rogers [18]) Let $(X,d)$ be metric space, $f : X \mapsto X$ a mapping such that for all $x,y \in X$,*

$$\begin{aligned} d(f(x),f(y)) \leq\ & ad(x,f(x)) + bd(y,f(y)) \\ & + cd(x,f(y)) + ed(y,f(x)) + fd(x,y), \end{aligned} \tag{1.8}$$

*where $a,b,c,e,f \geq 0$ and $\alpha = a+b+c+e+f$.*

*(i) If $(X,d)$ is complete metric space and $\alpha < 1$, then mapping $f$ has unique fixed point.*

*(ii) If, for compact metric space $(X,d)$ and continuous $f$, condition (1.8) is replaced with*

$$\begin{aligned} d(f(x),f(y)) < {} & ad(x,f(x))+bd(y,f(y)) \\ & +cd(x,f(y))+ed(y,f(x))+fd(x,y), \end{aligned} \tag{1.9}$$

*for any $x \neq y$, and $\alpha = 1$, then $f$ has unique fixed point.*

The following lemma is an integral part of proof of this theorem, but for the convenience of a reader, we state it separately.

**Lemma 1.2.26.** *Assume that (1.8) holds and $\alpha < 1$, then there exists $\beta < 1$, such that*

$$d(f(x),f^2(x)) \leq \beta d(x,f(x)). \tag{1.10}$$

*If $\alpha = 1$ and (1.9) holds, then*

$$x \neq f(x) \quad \Longrightarrow \quad d(f(x),f^2(x)) \leq \beta d(x,f(x)). \tag{1.11}$$

## 1.3 NON-ABSOLUTE INTEGRALS AND THEIR APPLICATIONS

Differential and integral calculus were independently discovered at the end of the seventeenth century by Gottfried Leibniz (1646–1716) and Isaac Newton (1642–1727). Integral calculus was used to calculate the area under a curve, and differential calculus was used to determine a curve's slope at a specific point. It determined that the relationship between differential and integral calculus was inverse for a large class of functions, such that integrating the derivative of a differentiable function might yield a differentiable function up to a constant. In fact, a function $F : [a,b] \to \mathbb{R}$ which is differentiable everywhere on $[a,b]$ except on a loosely stated small set $S \subseteq [a,b]$, could in certain cases be recovered by integrating a function $f : [a,b] \to \mathbb{R}$ which satisfies $F' = f$ on $[a,b] \setminus S$. Exactly how small $S$ had to be and how well-behaved $F$ had to be on $S$ was not known at the time. Many features of integration theory remained mysterious because the mathematics available at the time simply did not permit mathematicians to conduct research into descriptive characterizations of integrable functions. Mathematicians started to realize in the seventeenth century that the very basis of mathematical analysis was ill-defined and extremely unstable. This created a significant barrier to research when geometric intuition was no longer adequate. As a result, utilizing different epsilon-delta type definitions and proofs, mathematicians started creating a formal framework for mathematical analysis in the nineteenth century. Specifically, the formal approach to continuity served as the foundation for numerous important analytical advances. The modern treatment of continuity is generally credited to Augustin-Louis Cauchy (1789–1857), but numerous other mathematicians also made significant contributions in this area. Integration theorists started to obtain instruments during this period that would enable them to rigorously define and investigate various integrals. One of the first integrals to be formally defined was the Riemann integral, established in 1854 by Bernhard Riemann (1826–1866). A real-valued function defined on a compact interval is Riemann integrable if and only if it is bounded and discontinuous on a set of Lebesgue measure 0, as demonstrated by Henri Lebesgue (1875–1941) in 1904. Bounded derivatives, on the other hand, are not Riemann integrable when they are discontinuous on a set of positive Lebesgue measure. Vito Volterra (1860–1940) provided an example of such a function in 1881; it is developed in [37, Example 1.4.1]. Lebesgue introduced the Lebesgue integral in 1901, which fixed the drawback. As it turns out, the Lebesgue integral is a generalization of the Riemann integral and it is able to integrate all bounded derivatives restricted to compact intervals. Going back to our functions $F$ and $f$ where $F' = f$ on $[a,b] \setminus S$, if $[a,b]$ is compact then by Lebesgue's fundamental theorem of calculus we have that $f$ is Lebesgue integrable on $[a,b]$ and $F$ can be reconstructed by the Lebesgue integral of $f$ if and only if $F$ is absolutely continuous and $S$ has Lebesgue measure 0. Note that the absolute continuity of $F$ is a necessity here, thus it is easy to show that there are differentiable functions which cannot be recovered from their respective derivatives via the Lebesgue integral. If $F$ is not absolutely continuous and $S$ has Lebesgue measure 0, then in certain cases $f$ may still be Lebesgue integrable but $F$ can then not be recovered from the Lebesgue integral of $f$. For example, if $[a,b] = [0,1]$, $F$is the Cantor-Lebesgue function studied in [38], and $f = 0$ every-where on $[0,1]$, then $F' = f$ on $[0,1] \setminus S$, where $S$ has Lebesgue measure 0, yet $F$ cannot be reconstructed from $f$ via the Lebesgue integral. Unfortunately, the full power of the Lebesgue integral as it is known today was not understood by the mathematical community during this time and thus Lebesgue's work did not garner the immediate attention one perhaps would expect. On asecond thought this is not too surprising, since measure theory was still a very new topic and functional analysis had not yet been developed to a point where these two areas of mathematics could be researched in unity to produce a more complete theory. Perhaps this is a partial explanation to why Lebesgue was not appointed as professor until 1919, as mentioned in [39, Chapter VI.72]. However, later on in the twentieth century mathematicians incrementally developed the necessary tools to study the Lebesgue integral more thoroughly and eventually the Lebesgue integral became the standard integral in advanced mathematical analysis. Despite the power of the Lebesgue integral, there were still finite and unbounded derivatives that could not be integrated. In 1912, Arnaud Denjoy (1884–1974) presented a powerful integral which was able to integrate all finite derivatives and recover their primitive functions. As described in [41, Section 1.1], loosely stated, Denjoy constructed a transfinite sequence $(I_k)_k \leq \Omega$ of increasingly general integrals, where $\Omega$ is the first uncountable ordinal, $I_0$ is the Lebesgue integral and $I_\Omega$ is the socalled narrow Denjoy integral. After some modifications by Hans Bauer in 1915, Oskar Perron (1880–1975) presented an integral two years later in 1914 that could also integrate all finite derivatives and retrieve their original functions. The Hake-Aleksandrov-Looman theorem states that mathematicians discovered, after a considerable amount of time, that the narrow Denjoy integral and the Perron integral are comparable. As a result, the Denjoy-Perron integral is the current name for the previously described Denjoy and Perron integral.Many years later, in 1957, Jaroslav Kurzweil (born in 1926) presented a novel integral in a differential equations work he published. Ralph Henstock (1923–2007) presented the same integral as Kurzweil in a paper on integration theory that he published four years later, in 1961, despite being ignorant of Kurzweil's work. In the 1960s, Henstock developed a significant number of features of this integral in a series of publications. Since it closely resembles the Riemann integral and many of its features can be derived using Riemann sums and elementary epsilon-delta proofs, Kurzweil and Henstock's definition of this integral is fairly elegant. The Henstock-Kurzweil integral is the modern name for the integral of Henstock and Kurzweil. In fact, the Henstock-Kurzweil integral is identical to the Denjoy-Perron integral, as mathematicians have established later. Since many integration theorists had studied the Henstock-Kurzweil integral in great detail by the late 1990s, the theory surrounding this integral had advanced significantly. In the sense that all finite derivatives may be integrated and their corresponding primal functions can be reconstructed, the Henstock-Kurzweil integral leads to a comprehensive fundamental theorem of calculus. Compared to the calculus fundamental theorem for the Lebesgue integral, which also requires the indefinite integral to be perfectly continuous, this result is more generic. A monotone convergence theorem and a dominated convergence theorem are derived from the Henstock-Kurzweil (HK) integral. These theorems are more robust than their Lebesgue integral counterparts because they place less restrictions on the functional sequences under consideration. The integral that is derived from the Henstock-Kurzweil integral contains an incorrect one. This is an additional attribute that the Lebesgue integral lacks. Therefore, the Henstock-Kurzweil integral largely makes up for the theoretical flaws in the Lebesgue integral.

## 1.4 HOW WE REPLACE HENSTOCK-KURZWEIL INTEGRAL WITH LEBESGUE MEASURE?

An integral can be defined in two different ways. A descriptive description or an operational (or constructive) definition can be given. A descriptive definition does not provide a construction method; instead, it explains the integral in relation to its derivative.

**Definition 1.4.1.** *[40] We define the weak variation $V(F,E)$, and the strong variation, $V_*(F,E)$, by:*

$$V(F,E) = \sup\{\sum_{i=1}^{n} |F(b_i) - F(a_i)|\}$$

*and*

$$V_*(F,E) = \sup\{\sum_{i=1}^{n} w(F,[a_i,b_i])\},$$

*where the supremum is taken over all possible finite collections of nonoverlapping intervals that have end points in $E$.*

1) *We say that F is of bounded variation on* $E$, $(BV)$, *if* $V(F,E) < \infty$.
2) *We say that F is of restricted bounded variation on* $E, (BV_*)$, *if* $V_*(F,E) < \infty$.
3) *We say that F is absolutely continuous on E, (AC), if for each* $\varepsilon > 0$, *there exists a* $\delta > 0$ *such that, for every collection* $\{[a_i,b_i], 1 \leq i \leq n\}$, *of nonoverlapping intervals with end points in E and* $\sum_{i=1}^{n}(b_i - a_i) < \delta$, *then* $\sum_{i=1}^{n}|F(b_i) - F(a_i)| < \varepsilon$.
4) *We say that F is absolutely continuous on E in the restricted sense,* $(AC)_*$, *if for each* $\varepsilon > 0$, *there exists a* $\delta > 0$ *such that, for every collection* $\{[a_i,b_i], 1 \leq i \leq n\}$, *of nonoverlapping intervals with end points in E and* $\sum_{i=1}^{n}(b_i - a_i) < \delta$, *then*

$$\sum_{i=1}^{n} w(F,[a_i,b_i]) < \varepsilon.$$

5) *We say that F is generalized absolutely continuous on E, (ACG), if* $F|_E$ *is continuous and E is a countable union of sets* $\{E_i\}$ *such that F is (AC) on each* $E_i$.
6) *We say that F is generalized absolutely continuous in the restricted sense in* $E$, $(ACG)_*$, *if* $F|_E$ *is continuous and E is a countable union of sets* $\{E_i\}$ *such that F is* $(AC)_*$ *on each* $E_i$.

One thing that the Riemann approach in a first-year analysis course cannot accomplish is to provide a direct transfer to measure theory in the semester that follows. Still, there's more going on here than this casual comment would imply. Each Riemann approach has a set of natural metric measures associated with it because of a Henstock notion (created, in particular, by Thomson) that, upon analysis, is discovered to provide more exact information about the primitives than the classical conceptions of $ACG_*$ etc. The notion of variation, in particular, provides an additional definition for the different Riemann integrals. Though they are part of Thomson's great reworking of trigonometric integrals, these concepts are a bit too sophisticated to be covered in detail here.

**Definition 1.4.2.** *Descriptive Definitions: Let* $E$ *be a measurable subset of* $\mathbb{R}$ *and* $\mu(E)$ *denote the Lebesgue measure of* $E$. *Let* $c \in \mathbb{R}$.

1. *We say that c is a point of density for E if*

$$d_c E = \lim_{h \to 0^+} \frac{\mu(E \cap (c-h, c+h))}{2h} = 1.$$

2. *We say that c is a point of dispersion for E if*

$$d_c E = \lim_{h \to 0^+} \frac{\mu(E \cap (c-h, c+h))}{2h} = 0.$$

3. *We say that a function* $F : [a,b] \to \mathbb{R}$ *is approximately continuous at* $c \in E \subset [a,b]$, *if c is a point of density for E and* $F|_E$ *is differentiable at c.*

**Theorem 1.4.3.** *[40] Let F be a function defined on* $[a,b]$ *with* $F(a) = 0$, *then the following holds:*

1. *If $F$ is (AC) on $[a,b]$, then $F'$ exists a.e. and, $F'$ is Lebesgue integrable, then* $\int_a^x F'(y)d\mu(y) = F(x)$.
2. *If $F$ is $(ACG_*)$ on $[a,b]$, then $F'$ exists a.e. and, $F'$ is Henstock-Kurzweil integrable, then* $\int_a^x F'(y)d\mu(y) = F(x)$.

**Example 1.4.4.** *If $F$ is any interval function and if $V_*F(E) = \inf_\delta \sup_P \sum |F(J_i)|$, where $\delta$ is a gauge and $P$ is a partition in $I$ that is anchored in the set $E$, then $V_*$ is a regular Borel measure.*

If $V_*$ is $AC$ then $F$ is the HK-integral of its derivative, which exists almost everywhere; further, if in addition, $V_*$ is finite, then $F$ is the Lebesgue integral of this derivative.

## 1.5 APPLICATION OF HENSTOCK-KURZWEIL INTEGRAL AND NON ABSOLUTE INTEGRABLE FUNCTION SPACES

One of the most effective integration techniques that mathematicians are currently studying is the HK integral. Gauges allow for a more localized evaluation of functions than is possible with the classical Riemann integral. It has shown that this seemingly little modification to the conventional definition of the Riemann integral has far-reaching effects. For instance, integration and differentiation become genuinely inverse processes when the HK integral is used. The HK integral is perfect for integrating functions that oscillate wildly since it is a non-absolutely convergent integral, which the Lebesgue integral does not always offer. Instead of being compelled to study the negative and nonnegative instances individually, as is frequently the case in Lebesgue integration theory, this enables one to see the integration process as a whole. This benefit does, however, come with certain disadvantages. As of yet, no appropriate norm for the space $HK([a,b])$. has been created. Even though the Henstock-Kurzweil integral has great potency, it is hardly ever utilized in undergraduate or graduate courses on mathematical analysis. This is due to a number of factors. Students usually learn the Darboux integral in elementary analysis courses; they often interpret the integral as defined by the Cauchy criterion for Darboux integrability. The Riemann integral and the Darboux integral are known to be equivalent. First-year students often pay little to no attention to further subtleties and instead think in terms of formulas and graph representations of very well-behaved functions. Thus, a textbook on analysis intended for first-year students does not belong to have a relatively formal discussion of the Henstock-Kurzweil integral. Peng Yee Lee (born 1938), who studied under Henstock from 1961 to 1965, really notes in [72, Page 678] that Henstock once attempted to instruct first-year students on the Henstock-Kurzweil integral. He apparently never tried it again because it was a complete failure. One could argue that the Henstock-Kurzweil integral ought to be presented later on because it is essentially a logical extension of the Riemann integral. However, the Lebesgue integral is typically the integral of choice in postgraduate analysis classes (and possibly in certain advanced undergraduate analysis courses as well). Although the Henstock-Kurzweil integral can be used to create a significant portion of the theory of the Lebesgue integral. There has been some research on integral transformations, partial differential equations, and the Henstock-Kurzweil integral; for example, see [73]. There are situations where we want to solve issues with minimal smoothness assumptions, and these problems may contain problematic functions that the Lebesgue theory is unable to handle. Here, the Henstock-Kurzweil integral might be useful in solving some of the issues. But for this to be the case, there must be more than three references to papers that appear to be at random. For the Henstock–Kurzweil integral to be more commonly encountered in postgraduate courses, a significant body of theory that is both novel to the field of differential equations and integral transformations and accessible to postgraduate students is required. The question of whether this is feasible is one that is much outside the purview of this argument. Given what we have shown in this thesis, it is difficult to claim that the Henstock-Kurzweil integral should be taught more in advanced analysis classes, even though many of the results are highly intriguing.

Highly oscillating or unique functions arise in a number of physical phenomena [30]. For certain strongly oscillating functions, the Lebesgue integral is insufficient, which raises the possibility that the integral on the right side of [31, Equation 2] does not exist. In this case, the variational problem (i.e., problem 2) of [31] would not be well-defined. Changing the type of integral to be considered is one way to tackle this difficulty; in this study, we will utilize the Henstock-Kurzweil integral. One way to solve this problem is to change the type of integral to be considered; in this work, we will use the Henstock-Kurzweil in- tegral. Different authors have studied differential equations involving Henstock-Kurzweil integrable functions. In [74] the authors use the Finite Element Method (FEM) for finding numerical solutions of elliptic problems with Henstock-Kurzweil integrable functions. They use open quadratures and Lobatto quadratures to approximate numerically the integrals that appear in the FEM. In [36] are given conditions to establish the existence of a solution to nonlinear second-order differential equations of type:

$$-D^2x = f(t,x) + g(t,x)Du$$

subject to the boundary conditions $x(0) = \beta Dx(0),\ Dx(1) + Dx(\nu) = 0$, where the derivatives are in the distributional sense, $x, u$ are regulated functions and $g$ is of bounded variation. In [31] the Henstock-Kurzweil-Stieltjes integral is used to transform the distributional differential equation into an integral equation, then the Leray-Schauder nonlinear alternative theorem is applied for finding a solution. In [50] the existence and uniqueness of the Shrödinger equation, $-y'' + qy = f$ a.e. on $[a,b]$ subject to arbitrary boundary values, is guaranteed for functions $f, q$ Henstock-Kurzweil integrable. Properties of the inverse of the Shrödinger operator are established, then the authors give conditions so that the solution of the differential equation can be expressed as a Fourier-type series.

Gill and Zachary in [40] discussed the space $\mathbb{KS}^p(\mathbb{R}^n)$ called Kuelbs-Steadman space. In Feynman path integral: The properties of $\mathbb{KS}^p[\mathbf{R}^n]$ derived earlier suggest

that $\mathbb{KS}^2[\mathbf{R}^n]$ may be a better Hilbert space than $\mathbf{L}^2[\mathbf{R}^n]$ for the study of the path integral formulation of quantum theory developed by Feynman. Note that it is easy to prove that both the position and momentum operators have closed densely defined extensions to $\mathbb{KS}^2[\mathbf{R}^n]$. Furthermore, the extensions of Fourier, $\mathfrak{F}$ and Convolution, $\mathfrak{C}$ insure that all of the Schrödinger and Heisenberg theory has a faithful representation on $\mathbb{KS}^2[\mathbf{R}^n]$. These issues will be discussed more fully in another venue.
In operator theory on separable Hilbert spaces, the major problem with integration for operator-valued functions is that these functions need not have a Lebesgue (like) integral. However, they always have a HK-integral (see [42]).

## BIBLIOGRAPHY


1. N. H. Abel, Untersuchungen über die Reihe $1+mx+\frac{m(m-1)}{x}x^2+\ldots$, Journal für die reine und angewandte Mathematik., **1**, (1826), 311–339.
2. S. Banach, Sur les opérations dans les ensembles abstraits et leur application aux équations inté grales, Fund. Math., **3**(1), (1922), 133–181.
3. S. Banach, *Théorie des Operations Lineaires*, Warszawa, (1932).
4. J. Banaś, M. Mursaleen, *Sequence Spaces and Measures of Noncompactness with Applications to Diferential and Integral Equations*, Springer, (2014).
5. F. Başar, *Summability Theory and its Applications*, Bentham Science Publishers, e-books, Monograph, Istanbul, (2012).
6. V.W. Bryant, A remark on a fixed point theorem for iterated mappings, Amer. Math. Monthly, **75**, (1968), 399–400.
7. M. Cicchese, Questioni di completezza e contrazioni in spazi metrici generalizzati, Boll. Un. Mat. Ital., **13**-A(5), (1976), 175–179.
8. Lj. B. Ćirić, Generalized contractions and fixed point theorems, Publ. Inst. Math., **12** (26), (1971), 19–26.
9. Lj.B. Ćirić, A generalization of Banach's contraction principle, Proc. Amer. Math. Soc. **45**, (1974), 267–273.
10. S. Czerwik, Contraction Mappings in b-Metric Spaces, Acta Mathematica et Informatica Universitatis Ostraviensis, **1**(1), (1993), 5–11.
11. H. Fast, Sur la convergence statistique, Colloq. Math., **2**, (1951), 241–244.
12. M. Fréchet, Sur quelques points du calcul fonctionnel, Rendiconti del Circolo Matematico di Palermo., **22**, (1906), 1–74.
13. A. R. Freedman, J. J. Sember, Densities and summability, Pac. J. Math., **95**, (1981), 293–305.
14. J. A. Fridy, On statistical convergence, Analysis., **5**, (1985), 301–313.
15. J. A. Fridy, H. I. Miller, A matrix characterization of statistical convergence, Analysis., **11**(1), (1991), 59–66.
16. J. Gregory, *Vera Circuli et Hyperbalae Quadrature*, Padua., (1667).
17. G. H. Hardy, *Divergent Series*, Oxford University Press, Oxford, (1949).
18. G. E. Hardy, T. D. Rogers, A generalization of a fixed point theorem of Reich, Canadian Mathematical Bulletin., **16**, (1973), 201–206.
19. L. G. Huang, X. Zhang, Cone metric spaces and fixed point theorems of contractive mappings, J. Math. Anal. Appl. **332**, (2007), 1467–1475.
20. A. Latif, V. Parvaneh, P. Salimi, A. E. Al-Mazrooei, Various Suzuki type theorems in $b$-metric spaces, J. Nonlinear Sci. Appl., **8**(4), (2015), 363–377.

21. G. G. Lorentz, A contribution to the theory of divergent sequences, Acta Math., **80**, (1948), 167–190.
22. I. J. Maddox, *Elements of Functional Analysis*, The University Press, 2nd ed., Cambridge, (1988).
23. M. Mursaleen, O. H. H. Edely, Statistical convergence of double sequences, J. Math. Anal. Appl., **288**, (2003), 223–231.
24. Z. Mustafa, B. Sims, A new approach to generalized metric spaces, J. Nonlinear Convex Anal., **7** (2), (2006), 289–297.
25. T. Salat, On statistically convergent sequences of real numbers, Math. Slovaca., **30**, (1980), 139–150.
26. H. Steinhaus, Sur la convergence ordinaire et la convergence asymptotique, Colloq. Math., **2**, (1951), 73-74.
27. A. Sahiner, M. Gürdal, F. K. Düden, Triple sequences and their statistical convergence, Selcuk J. Appl. Math., **8**(2), (2007), 49–55.
28. A. Wilansky, *Summability through Functional Analysis*, North-Holland Mathematics Studies 85, Amsterdam New York Oxford, (1984).
29. K. Zeller, W. Beekmann, *Theorie der Limiierungsverfahren* (2. Aufl.), Springer, Berlin, (1970).
30. M. A. Hamed, B. Cummins, A numerical integration formula for the solution of the singular integral equation for classical crack problems in plane and antiplane elasticity, Journal of King Saud University - Engineering Sciences., **3**, (1991), 217–230.
31. T. P. Becerra, S. S. Perales, J. J. Oliveros, The HK-Sobolev space and applications to one-dimensional boundary value problems, Journal of King Saud University., **32**(6), (2020), 2790–2796.
32. Judith V. Grabiner, Who Gave You the Epsilon? Cauchy and the Origins of Rigorous Calculus. The American Mathematical Monthly, **90**(3), (1983), 185–194.
33. Brian S. Thomson, Judith B. Bruckner, Andrew M. Bruckner. *Elementary Real Analysis*, Prentice Hall, Second Edition, (2008).
34. H. Lebesgue. Intégrale, longueur, aire. Annali di Matematica Pura ed Ap-plicata, **7**(1), (1902), 231–359.
35. H. Lebesgue, *Leçons sur l'intégration et la recherche des fonctions primi-tives*, Gauthier-Villars, Paris, (1904).
36. W. Liu, G. Ye, D. Zhao, D. F. M. Torres, Existence theorems for a nonlinear second-order distributional differential equation, Journal of King Saud University-Science., **30**, (2018), 527–530.
37. Inder K. Rana, *An Introduction to Measure and Integration*, American Math-ematical Society, second edition, Providence, RI, (2002).
38. O. Dovgoshey, O. Martio, V. I. Ryazanov, M. Vuorinen, The Cantor Function, Expositiones Mathematicae, **24**(1), (2006), 1–37.
39. T. Gowers, J. B. Green, I. Leader, *The Princeton Companion to Mathematics.*, Princeton University Press, New Jersey, (2008).
40. T. L. Gill, W. W. Zachary, *Functional Analysis and the Feynman operator Calculus*, Springer, New York, (2016).
41. H. Hoffmann, *Descriptive Characterisation of the Variational Henstock-Kurzweil-Stieltjes Integral and Applications*, Karlsruhe Institute of Technology,Faculty of Mathematics, Institute for Analysis, (2014).
42. E. Hille, R. S. Phillips, *Functional Analysis and Semigroups*. American Mathematical Society Colloquium Publications, **31**, AMS (1957).

43. J. Mehra, H. Rechenberg, *The Historical Development of Quan-tum Theory*, Springer-Verlag, New York, (1982).
44. P. Bullen, *Nonabsolute Integration in the Twentieth Century*, American Mathematical Society, AMS Special Session on Nonabsolute Integration, September 23-24, (2000).
45. J. Andersson. *Course Plan SF3626, Analysis for PhD Students*, KTH Royal Institute of Technology, (2016).
46. J. Bell, *Vitali Coverings on the Real Line*, University of Toronto, Deptof Math, March 9, (2016).
47. D. Larsson, *Generalized Riemann Integration: Killing Two Birds with OneStone?*, Linköping University, Dept of Math, June, (2013).
48. P. Y. Lee, Z. Dongsheng, Upper and Lower Henstock Integrals. RealAnalysis Exchange, **22**(2), (1996-97), 734–739.
49. L. Paxton, A Sequential Approach to the Henstock Integral.arXiv:1609.05454v1 [math.CA], 2016-09-18.
50. S. S. Perales, F. J. Mendoza-Torres, Boundary value problems for the Schrödinger equation involving the Henstock-Kurzweil integral, Czechoslovak Math. **70**, (2020), 519–537.
51. J. Wells, *Generalizations of the Riemann Integral: An Investigation of the Henstock Integral*, Whitman College, May 15, (2011).
52. T. Y. Lee, *Henstock-Kurzweil Integration on Euclidean Spaces*, World Scientific Publishing, Singapore, (2011).
53. T. Y. Lee, Every Absolutely Integrable Function is McShane Integrable: An Alternative Proof. Rocky Mountain Journal of Mathematics, **34**(4), (2004), 1353–1365.
54. Juan A. Escamilla Reyna, F. J. Mendoza Torres, N. CamposLeón, The Completion of the Henstock-Kurzweil Space: An Alternative Proof.International Journal of Mathematical Analysis, **6**(5), (2012), 209–214.
55. R. Výborný, Some Applications of Kurzweil-Henstock Integration.Mathematica Bohemica, **118**(4), (1993), 425–441.
56. R. G. Bartle, The Concept of "Negligible Variation", Real Analysis Ex-change, **23**(1), (1997-98), 47–48.
57. V. Rutherfoord, Y. Sagher, Negligable Variation and the Change of Variables Theorem, arXiv:1212.4574v1 [math.CA], 2012–12–19.
58. R. Dimartino, W. O. Urbina, Excursions on Cantor-like Sets.arXiv:1411.7110v2 [math.CA], 04–06–2017. References69
59. R. Dimartino, W. O. Urbina, On Cantor-like sets and Cantor-Lebesgue Singular Functions. arXiv:1403.6554v1 [math.CA], 26–03–2014.
60. B. Szokefalvi Nagy, *Introduction to Real Functions and Orthogonal Expan-sions*, Oxford University Press, New York, (1965).
61. H. Okamoto, M. Wunsch, A Geometric Construction of Continuous, Strictly Increasing Singular Functions. Proceedings of the Japan Academy,ser. A., **83**(7), (2007), 114–118.
62. John D. De Pree, Charles W. Swartz, *Introduction to Real Analysis*, John Wiley and Sons, New York, (1988).
63. M. Al Ummam, C. Rini Indrati, Basic Properties and Some Convergence Theorems of the DL-Integral, AIP Conference Proceedings 2192,050004, (2019).
64. E. Pap, *Handbook of Measure Theory*, North Holland, **1**, Novi Sad,Yugoslavia, (2002).
65. A. Julia, *Stoke's Theorem and Integration on Integral Currents*, Paris Diderot University, Sorbonne Paris Cité University, (2018).
66. E. Mema, Equiintegrability and Controlled Convergence for the Henstock-Kurzweil Integral, International Mathematical Forum, **8**(19), (2013), 913–919.

67. P. Mohanty, Erik Talvila, A Product Convergence Theoremfor Henstock-Kurzweil Integrals, Real Analysis Exchange, **29**(1), (2003-24), 199–204.
68. P. Y. Lee, On $ACG_*$ Functions, Real Analysis Exchange, **15**(2), (1989-90), 754–759.
69. B. Bongiorno, L. Di Piazza, V. Skvortsov, A New FullDescriptive Characterization of Denjoy-Perron Integral, Real Analysis Ex-change, **21**(2), (1995-96), 656–663.
70. S. Saks, *Theory of the Integral*, Hafner Publishing Company, secondrevised edition, New York, (1937).
71. R. A. Gordon, *The Integrals of Lebesgue, Denjoy, Perron and Henstock*, American Mathematical Society, Providence, RI, (1994). References70
72. P. Y. Lee, The Integral á la Henstock, Scientiae Mathematicae Japonicae, **20**, (2007), 763–771.
73. T. S. Chew, B. van Brunt, G. C. Wake, First-Order Partial Differential Equations and Henstock-Kurzweil Integrals, Differential Integral Equations, **10**(5), (1997), 947–960.
74. D. A. L. Velasco, M. M. Moriń-Castillo, J. J. Oliveros-Oliveros, T. P. Becerra, J. A. Escamilla-Reyna, Numerical solution of some differential equations with Henstock-Kurzweil functions, J. Funct. Spaces, **9**, (2019), 1–9.

# 2 $q-$Bell Sequence Spaces and Associated Matrix Transformation

*Koray Ibrahim Atabey, Muhammed Cinar, Mikail Et*

In this work, we build $q-$analogue of the $q-$Bell matrix $\widehat{B}_q = \widehat{B}_{nk}(q)$ defined by

$$\widehat{B}_q = \widehat{B}_{nk}(q) = \begin{cases} \dfrac{\binom{n}{k} q^k B_q(k)}{B_q(n+1)}, & if\ 0 \leq k \leq n\ (k,n \in \mathbb{N}) \\ 0 & \text{, otherwise.} \end{cases}$$

Later, we define the sequence spaces $c_0(\widehat{B}_q)$, $c(\widehat{B}_q)$, $\ell_\infty(\widehat{B}_q)$, $\ell_p(\widehat{B}_q)$ $(1 \leq p < \infty)$ and study some topological properties and give some inclusion relations for these spaces. In addition, we build a base for the space $\ell_p(\widehat{B}_q)$, compute $\alpha$-, $\beta$-, $\gamma$-duals of the same space, characterize some matrix classes and give some geometric properties.

## 2.1 INTRODUCTION

Bell numbers [6] are a sequence of numbers that play an important role in combinatorics. Bell numbers are used to represent the numbers of many objects. They express the number of bins in a cluster containing a certain number of elements.

The reason why $q$ is used so frequently instead of $x$ is that if $q$ is understood as a power of a prime number, then these $q-$generalizations commonly count useful properties of finite dimensional vector spaces over a finite field of order $q$. When $q = 1$, the $q-$generalizations are simply polynomials that take the value of the classical number.

$q-$Bell numbers are regarded as Bell numbers' $q-$deformations. In the same way that ordinary Bell numbers are sums of ordinary Stirling numbers, $q-$Bell numbers are just sums of $q-$Stirling numbers.

$q-$Bell numbers are used in many mathematical fields, especially in fields such as quantum mathematics, statistical mechanics, and combinatorics and also play an important role in quantum mechanics in topics such as matrix theory, quantum groups and quantum field theory.

DOI: 10.1201/9781003596578-2

Several scientists have worked on this topic, including Carlitz [8], Balogh and Schlosser [4], Wagner [37, 38], Andrews [2], Kac and Cheung [21], Atabey et al. [3], Karakaş [26].

Now let's start with some notations.
$q-$analogue of $k \in \mathbb{N}$ is defined to be $[k]_q = k$ and

$$[k]_q = [k] = \frac{1-q^k}{1-q} = 1+q+q^2+\ldots+q^{k-1}$$

for $q = 1$ and $q \in \mathbb{R}^+ \backslash \{1\}$, respectively.

$$[k]! = \begin{cases} [1][2]\ldots[k], & n > 0 \\ 1 & , k = 0 \end{cases} \quad \text{and} \quad \begin{bmatrix} k \\ j \end{bmatrix} = \frac{[k]!}{[j]![k-j]!},$$

define $q-$factorial and $q-$combination, respectively. Two Pascal rules with $q$ are defined to be

$$\begin{bmatrix} k \\ j \end{bmatrix} = \begin{bmatrix} k-1 \\ j-1 \end{bmatrix} + q^j \begin{bmatrix} k-1 \\ j \end{bmatrix} \quad \text{and} \quad \begin{bmatrix} k \\ j \end{bmatrix} = q^{k-j} \begin{bmatrix} k-1 \\ j-1 \end{bmatrix} + \begin{bmatrix} k-1 \\ j \end{bmatrix},$$

where $0 \leq j \leq k-1$.

Carlitz [[8], Equation (3.3)] provided the explicit formula that follows:

$$S_q(n,k) = \frac{1}{[k]!} \sum_{j=0}^{k} (-1)^j q^{\binom{j}{2}} \begin{bmatrix} k \\ j \end{bmatrix} [k-j]_q^n.$$

Similarly, the $q-$Bell numbers are defined [37] by

$$B_q(n) = \sum_{k=0}^{n} S_q(n,k)$$

and the recurrence of $B_n(q)$ are defined to be

$$B_q(n+1) = \sum_{j=0}^{n} \binom{n}{j} q^j B_q(j) \qquad \forall\, n \in \mathbb{N}. \quad [37]$$

Authors like Cinar and Et [9], Demiriz and Şahin [10], Yaying et al. [34, 35, 36], Selmanogullari et al. [32], Aktuglu and Bekar [1], Mursaleen et al. [20], Bekar [7] have all used $q-$numbers in summability theory.

Let $\omega$ denote the set of all sequence spaces. The subspaces $\ell_\infty, c, c_0$ and $\ell_p$ of $\omega$ are defined respectively as bounded, convergent, null and $p-$absolutely summable sequences space. The spaces $\ell_\infty, c, c_0$ are Banach spaces under normed by

$$\|x\|_\infty = \sup_{k \in \mathbb{N}} |x_k| \quad \text{for } k \in \mathbb{N}$$

and the $\ell_p$ $(1 \leq p < \infty)$ is Banach space normed by

$$\|x\|_p = \left( \sum_k |x_k|^p \right)^{\frac{1}{p}}.$$

Furthermore, we denote by $\ell_1$, $cs$, $bs$ and $bv_p$ the spaces of all absolutely convergent series, convergent series, bounded series and $p-$bounded variation, respectively.

Let $D,E \subset \omega$ and $S=(s_{nk})$ be a real infinite matrix. The matrix $S$ defines a matrix transformation from $D$ to $E$ if for every sequence $x \in D$,

$$Sx=(S_n(x))=\left(\sum_{k=1}^{\infty} s_{nk}x_k\right) \in E$$

for each $n \in \mathbb{N}$. $(D,E)$ represents the family of all matrices that map from $D$ to $E$.

$$D_S=\{x \in \omega : Sx \in D\} \tag{2.1}$$

is a sequence space that defines the $S$'s matrix domain $D_S$ in a sequence space $D$.

Moreover, the sequence space $D_S$ is a $BK-$space normed by $\|x\|_{D_S}=\|Sx\|_D$ if $S$ is a triangular matrix and $D$ is a $BK-$space.

## 2.2 MAIN RESULT

In this part, we present the sequence sets $c_0(\widehat{B}_q)$, $c(\widehat{B}_q)$, $\ell_\infty(\widehat{B}_q)$ and $\ell_p(\widehat{E}_q)$ for $(1 \le p < \infty)$ utilizing the $q-$analogue of the $q-$Bell triangular matrix, with $q>0$. In addition, we build the Schauder basis of the space $\ell_p(\widehat{B}_q)$ and prove some inclusion theorems.

For $n \in \mathbb{N}$ and $q>0$, $B_q(n)$ is the $n$th $q-$Bell number, then

$$\widehat{B}_q=\widehat{B}_{nk}(q)=\begin{cases} \dfrac{\binom{n}{k}q^kB_q(k)}{B_q(n+1)} & , 0 \le k \le n \\ 0 & , \text{otherwise} \end{cases}$$

$$=\begin{bmatrix} \frac{B_q(0)}{B_q(1)} & 0 & 0 & 0 & 0 & \cdots \\ \frac{B_q(0)}{B_q(2)} & \frac{qB_q(1)}{B_q(2)} & 0 & 0 & 0 & \cdots \\ \frac{B_q(0)}{B_q(3)} & \frac{2qB_q(1)}{B_q(3)} & \frac{q^2B_q(2)}{B_q(3)} & 0 & 0 & \cdots \\ \frac{B_q(0)}{B_q(4)} & \frac{3qB_q(1)}{B_q(4)} & \frac{3q^2B_q(2)}{B_q(4)} & \frac{q^3B_q(3)}{B_q(4)} & 0 & \cdots \\ \frac{B_q(0)}{B_q(5)} & \frac{4qB_q(1)}{B_q(5)} & \frac{6q^2B_q(2)}{B_q(5)} & \frac{4q^3B_q(3)}{B_q(5)} & \frac{q^4B_q(4)}{B_q(5)} & \cdots \\ \vdots & \vdots & \vdots & \vdots & \vdots & \ddots \end{bmatrix}$$

defines $q-$analogue of the $\widehat{B}_q$ Bell matrix.

Let $y_n=(\widehat{B}_q)_n(x)$ be the matrix transformation for $n \in \mathbb{N}$ represented as

$$y_n=\frac{1}{B_q(n+1)}\sum_{k=0}^{n}\binom{n}{k}q^kB_q(k)x_k \tag{2.2}$$

and define the sequence spaces $c_0(\widehat{B}_q)$, $c(\widehat{B}_q)$, $\ell_\infty(\widehat{B}_q)$ and $\ell_p(\widehat{B}_q)$ for $(1 \leq p < \infty)$ by

$$c_0(\widehat{B}_q) = \left\{ x = (x_n) \in \omega : \lim_{n\to\infty} (\widehat{B}_q)_n(x) = 0 \right\},$$

$$c(\widehat{B}_q) = \left\{ x = (x_n) \in \omega : \lim_{n\to\infty} (\widehat{B}_q)_n(x) \text{ exists} \right\},$$

$$\ell_\infty(\widehat{B}_q) = \left\{ x = (x_n) \in \omega : \sup_{n\in\mathbb{N}} \left| \frac{1}{B_q(n+1)} \sum_{k=0}^{n} \binom{n}{k} q^k B_q(k) x_k \right| < \infty \right\},$$

$$\ell_p(\widehat{B}_q) = \left\{ x = (x_n) \in \omega : \sum_n \left| \frac{1}{B_q(n+1)} \sum_{k=0}^{n} \binom{n}{k} q^k B_q(k) x_k \right|^p < \infty \right\}.$$

The sequence spaces $\ell_p(\widehat{B}_q)$, $\ell_\infty(\widehat{B}_q)$, $c_0(\widehat{B}_q)$ and $c(\widehat{B}_q)$ can be redefined by

$$\ell_p(\widehat{B}_q) = (\ell_p)_{\widehat{B}_q} \ (1 \leq p < \infty), \ \ell_\infty(\widehat{B}_q) = (\ell_\infty)_{\widehat{B}_q}, \tag{2.3}$$

$$c_0(\widehat{B}_q) = (c_0)_{\widehat{B}_q} \text{ and } c(\widehat{B}_q) = (c)_{\widehat{B}_q} \tag{2.4}$$

respectively, when the notation of (2.1) is taken into account.

**Theorem 2.2.1.** *$\ell_p(\widehat{B}_q)$ is a BK−space normed by*

$$\|(\widehat{B}_q)_n(x)\|_{\ell_p} = \|x\|_{\ell_p(\widehat{B}_q)} = \left( \sum_n \left| (\widehat{B}_q)_n(x) \right|^p \right)^{\frac{1}{p}}, \quad (1 \leq p < \infty)$$

*and $\ell_\infty(\widehat{B}_q)$, $c_0(\widehat{B}_q)$ and $c(\widehat{B}_q)$ are BK−spaces normed by*

$$\|(\widehat{B}_q)_n(x)\|_{\ell_\infty} = \|x\|_{\ell_\infty(\widehat{B}_q)} = \|x\|_{c_0(\widehat{B}_q)} = \|x\|_{c(\widehat{B}_q)} = \sup_{n\in\mathbb{N}} \left| (\widehat{B}_q)_n(x) \right|.$$

**Proof:** Because (2.3) and (2.4) hold, $\ell_\infty$ and $\ell_p$ are $BK-$spaces in terms of their natural norms, and the matrix $\widehat{B}_q$ is a triangle; Theorem 4.3.12 of [39, p. 63] gives the fact that the spaces $\ell_p(\widehat{B}_q)$ and $\ell_\infty(\widehat{B}_q)$ are $BK-$spaces with the given norms, where $(1 \leq p < \infty)$.

According to the Theorem 4.3.2 of [39, p. 61], the spaces $c_0(\widehat{B}_q)$ and $c(\widehat{B}_q)$ are $BK-$spaces with the given norms.

**Theorem 2.2.2.** *$\ell_p(\widehat{B}_q)$ $(1 \leq p \leq \infty)$ is linearly isomorphic to $\ell_p$.*

**Proof:** Proving that $V : \ell_p(\widehat{B}_q) \to \ell_p$, $(x \to y = Vx = \widehat{B}_q x \in \ell_p)$, is a bijection and linear transformation suffices for $(1 \leq p \leq \infty)$.

$V$ is obviously linear. In addition, $V$ is implied to be injective as it is evident that $x = 0$ whenever $Vx = 0$.

For surjective, let us consider $y = (y_n) \in \ell_p$. Since

$$y_n = \frac{1}{B_q(n+1)} \sum_{k=0}^{n} \binom{n}{k} q^k B_q(k) x_k,$$

we get

$$x_k = \frac{1}{q^k B_q(k)} \sum_{i=0}^{k} (-1)^{k-i} \binom{k}{i} B_q(i+1) y_i.$$

Now we have, for $(1 \leq p < \infty)$

$$\begin{aligned}
\|x\|_{\ell_p(\widehat{B}_q)} &= \left( \sum_n \left| (\widehat{B}_q)_n(x) \right|^p \right)^{\frac{1}{p}} = \left( \sum_n \left| \frac{1}{B_q(n+1)} \sum_{k=0}^{n} \binom{n}{k} q^k B_q(k) x_k \right|^p \right)^{\frac{1}{p}} \\
&= \left( \sum_n \left| \frac{1}{B_q(n+1)} \sum_{k=0}^{n} \binom{n}{k} q^k B_q(k) \left( \frac{1}{q^k B_q(k)} \sum_{i=0}^{k} (-1)^{k-i} \binom{k}{i} B_q(i+1) y_i \right) \right.\right. \\
&= \left( \sum_n |y_n|^p \right)^{\frac{1}{p}} = \|y\|_p < \infty
\end{aligned}$$

and for $p = \infty$,

$$\|x\|_{\ell_\infty(\widehat{B}_q)} = \|y\|_\infty = \sup_{n \in \mathbb{N}} \left| (\widehat{B}_q)_n(x) \right| < \infty.$$

Now the proof is finished.

**Theorem 2.2.3.** *The $c_0(\widehat{B}_q)$ and $c(\widehat{B}_q)$ are linearly isomorphic to the spaces $c_0$ and $c$, respectively.*

**Proof:** It is sufficient to show that $T : c_0(\widehat{B}_q) \to c_0$, where $(x \to y = Tx = \widehat{B}_q x \in c_0)$, is a bijection and linear transformation.

Clearly, $T$ is linear. Additionally, as it is obvious that $x = 0$ whenever $Tx = 0$, $T$ is implied to be injective.

Let us assume $y = (y_n) \in c_0$ for surjective. Let us consider

$$\lim_{n \to \infty} (\widehat{B}_q)_n(x) = \lim_{n \to \infty} \left( \frac{1}{B_q(n+1)} \sum_{k=0}^{n} \binom{n}{k} q^k B_q(k) x_k \right)$$

and let's write $x_k$

$$\begin{aligned}
&= \lim_{n \to \infty} \left( \frac{1}{B_q(n+1)} \sum_{k=0}^{n} \binom{n}{k} q^k B_q(k) \frac{1}{q^k B_q(k)} \sum_{i=0}^{k} (-1)^{k-i} \binom{k}{i} B_q(i+1) y_i \right) \\
&= \lim_{n \to \infty} (y_n) = 0.
\end{aligned}$$

This means $x \in c_0(\widehat{B}_q)$.

To prove "$c(\widehat{B}_q)$ are linearly isomorphic to the spaces $c$", it is sufficient to replace "0" with another the limit real value.

**Theorem 2.2.4.** *The inclusions* $c_0 \subset c_0(\widehat{B}_q)$ *and* $c \subset c(\widehat{B}_q)$ *hold for* $q \geq 1$ *and strictly hold* $0 < q < 1$*, respectively.*

**Proof:** Let's get $x \in c_0$, so it means that $x \to 0$. The Silverman-Toeplitz criteria are satisfied for each $q \geq 1$ by the matrix $\widehat{B}_q$;

$$\sup_{n\in\mathbb{N}} \sum_k \left|\widehat{B}_{nk}(q)\right| = \sup_{n\in\mathbb{N}} \left( \left| \frac{1}{B_q(n+1)} \sum_{k=0}^{n} \binom{n}{k} q^k B_q(k) \right| \right) = 1 < \infty,$$

$$\lim_{n\to\infty} \widehat{B}_{nk}(q) = 0,$$

$$\lim_{n\to\infty} \sum_k \widehat{B}_{nk}(q) = \lim_{n\to\infty} \left( \frac{1}{B_q(n+1)} \sum_{k=0}^{n} \binom{n}{k} q^k B_q(k) \right) = 1.$$

Since the method $\widehat{B}_q$ is regular, we get $x \in c_0(\widehat{B}_q)$ whenever $\widehat{B}_q x \to 0$.

The matrix $\widehat{B}_q$ is conservative for $0 < q < 1$. Then $\widehat{B}_q x \in c$ when $x \in c$.

To demonstrate how strict the inclusion is, we get $x = (x_k) = \left( \frac{(-1)^k}{q^k B_q(k)} \right)$ for $q \to 0^+$ and $n \in \mathbb{N}$. It is obvious that this sequence is divergent. As a result, $x \notin c$ and $x \notin c_0$, however,

$$\lim_{n\to\infty} \left( \frac{1}{B_q(n+1)} \sum_{k=0}^{n} \binom{n}{k} q^k B_q(k) \left( \frac{(-1)^k}{q^k B_q(k)} \right) \right) = 0.$$

This indicates that both $x \in c_0(\widehat{B}_q)$ and $x \in c(\widehat{B}_q)$.

**Theorem 2.2.5.** *The inclusion* $\ell_p \subset \ell_p(\widehat{B}_q)$ *holds for* $1 \leq p \leq \infty$ *and* $q \geq 1$.

**Proof:** It is sufficient to prove that there is a constant $K > 0$ that satisfies the inequality $\|x\|_{\ell_p(\widehat{B}_q)} \leq K\|x\|_p$ for every $x \in \ell_p$.

For $(1 < p < \infty)$ and $q \geq 1$, let's get $x \in \ell_p$. Applying From Hölder's inequality for $\forall n \in \mathbb{N}$, we have

$$\begin{aligned}
\sum_{n=0}^{\infty} \left|(\widehat{B}_q)_n(x)\right|^p &= \sum_{n=0}^{\infty} \left| \sum_{k=0}^{n} \frac{\binom{n}{k} q^k B_q(k)}{B_q(n+1)} x_k \right|^p \\
&\leq \sum_{n=0}^{\infty} \left( \sum_{k=0}^{n} \frac{\binom{n}{k} q^k B_q(k)}{B_q(n+1)} |x_k| \right)^p \\
&\leq \sum_{n=0}^{\infty} \left( \sum_{k=0}^{n} \frac{\binom{n}{k} q^k B_q(k)}{B_q(n+1)} |x_k|^p \right) \left( \sum_{k=0}^{n} \frac{\binom{n}{k} q^k B_q(k)}{B_q(n+1)} \right)^{p-1} \\
&= \sum_{n=0}^{\infty} \left( \sum_{k=0}^{n} \frac{\binom{n}{k} q^k B_q(k)}{B_q(n+1)} |x_k|^p \right) \\
&= \sum_{k=0}^{\infty} |x_k|^p \left( \binom{n}{k} q^k B_q(k) \sum_{n=k}^{\infty} \frac{1}{B_q(n+1)} \right).
\end{aligned}$$

So this means

$$\|x\|_{\ell_p(\widehat{B}_q)} \leq K\|x\|_p, \tag{2.5}$$

where $K = \sup_{k\in\mathbb{N}} \left( \sum_{n=k}^{\infty} \frac{\binom{n}{k} q^k B_q(k)}{B_q(n+1)} \right)$. Also for $p = \infty$, we take $x_k \in \ell_\infty$. Then, for all $k \in \mathbb{N}$, there exists a constant $K > 0$ such that $|x_k| \leq K$. Therefore, using the triangle inequality

$$|(\widehat{B}_q)_n(x)| \leq \sum_{k=0}^{n} \frac{\binom{n}{k} q^k B_q(k)}{B_q(n+1)} |x_k| \leq \sum_{k=0}^{n} \frac{\binom{n}{k} q^k B_q(k)}{B_q(n+1)} K = K.$$

So $x \in \ell_p(\widehat{B}_q)$.

Similarly, the inequality (2.5) is easily demonstrated for $p = 1$, hence we omit the details. Therefore, for $1 \leq p \leq \infty$, the inclusion $\ell_p \subset \ell_p(\widehat{B}_q)$ holds.

**Theorem 2.2.6.** *If $1 \leq p < s$, then $\ell_p(\widehat{B}_q) \subset \ell_s(\widehat{B}_q)$.*

**Proof:** Let's take $1 \leq p < s$ and $x \in \ell_p(\widehat{B}_q)$. Then, we can deduce from Theorem 2.2.1 that $y \in \ell_p$ as defined in transformation (2.2). Consequently, utilizing the established $\ell_p \subset \ell_s$ inclusion, we can determine that $y \in \ell_s$. As a result, $x \in \ell_s(\widehat{B}_q)$, demonstrating the validity of the inclusion $\ell_p(\widehat{B}_q) \subset \ell_s(\widehat{B}_q)$. This finalizes the proof.

**Theorem 2.2.7.** *For all $q \in \mathbb{R}^+$, the inclusion $c_0(\widehat{B}_q) \subset c(\widehat{B}_q)$ is strict.*

**Proof:** Let's take $x \in c_0(\widehat{B}_q)$. Then we have $\lim_{n\to\infty}(\widehat{B}_q)_n(x) = 0$. Since $\lim_{n\to\infty}(\widehat{B}_q)_n(x)$ exists, we can write $x \in c(\widehat{B}_q)$.

For strict of inclusion, let's consider the sequence $x = (x_k) = (1^k)$ and calculate the transformation sequence

$$\lim_{n\to\infty} \left( \frac{1}{B_q(n+1)} \sum_{k=0}^{n} \binom{n}{k} q^k B_q(k)(1^k) \right) = 1 \neq 0.$$

so this means $x \in c_0(\widehat{B}_q) \backslash c(\widehat{B}_q)$.

**Theorem 2.2.8.** *For all $q \in \mathbb{R}^+$, the inclusion $\ell_p(\widehat{B}_q) \subset \ell_\infty(\widehat{B}_q)$ is strict.*

**Proof:** Let's take $x = (x_k) \in \ell_p(\widehat{B}_q)$. Then we have $\widehat{B}_q x \in \ell_p$. Since $\ell_p \subset \ell_\infty$, we can conclude that $\widehat{B}_q x \in \ell_\infty$. So $x = (x_k) \in \ell_\infty(\widehat{B}_q)$ which means that $\ell_p(\widehat{B}_q) \subset \ell_\infty(\widehat{B}_q)$. Let's consider the sequence $x = (x_k) = (1^k)$ for strict of the inclusion. Since

$$\sup_{n\in\mathbb{N}} \left| \frac{1}{B_q(n+1)} \sum_{k=0}^{n} \binom{n}{k} q^k B_q(k)(1^k) \right| = 1,$$

we have $x \in \ell_\infty(\widehat{B}_q)$. But since

$$\sum_n \left| \frac{1}{B_q(n+1)} \sum_{k=0}^{n} \binom{n}{k} q^k B_q(k)(1^k) \right|^p = \sum_n |1|^p \to \infty$$

we have $x \notin \ell_p(\widehat{B}_q)$.

**Theorem 2.2.9.** *Considering for $p \in [1,\infty] \setminus \{2\}$, $\ell_p(\widehat{B}_q)$ doesn't represent a Hilbert space.*

**Proof:** For proof, we take the following sequences

$$v = (v_n) = \left( \frac{B_1(q)}{B_0(q)}, \frac{-B_1(q)+B_2(q)}{qB_1(q)}, \frac{B_1(q)-2B_2(q)}{q^2B_2(q)}, \frac{-B_1(q)+3B_2(q)}{q^3B_3(q)}, \dots \right)$$

and

$$u = (u_n) = \left( \frac{B_1(q)}{B_0(q)}, \frac{-B_1(q)-B_2(q)}{qB_1(q)}, \frac{B_1(q)+2B_2(q)}{q^2B_2(q)}, \frac{-B_1(q)-3B_2(q)}{q^3B_3(q)}, \dots \right).$$

These sequences have the following transformations $\widehat{B}_q$:

$$\widehat{B}_q v = (1,1,0,0,\dots) \text{ and } \widehat{B}_q u = (1,-1,0,0,\dots)$$

respectively.

In the other hand we have $\widehat{B}_q(v-u) = (0,2,0,0,\dots)$ and $\widehat{B}_q(v+u) = (2,0,0,0,\dots)$. As a result, we get the phrase for $p \in [1,\infty] \setminus \{2\}$ as follows

$$\|v+u\|^2_{\ell_p(\widehat{B}_q)} + \|v-u\|^2_{\ell_p(\widehat{B}_q)} = 8 \neq 2^{2+\frac{2}{p}} = 2\left( \|v\|^2_{\ell_p(\widehat{B}_q)} + \|u\|^2_{\ell_p(\widehat{B}_q)} \right).$$

This shows that the parallelogram equality cannot be satisfied by the $\ell_p(\widehat{B}_q)$ space's norm.

**Theorem 2.2.10.** *The space $\ell_p(\widehat{B}_q)$ $(1 \leq p \leq \infty)$ isn't absolute type.*

**Proof:** Let's select $x = (1,-1,0,0,\dots)$ to show that it is not an absolute type. After we compute $\widehat{B}_q x$ and $\widehat{B}_q|x|$ by

$$\widehat{B}_q x = \left( \frac{B_0(q)}{B_1(q)}, \frac{B_0(q)-qB_1(q)}{B_2(q)}, \frac{B_0(q)-2qB_1(q)}{B_3(q)}, \frac{B_0(q)-3qB_1(q)}{B_4(q)}, \dots \right)$$

and

$$\widehat{F}_q|x| = \left( \frac{B_0(q)}{B_1(q)}, \frac{B_0(q)+qB_1(q)}{B_2(q)}, \frac{B_0(q)+2qB_1(q)}{B_3(q)}, \frac{B_0(q)+3qB_1(q)}{B_4(q)}, \dots \right),$$

respectively. We have $\|x\|_{\ell_p(\widehat{B}_q)} \neq \||x|\|_{\ell_p(\widehat{B}_q)}$. The proof is completed.

We now present $\ell_p(\widehat{B}_q)$ $(1 \leq p < \infty)$'s a basis.

**Theorem 2.2.11.** *For $1 \leq p < \infty$ and each fixed $k \in \mathbb{N}$, define a sequence $\eth^{(k)} \in \ell_p(\widehat{B}_q)$ as*

$$(\eth^{(k)})_n = \begin{cases} \dfrac{(-1)^{n-k}\binom{n}{k} B_q(k+1)}{q^n B_q(n)} & ,k \leq n \\ 0 & ,k > n \end{cases} \quad (n \in \mathbb{N}). \tag{2.6}$$

*Next, $\{\eth^{(k)}\}_{k\in\mathbb{N}}$ is a Schauder basis for the space $\ell_p(\widehat{B}_q)$ and each $u \in \ell_p(\widehat{B}_q)$ has a unique representation of the form*

$$u = \sum_k (\widehat{B}_q)_k(u)\eth^{(k)} \tag{2.7}$$

*for each $k \in \mathbb{N}$.*

**Proof:** Let's consider $1 \leq p < \infty$. Afterwards, it is clear by (2.6) that $(\widehat{B}_q)(\eth^{(k)}) = e^{(k)} \in \ell_p$ and hence $\eth^{(k)} \in \ell_p(\widehat{B}_q)$.

Let we take $u \in \ell_p(\widehat{B}_q)$ and for each non-negative integer $m$ and all $k \in \mathbb{N}$ we set

$$u^{(m)} = \sum_k (\widehat{B}_q)_k(u)\eth^{(k)}.$$

Then we get

$$\widehat{B}_q(u^{(m)}) = \sum_{k=0}^{m} (\widehat{B}_q)_k(u)(\widehat{B}_q)(\eth^{(k)}) = \sum_{k=0}^{m} (\widehat{B}_q)_k(u)e^{(k)}$$

and hence

$$(\widehat{B}_q)_n(u - u^{(m)}) = \begin{cases} 0 & ,(0 \leq n \leq m) \\ (\widehat{B}_q)_n(u) & ,(n > m) \end{cases} \quad (n,m \in \mathbb{N}). \tag{2.8}$$

For each given $\varepsilon > 0$, there exist a $m_0 \in \mathbb{N}$ such that

$$\sum_{k=m_0+1}^{\infty} \left|(\widehat{B}_q)_n(u)\right|^p = \left(\frac{\varepsilon}{2}\right)^p.$$

Consequently, we have for each $m > m_0$

$$\begin{aligned} \|u - u^{(m)}\|_{\ell_p(\widehat{B}_q)} &= \left(\sum_{k=m+1}^{\infty} \left|(\widehat{B}_q)_n(u)\right|^p\right)^{\frac{1}{p}} \\ &\leq \left(\sum_{k=m_0+1}^{\infty} \left|(\widehat{B}_q)_n(u)\right|^p\right)^{\frac{1}{p}} \leq \frac{\varepsilon}{2} < \varepsilon, \end{aligned}$$

showing that $\lim_{m\to\infty} \|u - u^{(m)}\|_{\ell_p(\widehat{B}_q)} = 0$ and consequently, $u$ is expressed as in (2.7).

In order to demonstrate the expression's uniqueness, we assume that another variant exists (2.7) such as

$$u = \sum_k (\widehat{L}_q)_k(u)\eth^{(k)}.$$

By utilizing the continuous transform $V$ we have demonstrated its isomorphism in Theorem 2.2.2, the equation that follows can be written as

$$(\widehat{B}_q)_n(u) = \sum_k (\widehat{L}_q)_k(u)(\widehat{B}_q)_n(\eth^{(k)}) = \sum_k (\widehat{L}_q)_k(u)\delta_{nk} = (\widehat{L}_q)_n(u).$$

This demonstrates the uniqueness of form (2.7). The proof is now complete.

## 2.3 $\alpha-$, $\beta-$, $\gamma-$ DUAL SPACE OF $\ell_P(\widehat{B}_Q)$

The $\alpha$-, $\beta$-, and $\gamma$-duals of the $\ell_p(\widehat{B}_q)$ are given in this section. We shall look at the situation $1 < p \leq \infty$ since $p = 1$ may be proved by analogy. We provide Stieglitz and Tietz [33] lemmas in order to prove Theorem 2.3.5 and Theorem 2.3.6. A number of authors have used the domain of specific matrices to study matrix transforms, sequence and dual spaces in the literature, such as [5, 12, 13, 23, 16, 17, 24, 25, 27].

**Lemma 2.3.1.** $S = (s_{nk}) \in (\ell_p, \ell_1) \Leftrightarrow$ *for* $(1 < p \leq \infty)$ *and* $(1/p + 1/r) = 1$

$$\sup_{K\in F} \sum_k \left| \sum_{n\in K} s_{nk} \right|^r < \infty.$$

**Lemma 2.3.2.** $S = (s_{nk}) \in (\ell_p, c) \Leftrightarrow$

$$\lim_{n\to\infty} s_{nk} \text{ exists for each } k \in \mathbb{N} \tag{2.9}$$

$$\sup_{n\in\mathbb{N}} \sum_k |s_{nk}|^r < \infty \quad \text{for } (1/p + 1/r) = 1. \tag{2.10}$$

**Lemma 2.3.3.** $S = (s_{nk}) \in (\ell_\infty, c) \Leftrightarrow$

$$\lim_{n\to\infty} \sum_k |s_{nk}| = \sum_k \left| \lim_{n\to\infty} s_{nk} \right| \tag{2.11}$$

*and (2.9) holds.*

**Lemma 2.3.4.** $S = (s_{nk}) \in (\ell_p, \ell_\infty) \Leftrightarrow$ *with* $(1 < p \leq \infty)$, *(2.10) holds.*

**Theorem 2.3.5.** *For* $1 < p \leq \infty$, *the set*

$$d_1(q) = \left\{ s = (s_k) \in \omega : \sup_{K\in F} \sum_k \left| \sum_{n\in K} \frac{(-1)^{n-k}\binom{n}{k} B_q(k+1)}{q^n B_q(n)} s_n \right|^r < \infty \right\}$$

*is the* $\alpha-$*dual of* $\ell_p(\widehat{B}_q)$.

**Proof:** Let's get any sequence $s = (s_n) \in \omega$ and $1 < p \leq \infty$. We define the matrix $T = (t_{nk})$ as

$$t_{nk} = \begin{cases} \frac{(-1)^{n-k}\binom{n}{k} B_q(k+1)}{q^n B_q(n)} s_n, & if \;\; 0 \leq k \leq n \\ 0 & , if \;\; k > n. \end{cases}$$

Additionally, we write $y = \widehat{B}_q x$ for each $x = (x_n) \in \omega$. Next, it tracks by (2.2) that

$$s_n x_n = \sum_{k=0}^{n} \frac{(-1)^{n-k}\binom{n}{k} B_q(k+1)}{q^n B_q(n)} a_n y_k = T_n(y) \quad (n \in \mathbb{N}). \tag{2.12}$$

Therefore, we deduce by (2.12) that $sx = (s_n x_n) \in \ell_1$ whenever $x \in \ell_p(\widehat{B}_q)$ if and only if $Ty \in \ell_1$ whenever $y \in \ell_p$.

Lemma 2.3.1 allows us to see that

$$\sup_{K \in F} \sum_k \left| \sum_{n \in K} \frac{(-1)^{n-k}\binom{n}{k} B_q(k+1)}{q^n B_q.(n)} s_n \right|^r < \infty.$$

So $\left(\ell_p(\widehat{B}_q)\right)^\alpha = d_1(q)$.

**Theorem 2.3.6.** *Define the following sets $d_2(q)$, $d_3(q)$, $d_4(q)$ as:*

$$d_2(q) = \left\{ s = (s_k) \in \omega : \sum_{j=k}^{\infty} \frac{(-1)^{j-k}\binom{j}{k} B_q(k+1)}{q^j B_q(j)} s_j \quad exists, \; \forall \, k \in \mathbb{N} \right\},$$

$$d_3(q) = \left\{ s = (s_k) \in \omega : \sup_{n \in \mathbb{N}} \sum_{k=1}^{n} \left| \sum_{j=k}^{n} \frac{(-1)^{j-k}\binom{j}{k} B_q(k+1)}{q^j B_q(j)} s_j \right|^r < \infty \right\},$$

$$d_4(q) = \left\{ s = (s_k) \in \omega : \lim_{n \to \infty} \sum_{k=1}^{n} \left| \sum_{j=k}^{n} \frac{(-1)^{j-k}\binom{j}{k} B_q(k+1)}{q^j B_q(j)} s_j \right| \right.$$
$$\left. = \sum_k \left| \sum_{j=k}^{\infty} \frac{(-1)^{j-k}\binom{j}{k} B_q(k+1)}{q^j B_q(j)} s_j \right| < \infty \right\}.$$

*Then we have*

*a)* $d_2(q) \cap d_3(q) = \left(\ell_p(\widehat{B}_q)\right)^\beta$ *for* $1 < p \leq \infty$,

*b)* $d_2(q) \cap d_4(q) = \left(\ell_\infty(\widehat{B}_q)\right)^\beta$.

**Proof:** Let's get $s = (s_k) \in \omega$, we define matrix $M = (m_{nk})$ by

$$m_{nk} = \begin{cases} \sum_{j=k}^{n} \frac{(-1)^{j-k}\binom{j}{k}B_q(k+1)}{q^j B_q(j)} s_j, & if \ 1 \le k \le n \\ 0 & , if \ k > n. \end{cases}$$

Think about the following equality now

$$\begin{aligned} \sum_{k=1}^{n} s_k x_k &= \sum_{k=1}^{n} s_k \left( \sum_{j=1}^{n} \frac{(-1)^{j-k}\binom{j}{k}B_q(k+1)}{q^j B_q(j)} y_j \right) \\ &= \sum_{k=1}^{n} \left( \sum_{j=k}^{n} \frac{(-1)^{j-k}\binom{j}{k}B_q(k+1)}{q^j B_q(j)} s_j \right) y_k = M_n(y). \end{aligned} \tag{2.13}$$

Afterwards, we obtain from Lemma 2.3.2 using (2.2) that $My \in c$ whenever $y = (y_k) \in \ell_p$ if and only if $sx = (s_k x_k) \in cs$ whenever $x \in \ell_p(\widehat{B}_q)$. Hence, $(s_k) \in d_2(q)$ and $(s_k) \in d_3(q)$ are defined by (2.9) and (2.10), respectively if and only if $(s_k) \in \left(\ell_p(\widehat{B}_q)\right)^{\beta}$. Consequently, we reach to $\left(\ell_p(\widehat{B}_q)\right)^{\beta} = d_2(q) \cap d_3(q)$.

Lemma 2.3.3 can be used in place of Lemma 2.3.2 in the situation of $p = \infty$, and identical techniques can be applied.

**Theorem 2.3.7.** *For* $1 < p \le \infty$, $\left(\ell_p(\widehat{B}_q)\right)^{\gamma} = d_3(q)$.

**Proof:** Let the sequence $s = (s_k)$, the matrix $M = (m_{nk})$ and equality (8.1.4) be defined as in the proof of Theorem 13.

Then, we use (2.2) to derive from Lemma 2.3.4 that $My \in \ell_\infty$ whenever $y = (y_k) \in \ell_p$ if and only if $sx = (s_k x_k) \in bs$ whenever $x \in \ell_p(\widehat{B}_q)$. Hence, $(s_k) \in d_3(q)$ is defined by (2.10) if and only if $(s_k) \in \left(\ell_p(\widehat{B}_q)\right)^{\gamma}$. Consequently, we reach to $\left(\ell_p(\widehat{B}_q)\right)^{\gamma} = d_3(q)$.

## 2.4 MATRIX TRANSFORMATIONS RELATED TO THE SPACE $\ell_P(\widehat{B}_Q)$

We characterize the matrix classes $(\ell_p(\widehat{B}_q), D)$, where $1 < p \le \infty$ and $D \in \{\ell_\infty, \ell_1, c, c_0\}$ in this section. We use

$$\tilde{s}_{nk} = \sum_{j=k}^{\infty} \frac{(-1)^{j-k}\binom{j}{k}B_q(k+1)}{q^j B_q(j)} s_{nj}$$

for the purpose of brevity.

The following lemma serves as the foundation for our findings.

**Lemma 2.4.1.** *Let $\mu$ be an arbitrary subset of $\omega$, $U$ a triangular matrix, $V$ its inverse, and $\lambda$ a $FK-$space. Define $W^{(n)} = (w_{mk}^{(n)})$ and $W = (w_{nk})$ by*

$$W^{(n)} = w_{mk}^{(n)} = \begin{cases} \sum_{j=k}^{m} s_{nj} v_{jk}, & if \ \ 1 \le k \le m \\ 0 & , if \ \ k > m \end{cases} \quad , \quad W = (w_{nk}) = \sum_{j=k}^{\infty} s_{nj} v_{jk},$$

*respectively (see Theorem 4.1 of [28]).*

*After we obtain $S = (s_{nk}) \in (\lambda_U, \mu)$ if and only if $W^{(n)} = (w_{mk}^{(n)}) \in (\lambda, c)$ and $W = (w_{nk}) \in (\lambda, \mu)$.*

*Next, we list the conditions that need to be satisfied in order to accept the validity of our results:*

$$\sup_{m\in\mathbb{N}} \sum_{k=1}^{m} \left| \sum_{j=k}^{m} \frac{(-1)^{j-k}\binom{j}{k} B_q(k+1)}{q^j B_q(j)} s_{nj} \right|^r < \infty, \tag{2.14}$$

$$\lim_{m\to\infty} \sum_{j=k}^{m} \frac{(-1)^{j-k}\binom{j}{k} B_q(k+1)}{q^j B_q(j)} s_{nj} = \tilde{s}_{nk}, \ \ each \ k,n \in \mathbb{N}, \tag{2.15}$$

$$\lim_{m\to\infty} \sum_{k=1}^{m} \left| \sum_{j=k}^{m} \frac{(-1)^{j-k}\binom{j}{k} B_q(k+1)}{q^j B_q(j)} s_{nj} \right| = \sum_{k} |\tilde{s}_{nk}| \ \ each \ n \in \mathbb{N}, \tag{2.16}$$

$$\sup_{m\in\mathbb{N}} \sum_{k} |\tilde{s}_{nk}|^r < \infty, \tag{2.17}$$

$$\sup_{N\in F} \sum_{k} \left| \sum_{n\in\mathbb{N}} \tilde{s}_{nk} \right|^r < \infty, \tag{2.18}$$

$$\lim_{n\to\infty} \tilde{s}_{nk} = \tilde{\Lambda}_k; \quad k \in \mathbb{N}, \tag{2.19}$$

$$\lim_{n\to\infty} \sum_{k} |\tilde{s}_{nk}| = \sum_{k} |\tilde{\Lambda}_k|, \tag{2.20}$$

$$\lim_{n\to\infty} \sum_{k} \tilde{s}_{nk} = 0, \tag{2.21}$$

$$\sup_{n,k\in\mathbb{N}} |\tilde{s}_{nk}| < \infty, \tag{2.22}$$

$$\sup_{k,m\in\mathbb{N}} \left| \sum_{j=k}^{m} \frac{(-1)^{j-k}\binom{j}{k} B_q(k+1)}{q^j B_q(j)} s_{nj} \right| < \infty, \tag{2.23}$$

$$\sup_{k\in\mathbb{N}} \sum_{n} |\tilde{s}_{nk}| < \infty, \tag{2.24}$$

$$\sup_{N,K\in F} \left| \sum_{n\in N} \sum_{k\in K} \tilde{s}_{nk} \right| < \infty. \tag{2.25}$$

Hence, we may obtain the following findings from [33] using Lemma 2.4.1 and the given conditions.

**Theorem 2.4.2.** *a)* $S=(s_{nk})\in\left(\ell_1(\widehat{B}_q),\ell_\infty\right)\Leftrightarrow(2.15),(2.22)$ *and* $(2.23)$ *hold.*

*b)* $S=(s_{nk})\in\left(\ell_1(\widehat{B}_q),c\right)\Leftrightarrow(2.15),(2.19),(2.22)$ *and* $(2.23)$ *hold.*

*c)* $S=(s_{nk})\in\left(\ell_1(\widehat{B}_q),c_0\right)\Leftrightarrow(2.15)$*, with* $\tilde{\alpha}_k=0,(2.19),(2.22)$ *and* $(2.23)$ *hold.*

*d)* $S=(s_{nk})\in\left(\ell_1(\widehat{B}_q),\ell_1\right)\Leftrightarrow(2.15),(2.23)$ *and* $(2.24)$ *hold.*

**Theorem 2.4.3.** *For* $1<p<\infty$,

*a)* $S=(s_{nk})\in\left(\ell_p(\widehat{B}_q),\ell_\infty\right)\Leftrightarrow(2.14),(2.15)$ *and* $(2.17)$ *hold.*

*b)* $S=(s_{nk})\in\left(\ell_p(\widehat{B}_q),c\right)\Leftrightarrow(2.14),(2.15),(2.17)$ *and* $(2.19)$ *hold.*

*c)* $S=(s_{nk})\in\left(\ell_p(\widehat{B}_q),c_0\right)\Leftrightarrow(2.14),(2.15),(2.17)$ *and with* $\tilde{\alpha}_k=0$ $(2.19)$ *hold.*

*d)* $S=(s_{nk})\in\left(\ell_p(\widehat{B}_q),\ell_1\right)\Leftrightarrow(2.14),(2.15)$ *and* $(2.18)$ *hold.*

**Theorem 2.4.4.** *a)* $S=(s_{nk})\in\left(\ell_\infty(\widehat{B}_q),\ell_\infty\right)\Leftrightarrow(2.15),(2.16)$ *and in case* $r=1$ $(2.17)$ *hold.*

*b)* $S=(s_{nk})\in\left(\ell_\infty(\widehat{B}_q),c\right)\Leftrightarrow(2.15),(2.16),(2.19)$ *and* $(2.20)$ *hold.*

*c)* $S=(s_{nk})\in\left(\ell_\infty(\widehat{B}_q),c_0\right)\Leftrightarrow(2.15),(2.16)$ *and* $(2.21)$ *hold.*

*d)* $S=(s_{nk})\in\left(\ell_\infty(\widehat{B}_q),\ell_1\right)\Leftrightarrow(2.15),(2.16)$ *and* $(2.25)$ *hold.*

## 2.5 THE $\ell_P(\widehat{B}_Q)$'S CERTAIN GEOMETRIC CHARACTERISTICS

One of the important properties of Banach spaces in functional analysis is their geometric properties. We refer to [11, 15, 14, 18, 19, 22, 29, 30] for further information. We will discuss the $\ell_p(\widehat{B}_q)$'s some geometric properties for $1<p<\infty$ in this section.

If every bounded sequence $(x_n)$ in $D$ admits a subsequence $(r_n)$ such that the $\{t_k(r)\}$ is convergent in the norm in $D$, a Banach space $D$ has the Banach-Saks property (see [19]), where

$$\{t_k(r)\}=\frac{1}{k+1}(r_0+r_1+\ldots+r_k)\quad(k\in\mathbb{N}). \tag{2.26}$$

A Banach space $D$ has the weak Banach-Saks property for given any weakly null sequence $(s_n)\subset D$ if there exists a subsequence $(r_n)$ of $(s_n)$ such that the $\{t_k(r)\}$ is strongly convergent to zero.

The coefficient that Garcia-Falset provides in [14] is as follows:

$$R(D)=\sup\left\{\liminf_{n\to\infty}\|s_n-s\|:(s_n)\subset B(D),s_n\stackrel{w}{\to}s,s\in B(D)\right\}, \tag{2.27}$$

where $B(D)$ denotes the unit ball of $D$.

**Remark 2.5.1.** *For $R(D) < 2$, a Banach space $D$ has the weak fixed point property [15].*

For $\forall\, n \in \mathbb{N}$, $1 < p < \infty$ and some $M > 0$, if every weakly null sequence $(s_k)$ has a subsequence $(s_{k_l})$ such that

$$\left\| \sum_{l=0}^{n} s_{k_l} \right\| < M(n+1)^{1/p}, \tag{2.28}$$

a Banach space has the Banach-Saks type $p$ or the property $(BS)_p$ (see [29]).

The $\ell_p(\widehat{B}_q)$'s some geometric properties, where $1 < p < \infty$, allow us to calculate the following conclusions.

**Theorem 2.5.2.** *The $\ell_p(\widehat{B}_q)$ $1 < p < \infty$ has the Banach-Saks type $p$.*

**Proof:** Let we get $(\varepsilon_n)$ sequence such that $(\varepsilon_n) > 0$ for every $n \in \mathbb{N}$ and $\sum_{n=1}^{\infty} \varepsilon_n \leq \frac{1}{2}$, and additionally, we take a weakly null sequence $(s_n)$ in $B(\ell_p(\widehat{B}_q))$. Set $r_0 = s_0 = 0$ and $r_1 = s_{n_1} = s_1$. After, a $u_1 \in \mathbb{N}$ exists such that

$$\left\| \sum_{i=u_1+1}^{\infty} r_1(i) e^{(i)} \right\|_{\ell_p(\widehat{B}_q)} < \varepsilon_1. \tag{2.29}$$

There is an $n_2 \in \mathbb{N}$ such that

$$\left\| \sum_{i=0}^{u_1} s_n(i) e^{(i)} \right\|_{\ell_p(\widehat{B}_q)} < \varepsilon_1, \tag{2.30}$$

when $n \geq n_2$, because $(s_n)$ is a weakly null sequence implies $s_n \to 0$ coordinatewise. Set $r_2 = s_{n_2}$. Then there exists an $u_2 > u_1$ such that

$$\left\| \sum_{i=u_2+1}^{\infty} r_2(i) e^{(i)} \right\|_{\ell_p(\widehat{B}_q)} < \varepsilon_2. \tag{2.31}$$

Applying once more the understanding that $s_n \to 0$ coordinatwise, there exists an such that $n_3 > n_2$

$$\left\| \sum_{i=0}^{u_2} s_n(i) e^{(i)} \right\|_{\ell_p(\widehat{B}_q)} < \varepsilon_2, \tag{2.32}$$

when $n \geq n_3$.

Two increasing subsequences $(u_i)$ and $(n_i)$ may be identified if we carry out this procedure further, such that

$$\left\| \sum_{i=0}^{u_j} s_n(i) e^{(i)} \right\|_{\ell_p(\widehat{B}_q)} < \varepsilon_j, \tag{2.33}$$

for each $n \geq n_{j+1}$ and

$$\left\| \sum_{i=u_j+1}^{\infty} r_j(i)e^{(i)} \right\|_{\ell_p(\widehat{B}_q)} < \varepsilon_j. \tag{2.34}$$

where $r_j = s_{n_j}$. Thus,

$$\begin{aligned} \left\| \sum_{j=0}^{n} r_j \right\|_{\ell_p(\widehat{B}_q)} &= \left\| \sum_{j=0}^{n} \left( \sum_{i=0}^{u_{j-1}} r_j(i)e^{(i)} + \sum_{i=u_{j-1}+1}^{u_j} r_j(i)e^{(i)} + \sum_{i=u_j+1}^{\infty} r_j(i)e^{(i)} \right) \right\|_{\ell_p(\widehat{B}_q)} \\ &\leq \left\| \sum_{j=0}^{n} \left( \sum_{i=u_{j-1}+1}^{u_j} r_j(i)e^{(i)} \right) \right\|_{\ell_p(\widehat{B}_q)} + 2\sum_{j=0}^{n} \varepsilon_j. \end{aligned}$$

The other side, $\|s\|_{\ell_p(\widehat{B}_q)} \leq 1$ is apparent. Thus, we have that

$$\begin{aligned} &\left\| \sum_{j=0}^{n} \left( \sum_{i=u_{j-1}+1}^{u_j} r_j(i)e^{(i)} \right) \right\|_{\ell_p(\widehat{B}_q)}^p = \\ &= \sum_{j=0}^{n} \sum_{i=u_{j-1}+1}^{u_j} \left| \frac{1}{B_q(i+1)} \sum_{k=0}^{i} \binom{i}{k} q^k B_q(k).r_j(i) \right|^p \\ &\leq \sum_{j=0}^{n} \sum_{i=0}^{\infty} \left| \frac{1}{B_q(i+1)} \sum_{k=0}^{i} \binom{i}{k} q^k B_q(k).r_j(i) \right|^p \leq (n+1). \end{aligned}$$

Hence, it can be obtained that

$$\left\| \sum_{j=0}^{n} \left( \sum_{i=u_{j-1}+1}^{u_j} r_j(i)e^{(i)} \right) \right\|_{\ell_p(\widehat{B}_q)} \leq (n+1)^{\frac{1}{p}}.$$

Utilizing the knowledge that $1 \leq (n+1)^{\frac{1}{p}}$ for $1 < p < \infty$ and $\forall\, n \in \mathbb{N}$, we get

$$\left\| \sum_{j=0}^{n} r_j \right\|_{\ell_p(\widehat{B}_q)} \leq (n+1)^{\frac{1}{p}} + 1 \leq 2(n+1)^{\frac{1}{p}}.$$

As a consequence, $\ell_p(\widehat{B}_q)$ has the Banach-Saks type $p$. The proof is finished now.

**Remark 2.5.3.** *$R(\ell_p(\widehat{B}_q)) = R(\ell_p) = 2^{\frac{1}{p}}$, because $\ell_p(\widehat{B}_q)$ is linearly isomorphic to $\ell_p$.*

We derive the following theorem from Remarks 2.5.1 and 2.5.3.

**Theorem 2.5.4.** *The space $\ell_p(\widehat{B}_q)$ $(1 < p < \infty)$ satisfies the weak fixed point property.*

## CONCLUSION

The triangular matrix with $q-$Bell numbers is used in this article to define the sequence spaces $c_0(\widehat{B}_q)$, $c(\widehat{B}_q)$, $\ell_\infty(\widehat{B}_q)$ and $\ell_p(\widehat{B}_q)$ $(1 \leq p < \infty)$. The inclusion connections between these spaces were significantly influenced by the variability of $q$. Next, the $\ell_p(\widehat{B}_q)$ space's topological and a few geometric properties were investigated.

The $q-$Bell numbers, which play a significant role in algebra, were moved to the area of sequence spaces and summability, which is an invention. Furthermore, this work provides a platform for further sequence space research; for example, one may expand on the findings by utilizing $q-$Genocchi, $q-$Bernoulli and $(p,q)-$Fibonacci sequence spaces.

## REFERENCES

1. Aktuglu H., Bekar S., On $q-$Cesáro matrix and $q-$statistical convergence. Journal of Computational and Applied Mathematics, **235**(16), (2011), 4717–4723.
2. Andrews G. E., $q-$Catalan identities, In The legacy of Alladi Ramakrishnan in the mathematical sciences. Springer New York, (2010), 183–190.
3. Atabey K. I., Çinar M., Et M., $q-$Fibonacci sequence spaces and related matrix transformations, Journal of Applied Mathematics and Computing, **69**(2), (2023), 2135–2154.
4. Zsófia R. Kereskényiné Balogh., Michael J. Schlosser, $q-$Stirling numbers of the second kind and $q-$Bell numbers for graphs, Electronic notes in discrete mathematics, **54**, (2016), 361–366.
5. Basar F., Dutta H., *Summable spaces and their duals, matrix transformations and geometric properties*, CRC Press (2020).
6. Bell E. T., Exponential numbers, The American Mathematical Monthly, **41**, (1934), 411–419.
7. Bekar S., *$q-$Matrix summability methods*, PhD Thesis. Eastern Mediterranean University (EMU), (2011).
8. Carlitz L., Generalized Stirling numbers, Combinatorial Analysis Notes. Duke University. (1968), 8–15.
9. Çinar M., Et M., $q-$Double Cesaro Matrices and $q-$Statistical Convergence of Double Sequences, National Academy Science Letters, **43**(1), (2020), 73–76.
10. Demiriz S., Sahin A., $q-$Cesáro sequence spaces derived by $q-$analogues, Advances in Mathematics, **5**(2), (2016), 97–110.
11. Diestel J., *Sequences and series in Banach spaces*, Springer Science and Business Media., **92**, (2012).
12. Et M., Colak R., On some generalized difference sequence spaces, Soochow Journal of Mathematics, **21**(4), (1995), 377–386.
13. Et M., Esi A., On Köthe-Toeplitz duals of generalized difference sequence spaces, Bull. Malaysian Mathematical Sciences Society, **23**, (2000), 25–32.
14. Garcia-Falset J., Stability and fixed points for nonexpansive mappings. Houston Journal of Mathematics, **20**(3), (1994), 495–506.
15. Garcia-Falset J., The fixed point property in Banach spaces with the NUS-property. Journal of Mathematical Analysis and Applications, **215**(2), (1997), 532–542.
16. Gökce F., Sarigöl M. A., Series spaces derived from absolute Fibonacci summability and matrix transformations, Bollettino dell'Unione Matematica Italiana, **13**(1), (2020), 29–38.
17. Gökce F., Sarigöl M. A., Some matrix and compact operators of the absolute Fibonacci series spaces, Kragujevac Journal of Mathematics, **44**(2), (2020), 273–286.

18. Hudzik H., Karakaya V., Mursaleen M., Simsek N., Banach-Saks Type and Gurarii Modulus of Convexity of Some Banach Sequence Spaces, In Abstract and Applied Analysis., **2014**, (2014), 1–9.
19. Mursaleen M., Basar F., Altay B., On the Euler sequence spaces which include the spaces $\ell_p$ and $\ell_\infty$, Nonlinear Analysis: Theory, Methods and Applications, **65**(3), (2006), 707–717.
20. Mursaleen M., Tabassum S., Fatma R., On $q-$Statistical Summability Method and Its Properties, Iranian Journal of Science and Technology, Transactions A: Science, **46**(2), (2022), 455–460.
21. Kac V., Cheung P., *Quantum Calculus*, Springer. New York. (2002).
22. Kananthai A., Musarleen M., Sanhan W., Suantai S., On property $(H)$ and rotundity of difference sequence spaces, Journal of Nonlinear and Convex analysis, **3**(3), (2002), 401–410.
23. Kara E. E., Some topological and geometrical properties of new Banach sequence spaces, Journal of Inequalities and Applications, **2013**(1), (2013), 1–15.
24. Kara E. E., Ilkhan M., Some properties of generalized Fibonacci sequence spaces, Linear and Multilinear Algebra, **64**(11), (2016), 2208–2223.
25. Kara M. I., Kara E. E., Matrix transformations and compact operators on Catalan sequence spaces, Journal of Mathematical Analysis and Applications, **498**(1), (2021), 124925.
26. Karakas M., On the sequence spaces involving bell numbers. Linear and Multilinear Algebra, (2022), 1–12.
27. Khan V. A., Tuba U., On paranormed Ideal convergent sequence spaces defined by Jordan totient function, Journal of Inequalities and Applications, **2021**(1), (2021), 1–16.
28. M. Kirisci, Basar F., Some new sequence spaces derived by the domain of generalized difference matrix, Computers and Mathematics with Applications, **60**(5), (2010), 1299–1309.
29. Knaust H., Orlicz sequence spaces of Banach-Saks type. Archive for Mathematics, **59**(6), (1992), 562–565.
30. Savas E., Karakaya V., Simsek N., Some $\ell(p)-$type new sequence spaces and their geometric properties, Abstract and Applied Analysis, **2009**, (2009): 696971.
31. Schur I., Ein Beitrag zur Additiven Zahlentheorie, Sitzungsber, Akad. Wissensch. Berlin, Phys. Math. Klasse. (1917), 302–321.
32. Selmanogullari T., Savas E., Rhoades B. E., On $q-$Hausdorff matrices. Taiwanese Journal of Mathematics, **15**(6), (2011), 2429–2437.
33. Stieglitz M., Tietz H., Matrix transformationen von Folgenräumen eine Ergebnisübersicht. Mathematische Zeitschrift, **154**, (1977), 1–16.
34. Yaying Y., Hazarika B., Mursaleen M., On sequence space derived by the domain of $q-$Cesáro matrix in $\ell_p$ space and the associated operator ideal. Journal of Mathematical Analysis and Applications, **493**(1), (2021), 1–17.
35. Yaying Y., Hazarika B., Mursaleen M., On generalized $(p,q)-$Euler matrix and associated sequence spaces. Journal Function Spaces, **2021**(1), (2021).
36. Yaying T., Hazarika B., Tripathy B. C., Mursaleen M., The Spectrum of Second Order Quantum Difference Operator, Symmetry, **557**, (2022), 14-3 (2022), 1–14.
37. Wagner G. C., Generalized Stirling and Lah numbers, Discrete Math, **160**, (1996), 199–218.
38. Wagner G. C., Partition statistics and $q-$Bell numbers $(q=-1)$, Journal of Integer Sequences, **7**(1), (2004), 2–3.
39. Wilansky A., *Summability through functional analysis*, Elsevier, (2000).

# 3 Almost Convergence via Bell Matrix

*Muhammet Cihat Dagli, Murat Karakas*

In this chapter, we introduce a new almost convergent sequence space derived from Bell matrix. We show that this new sequence space is linearly isomorphic to the space of all almost convergent sequences. Also, we express the $\beta$-dual of this new space. Furthermore, after defining the Bell core of a complex-valued sequence, we give several inclusion theorems for this core.

## 3.1 INTRODUCTION

A sequence space is a linear subspace of the space $\omega$ of all real-valued sequences. We denote the spaces of all bounded, convergent, null and absolutely $p$-summable real-valued sequences by $\ell_\infty, c, c_0$ and $\ell_p$, respectively. It is well known that these are Banach spaces according to the following norms

$$\|z\|_{\ell_\infty} = \|z\|_c = \|z\|_{c_0} = \sup_{k\in\mathbb{N}} |z_k|$$

and

$$\|z\|_{\ell_p} = \left(\sum_{k=0}^{\infty} |z_k|^p\right)^{1/p}.$$

Let $T = (t_{nk})$ be an infinite matrix and $T_n$ be the sequence in the $n$th row of the matrix $T$ for each $n \in \mathbb{N}$. Then, for a given $z = (z_k) \in \omega$ the sequence $Tz = \{(Tz)_n\} = \left\{\sum_k t_{nk} z_k\right\}$ is referred to as the $T-$transform of $z$ subject to the condition that the convergence of series is satisfied for each $n \in \mathbb{N}$. In the rest of the chapter, the notation $\sum_k$ will be used instead of $\sum_{k=0}^{\infty}$. $T$ is called as a matrix mapping from a sequence space $\Gamma$ to a sequence space $\Theta$ if the sequence $Tz$ exists and $Tz \in \Theta$ for all $z \in \Gamma$. Denote $(\Gamma, \Theta)$ as the family of all infinite matrices from $\Gamma$ to $\Theta$. If $Tz$ converges to $L$, then, we say that $z$ is $T-$summable to $L$ and so $L$ is $T-$limit of $z$. A matrix $T$ is a regular mapping from $\Gamma$ into $\Theta$, denoted by $T \in (\Gamma, \Theta)_{reg}$ if $T \in (\Gamma, \Theta)$ and $\lim_n (Tz)_n = \lim_n z_n$ for all $z = (z_n) \in \Gamma$.

DOI: 10.1201/9781003596578-3

Since Fast [26] and also Steinhaus [51] initiated the almost convergence as generalization of convergence of a sequence, this concept has been widely discussed by researchers and lots of substantial papers have surfaced in the literature.

A sequence $z = (z_n)$ is called as almost convergent to $L$ iff

$$\lim_{r\to\infty}\sum_{n=0}^{r}\frac{z_{m+n}}{r+1} = L, \text{ uniformly in } m$$

and is denoted by $h' - \lim_n z_n = L$. By $h'$ and $h'_0$, we denote the spaces of all almost convergent and almost null sequences, respectively, defined by

$$h' = \left\{ z = (z_m) \in \omega : \exists L \in \mathbb{C} \ni \lim_{r\to\infty}\sum_{n=0}^{r}\frac{z_{m+n}}{r+1} = L \text{ uniformly in } m \right\}$$

and

$$h'_0 = \left\{ z = (z_m) \in \omega : \lim_{r\to\infty}\sum_{n=0}^{r}\frac{z_{m+n}}{r+1} = 0 \text{ uniformly in } m \right\},$$

where $\mathbb{C}$ is the set of all complex numbers.

The matrix domain $h_T$ of an infinite matrix $T$ in the sequence space $h$ is defined by

$$h_T = \{z \in \omega : Tz \in h\},$$

which is a sequence space. For detailed discussions on almost convergence via matrix operators, one can refer to the studies [6, 9, 10, 12, 13, 14, 25, 37, 39, 46, 53].

The concept of Knopp Core (in short $\mathcal{K}$−core) of a sequence $z = (z_n) \in \omega$ is the intersection of all $Q_n$, the least convex closed region of the complex plane including $z_n, z_{n+1}, \ldots$ for $n = 1, 2, \ldots$ (cf. [16, p. 137]). The following relation is established in [47]

$$\mathcal{K} - core\,(z) = \cap_{u\in\mathbb{C}} M_z(u),$$

where $M_z(u)$ is the set given by

$$M_z(u) = \left\{ w \in \mathbb{C} : |w - u| \le \limsup_n |z_n - u| \right\},$$

The notion of natural density of $E \subset \mathbb{N}$, denoted as $\delta(E)$ can be defined by

$$\delta(E) = \lim_n \frac{1}{n}|k \le n : k \in E|,$$

in which $|A|$ is the number of elements of $A$. Given any $\varepsilon > 0$ such that

$$\delta(\{k : |z_k - z_0| \ge \varepsilon\}) = 0,$$

then, the sequence $z = (z_k)$ is referred to as statistically convergent to $z_0$. In this case, we write $sc - \lim z = z_0$ (see [52]). We remark that the spaces of all statistically convergent and statistically null convergent sequences will be denoted by $sc$ and $sc_0$, respectively.

Some applications of statistical convergence concept for approximating functions by certain types of Bernstein operators can be found in the nice papers [3, 44, 45, 50].

Fridy and Orhan [27] presented the notion of the statistically core and derived the following relation for any statistically bounded sequence $z$,

$$sc - core(z) = \cap_{u \in \mathbb{C}} D_z(u),$$

where $D_z(u) = \{w \in \mathbb{C} : |w-u| \leq sc - \limsup_n |z_n - u|\}$. For additional studies concerning with core theorems, the reader may consult to [15, 17, 18, 19] and cited references therein. Also, the monographs [7] and [41] and the papers [2, 11, 40, 42] of this research area can be investigated.

In recent years, several important sequence spaces have been developed as the domain of some special conservative/regular matrices involving special number sequences, for instance, Fibonacci numbers [34, 35, 43], Tribonacci number [54, 55], Catalan number [29, 33, 36], Schröder number [20, 21, 22] and certain functions [24, 28, 30, 31, 32]. Moreover, various significant features such as almost convergence and core theorems, matrix transformations, compact operators, special duals have been discussed.

In a recent paper [38], another noteworthy number, referred as Bell number $B_n$ is considered. These numbers describe the number of ways a set of $n$ elements can be partitioned into non-empty subsets. For instance, there are five ways the numbers $\{1,2,3\}$ can be partitioned:

$$\{1\} \cup \{2\} \cup \{3\}, \ \{1,2\} \cup \{3\}, \ \{1,3\} \cup \{2\}, \ \{1\} \cup \{2,3\}, \ \{1,2,3\}.$$

The first few Bell numbers $1,2,5,15,52,203,877,4140,...$ with the initial condition $B_0 = 1$. They can be defined as

$$B_{n+1} = \sum_{k=0}^{n} \binom{n}{k} B_k$$

and

$$B_n = \sum_{k=0}^{n} S_2(n,k),$$

where $S_2(n,k)$ is the Stirling numbers of the second kind, counts the set partitions of $\{1,2,3,...,n\}$ which consists of exactly $k$ subsets or parts. In [38], he introduced a new matrix $\tilde{B} = (\tilde{b}_{nk})$, whose entries are Bell numbers as

$$\tilde{b}_{nk} = \begin{cases} \binom{n}{k} \dfrac{B_k}{B_{n+1}}, & \text{if } 0 \leq k \leq n; \\ 0, & \text{if } k > n; \end{cases} \tag{3.1}$$

and observed that this matrix is regular. Also, the author offered the sequence spaces $\ell_p(\tilde{B})$ for $1 \leq p < \infty$ and $\ell_\infty(\tilde{B})$ as the set of all sequences whose $\tilde{B}-$transforms (see eq. 3.2, below) are in the spaces $\ell_p$ and $\ell_\infty$. In other words,

$$\ell_p\left(\tilde{B}\right) = \left\{z = (z_k) \in \omega : \tilde{B}z \in \ell_p\right\} \text{ for } 1 \leq p < \infty$$

and

$$\ell_\infty\left(\tilde{B}\right) = \left\{z = (z_k) \in \omega : \tilde{B}z \in \ell_\infty\right\}.$$

Besides, certain topological properties such as giving Schauder basis, determining the $\alpha-$, $\beta-$ and $\gamma-$duals, characterizing some matrix classes on $\ell_p\left(\tilde{B}\right)$ and geometric properties like uniform convexity, strict convexity, super reflexivity for the resulting spaces were given.

The sequence $y_n$ represents the $\tilde{B}-$transform of a sequence $z = (z_k)$ as

$$y_n = \left(\tilde{B}z\right)_n = \frac{1}{B_{n+1}} \sum_{k=0}^{n} \binom{n}{k} B_k z_k, \tag{3.2}$$

for all $n \in \mathbb{N}$ as the relation

$$z_k = \left(\tilde{B}^{-1}z\right)_k = \sum_{j=0}^{k} (-1)^{k-j} \binom{k}{j} \frac{B_{j+1}}{B_k} y_j \tag{3.3}$$

holds for all $k \in \mathbb{N}$.

The present chapter deals with a new almost convergent sequence space constructed by Bell matrix (3.1). We also demostrate that this novel space is linearly isomorphic to almost convergent sequence space and we offer its $\beta-$dual. Furthermore, after defining the Bell core of a complex-valued sequence, we characterize the class of matrices for which $\tilde{B}-core\,(Tz) \subseteq \mathcal{K}-core\,(z)$ and $\tilde{B}-core\,(Tz) \subseteq sc-core\,(z)$ for all $z \in \ell_\infty$.

## 3.2 A NEW ALMOST CONVERGENT SEQUENCE SPACE

Denote the Bell almost convergent sequence space as

$$B'\left(\tilde{B}\right) = \left\{ z = (z_m) \in \omega : \exists\, L \in \mathbb{C} \ni \lim_{r\to\infty} \sum_{n=0}^{r} \frac{y_{m+n}}{r+1} = L \text{ uniformly in } m \right\}.$$

**Theorem 3.2.1.** *The sequence space $B'\left(\tilde{B}\right)$ is linearly isomorphic to $h'$.*

**Proof:** Let the transformation $F : B'\left(\tilde{B}\right) \to h'$ be defined by $z \to y$ such that $Fz = \tilde{B}z$. Firstly, the linearity of $F$ is obvious due to the linearity of any matrix transformation. As well, it is satisfied that $z = \theta$ while $Fz = \theta$, i.e. $F$ is injective. For any sequence $y = (y_n) \in h'$, if we apply (3.3), i.e.,

$$z_n = \sum_{k=0}^{n} \left( (-1)^{n-k} \binom{n}{k} \frac{B_{k+1}}{B_n} \right) y_k,$$

then, we can write

$$\begin{aligned}
&\frac{1}{B_{n+j+1}}\sum_{k=0}^{n+j}\binom{n+j}{k}B_k z_k \\
&=\frac{1}{B_{n+j+1}}\sum_{k=0}^{n+j}\binom{n+j}{k}B_k\sum_{i=0}^{k}\left((-1)^{k-i}\binom{k}{i}\frac{B_{i+1}}{B_k}\right)y_i \\
&=\frac{1}{B_{n+j+1}}\sum_{i=0}^{n+j}\left(\sum_{k=0}^{n+j-i}\binom{n+j}{k+i}(-1)^k\right)B_{i+1}y_i \\
&=y_{n+j}.
\end{aligned}$$

Thus,

$$\frac{1}{r+1}\sum_{j=0}^{r}\frac{1}{B_{n+j+1}}\sum_{k=0}^{n+j}\binom{n+j}{k}B_k z_k=\frac{1}{r+1}\sum_{j=0}^{r}y_{n+j},$$

which implies that the following limit exists uniformly in $n$:

$$\lim_{r\to\infty}\frac{1}{r+1}\sum_{j=0}^{r}\left(\tilde{B}z\right)_{n+j}=\lim_{r\to\infty}\frac{1}{r+1}\sum_{j=0}^{r}y_{n+j}.$$

So that $z\in B'\left(\tilde{B}\right)$ and $Fz=y$, i.e. $F$ is surjective. The proof is complete.

Now, we recall the definition of $\beta-$dual. The $\beta-$dual of a sequence space $\Gamma$ can be defined as in the following

$$\Gamma^{\beta}=\{u=(u_n)\in\omega : zu=(z_n u_n)\in cs \text{ for all } z=(z_n)\in\Gamma\},$$

where $cs$ denotes the space of convergent series.

The following is necessary in order to state $\beta-$dual of the space $B'\left(\tilde{B}\right)$.

**Lemma 3.2.2.** *([48]) $T=(t_{nk})\in(h',c)$ iff*

$$\sup_{n\in\mathbb{N}}\sum_k|t_{nk}|<\infty$$

*is valid and there exists $\lambda_k\in\mathbb{R}$ and $\lambda\in\mathbb{R}$ such that*

$$\lim_{n\to\infty}t_{nk}=\lambda_k$$

*for each $k\in\mathbb{N}$ and*

$$\lim_{n\to\infty}\sum_k t_{nk}=\lambda$$

*and*

$$\lim_{n\to\infty}\sum_k|\Delta(t_{nk}-\lambda_k)|=0,\quad (\text{for each } k\in\mathbb{N})$$

*where $\Delta(t_{nk}-\lambda_k)=t_{nk}-\lambda_k-\left(t_{n,k+1}-\lambda_{k+1}\right)$.*

**Theorem 3.2.3.** *Define the sets* $\mu_1, \mu_2, \mu_3, \mu_4$ *and* $\mu_5$ *as*

$$\mu_1 = \left\{ b = (b_k) \in \omega : \sup_{r \in \mathbb{N}} \sum_{m=0}^{r} \left| \sum_{k=m}^{r} (-1)^{k-m} \binom{k}{m} \frac{B_{m+1}}{B_k} b_k \right| < \infty \right\},$$

$$\mu_2 = \left\{ b = (b_k) \in \omega : \lim_{r \to \infty} \sum_{k=m}^{r} (-1)^{k-m} \binom{k}{m} \frac{B_{m+1}}{B_k} b_k \text{ exists} \right\},$$

$$\mu_3 = \left\{ b = (b_k) \in \omega : \lim_{r \to \infty} \sum_{m=0}^{r} \left( \sum_{k=m}^{r} (-1)^{k-m} \binom{k}{m} \frac{B_{m+1}}{B_k} b_k \right) \text{ exists} \right\},$$

$$\mu_4 = \left\{ b = (b_k) \in \omega : \lim_{r \to \infty} \sum_{m=0}^{r} \left| \sum_{k=m}^{r} (-1)^{k-m} \binom{k}{m} \frac{B_{m+1}}{B_k} b_k \right| = 0 \right\},$$

$$\mu_5 = \left\{ b = (b_k) \in \omega : \lim_{r \to \infty} \sum_{m} \left| \Delta \left( \sum_{k=m}^{r} (-1)^{k-m} \binom{k}{m} \frac{B_{m+1}}{B_k} b_k - \lambda_m \right) \right| = 0 \right\}.$$

*Then, we have* $\left(B'\left(\tilde{B}\right)\right)^{\beta} = \mu_1 \cap \mu_2 \cap \mu_3 \cap \mu_4 \cap \mu_5$.

**Proof:** For any $b = (b_n) \in \omega$ and for all $r \in \mathbb{N}$ , it follows from (3.3) that

$$\begin{aligned} \sum_{m=0}^{r} b_m z_m &= \sum_{m=0}^{r} b_m \left( \sum_{n=0}^{m} (-1)^{m-n} \binom{m}{n} \frac{B_{n+1}}{B_m} y_n \right) \\ &= \sum_{m=0}^{r} \left( \sum_{n=m}^{r} (-1)^{n-m} \binom{n}{m} \frac{B_{m+1}}{B_n} b_n \right) y_m = (\mathscr{A} y)_r, \end{aligned}$$

where $\mathscr{A} = (a_{rm})$ is the matrix, defined by

$$a_{rm} = \begin{cases} \sum_{n=m}^{r} (-1)^{n-m} \binom{n}{m} \frac{B_{m+1}}{B_n} b_n, & \text{if } 0 \leq m \leq r; \\ 0, & \text{otherwise;} \end{cases}$$

for all $r, m \in \mathbb{N}$. Thus, we disclose that $bz = (b_m z_m) \in cs$ when $z = (z_m) \in B'\left(\tilde{B}\right)$ iff $\mathscr{A} y \in c$ for all $y = (y_m) \in h'$. Hence, $b = (b_m) \in \left(B'\left(\tilde{B}\right)\right)^{\beta}$ iff $\mathscr{A} \in (h', c)$. Applying Lemma 3.2.2 gives the result.

## 3.3 THE BELL CORE

In this section, we aim to define Bell core (or say $\tilde{B} - core$) of a complex-valued sequence and characterize the class of matrices that has the relation $\tilde{B} - core(Tz) \subseteq \mathscr{K} - core(z)$ and $\tilde{B} - core(Tz) \subseteq sc - core(z)$ for all bounded sequences $z$.

The concept of Knopp Core Theorem [16, p. 138] clarifies that $\mathscr{K} - core(Tz) \subseteq \mathscr{K} - core(z)$ for a positive matrix $T \in (c, c)_{reg}$ and all real-valued sequences $z$. One can suggest the the monograph [7] and the papers [4, 5, 8] for comprehensive background of this notion.

Before mentioning our results, we recall the notion of $\tilde{B} - core$ of $z$. Let $H_n$ be the least closed convex hull containing $\left(\tilde{B}z\right)_n, \left(\tilde{B}z\right)_{n+1}, \ldots$. Then, the $\tilde{B} - core$ of $z$ is given by

$$\tilde{B} - core(z) = \cap_{n=1}^{\infty} H_n.$$

In fact, the definition of the $\tilde{B}-core$ of $z$ can also be expressed by means of $\mathscr{K}-core$ of the sequence $\left(\tilde{B}z\right)_n$ as

$$\tilde{B}-core\left(z\right)=\cap_{p\in\mathbb{C}}\mathscr{V}_z\left(p\right),\quad (\text{for } z\in\ell_\infty),$$

in which

$$\mathscr{V}_z\left(p\right)=\left\{w\in\mathbb{C}:|w-p|\leq\limsup_n\left|\left(\tilde{B}z\right)_n-p\right|\right\}$$

for any complex number $p$.

Consider a matrix $\tilde{T}=(\tilde{t}_{nk})$ via $T=(t_{nk})$ as

$$\tilde{t}_{nk}=\frac{1}{B_{n+1}}\sum_{k=0}^{n}\binom{n}{k}B_kt_{nk},\quad \text{for all } n,k\in\mathbb{N}.$$

Let the spaces $c_0\left(\tilde{B}\right)$ and $c\left(\tilde{B}\right)$ consist of all sequences whose $\tilde{B}-$transforms are in the spaces $c_0$ and $c$, respectively. We require several lemmas in order to achieve the proofs of our theorems,

**Lemma 3.3.1.** $T=(t_{nk})\in\left(\ell_\infty,c\left(\tilde{B}\right)\right)$ *iff*

$$\left\|\tilde{T}\right\|=\sup_n\sum_k|\tilde{t}_{nk}|<\infty, \tag{3.4}$$

$$\lim_n\tilde{t}_{nk}=\tau_k,\ \textit{for each } k\in\mathbb{N} \tag{3.5}$$

*and*

$$\lim_n\sum_k|\tilde{t}_{nk}-\tau_k|=0. \tag{3.6}$$

**Lemma 3.3.2.** $T=(t_{nk})\in\left(c,c\left(\tilde{B}\right)\right)_{reg}$ *iff (3.4) is valid and (3.5) is also valid with* $\tau_k=0$ *for all* $k\in\mathbb{N}$ *and the following limit exists:*

$$\lim_n\sum_k\tilde{t}_{nk}=1. \tag{3.7}$$

**Lemma 3.3.3.** $T=(t_{nk})\in\left(sc\cap\ell_\infty,c\left(\tilde{B}\right)\right)_{reg}$ *if and only if* $T\in\left(c,c\left(\tilde{B}\right)\right)_{reg}$ *and*

$$\lim_n\sum_{k\in E,\ \delta(E)=0}|\tilde{t}_{nk}|=0. \tag{3.8}$$

**Proof:** Firstly, it follows from $c\subset sc\cap\ell_\infty$ that $T\in\left(c,c\left(\tilde{B}\right)\right)_{reg}$. For $z\in\ell_\infty$ and $E\subset\mathbb{N}$ with $\delta\left(E\right)=0$, take the sequence $\tilde{z}=(\tilde{z}_k)$ such that

$$\tilde{z}_k=\begin{cases}z_k, & \text{if } k\in E;\\ 0, & \text{if } k\notin E.\end{cases}$$

Observe that $T\tilde{z}\in c_0\left(\tilde{C}\right)$ whenever $\tilde{z}\in sc_0$. As well, from the relation

$$\sum_k\tilde{t}_{nk}\tilde{z}_k=\sum_{k\in E}\tilde{t}_{nk}z_k,$$

the matrix $W = (W_{nk})$ belongs to the class $\left(\ell_\infty, c\left(\tilde{B}\right)\right)$, where $W_{nk} = \tilde{t}_{nk}$ if $k \in E$ and $W_{nk} = 0$ if $k \notin E$. Therefore, employing Lemma 3.3.1 yields (3.8).

Conversely, if $z \in sc \cap \ell_\infty$ and $sc - \lim z = \kappa$, then, we have for any $\varepsilon > 0$,

$$\delta(E) = \delta(\{k : |z_k - \kappa| \geq \varepsilon\}) = 0.$$

Consider that

$$\sum_k \tilde{t}_{nk} z_k = \sum_k \tilde{t}_{nk}(z_k - \kappa) + \kappa \sum_k \tilde{t}_{nk}. \tag{3.9}$$

Then, in the light of the inequality

$$\sum_k \tilde{t}_{nk}(z_k - \kappa) \leq \|z\| \sum_{k \in E} |\tilde{t}_{nk}| + \varepsilon \left\|\tilde{T}\right\|,$$

passing the limit $n \to \infty$ in (3.9), and using (3.7) and (3.8) give that

$$\lim_n \sum_k \tilde{t}_{nk} z_k = \kappa,$$

from which we get $T \in \left(sc \cap \ell_\infty, c\left(\tilde{B}\right)\right)_{reg}$, as desired. So, the proof is complete.

**Lemma 3.3.4.** *([49]) Let $T = (t_{nk})$ be a matrix, satisfying that $\sum_k |t_{nk}| < \infty$ and $\lim_n t_{nk} = 0$. Then, there exists $z \in \ell_\infty$ with $\|z\| \leq 1$ such that*

$$\limsup_n \sum_k t_{nk} z_k = \limsup_n \sum_k |t_{nk}|.$$

We are ready to give some inclusion theorems.

**Theorem 3.3.5.** *Suppose that $T \in \left(c, c\left(\tilde{B}\right)\right)_{reg}$. For all $z \in \ell_\infty$, we have $\tilde{B} - core(Tz) \subseteq \mathscr{K} - core(z)$ if and only if*

$$\lim_n \sum_k |\tilde{t}_{nk}| = 1. \tag{3.10}$$

**Proof:** Since the matrix satisfies the conditions of Lemma 3.3.4, there exists $z \in \ell_\infty$ with $\|z\| \leq 1$ such that

$$\left\{l \in \mathbb{C} : |l| \leq \limsup_n \sum_k \tilde{t}_{nk} z_k\right\} = \left\{l \in \mathbb{C} : |l| \leq \limsup_n \sum_k |\tilde{t}_{nk}|\right\}.$$

Also, since $\mathscr{K} - core(z) \subset T_1(0)$, by

$$\left\{l \in \mathbb{C} : |l| \leq \limsup_n \sum_k |\tilde{t}_{nk}|\right\} \subseteq T_1(0) = \{l \in \mathbb{C} : |l| \leq 1\},$$

which gives (3.10). On the contrary, for $l \in \mathcal{K}-core(Tz)$, one arrives the inequality

$$\begin{aligned} |l-p| &\leq \limsup_n |\tilde{B}_n(Tz)-p| \\ &= \limsup_n \left|p-\sum_k \tilde{t}_{nk}z_k\right| \\ &\leq \limsup_n \left|\sum_k \tilde{t}_{nk}(p-z_k)\right| + \limsup_n |p| \left|1-\sum_k \tilde{t}_{nk}\right| \\ &= \limsup_n \left|\sum_k \tilde{t}_{nk}(p-z_k)\right|, \quad (\text{for } p \in \mathbb{C}). \end{aligned} \tag{3.11}$$

Now, if $\limsup_k |z_k - p| = \kappa$, then, for any $\varepsilon > 0$, we have $|z_k - p| \leq \kappa + \varepsilon$ whenever $k \geq k_0$. So, we have

$$\begin{aligned} \left|\sum_k \tilde{t}_{nk}(p-z_k)\right| &= \left|\sum_{k<k_0} \tilde{t}_{nk}(p-z_k) + \sum_{k\geq k_0} \tilde{t}_{nk}(p-z_k)\right| \\ &\leq \sup_k |p-z_k| \sum_{k<k_0} |\tilde{t}_{nk}| + (\kappa+\varepsilon) \sum_{k\geq k_0} |\tilde{t}_{nk}| \\ &\leq \sup_k |p-z_k| \sum_{k<k_0} |\tilde{t}_{nk}| + (\kappa+\varepsilon) \sum_k |\tilde{t}_{nk}|. \end{aligned} \tag{3.12}$$

If we combine (3.11) and (3.12), then, we have

$$|l-p| \leq \limsup_n \left|\sum_k \tilde{t}_{nk}(p-z_k)\right| \leq \kappa+\varepsilon,$$

from which we conclude that $l \in \mathcal{K}-core(z)$. Thus, the proof is complete.

**Theorem 3.3.6.** *Suppose that $T \in \left(sc \cap \ell_\infty, c\left(\tilde{B}\right)\right)_{reg}$. Then, $\tilde{B}-core(Tz) \subseteq sc-core(z)$ for all $z \in \ell_\infty$ if and only if (3.10) holds.*

**Proof:** Since $sc-core(z) \subseteq \mathcal{K}-core(z)$, it follows from our previous theorem that (3.10) holds. Conversely, consider $l \in \tilde{B}-core(Tz)$. If $sc-\limsup |z_k - p| = \kappa'$, then, it is aware from [23] that $\delta(\{k : |z_k - p| > \kappa' + \varepsilon\}) = 0$. Thus,

$$\begin{aligned} &\left|\sum_k \tilde{t}_{nk}(p-z_k)\right| \\ &= \left|\sum_{k\in E} \tilde{t}_{nk}(p-z_k) + \sum_{k\notin E} \tilde{t}_{nk}(p-z_k)\right| \\ &\leq \sup_k |p-z_k| \sum_{k\in E} |\tilde{t}_{nk}| + \left(\kappa'+\varepsilon\right) \sum_{k\notin E} |\tilde{t}_{nk} \\ &\leq \sup_k |p-z_k| \sum_{k\in E} |\tilde{t}_{nk}| + \left(\kappa'+\varepsilon\right) \sum_k |\tilde{t}_{nk}|. \end{aligned}$$

Applying (3.8) and (3.10) gives that

$$\limsup_n \left|\sum_k \tilde{t}_{nk}(p-z_k)\right| \leq \kappa'+\varepsilon \tag{3.13}$$

and so combining (3.11) and (3.13) yields that

$$|l-p| \leq sc-\limsup_{k}|z_k-p|.$$

Consequently, we have $l \in sc-core(z)$, as required.

## BIBLIOGRAPHY


1. Altay B., Basar F., On some Euler sequence spaces of non-absolute type, Ukrainian Math. J., **57**(1), (2005), 1–17.
2. Altay B., Basar F., Malkowsky E., Matrix transformations on some sequence spaces related to strong Cesáro summability and boundedness, Appl. Math. Comput., **211**(2009), 255–264.
3. Ansari K. J., Karaklic S., Özger F., Bivariate Bernstein-Kantorovich operators with a summability method and related GBS operators, Filomat, **36**(19), (2022), 6751–6765.
4. Aydin C., Basar F., On the new sequence spaces which include the spaces $c_0$ and $c$, Hokkaido Math. J., **33**(2), (2004), 383–398.
5. Aydin C., Basar F., Some new difference sequence spaces, Appl. Math. Comput., **157**(3), (2004), 677–693.
6. Basar F., $f-$conservative matrix sequences, Tamkang J. Math., **22**, (1991), 205–212.
7. Basar F., *Summability Theory and Its Applications*, 2nd ed., CRC Press/Taylor & Francis Group, Boca Raton · London · New York, (2022).
8. Basar F., Altay B., On the space of sequences of $p-$bounded variation and related matrix mappings, (English, Ukrainian summary) Ukrain. Mat. Zh., **55**(1), (2003), 108–118.
9. Basar F., Colak R., Almost-conservative matrix transformations, Turkish J. Math., **13**, (1989), 91–100.
10. Basar F., Kirisci M., Almost convergence and generalized difference matrix, Comput. Math. App, **61**(3), (2011), 602–611.
11. Basar F., Malkowsky E., Altay B., Matrix transformations on the matrix domains of triangles in the spaces of strongly $C_1-$summable and bounded sequences, Publ. Math. Debrecen., **73**(1-2), (2008), 193–213.
12. Basar F., Solak I., Almost-coercive matrix transformations, Rend. Mat. Appl., **11–2**, (1991), 249–256.
13. Candan M., Vector-valued FK-spaces defined by a modulus function and an infinite matrix, Thai J. Math., **12**(1), (2014), 155–165.
14. Candan M., Almost convergence and double sequential band matrix, Acta Math. Sci., **34**(2), (2014), 354–366.
15. Connor J., Fridy J. A., Orhan C., Core equality results for sequences, J. Math. Anal. Appl., **321**(2006), 515–523.
16. Cooke R. G., *Infinite matrices and sequence spaces*, MacMillan, New York, (1950).
17. Cakan C., Coskun H., Some new inequalities related to the invariant means and uniformly bounded function sequences, Appl. Math. Lett., **20**, (2007), 605–609.
18. Coskun H., Cakan C., A class of statistical and $\sigma-$conservative matrices, Czechoslovak Math. J., **55**, (2005), 791–801.
19. Coskun H., Cakan C., Mursaleen M., On the statistical and $\sigma-$cores, Stud. Math., **154**, (2003), 29–35.
20. Dagli M. C., A novel conservative matrix arising from Schröder numbers and its properties, Linear Multilinear Algebra, **71**(8), (2023), 1338–1351.

21. Dagli M. C., A new almost convergent sequence space defined by Schröder matrix, Linear Multilinear Algebra, (2022).
22. Dagli M. C., Matrix mappings and compact operators for Schröder sequence spaces, Turkish J. Math., **46**(6), (2022), 2304–2320.
23. Demirci K., A-statistical core of a sequence, Demonstr. Math., **33**, (2000), 43–51.
24. Demiriz S., Ilkhan M., Kara E. E., Almost convergence and Euler totient matrix, Ann. Funct. Anal., **11**, (2020), 604–616.
25. Duran J. P., Infinite matrices and almost convergence, Math. Z., **128**, (1972), 75–83.
26. Fast H., Sur la convergence statistique, Colloq. Math., **2**, (1951), 241–244.
27. Fridy A., Orhan C., Statistical core theorems, J. Math. Anal. Appl., **208**, (1997), 520–527.
28. Ilkhan M., Matrix domain of a regular matrix derived by Euler totient function in the spaces $c_0$ and $c$, Mediterr. J. Math., **17**(1), (2020), 215–227.
29. Ilkhan M., A new conservative matrix derived by Catalan numbers and its matrix domain in the spaces $c$ and $c_0$, Linear Multilinear Algebra., **68**(2), (2020), 417–434.
30. Ilkhan M., Kara E. E., A new Banach space defined by Euler totient matrix operator, Oper. Matrices., **13**(2), (2019), 527–544.
31. Ilkhan M., Kara E. E., Usta F., Compact operators on the Jordan totient sequence spaces, Math. Methods Appl. Sci., **44**(9), (2021), 7666–7675.
32. Ilkhan M., Simsek N., Kara E. E., A new regular infinite matrix defined by Jordan totient function and its matrix domain in $\ell_p$, Math. Methods Appl. Sci., **44**(9), (2020), 7622–7633.
33. Jasrotia S., Singh U. P., Raj K., Some new observations on Catalan almost convergent sequence spaces and the Catalan core, Acta Sci. Math. (Szeged), **87**(1-2), (2021), 153–163.
34. Kara E. E., Demiriz S., Some new paranormed difference sequence spaces derived by Fibonacci numbers, Miskolc Math. Notes., **16**(2), (2015), 907–923.
35. Kara E. E., Ilkhan M., Some properties of generalized Fibonacci sequence spaces, Linear Multilinear Algebra., **64**(11), (2016), 2208–2223.
36. Kara M. I., Kara E. E., Matrix transformations and compact operators on Catalan sequence spaces, J. Math. Anal. Appl., **498**(1), (2021) Article no: 124925.
37. Karaisa A., Özger F., On almost convergence and difference sequence spaces of order $m$ with core theorems, Gen. Math. Notes., **26**(1), (2015), 102–125.
38. Karakas M., On the sequence spaces involving bell numbers, Linear Multilinear Algebra, (2022), 2298–2309
39. Kilinc G., Candan M., Some generalized Fibonacci difference spaces defined by a sequence of modulus functions, Facta Univ. Ser. Math. Inf., **32**, (2017), 95–116.
40. Kirisci M., Basar F., Some new sequence spaces derived by the domain of generalized difference matrix, Comput. Math. Appl., **60**(5), (2010), 1299–1309.
41. Mursaleen M., Basar F., *Sequence Spaces: Topics in Modern Summability Theory*, CRC Press/Taylor & Francis Group, Series: Mathematics and Its Applications, Boca Raton · London · New York, (2020).
42. Mursaleen M., Basar F., Altay B., On the Euler sequence spaces which include the spaces $\ell_p$ and $\ell_\infty$, II. Nonlinear Anal., **65**(3), (2006), 707–717.
43. Ogur O., Superposition operators on sequence spaces $\ell_p(F)$ derived by using matrix of Fibonacci numbers, Linear Multilinear Algebra,**68**(10), (2020), 2087–2098.
44. Özger F., Aljimi E., Temizer Ersoy M., Rate of Weighted Statistical Convergence for Generalized Blending-Type Bernstein-Kantorovich Operators, Mathematics., **10**(12), (2022), 1–21.

45. Özger F., Ansari K. J., Statistical convergence of bivariate generalized bernstein operators via four-dimensional infinite matrices, Filomat., **36**(2), (2022), 507–525.
46. Özger F., Basar F., Domain of the double sequential band matrix $B(\tilde{r},\tilde{s})$ on some Maddox's spaces, Acta Math. Sci., **34**, (2014), 394–408.
47. Shcherbakov A. A., Kernels of sequences of complex numbers and their regular transformations, Math. Notes, **22**, (1977), 948–953.
48. Siddiqi J. A., Infinite matrices summing every almost periodic sequences, Pacific J. Math., **39**, (1971), 235–251.
49. Simons S., Banach limits, infinite matrices and sublinear functionals, J. Math. Anal. Appl., **26**, (1969), 640–655.
50. Srivastava H. M., Ansari K. J., Özger F., Ödemis Z., A Link between Approximation Theory and Summability Methods via Four-Dimensional Infinite Matrices, Mathematics, **9**(16), (2021), 1–18.
51. Steinhaus H., Sur la convergence ordinaire et la convergence asymptotique, Colloq. Math., **2**, (1951), 73–74.
52. Steinhaus H., Quality control by sampling, Colloq. Math., **2**, (1951), 98–108.
53. Qamaruddin Q., Mohuiddine S. A., Almost convergence and some matrix transformations, Filomat., **21**, (2007), 261–266.
54. Yaying T., Hazarika B., On sequence spaces defined by the domain of a regular Tribonacci matrix, Math. Slovaca., **70**(3), (2020), 697–706.
55. Yaying T., Kara M. I., On sequence spaces defined by the domain of tribonacci matrix in $c_0$ and $c$, Korean J. Math., **29**(1), (2021), 25–40.

# 4 On Ideal Lacunary Statistical Convergence of Sequence in Gradual Normed Linear Spaces Order $\alpha$

*Rabia Savas*

The main aim of this chapter is to present the concepts of the notions of ideal lacunary statistical convergence of sequences in gradual normed linear spaces order $\alpha$. To accomplish this goal, we will investigate some fundamental properties of the newly introduced notion and its relationship with ideal statistical convergence in gradual normed linear spaces. Additionally, some inclusion results are presented.

## 4.1 INTRODUCTION

The idea of statistical convergence was introduced by Fast [15] in 1951 and some basic features of statistical convergence were presented by Schoenberg [54] in 1959. Active research on this topic started after the paper of Fridy [17] and since then a large collection of literature has appeared. In the last quarter of the 20th century, statistical convergence has been discussed and captured important aspects in creating the basis of several investigations conducted in main branches of mathematics such as the theory of number [13], measure theory [30], trigonometric series [58], probability theory [10], and approximation theory [22]. In addition, it was further investigated from the sequence space point of view and linked with summability theory by Connor [7], Et at. al. [21], Mursaleen [31], Móricz [34], Kolk [26], Fridy and Orhan [18], Kumar and Mursaleen [28], Rath and Tripathy [36], Salat [38], Bhardwarj [2], Savas and Gurdal [40], and many others made substantial contributions to the theory. Let us consider the following notations:

The density of $E$ is defined by

$$\delta(E) = \lim_{w} \frac{1}{w} \left|\{m \leq w : m \in E\}\right|,$$

DOI: 10.1201/9781003596578-4

where the limit exists. The number sequence $\psi = (\psi_m)$ is said to be statistically convergent to the number $\varpi$ if for each $\varepsilon > 0$,

$$\lim_{w\to\infty} \frac{1}{w} |\{m \leq w : |\psi_m - \varpi| \geq \varepsilon\}| = 0.$$

In this case, we write $S - \lim \psi = \varpi$ or $\xi_m \to \varpi\,(S)$ and $S$ denotes the set of all statistically convergent sequences.

The statistical convergence with degree $0 < \beta < 1$ was given by Gadjiev and Orhan [22] for a number sequence, and then it was generalized for the method of $A-$statistical convergence by Duman et al. [12]. After then the statistical convergence of order $\alpha$ and strong $p-$Cesàro summability of order $\alpha$ were studied by Çolak [5] for number sequences.

Let $0 < \alpha \leq 1$ be given. The sequence $(\psi_m)$ is said to be statistically convergent of order $\alpha$ if there is a complex number $\varpi$ such that

$$\lim_{w\to\infty} \frac{1}{w^\alpha} |\{m \leq w : |\psi_m - \varpi| \geq \varepsilon\}| = 0$$

for every $\varepsilon > 0$. In this case, we write $S^\alpha - \lim \psi_m = \varpi$. The set of all statistically convergent sequences of order $\alpha$ will be denoted by $S^\alpha$. This definition is the same as Definition 4 in [22] if we take $\alpha = 1 - \beta$, $0 < \beta < 1$ and, when $\alpha = 1$ the statistical convergence of order $3b1$ reduces to the statistical convergence. By considering this way, related studies were presented in [6], [9], [46], [47], [48], [49], [55].

Following this concept, in 1993 Fridy and Orhan [18] introduced the concept of lacunary statistically convergence as follows:

A lacunary sequence is an increasing integer sequence $\theta = (\sigma_w)_{w\in\mathbb{N}\cup\{0\}}$ satisfying

$$\sigma_0 = 0 \text{ and } \zeta_w = \sigma_w - \sigma_{w-1} \to \infty, \text{ as } w \to \infty.$$

Throughout this article, the intervals determined by $\theta$ will be denoted by $J_w = (\sigma_{w-1}, \sigma_w]$ and the ratio $\frac{\sigma_w}{\sigma_{w-1}}$ will be abbreviated by $q_r$. A real-valued sequence $(\psi_m)$ is said to be lacunary statistical convergent to a real number $\varpi$, if for any $\varepsilon > 0$,

$$\lim_w \frac{1}{\zeta_w} |\{m \in J_w : |\psi_m - \varpi| \geq \varepsilon\}| = 0.$$

In this case, $\varpi$ is called the lacunary statistical limit of the sequence $(\psi_m)$ and it was written as $(\psi_m) \longrightarrow \varpi\,(S_\theta)$. For more information about lacunary statistical convergence and its generalizations, [11], [20], [23], [24], [25], [32], [33], [35], [42], [50], [51], [56] can be addressed where many more references can be found.

In 1970, Bernstein [1] introduced the convergence of sequences with respect to a filter $\mathscr{F}$ on $\mathbb{N}$. Additionally, in 2001, Kostyrko et al. [27] presented the concept of $\mathscr{I}-$convergence that is a generalization form of statistical convergence.

Recall that A collection $\mathscr{I}$ of subsets of $\mathbb{N}$ is called an ideal provided that $\emptyset \in \mathscr{I}$, $A, B \in \mathscr{I}$ implies $A \cup B \in \mathscr{I}$, and $B \in \mathscr{I}$, $A \subset B$ implies $A \in \mathscr{I}$. A non-empty family

$\mathscr{F} \subset 2^{\mathbb{N}}$ is said to be a filter of $\mathbb{N}$ if the following conditions hold; $\emptyset \notin \mathscr{F}$, $A, B \in \mathscr{F}$ implies $A \cap B \in \mathscr{F}$, and $A \in \mathscr{F}$, $A \subset B$ implies $B \in \mathscr{F}$. If $\mathscr{I}$ is a proper ideal of $\mathbb{N}$ (i.e., $\mathbb{N} \notin \mathscr{I}$), then the family of sets

$$\mathscr{F}(\mathscr{I}) = \{M \subset \mathbb{N} : \exists A \in \mathscr{I} : M = \mathbb{N} \backslash A\}$$

is a filter of $\mathbb{N}$. It is called the filter associated with the ideal. Also, a proper ideal $\mathscr{I}$ is said to be admissible if $\{n\} \in \mathscr{I}$ for each $n \in \mathbb{N}$. Let $\mathscr{I}$ be an admissible ideal of subsets of $\mathbb{N}$, the sequence $\psi = (\psi_m)_{m \in \mathbb{N}}$ of elements of $\mathbb{R}$ is said to be $\mathscr{I}-$convergent to $\varpi \in \mathbb{R}$ if for $\varepsilon > 0$ the set $A(\varepsilon) = \{m \in \mathbb{N} : |\psi_m - \varpi| \geq \varepsilon\} \in \mathscr{I}$. An ideal $\mathscr{I}$ is called non-trivial if $\mathscr{I} \neq \emptyset$ and $\check{N} \notin \mathscr{I}$. The filter $F(\mathscr{I}) = \{\check{N} \setminus A : A \in \mathscr{I}\}$ is called the filter associated with the ideal $\mathscr{I}$. A non-trivial ideal $\mathscr{I} \subset P(\check{N})$ is called an admissible ideal in real vector space $\check{N}$ if and only if $\mathscr{I} \supset \{\{\psi\} : \psi \in \check{N}\}$.

In a natural way, more general summability methods, namely, $\mathscr{I}-$statistical convergence was introduced Das and Savas [9] as follows:

A sequence $\psi = (\psi_m)$ is said to be $\mathscr{I}-$statistical convergent to the number $\varpi \in \mathbb{N}$ if, for every $\varepsilon > 0$ and $\delta > 0$

$$\left\{m \in \mathbb{N} : \frac{1}{w} |\{m \leq w : |\psi_m - \varpi| \geq \varepsilon\}| \geq \delta\right\} \in \mathscr{I}.$$

For $\mathscr{I} = \mathscr{I}_{fin}$, $\mathscr{I}-$statistical convergence coincides with statistical convergence. Later it was further studied by [8], [39], [41], [43], [44], [45], [52], [53], and many others. During this paper, we use 0 to denote the $m-tuple$ $(0,0,...,0,0)$ and $\mathscr{I}$ stands for a non-trivial admissible ideal in $\mathbb{N}$.

On the other hand, the notion of fuzzy sets was introduced by Zadeh [57] in 1965 which was an extension of the classical set-theoretical concept. The term "fuzzy number" plays an important role in the study of fuzzy set theory. Fuzzy numbers were basically the generalization of intervals, not numbers. Therefore, this term is debatable to many authors due to its different behaviour. The term "fuzzy intervals" is often used by many authors instead of fuzzy numbers. To overcome the confusion among the researchers, in 2008 Fortin et al. [19] presented the notion of gradual real numbers as elements of fuzzy intervals. Gradual real numbers are mainly known by their respective assignment function which is defined in the interval $(0, 1]$. Hence, in some sense, every real number can be viewed as a gradual number with a constant assignment function. Furthermore, in 2011, Sadeqi and Azari [37] first introduced the concept of gradual normed space and the usual convergence of sequences in gradual normed linear spaces has been introduced by Ettefagh et al. [14]. Recently, Choudhury and Debnath [3] have presented $\mathscr{I}-$statistical convergence in these spaces. Further improvement in this direction has been appeared in [4]. By considering above mentioned studies, it should be mentioned that the concept of ideal lacunary statistical convergence of sequences in gradual normed linear spaces order $\alpha$ has also not been studied till now. Therefore, the main goal of this paper is to present gradual $\mathscr{I}-$lacunary statistical convergence of sequences of order $\alpha$ and gradually $\mathscr{I}-$strongly summable sequences spaces of order $\alpha$. We also investigate their relationship and make some observations about these classes and in this way try to present an answer to an open problem in the literature.

## 4.2 DEFINITIONS AND PRELIMINARIES

Before we present some definitions.

**Definition 4.2.1.** *[19] A gradual real number $\widetilde{\tau}$ is defined by an assignment function $A_{\widetilde{\tau}}:(0,1]\to\mathbb{R}$. The class of all gradual real numbers is shown by $G(\mathbb{R})$. A gradual real number $\widetilde{\tau}$ is said to be non-negative, if for every $\varphi\in(0,1]$, $A_{\widetilde{\tau}}(\varphi)\geq 0$. The class of all non- negative gradual real numbers is denoted by $G^{*}(\mathbb{R})$.*

In [19], the gradual operations between the elements of $G(\mathbb{R})$ was defined as follows:

**Definition 4.2.2.** *Let $*$ be any operation in $\mathbb{R}$ and suppose $\widetilde{\tau}_1,\widetilde{\tau}_2\in G(\mathbb{R})$ with assignment functions $A_{\widetilde{\tau}_1}$ and $A_{\widetilde{\tau}_2}$ respectively. Then $\widetilde{\tau}_1*\widetilde{\tau}_2\in G(\mathbb{R})$ is defined with the assignment function $A_{\widetilde{\tau}_1*\widetilde{\tau}_2}$ given by $A_{\widetilde{\tau}_1*\widetilde{\tau}_2}(\xi)=A_{\widetilde{\tau}_1}(\xi)*A_{\widetilde{\tau}_2}(\xi)$, $\forall\xi\in(0,1]$.Then the gradual addition $\widetilde{\tau}_1+\widetilde{\tau}_2$ and the gradual scalar multiplication $c\widetilde{\tau}\,(c\in\mathbb{R})$ are defined by*

$$A_{\widetilde{\tau}_1+\widetilde{\tau}_2}(\xi)=A_{\widetilde{\tau}_1}(\xi)+A_{\widetilde{\tau}_2}(\xi)$$

*and*

$$A_{c\widetilde{\tau}}(\xi)=cA_{\widetilde{\tau}}(\xi),\ \forall\,\xi\in(0,1].$$

*For any real number $p\in\mathbb{R}$, the constant gradual real number $\widetilde{p}$ is defined by the constant assignment function $A_{\widetilde{p}}(\xi)=p$ for any $\xi\in(0,1]$. In particular, $\widetilde{0}$ and $\widetilde{1}$ are the constant fradyal numbers defined by $A_{\widetilde{0}}(\xi)=0$ and $A_{\widetilde{1}}(\xi)=1$ respectively. It can be easily seen that $G(\mathbb{R})$ with the gradual addition and gradual scalar multiplication forms a real vector space [19].*

**Definition 4.2.3.** *[37] Let $\check{N}$ be a real vector space. The function $\|.\|_G:\check{N}\to G^{*}(\mathbb{R})$ is said to be a gradual norm on $\check{N}$, if for every $\xi\in(0,1]$, the following conditions hold for any $\psi,\varsigma\in\check{N}$:*

1. $A_{\|\psi\|_G}(\xi)=A_{\widetilde{0}}(\xi)$ if and only if $\xi=0$;
2. $A_{\|c\psi\|_G}(\xi)=|c|A_{\|\psi\|_G}(\varphi)$ for any $c\in\mathbb{R}$;
3. $A_{\|\psi+\varsigma\|_G}(\xi)\leq A_{\|\psi\|_G}(\xi)+A_{\|\varsigma\|_G}(\xi)$.

The pair $(\check{N},\|.\|_G)$ is called a gradual normed linear space $(GNLS)$.

**Definition 4.2.4.** *[14] Let $(\psi_m)$ be a sequence in the GNLS $(\check{N},\|.\|_G)$. Then, $(\psi_m)$ is said to be gradually bounded if for every $\xi\in(0,1]$, there exists $B=B(\xi)>0$ such that $A_{\|\psi_m\|_G}\leq B$ for all $m\in\mathbb{N}$.*

**Definition 4.2.5.** *[37] Let $(\psi_m)$ be a sequence in the GNLS $(\check{N}, ||.||_G)$. Then $(\psi_m)$ is said to be gradually convergent to $\varpi \in \check{N}$, if for every $\xi \in (0,1]$ and $\varepsilon > 0$, there exists $N(= N_\varepsilon(\xi)) \in \mathbb{N}$ such that $A_{||\psi_m - \psi||_G}(\xi) < \varepsilon$, $\forall\, m \geq N$.*

**Example 4.2.6.** *[37] Let $\check{N} = \mathbb{R}^m$ and for $\psi_m = (\psi_1, \psi_2, ..., \psi_m) \in \mathbb{R}^m$, $\xi \in (0,1]$, define $||.||_G$ by*

$$A_{||\psi||_G}(\xi) = e^{\xi} \sum_{i=1}^{m} |\psi_i|.$$

*Then, $||.||_G$ is a gradual norm on $\mathbb{R}^m$, and $(\mathbb{R}^m, ||.||_G)$ is a GNLS.*

Additionally, the notion of gradually lacunary statistically convergence sequence spaces was presented as follows:

**Definition 4.2.7.** *[4] Let $(\psi_m)$ be a sequence in the GNLS $(\check{N}, ||.||_G)$ and $\theta$ be any lacunary sequence. Then $(\psi_m)$ is said to be gradually lacunary statistical convergent to $\varpi \in \check{N}$ if for every $\varepsilon > 0$ and $\xi \in (0,1]$,*

$$\lim_{w \to \infty} \frac{1}{\sigma_w} \left| \left\{ m \in I_w : A_{||\psi_m - \psi||_G(\xi)} \geq \varepsilon \right\} \right| = 0.$$

*In this case, we write $S_\theta(G) - \lim \psi_m = \psi$.*

**Example 4.2.8.** *Let $\check{N} = \mathbb{R}^m$ and $||.||_G$ be the norm defined in Example 2.1. Consider $\theta = (\sigma_w)$ defined by*

$$\sigma_w = \begin{cases} 0, & w = 0; \\ 5^w, & w \geq 1. \end{cases}$$

*Then, the sequence $(\psi_m)$ in $\mathbb{R}^m$ is defined as*

$$(\psi_m) = \begin{cases} (0,0,0,...,0,q), & \text{if } w = s^2, s \in \mathbb{N} \\ (0,0,0,...,0,0), & \text{otherwise} \end{cases}$$

*is gradually lacunary statistical convergent to 0 in $\mathbb{R}^m$. To prove that let us consider the following:*

$$\begin{aligned} & \lim_{w} \frac{1}{\zeta_w} \left| \left\{ m \in I_w : A_{||\psi_m - 0||_G(\xi)} \geq \varepsilon \right\} \right| \\ = \; & 5 \lim_{w} \frac{1}{5^w} \left| \left\{ m \in (5^{w-1}, 5^w] : A_{||\psi_m - 0||_G}(\xi) \geq \varepsilon \right\} \right| \\ \leq \; & 5 \lim_{w} \frac{1}{5^w} \left| \left\{ m \leq 5^w : A_{||\psi_m - 0||_G}(\xi) \geq \varepsilon \right\} \right| \\ \leq \; & 5 \lim_{w} \frac{[\sqrt{w}]}{w} \\ = \; & 0, \end{aligned}$$

*where $[s]$ denotes the largest integer $\leq s$. Consequently, $S_\theta(G) - \lim \psi_m = 0$.*

**Example 4.2.9.** *Let $\check{N} = \mathbb{R}$ and for any $\psi \in \mathbb{R}$, let $||.||_G$ be the norm defined as $A_{||\psi||_G} = e^{\xi}|\psi|$. Let us consider the sequence $\theta = (\sigma_w)$ defined by*

$$(\sigma_w) = \begin{cases} 0, & if \ w = 0; \\ 2^{w,} & if \ w \geq 1. \end{cases}$$

*Then the sequence $(\psi_m)$ in $\check{N}$ defined as $\psi_m = m^2$ is not gradually lacunary statistical convergent. To show that for any $\psi \in \mathbb{R}$, we obtain $\psi \leq 0$ or $\psi > 0$. Firstly, if $\psi \leq 0$, we choose $\varepsilon = \frac{1}{2}e^{\xi}$. Then we are granted the following:*

$$\begin{aligned} & \lim_w \frac{1}{\zeta_w}\left|\left\{m \in I_w : A_{||\psi_m - \psi||_G(\xi)} \geq \varepsilon\right\}\right| \\ = \ & \lim_w \frac{2}{w}\left|\left\{m \in (2^{w-1}, 2^w] : A_{||m^2-\psi||_G}(\xi) \geq \frac{1}{2}e^{\xi}\right\}\right| \\ = \ & 1. \end{aligned}$$

*If $\psi > 0$, then there exists $m_0 \in \mathbb{N}$ such that $\psi_{m_0-1} \leq \psi \leq \psi_{m_0}$. In this case, we need to consider the following cases; if $0 < \psi < 1$, then choose $\varepsilon = \frac{e^{\xi}}{2}\min\{\psi, 1-\psi\}$. Then, it is easy to verify that*

$$\lim_w \frac{1}{\zeta_w}\left|\left\{m \in I_w : A_{||\psi_m - \psi||_G(\xi)} \geq \varepsilon\right\}\right| = 1.$$

*And if $\psi \geq 1$, and $\varepsilon = \frac{e^{\xi}}{2}\min\{\psi - \psi_{m_0-1}, \psi_{m_0} - \psi\}$ then it is clear that*

$$\lim_w \frac{1}{\zeta_w}\left|\left\{m \in I_w : A_{||\psi_m - \psi||_G(\xi)} \geq \varepsilon\right\}\right| = 1.$$

*From the above results, $(\psi_m)$ is not gradually lacunary statistical convergent.*

**Definition 4.2.10.** *[4] Let $(\psi_m)$ be a sequence in the GNLS $(\check{N}, ||.||_G)$. Then, $(\psi_m)$ is said to be gradually $\mathscr{I}-$convergent to $\psi \in \check{N}$, if for every $\xi \in (0,1]$ and $\varepsilon > 0$, the set $B(\xi,\varepsilon) = \left\{m \in \mathbb{N} : A_{||\psi_m - \psi||_G}(\xi) \geq \varepsilon\right\} \in \mathscr{I}$. In this case, we write $\psi_m \to \psi(\mathscr{I}(G))$. The class of all gradually $\mathscr{I}-$convergent sequences will be denoted by $\mathscr{I}(G)^{\alpha}$.*

## 4.3 MAIN RESULTS

In this section, we present the main results of this paper.

**Definition 4.3.1.** *Let $(\psi_m)$ be a sequence in the GNLS $(\check{N}, ||.||_G)$. Then, $(\psi_m)$ is said to be gradually $\mathscr{I}-$statistical convergent of order $\alpha$ to $\psi \in \check{N}$ or $\mathscr{I}^{st}(G)^{\alpha}-$convergent to $\xi \in \check{N}$, $\alpha \in (0,1]$ for any real number, if for every $\xi \in (0,1]$ and $\varepsilon > 0$, $\delta > 0$,*

$$\left\{w \in \mathbb{N} : \frac{1}{w^{\alpha}}\left|\left\{m \leq w : A_{||\psi_m - \psi||_G}(\xi) \geq \varepsilon\right\}\right| \geq \delta\right\} \in \mathscr{I}.$$

*In this case, we write* $\psi_m \to \psi\left(\mathscr{I}^{st}(G)^{\alpha}\right)$*. The class of all gradually* $\mathscr{I}-$*statistical convergent sequences of order* $\alpha$ *will be denoted by* $\mathscr{I}^{st}(G)^{\alpha}$.

**Example 4.3.2.** *Let* $\check{N} = \mathbb{R}$ *and* $||.||_G$ *be the norm defined by* $A_{||\psi||_G}(\xi) = e^{\varphi}\sum_{i=1}^{m}|\psi_i|$*. Let us also consider the ideal* $\mathscr{I} = \mathscr{I}_d$*, ideal consisting of all subsets of* $\mathbb{N}$ *having natural density zero. Additionally, let us define the following sequence*

$$(v_w) = \begin{cases} 1, & 1 \leq w \leq 5; \\ w-5, & w \geq 5 \end{cases}$$

*and* $S = \{1^2, 2^2, 3^2, ...\}$*.Then, the sequence* $(\psi_m)$ *defined by*

$$(\psi_m) = \begin{cases} (0,0,0,...,0,m), \ w-[\sqrt{v_w}]+1 \leq m \leq w, \ w \notin S \\ (0,0,0,...,0,m), \ w-v_w+1 \leq m \leq w, \ w \in S \\ 0, \ otherwise \end{cases}$$

*is* $\mathscr{I}^{st}(G)^{\alpha}$ *convergent to* 0*. To show this for any* $\varepsilon > 0$ $(0 < \varepsilon < 1)$*, since*

$$\frac{1}{v_w^{\alpha}}\left|\left\{w - v_w + 1 \leq m \leq w : A_{||\psi_m - 0||_G}(\xi) \geq \varepsilon\right\}\right| = \frac{\sqrt{v_w^{\alpha}}}{v_w^{\alpha}} \to \infty$$

*as* $w \to \infty$ *and* $w \notin S$*; therefore, for every* $\delta > 0$*,*

$$\left\{w \in \mathbb{N} : \frac{1}{v_w^{\alpha}}\left|\left\{w - v_w + 1 \leq m \leq w : A_{||\psi_m - 0||_G}(\xi) \geq \varepsilon\right\}\right| \geq \delta\right\} \tag{4.1}$$
$$\subset S \cup \{1, 2, ..., m_1\}$$

*for some* $m_1 \in \mathbb{N}$*. Now let* $\delta > 0$ *be given. Then, the fact* $\lim_w \left(1 - \frac{v_w}{w}\right) = 0$ *enables us to choose* $m_2 \in \mathbb{N}$ *such that* $1 - \frac{v_w}{w} < \frac{\delta}{2}$ *for every* $w \geq m_2$*. Thus for the above* $\varepsilon > 0$*, we obtain*

$$\begin{aligned} & \frac{1}{w^{\alpha}}\left|\left\{m \leq w : A_{||\psi_m - 0||_G}(\xi) \geq \varepsilon\right\}\right| \\ = & \frac{1}{w^{\alpha}}\left|\left\{m \leq w - v_w : A_{||\psi_m - 0||_G}(\xi) \geq \varepsilon\right\}\right| \\ & + \frac{1}{w^{\alpha}}\left|\left\{w - v_w + 1 \leq m \leq w : A_{||\psi_m - 0||_G}(\xi) \geq \varepsilon\right\}\right| \\ \leq & 1 - \frac{v_w^{\alpha}}{w} + \frac{1}{w}\left|\left\{w - v_w + 1 \leq m \leq w : A_{||\psi_m - 0||_G}(\xi) \geq \varepsilon\right\}\right| \\ \leq & \frac{\delta}{2} + \frac{1}{v_w^{\alpha}}\left|\left\{w - v_w + 1 \leq m \leq w : A_{||\psi_m - 0||_G}(\xi) \geq \varepsilon\right\}\right| \end{aligned}$$

*for all* $w \geq m_2$. *Thus,*

$$\begin{aligned}&\left\{w \in \mathbb{N} : \frac{1}{w^{\alpha}}\left|\left\{m \leq w : A_{||\psi_m - 0||_G}(\xi) \geq \varepsilon\right\}\right| \geq \delta\right\}\\ &\subset \left\{w \in \mathbb{N} : \frac{1}{\nu_w^{\alpha}}\left|\left\{w - \nu_w + 1 \leq m \leq w : A_{||\psi_m - 0||_G}(\xi) \geq \varepsilon\right\}\right| \geq \frac{\delta}{2}\right\}\\ &\cup \{1,2,3,...,m_2\}\\ &\subset S \cup \{1,2,3,...,m_3\},\end{aligned}$$

*from* (3.1) *where* $m_3 = \max\{m_1, m_2\}$. *Now, since* $\delta(S) = 0$, *from the above conclusion, we are granted*

$$\left\{w \in \mathbb{N} : \frac{1}{w^{\alpha}}\left|\left\{m \leq w : A_{||\psi_m - 0||_G}(\xi) \geq \varepsilon\right\}\right| \geq \delta\right\} \in \mathcal{I},$$

*providing that* $\psi_m \to 0\left(\mathcal{I}^{st}(G)^{\alpha}\right)$.

**Example 4.3.3.** *Let* $\check{N} = \mathbb{R}$ *and* $||.||_G$ *be the norm defined by* $A_{||\psi||_G}(\xi) = e^{\xi}\sum_{i=1}^{m}|\psi_i|$. *Let us consider the ideal* $\mathcal{I} = \mathcal{I}_{fin}$, *ideal consisting of all finite subsets of* $\mathbb{N}$. *Define the sequence* $(\psi_m)$ *as follows:*

$$(\psi_m) = \begin{cases} 0, & m = s^2, s \in \mathbb{N} \\ (0,0,...,0,1), & \textit{otherwise} \end{cases}$$

*then* $(\psi_m)$ *is* $\mathcal{I}^{st}(G)^{\alpha}$ *convergent to* $(0,0,...,0,1)$, *but not* $\mathcal{I}(G)^{\alpha}$ *convergent to* $(0,0,...,0,1)$.

**Remark 4.3.4.** *For* $\mathcal{I} = \mathcal{I}_{fin}$, $\mathcal{I}^{st}(G)^{\alpha}$ *convergence coincides with gradually statistical convergent of order* $\alpha$. *For arbitrary ideal* $\mathcal{I}$ *and for* $\alpha = 1$ *it coincides with the notion of gradually* $\mathcal{I}-$*statistical convergence that was defined in [3]. When* $\mathcal{I} = \mathcal{I}_{fin}$ *and* $\alpha = 1$ *it reduces to the notion of gradual statistical convergence that was seen in [3].*

**Definition 4.3.5.** *Let* $\theta$ *be a lacunary sequence,* $(\psi_m)$ *be a sequence in the GNLS* $(\check{N}, ||.||_G)$ *and* $\alpha \in (0,1]$ *be any real number.* $(\psi_m)$ *is said to be gradually* $\mathcal{I}-$*lacunary statistical convergent of order* $\alpha$ *to* $\psi \in \check{N}$ *or* $\mathcal{I}_{\theta}^{st}(G)^{\alpha}-$*convergent to* $\psi$ *if for every* $\xi \in (0,1]$ *and* $\varepsilon > 0$, $\delta > 0$,

$$\left\{w \in \mathbb{N} : \frac{1}{\zeta_w^{\alpha}}\left|\left\{m \in J_w : A_{||\psi_m - \psi||_G}(\xi) \geq \varepsilon\right\}\right| \geq \delta\right\} \in \mathcal{I}.$$

*In this case, we write* $\psi_m \to \psi\left(\mathcal{I}_{\theta}^{st}(G)^{\alpha}\right)$. *The class of all gradually* $\mathcal{I}-$*lacunary statistical convergent sequences of order* $\alpha$ *will be denoted by* $\mathcal{I}_{\theta}^{st}(G)^{\alpha}$.

**Remark 4.3.6.** *For* $\alpha = 1$ *the below definition coincides with gradually* $\mathcal{I}-$*lacunary statistical convergence. Additionally, it must be noted in this context* *that gradual lacunary statistical convergence of order $\alpha$ has not been studied till now. It is easy to say that gradual lacunary statistical convergence of order $\alpha$ is a special case of gradually $\mathscr{I}-$lacunary statistical convergence of order $\alpha$ when we consider $\mathscr{I} = \mathscr{I}_{fin}$. Therefore, properties of gradually lacunary statistical convergence of order $\alpha$ can be clearly granted from our results with obvious modifications.*

**Definition 4.3.7.** *Let $\theta$ be a lacunary sequence, $(\psi_m)$ be a sequence in the GNLS $(\check{N}, ||.||_G)$ and $\alpha \in (0,1]$ be any real number. A sequence $(\psi_m)$ is said to be gradually strongly $N_\theta^G(\mathscr{I})^\alpha-$summable to $\psi \in \check{N}$ (or strongly $N_\theta^G(\mathscr{I})-$summable of order $\alpha$), if for every $\xi \in (0,1]$ and $\varepsilon > 0$,*

$$\left\{ w \in \mathbb{N} : \frac{1}{\zeta_w^\alpha} \left( \sum_{m \in I_w} A_{||\psi_m - \psi||_G}(\xi) \geq \varepsilon \right) \right\} \in \mathscr{I}.$$

*Under these circumstances, we write $\psi_m \longrightarrow \psi \left(N_\theta^G(\mathscr{I})^\alpha\right)$. The set of all gradually strongly $N_\theta^G(\mathscr{I})^\alpha-$summable sequences will be denoted by $N_\theta^G(\mathscr{I})^\alpha$.*

**Theorem 4.3.8.** *Let $(\psi_m)$ be a sequence in the GNLS $(\check{N}, ||.||_G)$ such that $\psi_m \to \psi\left(\mathscr{I}_\theta^{st}(G)^\alpha\right)$. Then, $\psi$ is uniquely determined.*

**Proof:** If we assume $\psi_m \to \psi\left(\mathscr{I}_\theta^{st}(G)^\alpha\right)$ and $\psi_m \to \eta\left(\mathscr{I}_\theta^{st}(G)^\alpha\right)$ hold for $\psi, \eta \in \check{N}$ with $\psi \neq \eta$, then, for any $\varepsilon > 0$, $\delta > 0$ and $\xi \in (0,1]$, we have $B_1 = B_1(\xi, \varepsilon, \delta), B_2 = B_2(\xi, \varepsilon, \delta) \in F(\mathscr{I})$, where

$$E_1 = \left\{ w \in \mathbb{N} : \frac{1}{\zeta_w^\alpha} \left| \left\{ m \in J_w : A_{||\psi_m - \psi||_G}(\varphi) \geq \varepsilon \right\} \right| < \delta \right\}$$

and

$$E_2 = \left\{ w \in \mathbb{N} : \frac{1}{\zeta_w^\alpha} \left| \left\{ m \in J_w : A_{||\psi_m - \psi||_G}(\varphi) \geq \varepsilon \right\} \right| < \delta \right\}.$$

Obviously, $E_1 \cap E_2 \in F(\mathscr{I})$ and is non-empty. Let us choose $z \in E_1 \cap E_2$ and take $\varepsilon = A_{||\frac{\psi - \eta}{3}||_G}(\xi) > 0$. Then,

$$\frac{1}{\zeta_z^\alpha} \left| \left\{ m \in J_z : A_{||\psi_m - \psi||_G}(\xi) \geq \varepsilon \right\} \right| < \delta$$

and

$$\frac{1}{\zeta_z^\alpha} \left| \left\{ m \in J_z : A_{||\psi_m - \psi||_G}(\xi) \geq \varepsilon \right\} \right| < \delta.$$

Now choosing $\delta$ sufficiently small, we can say most of the $m \in J_z$ will satisfy $A_{||\psi_m - \psi||_G}(\xi) < \varepsilon$ and $A_{||\psi_m - \eta||_G}(\xi) < \varepsilon$. Therefore, we have

$$E = \left\{ m \in J_z : A_{||\psi_m - \psi||_G}(\xi) < \varepsilon \right\} \cap \left\{ m \in J_z : A_{||\psi_m - \eta||_G}(\xi) < \varepsilon \right\} \neq \emptyset.$$

If we take $p \in B$, then we obtain the following

$$\begin{aligned} \varepsilon &= A_{\|\frac{\psi-\eta}{3}\|_G}(\xi) \leq \frac{1}{3}\left(A_{\|\psi_p-\psi\|_G}(\xi)+A_{\|\psi_p-\eta\|_G}(\xi)\right) \\ &< \frac{1}{3}(\varepsilon+\varepsilon)=\frac{2\varepsilon}{3}, \end{aligned}$$

which is a contradiction.

**Theorem 4.3.9.** *Let $(\psi_m)$ and $(\eta_m)$ be two sequences in the GNLS $(\check{N},\|.\|_G)$ such that $\psi_m \to \psi\left(\mathscr{I}_\theta^{st}(G)^\alpha\right)$ and $\eta_m \to \eta\left(\mathscr{I}_\theta^{st}(G)^\alpha\right)$. Then,*

**(i)** $\psi_m+\eta_m \to \psi+\eta\left(\mathscr{I}_\theta^{st}(G)^\alpha\right)$.
**(ii)** $c\psi_m \to c\psi\left(\mathscr{I}_\theta^{st}(G)^\alpha\right)$, where $c \in \mathbb{R}$.

**Proof:** (i) From the hypothesis, we can say that for every $\xi \in (0,1]$ and $\varepsilon>0$, $\delta>0$, the two sets $Q_1=Q_1(\xi,\varepsilon,\delta)$, $Q_2=Q_2(\xi,\varepsilon,\delta) \in \mathscr{I}$, where

$$Q_1=\left\{w \in \mathbb{N}: \frac{1}{\zeta_w^\alpha}\left|\left\{m \in J_w: A_{\|\psi_m-\psi\|_G}(\xi) \geq \frac{\varepsilon}{2}\right\}\right|<\frac{\delta}{2}\right\}$$

and

$$Q_2=\left\{w \in \mathbb{N}: \frac{1}{\zeta_w^\alpha}\left|\left\{m \in J_w: A_{\|\eta_m-\eta\|_G}(\xi) \geq \frac{\varepsilon}{2}\right\}\right|<\frac{\delta}{2}\right\}.$$

Then, $(\mathbb{N} \setminus Q_1) \cap (\mathbb{N} \setminus Q_2) \in F(\mathscr{I})$ and so $(\mathbb{N} \setminus Q_1) \cap (\mathbb{N} \setminus Q_2) \neq \emptyset$. Choose $w \in (\mathbb{N} \setminus Q_1) \cap (\mathbb{N} \setminus Q_2)$. Then, the following inequality

$$\begin{aligned} & \frac{1}{\zeta_w^\alpha}\left|\left\{m \in J_w: A_{\|(\psi_m+\eta_m)-(\psi+\eta)\|_G}(\xi) \geq \varepsilon\right\}\right| \\ \leq\ & \frac{1}{\zeta_w^\alpha}\left|\left\{m \in J_w: A_{\|\psi_m-\psi\|_G}(\xi) \geq \frac{\varepsilon}{2}\right\}\right| \\ & +\frac{1}{\zeta_w^\alpha}\left|\left\{m \in J_w: A_{\|\eta_m-\eta\|_G}(\xi) \geq \frac{\varepsilon}{2}\right\}\right|, \end{aligned}$$

yields the following inclusion

$$\begin{aligned} & (\mathbb{N} \setminus Q_1) \cap (\mathbb{N} \setminus Q_2) \\ \subseteq\ & \left\{w \in \mathbb{N}: \frac{1}{\zeta_w^\alpha}\left|\left\{m \in J_w: A_{\|(\psi_m+\eta_m)-(\psi+\eta)\|_G}(\xi) \geq \varepsilon\right\}\right|<\delta\right\}. \end{aligned} \tag{4.2}$$

Now as $(\mathbb{N} \setminus Q_1) \cap (\mathbb{N} \setminus Q_2) \in F(\mathscr{I})$, so the set on the right-hand side of (3.2) also belongs to $F(\mathscr{I})$ which means that $\psi_m+\eta_m \to \psi+\eta\left(\mathscr{I}_\theta^{st}(G)^\alpha\right)$.

(ii) For $c = 0$, there is nothing to prove. So let $c \neq 0$. Then, for every $\xi \in (0,1]$ and $\varepsilon > 0$, $0 < \alpha \leq 1$, we obtain the following inequality

$$\begin{aligned} &\frac{1}{\zeta_w^\alpha}\left|\left\{m \in J_w : A_{\|c\psi_m - c\psi\|_G}(\xi) \geq \varepsilon\right\}\right| \\ = &\frac{1}{\zeta_w^\alpha}\left|\left\{m \in J_w : |c| A_{\|\psi_m - \psi\|_G}(\xi) \geq \varepsilon\right\}\right| \\ \leq &\frac{1}{\zeta_w^\alpha}\left|\left\{m \in J_w : A_{\|\psi_m - \psi\|_G}(\xi) \geq \frac{\varepsilon}{|c|}\right\}\right|. \end{aligned}$$

Consequently, we obtain $c\psi_m \to c\psi\left(\mathscr{I}_\theta^{st}(G)^\alpha\right)$, where $c \in \mathbb{R}$.

**Theorem 4.3.10.** *Let $(\psi_m)$ be a sequence in the GNLS $\left(\check{N}, ||\cdot|_G|\right)$ and $\theta$ be a lacunary sequence. Then,*
*(i) If $\psi_m \to \psi\left(N_\theta^G(\mathscr{I})^\alpha\right)$ then $\psi_m \to \psi\left(\mathscr{I}_\theta^{st}(G)^\alpha\right)$; however, the reverse is not necessarily true.*
*(ii) The reverse of (i) holds if $(\psi_m)$ is gradually bounded.*

**Proof** (i) Let $\varepsilon > 0$ and $\psi_m \to \psi\left(N_\theta^G(\mathscr{I})^\alpha\right)$. Then, we are granted

$$\begin{aligned} \sum_{m \in J_w} A_{\|\psi_m - \psi\|_G}(\xi) &\geq \sum_{m \in J_w,\, A_{\|\psi_m - \psi\|_G}(\varphi) \geq \varepsilon} A_{\|\psi_m - \psi\|_G}(\xi) \\ &\geq \varepsilon\left|\left\{m \in J_w : A_{\|\psi_m - \psi\|_G}(\xi) \geq \varepsilon\right\}\right|. \end{aligned}$$

This implies that

$$\frac{1}{\varepsilon\zeta_w^\alpha}\sum_{m \in J_w} A_{\|\psi_m - \psi\|_G}(\xi) \geq \frac{1}{\zeta_w^\alpha}\left|\left\{m \in J_w : A_{\|\psi_m - \psi\|_G}(\xi) \geq \varepsilon\right\}\right|.$$

As a result for any $\delta > 0$, the following inclusion is obtained

$$\begin{aligned} &\left\{w \in \mathbb{N} : \frac{1}{\zeta_w^\alpha}\left|\left\{m \in J_w : A_{\|\psi_m - \psi\|_G}(\xi) \geq \varepsilon\right\}\right| \geq \delta\right\} \\ \subseteq &\left\{w \in \mathbb{N} : \frac{1}{\zeta_w^\alpha}\left(\sum_{m \in J_w} A_{\|\psi_m - \psi\|_G}(\xi)\right) \geq \varepsilon\delta\right\}. \end{aligned}$$

Now for the converse part, we construct a counterexample by considering the gradual normed space $(\mathbb{R}^m, \|\cdot\|_G)$, where $\|\cdot\|_G$ is the norm defined as

$$A_{\|\psi_m\|_G}(\xi) = e^\xi \sum_{i=1}^{m} |\psi_i|$$

$\xi \in (0,1]$. Let $\theta$ be given and $\psi_m$ be the sequence defined as $(0,0,...,1)$, $(0,0,...,2)$, ..., $\left(0,0,...,\left[\sqrt{\zeta_w^\alpha}\right]\right)$ at the first $\left[\sqrt{\zeta_w^\alpha}\right]$ integers in $J_w$, and $\psi_m = 0$ otherwise. Then, we have for any $\varepsilon > 0$ with $0 < \varepsilon e^\xi \leq 1$,

$$\frac{1}{\zeta_w^\alpha}\left|\left\{m \in J_w : A_{||\psi_m - \psi||_G}(\xi) \geq \varepsilon\right\}\right| \leq \frac{\left[\sqrt{\zeta_w^\alpha}\right]}{\zeta_w^\alpha}.$$

Thus, for any $\delta > 0$, the following inclusion holds

$$\left\{w \in \mathbb{N} : \frac{1}{\zeta_w^\alpha}\left|\left\{m \in J_w : A_{||\psi_m - \psi||_G}(\xi) \geq \varepsilon\right\}\right| \geq \delta\right\} \subseteq \left\{w \in \mathbb{N} : \frac{\left[\sqrt{\zeta_w^\alpha}\right]}{\zeta_w^\alpha} \geq \delta\right\}.$$

Now since $\mathscr{I}$ is admissible, so the set on the right-hand side belongs to $\mathscr{I}$ and as a consequence, the set on the left-hand side also belongs to $\mathscr{I}$. This means that $\psi_m \to 0\left(\mathscr{I}_\theta^{st}(G)^\alpha\right)$. On the other hand, since

$$\frac{1}{\zeta_w^\alpha}\left(\sum_{m \in J_w} A_{||\psi_m - \psi||_G}(\xi)\right) = \frac{\left[\sqrt{\zeta_w^\alpha}\right]\left(\left[\sqrt{\zeta_w^\alpha}\right]+1\right)}{2\zeta_w^\alpha}$$

holds, so we must have

$$\begin{aligned}
&\left\{w \in \mathbb{N} : \frac{1}{\zeta_w^\alpha}\left(\sum_{m \in J_w} A_{||\psi_m - 0||_G}(\varphi)\right) \geq \frac{1}{4}\right\} \\
= &\left\{w \in \mathbb{N} : \frac{\left[\sqrt{\zeta_w^\alpha}\right]\left(\left[\sqrt{\zeta_w^\alpha}\right]+1\right)}{2\zeta_w^\alpha} \geq \frac{1}{2}\right\} \\
= &\{w_0, w_0+1, w_0+2, ...\}
\end{aligned}$$

for some $w_0 \in \mathbb{N}$. Since $\mathscr{I}$ is admissible, so

$$\left\{w \in \mathbb{N} : \frac{1}{\zeta_w^\alpha}\left(\sum_{m \in J_w} A_{||\psi_m - 0||_G}(\xi)\right) \geq \frac{1}{4}\right\} \in F(\mathscr{I}),$$

which means that $\psi_m \nrightarrow 0\left(N_\theta^G(\mathscr{I})^\alpha\right)$.

(ii) Suppose $\psi_m \to \psi\left(\mathscr{I}_\theta^{st}(G)^\alpha\right)$ and $(\psi_m)$ is gradually bounded. Then, gradual boundedness of $(\psi_m)$ implies the existence of $B = B(\xi) > 0$ such that $A_{|\psi_m - \psi||_G} \leq B$

for all $m \in \mathbb{N}$. Then, for given $\varepsilon > 0$, we have

$$\begin{aligned}
&\frac{1}{\zeta_w^\alpha}\left(\sum_{m\in J_w} A_{\|\psi_m-\psi\|_G}(\xi)\right)\\
= &\frac{1}{\zeta_w^\alpha}\left(\sum_{A_{\|\psi_m-\psi\|_G}(\xi)\geq\frac{\varepsilon}{2}} A_{\|\psi_m-\psi\|_G}(\xi)\right)\\
&+\frac{1}{\zeta_w^\alpha}\left(\sum_{A_{\|\psi_m-\psi\|_G}(\xi)<\frac{\varepsilon}{2}} A_{\|\psi_m-\psi\|_G}(\xi)\right)\\
\leq &\frac{B}{\zeta_w^\alpha}\left|\left\{m \in J_w : A_{\|\psi_m-\psi\|_G}(\xi)\geq\frac{\varepsilon}{2}\right\}\right| + \frac{\varepsilon}{2}.
\end{aligned}$$

From the above inequality, we can conclude that

$$\begin{aligned}
&\left\{w \in \mathbb{N} : \frac{1}{\zeta_w^\alpha}\left(\sum_{m\in J_w} A_{\|\psi_m-\psi\|_G}(\xi)\right)\geq\varepsilon\right\}\\
\subseteq &\left\{w\in\mathbb{N} : \frac{1}{\zeta_w^\alpha}\left|\left\{m\in J_w : A_{\|\psi_m-\psi\|_G}(\xi)\geq\frac{\varepsilon}{2}\right\}\right|\geq\frac{\varepsilon}{2B}\right\}\in\mathscr{I}.
\end{aligned}$$

Hence, $\psi_m \to \psi\left(N_\theta^G(\mathscr{I})^\alpha\right)$.

**Theorem 4.3.11.** *Let $(\psi_m)$ be a sequence in the GNLS $\left(\check{N}, \|\cdot\|_G\right)$. Then, for any lacunary sequence $\theta$, $\psi_m \to \psi\left(\mathscr{I}^{st}(G)^\alpha\right)$ implies $\psi_m \to \psi\left(\mathscr{I}_\theta^{st}(G)^\alpha\right)$ if and only if*

$$\liminf_w q_w^\alpha > 1.$$

*Further, if*

$$\liminf_w q_w^\alpha = 1,$$

*then there exists a gradually bounded sequence which is $\mathscr{I}^{st}(G)^\alpha$ −convergent but not $\mathscr{I}_\theta^{st}(G)^\alpha$ −convergent.*

**Proof:** Let us first assume $\liminf_w q_w > 1$. Then, for sufficiently large $w$, there exists an $v > 0$ such that $q_w^\alpha > 1 + v$ for sufficiently large $w$ which implies that

$$\frac{\sigma_w^\alpha}{\zeta_w^\alpha} \leq \frac{1+v}{v}.$$

Now as $\psi_m \to \psi\left(\mathscr{I}^{st}(G)^\alpha\right)$, then for every $\varepsilon > 0$ and for sufficiently large $w$, we obtain the following

$$\begin{aligned}
\frac{1}{\zeta_w^\alpha}\left|\left\{m\leq I_w : A_{\|\psi_m-\psi\|_G}(\xi)\geq\varepsilon\right\}\right| &\geq \frac{\sigma_w^\alpha}{\zeta_w^\alpha}\frac{1}{\sigma_w^\alpha}\left|\left\{m\in J_w : A_{\|\psi_m-\psi\|_G}(\xi)\geq\varepsilon\right\}\right|\\
&\geq \frac{1+v}{v}\frac{1}{\sigma_w^\alpha}\left|\left\{m\in J_w : A_{\|\psi_m-\psi\|_G}(\xi)\geq\varepsilon\right\}\right|.
\end{aligned}$$

For any $\delta > 0$,

$$\left\{w \in \mathbb{N} : \frac{1}{\zeta_w^\alpha}\left|\left\{m \in J_w : A_{||\psi_m - \psi||_G}(\xi) \geq \varepsilon\right\}\right| \geq \delta\right\}$$

$$\subseteq \left\{w \in \mathbb{N} : \frac{1}{\sigma_w^\alpha}\left|\left\{m \leq \sigma_w : A_{||\psi_m - \psi||_G}(\xi) \geq \varepsilon\right\}\right| \geq \frac{v\delta}{1+v}\right\} \in \mathscr{I}.$$

This proves the sufficiency.

For the opposite part, let us consider $\liminf_w q_w^\alpha = 1$. We assume to form a sequence which is $\mathscr{I}^{st}(G)^\alpha$ –convergent but is not $\mathscr{I}_\theta^{st}(G)^\alpha$ –convergent. Proceeding as in ([16], page-510 and [18], page-46). Let us select a subsequence $(\sigma_{w_j})$ of the lacunary sequence $\theta$ satisfying

$$\frac{\sigma_{w_j-1}^\alpha}{\sigma_{w_j}^\alpha} > \frac{j}{1+j}$$

and

$$\frac{\sigma_{w_{j-1}}^\alpha}{\sigma_{w_j}^\alpha - 1} < j,$$

where $w_j - w_{j-1} \geq 2$. Now we define a gradually bounded sequence $(\psi_m)$ in $(\mathbb{R}^m, ||\cdot||_G)$ as follows:

$$(\psi_m) = \begin{cases} (0,0,...,0,1) & m \in J_{w_j},\ j = 1,2,3,... \\ 0, & \text{otherwise} \end{cases}$$

Then, for any $\psi \in \mathbb{R}^m$, we obtain

$$\frac{1}{\zeta_{w_j}^\alpha}\left(\sum_{m \in J_{w_j}} A_{||\psi_m - \psi||_G}(\xi)\right) = A_{||(0,0,...,0,1) - \psi||_G}(\xi), \quad j = 1,2,3,...$$

and

$$\frac{1}{\zeta_{w_j}^\alpha}\left(\sum_{m \in J_w} A_{||\psi_m - \psi||_G}(\xi)\right) = A_{||\psi||_G}(\xi), \quad for\ w \neq w_j.$$

Hence, $\psi_m \nrightarrow \psi\left(N_\theta^G(\mathscr{I})^\alpha\right)$ and since $(\psi_m)$ is gradually bounded, we obtain $\psi_m \nrightarrow \psi\left(\mathscr{I}_\theta^{st}(G)^\alpha\right)$. However, $(\psi_m)$ is $\mathscr{I}^{st}(G)^\alpha$ –convergent, since if $p$ is an integer that is sufficiently large, we can have a unique $j$ satisfying $\sigma^{\alpha}_{w_j-1} < p \leq \sigma^{\alpha}_{w_{j+1}-1}$ and then

$$\begin{aligned}\frac{1}{p}\left|\left\{m \leq p : A_{||\psi_m-\psi||_G}(\xi) \geq \varepsilon\right\}\right| &\leq \frac{1}{p}\sum_{m=1}^{p} A_{||\psi_m||_G}(\xi) \\ &\leq \frac{\sigma^{\alpha}_{w_{j-1}} + \zeta^{\alpha}_{w_j}}{\sigma^{\alpha}_{w_j-1}} \\ &\leq \frac{1}{j} + \frac{1}{j} = \frac{2}{j}\end{aligned}$$

as $p \to \infty$, it follows that $j \to \infty$. Now by applying the technique of proof of Theorem 2.1 of [7], one can easily show that $(\psi_m)$ is $\mathscr{I}^{st}(G)^{\alpha}$ −convergent.

For the following result, we presume that the lacunary sequence $\theta$ satisfies the condition that for any set $\widetilde{Q} \in F(\mathscr{I})$, $\cup\left\{m : \sigma_{w-1} < m < \sigma_w, w \in \widetilde{Q}\right\} \in F(\mathscr{I})$.

**Theorem 4.3.12.** *For any lacunary sequence $\theta$ satisfying the above condition, gradual $\mathscr{I}$−lacunary statistical convergence of order $\alpha$ implies gradual $\mathscr{I}$−statistical convergence of order $\alpha$, $0 < \alpha < 1$, if*

$$\sup_n \sum_{i=0}^{n-1} \frac{\zeta^{\alpha}_{i+1}}{(\sigma_{w-1})^{\alpha}} = B < \infty.$$

**Proof:** Assume that $\psi_m \to \psi\left(\mathscr{I}^{st}(G)^{\alpha}\right)$ and for $\varepsilon, \delta, \delta_1 > 0$ define the sets

$$C = \left\{w \in \mathbb{N} : \frac{1}{\zeta^{\alpha}_w}\left|\left\{m \in J_w : A_{||\psi_m-\psi||_G}(\xi) \geq \varepsilon\right\}\right| < \delta\right\}$$

and

$$T = \left\{w \in \mathbb{N} : \frac{1}{w^{\alpha}}\left|\left\{m \leq w : A_{||\psi_m-\psi||_G}(\xi) \geq \varepsilon\right\}\right| \geq \delta_1\right\}.$$

It is obvious from the assumption that $C \in F(\mathscr{I})$, the filter associated with the ideal $\mathscr{I}$. Further, observe that

$$A_j = \frac{1}{\zeta^{\alpha}_j}\left|\left\{m \in J_j : A_{||\psi_m-\psi||_G}(\xi) \geq \varepsilon\right\}\right| < \delta$$

for all $j \in C$. Let $w \in \mathbb{N}$ be such that $\sigma_{w-1} < m < \sigma_w$ for some $w \in C$. Now

$$\begin{aligned}
& \frac{1}{w^\alpha}\left|\left\{m \leq w : A_{\|\psi_m-\psi\|_G}(\xi) \geq \varepsilon\right\}\right| \\
\leq\; & \frac{1}{\sigma_{w-1}^\alpha}\left|\left\{m \leq \sigma_w : A_{\|\psi_m-\psi\|_G}(\xi) \geq \varepsilon\right\}\right| \\
=\; & \frac{1}{\sigma_{w-1}^\alpha}\left|\left\{m \in J_w : A_{\|\psi_m-\psi\|_G}(\xi) \geq \varepsilon\right\}\right| \\
& + \cdots + \frac{1}{\sigma_{w-1}^\alpha}\left|\left\{m \in J_w : A_{\|\psi_m-\psi\|_G}(\xi) \geq \varepsilon\right\}\right| \\
& + \frac{\sigma_1^\alpha}{\sigma_{w-1}^\alpha}\frac{1}{\zeta_1^\alpha}\left|\left\{m \in J_1 : A_{\|\psi_m-\psi\|_G}(\xi) \geq \varepsilon\right\}\right| \\
& + \frac{(\sigma_2-\sigma_1)^\alpha}{\sigma_{w-1}^\alpha}\frac{1}{\zeta_2^\alpha}\left|\left\{m \in J_2 : A_{\|\psi_m-\psi\|_G}(\xi) \geq \varepsilon\right\}\right| \\
& + \cdots + \frac{(\sigma_2-\sigma_1)^\alpha}{\sigma_{w-1}^\alpha}\frac{1}{\zeta_w^\alpha}\left|\left\{m \in J_w : A_{\|\psi_m-\psi\|_G}(\xi) \geq \varepsilon\right\}\right| \\
=\; & \frac{\sigma_1^\alpha}{\sigma_{n-1}^\alpha}A_1 + \frac{(\sigma_2-\sigma_1)^\alpha}{\sigma_{w-1}^\alpha}A_2 + \cdots \frac{(\sigma_w-\sigma_{w-1})^\alpha}{\sigma_{w-1}^\alpha}A_w \\
\leq\; & \sup_{j\in C} A_j . \sup_w \sum_{i=0}^{w-1} \frac{(\sigma_{i+1}-\sigma_i)^\alpha}{\sigma_{w-1}^\alpha} \\
<\; & B\delta.
\end{aligned}$$

When we choose $\delta_1 = \frac{\delta}{2}$ and in view of the fact that $\cup\{m : \sigma_{w-1} < m < \sigma_w : w \in C\} \subset T$ where $C \in F(\mathcal{I})$, it follows from our assumption on $\theta$ that the set $T$ also belongs to $F(\mathcal{I})$ and this completes the proof of the theorem.

## 4.4 CONCLUSION

Some summability methods in ideal lacunary statistical convergence sequence spaces were studied by several authors. Additionally, Sadeqi and Azari [37] introduced the concept of gradual normed space and the usual convergence of sequences in gradual normed linear spaces has been introduced by Ettefagh et al. [14]. Recently, Choudhury and Debnath [3] have presented $\mathcal{I}-$statistical convergence in these spaces. However, the concept of the notions of ideal lacunary statistical convergence of sequences in gradual normed linear spaces order $\alpha$ has not been studied so far. Therefore, the present paper is filled up with a big gab in the existing literature. Also, These concepts for double sequences are still open questions for future studies.

## BIBLIOGRAPHY

1. Bernstein A. I., A new kind of compactness for topological spaces, Fund. Math., **66**, (1970), 185–193.

2. Bhardwaj V. K., Bala I., On weak statistical convergence, Int. J. Math and Math. Sci, 2007, Article ID 38530.
3. Choudhury C., Debnath S., On $I-$statistical convergence of sequences in gradual normed linear spaces, Math. Vesnik., **74**(3), (2022), 218–228.
4. Choudhury C., Debnath S., On lacunary statistical convergence of sequences in gradual normed linear spaces, An. Univ. Craiova Ser. Mat. Inform., **36**(3), (2021), 595–604.
5. Colak R., Statistical convergence of order $\alpha$, Modern Methods in Analysis and Its Applications, Anamaya Publ., New Delhi, India (2010), 121–129.
6. Colak R., Bektas C. A., 3bb-Statistical convergence of order $\alpha$, Acta Math. Sci. Ser. B Eng. Ed., **31**(3), (2011), 953–959.
7. Connor J., The statistical and strong p-Cesaro convergence of sequences, Analysis, **8**(1-2), (1988), 47–63.
8. Das P., Savas E., Ghosal S. K., On generalizations of certain summability methods using ideals, Appl. Math. Lett., **24**(9), (2011), 1509–1514.
9. Das P., Savas E., On I-statistical and I-lacunary statistical convergence of order $\alpha$, Bull. Iran. Math. Soc., **40**(2), (2014), 459–472
10. Das P., Ghosal S., Sam S., Statistical Convergence of order $\alpha$ in probability, Arab. J. Math. Sci., **21**, (2015), 253–265.
11. Demirci I. A., Gurdal M., On lacunary generalized statistical convergent complex uncertain triple sequence, J. Intell. Fuzzy Systems, **41**(1), (2021), 1021–1029.
12. Duman O., Khan M. K., Orhan C., A-statistical convergence of approximating operators, Math. Inequal. Appl., **6**(4), (2003), 689–699.
13. Erdös R., Tenenbaum G., Surles densities de certaines suites d'entiers., Proc. Lond. Math. Soc., **3**(59), (1989), 417–438.
14. Ettefagh M., Etemad S., Azari F. Y., On some properties of sequences in gradual normed spaces. Asian-Eur. J. Math., **13**(4), (2020), 2050085.
15. Fast H., Sur la convergence statistique, Colloq. Math., **2**, (1951), 241–244.
16. Freedman A. R., Sember J. J., Raphael M., Some Cesáro type summability spaces, Proc Lond. Math. Soc., **37**(3), (1978), 508–520.
17. Fridy J. A., On statistical convergence, Analysis., **5**, (1985), 301–313.
18. Fridy J. A., Orhan C., Lacunary Statistical Convergence, Pacific J. Math., **160**, (1993), 43–51.
19. Fortin J., Dubois D., Fargier H., Gradual numbers and their application to fuzzy interval analysis. IEEE Trans. Fuzzy Syst., **16**(2), (2008), 388–402.
20. Esi A., Acikgoz M., On some double lacunary strong Zweier convergent sequence spaces, An. Univ. Craiova. Ser. Mat. Inform., **40**(2), (2013), 121–127.
21. Et M., Nuray F., $m$-statistical convergence, Indian J. Pure Appl. Math., **32**(6), (2001), 961–969.
22. Gadjiev A. D., Orhan C., Some approximation theorems via statistical convergence, Rocky Mountain. J. Math., **32**(2002), 129–138.
23. Hazarika B., Esi A., On fuzzy real valued asymptotically equivalent sequences and lacunary ideal convergence, An. Univ. Craiova. Ser. Mat. Inform., **41**(2), (2014), 209–225.
24. Hazarika B., Esi A., Lacunary ideal quasi Cauchy sequences, An. Univ. Craiova. Ser. Mat. Inform., **45**(2), (2018), 220–231.
25. Huban M. B., Gurdal M., Wijsman lacunary invariant statistical convergence for triple sequences via Orlicz function, J. Classical Analy., **17**(2), (2021), 119–128.
26. Kolk E., The statistical convergence in Banach space, Acta Comment. Univ. Tartu Math., **928**, (1991), 41–52.

27. Kostyrko P., Salat T., Wilczynski W., $I-$convergence, Real Anal. Exch., **26**(2) (2000-2001), 669–686.
28. Kumar V., Mursaleen M., On $(\lambda,\mu)$-statistical convergence of double sequences on intuitionistic fuzzy normed spaces., Filomat., **25**(2) (2011), 109–120.
29. Lorentz G. G., A contribution to the theory of divergent sequences, Acta Math,, **80**, (1948), 167–190.
30. Miller H. I., A measure theoretical subsequence characterization of statistical convergence, Trans. Amer. Math. Soc., **347**, (1995), 1811–1819.
31. Mursaleen M., $\lambda-$ Statistical Convergence, Math. Slovaca., **50**(11), (2000), 111–115.
32. Mursaleen M., Belen C., On statistical lacunary summability of double sequences, Filomat., **28**(2), (2014), 231–239.
33. Mursaleen M., Mouhiuddine S. A., On lacunary statistical convergence with respect to the intuitionistic fuzzy normed space, J. Comput. Appl. Math., **233**(2), (2009), 142–149.
34. Móricz F., Statistical convergence of multiple sequences, Arch. Math., **81**, (2003), 82–89.
35. Pehlivan S., Gurdal M., Fisher B., Lacunary statistical cluster points of sequences, Math. Commun., **11**(1), (2006), 39–46.
36. Rath D., Tripathy B. C., On statistically convergent and statistically Cauchy sequences., Indian J. Pure Appl. Math., **25**(4), (1994), 381–386.
37. Sadeqi I., Azari F. Y., Gradual normed linear space, Iran. J. Fuzzy Syst., **8**(5), (2011), 131–139.
38. Salat T., On statistically convergent sequences of real numbers, Math. Slovaca., **30**, (1980), 139–150.
39. Savas E., On $I-$lacunary statistical convergence of weight of sequence of sets, Filomat., **31**(16), (2017), 5315–5322.
40. Savas E., Gurdal M., A generalized statistical convergence in intuitionistic fuzzy normed spaces, Science Asia., **41**(4), (2015), 289–294.
41. Savas E., Gurdal M., I-statistical convergence in probabilistic normed spaces, Politehn. Univ. Bucharest Sci. Bull. Ser. A Appl. Math. Phys., **77**(4), (2015), 195–204.
42. Savas E., Borgohain S., On strongly almost lacunary statistical $A-$convergence defined by a Musielak-Orlicz function, Filomat., **30**(3), (2016), 689–697.
43. Savas E., Gumus H., A generalization on $I-$asymptotically lacunary statistical equivalent sequences, J. Inequal. Appl., **270**(2013), (2013), 1–11.
44. Savas E., Das P., A generalized statistical convergence via ideals, Appl. Math. Lett., **24**(6), (2011), 826–830.
45. Savas E., On asymptotically I-statistical equivalent sequences of order $\alpha$, Indian J. Math., **56**(2), (2014), 1–9.
46. Savas R., Patterson R. F., Generalization of Statistically Convergent, Methods of Functional Analysis and Topology., **26**(4), (2020), 384–389.
47. Savas R., On Some Multiple Statistical Convergence of order 3b8 Defined by Almost Lacunary Sequences, J. Ana. Num. Theory., **8**(1), (2020), 7–10.
48. Savas R., Multidimensional Strongly Deferred Invariant Convergence, Numerical Functional Analysis and Optimization., **42**(11), (2021), 1323–133.
49. Savas R., On Asymptotically Deferred Statistical Equivalent Measurable Functions, Journal of Classical Analysis, **16**(2), (2020), 141–147.
50. Savas R., Lacunary statistical boundedness of measurable functions, Math. Nat. Sci., **6**, (2020), 15–19.

51. Savas R., Patterson R. F., $I_2-$lacunary strongly summability for multidimensional measurable functions, Publications de l'Institut Mathematique, **107**(121), (2020), 93–107.
52. Savas R., $I-$Almost Lacunary vector valued sequence spaces in 2-normed spaces, Maltepe Journal of Mathematics., **2**(2), (2020), 76–81.
53. Savas R., Patterson R. F., Multidimensional Matrix Characterization of Asymptotic $I_2-$Equivalent and Ideal for Double Sequences, Numerical Functional Analysis and Optimization., **40**(6), (2019), 654–669.
54. Schoenberg I. J., The integrability of certain functions and related summability methods, The Amer, Math. Monthly., **66**, (1959), 361–375.
55. Sengul H., Et M., On $I-$lacunary statistical convergence of order $\alpha$ of sequences of sets, Filomat., **31**(8), (2017), 2403–2412.
56. Tripathy B. C., Hazarika B., Choudhary B., Lacunary $I-$ convergent sequences, Kyungpook Math. J., **52**(4), (2012), 473–482.
57. Zadeh L. A., Fuzzy sets, Inf. Control., **8**(3), (1965), 338–353.
58. Zygmund A., *Trigonometric Series*, Cambridge University Press, New York, NY, USA, 1st edition, (1959).

# 5 On the Four-Dimensional Forward Difference Matrix

*Orhan Tug, Eberhard Malkowsky, Vladimir Rakocevic, Bipan Hazarika*

In this chapter, we investigate the spaces $\mathcal{M}_u(\Delta)$ and $\mathcal{C}_\vartheta(\Delta)$ of double sequences as domain of the four-dimensional forward difference matrix in the double sequence spaces $\mathcal{M}_u$ and $\mathcal{C}_\vartheta$, respectively, where $\vartheta \in \{bp, bp0, r, r0\}$. Then, we present some topological and algebraic properties of the new double sequence spaces $\mathcal{M}_u(\Delta)$ and $\mathcal{C}_\vartheta(\Delta)$. Moreover, we determine the $\alpha$-, $\beta(\vartheta)$-, and $\gamma$-duals of the new double sequence spaces $\mathcal{M}_u(\Delta)$ and $\mathcal{C}_\vartheta(\Delta)$ after proving some significant lemmas. Finally, we characterize some four-dimensional matrix classes $(\lambda(\Delta), \mu)$ and $(\mu, \lambda(\Delta))$, where $\lambda, \mu = \{\mathcal{M}_u, \mathcal{C}_\vartheta\}$. The main purpose of this article is to define some new double sequence spaces by calculating the four-dimensional forward difference matrix domain in the double sequence spaces $\mathcal{M}_u$ and $\mathcal{C}_\vartheta$, respectively, where $\vartheta \in \{bp, bp0, r, r0\}$.

## 5.1 INTRODUCTION

Summability theory is the theory of the assignment of the limit of the single or double sequences and series, which is one of the fundamentals in modern functional analysis and its applications. In the classical and modern methodology of summability, the matrix method is the classical method concerning infinite matrices and their domains and the concepts of convergence generated by them. The special summability methods are given by some special triangular matrices named by Cesàro mean, Hölder methods, Weighted mean, Riesz methods, Nörlund methods, Hausdorff methods, methods of function theoretical type and summability method defined by power series.

The four-dimensional backward difference matrix domain on the double sequence spaces $\mathcal{M}_u$, $\mathcal{C}_p$, $\mathcal{C}_{0p}$, $\mathcal{C}_r$, and $\mathcal{L}_q$ has been studied by Demiriz and Duyar [1]. Right after that, the four-dimensional generalized backward difference matrix was defined by Başar and Tug [2], and its domain in some double sequence spaces were studied by Tug [3] and [4]. Then the four-dimensional generalized backward difference matrix domain on almost and strongly almost convergent sequence spaces were studied by Tuğ [5], [6], [7], [8] and [9]. Moreover, Tuğ at al. [10] defined the sequentially defined four-dimensional backward difference matrix and studied its domain on some double sequence spaces. Furthermore, Çapan and Başar [11] studied the

DOI: 10.1201/9781003596578-5

four-dimensional backward difference matrix domain on almost and strongly almost convergent double sequence spaces. Most recently, Tug [12] calculated the sequentially defined four-dimensional backward difference matrix domain on almost and strongly almost convergent double sequence spaces.

Before presenting the main results, we recall some well-known terminology and notations.

We denote all complex-valued double sequences by the set $\Omega := \{x = (x_{mn}) : x_{mn} \in \mathbb{C}, \ \forall\, m,n \in \mathbb{N}\}$ which is a vector space with coordinatewise addition and scalar multiplication, where $\mathbb{C}$ denotes the complex field and $\mathbb{N} = \{0,1,2,...\}$. Any vector subspace of $\Omega$ is called a double sequence space. A double sequence $x = (x_{mn})$ is called convergent in Pringsheim's sense to a limit point $L$ if for every $\varepsilon > 0$ there exists a natural number $n_0 = n_0(\varepsilon)$ and $L \in \mathbb{C}$ such that $|x_{mn} - L| < \varepsilon$ for all $m,n > n_0$. Therefore, we write $L = p - \lim_{m,n\to\infty} x_{mn}$. By $\mathscr{C}_p$, we denote the space of all convergent double sequences in Pringsheim's sense, that is,

$$\mathscr{C}_p := \{x = (x_{mn}) \in \Omega : \forall\, \varepsilon > 0,\ \exists\, n_0 \in \mathbb{N}\ and\ L \in \mathbb{C} \ni, |x_{mn} - L| < \varepsilon\}$$

for all $m,n > n_0$. Móricz [13] proved that the double sequence space $\mathscr{C}_p$ is a complete semi-normed space with the seminorm

$$\|x\|_\infty = \lim_{N\to\infty} \sup_{m,n\geq N} |x_{mn}|.$$

The space of all null double sequences in Pringsheim's sense is denoted by $\mathscr{C}_{p0}$.

A double sequence $x = (x_{mn})$ of complex numbers is called bounded if $\|x\|_\infty = \sup_{m,n\in\mathbb{N}} |x_{mn}| < \infty$, where $\mathbb{N} = \{0,1,2,\cdots\}$, and the space of all bounded double sequences is denoted by $\mathscr{M}_u$, that is,

$$\mathscr{M}_u := \{x = (x_{mn}) \in \Omega : \|x\|_\infty = \sup_{m,n\in\mathbb{N}} |x_{m,n}| < \infty\};$$

it is a Banach space with the norm $\|\cdot\|_\infty$.

Unlike in the case of single sequences, there are double sequences that are convergent in Pringsheim's sense but unbounded. Therefore, the set $\mathscr{C}_p \setminus \mathscr{M}_u$ is not empty. Boos [14] defined the sequence $x = (x_{mn})$ by

$$x_{mn} = \begin{cases} n & ,\ if \quad m = 0, n \in \mathbb{N} \\ 0 & , \quad if\ m \geq 1, n \in \mathbb{N}, \end{cases}$$

which is obviously in $\mathscr{C}_p$, i.e., $p - \lim_{m,n\to\infty} x_{mn} = 0$, but not in the set $\mathscr{M}_u$, i.e., $\|x\|_\infty = \sup_{m,n\in\mathbb{N}} |x_{mn}| = \infty$. Thus, $x \in \mathscr{C}_p \setminus \mathscr{M}_u$.

By $\mathscr{C}_{bp}$, we denote the set of all convergent in Pringsheim's sense and bounded double sequences, that is,

$$\mathscr{C}_{bp} := \mathscr{C}_p \cap \mathscr{M}_u = \left\{ x = (x_{mn}) \in \mathscr{C}_p : \|x\|_\infty = \sup_{m,n\in\mathbb{N}} |x_{mn}| < \infty \right\}.$$

The set $\mathscr{C}_{bp}$ is a Banach space with the norm

$$\|x\|_\infty = \sup_{m,n\in\mathbb{N}} |x_{mn}| < \infty.$$

Hardy [15] defined that a sequence in the space $\mathscr{C}_p$ is called regularly convergent if it is a single convergent sequence concerning each index. We denote the set of such double sequences by $\mathscr{C}_r$, that is,

$$\mathscr{C}_r := \{x = (x_{mn}) \in \mathscr{C}_p : \forall_{m\in\mathbb{N}} (x_{mn})_m \in c, \textit{ and } \forall_{n\in\mathbb{N}} (x_{mn})_n \in c\},$$

where $c$ denotes the set of all convergent single sequences of complex numbers. Regular convergence requires the boundedness of double sequences; this is the main difference between regular convergence and convergence in Pringsheim's sense. We also use the notations $\mathscr{C}_{bp0} = \mathscr{M}_u \cap \mathscr{C}_{p0}$ and $\mathscr{C}_{r0} = \mathscr{C}_r \cap \mathscr{C}_{p0}$.

Throughout the text, unless otherwise stated we mean by the summation $\sum_{kl} x_{kl}$ without limits run from 0 to $\infty$ is $\sum_{k,l=0}^{\infty} x_{kl}$.

The space $\mathscr{L}_q$ of all absolutely $q-$summable double sequences was introduced by Başar and Sever [16] as follows

$$\mathscr{L}_q := \left\{ x = (x_{kl}) \in \Omega : \sum_{k,l} |x_{kl}|^q < \infty \right\}, \quad (1 \le q < \infty).$$

It is a Banach space with the norm $\|\cdot\|_q$ defined by

$$\|x\|_q = \left( \sum_{k,l} |x_{kl}|^q \right)^{1/q}.$$

Moreover, Zeltser [17] introduced the space $\mathscr{L}_u$ as a special case of the space $\mathscr{L}_q$ for $q = 1$.

The double series spaces $\mathscr{BS}, \mathscr{CS}_\vartheta$, where $\vartheta \in \{p, bp, r\}$, and $\mathscr{BV}$ were introduced by Altay and Başar [18]. The set $\mathscr{BS}$ of all double series whose sequences of partial sums are bounded is defined by

$$\mathscr{BS} = \left\{ x = (x_{kl}) \in \Omega : \sup_{m,n\in\mathbb{N}} |s_{mn}| < \infty \right\}$$

where the sequence $s_{mn} = \sum_{k,l=0}^{m,n} x_{kl}$ is the $(m,n)-th$ partial sum of the series. The series space $\mathscr{BS}$ is a Banach space with a norm defined as

$$\|x\|_{\mathscr{BS}} = \sup_{m,n\in\mathbb{N}} \left| \sum_{k,l=0}^{m,n} x_{kl} \right|. \tag{5.1}$$

Moreover, the series space $\mathscr{BS}$ is linearly isomorphic to the sequence space $\mathscr{M}_u$. The set $\mathscr{CS}_\vartheta$ of all series whose sequences of partial sums are $\vartheta-$convergent in Pringsheim's sense is defined by

$$\mathscr{CS}_\vartheta = \{x = (x_{kl}) \in \Omega : (s_{mn}) \in \mathscr{C}_\vartheta\}$$

where $\vartheta \in \{p, bp, r\}$. The space $\mathscr{CS}_p$ is a complete semi-normed space with the seminorm defined by

$$\|x\|_\infty = \lim_{n\to\infty}\left(\sup_{k,l\geq n}\left|\sum_{i,j=0}^{k,l} x_{ij}\right|\right).$$

Moreover, the series space $\mathscr{CS}_p$ is isomorphic to the sequence space $\mathscr{C}_p$. Furthermore, the sets $\mathscr{CS}_{bp}$ and $\mathscr{CS}_r$ are also Banach spaces with the norm (5.1) and the inclusion $\mathscr{CS}_r \subset \mathscr{CS}_{bp}$ holds. The set $\mathscr{BV}$ of all double sequences of bounded variation is defined by

$$\mathscr{BV} = \left\{x = (x_{kl}) \in \Omega : \sum_{k,l}\left|x_{kl} - x_{k-1,l} - x_{k,l-1} + x_{k-1,l-1}\right| < \infty\right\}.$$

The space $\mathscr{BV}$ is Banach space with the norm defined by

$$\|x\|_{\mathscr{BV}} = \sum_{k,l}\left|x_{kl} - x_{k-1,l} - x_{k,l-1} + x_{k-1,l-1}\right|,$$

that is linearly isomorphic to the space $\mathscr{L}_u$ of absolutely convergent double series. Moreover, the inclusions $\mathscr{BV} \subset \mathscr{C}_\vartheta$ and $\mathscr{BV} \subset \mathscr{M}_u$ strictly hold.

Let $E$ be any double sequence space. Then,

$$dE := \left\{x = (x_{kl}) \in \Omega : \left\{\frac{1}{kl}x_{kl}\right\}_{k,l\in\mathbb{N}} \in E\right\},$$

$$\int E := \left\{x = (x_{kl}) \in \Omega : \{klx_{kl}\}_{k,l\in\mathbb{N}} \in E\right\},$$

$$E^{\beta(\vartheta)} := \left\{a = (a_{kl}) \in \Omega : \{a_{kl}x_{kl}\} \in \mathscr{CS}_\vartheta,\ for\ every\ x = (x_{kl}) \in E\right\},$$

$$E^{\alpha} := \left\{a = (a_{kl}) \in \Omega : \{a_{kl}x_{kl}\} \in \mathscr{L}_u,\ for\ every\ x = (x_{kl}) \in E\right\},$$

$$E^{\gamma} := \left\{a = (a_{kl}) \in \Omega : \{a_{kl}x_{kl}\} \in \mathscr{BS},\ for\ every\ x = (x_{kl}) \in E\right\}.$$

Therefore, let $E_1$ and $E_2$ are arbitrary double sequences with $E_2 \subset E_1$ then the inclusions $E_1^\alpha \subset E_2^\alpha$, $E_1^\gamma \subset E_1^\alpha$ and $E_1^{\beta(\vartheta)} \subset E_1^\alpha$ hold. But the inclusion $E_1^\gamma \subset E_1^{\beta(\vartheta)}$ does not hold, since $\mathscr{C}_p \setminus \mathscr{M}_u$ is not empty.

Let $A = (a_{mnkl})_{m,n,k,l\in\mathbb{N}}$ be an infinite four-dimensional matrix and $E_1$, $E_2 \in \Omega$. We write

$$y_{mn} = A_{mn}(x) = \vartheta - \sum_{k,l} a_{mnk}x_{kl} \text{ for each } m,n \in \mathbb{N}. \tag{5.2}$$

We say that $A$ defines a matrix transformation from $E_1$ to $E_2$ if

$$A(x) = (A_{mn}(x))_{m,n} \in E_2 \text{ for all } x \in E_1. \tag{5.3}$$

The $\vartheta-$summability domain $E_A^{(\vartheta)}$ of a four-dimensional infinite matrix $A$ in a double sequence space $E$ is defined by

$$E_A^{(\vartheta)} = \left\{ x = (x_{kl}) \in \Omega : Ax = \left( \vartheta - \sum_{k,l} a_{mnkl} x_{kl} \right)_{m,n\in\mathbb{N}} \textit{ exists and is in } E \right\},$$

which is a sequence space. The above notation (5.3) says that $A = (a_{mnkl})_{m,n,k,l\in\mathbb{N}}$ maps the space $E_1$ into the space $E_2$ if $E_1 \subset (E_2)_A^{(\vartheta)}$ and we denote the set of all four-dimensional matrices that map the space $E_1$ into the space $E_2$ by $(E_1 : E_2)$. Thus, $A \in (E_1 : E_2)$ if and only if the double series on the right side of (5.3) $\vartheta-$converges for each $m, n \in \mathbb{N}$, i.e, $A_{mn} \in (E_1)^{\beta(\vartheta)}$ for all $m, n \in \mathbb{N}$ and we have $Ax \in E_2$ for all $x \in E_1$.

Adams [19] defined that the four-dimensional infinite matrix $A = (a_{mnkl})$ is a triangular matrix if $a_{mnkl} = 0$ for $k > m$ or $l > n$ or both. We also say by [19] that a triangular matrix $A = (a_{mnkl})$ is called a triangle if $a_{mnmn} \neq 0$ for all $m, n \in \mathbb{N}$. One can observe easily that if $A$ is a triangle, then $E_A^{(\vartheta)}$ and $E$ are linearly isomorphic.

Wilansky [20, Theorem 4.4.2, p. 66] defined that if $E$ is a sequence space, then the continuous dual $E_A^*$ of the space $E_A$ is given by

$$E_A^* = \{f : f = g \circ A, g \in E^*\}.$$

Zeltser [21] stated the notations of the double sequences $e^{kl} = (e^{kl}_{mn}), e^1, e_k$ and $e$ by

$$e^{kl}_{mn} = \begin{cases} 1, & if\ (k,l) = (m,n); \\ 0, & otherwise. \end{cases}$$

$e^1 = \sum_k e^{kl}$; *the double sequence that all terms of l-th column are one and other terms are zero,*
$e_k = \sum_l e^{kl}$; *the double sequence that all terms of k-th row are one and other terms are zero,*
$e = \sum_{kl} e^{kl}$; *the double sequence that all terms are one*

for all $k, l, m, n \in \mathbb{N}$.

The four-dimensional forward difference matrix $\Delta = (\delta_{mnkl})$ is defined by

$$\delta_{mnkl} := \begin{cases} (-1)^{m+n-k-l} & , \quad m \le k \le m+1,\ n \le l \le n+1, \\ 0 & , \quad \text{otherwise} \end{cases}$$

for all $m,n,k,l \in \mathbb{N}$. The $\Delta-$transform of a double sequence $x=(x_{mn})$ is given by

$$y_{mn} := \{\Delta x\}_{mn} = x_{mn} - x_{m+1,n} - x_{m,n+1} + x_{m+1,n+1}$$

for all $m,n \in \mathbb{N}$. We shall briefly discuss $\Delta^{-1}$ which is the inverse of four-dimensional forward difference matrix $\Delta$, where $(\Delta^{-1}\Delta)(x_{kl}) = x_{kl}$. Let $\Delta^{-1}y_{kl} = x_{kl}$. Then we can show that $x_{kl}$ is a finite summation of the original double sequence $y_{kl}$.

$$\Delta(\Delta^{-1}y_{kl}) = \Delta x_{kl} = x_{kl} - x_{k+1,l} - x_{k,l+1} + x_{k+1,l-1}. \tag{5.4}$$

If we write the equation (5.4) for $y_{00}, y_{01}, y_{10}, ..., y_{kl}$

$$\begin{aligned}
\Delta(\Delta^{-1}y_{00}) &= \Delta x_{00} = x_{00} - x_{10} - x_{01} + x_{11} \\
\Delta(\Delta^{-1}y_{01}) &= \Delta x_{01} = x_{01} - x_{11} - x_{02} + x_{12} \\
\Delta(\Delta^{-1}y_{10}) &= \Delta x_{10} = x_{10} - x_{20} - x_{11} + x_{21} \\
\Delta(\Delta^{-1}y_{11}) &= \Delta x_{11} = x_{11} - x_{21} - x_{12} + x_{22} \\
&\vdots \\
\Delta(\Delta^{-1}y_{kl}) &= \Delta x_{kl} = x_{kl} - x_{k+1,l} - x_{k,l+1} + x_{k+1,l+1}.
\end{aligned}$$

Then we add the left-hand sides up to $y_{00} + y_{01} + y_{10} + ... + y_{kl}$

$$\sum_{i,j=0}^{k,l} y_{i,j} = x_{k+1,l+1} + x_{00} - x_{k+1,0} - x_{0,l+1}$$

for all $k,l \in \mathbb{N}$. To be able to have $x_{kl}$ instead of having $x_{k+1,l+1}$, we must write it as

$$x_{kl} = \sum_{i,j=0}^{k-1,l-1} y_{i,j} - x_{00} + x_{k,0} + x_{0,l} \tag{5.5}$$

for all $k,l \in \mathbb{N}$. With this result, we can introduce the role of inverse four-dimensional forward difference operator $\Delta^{-1}$ on the double sequence $y_{kl}$, where $x_{kl} = \Delta^{-1}y_{kl}$, as the $(k-1,l-1)^{th}-$partial sum of the double sequence $y_{kl}$ plus arbitrary constants on the first row and the first column of the double sequence $x=(x_{kl})$.

## 5.2 NEW DOUBLE SEQUENCE SPACES

In this section, we introduce new double sequence spaces $\mathcal{M}_u(\Delta)$, $\mathcal{C}_\vartheta(\Delta)$, where $\vartheta \in \{bp, r\}$, as the matrix domains of the four-dimensional matrix of the forward differences in the sequence spaces $\mathscr{M}_u$ and $\mathscr{C}_\vartheta$ as follow;

$$\mathscr{M}_u(\Delta) := \left\{ x = (x_{kl}) \in \Omega : \sup_{k,l \in \mathbb{N}} |y_{kl}| < \infty \right\},$$

$$\mathscr{C}_\vartheta(\Delta) := \left\{ x = (x_{kl}) \in \Omega : \exists\, L \in \mathbb{C} \ni \vartheta - \lim_{k,l\to\infty} |y_{kl} - L| = 0 \right\},$$

$$\mathscr{C}_{\vartheta 0}(\Delta) := \left\{ x = (x_{kl}) \in \Omega : \vartheta - \lim_{k,l\to\infty} |y_{kl}| = 0 \right\},$$

where $y_{kl} = \Delta x_{kl} = (x_{kl} - x_{k+1,l} - x_{k,l+1} + x_{k+1,l+1})$ for all $k,l \in \mathbb{N}$.

**Theorem 5.2.1.** *The spaces $\mathscr{M}_u(\Delta)$ and $\mathscr{C}_\vartheta(\Delta)$, where $\vartheta \in \{bp, bp0, r, r0\}$ are Banach spaces with the norm*

$$\begin{aligned} \|x\|_{\mathscr{M}_u(\Delta)} &:= |x_{k,0} + x_{0,l} - x_{00}| + \|\Delta x\|_{\mathscr{M}_u} \\ &:= |x_{k,0} + x_{0,l} - x_{00}| + \sup_{k,l\in\mathbb{N}} \left| x_{kl} - x_{k+1,l} - x_{k,l+1} + x_{k+1,l+1} \right|. \end{aligned} \tag{5.6}$$

**Proof:** The linearity of those spaces is clear. Suppose that $x^i = (x^i_{kl})$ is a Cauchy sequence in the space $\mathscr{M}_u(\Delta)$ for all $k,l \in \mathbb{N}$. Then

$$\begin{aligned} \|x^i - x^j\|_{\mathscr{M}_u(\Delta)} &= |(x^i_{k,0} - x^j_{k,0}) + (x^i_{0,l} - x^j_{0,l}) - (x^i_{00} - x^j_{00})| \\ &\quad + \sup_{k,l\in\mathbb{N}} |\Delta(x^i_{kl} - x^j_{kl})| \to 0 \end{aligned}$$

as $i,j \to \infty$. Thus, we obtain $|x^i_{kl} - x^j_{kl}| \to 0$ for $i,j \to \infty$ and for every $k,l \in \mathbb{N}$. Hence $x^i = (x^i_{kl})$ is a Cauchy sequence in $\mathbb{C}$ for each $k,l \in \mathbb{N}$. Since $\mathbb{C}$ is complete, then it converges to a sequence $x = (x_{kl})$, i.e., we have

$$\lim_{i\to\infty} x^i_{kl} = x_{kl}$$

for each $k,l \in \mathbb{N}$. Therefore, for every $\varepsilon > 0$, there exists a natural number $N(\varepsilon)$, such that for all $i,j \geq N(\varepsilon)$, and for all $k,l \in \mathbb{N}$ we have

$$|x^i_{k,0} - x^j_{k,0}| < \frac{\varepsilon}{4},\ |x^i_{0,l} - x^j_{0,l}| < \frac{\varepsilon}{4},\ |x^i_{0,0} - x^j_{0,0}| < \frac{\varepsilon}{4},\ |\Delta(x^i_{kl} - x^j_{kl})| < \frac{\varepsilon}{4}.$$

Moreover,

$$\begin{aligned} \lim_{j\to\infty} |x^i_{k,0} - x^j_{k,0}| &= |x^i_{k,0} - x_{k,0}| < \frac{\varepsilon}{4}, \\ \lim_{j\to\infty} |x^i_{0,l} - x^j_{0,l}| &= |x^i_{0,l} - x_{0,l}| < \frac{\varepsilon}{4}, \\ \lim_{j\to\infty} |x^i_{0,0} - x^j_{0,0}| &= |x^i_{0,0} - x_{0,0}| < \frac{\varepsilon}{4}, \\ \lim_{j\to\infty} |\Delta(x^i_{kl} - x^j_{kl})| &= |\Delta(x^i_{kl} - x_{kl})| < \frac{\varepsilon}{4} \end{aligned}$$

for all $i \geq N(\varepsilon)$. Hence, we obtain that

$$\begin{aligned}\|x^i - x\|_{\mathcal{M}_u(\Delta)} &= |(x^i_{k,0} - x_{k,0}) + (x^i_{0,l} - x_{0,l}) - (x^i_{00} - x_{00})| \\ &\quad + \sup_{k,l\in\mathbb{N}} |\Delta(x^i_{kl} - x_{kl})| \\ &\leq |x^i_{k,0} - x_{k,0}| + |x^i_{0,l} - x_{0,l}| + |x^i_{00} - x_{00}| \\ &\quad + \sup_{k,l\in\mathbb{N}} |\Delta(x^i_{kl} - x_{kl})| < \varepsilon.\end{aligned}$$

Now we must show that $x \in \mathcal{M}_u(\Delta)$.

$$\begin{aligned}\sup_{k,l\in\mathbb{N}} |\Delta x_{kl}| &= \sup_{k,l\in\mathbb{N}} |x_{kl} - x_{k+1,l} - x_{k,l+1} + x_{k+1,l+1}| \\ &= \sup_{k,l\in\mathbb{N}} \big|x_{kl} - x^i_{kl} + x^i_{kl} - x_{k+1,l} + x^i_{k+1,l} - x^i_{k+1,l} - x_{k,l+1} + x^i_{k,l+1} - x^i_{k,l+1} \\ &\quad + x_{k+1,l+1} - x^i_{k+1,l+1} + x^i_{k+1,l+1}\big| \\ &\leq \sup_{k,l\in\mathbb{N}} |\Delta x^i_{kl}| + \sup_{k,l\in\mathbb{N}} |\Delta x^i_{kl} - \Delta x_{kl}| < \infty.\end{aligned}$$

Hence $x = (x_{kl}) \in \mathcal{M}_u(\Delta)$. This completes the proof.

Let $\vartheta = \{bp, bp0, r, r0\}$. We define the operator $P$ form $\lambda(\Delta)$ into itself, where $\lambda \in \{\mathcal{M}_u, \mathcal{C}_\vartheta\}$ as

$$\begin{aligned}P : \lambda(\Delta) &\rightarrow \lambda(\Delta) \\ x &\rightarrow Px = \begin{bmatrix} 0 & 0 & 0 & 0 & \cdots \\ 0 & x_{11} & x_{12} & x_{13} & \cdots \\ 0 & x_{21} & x_{22} & x_{23} & \cdots \\ 0 & x_{31} & x_{32} & x_{33} & \cdots \\ \vdots & \vdots & \vdots & \vdots & \ddots \end{bmatrix}\end{aligned}$$

for all $x = (x_{kl}) \in \lambda(\Delta)$. Clearly, $P$ is a linear and bounded operator on $\lambda(\Delta)$.

Now we show that the four-dimensional forward difference operator $\Delta$ is a linear homeomorphism.

$$\begin{aligned}\Delta : P(\lambda(\Delta)) &\rightarrow \lambda \\ x &\rightarrow \Delta x = y = (x_{kl} - x_{k+1,l} - x_{k,l+1} + x_{k+1,l+1}),\end{aligned} \tag{5.7}$$

where the set $P(\lambda(\Delta))$ is defined by

$$\begin{aligned}P(\lambda(\Delta)) &:= \{x = (x_{kl}) \in \mathbb{C} : x \in \lambda(\Delta) \text{ and } x_{00} = x_{k,0} = x_{0,l} = 0, \forall k, l \in \mathbb{N}\} \\ &\subset \lambda(\Delta)\end{aligned}$$

and

$$\|x\|_{P(\lambda(\Delta))} = \|\Delta x\|_\lambda.$$

Therefore, the spaces $P(\lambda(\Delta))$ and $\lambda$ are equivalent as topological spaces, and the $\Delta$ and $\Delta^{-1}$ are norm preserving and $\|\Delta\| = \|\Delta^{-1}\| = 1$. We prove the following Lemma 5.2.2 for the case $\lambda = \mathscr{C}_{r0}$ by using the results in [13, Theorem 5, Remark 3, p.132]. Since the proofs of the other cases are similar to that of following Lemma 5.2.2, we left them as an exercise for the reader.

**Lemma 5.2.2.** *A linear functional $f_\Delta$ on $P(\mathscr{C}_{r0}(\Delta))$ is continuous if and only if there exists a double sequence $a = (a_{kl})_{k,l\geq 1} \in \mathscr{L}_u$ such that*

$$f_\Delta(x) = \sum_{k,l=1}^{\infty} a_{kl}(\Delta x)_{kl} \tag{5.8}$$

*for all $x \in P(\mathscr{C}_{r0}(\Delta))$.*

**Proof:** First we show that $\Delta : P(\mathscr{C}_{r0}(\Delta)) \to \mathscr{C}_{r0}$, $\Delta x_{kl} = x_{kl} - x_{k+1,l} - x_{k,l+1} + x_{k+1,l+1}$ with $x_{00} = x_{k,0} = x_{0,l} = 0$ for each $k,l \in \mathbb{N}$ is an isometric linear isomorphism, that is, we prove that $\Delta$ is a bijection between $P(\mathscr{C}_{r0}(\Delta))$ and $\mathscr{C}_{r0}$ by $\Delta x_{kl} = x_{kl} - x_{k+1,l} - x_{k,l+1} + x_{k+1,l+1}$ with $x_{00} = x_{k,0} = x_{0,l} = 0$ for each $k,l \in \mathbb{N}$. Linearity is clear. Moreover, $x = 0$ whenever $\Delta x = 0$, and hence $\Delta$ is injective. Now suppose that $y = (y_{kl}) \in \mathscr{C}_{r0}$, we define the sequence $x = (x_{kl})$ by $x_{kl} = \sum_{i,j=0}^{k-1,l-1} y_{ij}$ with $x_{00} = x_{k,0} = x_{0,l} = 0$ for each $k,l \in \mathbb{N}$. Then we have,

$$\begin{aligned}
\|x\|_{P(\mathscr{C}_{r0}(\Delta))} &= \sup_{k,l\in\mathbb{N}} |\Delta x_{kl}| \\
&= \sup_{k,l\in\mathbb{N}} \left|\Delta\left(\sum_{i,j=0}^{k-1,l-1} y_{ij}\right)\right| \\
&= \sup_{k,l\in\mathbb{N}} \left|\sum_{i,j=0}^{k-1,l-1} y_{ij} - \sum_{i,j=0}^{k,l-1} y_{ij} - \sum_{i,j=0}^{k-1,l} y_{ij} + \sum_{i,j=0}^{k,l} y_{ij}\right| \\
&= \sup_{k,l\in\mathbb{N}} \left|\sum_{i,j=0}^{k-1,l-1} y_{ij} - \left(\sum_{j=0}^{l-1} y_{kj} + \sum_{i,j=0}^{k-1,l-1} y_{ij}\right)\right. \\
&\quad - \left(\sum_{i=0}^{k-1} y_{il} + \sum_{i,j=0}^{k-1,l-1} y_{ij}\right) \\
&\quad \left. + \left(\sum_{j=0}^{l-1} y_{kj} + \sum_{i=0}^{k-1} y_{il} + \sum_{i,j=0}^{k-1,l-1} y_{ij} + y_{kl}\right)\right| \\
&= \sup_{k,l\in\mathbb{N}} |y_{kl}| = \|y\|_\infty < \infty.
\end{aligned}$$

It shows that $x \in P(\mathscr{C}_{r0}(\Delta))$ and consequently $\Delta$ is surjective and norm preserving. It completes the first part of the proof.

Now suppose that $f_\Delta$ is a linear functional on $P(\mathscr{C}_{r0}(\Delta))$. If $f_\Delta$ is continuous, then $f_\Delta \circ \Delta^{-1}$ is a continuous linear functional on $\mathscr{C}_{r0}$. Then by [13, Remark 3] there exists a double sequence $a = (a_{kl})_{k,l \geq 1} \in \mathscr{L}_u$ such that

$$f_\Delta \circ \Delta^{-1}(y) = \sum_{k,l=0}^{\infty} a_{kl} y_{kl}$$

for all $y \in \mathscr{C}_{r0}$. It gives

$$f_\Delta(x) = \left(f_\Delta \circ \Delta^{-1}\right)(\Delta x) = \sum_{k,l=0}^{\infty} a_{kl} (\Delta x)_{kl}$$

for all $x \in P(\mathscr{C}_{r0}(\Delta))$. Conversely, if $f_\Delta(x) = \sum_{k,l=1}^{\infty} a_{kl}(\Delta x)_{kl}$ for all $x \in P(\mathscr{C}_{r0}(\Delta))$ and for some $a = (a_{kl}) \in \mathscr{L}_u$, then

$$\begin{aligned} |f_\Delta(x)| = \left| \sum_{k,l=0}^{\infty} a_{kl} (\Delta x)_{kl} \right| &\leq \sum_{k,l=1}^{\infty} |a_{kl}| |(\Delta x)_{kl}| \\ &\leq \|x\|_{P(\mathscr{C}_{r0}(\Delta))} \sum_{k,l=0}^{\infty} |a_{kl}| \\ &= \|x\|_{P(\mathscr{C}_{r0}(\Delta))} \|a\|_{\mathscr{L}_u}. \end{aligned}$$

Therefore, $\|f_\Delta\| \leq \|a\|_{\mathscr{L}_u}$ and then we see that $f_\Delta$ is a bounded(continuous) linear functional on $P(\mathscr{C}_{r0}(\Delta))$. This completes the proof.

**Definition 5.2.3.** *Let $X$ and $Y$ be Banach spaces, and $\mathscr{B}(X,Y)$ be the space of bounded linear operators from $X$ into $Y$. An operator $T \in \mathscr{B}(X,Y)$ is called an isometry if $\|Tx\| = \|x\|$ for all $x \in X$.*

Now we denote the continuous duals of $P(\lambda(\Delta))$ and $\lambda$ by $[P(\lambda(\Delta))]^*$ and $\lambda^*$, respectively. We may now show that the operator

$$\begin{aligned} T : [P(\lambda(\Delta))]^* &\rightarrow \lambda^* \\ f_\Delta &\rightarrow f = f_\Delta \circ (\Delta^{-1}) \end{aligned}$$

is a linear isometry. Hence, $[P(\mathscr{M}_u(\Delta))]^* \cong \mathscr{M}_u^*$, by [13, Remark 3] we have $[P(\lambda(\Delta))]^* \cong \lambda^* \cong \mathscr{L}_u$, where $\lambda \in \{\mathscr{C}_r, \mathscr{C}_{r0}\}$, by [13, Theorem 8] we have $[P(\mu(\Delta))]^* \cong \mu^* \cong \ell_1(\ell_\infty^*)$, where $\mu \in \{\mathscr{C}_{bp}, \mathscr{C}_{bp0}\}$, and the sets $\ell_1$ and $\ell_\infty$ represent absolutely summable and bounded single sequence spaces, respectively.

Now we prove the following Theorem only for the case $\lambda = \mathscr{C}_{r0}$.

**Theorem 5.2.4.** *The continuous dual $[P(\mathscr{C}_{r0}(\Delta))]^*$ is isometrically isomorphic to $\mathscr{C}_{r0}^* \cong \mathscr{L}_u$.*

**Proof:** Let us define an operator

$$T : [P(\mathscr{C}_{r0}(\Delta))]^* \rightarrow \mathscr{C}_{r0}^* \cong \mathscr{L}_u$$

with $T(f_\Delta) = \left(f_\Delta(e^{kl})\right)_{k,l\geq 1}$,

$$T(f_\Delta(x)) = T\left(\left(f_\Delta \circ \Delta^{-1}\right)(\Delta x)\right) = \sum_{k,l=1}^{\infty} a_{kl} T((\Delta x)_{kl}),$$

where $a = (a_{kl}) \in \mathcal{L}_u$. Therefore, $T$ is a surjective linear map by Lemma 5.2.2. Moreover, since $T(f_\Delta(e^{kl})) = 0 = (0,0,0,...)$ implies $f_\Delta = 0$, where $(x_{kl}) = e^{kl}$ is Schauder basis for $\mathscr{C}_{r0}$ by the definition of double Schauder basis [22, Definition 4.2, p. 14], $T$ is injective. Let $f_\Delta \in [P(\mathscr{C}_{r0}(\Delta))]^*$ and $x \in P(\mathscr{C}_{r0}(\Delta))$. Then we have

$$\begin{aligned}
|f_\Delta(x)| = \left| f_\Delta\left(\sum_{k,l=1}^{\infty} (\Delta x)_{kl} e^{kl}\right)\right| &= \left|\sum_{k,l=1}^{\infty} (\Delta x)_{kl} f_\Delta(e^{kl})\right| \\
&\leq \sum_{k,l=1}^{\infty} \left|f_\Delta(e^{kl})\right| |(\Delta x)_{kl}| \\
&\leq \sup_{k,l\in\mathbb{N}} |(\Delta x)_{kl}| \sum_{k,l=1}^{\infty} \left|f_\Delta(e^{kl})\right| \\
&\leq \|x\|_{P(\mathscr{C}_{r0}(\Delta))} \|T(f_\Delta)\|_{\mathcal{L}_u}.
\end{aligned}$$

Then we obtain

$$\|f_\Delta\|_\infty \leq \|T(f_\Delta)\|_{\mathcal{L}_u}. \tag{5.9}$$

Furthermore, since $\left|f_\Delta(e^{kl})\right| \leq \|f_\Delta\|_\infty \|e^{kl}\|_{P(\mathscr{C}_{r0}(\Delta))} = \|f_\Delta\|_\infty$, then we have

$$\|T(f_\Delta)\|_{\mathcal{L}_u} = \sup_{k,l\in\mathbb{N}} \left|f_\Delta(e^{kl})\right| \leq \|f_\Delta\|_\infty. \tag{5.10}$$

We obtain by (5.9) and (5.10) that $\|T(f_\Delta)\|_{\mathcal{L}_u} = \|f_\Delta\|_\infty$. This completes the proof.

## 5.3 DUAL SPACES OF THE NEW DOUBLE SEQUENCE SPACES

In this section, we determine the $\alpha-$, $\beta(\vartheta)-$ and $\gamma-$duals of our new double sequence spaces. First, we begin with some lemmas to determine the $\alpha-$, $\beta(\vartheta)-$ and $\gamma-$duals of the spaces $\mathscr{M}_u(\Delta)$, $\mathscr{C}_\vartheta(\Delta)$, where $\vartheta \in \{bp, r\}$.

**Lemma 5.3.1.** *We have* $\sup_{k,l\in\mathbb{N}} |\Delta x_{kl}| < \infty$ *if and only if*

*(i)* $\sup_{k,l\in\mathbb{N}} \frac{1}{kl}|x_{kl}| < \infty$,
*(ii)* $\sup_{k,l\in\mathbb{N}} \left|kl\Delta\left(\frac{1}{kl}x_{kl}\right)\right| < \infty$.

**Proof:** Suppose that there exists a positive real number $M$ such that

$$\sup_{k,l\in\mathbb{N}} |x_{kl} - x_{k+1,l} - x_{k,l+1} + x_{k+1,l+1}| \leq M.$$

Then

$$|x_{kl}| = |x_{k,0} + x_{0,l} - x_{00} + x_{kl}| = \left|\sum_{i,j=0}^{k-1,l-1} \Delta x_{ij}\right| \leq \sum_{i,j=0}^{k-1,l-1} |\Delta x_{ij}| \leq M(kl).$$

It is seen that (i) is necessary. Moreover, by considering the condition (i) there exist positive real numbers $N_1, N_2, N_3$ such that

$$\sup_{k,l\in\mathbb{N}} \frac{1}{(k+1)l}|x_{k+1,l}| \leq N_1, \tag{5.11}$$

$$\sup_{k,l\in\mathbb{N}} \frac{1}{k(l+1)}|x_{k,l+1}| \leq N_2, \tag{5.12}$$

$$\sup_{k,l\in\mathbb{N}} \frac{1}{(k+1)(l+1)}|x_{k+1,l+1}| \leq N_3. \tag{5.13}$$

Then we have

$$\begin{aligned}
kl\left|\Delta\left(\frac{1}{kl}x_{kl}\right)\right| &= kl\left|\frac{1}{kl}x_{kl} - \frac{1}{(k+1)l}x_{k+1,l} - \frac{1}{k(l+1)}x_{k,l+1}\right.\\
&\quad \left. + \frac{1}{(k+1)(l+1)}x_{k+1,l+1}\right|\\
&= kl\left|\frac{1}{kl}\Delta x_{kl} + \left(\frac{1}{kl(k+1)}x_{k+1,l} + \frac{1}{kl(l+1)}x_{k,l+1}\right.\right.\\
&\quad \left.\left. - \frac{(k+l+1)}{kl(k+1)(l+1)}x_{k+1,l+1}\right)\right|\\
&\leq kl\left(\left|\frac{1}{kl}\Delta x_{kl}\right| + \left|\frac{1}{kl(k+1)}x_{k+1,l}\right| + \left|\frac{1}{kl(l+1)}x_{k,l+1}\right|\right.\\
&\quad \left. + \left|\frac{(k+l+1)}{kl(k+1)(l+1)}x_{k+1,l+1}\right|\right)\\
&\leq M',
\end{aligned}$$

where $M' = M + N_1 + N_2 + N_3$. So it gives the necessity of (ii).

Now let us suppose that the conditions (i) and (ii) hold. By only considering the following inequality

$$
\begin{aligned}
kl\left|\Delta\left(\frac{1}{kl}x_{kl}\right)\right| &= \left|\frac{kl}{kl}x_{kl}-\frac{kl}{(k+1)l}x_{k+1,l}-\frac{kl}{k(l+1)}x_{k,l+1}\right. \\
&\quad \left.+\frac{kl}{(k+1)(l+1)}x_{k+1,l+1}\right| \\
&= kl\left|\frac{1}{kl}\Delta x_{kl}-\left(\frac{1}{kl(k+1)}x_{k+1,l}+\frac{1}{kl(l+1)}x_{k,l+1}\right.\right. \\
&\quad \left.\left.-\frac{(k+l+1)}{kl(k+1)(l+1)}x_{k+1,l+1}\right)\right| \\
&\geq |\Delta x_{kl}|-\left|-\frac{1}{(k+1)}x_{k+1,l}-\frac{1}{(l+1)}x_{k,l+1}\right. \\
&\quad \left.+\frac{(k+l+1)}{(k+1)(l+1)}x_{k+1,l+1}\right|
\end{aligned}
$$

we can see the necessity of $\sup_{k,l\in\mathbb{N}}|\Delta x_{kl}|<\infty$.

**Lemma 5.3.2.** *Let* $\Delta x_{kl}=y_{kl}$. *If*

$$\sup_{m,n\in\mathbb{N}}\left|\sum_{k,l=1}^{m,n}y_{kl}\right|<\infty,$$

*then*

$$\sup_{m,n\in\mathbb{N}}\left((m+1)(n+1)\left|\sum_{k,l=1}^{\infty}\frac{y_{m+k-1,n+l-1}}{(m+k)(n+l)}\right|\right)<\infty$$

**Proof:** Let us consider Abel's double partial summation on the $(s,t)^{th}-$ partial sum of the series $\sum_{k,l=1}^{\infty}\frac{y_{m+k+1,n+l+1}}{(m+k)(n+l)}$ as in the following equation.

$$
\begin{aligned}
\sum_{k,l=1}^{s,t}\frac{y_{m+k-1,n+l-1}}{(m+k)(n+l)} &= \sum_{k,l=1}^{s,t}y_{m+k-1,n+l-1}\left(\frac{1}{(m+k)(n+l)}\right) \qquad (5.14)\\
&= \sum_{k,l=1}^{s-1,t-1}\left(\sum_{i,j=1}^{k,l}y_{m+i-1,n+j-1}\right)\Delta_{11}^{kl}\left(\frac{1}{(m+k)(n+l)}\right) \\
&\quad +\sum_{k=1}^{s-1}\left(\sum_{i,j=1}^{k,t}y_{m+i-1,n+j-1}\right)\Delta_{10}^{kl}\left(\frac{1}{(m+k)(n+t)}\right) \\
&\quad +\sum_{l=1}^{t-1}\left(\sum_{i,j=1}^{s,l}y_{m+i-1,n+j-1}\right)\Delta_{01}^{kl}\left(\frac{1}{(m+s)(n+l)}\right) \\
&\quad +\sum_{i,j=1}^{s,t}y_{m+i-1,n+j-1}\left(\frac{1}{(m+s)(n+t)}\right),
\end{aligned}
$$

where for the double sequence $a_{kl} = \frac{1}{(m+k)(n+l)}$

$$\Delta_{10}^{kl} a_{kl} = a_{kl} - a_{k+1,l}$$
$$\Delta_{01}^{kl} a_{kl} = a_{kl} - a_{k,l+1}$$
$$\Delta_{11}^{kl} a_{kl} = \Delta_{10}^{kl}(\Delta_{01}^{kl} a_{kl}) = \Delta_{01}^{kl}(\Delta_{10}^{kl} a_{kl}) = a_{kl} - a_{k+1,l} - a_{k,l+1} + a_{k+1,l+1}.$$

Since there exists a positive real number $M$ such that

$$\sup_{m,n\in\mathbb{N}} \left| \sum_{k,l=1}^{m,n} y_{kl} \right| \le M, \tag{5.15}$$

the equation (5.14) is written as

$$\begin{aligned} \sum_{k,l=1}^{s,t} \frac{y_{m+k-1,n+l-1}}{(m+k)(n+l)} &\le M\Bigg[\sum_{k,l=1}^{s-1,t-1}\left(\frac{1}{(m+k)(n+l)} - \frac{1}{(m+k+1)(n+l)}\right.\\ &\quad \left. - \frac{1}{(m+k)(n+l+1)} + \frac{1}{(m+k+1)(n+l+1)}\right)\\ &\quad + \sum_{k=1}^{s-1} \frac{1}{(n+t)}\left(\frac{1}{(m+k)} - \frac{1}{(m+k+1)}\right)\\ &\quad + \sum_{l=1}^{t-1} \frac{1}{(m+s)}\left(\frac{1}{(n+l)} - \frac{1}{(n+l+1)}\right)\\ &\quad + \frac{1}{(m+s)(n+t)}\Bigg]\\ &= \frac{M}{(m+1)(n+1)}. \end{aligned}$$

Therefore by passing to $\vartheta-$limit as $s,t \to \infty$, where $\vartheta = \{bp,r\}$, and taking supremum over $m,n \in \mathbb{N}$, then the condition

$$\sup_{m,n\in\mathbb{N}} \left( (m+1)(n+1) \left| \sum_{k,l=1}^{\infty} \frac{y_{m+k-1,n+l-1}}{(m+k)(n+l)} \right| \right) < \infty$$

is immediate.

**Lemma 5.3.3.** *Let $\vartheta \in \{bp,r\}$. If the series $\sum_{k,l=1}^{\infty} \Delta x_{kl}$ is $\vartheta-$convergent, then*

$$\vartheta - \lim_{m,n\to\infty} \left( (m+1)(n+1) \left| \sum_{k,l=1}^{\infty} \frac{y_{m+k-1,n+l-1}}{(m+k)(n+l)} \right| \right) = 0.$$

**Proof:** Since the partial sum of the series $\sum_{k,l=1}^{\infty} \Delta x_{kl}$ is $\vartheta-$convergent, where $\vartheta \in \{bp,r\}$, we have

$$\left| \sum_{i,j=1}^{k,l} y_{m+i-1,n+j-1} \right| = \left| \sum_{i,j=m,n}^{m+k-1,n+l-1} y_{ij} \right| = O(1).$$

Then by using the equality (5.14) we write

$$(m+1)(n+1)\left|\sum_{k,l=1}^{\infty}\frac{y_{m+k-1,n+l-1}}{(m+k)(n+l)}\right| = O(1).$$

If we let $\vartheta-$limit as $m,n\to\infty$, we reach the proof.

**Corollary 5.3.4.** *Let $\vartheta\in\{bp,r\}$ and $a=(a_{kl})$ be any double sequence. Then*

*(i) If $\sup_{m,n\in\mathbb{N}}\left|\sum_{k,l=1}^{m,n} kla_{kl}\right| < \infty$, then*

$$\sup_{m,n\in\mathbb{N}}\left|mn\sum_{k,l=m+1,n+1}^{\infty} a_{kl}\right| < \infty.$$

*(ii) If $\sum_{k,l=1}^{\infty} kla_{kl}$ is $\vartheta-$convergent, then*

$$\vartheta-\lim_{m,n\to\infty}\left(mn\sum_{k,l=m+1,n+1}^{\infty} a_{kl}\right) = 0.$$

*(iii) $\sum_{k,l=1}^{\infty} kla_{kl}$ is $\vartheta-$convergent if and only if*

$$\sum_{k,l=1}^{\infty} R_{kl} \text{ is } \vartheta-\text{convergent with } mnR_{mn} = O(1),$$

*where $R_{mn}=\sum_{k,l=m+1,n+1}^{\infty} a_{kl}$.*

**Proof:** The proof of (i) and (ii) can be easily seen by writing $kla_{kl}$ instead of $y_{kl}$ in Lemma 5.3.2, and writing $(k+1)(l+1)a_{k+1,l+1}$ instead of $y_{kl}$ in Lemma 5.3.3, respectively.

To prove the corollary (iii), the following $(s,t)^{th}-$ partial sum can be written by using Abel's double summation formula that

$$\begin{aligned}\sum_{k,l=1}^{s,t} kla_{kl} &= \sum_{k,l=1}^{s-1,t-1}\left(\sum_{i,j=0}^{k,l} a_{ij}\right)\Delta_{11}^{kl}(kl)+\sum_{k=1}^{s-1}\left(\sum_{i,j=0}^{k,t} a_{ij}\right)\Delta_{10}^{kl}(kl)\\ &\quad +\sum_{i,j=0}^{s,t} a_{ij}(st)\\ &= \sum_{k,l=1}^{s,t}\left(\sum_{i,j=k,l}^{s,t} a_{ij}\right)+st\sum_{k,l=s+1,t+1}^{\infty} a_{kl}.\end{aligned}$$

Letting $\vartheta-$limit as $s,t\to\infty$, we obtain the statement in Part (iii).

Let us define the following sets to be able to define the dual spaces of $\lambda(\Delta)$.

$$D_1 := \int \mathscr{L}_u := \left\{ a = (a_{kl}) \in \Omega : \sum_{k,l=1}^{\infty} kl|a_{kl}| < \infty \right\},$$

$$D_2 := \int \mathscr{C}\mathscr{S}_{\vartheta} := \left\{ a = (a_{kl}) \in \Omega : \sum_{k,l=1}^{\infty} kla_{kl} \ \ is \ \ \vartheta - convergent \right\},$$

$$D_3 := \int \mathscr{B}\mathscr{S} := \left\{ a = (a_{kl}) \in \Omega : \sum_{m,n} \left| \sum_{k,l=1}^{m,n} kla_{kl} \right| < \infty \right\},$$

$$D_4 := \left\{ a = (a_{kl}) \in \Omega : \sum_{k,l=1}^{\infty} \left| \sum_{i,j=k,l}^{\infty} a_{ij} \right| < \infty \right\}.$$

**Theorem 5.3.5.** *Let $\lambda \in \{\mathscr{M}_u, \mathscr{C}_{bp}, \mathscr{C}_r\}$. Then $[P(\lambda(\Delta))]^{\alpha} = D_1$.*

**Proof:** We need to prove the existence of the inclusion relations $D_1 \subset [P(\lambda(\Delta))]^{\alpha}$ and $[P(\lambda(\Delta))]^{\alpha} \subset D_1$.

Suppose that $a = (a_{kl}) \in D_1$, i.e., $\sum_{k,l=1}^{\infty} kl|a_{kl}| < \infty$. Then by using Lemma 5.3.1 we have

$$\sum_{k,l=1}^{\infty} |a_{kl}x_{kl}| = \sum_{k,l=1}^{\infty} kl|a_{kl}| \left( \frac{|x_{kl}|}{kl} \right) < \infty$$

for all $x = (x_{kl}) \in P(\lambda(\Delta))$. This shows that $a = (a_{kl}) \in [P(\lambda(\Delta))]^{\alpha}$. Hence, the inclusion $D_1 \subset [P(\lambda(\Delta))]^{\alpha}$ holds.

Now suppose that $a = (a_{kl}) \in [P(\lambda(\Delta))]^{\alpha}$, i.e., $\sum_{k,l=1}^{\infty} |a_{kl}x_{kl}| < \infty$ for all $x = (x_{kl}) \in P(\lambda(\Delta))$. If we consider the double sequence $x = (x_{kl})$ as

$$x_{kl} := \begin{cases} 0, & k = 0, l \geq 0 \\ 0, & l = 0, k \geq 0 \\ kl, & k \geq 1, l \geq 1. \end{cases} \tag{5.16}$$

Then we have

$$\sum_{k,l=1}^{\infty} |a_{kl}x_{kl}| = \sum_{k,l=1}^{\infty} kl|a_{kl}| < \infty,$$

which says $a = (a_{kl}) \in D_1$. Hence, the inclusion $[P(\lambda(\Delta))]^{\alpha} \subset D_1$ holds. This concludes the proof.

**Theorem 5.3.6.** *Let $\lambda \in \{\mathscr{M}_u, \mathscr{C}_{bp}, \mathscr{C}_r\}$. Then $[P(\lambda(\Delta))]^{\beta(\vartheta)} = D_2 \cap D_4$.*

**Proof:** We should show the validity of the inclusions $D_2 \cap D_4 \subset [P(\lambda(\Delta))]^{\beta(\vartheta)}$ and $[P(\lambda(\Delta))]^{\beta(\vartheta)} \subset D_2 \cap D_4$.

Suppose that the double sequence $a = (a_{kl}) \in D_2 \cap D_4$ and the sequence $x = (x_{kl}) \in P(\lambda(\Delta))$ are defined with the relation (5.7) between the terms of the sequence $x = (x_{kl})$ and $y = (y_{kl})$ as

$$x_{kl} = \sum_{i,j=1}^{k,l} y_{i-1,j-1}, \tag{5.17}$$

where $y = (y_{kl}) \in \lambda$ which is defined as

$$y_{kl} := \begin{cases} x_{11} & , \quad k=0, l=0 \\ -x_{11}+x_{12} & , \quad k=0, l=1 \\ -x_{11}+x_{21} & , \quad k=1, l=0 \\ x_{kl}-x_{k+1,l}-x_{k,l+1}+x_{k+1,l+1}, & \quad k \geq 1, l \geq 1. \end{cases} \tag{5.18}$$

Then, we have the following $(s,t)^{th}$−partial sum of the series $\sum_{k,l} a_{kl}x_{kl}$ that

$$\begin{aligned} \sum_{k,l=1}^{s,t} a_{kl}x_{kl} &= \sum_{k,l=1}^{s,t} a_{kl}\left(\sum_{i,j=1}^{k,l} y_{i-1,j-1}\right) \\ &= \sum_{k,l=1}^{s-1,t-1}\left(\sum_{i,j=k,l}^{s-1,t-1} a_{ij}\right) y_{kl} \\ &= \sum_{k,l=1}^{s-1,t-1}\left(\sum_{i,j=k,l}^{\infty} a_{ij}\right) y_{kl} - \sum_{k,l=1}^{s-1,t-1}\left(\sum_{i,j=s,t}^{\infty} a_{ij}\right) y_{kl} \\ &= \sum_{k,l=1}^{s-1,t-1} R_{kl}y_{kl} - R_{st}\sum_{k,l=1}^{s-1,t-1} y_{kl}. \end{aligned}$$

Now, by the Corollary 5.3.4(iii), we can say that the sequence $\sum_{k,l=1}^{s,t} a_{kl}x_{kl}$ is $\vartheta-$ convergent for every $x = (x_{kl}) \in P(\lambda(\Delta))$, since $\sum_{k,l=1}^{s-1,t-1} R_{kl}y_{kl}$ is $\vartheta-$ convergent with $x_{st}R_{st} \to 0$ as $s,t \to \infty$. This yields that $a = (a_{kl}) \in [P(\lambda(\Delta))]^{\beta(\vartheta)}$ and the inclusion $D_2 \cap D_4 \subset [P(\lambda(\Delta))]^{\beta(\vartheta)}$ holds.

Now, suppose that $a = (a_{kl}) \in [P(\lambda(\Delta))]^{\beta(\vartheta)}$. Then the series $\sum_{k,l=1}^{\infty} a_{kl}x_{kl}$ is $\vartheta-$convergent for every $x = (x_{kl}) \in P(\lambda(\Delta))$. If we consider the sequence $x = (x_{kl})$ defined in (5.16) Then, we can observe that

$$\sum_{k,l=1}^{\infty} a_{kl}x_{kl} = \sum_{k,l=1}^{\infty} kl a_{kl}$$

and by the equality $y = \Delta x$ we have the following series

$$\begin{aligned} \sum_{k,l=1}^{s,t} kl a_{kl} &= \sum_{k,l=1}^{s-1,t-1}\left(\sum_{i,j=k,l}^{\infty} a_{ij}\right) - \sum_{k,l=1}^{s-1,t-1}\left(\sum_{i,j=s,t}^{\infty} a_{ij}\right) \\ &= \sum_{k,l=1}^{s-1,t-1} R_{kl} - stR_{st}, \end{aligned}$$

which is $\vartheta-$convergent as $s,t \to \infty$. Thus, $a = (a_{kl}) \in D_2$. Moreover, by Corollary 5.3.4 (ii) we can write that $stR_{st} \to 0$ as $s,t \to \infty$ for every $y = (y_{kl}) \in \lambda$, and $\sum_{k,l=1}^{\infty} R_{kl} < \infty$. Therefore, $a = (a_{kl}) \in D_4$. Hence the inclusion $[P(\lambda(\Delta))]^{\beta(\vartheta)} \subset D_2 \cap D_4$ holds. This completes the proof.

**Theorem 5.3.7.** *Let $\lambda \in \{\mathcal{M}_u, \mathcal{C}_\vartheta\}$. Then $[P(\lambda(\Delta))]^\gamma = D_3 \cap D_4$, where $\vartheta \in \{bp, r\}$.*

**Proof:** The proof can be done with a similar path as above by considering Corollary 5.3.4 (i). So, we omit the repetition.

## 5.4 MATRIX TRANSFORMATIONS

In this section, we characterize the four-dimensional matrix mapping from the sequence space $\lambda(\Delta)$ to $\mu$ and vice-versa. Then we conclude the section with some significant results.

**Theorem 5.4.1.** *The four-dimensional matrix $A = (a_{mnkl}) \in (\lambda(\Delta) : \mu)$ if and only if*

$$A_{mn} = (a_{mnkl})_{k,l\in\mathbb{N}} \in (\lambda(\Delta))^{\beta(\vartheta)} \text{ for all } m,n \in \mathbb{N}, \tag{5.19}$$

$$A_{mn}(kl) = \sum_{k,l=1}^{\infty} kl a_{mnkl} \in \mu, \tag{5.20}$$

$$B = (b_{mnkl}) \in (\lambda : \mu), \tag{5.21}$$

*where the four-dimensional matrix*

$$B = (b_{mnkl}) = \sum_{i,j=k,l}^{\infty} a_{mnij} \text{ for all } m,n,k,l \in \mathbb{N}. \tag{5.22}$$

**Proof:** Suppose that $A = (a_{mnkl}) \in (\lambda(\Delta) : \mu)$. Then, $A_{mn}(x)$ exists for every $x = (x_{kl}) \in \lambda(\Delta)$ and is in $\mu$ for all $m,n \in \mathbb{N}$. If we define the sequence $x = (x_{kl})$ by

$$x_{kl} := \begin{cases} 1 & , \quad k = l \\ 0 & , \quad otherwise \end{cases} \tag{5.23}$$

for all $k,l \in \mathbb{N}$, then the necessity of (5.19) is clear. If we define the sequence $x = (x_{kl})$ as $x_{kl} = kl$ for all $k,l \in \mathbb{N}$, then the necessity of (5.20) is also clear by Theorem 5.3.6. Moreover, by Theorem 5.3.6, we have $\sum_{k,l=1}^{\infty} |a_{mnkl}| < \infty$ for each $m,n \in \mathbb{N}$.

Now suppose that $x = (x_{kl}) \in P(\lambda(\Delta)) \subset \lambda(\Delta)$ let us consider the $(s,t)^{th}-$partial sum of the series $\sum_{k,l=1}^{\infty} a_{mnkl}x_{kl}$ by considering the relation $x_{kl} = \sum_{i,j=0}^{k-1,l-1} y_{ij}$

between terms of the sequences $x=(x_{kl})$ and $y=(y_{kl})$ as in the following

$$\begin{aligned}
A_{mn}^{st}(x) &= \sum_{k,l=1}^{s,t} a_{mnkl}x_{kl} \\
&= \sum_{k,l=1}^{s,t} a_{mnkl}\left(\sum_{i,j=0}^{k-1,l-1} y_{ij}\right) \\
&= \sum_{k,l=1}^{s-1,t-1}\left(\sum_{i,j=k,l}^{s-1,t-1} a_{mnij}\right)y_{kl} \\
&= \sum_{k,l=1}^{s-1,t-1}\left(\sum_{i,j=k,l}^{\infty} a_{mnij}\right)y_{kl} - \sum_{k,l=1}^{s-1,t-1}\left(\sum_{i,j=s,t}^{\infty} a_{mnij}\right)y_{kl} \\
&= \sum_{k,l=1}^{s-1,t-1} b_{mnkl}y_{kl} - b_{mnst}\sum_{k,l=1}^{s-1,t-1} y_{kl},
\end{aligned}$$

where $y\in\lambda$. We obtain by letting $\vartheta-$limit as $s,t\to\infty$ and by considering the Corollary 5.3.4(iii) that $A_{mn}(x)=\sum_{k,l=1}^{\infty} b_{mnkl}y_{kl}$, that is $Ax=By$ for each $m,n\in\mathbb{N}$. Therefore, $A=(a_{mnkl})\in(\lambda(\Delta):\mu)$ implies that $B=(b_{mnkl})\in(\lambda:\mu)$.

Now suppose that the conditions (5.19)–(5.21) hold. Let us take a sequence $x=(x_{kl})\in\lambda(\Delta)$ defined by

$$x_{kl} := \begin{cases} x_{k,1}, & k\geq 1, l=1 \\ x_{1,l}, & k=l, l\geq 1 \\ \widetilde{x_{kl}}, & k>l, l>1, \end{cases}$$

where $\widetilde{x}=(\widetilde{x_{kl}})\in P(\lambda(\Delta))$. Then, if we write again the above $(s,t)^{th}-$partial sum of the series $\sum_{k,l=1}^{\infty} a_{mnkl}x_{kl}$, we have

$$\begin{aligned}
A_{mn}^{st}(x) &= \sum_{k,l=1}^{s,t} a_{mnkl}x_{kl} \\
&= a_{mn11}x_{11} + \sum_{l=2}^{t} a_{mn,1,l}x_{1,l} + \sum_{k=2}^{s} a_{mn,k,1}x_{k,1} + \sum_{k,l=2}^{s,t} a_{mnkl}\widetilde{x_{kl}} \\
&= a_{mn11}x_{11} + \sum_{k=2}^{s-1} b_{mnk,1}y_{k,1} + \sum_{l=2}^{t-1} b_{mn,1,l}y_{1,l} \\
&\quad + \sum_{k,l=1}^{s-1,t-1} b_{mnkl}y_{kl} - b_{mnst}\sum_{k,l=1}^{s-1,t-1} y_{kl}.
\end{aligned}$$

Therefore, we obtain by letting limit as $s,t\to\infty$ that

$$A_{mn}(x) = a_{mn11}x_{11} + \sum_{k=2}^{\infty} b_{mnk,1}y_{k,1} + \sum_{l=2}^{\infty} b_{mn,1,l}y_{1,l} + \sum_{k,l=1}^{\infty} b_{mnkl}y_{kl}.$$

Thus, $A_{mn}(x)$ exists for each $x = (x_{kl}) \in \lambda(\Delta)$ and is in $\mu$ since $B \in (\lambda : \mu)$. This completes the proof.

We list some four-dimensional matrix classes from and into the sequence spaces $\lambda, \mu = \{\mathscr{M}_u, \mathscr{C}_{bp}, \mathscr{C}_r\}$ as in the following table, which have been characterized in some distinguished papers (see [3, Theorem 3.5], [23, Lemma 3.2], [24, Theorem 2.2], [25, Theorem 3.2]).

$$\sup_{m,n\in\mathbb{N}} \sum_{k,l} |a_{mnkl}| < \infty, \tag{5.24}$$

$$\exists\, a_{kl} \in \mathbb{C} \ni \vartheta - \lim_{m,n\to\infty} a_{mnkl} = a_{kl} \text{ for all } k,l \in \mathbb{N}, \tag{5.25}$$

$$\exists\, l \in \mathbb{C} \ni \vartheta - \lim_{m,n\to\infty} \sum_{k,l} a_{mnkl} = l \text{ exists }, \tag{5.26}$$

$$\exists\, k_0 \in \mathbb{N} \ni \vartheta - \lim_{m,n\to\infty} \sum_{l} |a_{mnk_0l} - a_{k_0l}| = 0, \tag{5.27}$$

$$\exists\, l_0 \in \mathbb{N} \ni \vartheta - \lim_{m,n\to\infty} \sum_{k} |a_{mnkl_0} - a_{kl_0}| = 0, \tag{5.28}$$

$$\exists\, l_0 \in \mathbb{N} \ni \vartheta - \lim_{m,n\to\infty} \sum_{k} a_{mnkl_0} = u_{l_0}, \tag{5.29}$$

$$\exists\, k_0 \in \mathbb{N} \ni \vartheta - \lim_{m,n\to\infty} \sum_{l} a_{mnk_0l} = v_{k_0}, \tag{5.30}$$

$$\exists\, a_{kl} \in \mathbb{C} \ni bp - \lim_{m,n\to\infty} \sum_{k,l} |a_{mnkl} - a_{kl}| = 0, \tag{5.31}$$

$$bp - \lim_{m,n\to\infty} \sum_{l=0}^{n} a_{mnkl} \text{ exists for each } k \in \mathbb{N}, \tag{5.32}$$

$$bp - \lim_{m,n\to\infty} \sum_{k=0}^{m} a_{mnkl} \text{ exists for each } l \in \mathbb{N}, \tag{5.33}$$

$$\sum_{k,l} |a_{mnkl}| \text{ converges.} \tag{5.34}$$

**Theorem 5.4.2.** *The four-dimensional matrix $A = (a_{mnkl}) \in (\mu : \lambda(\Delta))$ if and only if*

$$A_{mn} \in \mu^{\beta(\vartheta)}, \tag{5.35}$$

$$F = (f_{mnkl}) \in (\mu : \lambda), \tag{5.36}$$

*where the four-dimensional matrix*

$$F = (f_{mnkl}) = \Delta_{11}^{mn} a_{mnij} = a_{mnij} - a_{m+1,nij} - a_{m,n+1,ij} + a_{m+1,n+1,ij}. \tag{5.37}$$

**Proof:** Suppose that $A = (a_{mnkl}) \in (\mu : \lambda(\Delta))$. Then, $A_{mn}(x)$ exists for every $x = (x_{kl}) \in \mu$ and is in $\lambda(\Delta)$ for all $m,n \in \mathbb{N}$. Thus, the necessity of (5.35) is immediate. Since $A_{mn}(x) \in \lambda(\Delta)$, then $\Delta A \in \lambda$ for every $x = (x_{kl}) \in \mu$. $\Delta A$ is the matrix $F$. Hence, the necessity of the condition $F = (f_{mnkl}) \in (\lambda : \mu)$ can be clearly seen. The rest of the theorem can be followed by a similar path as in Theorem 5.4.1. We omit the details.

## 5.5 CONCLUSION

In this work, we introduced the new double sequence spaces $\mathscr{M}_u(\Delta), \mathscr{C}_\vartheta(\Delta)$, where $\vartheta \in \{bp, r\}$ derived by the domain of four-dimensional forward difference matrix $\Delta$ in the double sequence spaces $\mathscr{M}_u, \mathscr{C}_\vartheta$, where $\vartheta \in \{bp, bp0, r, r0\}$. Then we presented some topological and algebraic properties, determined $\alpha-$, $\beta(\vartheta)-$, and $\gamma-$duals, and characterized some four-dimensional matrix classes related to these new double sequence spaces.

The chapter contributes significant nonstandard results and represents some useful contributions to the theory of double sequences. As a natural continuation of this paper, the four-dimensional forward difference matrix domain in the double sequence spaces $\mathscr{C}_p$ and $\mathscr{L}_q$, where $0 < q < \infty$ can be calculated as an open problem. Moreover, the four-dimensional forward difference matrix domain in the spaces $\mathscr{C}_f$, $\mathscr{BS}$, $\mathscr{CS}$ and $\mathscr{BV}$ can be calculated. Furthermore, Hahn double sequence space can be defined and studied by using some significant results stated in this work.

## REFERENCES

1. Demiriz S., Duyar O., Domain of difference matrix of order one in some spaces of double sequences. arXiv preprint arXiv:1501.01113 (2015).
2. Tug O., Basar F., Four-dimensional generalized difference matrix and some double sequence spaces. AIP Conference Proceedings, **1759**(1), (2016).
3. Tug O., Four-dimensional generalized difference matrix and some double sequence spaces. J. Inequal. Appl., **149**(1), (2017).
4. Tug O., On the Matrix Domain in the Sequence $\mathscr{L}_u$, Eurasian J. Sci. Eng., **3**(1), (2017), 204–211.
5. Tug O., On almost B-summable double sequence spaces, J. Inequal. Appl., **1**(9), (2018), 1–9.
6. Tug O., On the characterization of some classes of four-dimensional matrices and almost $B-$summable double sequences, J. Math., **2018**, (2018), Article ID 1826485, 7 pages.
7. Tug O., Four-Dimensional Generalized Difference Matrix and Almost Convergent Double Sequence Spaces. In: Kalmenov T., Nursultanov E., Ruzhansky M., Sadybekov M. (eds) Functional Analysis in Interdisciplinary Applications. FAIA 2017. Springer Proceedings in Mathematics & Statistics, vol 216. Springer, Cham, (2017).
8. Tug O., *$B(r,s,t,u)$-summable Double Sequence Spaces and Matrix Transformations*, Ph.D. Dissertation, University of Nis; Nis, Serbia, (2019).
9. Tug O., The spaces of $B(r,s,t,u)$ strongly almost convergent double sequences and matrix transformations, Bull. Sci. Math., **169**, (2021), 1–12.
10. Tug O., Rakovevic V., Malkowsky E., On the Domain of the Four-Dimensional Sequential Band Matrix in Some Double Sequence Spaces, Mathematics, **8**(789), (2020), 23–43.
11. Capan H., Basar F., On the difference spaces of almost convergent and strongly almost convergent double sequences, Positivity, **23**(2), (2019), 493–506.
12. Tug O., Almost and Strongly Almost $B(\tilde{r},\tilde{s},\tilde{t},\tilde{u})-$Summable Double Sequences, Eurasian J. Sci. Eng., **7**(2), (2021), 71–89.
13. Móricz F., Extensions of the spaces $c$ and $c_0$ from single to double sequence, Acta Math. Hungar., **57**, (1991), 129–136.

14. Boos J., *Classical and Modern Methods in Summability*, Oxford University Press, New York, (2000).
15. Hardy G. H., On the convergence of certain multiple series, Math. Proc. Cambridge Philos. Soc., **19**, (1916-1919), 86–95.
16. Basar F., Sever Y., The space $\mathscr{L}_q$ of double sequences, Math. J. Okayama Univ., **51**, (2009), 149–157.
17. Zeltser M., On conservative matrix methods for double sequence spaces, Acta Math. Hungar., **95**(3), (2002), 225–242.
18. Altay B., Basar F., Some new spaces of double sequences, J. Math. Anal. Appl., **309**(1), (2005), 70–90.
19. Adams C. R., On non-factorable transformations of double sequences. Proceedings of the National Academy of Sciences of the United States of America, **19**(5), (1933), 564.
20. Wilansky A., *Summability through Functional Analysis*, North-Holland Mathematics Studies 85, Amsterdam-New York-Oxford, (1984).
21. Zeltser M., *Investigation of double sequence spaces by soft and hard analytical methods*, Tartu, Tartu Univ. Press, (2001).
22. Loganathan S., Moorthy C. G., A net convergence for Schauder double bases, Asian-European J. Math., **9**(01), (2016), 1650010.
23. Cakan C., Altay B., Mursaleen M., The $\sigma$-convergence and $\sigma$-core of double sequences. Appl. Math. Lett., **2006**(19), (2016), 387–399.
24. Zeltser M., Mursaleen M., Mohiuddin S. A., On almost conservative matrix methods for double sequence spaces, Publ. Math., **75**, (2009), 387–399.
25. Yesilkayagil M., Basar F., Mercerian theorm for four-dimensional matrices, Commun. Fac. Sci. Univ. Ank. Ser., **65**(2016), (2016), 147–155.

# 6 Absolute and Strong $\sigma$–Convergence defined by Modulus Function

*Ekrem Savas*

In this chapter, we generalize the spaces $l^{\sigma}$ and $m^{\sigma}$ simultaneously in two directions: by using the modulus function and index. We also prove some inclusion theorems.

## 6.1 INTRODUCTION

Let $l_{\infty}$ and $c$ be the Banach spaces of bounded and convergent sequences $x = (x_k)$ with the usual norm $||x|| = sup_k|x_k|$.

Let us observe the outline of almost convergence. A linear functional $\xi$ on $l_{\infty}$ is said to be a Banach limit [2] if it has the following properties:

1. $\xi(x) \geq 0$ if $n \geq 0$ (i.e. $x_n \geq 0$ for all $n$),
2. $\xi(e) = 1$ where $e = (1, 1, \ldots)$,
3. $\xi(Dx) = \xi(x)$, where the shift operator $D$ is defined by $D(x_n) = \{x_{n+1}\}$.

Let $B$ be the set of all Banach limits on $l_{\infty}$. A sequence $x \in \ell_{\infty}$ is said to be almost convergent if all Banach limits of $x$ coincide. We denote the set of all almost convergent sequences by $\hat{c}$. A new sequence space $\hat{l}$ of absolute almost convergence was introduced by Das, Kuttner and Nanda [4].

For any sequence $x$, write

$$d_{mn} = d_{mn}(x) = \frac{1}{m+1}\sum_{i=0}^{m} x_{n+i}.$$

Lorentz [6] proved that $x \in \hat{c}$ if and only if $d_{mn} \to a$ limit $s$ as $m \to \infty$ uniformly in $n$; and in this case we write $\hat{c} - \lim x = s$. It is the characterization of $\hat{c}$ which motivates to define an absolute analogue of almost convergence.
Given an infinite series

$$\sum a_n$$

DOI: 10.1201/9781003596578-6

which we denote by $a$, let

$$x_n = a_0 + a_1 + \cdots + a_n. \tag{6.1}$$

We will suppose throughout that $a$ and $x$ are related by (1.1) (where limits of summation are not stated, sums are to be treated from $0 \to \infty$).
It was extended the definition of $d_{mn}$ to $m = -1$ by taking

$$d_{-1,n} = x_{n-1}.$$

We then write, for $m, n \geq 0$,

$$\phi_{mn} = \phi_{mn}(a) = d_{mn} - d_{m-1,n}.$$

A straightforward calculation then shows that

$$\phi_{on} = a_n,$$

$$\phi_{mn} = \frac{1}{m(m+1)} \sum_{v=1}^{m} v a_{n+v}.$$

The series $a$ (or the sequence $x$) is said to be absolutely almost convergent if $\sum_m |\varphi_{mn}|$ converges uniformly in $n$. Das and Mishra [5] extended this concept and defined the space of absolutely almost convergent sequences with index $k \geq 1$ as follows.

$$\sum_{m=1}^{\infty} m^{k-1} |\phi_{mn}|^k < \infty$$

uniformly in $n$. Note that the case $k = 1$ is the same as the definition of absolute almost convergence.

Let $\sigma$ be a one-to-one mapping of the set of positive integers into itself. A continuous linear functional $\varphi$ on $l_\infty$ is said to be an invariant mean or a $\sigma$- mean if and only if

1. $\varphi \geq 0$ when the sequence $x = (x_n)$ has $x_n \geq 0$ for all $n$.
2. $\varphi(e) = 1$, where $e = (1, 1, \ldots)$ and
3. $\varphi\left(x_{\sigma(n)}\right) = \varphi(x)$ for all $x \in l_\infty$.

For certain kinds of mapping $\sigma$ every invariant mean $\varphi$ extends the limit functional on space $c$, in the sense that $\varphi(x) = \lim x$ for all $x \in c$. Consequently, $c \subset V_\sigma$ where $V_\sigma$ is the bounded sequences all of whose $\sigma$-means are equal, (see, [18]).

In case $\sigma$ is the translation mapping $n \to n+1$, a $\sigma$-mean is often called a Banach limit and $V_\sigma$ is the set of almost convergent sequences (Lorentz [6]).

If $x = (x_n)$, set $Tx = (Tx_n) = \left(x_{\sigma(n)}\right)$. It can be shown (Schaefer, [18]) that

$$V_\sigma = \{x : \lim_m t_{mn} = L \text{ uniformly in n}, L = \sigma - \lim x\}, \tag{6.2}$$

where,

$$t_{mn} = (x_n + Tx_n + \cdots + T^m x_n)/(m+1)$$

and

$$t_{-1,n}(x) = 0.$$

Recently Mursaleen [7] defined strongly $\sigma$-convergent sequences, replacing the Banach limits by $\sigma$-means , in the following manner:

A bounded sequence $x = (x_k)$ is said to be strongly $\sigma$-convergent to a number $L$ if and only if,

$$\lim_m \frac{1}{m} \sum_{i=1}^{m} |x_{\sigma^i(n)} - L| = 0, \tag{6.3}$$

uniformly in $n$. By $[V_\sigma]$ we denote the set of all strongly $\sigma$-convergent sequences. When (1.3) holds, we write $[\sigma] - \lim x = \text{L}$.

$\sigma$-convergent sequences are studied by Mursaleen [7], Savaş ([11]–[14]) and others.

In [16], Savas extended the set $[V_\sigma]$ to $[V_\sigma]_{(p_k)}$, where $p = (p_k)$ is a sequence of strictly positive real numbers and introduced the spaces of sequences strongly $\sigma-$convergent to zero and strongly $\sigma$-bounded.

Let $p = (p_k)$ be a sequence of strictly positive real numbers. We define,

$$[V_\sigma]_{(p_k)} = \left\{ x : \lim_m \frac{1}{m} \sum_{k=1}^{m} |x_{\sigma^k(n)} - L|^{p_k} = 0 \text{ uniformly in n} \right\},$$

$$[V_\sigma]^o_{(p_k)} = \left\{ x : \lim_m \frac{1}{m} \sum_{k=1}^{m} |x_{\sigma^k(n)}|^{p_k} = 0 \text{ uniformly in n} \right\},$$

$$[V_\sigma]^*_{(p_k)} = \left\{ x : \sup_{m,n} \frac{1}{m} \sum_{k=1}^{m} |x_{\sigma^k(n)}|^{p_k} < \infty \right\}.$$

If $p = (p_k)$ be a constant sequence, that is, $p_k = p$ for all $k$, then write $[V_\sigma]_{(p_k)} = [V_\sigma]_p$ and if $p_k = p = 1$ for all $k$, we have $[V_\sigma]_1 = [V_\sigma]$.
If $\sigma(n) = n+1$ we have (see, Nanda [9]),

$$[V_\sigma]_{p_k} = [\hat{c}, p], [V_\sigma]^o_{(p_k)} = [\hat{c}, p]_o \text{ and } [V_\sigma]^\infty_{(p_k)} = [\hat{c}, p]_\infty.$$

A mapping $f : [0, \infty) \to [0, \infty)$ is called a modulus if

i) $f(x) = 0 \iff x = 0$,
ii) $f(x+y) \leq f(x) + f(y), \forall x \geq 0, y \geq 0$,
iii) $f$ is increasing,
iv) $f$ is continuous from the right at 0.

Since $|f(x)-f(y)| \leq f(|x-y|)$, it follows from condition $(iv)$ that $f$ is continuous on $[0,\infty)$.

A modulus may be bounded or unbounded. For example, $f(x)=x^p$, for $0<p\leq 1$ is unbounded, but $f(x)=\dfrac{x}{1+x}$ is bounded.

Ruckle [10] used the concept of modulus function $f$ to define the sequence space.

$$L(f)=\{x:\sum_{k=1}^{\infty} f(|x_k|)<\infty\}.$$

If we take $f(x)=x^p$, then $L(f)$ reduces to the familiar sequence space $l_p$. In particular, if $f(x)=x$, then $L(f)=l_1$.

We now extend $[V_\sigma]_{(p_k)}$ to $[V_\sigma, f]_{(p_k)}$ by using modulus as follows.

$$[V_\sigma, f]_{(p_k)}=\left\{x:\lim_m \frac{1}{m}\sum_{k=1}^{m} f(|x_{\sigma^k(n)}-L|)^{p_k}=0 \text{ uniformly in n}\right\}.$$

We write the following definition. If $X$ is a linear space over the field $C$, then a paranorm on $X$ is a function $g: X\to R$ which satisfies the following axiom: for $x,y\in X$,

1. $g(0)=0$,
2. $g(x)=g(-x)$,
3. $g(x+y)\leq g(x)+g(y)$,
4. $\lambda\to\lambda_0, x\to x_0$ imply $\lambda x\to\lambda_0 x_0$, where $\lambda,\lambda_0\in C$ and $x,x_0\in X$; in other words, $|\lambda-\lambda_0|\to 0$, $g(x-x_0)\to 0$ imply $g(\lambda x-\lambda_0 x_0)\to 0$. A paranormed space is a linear space $X$ with a paranorm $g$ and is written as $(X,g)$.

We now have the following result.

**Theorem 6.1.1.** *Let $(p_k)$ be bounded. Then $[V_\sigma, f]_{(p_k)}$ and $[V_\sigma, f]^o_{(p_k)}$ are linear spaces over the complex field.*

**Theorem 6.1.2.** *Let $(p_k)$ be bounded. Then $[V_\sigma, f]^o_{(p_k)}$ is a complete linear topological space paranormed by g defined by $g(x)=\sup f(|x_k|)^{p_k/M}$ where $M=\max(1,H=\sup p_k)$. Let $(p_k)$ be bounded and $\inf p_k>0$. Then $[V_\sigma, f]^o_{(p_k)}$ is a complete paranormed space paranormed by g.*

The proofs are a routine verification by using standard techniques and therefore we omit it.

It is evident that

$$[V_\sigma, f]^o_{(p_k)}\subset[V_\sigma, f]_{(p_k)}\subset[V_\sigma, f]^\infty_{(p_k)}.$$

We now write inclusion relation between $[V_\sigma, f]_{(p_k)}$ spaces.

**Theorem 6.1.3.** *Let $p_k=p$ for all $k, q_k=q$ for all $k$ and $0<p<q$. Then $[V_\sigma, f]_q$ is a closed subspace of $[V_\sigma, f]_p$.*

**Proof:** Let $r_1 = \frac{q}{p} > 1$ and $r_q > 0$ s.t $\frac{1}{r_1} + \frac{1}{r_2} = 1$. If $x \in [V_\sigma, f]_q$, then for some $l \in C$

$$\frac{1}{m}\sum_{k=1}^{m} f(|x_{\sigma^k(n)} - l|)^p \leq \left(\sum_{k=1}^{m}(|x_{\sigma^k(n)} - l|^p)^{r_1}\right)^{1/r_1}\left(\sum_{k=1}^{m}(\frac{1}{m})^{r_2}\right)^{1/r_2}$$

$$= \left(\sum_{k=1}^{m}(f(|x_{\sigma^k(n)} - l|)^q)\right)^{1/r_1}\frac{1}{m^{1-1/r_2}}$$

$$= \left(\frac{1}{m}\sum_{k=1}^{m}(f(|[V_\sigma, f]^o_{(p_k)} - l|)^q)\right)^{1/r_1}$$

and therefore $[V_\sigma, f]_p \subset [V_\sigma, f]_q$. Also note that $[V_\sigma, f]_p$ and $[V_\sigma, f]_q$ are complete linear topological spaces by Theorem 1.1. (by taking $p = p_k$ for all $k, q = q_k$ for all $k$). Now, if $(x_n)$ is a Cauchy sequence in $[V_\sigma, f]_q$ then it follows that $(x_n)$ is also Cauchy sequence in $[V_\sigma, f]_p$. This proves that $[V_\sigma, f]_q$ is a closed subspace of $[V_\sigma, f]_p$.

If $0 < p_k \leq q_k$ and $(p_k)$ and $(q_k)$ are not constant sequences we are able to prove "$[V_\sigma, f]_{(q_k)} \subset [V_\sigma, f]_{(p_k)}$" only under additional conditions.

**Theorem 6.1.4.** *Let $0 < p_k \leq q_k$ and let $\frac{q_k}{p_k}$ be bounded. Then $[V_\sigma, f]_{(q_k)} \subset [V_\sigma, f]_{(p_k)}$.*

**Proof:** Let $x \in [V_\sigma, f]_{(q_k)}$. Write $w_k = f(|x_k - l|)^{q_k}$ and $\lambda_k = \frac{p_k}{q_k}$, so that $0 < \lambda < \lambda_k < 1$.
Define $u_k = w_k(w_k \geq 1), = 0(w_k < 1)$. and $v_k = 0(w_k \geq 1), = w_k(w_k < 1)$. So $w_k = u_k + v_k$ , $w_k^{\lambda k} = u_k^{\lambda k} + v_k^{\lambda k}$. Now it follows that $u_k^{\lambda_k} \leq u_k \leq w_k, v_k^{\lambda_k} \leq v_k^{\lambda}$
We have(Savas [16]),

$$\frac{1}{m}\sum_{k=1}^{m} w_{\sigma^k(n)}^{\lambda_k} \leq \frac{1}{m}\sum_{k=1}^{m} w_{\sigma^k(n)} + \{\frac{1}{m}\sum_{k=1}^{m} v_{\sigma^k(n)}\}.$$

Now it follows that $x \in [V_\sigma, f]_{(p_k)}$ and this completes the proof.

**Theorem 6.1.5.** *(i) Let $0 < \inf p_k \leq p_k \leq 1$. Then $[V_\sigma, f, p] \subset [V_\sigma, f]$.*
*(i) Let $1 \leq p_k \leq \sup p_k < \infty$, Then $[V_\sigma, f] \subset [V_\sigma, f, p]$*

**Proof:** (i) Let $x \in [V_\sigma, f, p]$, since $0 < \inf p_k \leq 1$, we get

$$\frac{1}{m}\sum_{k=1}^{m} f(|x_{\sigma^k(n)} - le|) \leq \frac{1}{m}\sum_{k=1}^{m} f(|x_{\sigma^k(n)} - l|)^{p_k} \tag{6.4}$$

and hence $x \in [V_\sigma, f]$.
(ii) Let $p_k \geq 1$ for each $k$, and $\sup p_k < \infty$. Let $x \in [V_\sigma, f]$. Then for each $k, 1 > \varepsilon > 0$ there exists a positive integer $N$ such that

$$\frac{1}{m}\sum_{k=1}^{m} f(|x_{\sigma^k(n)} - l|) \leq \varepsilon < 1$$

for all $m \geq N$. This implies that

$$\frac{1}{m}\sum_{k=1}^{m} f(|x_{\sigma^k(n)} - l|)^{p_k} \leq \frac{1}{m}\sum_{k=1}^{m} f(|x_{\sigma^k(n)} - l|). \tag{6.5}$$

Therefore, $x \in [V_\sigma, f, p]$. This completes the proof.

**Theorem 6.1.6.** *(i) Let* $\liminf p_k = r > 0$. *Then* $x_k \to l$ *implies* $x_k \to l[V_\sigma, f, p]$.
*(ii) Let* $\liminf p_k = r > 0$. *If* $x_k \to l[V_\sigma, f, p]$, *then* $l$ *is unique.*

**Proof:** (i) Let $x_k \to l$. By definition of $f$ we have $f(|x_{\sigma^k(n)} - l|) \to 0$. Since $\liminf p_k > 0$, it follows $f(|x_{\sigma^k(n)} - l|)^{p_k} \to 0$ and consequently, $x_k \to l[V_\sigma, f, p]$.
(ii) Let $\liminf p_k = r > 0$. Suppose that $x_k \to l_1[V_\sigma, f, p]$, $x_k \to l_2[V_\sigma, f, p]$ and $|l_1 - l_r| = a > 0$. Then from the definition of $f$, we have

$$\frac{1}{m}\sum_{k=1}^{m} f(|l_1 - l_2|)^{p_k} \leq \frac{C}{m}\sum_{k=1}^{m} f(|x_{\sigma^k(n)} - l_1|)^{p_k} + \frac{C}{m}\sum_{k=1}^{m} f(|x_{\sigma^k(n)} - l_2|)^{p_k}. \tag{6.6}$$

Thus, $\lim_m \frac{1}{m}\sum_{k=1}^{m} f(|l_1 - l_2|)^{p_k} = 0$.
Also, $f(|l_1 - l_2|)^{p_k} \to f(a)^r$ as $k \to \infty$ and therefore

$$\lim_m \frac{1}{m}\sum_{k=1}^{m} f(|l_1 - l_2|)^{p_k} = f(a)^r. \tag{6.7}$$

From (4) and (5), it follows that $f(0) = 0$ and by definition of $f$, we have $a = 0$. Hence, $l_1 = l_2$ and this completes the proof.

## 6.2 ABSOLUTELY $\sigma$–CONVERGENCE WITH INDICES

Let $x_n = z_0 + z_1 + \cdots + z_n$ and $\sigma$ be monotonicaly increasing. Write

$$\Phi_{mn}(z) = \Phi_{mn}(z) = t_{mn} - t_{m-1,n} = \frac{1}{m+1}\sum_{i=0}^{m} T^1 x_n - \frac{1}{m}\sum_{i=0}^{m-1} T^i x_n.$$

Therefore,

$$\begin{aligned}
\Phi_{mn} &= \frac{1}{m(m+1)}[mT^m x_n - T^0 x_n - T^1 x_n - \cdots - T^{m-1} x_n] \\
&= \frac{1}{m(m+1)}[z_{\sigma^0(n)+1} + z_{\sigma^0(n)+2} + \cdots + z_{\sigma^1(n)} \\
&= 2(z_{\sigma^1(n)+1} + z_{\sigma^1(n)+2} + \cdots + z_{\sigma^2(n)}) + \cdots + \\
&\quad m(z_{\sigma^{m-1}(n)+1} + z_{\sigma^{m-1}(n)+2} + \cdots + z_{\sigma^m(n)})] \\
&= \frac{1}{m(m+1)}\sum_{j=1}^{m} j(\sum_{i=d_j}^{h_j} z_i), m \geq 1,
\end{aligned}$$

and $\phi_{0n}(z) = z_n$; with $d_j = \sigma^{j-1}(n) + 1$, $h_j = \sigma^j(n)$, (see, [1]).

The spaces of absolutely $\sigma-$ convergent and absolute $\sigma-$ bounded sequences were respectively introduced by Savas [17], Ahmad, Mursaleen and Khan [1] as follows:

$$l^{\sigma} := \{z : \sum_{m=0}^{\infty} |\Phi_{mn}(z)|, \text{converges uniformly in n}\}$$

and

$$m^{\sigma} := \{z : \sup_{n} \sum_{m=0}^{\infty} |\Phi_{mn}(z)| < \infty\}.$$

Let $p = (p_m)$ be a sequence of positive real numbers such that $\sup_m p_m \leq \infty$. We further extend the above sequence spaces to $l_{\sigma}(f,p)$ and $m_{\sigma}(f,p)$ as follows.

$$l^{\sigma}(f,p) = \left\{z : \sum_{m} f(|\Phi_{mn}(z)|)^{p_m} \text{ converges uniformly in n}\right\},$$

$$m^{\sigma}(f,p) = \left\{z : \sup_{n} \sum_{m} f(|\Phi_{mn}(z)|)^{p_m} < \infty\right\}.$$

**Remark 6.2.1.** *(i) If $f(x) = x$, and $p_m = 1$ for all m, we write $l^{\sigma}$ and $m^{\sigma}$ in place of $l^{\sigma}(f,p)$ and $m^{\sigma}(f,p)$ respectively.*
*(ii) If $f(x) = x$ and $\sigma(n) = n+1, l^{\sigma}(f,p) = \hat{l}(p)$ and $m^{\sigma}(f,p) = \hat{l}(p)$, (see, [4]).*

**Theorem 6.2.2.** *$l^{\sigma}(f,p) \subset m^{\sigma}(f,p)$.*

**Proof:** Suppose that $z \in l^{\sigma}(f,p)$. Therefore, for an integer $M$

$$\sum_{m \geq M} f(|\Phi_{mn}(z)|)^{p_m} \leq 1,$$

which gives

$$f(|\Phi_{mn}(z)|) \leq 1$$

for $m \geq M$ and all $n$. But, if $m \geq 1$,

$$\sum_{i=d_m}^{h_m} z_i = (m+1)\Phi_{mn} - (m-1)\Phi_{m-1,n}$$

and we get each $f(|z_i|)$ bounded for any fixed $m \geq M+1$. Hence $f(|\Phi_{mn}|)$ is bounded for all $m,n$. Hence we get the result.

**Theorem 6.2.3.** *$l^\sigma(f,p)$ and $m^\sigma(f,p)$ are complete linear spaces paranormed by*

$$g_p(z) = \sup_n \left( \sum_m f(|\Phi_{mn}(z)|) \right)^{1/H},$$

*where $H = \max(1, \sup p_m)$.*

The proof is a routine verification by using standard arguments and therefore omitted.

**Theorem 6.2.4.** *Suppose that for all $m$, $p_m \leq q_m$. Then*
*(i) $l^\sigma(f,p) \subset l^\sigma(f,q)$, and*
*(ii) $m^\sigma(f,p) \subset m^\sigma(f,q)$.*

**Proof:** (i) Let $z \in l^\sigma(f,p)$. Therefore, there exists an integer $M$ such that, for all $n$

$$\sum_{m=M}^{\infty} f(|\Phi_{mn}(z)|)^{p_m} \leq 1.$$

Hence

$$f(|\Phi_{mn}(z)|) \leq 1$$

for $m \geq M$ and all $n$ so that

$$f(|\Phi_{mn}|)^{q_m} \leq f(|\Phi_{mn}|)^{p_m}$$

and the uniform convergence of $\sum f(|\Phi_{mn}|)^{q_m}$ follows from that of $\sum_m f(|\Phi_{mn}|)^{p_m}$.
(ii) Let $z \in m_\sigma(p)$, then

$$\sup_n \sum_m f(|\Phi_{mn}|)^{p_m} \leq \infty.$$

Hence, $f(|\Phi_{mn}(z)|)$ is bounded for all $m,n$. Therefore there exists $K \geq 1$ such that

$$f(|\Phi_{mn}|) \leq K$$

so that

$$\sum_m f(|\Phi_{mn}|)^{q_m} \leq \sum_m K^{q_m - p_m} f(|\Phi_{mn}|)^{p_m} \leq K^H \sum_m f(|\Phi_{mn}|)^{p_m},$$

where $H = \sup_m q_m$.
Hence, $z \in m^\sigma(f,q)$.
This completes the proof of the theorem.

Savas [16] generalized the spaces $l^\sigma$ and $w^\sigma$ with the help of indices and also studied some properties of those spaces.

Write,

$$S_p^\sigma = \left\{ x : \sum_{m=1}^{\infty} m^{p-1} |\Phi_{m,n}(x)|^p \text{ converges uniformly in n} \right\} \tag{6.8}$$

for $p > 0$. In particular if $p = 1$ we get $S_p = l^\sigma$, the space of absolute $\sigma$−convergent sequences. If we take $\sigma(n) = n+1$, the above definition reduces to the definition of Das and Mishra.

We now generalize the spaces $l^\sigma$ and $w^\sigma$ simultaneously by using indices and also modulus functions as follows:

$$l^\sigma(f) := \{z : \sum_{m=0}^{\infty} m^{p-1} f(|\Phi_{mn}(z)|), \text{converges uniformly in n}\}$$

and

$$m^\sigma(f) := \{z : \sup_n \sum_{m=0}^{\infty} m^{p-1} f(|\Phi_{mn}(z)|) < \infty\}.$$

We now write,

$$w^\sigma(f) = \{x : \frac{1}{m+1} \sum_{i=0}^{m} m^{p-1} f(|x_{\sigma^i(n)} - L|) \text{ as } m \to \infty \text{ uniformly in n}\}.$$

If $\sigma(n) = n+1$, the above sequence spaces reduce to the following sequence spaces, respectively, which are defined by Nanda [8].

$$\hat{L}(f) = \{a : \sum_m m^{p-1} f(|\phi_{mn}|) \text{ converges uniformly in n}\},$$

$$\hat{\hat{L}}(f) = \{a : \sup_n \sum_m m^{p-1} f(|\phi_{mn}|) \leq \infty\},$$

$$\hat{W}(f) = \{x : \frac{1}{m+1} \sum_{i=0}^{m} m^{p-1} f(|x_{n+i} - s|) \to 0 \text{ as } m \to \infty \text{ uniformly in n}\}.$$

If $f(x) = x$, we write $l^\sigma(f) = l^\sigma$, $m_\sigma(f) = m_\sigma$, and $w^\sigma(f) = w^\sigma$ where $l^\sigma$ and $m_\sigma$ are, respectively, defined above. If we take $\sigma(n) = n+1, l_\sigma$ and $m_\sigma$ reduce to $\hat{l}$ and $\hat{\hat{l}}$ respectively, (see, [3, 4]). We now have

**Theorem 6.2.5.** *If $f$ is absolutely homogenous, then $l^\sigma(f) \subset m^\sigma(f)$.*

**Proof:** Let $z \in l^\sigma(f)$. Then there exists a constant $M > 0$ such that for all $n$,

$$\sum_{m \geq M} m^{p-1} f(|\Phi_{mn}|) = O(1), \tag{6.9}$$

that is enough to show that for all $m \geq M$ and for all $n$,

$$f(|\Phi_{mn}|)$$

is bounded. From (2.1) it follows that for all $m \geq M$ and for all $n$,

$$f(|\Phi_{mn}|) \leq \frac{1}{m^{p-1}} \leq 1. \tag{6.10}$$

But if $m \geq 1$,

$$\sum_{i=d_j}^{h_j} z_i = (m+1)\Phi_{mn} - (m-1)\Phi_{m-1,n}.$$

So

$$f(|\sum_{i=d_j}^{h_j} z_i|) = f[|(m+1)\Phi_{mn} - (m-1)\Phi_{m-1,n}|]$$

$$\leq f(|(m+1)\Phi_{mn}|) + f(|(m-1)\Phi_{m-1,n}|]$$

$$\leq (m+1)f(|\Phi_{mn}|) + |(m-1)f(|hi_{m-1,n}|).$$

Applying this with any fixed $m \geq M+1$, we deduce that $f(|\sum_{i=d_j}^{h_j} z_i|)$ is bounded. Now

$$f(|\Phi_{mn}|) \leq \frac{1}{m(m+1)} \sum_{j=1}^{m} jf(|\sum_{i=d_j}^{h_j} z_i|)$$

$$= \frac{1}{m(m+1)} O(1) \sum_{j=1}^{m} j$$

$$= \frac{1}{m(m+1)} O(1),$$

and this completes the proof.

In order to prove the next theorem we, first prove the following lemma.

**Lemma 6.2.6.** $V_\sigma(f) - \lim x = s \iff$

*i)* $\sigma - \lim x = s$,
*ii)* $\frac{1}{m+1}\sum_{r=0}^{m} f(|x_{\sigma^r(n)} - t_{r,n}(x)|) \to 0$ *as* $m \to \infty$ *uniformly in n.*

**Proof:** Let $(i)$ and $(ii)$ hold. We have

$$\frac{1}{m+1}\sum_{i=0}^{m} m^{p-1} f(|x_{\sigma^i(n)} - L|)$$

$$\leq \frac{1}{m+1}\sum_{i=0}^{m} m^{p-1} f(|x_{\sigma^i(n)} - t_{i,n}|) + \frac{1}{m+1}\sum_{i=0}^{m} m^{p-1} f(|t_{i,n} - L|) \to 0$$

as $m \to \infty$ uniformly in $n$ and hence $x \in w^\sigma(f)$. Conversely, let $x \in w^\sigma(f)$. We have

$$|\frac{1}{m+1}\sum_{i=0}^{m} m^{p-1} f(x_{\sigma^i(n)} - L)|$$

$$\leq \frac{1}{m+1}\sum_{i=0}^{m} m^{p-1} |f(x_{\sigma^i(n)} - L)| + \frac{1}{m+1}\sum_{i=0}^{m} m^{p-1} |f(x_{\sigma^i(n)} - L)|.$$

Since $f$ is continuous, $x \in V_\sigma$ and $\sigma - \lim x = L$. Further

$$\frac{1}{m+1}\sum_{i=0}^{m} m^{p-1} f(|x_{\sigma^i(n)} - t_{i,n}|)$$

$$= \frac{1}{m+1}\sum_{i=0}^{m} m^{p-1} f((|x_{\sigma^i(n)} - L) + (L - t_{i,n}|))$$

$$\leq \frac{1}{m+1}\sum_{i=0}^{m} m^{p-1} f(|x_{\sigma^i(n)} - L|) + \frac{1}{m+1}\sum_{i=0}^{m} m^{p-1} f(|t_{i,n} - L|) = \sum_1 + \sum_2$$

say.
Since $x \in w^\sigma(f)$, $\sum_1 = O(1)$ as $m \to \infty$ uniformly in $n$. Again since $V_\sigma - \lim x = L$, it follows that $t_{i,n} \to L$ as $i \to \infty$ as $i \to \infty$ uniformly in $n$ and so its weighted arithmetic average $\sum_2 \to 0$ as $m \to \infty$ uniformly in $n$. This completes the proof.

**Theorem 6.2.7.** *If $f$ is absolutely homogenous, then $l^\sigma(f) \subset w^\sigma(f)$.*

**Proof:** Suppose $x \in l^\sigma(f)$.
Hence

$$V_{r,n}(x) = \sum_{m=r}^{\infty} m^{p-1} f(|t_{mn} - t_{m-1,n}|)$$

exists for each $r \geq 0$ and

$$V_{r,n}(x) \to 0 \text{ as } r \to \infty \text{ uniformly in n.}$$

Now in order to prove the theorem, it is enough to show (by Lemma 2.7) that

$$\frac{1}{m+1}\sum_{r=0}^{m} m^{p-1} f(|x_{\sigma^r(n)} - t_{r,n}| \to 0$$

as $m \to \infty$ uniformly in n. Now since,

$$x_{\sigma^i(n)} - t_{r,n} = r(t_{r,n} - t_{r-1,n}),$$

we have

$$\frac{1}{m+1}\sum_{r=0}^{m} m^{p-1} f(|x_{\sigma^r(n)} - t_{r,n}|$$

$$= \frac{1}{m+1}\sum_{r=0}^{m} m^{p-1} f(r|t_{r,n} - t_{r-1,n}|$$

$$= \frac{1}{m+1}\sum_{r=0}^{m} m^{p-1} r f(|t_{r,n} - t_{r-1,n}|$$

$$= \frac{1}{m+1}\sum_{r=0}^{m} m^{p-1} r f(|V_{r+n} - V_{r+1,n}|$$

$$= \frac{V_{1,p}}{m+1} - V_{m+1,n} + \frac{1}{m+1}\sum_{r=0}^{m} V_{r+1,n} = o(1)$$

as $m \to \infty$ uniformly in $n$ and this completes the proof.

Let X and Y be any two nonempty subsets of s. If $x = (x_k) \in X$ implies that $Ax = (A_n(x)) \in Y$, we say that A defines a matrix transformation from X into Y and we denote it by $A : X \to Y$. By $(X,Y)$ we mean the class of matrices A such that $A : X \to Y$.

$$(A, f, p)_\infty = \left\{x : \sup_n f(|A_n(x)|)^{p_n} < \infty\right\},$$

$$(\bar{A}, f, p) = \left\{x : \sum_m f(|t_{mn}(Ax)|)^{p_m} \text{ converges uniformly in } n\right\},$$

$$(\bar{\bar{A}}, f, p) = \left\{x : \sup_n \sum_m f(|t_{mn}(Ax)|)^{p_m} < \infty\right\},$$

where

$$t_{mn}(Ax) = \sum_k a(n,k,m)x_k$$

is such that

$$a(n,k,m) = \frac{1}{m+1}\sum_{j=0}^{m} a\left(\sigma^j(n),k\right).$$

If $p_n = p$ for all $n$. We write $(A, f)_\infty$ for $(A, f, p)_\infty$. If $p_m = p$ for all $m$ we write $(\bar{A}, f)_p$ and $(\bar{\bar{A}}, f)_p$ for $(\bar{A}, f, p)$ and $(\bar{\bar{A}}, f, p)$ respectively.

**Theorem 6.2.8.** $(\bar{A}, f, p) \subset (\bar{\bar{A}}, f, p)$.

**Proof:** Let $x \in (\bar{A}, f, p)$. Then there is an integer $M$ such that

$$\sum_{m \geq M} f(\left|\sum_k a(n,k,m)x_k\right|)^{p_m} \leq 1.$$

Hence it is enough to show that, for fixed $m$, $f(|\sum_k a(n,k,m)x_k|)$ is bounded. It follows from (2.1) that

$$f(\left|\sum_k a(n,k,m)x_k\right|) \leq 1 \text{ for } m \geq M \text{ and for all } n.$$

But if $m \geq 1$

$$(m+1)\sum_k a(n,k,m)x_k - m\sum_k a(n,k,m-1)x_k = \sum_k a\left(\sigma^m(n),k\right)x_k.$$

Applying this with any fixed $m \geq M+1, f(|\sum_k a(n,k,m)x_k|$ is bounded. Therefore $f(|\sum_k a(n,k,m)x_k|$ is bounded for all $m,n$ and this completes the proof.

**Theorem 6.2.9.** $(\bar{\bar{A}}, f)_p \subset (A, f)_\infty$.

**Proof:** Note that

$$\sup_{m,n} f(\left|\sum_k a(n,k,m)x_k\right|) \leq \sup_n \left(\sum_m f(\left|\sum_k a(n,k,m)x_k\right|)^p\right)^{1/p}$$

$\sup_{n,m} f(|\sum_k a(n,k,m)x_k|) \geq \sup_n f(|\sum_k a(n,k,o)x_k|) = \sup_n f(|\sum_k a_{nk}x_k|)$

$$\begin{aligned}\sup_{n,m} f(\left|\sum_k a(n,k,m)x_k\right|) &= \sup_{n,m} f(\left|\sum_k \frac{1}{m+1}\sum_{j=0}^m a\left(\sigma^j(n),k\right)x_k\right|)\\ &\leq \sup_{n,j} f(\left|\sum_k a\left(\sigma^j(n),k\right)x_k\right|)\\ &= \sup_n f(\left|\sum_k a_{nk}x_k\right|).\end{aligned}$$

Now the result follows from (2.3), (2.4) and (2.5).

**Theorem 6.2.10.** *Let inf $p_n > 0$. Then $(A, p)_\infty$ is a linear topological space paranormed by*

$$f(x) = \sup_n \left|\sum_k a_{nk}x_k\right|^{p_n/M},$$

*where $M = \max(1, \sup p_n) \cdot (\bar{A}, p)$ is paranormed by*

$$g(x) = \sup_n \left(\sum_m \left|\sum_k a(n,k,m)x_k\right|^{p_m}\right)^{1/M}.$$

*$(\bar{A}, p)$ is paranormed by (2.6) if inf $p_m > 0$.*

**Proof:** The proof is a routine verification by using standard techniques and therefore we omit it.

**Lemma 6.2.11.** *The following two conditions are equivalent.*

$$\sup_n \sum_k f(|a_n k|) < \infty,$$

$$\sup_{m,n} \sum_k f(|a(n,k,m)|) < \infty.$$

**Proof:** We have

$$\sup_{m,n} \sum_k f(|a(n,k,m)|) \geq \sup_n \sum_k f(|a(n,k,0)|) = \sup_n \sum_k f(|a_{nk}|)$$

and

$$\sup_{m,n} \sum_k f(|a(n,k,m)|) \leq \sup_{j,n} \sum_k f(|a(\sigma^j(n),k)|) \sup_m \frac{1}{m+1} \sum_{j=0}^{m} 1$$
$$= \sup_n \sum_k f(|a_{nk}|).$$

Therefore the result follows.

## BIBLIOGRAPHY


1. Ahmad Z. U., Mursaleen M., Khan Q. A., Invariant means and some matrix transformations, Southeast Asian Bul. Math., **23**(4), (1999), 533–539.
2. Banach S., *Theorie des Operations linearies*, Warszawa, (1932).
3. Das G., Kuttner B., Nanda S., On absolute almost convergence, J. Math. Anal. Appl., **161**(1), (1991), 50–56.
4. Das G., Kuttner B., Nanda S., Some sequence spaces and absolute almost convergence, Tran. Amer. Math. Soc., **283**(2), (1984), 729–739.
5. Das G., Mishra S. K., Absolute almost convergence with an index, Rendiconti di Mat. Serie VII., **8**, (1988), 501–510.
6. Lorentz G. G., A contribution to the theory of divergent sequences, Acta math., **80**, (1948), 167–190.
7. Mursaleen M., On some new invariant matrix methods of summability, Quarterly J. Math. Oxford, 34–77.
8. Nanda S., Absolute and strong almost convergence with indices defined by moduli, Journal of the Indian Math. Soc., **65**(1-4), (1998), 191–196.
9. Nanda S., Strong almost convergent sequnces, Bull. Cal. Math. Soc., **76**, (1984), 236–250.
10. Ruckle W. H., FK spaces in which sequences of coordinate vectors is bounded, Canad. J. Math., **25**, (1973), 973–978.
11. Savas E., Some Sequence Spaces Involving Invariant Means, Indian J. Math., **31**, (1989).
12. Savas E., A note on Some Sequence Spaces, Doga Türk.. J. Math., **15**, (1991).
13. Savas E., Invariant Means and Generalization of a Theorem of S. Mishra, Doga Türk. J. Math., **14**, (1989).
14. Savas E., Invariant Coregular and Conull Matrices of Operators, Hacettepe Bull. Math. Sci. and Eng., **19**, (1990).

15. Savas E., On some generalized sequence spaces defined by a modulus, Indian J. Pur. Appl. Math., **30**(5), (1999), 459–464.
16. Savas E., Strongly $\sigma-$ convergent sequences, Bull. Cal. Math. Soc., **81**, (1989), 295–300.
17. Savas E., On strong $\sigma-$ convergence, J. Orissa Math., **5**(2), (1986), 45–53.
18. Schaefer P., Infinite matrices and invariant means, Proc. Amer. Math. Soc., **36**, (1972), 104–110.

# 7 Remarks on Ideal Variations of the Star-Hurewicz and Related Properties

*Debraj Chandra, Nur Alam*

This chapter is a continuation of the study of variants of $\mathcal{I}$-Hurewicz property done in (Das et al., Topology Appl., 2018). We follow the approach of neighbourhood assignments to study the star-$\mathcal{I}$-Hurewicz and strongly star-$\mathcal{I}$-Hurewicz properties. We obtain certain results in the context of Alexandroff duplicate, groupability and Alster covers which have not been investigated in prior works. A number of observations concerning the cardinality $\mathfrak{b}(\mathcal{I})$ are presented. Few related games are also studied.

## 7.1 INTRODUCTION

Let $\mathscr{A}$ and $\mathscr{B}$ be collections of families of subsets of an infinite space $X$. In [26] (see also [19]), Scheepers began the systematic study of selection principles in general topology by introducing the following.

$S_1(\mathscr{A},\mathscr{B})$: For each sequence $(\mathscr{U}_n)$ of elements of $\mathscr{A}$ there exists a sequence $(V_n)$ such that for each $n$ $V_n \in \mathscr{U}_n$ and $\{V_n : n \in \mathbb{N}\} \in \mathscr{B}$.

$S_{fin}(\mathscr{A},\mathscr{B})$: For each sequence $(\mathscr{U}_n)$ of elements of $\mathscr{A}$ there exists a sequence $(\mathscr{V}_n)$ such that for each $n$ $\mathscr{V}_n$ is a finite subset of $\mathscr{U}_n$ and $\cup_{n\in\mathbb{N}}\mathscr{V}_n \in \mathscr{B}$.

$U_{fin}(\mathscr{A},\mathscr{B})$: For each sequence $(\mathscr{U}_n)$ of elements of $\mathscr{A}$ there exists a sequence $(\mathscr{V}_n)$ such that for each $n$ $\mathscr{V}_n$ is a finite subset of $\mathscr{U}_n$ and $\{\cup\mathscr{V}_n : n \in \mathbb{N}\} \in \mathscr{B}$ or $\cup\mathscr{V}_n = X$ for some $n$.

For a subset $A$ of a space $X$ and a collection $\mathscr{P}$ of subsets of $X$, $St(A,\mathscr{P})$ denotes the star of $A$ with respect to $\mathscr{P}$, that is the set $\cup\{B \in \mathscr{P} : A \cap B \neq \emptyset\}$. For $A = \{x\}$, $x \in X$, we write $St(x,\mathscr{P})$ instead of $St(\{x\},\mathscr{P})$ [16]. In [20], Kocinac introduced star selection principles in the following way.

$S_1^*(\mathscr{A},\mathscr{B})$: For each sequence $(\mathscr{U}_n)$ of elements of $\mathscr{A}$ there exists a sequence $(V_n)$ such that for each $n$ $V_n \in \mathscr{U}_n$ and $\{St(V_n,\mathscr{U}_n) : n \in \mathbb{N}\} \in \mathscr{B}$.

DOI: 10.1201/9781003596578-7

$S^*_{fin}(\mathscr{A},\mathscr{B})$: For each sequence $(\mathscr{U}_n)$ of elements of $\mathscr{A}$ there exists a sequence $(\mathscr{V}_n)$ such that for each $n$ $\mathscr{V}_n$ is a finite subset of $\mathscr{U}_n$ and $\cup_{n\in\mathbb{N}}\{St(V,\mathscr{U}_n) : V \in \mathscr{V}_n\} \in \mathscr{B}$.
$U^*_{fin}(\mathscr{A},\mathscr{B})$: For each sequence $(\mathscr{U}_n)$ of elements of $\mathscr{A}$ there exists a sequence $(\mathscr{V}_n)$ such that for each $n$ $\mathscr{V}_n$ is a finite subset of $\mathscr{U}_n$ and $\{St(\cup\mathscr{V}_n,\mathscr{U}_n) : n \in \mathbb{N}\} \in \mathscr{B}$ or there is some $n$ such that $St(\cup\mathscr{V}_n,\mathscr{U}_n) = X$.
$SS^*_{fin}(\mathscr{A},\mathscr{B})$: For each sequence $(\mathscr{U}_n)$ of elements of $\mathscr{A}$ there exists a sequence $(x_n)$ of elements of $X$ such that $\{St(x_n,\mathscr{U}_n) : n \in \mathbb{N}\} \in \mathscr{B}$.
$SS^*_{fin}(\mathscr{A},\mathscr{B})$: For each sequence $(\mathscr{U}_n)$ of elements of $\mathscr{A}$ there exists a sequence $(F_n)$ of finite subsets of $X$ such that $\{St(F_n,\mathscr{U}_n) : n \in \mathbb{N}\} \in \mathscr{B}$.

A family $\mathscr{I} \subseteq 2^X$ of subsets of a nonempty set $X$ is said to be an ideal in $X$ if $(i)$ $A,B \in \mathscr{I}$ imply $A\cup B \in \mathscr{I}$ and $(ii)$ $A \in \mathscr{I}$ and $B \subseteq A$ imply $B \in \mathscr{I}$, while an admissible ideal $\mathscr{I}$ of $X$ further satisfies $\{x\} \in \mathscr{I}$ for each $x \in X$. Such ideals are also called free ideals. If $\mathscr{I}$ is a proper ideal in $X$ (i.e. $X \notin \mathscr{I}$, $\mathscr{I} \neq \{\emptyset\}$), then the family of sets $\mathscr{F}(\mathscr{I}) = \{M \subseteq X : \text{there exists } A \in \mathscr{I} \text{ such that } M = X\setminus A\}$ is a filter in $X$ and $\mathscr{I}^+$ is the coideal, i.e. the family of all subsets of $X$ outside $\mathscr{I}$.

The study of classical selection principles, star selection principles and corresponding games has renowned history, see [1, 4, 6, 7, 19, 20, 21, 22, 26] and references therein for recent development. In [10], Das initiated a more general approach to study certain types of open covers and selection principles using ideals. Many studies have been done on classical selection principles as well as star selection principles using ideals. Readers interested in the study of selection principles using ideals can consult the papers [10, 11, 12, 13, 14, 15, 22] where more basic references can be found.

This chapter is a continuation of the study of star-$\mathscr{I}$-Hurewicz property [12] and strongly star-$\mathscr{I}$-Hurewicz property [12] done by Das et al. in [12]. We follow the approach of neighbourhood assignments to study the star-$\mathscr{I}$-Hurewicz and strongly star-$\mathscr{I}$-Hurewicz properties. We obtain certain results in the context of Alexandroff duplicate, groupability and Alster covers which have not been investigated in prior works. A number of observations concerning the cardinality $\mathfrak{b}(\mathscr{I})$ are presented. Few game theoretic observations are also obtained. In particular, the games $G_{ufin}(\mathscr{O},\mathscr{I}\text{-}\Gamma)$, $G^*_{ufin}(\mathscr{O},\mathscr{I}\text{-}\Gamma)$ and $SG^*_{fin}(\mathscr{O},\mathscr{I}\text{-}\Gamma)$ are investigated.

## 7.2 DEFINITIONS AND TERMINOLOGIES

By a space, we always mean a topological space. For undefined notions and terminologies, see [16]. Throughout the chapter, $\mathscr{I}$ stands for a proper admissible ideal on $\mathbb{N}$. The symbol $fin$ denotes the ideal of all finite subsets of $\mathbb{N}$.

Let $X$ be a space and $\mathscr{I}$ be an ideal on $\mathbb{N}$. We use the following notions of covers throughout the chapter (see [11, 12, 13, 14, 15, 19, 9, 23, 26].

1. $(\mathscr{O})$ The collection of all open covers of $X$.
2. $(\Lambda)$ The collection of all large covers of $X$. An open cover $\mathscr{U}$ of $X$ is said to be a large cover if for each $x \in X$, $\{U \in \mathscr{U} : x \in U\}$ is infinite.
3. $(\Omega)$ The collection of all $\omega$-covers of $X$. An open cover $\mathscr{U}$ of $X$ is said to be an $\omega$-cover if $X$ does not belong to $\mathscr{U}$ and for each finite subset $F$ of $X$ there is a set $U \in \mathscr{U}$ such that $F \subseteq U$.

4. ($\Gamma$) The collection of all $\gamma$-covers of $X$. An open cover $\mathscr{U}$ of $X$ is said to be a $\gamma$-cover if it is infinite and each element of $X$ does not belong to at most finitely many members of $\mathscr{U}$.
5. ($\mathscr{O}^{gp}$) The collection of all groupable covers of $X$. An open cover $\mathscr{U}$ of $X$ is said to be groupable if it can be expressed as a countable union of finite, pairwise disjoint subfamilies $\mathscr{U}_n$, $n \in \mathbb{N}$, such that each $x \in X$ belongs to $\cup\mathscr{U}_n$ for all but finitely many $n$.
6. ($\Lambda^{gp}$) $\Lambda^{gp} = \Lambda \cap \mathscr{O}^{gp}$.
7. ($\mathscr{I}$-$\Lambda$) The collection of all $\mathscr{I}$-large covers of $X$. A countable open cover $\mathscr{U}$ of $X$ is said to be $\mathscr{I}$-large if there exists an enumeration $\mathscr{U} = \{U_n : n \in \mathbb{N}\}$ such that for each $x \in X$, $\{n \in \mathbb{N} : x \in U_n\} \in \mathscr{I}^+$.
8. ($\mathscr{I}$-$\mathscr{O}^{gp}$) The collection of all $\mathscr{I}$-groupable covers of $X$. An open cover $\mathscr{U}$ of $X$ is said to be $\mathscr{I}$-groupable if it can be represented in the form $\mathscr{U} = \cup_{n\in\mathbb{N}}\mathscr{U}_n$, where $\mathscr{U}_n$'s are finite, pairwise disjoint and for each $x \in X$, $\{n \in \mathbb{N} : x \notin \cup\mathscr{U}_n\} \in \mathscr{I}$.
9. ($\mathscr{I}$-$(\mathscr{I}$-$\Lambda)^{gp}$) $\mathscr{I}$-$(\mathscr{I}$-$\Lambda)^{gp} = \mathscr{I}$-$\Lambda \cap \mathscr{I}$-$\mathscr{O}^{gp}$.
10. ($\mathscr{I}$-$(\mathscr{O}^{gp})^*$) The collection of all star-$\mathscr{I}$-groupable covers of $X$. An open cover $\mathscr{U}$ of $X$ is said to be star-$\mathscr{I}$-groupable if it can be represented in the form $\mathscr{U} = \cup_{n\in\mathbb{N}}\mathscr{U}_n$, where $\mathscr{U}_n$'s are finite, pairwise disjoint and for each $x \in X$, $\{n \in \mathbb{N} : x \notin St(\cup\mathscr{U}_n, \mathscr{U})\} \in \mathscr{I}$.

Clearly $Fin$-$\Lambda = \Lambda$, $Fin$-$\mathscr{O}^{gp} = \mathscr{O}^{gp}$ and $Fin$-$(fin$-$\Lambda)^{gp} = \Lambda^{gp}$.

A space $X$ is said to have the Hurewicz [26] (respectively, strongly star-Hurewicz [4], star-Hurewicz [4]) property if it satisfies the selection principle $U_{fin}(\mathscr{O},\Gamma)$ (respectively, $SS^*_{fin}(\mathscr{O},\Gamma)$, $U^*_{fin}(\mathscr{O},\Gamma)$). A space $X$ is said to have the $\mathscr{I}$-Hurewicz property (in short, *IH*) [10] (see also [15]) if for each sequence $(\mathscr{U}_n)$ of open covers of $X$ there exists a sequence $(\mathscr{V}_n)$ such that for each $n$ $\mathscr{V}_n$ is a finite subset of $\mathscr{U}_n$ and for each $x \in X$, $\{n \in \mathbb{N} : x \notin \cup\mathscr{V}_n\} \in \mathscr{I}$. A space $X$ is said to have the star-$\mathscr{I}$-Hurewicz property (in short, *SIH*) [12] if for each sequence $(\mathscr{U}_n)$ of open covers of $X$ there exists a sequence $(\mathscr{V}_n)$ such that for each $n$ $\mathscr{V}_n$ is a finite subset of $\mathscr{U}_n$ and for each $x \in X$, $\{n \in \mathbb{N} : x \notin St(\cup\mathscr{V}_n, \mathscr{U}_n)\} \in \mathscr{I}$. A space $X$ is said to have the strongly star-$\mathscr{I}$-Hurewicz property (in short, SS$\mathscr{I}$H) [12] if for each sequence $(\mathscr{U}_n)$ of open covers of $X$ there exists a sequence $(F_n)$ of finite subsets of $X$ such that for each $x \in X$, $\{n \in \mathbb{N} : x \notin St(F_n, \mathscr{U}_n)\} \in \mathscr{I}$. A space $X$ is said to be star-Lindelöf (respectively, strongly star-Lindelöf) if for every open cover $\mathscr{U}$ of $X$ there exists a countable $\mathscr{V} \subseteq \mathscr{U}$ (respectively, $A \subseteq X$) such that $St(\cup\mathscr{V}, \mathscr{U}) = X$ (respectively, $St(A, \mathscr{U}) = X$). A space $X$ is called an $\varepsilon$-space if each $\omega$-cover of $X$ has a countable subset which is an $\omega$-cover [18].

The eventual dominance relation $\leq^*$ on the Baire space $\mathbb{N}^\mathbb{N}$ is defined by $f \leq^* g$ if and only if $f(n) \leq g(n)$ for all but finitely many $n$. A subset $A$ of $\mathbb{N}^\mathbb{N}$ is said to be bounded if there is a $g \in \mathbb{N}^\mathbb{N}$ such that $f \leq^* g$ for all $f \in A$. Let $\mathfrak{b}$ be the minimum cardinality of an unbounded subset of $\mathbb{N}^\mathbb{N}$ and $\mathfrak{c}$ be the cardinality of the set of reals. It is well known that $\omega_1 \leq \mathfrak{b} \leq \mathfrak{c}$. Let $\mathscr{I}$ be an ideal on $\mathbb{N}$. In [17], the authors introduced an ideal version of the bounding number $\mathfrak{b}(\mathscr{I}) = \min\{|A| : A \subseteq \mathbb{N}^\mathbb{N}$ and for each $g \in \mathbb{N}^\mathbb{N}$ there is a $f \in A$ such that $\{n \in \mathbb{N} : f(n) \geq g(n)\} \in \mathscr{I}^+\}$. Note that $\mathfrak{b} \leq \mathfrak{b}(\mathscr{I}) \leq \mathfrak{c}$ and $\mathfrak{b}(Fin) = \mathfrak{b}$ (see [17, 27]).

A family $\mathscr{A} \subseteq P(\mathbb{N})$ is said to be an almost disjoint family if each $A \in \mathscr{A}$ is infinite and for any two distinct elements $B, C \in \mathscr{A}$, $|B \cap C| < \omega$. For an almost disjoint family $\mathscr{A}$, let $\Psi(\mathscr{A}) = \mathscr{A} \cup \mathbb{N}$ be the Isbell-Mrówka space (or, $\Psi$-space) (see [24]). It is well known that $\Psi(\mathscr{A})$ is pseudocompact if and only if $\mathscr{A}$ is a maximal almost disjoint family.

Recall that a neighbourhood assignment for a space $(X, \tau)$ is a function $N : X \to \tau$ such that $x \in N(x)$ for each $x \in X$.

**Lemma 7.2.1.** *1. A space X is IH if and only if for each sequence* $(\mathbb{N}_n)$ *of neighbourhood assignments of X there exists a sequence* $(F_n)$ *of finite subsets of X such that for each* $x \in X$, $\{n \in \mathbb{N} : x \notin \cup \mathbb{N}_n(F_n)\} \in \mathscr{I}$.

2. *A space X is SIH if and only if for each sequence* $(\mathbb{N}_n)$ *of neighbourhood assignments of X there exists a sequence* $(F_n)$ *of finite subsets of X such that for each* $x \in X$, $\{n \in \mathbb{N} : x \notin St(\cup \mathbb{N}_n(F_n), \mathbb{N}_n(X))\} \in \mathscr{I}$.
3. *A space X is SSIH if and only if for each sequence* $(\mathbb{N}_n)$ *of neighbourhood assignments of X there exists a sequence* $(F_n)$ *of finite subsets of X such that for each* $x \in X$, $\{n \in \mathbb{N} : x \notin St(F_n, \mathbb{N}_n(X))\} \in \mathscr{I}$.

*Proof.* (1). Let $(\mathbb{N}_n)$ be a sequence of neighbourhood assignments of $X$. For each $n$ $\mathscr{U}_n = \{\mathbb{N}_n(x) : x \in X\}$ is an open cover of $X$. Since $X$ is $IH$, there exists a sequence $(\mathscr{V}_n)$ such that for each $n$ $\mathscr{V}_n$ is a finite subset of $\mathscr{U}_n$ and for each $x \in X$, $\{n \in \mathbb{N} : x \notin \cup \mathscr{V}_n\} \in \mathscr{I}$. Thus we get a sequence $(F_n)$ of finite subsets of $X$ satisfying $\mathscr{V}_n = \{\mathbb{N}_n(x) : x \in F_n\}$ for each $n$. Observe that for each $x \in X$, $\{n \in \mathbb{N} : x \notin \cup \mathbb{N}_n(F_n)\} \in \mathscr{I}$.

Conversely, let $(\mathscr{U}_n)$ be a sequence of open covers of $X$. For each $x \in X$ there is a $U_x^{(n)} \in \mathscr{U}_n$ with $x \in U_x^{(n)}$. Define a sequence $(\mathbb{N}_n)$ of neighbourhood assignments of $X$ by $\mathbb{N}_n(x) = U_x^{(n)}$ for each $n$ and each $x \in X$. Then we obtain a sequence $(F_n)$ of finite subsets of $X$ such that for each $x \in X$, $\{n \in \mathbb{N} : x \notin \cup \mathbb{N}_n(F_n)\} \in \mathscr{I}$. For each $n$ let $\mathscr{V}_n = \{\mathbb{N}_n(x) : x \in F_n\}$. Then the sequence $(\mathscr{V}_n)$ guarantees for $(\mathscr{U}_n)$ that $X$ is $IH$.

(2). Let $(\mathbb{N}_n)$ be a sequence of neighbourhood assignments of $X$. For each $n$ $\mathscr{U}_n = \{\mathbb{N}_n(x) : x \in X\}$ is an open cover of $X$. Since $X$ is $SIH$, there exists a sequence $(\mathscr{V}_n)$ such that for each $n$ $\mathscr{V}_n$ is a finite subset of $\mathscr{U}_n$ and for each $x \in X$, $\{n \in \mathbb{N} : x \notin St(\cup \mathscr{V}_n, \mathscr{U}_n)\} \in \mathscr{I}$. Thus we get a sequence $(F_n)$ of finite subsets of $X$ satisfying $\mathscr{V}_n = \{\mathbb{N}_n(x) : x \in F_n\}$ for each $n$. Observe that for each $x \in X$, $\{n \in \mathbb{N} : x \notin St(\cup \mathbb{N}_n(F_n), \mathbb{N}_n(X))\} \in \mathscr{I}$.

Conversely, let $(\mathscr{U}_n)$ be a sequence of open covers of $X$. For each $x \in X$ there is a $U_x^{(n)} \in \mathscr{U}_n$ with $x \in U_x^{(n)}$. Define a sequence $(\mathbb{N}_n)$ of neighbourhood assignments of $X$ by $\mathbb{N}_n(x) = U_x^{(n)}$ for each $n$ and each $x \in X$. Then we obtain a sequence $(F_n)$ of finite subsets of $X$ such that for each $x \in X$, $\{n \in \mathbb{N} : x \notin St(\cup \mathbb{N}_n(F_n), \mathbb{N}_n(X))\} \in \mathscr{I}$. For each $n$ let $\mathscr{V}_n = \{\mathbb{N}_n(x) : x \in F_n\}$. Now the sequence $(\mathscr{V}_n)$ guarantees for $(\mathscr{U}_n)$ that $X$ is $SIH$.

(3). Let $(\mathbb{N}_n)$ be a sequence of neighbourhood assignments of $X$. For each $n$ $\mathscr{U}_n = \{\mathbb{N}_n(x) : x \in X\}$ is an open cover of $X$. Since $X$ is $SSIH$, there exists a sequence $(F_n)$ of finite subsets of $X$ such that for each $x \in X$, $\{n \in \mathbb{N} : x \notin St(F_n, \mathscr{U}_n)\} \in \mathscr{I}$, i.e. $\{n \in \mathbb{N} : x \notin St(F_n, \mathbb{N}_n(X))\} \in \mathscr{I}$.

Conversely, let $(\mathscr{U}_n)$ be a sequence of open covers of $X$. For each $x \in X$ there is a $U_x^{(n)} \in \mathscr{U}_n$ with $x \in U_x^{(n)}$. Define a sequence $(\mathbb{N}_n)$ of neighbourhood assignments of $X$ by $\mathbb{N}_n(x) = U_x^{(n)}$ for each $n$ and each $x \in X$. Then we obtain a sequence $(F_n)$ of finite subsets of $X$ such that for each $x \in X$, $\{n \in \mathbb{N} : x \notin St(F_n, \mathbb{N}_n(X))\} \in \mathscr{I}$. Thus for each $x \in X$, $\{n \in \mathbb{N} : x \notin St(F_n, \mathscr{U}_n)\} \in \mathscr{I}$ and hence $X$ is *SSIH*. □

**Lemma 7.2.2.** *1. A space X is Lindelöf if and only if for each neighbourhood assignment* $\mathbb{N}$ *of X there exists a countable subset A of X such that* $\cup N(A) = X$.

2. *A space X is star-Lindelöf if and only if for each neighbourhood assignment* $\mathbb{N}$ *of X there exists a countable subset A of X such that* $St(\cup N(A), N(X)) = X$.
3. *A space X is strongly star-Lindelöf if and only if for each neighbourhood assignment* $\mathbb{N}$ *of X there exists a countable subset A of X such that* $St(A, \mathbb{N}(X)) = X$.

**Proof:** $(a)$. Let $\mathbb{N}$ be a neighbourhood assignment of $X$. Then $\mathscr{U} = \{\mathbb{N}(x) : x \in X\}$ is an open cover of $X$. Since $X$ is Lindelöf, there exists a countable $A \subseteq X$ such that $\{\mathbb{N}(x) : x \in A\}$ covers $X$. It follows that $\cup\mathbb{N}(A) = X$.

Conversely, let $\mathscr{U}$ be an open cover of $X$. For each $x \in X$ pick a $U_x \in \mathscr{U}$ with $x \in U_x$. Define a neighbourhood assignment $\mathbb{N}$ of $X$ by $\mathbb{N}(x) = U_x$ for each $x \in X$. Then we get a countable $A \subseteq X$ such that $\cup\mathbb{N}(A) = X$. Choose $\mathscr{V} = \{\mathbb{N}(x) : x \in A\}$. Now $\mathscr{V}$ guarantees for $\mathscr{U}$ that $X$ is Lindelöf.

$(b)$. Let $\mathbb{N}$ be a neighbourhood assignment of $X$. Then $\mathscr{U} = \{\mathbb{N}(x) : x \in X\}$ is an open cover of $X$. Since $X$ is star-Lindelöf, there exists a countable $A \subseteq X$ such that $St(\cup\{\mathbb{N}(x) : x \in A\}, \mathscr{U}) = X$, i.e. $St(\cup\mathbb{N}(A), \mathbb{N}(X)) = X$.

Conversely, let $\mathscr{U}$ be an open cover of $X$. For each $x \in X$ pick a $U_x \in \mathscr{U}$ with $x \in U_x$. Define a neighbourhood assignment $\mathbb{N}$ of $X$ by $\mathbb{N}(x) = U_x$ for each $x \in X$. Then we get a countable $A \subseteq X$ such that $St(\cup\mathbb{N}(A), \mathbb{N}(X)) = X$. Choose $\mathscr{V} = \{\mathbb{N}(x) : x \in A\}$. Now $\mathscr{V}$ guarantees for $\mathscr{U}$ that $X$ is star-Lindelöf.

$(c)$. Let $\mathbb{N}$ be a neighbourhood assignment of $X$. Then $\mathscr{U} = \{\mathbb{N}(x) : x \in X\}$ is an open cover of $X$. Since $X$ is strongly star-Lindelöf, there exists a countable $A \subseteq X$ such that $St(A, \mathscr{U}) = X$, i.e. $St(A, \mathbb{N}(X)) = X$.

Conversely, let $\mathscr{U}$ be an open cover of $X$. For each $x \in X$ pick a $U_x \in \mathscr{U}$ with $x \in U_x$. Define a neighbourhood assignment $\mathbb{N}$ of $X$ by $\mathbb{N}(x) = U_x$ for each $x \in X$. Then we get a countable $A \subseteq X$ such that $St(A, \mathbb{N}(X)) = X$, i.e. $St(A, \mathscr{U}) = X$. Thus $X$ is strongly star-Lindelöf.

## 7.3 MAIN RESULTS

The Alexandroff duplicate $AD(X)$ of a space $X$ (see [5, 16]) is defined as follows. $AD(X) = X \times \{0,1\}$; each point of $X \times \{1\}$ is isolated and a basic neighbourhood of $(x,0) \in X \times \{0\}$ is a set of the form $(U \times \{0\}) \cup ((U \times \{1\}) \setminus \{(x,1)\})$, where $U$ is a neighbourhood of $x$ in $X$.

**Example 7.3.1.** There exist a space $X$ and an ideal $\mathscr{I}$ such that $X$ is *SIH* (respectively, *SSIH*) but $AD(X)$ is not *SIH* (respectively, not *SSIH*).

*Let $\mathcal{I}$ be an ideal such that $\mathfrak{b} < \mathfrak{b}(\mathcal{I})$ and $X = \Psi(\mathscr{A})$ with $|\mathscr{A}| = \mathfrak{b}$. By Corollary 4a, X is SSIH and hence X is SIH. Observe that $AD(X)$ is not SIH and so $AD(X)$ is not SSIH. Indeed, if possible suppose that $AD(X)$ is SIH. Since $\mathscr{A} \times \{1\}$ is a clopen subset of $AD(X)$, $\mathscr{A} \times \{1\}$ is SIH. Which is a contradiction because $\mathscr{A} \times \{1\}$ is a discrete subset of $AD(X)$ with $|\mathscr{A} \times \{1\}| = \mathfrak{b}$.*

**Theorem 7.3.2.** *1. If $AD(X)$ is SIH, then X is SIH.*
*2. If $AD(X)$ is SSIH, then X is SSIH.*

**Proof:** (a). Let $(\mathbb{N}_n)$ be a sequence of neighbourhood assignments of $X$. We define a sequence $(\mathbb{N}'_n)$ of neighbourhood assignments of $AD(X)$ by $\mathbb{N}'_n(x,y) = \mathbb{N}_n(p(x,y)) \times \{0,1\}$, where $p : X \times \{0,1\} \to X$ is the projection mapping. Subsequently we get a sequence $(F_n)$ of finite subsets of $AD(X)$ such that for each $z \in AD(X)$, $\{n \in \mathbb{N} : z \notin St(\cup \mathbb{N}'_n(F_n), \mathbb{N}'_n(AD(X)))\} \in \mathcal{I}$ since $AD(X)$ is $SIH$. For each $n$ let $A_n = p(F_n)$. Thus the sequence $(A_n)$ witnesses for $(\mathbb{N}_n)$ that $X$ is $SIH$.

(b). Consider a sequence $(\mathbb{N}_n)$ of neighbourhood assignments of $X$. Let us define a sequence $(\mathbb{N}'_n)$ of neighbourhood assignments of $AD(X)$ by $\mathbb{N}'_n(x,y) = \mathbb{N}_n(p(x,y)) \times \{0,1\}$, where $p : X \times \{0,1\} \to X$ is the projection mapping. Since $AD(X)$ is $SSIH$, there exists a sequence $(F_n)$ of finite subsets of $AD(X)$ such that for each $z \in AD(X)$, $\{n \in \mathbb{N} : z \notin St(F_n, \mathbb{N}'_n(AD(X)))\} \in \mathcal{I}$. For each $n$ choose $A_n = p(F_n)$. Observe that the sequence $(A_n)$ witnesses for $(\mathbb{N}_n)$ that $X$ is $SSIH$.

**Theorem 7.3.3.** *If $f$ is an open perfect mapping from X onto Y, then X is SIH whenever Y is SIH.*

**Proof:** Let $(\mathbb{N}_n)$ be a sequence of neighbourhood assignments of $X$ and $y \in Y$. Since $f^{-1}(y)$ is compact, for each $n$ there exists a finite subset $F_n^{(y)}$ of $X$ such that $f^{-1}(y) \subseteq \cup N_n(F_n^{(y)})$ and $f^{-1}(y) \cap \mathbb{N}_n(x) \neq \emptyset$ for each $x \in F_n^{(y)}$. Since $f$ is closed, for each $n$ there exists an open set $U_y^{(n)}$ in $Y$ containing $y$ such that $f^{-1}(U_y^{(n)}) \subseteq \cup N_n(F_n^{(y)})$. Then by the openness of $f$, for each $n$ we get an open set $V_y^{(n)}$ in $Y$ containing $y$ such that $V_y^{(n)} \subseteq \cap\{f(N_n(x)) : x \in F_n^{(y)}\}$ and $f^{-1}(V_y^{(n)}) \subseteq f^{-1}(U_y^{(n)})$. We define a sequence $(N'_n)$ of neighbourhood assignments of $Y$ as $N'_n(y) = V_y^{(n)}$ for all $y \in Y$. Since $Y$ is $SIH$, there exists a sequence $(A_n)$ of finite subsets of $Y$ such that for each $y \in Y$, $\{n \in \mathbb{N} : y \notin St(\cup \mathbb{N}'_n(A_n), \mathbb{N}'_n(Y))\} \in \mathcal{I}$. For each $n$ let $F_n = \cup_{y \in A_n} F_n^{(y)}$. We claim that the sequence $(F_n)$ guarantees for $(\mathbb{N}_n)$ that $X$ is $SIH$. Pick a $x \in X$. We now show that $\{n \in \mathbb{N} : x \notin St(\cup \mathbb{N}_n(F_n), \mathbb{N}_n(X))\} \in \mathcal{I}$. Choose $y = f(x)$ and then we get $\{n \in \mathbb{N} : y \notin St(\cup \mathbb{N}'_n(A_n), \mathbb{N}'_n(Y))\} \in \mathcal{I}$. Since for each $y \in A_n$, $f^{-1}(V_y^{(n)}) \subseteq \cup \mathbb{N}_n(F_n^{(y)})$, we have $f^{-1}(\cup \mathbb{N}'_n(A_n)) \subseteq \cup \mathbb{N}_n(F_n)$ for each $n$. It follows that $\{n \in \mathbb{N} : x \notin St(\cup \mathbb{N}_n(F_n), \mathbb{N}_n(X))\} \subseteq \{n \in \mathbb{N} : y \notin St(\cup \mathbb{N}'_n(A_n), \mathbb{N}'_n(Y))\}$ and hence $\{n \in \mathbb{N} : x \notin St(\cup \mathbb{N}_n(F_n), \mathbb{N}_n(X))\} \in \mathcal{I}$. Hence the result.

**Corollary 7.3.4.** *If X is SIH and Y is $\sigma$-compact, then $X \times Y$ is SIH.*

We now observe that Theorem 7.3.3 may not hold if $Y$ is $SSIH$. Which contradicts the Proposition 4.6 of [12].

**Example 7.3.5.** There exists a *SSIH* space whose preimage under an open perfect mapping is not *SSIH*.

*Consider $X = \Psi(\mathscr{A})$ with $|\mathscr{A}| = \mathfrak{b}$. Let $\mathscr{I}$ be an ideal such that $\mathfrak{b} < \mathfrak{b}(\mathscr{I})$. By Corollary 4a, X is SSIH. Let $D = \{d_\alpha : \alpha < \mathfrak{b}\}$ be the discrete space of cardinality $\mathfrak{b}$ and Y be the one-point compactification of it. Obviously, the projection mapping $p : X \times Y \to X$ is open perfect. We claim that $X \times Y$ is not SSIH. If possible suppose that $X \times Y$ is SSIH. Since $|\mathscr{A}| = \mathfrak{b}$, enumerate $\mathscr{A}$ as $\{x_\alpha : \alpha < \mathfrak{b}\}$. Define a neighbourhood assignment $\mathbb{N}$ of $X \times Y$ by*

$$\mathbb{N}(x,y) = \begin{cases} (\{x\} \cup x) \times (Y \setminus \{y\}), & \text{if } (x,y) \in \{x_\alpha\} \times (Y \setminus \{d_\alpha\}) \text{ for } \alpha < \mathfrak{b} \\ X \times \{y\}, & \text{if } (x,y) = (x_\alpha, d_\alpha) \text{ for } \alpha < \mathfrak{b} \\ \{x\} \times Y, & \text{otherwise.} \end{cases}$$

*For each n choose $\mathbb{N}_n = \mathbb{N}$. Then we get a sequence $(F_n)$ of finite subsets of $X \times Y$ such that for each $z \in X \times Y$, $\{n \in \mathbb{N} : z \notin St(F_n, \mathbb{N}_n(X \times Y))\} = \{n \in \mathbb{N} : z \notin St(F_n, \mathbb{N}(X \times Y))\} \in \mathscr{I}$. Since $\cup_{n \in \mathbb{N}} F_n$ is a countable subset of $X \times Y$, we can find a $\beta < \mathfrak{b}$ such that $(\cup_{n \in \mathbb{N}} F_n) \cap (X \times \{d_\alpha\}) = \emptyset$ for each $\alpha > \beta$. It follows that $F_n \cap (X \times \{d_\alpha\}) = \emptyset$ for each n and each $\alpha > \beta$. Choose $\gamma > \beta$. Thus $(x_\gamma, d_\gamma) \notin St(F_n, \mathbb{N}(X \times Y))$ for each n as $X \times \{d_\gamma\}$ is the only member of $\mathbb{N}(X \times Y)$ containing the point $(x_\gamma, d_\gamma)$. Consequently, $\mathbb{N} = \{n \in \mathbb{N} : (x_\gamma, d_\gamma) \notin St(F_n, \mathbb{N}(X \times Y))\} \in \mathscr{I}$, which is absurd as $\mathscr{I}$ is a proper ideal on $\mathbb{N}$. Hence $X \times Y$ is not SSIH.*

From this example, it is also clear that Corollary 7.3.4 is not true for the SS$\mathscr{I}$H property.

**Theorem 7.3.6.** *If f is an open, closed and finite-to-one continuous mapping from X onto Y, then X is SSIH whenever Y is SSIH.*

**Proof:** Let $(\mathbb{N}_n)$ be a sequence of neighbourhood assignments of $X$ and $y \in Y$. Since $f^{-1}(y)$ is finite, for each $n$ there exists a finite subset $F_n^{(y)}$ of $X$ such that $f^{-1}(y) \subseteq \cup \mathbb{N}_n(F_n^{(y)})$ and $f^{-1}(y) \cap \mathbb{N}_n(x) \neq \emptyset$ for each $x \in F_n^{(y)}$. Since $f$ is closed, for each $n$ there exists an open set $U_y^{(n)}$ in $Y$ containing $y$ such that $f^{-1}(U_y^{(n)}) \subseteq \cup \mathbb{N}_n(F_n^{(y)})$. Then by the openness of $f$, for each $n$ we get an open set $V_y^{(n)}$ in $Y$ containing $y$ such that $V_y^{(n)} \subseteq \cap \{f(\mathbb{N}_n(x)) : x \in F_n^{(y)}\}$ and $f^{-1}(V_y^{(n)}) \subseteq f^{-1}(U_y^{(n)})$. We define a sequence $(\mathbb{N}'_n)$ of neighbourhood assignments of $Y$ as $\mathbb{N}'_n(y) = V_y^{(n)}$ for all $y \in Y$. Since $Y$ is *SSIH*, there exists a sequence $(F_n)$ of finite subsets of $Y$ such that for each $y \in Y$, $\{n \in \mathbb{N} : y \notin St(F_n, \mathbb{N}'_n(Y))\} \in \mathscr{I}$. Clearly $(f^{-1}(F_n))$ is a sequence of finite subsets of $X$. We now show that for each $x \in X$, $\{n \in \mathbb{N} : x \notin St(f^{-1}(F_n), \mathbb{N}_n(X))\} \in \mathscr{I}$. It is easy to see that $f^{-1}(St(F_n, \mathbb{N}'_n(Y))) \subseteq St(f^{-1}(F_n), \mathbb{N}_n(X))$ for each $n$. Let $x \in X$. Observe that $\{n \in \mathbb{N} : x \notin St(f^{-1}(F_n), \mathbb{N}_n(X))\} \subseteq \{n \in \mathbb{N} : f(x) \notin St(F_n, \mathbb{N}'_n(Y))\}$ because $f(x) \in St(F_n, \mathbb{N}'_n(Y))$ implies that $x \in St(f^{-1}(F_n), \mathbb{N}_n(X))$. Thus $\{n \in \mathbb{N} : x \notin St(f^{-1}(F_n), \mathbb{N}_n(X))\} \in \mathscr{I}$. Hence $X$ is *SSIH*.

### 7.3.1 GROUPABILITY AND ALSTER COVERS

**Definition 7.3.7.** *A countable cover $\mathscr{U} = \{U_n : n \in \mathbb{N}\}$ of a space $X$ is said to be strongly $\mathscr{I}$-groupable if there exists a sequence of positive integers $n_1 < n_2 < \cdots$ such that $\mathscr{U} = \cup_{k\in\mathbb{N}}\{U_i : n_k \leq i < n_{k+1}\}$ and for each $x \in X$, $\{k \in \mathbb{N} : x \notin \cup\{U_i : n_k \leq i < n_{k+1}\}\} \in \mathscr{I}$. We denote the collection of all strongly $\mathscr{I}$-groupable covers by $\mathscr{I}$-$\mathscr{O}^{sgp}$.*

A strongly Fin-groupable cover is said to be strongly groupable. Clearly $\mathscr{O}^{sgp} \subseteq \mathscr{O}^{gp}$, where $\mathscr{O}^{sgp}$ denotes the collection of all strongly groupable covers.

**Remark 7.3.8.** *1. The notion of $\mathscr{I}$-groupability of a cover is independent of enumeration, i.e. if $\mathscr{U} = \{U_n : n \in \mathbb{N}\}$ is $\mathscr{I}$-groupable, then for any bijection $\sigma$ on $\mathbb{N}$, $\{U_{\sigma(n)} : n \in \mathbb{N}\}$ is also $\mathscr{I}$-groupable.*

*2. In contrast to the above, a strongly $\mathscr{I}$-groupable cover $\mathscr{U} = \{U_n : n \in \mathbb{N}\}$ may not remain a strongly $\mathscr{I}$-groupable cover under a changed enumeration.*

Also note that $\mathscr{I}$-$\mathscr{O}^{sgp} \subseteq \mathscr{I}$-$\mathscr{O}^{gp}$ and a $\mathscr{I}$-groupable cover need not be strongly $\mathscr{I}$-groupable. However, we can prove the following.

**Proposition 7.3.9.** *Every $\mathscr{I}$-groupable cover $\mathscr{U}$ of a space $X$ is strongly $\mathscr{I}$-groupable (with respect to some enumeration of $\mathscr{U}$).*

**Proof:** Since $\mathscr{U}$ is $\mathscr{I}$-groupable, there is a sequence $(\mathscr{U}_n)$ of finite, pairwise disjoint subsets of $\mathscr{U}$ such that $\mathscr{U} = \cup_{n\in\mathbb{N}}\mathscr{U}_n$ and for each $x \in X$, $\{n \in \mathbb{N} : x \notin \cup\mathscr{U}_n\} \in \mathscr{I}$. For each $k \in \mathbb{N}$ we enumerate $\mathscr{U}_k$ by $\mathscr{U}_k = \{U_i : n_k \leq i < n_{k+1}\}$, where $n_1 < n_2 < \cdots < n_k < \cdots$ is a sequence of positive integers with $n_1 = 1$. We now fix the enumeration of $\mathscr{U}$ as $\{U_n : n \in \mathbb{N}\}$. It follows that $\mathscr{U} = \{U_n : n \in \mathbb{N}\}$ is strongly $\mathscr{I}$-groupable.

**Corollary 7.3.10.** *Every groupable cover $\mathscr{U}$ of a space $X$ is strongly groupable (with respect to some enumeration of $\mathscr{U}$).*

For any families $\mathscr{U}$ and $\mathscr{V}$ of subsets of $X$, $\mathscr{U} \wedge \mathscr{V}$ denotes the collection $\{U \cap V : U \in \mathscr{U}$ and $V \in \mathscr{V}\}$.

**Lemma 7.3.11.** *A space $X$ satisfies $U^*_{fin}(\mathscr{O}, \mathscr{I}\text{-}\mathscr{O}^{sgp})$ if and only if for each sequence $(\mathrm{N}_n)$ of neighbourhood assignments of $X$ there exists a sequence $(F_n)$ of finite subsets of $X$ such that $\{St(\cup \mathrm{N}_n(F_n), N_n(X)) : n \in \mathbb{N}\} \in \mathscr{I}$-$\mathscr{O}^{sgp}$.*

**Proof:** Let $(\mathrm{N}_n)$ be a sequence of neighbourhood assignments of $X$. For each $n$ $\mathscr{U}_n = \{\mathrm{N}_n(x) : x \in X\}$ is an open cover of $X$. Since $X$ satisfies $U^*_{fin}(\mathscr{O}, \mathscr{I}\text{-}\mathscr{O}^{sgp})$, there exists a sequence $(\mathscr{V}_n)$ such that for each $n$ $\mathscr{V}_n$ is a finite subset of $\mathscr{U}_n$ and $\{St(\cup\mathscr{V}_n, \mathscr{U}_n) : n \in \mathbb{N}\} \in \mathscr{I}$-$\mathscr{O}^{sgp}$. Now we can obtain a sequence $(F_n)$ of finite subsets of $X$ such that for each $n$ $\mathscr{V}_n = \{N_n(x) : x \in F_n\}$. Clearly $\{St(\cup N_n(F_n), \mathrm{N}_n(X)) : n \in \mathbb{N}\} \in \mathscr{I}$-$\mathscr{O}^{sgp}$.

Conversely, let $(\mathscr{U}_n)$ be a sequence of open covers of $X$. For each $x \in X$ there exists a $U_x^{(n)} \in \mathscr{U}_n$ such that $x \in U_x^{(n)}$. We define a sequence $(N_n)$ of neighbourhood assignments of $X$ by $N_n(x) = U_x^{(n)}$ for each $n$ and each $x \in X$. By the given hypothesis we obtain a sequence $(F_n)$ of finite subsets of $X$ such that $\{St(\cup \mathbb{N}_n(F_n), N_n(X)) : n \in \mathbb{N}\} \in \mathscr{I}\text{-}\mathscr{O}^{sgp}$. For each $n$ let $\mathscr{V}_n = \{N_n(x) : x \in F_n\}$. Now the sequence $(\mathscr{V}_n)$ witnesses for $(\mathscr{U}_n)$ that $X$ satisfies $U^*_{fin}(\mathscr{O}, \mathscr{I}\text{-}\mathscr{O}^{sgp})$.

**Theorem 7.3.12.** *If $X$ satisfies $U^*_{fin}(\mathscr{O}, \mathscr{I}\text{-}\mathscr{O}^{sgp})$, then $X$ is SIH.*

**Proof:** Let $(\mathbb{N}_n)$ be a sequence of neighbourhood assignments of $X$. We define a sequence $(\mathbb{N}'_n)$ of neighbourhood assignments of $X$ by $\mathbb{N}'_n(x) = \wedge_{i \leq n} \mathbb{N}_i(x)$. Apply $U^*_{fin}(\mathscr{O}, \mathscr{I}\text{-}\mathscr{O}^{sgp})$ to $(\mathbb{N}'_n)$ to obtain a sequence $(F_n)$ of finite subsets of $X$ such that $\{St(\cup \mathbb{N}'_n(F_n), \mathbb{N}'_n(X)) : n \in \mathbb{N}\} \in \mathscr{I}\text{-}\mathscr{O}^{sgp}$. Consider a sequence $n_1 < n_2 < \cdots$ of positive integers which witnesses that $\{St(\cup \mathbb{N}'_n(F_n), \mathbb{N}'_n(X)) : n \in \mathbb{N}\} \in \mathscr{I}\text{-}\mathscr{O}^{sgp}$, i.e. for each $x \in X$, $\{k \in \mathbb{N} : x \notin \cup\{St(\cup \mathbb{N}'_i(F_i), \mathbb{N}'_i(X)) : n_k \leq i < n_{k+1}\}\} \in \mathscr{I}$. For each $k \in \mathbb{N}$ let $A_k = \cup_{n_k \leq i < n_{k+1}} F_i$. We claim that the sequence $(A_n)$ witnesses for $(\mathbb{N}_n)$ that $X$ is *SIH*. Let $x \in X$. Choose a $m \in \mathbb{N}$ so that $x \notin St(\cup \mathbb{N}_m(A_m), \mathbb{N}_m(X))$. Suppose that $x \in \cup\{St(\cup \mathbb{N}'_i(F_i), \mathbb{N}'_i(X)) : n_m \leq i < n_{m+1}\}$. Then $x \in St(\cup \mathbb{N}'_j(F_j), \mathbb{N}'_j(X))$ for some $n_m \leq j < n_{m+1}$ and hence $x \in \mathsf{N}'_j(y)$ for some $y \in X$ and $\mathbb{N}'_j(y) \cap (\cup \mathbb{N}'_j(F_j)) \neq \emptyset$. Note that $j \geq m$ since $n_m \geq m$. It follows that $x \in \mathbb{N}_m(y)$ and $\mathbb{N}_m(y) \cap (\cup \mathbb{N}_m(A_m)) \neq \emptyset$. So $x \in St(\cup \mathbb{N}_m(A_m), \mathbb{N}_m(X))$, which is a contradiction. Thus $x \notin \cup\{St(\cup \mathbb{N}'_i(F_i), \mathbb{N}'_i(X)) : n_m \leq i < n_{m+1}\}$ and consequently $m \in \{k \in \mathbb{N} : x \notin \cup\{St(\cup \mathsf{N}'_i(F_i), \mathbb{N}'_i(X)) : n_k \leq i < n_{k+1}\}\}$. Accordingly we have $\{n \in \mathbb{N} : x \notin St(\cup \mathbb{N}_n(A_n), \mathbb{N}_n(X))\} \subseteq \{k \in \mathbb{N} : x \notin \cup\{St(\cup \mathbb{N}'_i(F_i), \mathbb{N}'_i(X)) : n_k \leq i < n_{k+1}\}\}$ and hence $\{n \in \mathbb{N} : x \notin St(\cup \mathbb{N}_n(A_n), \mathbb{N}_n(X))\} \in \mathscr{I}$. The proof is now complete.

**Lemma 7.3.13.** *A space $X$ satisfies $SS^*_{fin}(\mathscr{O}, \mathscr{I}\text{-}\mathscr{O}^{sgp})$ if and only if for each sequence $(\mathbb{N}_n)$ of neighbourhood assignments of $X$ there exists a sequence $(F_n)$ of finite subsets of $X$ such that $\{St(F_n, \mathbb{N}_n(X)) : n \in \mathbb{N}\} \in \mathscr{I}\text{-}\mathscr{O}^{sgp}$.*

**Proof:** Let $(\mathbb{N}_n)$ be a sequence of neighbourhood assignments of $X$. For each $n$ $\mathscr{U}_n = \{\mathbb{N}_n(x) : x \in X\}$ is an open cover of $X$. Since $X$ satisfies $SS^*_{fin}(\mathscr{O}, \mathscr{I}\text{-}\mathscr{O}^{sgp})$, there exists a sequence $(F_n)$ of finite subsets of $X$ such that $\{St(F_n, \mathscr{U}_n) : n \in \mathbb{N}\} \in \mathscr{I}\text{-}\mathscr{O}^{sgp}$, i.e. $\{St(F_n, \mathbb{N}_n(X)) : n \in \mathbb{N}\} \in \mathscr{I}\text{-}\mathscr{O}^{sgp}$.

Conversely, let $(\mathscr{U}_n)$ be a sequence of open covers of $X$. For each $x \in X$ there exists a $U_x^{(n)} \in \mathscr{U}_n$ such that $x \in U_x^{(n)}$. We now define a sequence $(\mathsf{N}_n)$ of neighbourhood assignments of $X$ by $\mathsf{N}_n(x) = U_x^{(n)}$ for each $n$ and each $x \in X$. By the given hypothesis we get a sequence $(F_n)$ of finite subsets of $X$ such that $\{St(F_n, \mathsf{N}_n(X)) : n \in \mathbb{N}\} \in \mathscr{I}\text{-}\mathscr{O}^{sgp}$, i.e. $\{St(F_n, \mathscr{U}_n) : n \in \mathbb{N}\} \in \mathscr{I}\text{-}\mathscr{O}^{sgp}$. Thus $X$ satisfies $\mathsf{SS}^*_{\mathrm{fin}}(\mathscr{O}, \mathscr{I}\text{-}\mathscr{O}^{sgp})$.

**Theorem 7.3.14.** *If $X$ satisfies $SS^*_{fin}(\mathscr{O}, \mathscr{I}\text{-}\mathscr{O}^{sgp})$, then $X$ is* SS$\mathscr{I}$H.

**Proof:** Let $(N_n)$ be a sequence of neighbourhood assignments of $X$. We define a sequence $(N'_n)$ of neighbourhood assignments of $X$ by $N'_n(x) = \wedge_{i \leq n} \mathbb{N}_n(x)$. Apply $SS^*_{fin}(\mathscr{O}, \mathscr{I}\text{-}\mathscr{O}^{sgp})$ to $(\mathbb{N}'_n)$ to obtain a sequence $(A_n)$ of finite subsets of $X$ such that

$\{St(A_n, \mathbb{N}'_n(X)) : n \in \mathbb{N}\} \in \mathscr{I}\text{-}\mathscr{O}^{sgp}$. Consider a sequence $n_1 < n_2 < \cdots$ of positive integers which witnesses that $\{St(A_n, \mathbb{N}'_n(X)) : n \in \mathbb{N}\} \in \mathscr{I}\text{-}\mathscr{O}^{sgp}$, i.e. for each $x \in X$, $\{k \in \mathbb{N} : x \notin \cup\{St(A_i, \mathbb{N}'_i(X)) : n_k \leq i < n_{k+1}\}\} \in \mathscr{I}$. For each $k \in \mathbb{N}$ let $F_k = \cup_{n_k \leq i < n_{k+1}} A_i$. Observe that the sequence $(F_n)$ guarantees for $(\mathbb{N}_n)$ that $X$ is $SSIH$. Let $x \in X$. Pick a $m \in \{n \in \mathbb{N} : x \notin St(F_n, \mathbb{N}_n(X))\}$ and so $x \notin St(F_m, \mathbb{N}_m(X))$. If possible suppose that $x \in \cup\{St(A_i, \mathbb{N}'_i(X)) : n_m \leq i < n_{m+1}\}$. This implies that $x \in St(A_j, \mathbb{N}'_j(X))$ for some $n_m \leq j < n_{m+1}$ and then $x \in \mathbb{N}'_j(y)$ and $\mathbb{N}'_j(y) \cap A_j \neq \emptyset$ for some $y \in X$. Now $j \geq m$ as $n_m \geq m$ and so $\mathbb{N}'_j(y) \subseteq \mathbb{N}_m(y)$. Since $A_j \subseteq F_m$, we obtain $x \in St(F_m, \mathbb{N}_m(X))$, which is a contradiction. Thus $x \notin \cup\{St(A_i, \mathbb{N}'_i(X)) : n_m \leq i < n_{m+1}\}$ and consequently $m \in \{k \in \mathbb{N} : x \notin \cup\{St(A_i, \mathbb{N}'_i(X)) : n_k \leq i < n_{k+1}\}\}$. It follows that $\{n \in \mathbb{N} : x \notin St(F_n, \mathbb{N}_n(X))\} \subseteq \{k \in \mathbb{N} : x \notin \cup\{St(A_i, \mathbb{N}'_i(X)) : n_k \leq i < n_{k+1}\}\}$ and hence $\{n \in \mathbb{N} : x \notin St(F_n, \mathbb{N}_n(X))\} \in \mathscr{I}$. The proof is now complete.

We now recall the following classes of covers (see [3, 2, 12]).

1. $(\mathscr{G})$ The family of all covers $\mathscr{U}$ of a space $X$ for which each element of $\mathscr{U}$ is a $G_\delta$ set.
2. $(\mathscr{G}_K)$ The family of all covers $\mathscr{U}$ of a space $X$ not containing $X$, each element of $\mathscr{U}$ is a $G_\delta$ set and for each compact set $C \subseteq X$ there is a $U \in \mathscr{U}$ such that $C \subseteq U$. Such type of covers are called Alster covers.
3. $(\mathscr{G}_\Gamma)$ The family of all covers $\mathscr{U} \in \mathscr{G}$ which are infinite and each infinite subset of $\mathscr{U}$ is a cover of $X$.
4. $(\mathscr{G}_{\mathscr{I}\text{-}\Gamma})$ The family of all countable covers $\mathscr{U} \in \mathscr{G}$, each of which has an enumeration $\mathscr{U} = \{U_n : n \in \mathbb{N}\}$ such that for each $x \in X$, $\{n \in \mathbb{N} : x \notin U_n\} \in \mathscr{I}$.

If $\mathscr{I} = Fin$, then $\mathscr{G}_{\mathscr{I}\text{-}\Gamma} \subseteq \mathscr{G}_\Gamma$. Also if we restrict ourself to countable covers, then $\mathscr{G}_\Gamma \subseteq \mathscr{G}_{\mathscr{I}\text{-}\Gamma}$ for any ideal $\mathscr{I}$ on $\mathbb{N}$. Thus under this assumption and if $\mathscr{I} = Fin$, then $S_1^*(\mathscr{G}_K, \mathscr{G}_{\mathscr{I}\text{-}\Gamma}) = S_1^*(\mathscr{G}_K, \mathscr{G}_\Gamma)$ and $\mathrm{SS}_1^*(\mathscr{G}_K, \mathscr{G}_{\mathscr{I}\text{-}\Gamma}) = SS^*_{fin}(\mathscr{G}_K, \mathscr{G}_\Gamma)$.

Let $\mathfrak{G}_\delta$ denote the collection of all $G_\delta$ subsets of $X$. A neighbourhood $g\delta$-assignment is a function $\mathbb{N} : X \to \mathfrak{G}_\delta$ such that for each $x \in X$, $x \in \mathbb{N}(x)$. A neighbourhood $g\delta$-assignment $\mathbb{N}$ of $X$ is said to be Alster-neighbourhood $g\delta$-assignment if $\mathbb{N}(X)$ is a Alster cover of $X$. For the rest of this subsection we assume that all considered neighbourhood assignments are closed under finite unions (a neighbourhood assignment $\mathbb{N}$ of $X$ is said to be closed under finite unions if for each finite subset $F$ of $X$, $\cup\mathbb{N}(F) \in \mathbb{N}(X)$).

**Lemma 7.3.15.** *A space $X$ satisfies $\mathrm{S}_1^*(\mathscr{G}_K, \mathscr{G}_{\mathscr{I}\text{-}\Gamma})$ if and only if for each sequence $(\mathsf{N}_n)$ of Alster-neighbourhood $g\delta$-assignments of $X$ there exists a sequence $(x_n)$ of elements of $X$ such that $\{St(\mathsf{N}_n(x_n), \mathsf{N}_n(X)) : n \in \mathbb{N}\} \in \mathscr{G}_{\mathscr{I}\text{-}\Gamma}$.*

**Proof:** Let $(N_n)$ be a sequence of Alster-neighbourhood $g\delta$-assignments of $X$. For each $n$ $\mathscr{U}_n = \{N_n(x) : x \in X\}$ is an Alster cover of $X$. Since $X$ satisfies $S_1^*(\mathscr{G}_K, \mathscr{G}_{\mathscr{I}\text{-}\Gamma})$, there exists a sequence $(U_n)$ such that for each $n$ $U_n \in \mathscr{U}_n$ and $\{St(U_n, \mathscr{U}_n) : n \in \mathbb{N}\} \in \mathscr{G}_{\mathscr{I}\text{-}\Gamma}$. Now we can obtain a sequence $(x_n)$ of elements of $X$ such that for each $n$ $U_n = N_n(x_n)$. Clearly $\{St(N_n(x_n), N_n(X)) : n \in \mathbb{N}\} \in \mathscr{G}_{\mathscr{I}\text{-}\Gamma}$.

Conversely, let $(\mathscr{U}_n)$ be a sequence of Alster covers of $X$. For each $x \in X$ there exists a $U_x^{(n)} \in \mathscr{U}_n$ with $x \in U_x^{(n)}$. Define a sequence $(N_n)$ of Alster-neighbourhood $g\delta$-assignments of $X$ by $N_n(x) = U_x^{(n)}$ for each $n$ and each $x \in X$. Using the given hypothesis we obtain a sequence $(x_n)$ of elements of $X$ such that $\{St(\cup N_n(x_n), N_n(X)) : n \in \mathbb{N}\} \in \mathscr{G}_{\mathscr{I}}\text{-}\Gamma$. For each $n$ let $U_n = N_n(x_n)$. Observe that $\{St(U_n, \mathscr{U}_n) : n \in \mathbb{N}\} \in \mathscr{G}_{\mathscr{I}}\text{-}\Gamma$. Thus $X$ satisfies $S^*_{fin}(\mathscr{G}_K, \mathscr{G}_{\mathscr{I}}\text{-}\Gamma)$.

**Lemma 7.3.16.** *Let $(N_n)$ be sequence of neighbourhood assignments of $X \times Y$. Then for each $n$ and each $K \times \{y\} \subseteq X \times Y$, where $K$ is compact there are a $z \in X \times Y$ and a $G_\delta$ subset $\Phi(K)$ of $X$ containing $K$ such that $\Phi(K) \times \{y\} \subseteq N_n(z)$.*

**Proof:** Let $K$ be a compact subset of $X$. For each $y \in Y$ and each $n$ there exists a $z \in X \times Y$ such that $K \times \{y\} \subseteq N_n(z)$. Let $V$ be an open set in $X$ containing $K$ such that $K \times \{y\} \subseteq V \times \{y\} \subseteq N_n(z)$. If $(N'_n)$ is a sequence of neighbourhood assignments of $X$, then we get a sequence $(F_n)$ of finite subsets of $X$ such that $K \subseteq \cup N'_n(F_n) = V_n$ (say). It follows that $K \subseteq V_n \cap V$ for all $n$. Choose $\Phi(K) = \cap_{n \in \mathbb{N}}(V_n \cap V)$. Thus for each compact subset $K$ of $X$ we obtain a $G_\delta$ subset $\Phi(K)$ of $X$ such that $K \subseteq \Phi(K)$. Also we can say that for each compact subset $K$ of $X$, each $y \in Y$ and each $n$ there are a $z \in X \times Y$ and a $G_\delta$ subset $\Phi(K)$ of $X$ containing $K$ such that $\Phi(K) \times \{y\} \subseteq N_n(z)$.

**Theorem 7.3.17** ([12, Theorem 4.3]). *If $X$ satisfies $S^*_{fin}(\mathscr{G}_K, \mathscr{G}_{\mathscr{I}}\text{-}\Gamma)$ and $Y$ is SIH, then $X \times Y$ is SIH.*

**Lemma 7.3.18.** *A space $X$ satisfies $SS^*_{fin}(\mathscr{G}_K, \mathscr{G}_{\mathscr{I}}\text{-}\Gamma)$ if and only if for each sequence $(N_n)$ of Alster-neighbourhood $g\delta$-assignments of $X$ there exists a sequence $(x_n)$ of elements of $X$ such that $\{St(x_n, N_n(X)) : n \in \mathbb{N}\} \in \mathscr{G}_{\mathscr{I}}\text{-}\Gamma$.*

**Proof:** Let $(N_n)$ be a sequence of Alster-neighbourhood $g\delta$-assignments of $X$. For each $n$ $\mathscr{U}_n = \{N_n(x) : x \in X\}$ is an Alster cover of $X$. Since $X$ satisfies $SS^*_{fin}(\mathscr{G}_K, \mathscr{G}_{\mathscr{I}}\text{-}\Gamma)$, there exists a sequence $(x_n)$ of elements of $X$ such that $\{St(x_n, \mathscr{U}_n) : n \in \mathbb{N}\} \in \mathscr{G}_{\mathscr{I}}\text{-}\Gamma$, i.e. $\{St(x_n, N_n(X)) : n \in \mathbb{N}\} \in \mathscr{G}_{\mathscr{I}}\text{-}\Gamma$.

Conversely, let $(\mathscr{U}_n)$ be a sequence of Alster covers of $X$. For each $x \in X$ there exists a $U_x^{(n)} \in \mathscr{U}_n$ with $x \in U_x^{(n)}$. Define a sequence $(N_n)$ of Alster-neighbourhood $g\delta$-assignments of $X$ by $N_n(x) = U_x^{(n)}$ for each $n$ and each $x \in X$. Using the given hypothesis we obtain a sequence $(x_n)$ of elements of $X$ such that $\{St(x_n, N_n(X)) : n \in \mathbb{N}\} \in \mathscr{G}_{\mathscr{I}}\text{-}\Gamma$. Clearly $\{St(x_n, \mathscr{U}_n) : n \in \mathbb{N}\} \in \mathscr{G}_{\mathscr{I}}\text{-}\Gamma$. Thus $X$ satisfies $SS^*_{fin}(\mathscr{G}_K, \mathscr{G}_{\mathscr{I}}\text{-}\Gamma)$.

**Theorem 7.3.19.** *If $X$ satisfies $SS^*_1(\mathscr{G}_K, \mathscr{G}_{\mathscr{I}}\text{-}\Gamma)$ and $Y$ is SSIH, then $X \times Y$ is SSIH.*

**Proof:** Let $(N_n)$ be a sequence of neighbourhood assignments of $X \times Y$. By Lemma 7.3.16, for each $n$ and each $K \times \{y\} \subseteq X \times Y$, where $K$ is compact there are a $z \in X \times Y$ and a $G_\delta$ subset $\Phi(K)$ of $X$ containing $K$ such that $\Phi(K) \times \{y\} \subseteq N_n(z)$.

Let $(N'_n)$ be a sequence of Alster-neighbourhood $g\delta$-assignments of $X$ such that $N'_n(X) \subseteq \{\Phi(K) : K \text{ is a compact subset of } X\}$. Since $X$ satisfies $SS^*_1(\mathscr{G}_K, \mathscr{G}_{\mathscr{I}}\text{-}\Gamma)$, there exists a sequence $(x_n)$ of elements of $X$ such that $\{St(x_n, N'_n(X)) : n \in \mathbb{N}\} \in$

$\mathcal{G}_{\mathcal{I}}$-$\Gamma$. Define a sequence $(N_n^*)$ of neighbourhood assignments of $Y$ such that for each $y \in Y$, $\{x_n\} \times N_n^*(y) \subseteq N_n(z)$ for some $z \in X \times Y$. Since $Y$ is *SSIH*, there is a sequence $(A_n)$ of finite subsets of $Y$ such that for each $y \in Y$, $\{n \in \mathbb{N} : y \notin St(A_n, N_n^*(Y))\} \in \mathcal{I}$. For each $n$ and each $y \in A_n$ choose a $z_y \in X \times Y$ such that $\{x_n\} \times N_n^*(y) \subseteq N_n(z_y)$. Consider $F_n = \{z_y : y \in A_n\}$. We now show that the sequence $(F_n)$ witnesses for $(N_n)$ that $X \times Y$ is *SSIH*. Let $(x,y) \in X \times Y$. Clearly $\{n \in \mathbb{N} : x \notin St(x_n, N_n'(X))\} \in \mathcal{I}$ because $\{St(x_n, N_n'(X)) : n \in \mathbb{N}\} \in \mathcal{G}_{\mathcal{I}}$-$\Gamma$. Also since $\{n \in \mathbb{N} : y \notin St(A_n, N_n^*(Y))\} \in \mathcal{I}$, $\{n \in \mathbb{N} : x \notin St(x_n, N_n'(X))\} \cup \{n \in \mathbb{N} : y \notin St(A_n, N_n^*(Y))\} \in \mathcal{I}$.

It now remains to show that $\{n \in \mathbb{N} : (x,y) \notin St(F_n, N_n(X \times Y))\} \subseteq \{n \in \mathbb{N} : x \notin St(x_n, N_n'(X))\} \cup \{n \in \mathbb{N} : y \notin St(A_n, N_n^*(Y))\}$. If $k \in \{n \in \mathbb{N} : (x,y) \notin St(F_n, N_n(X \times Y))\}$, then $(x,y) \notin St(F_k, N_k(X \times Y))$. If possible assume that $x \in St(x_k, N_k'(X))$ and $y \in St(A_k, N_k^*(Y))$. Accordingly, we obtain a $p \in X$, $q \in Y$ and $r \in A_k$ such that $x, x_k \in N_k'(p)$ and $y, r \in N_k^*(q)$. Choose a $z_r \in F_k$ such that $\{x_k\} \times N_k^*(r) \subseteq N_k(z_r)$. Also we get a $z_1 \in X \times Y$ such that $N_k'(p) \times \{y\} \subseteq N_k(z_1)$. Since $N_k$ is closed under finite unions, we can say that there exists a $z \in X \times Y$ such that $N_k(z) = \cup N_k(\{z_r, z_1\})$ and so $(x,y) \in N_k(z)$ with $N_k(z) \cap F_k \neq \emptyset$. This implies that $(x,y) \in St(F_k, N_k(X \times Y))$, which is absurd. Consequently either $x \notin St(x_k, N_k'(X))$ or $y \notin St(A_k, N_k^*(Y))$. Thus $k \in \{n \in \mathbb{N} : x \notin St(x_n, N_n'(X))\} \cup \{n \in \mathbb{N} : y \notin St(A_n, N_n^*(Y))\}$. Hence $\{n \in \mathbb{N} : (x,y) \notin St(F_n, N_n(X \times Y))\} \in \mathcal{I}$. This completes the proof.

### 7.3.2 THE CARDINAL $\mathfrak{B}(\mathcal{I})$

Let $Y$ be a subspace of $X$. We say that $Y$ is star-$\mathcal{I}$-Hurewicz in $X$ (in short, S$\mathcal{I}$H in $X$) (respectively, strongly star-$\mathcal{I}$-Hurewicz in $X$ (in short, SS$\mathcal{I}$H in $X$)) if for each sequence $(n_n)$ of neighbourhood assignments of $X$ there exists a sequence $(F_n)$ of finite subsets of $X$ such that for each $x \in Y$, $\{n \in \mathbb{N} : x \notin St(\cup \mathrm{N}_n(F_n), \mathrm{N}_n(X))\} \in \mathcal{I}$ (respectively, $\{n \in \mathbb{N} : x \notin St(F_n, \mathrm{N}_n(X))\} \in \mathcal{I}$). Clearly if $Y$ is a S$\mathcal{I}$H (respectively, SS$\mathcal{I}$H) subspace of $X$, then $Y$ is S$\mathcal{I}$H (respectively, SS$\mathcal{I}$H) in $X$. $X$ is S$\mathcal{I}$H (respectively, SS$\mathcal{I}$H) if and only if $X$ is S$\mathcal{I}$H (respectively, SS$\mathcal{I}$H) in $X$. If $Y$ is S$\mathcal{I}$H (respectively, SS$\mathcal{I}$H) in $X$, then so is any subset of $Y$. If $Y$ is SS$\mathcal{I}$H in $X$, then $Y$ is S$\mathcal{I}$H in $X$. Also if for each $n \in \mathbb{N}$, $X_n$ is S$\mathcal{I}$H (respectively, SS$\mathcal{I}$H) in $X$, then so is $\cup_{n\in\mathbb{N}} X_n$.

**Example 7.3.20.** There exists a space $X$ and an ideal $\mathcal{I}$ such that $X$ has a subspace $Y$ which is *SSIH* in $X$ (hence *SIH* in $X$) but $Y$ is not S$\mathcal{I}$H (hence not *SSIH*).
*Let $\mathcal{I}$ be an ideal on $\mathbb{N}$ such that $\omega_1 < \mathfrak{b}(\mathcal{I})$. Let $X = \Psi(\mathcal{A})$ with $|\mathcal{A}| = \omega_1$ and $Y = \mathcal{A}$. By Corollary 4a, $Y$ is SSIH in $X$. Since $Y$ is a discrete space with cardinality $\omega_1$, $Y$ is not SIH.*

**Example 7.3.21.** There exists a space $X$ and an ideal $\mathcal{I}$ such that $X$ has a subspace $Y$ which is *SIH* in $X$ but not *SSIH* in $X$.
*Let $X = \Psi(\mathcal{A})$ with $|\mathcal{A}| = \mathfrak{b}$ and $\mathcal{I}$ be an ideal on $\mathbb{N}$ with $\mathfrak{b} < \mathfrak{b}(\mathcal{I})$. By Corollary 4a, $\mathcal{A}$ is SSIH in $X$. Let $Y$ be the one-point compactification of the discrete space $D = \{d_\alpha : \alpha < \mathfrak{b}\}$ of cardinality $\mathfrak{b}$. Clearly, $\mathcal{A} \times Y$ is SIH in $X \times Y$ by Corollary 7.3.4. We now show that $\mathcal{A} \times Y$ is not SSIH in $X \times Y$. If possible suppose that $\mathcal{A} \times Y$ is*

*$S$ in $SHX \times Y$. We enumerate $\mathscr{A}$ as $\{x_\alpha : \alpha < \mathfrak{b}\}$. Let us define a neighbourhood assignment $\mathsf{N}$ of $X \times Y$ by*

$$\mathsf{N}(x,y) = \begin{cases} (\{x\} \cup x) \times (Y \setminus \{y\}), & \text{if } (x,y) \in \{x_\alpha\} \times (Y \setminus \{d_\alpha\}) \text{ for } \alpha < \mathfrak{b} \\ X \times \{y\}, & \text{if } (x,y) = (x_\alpha, d_\alpha) \text{ for } \alpha < \mathfrak{b} \\ \{x\} \times Y, & \text{otherwise.} \end{cases}$$

*For each $n$ choose $N_n = N$. Then we get a sequence $(F_n)$ of finite subsets of $X \times Y$ such that for each $z \in \mathscr{A} \times Y$, $\{n \in \mathbb{N} : z \notin St(F_n, N_n(X \times Y))\} = \{n \in \mathbb{N} : z \notin St(F_n, N(X \times Y))\} \in \mathscr{I}$. Since $\cup_{n \in \mathbb{N}} F_n$ is a countable subset of $X \times Y$, we can find a $\beta < \mathfrak{b}$ such that $(\cup_{n \in \mathbb{N}} F_n) \cap (X \times \{d_\alpha\}) = \emptyset$ for each $\alpha > \beta$. It follows that $F_n \cap (X \times \{d_\alpha\}) = \emptyset$ for each $n$ and each $\alpha > \beta$. Choose $\gamma > \beta$. Thus $(x_\gamma, d_\gamma) \notin St(F_n, N(X \times Y))$ for each $n$ as $X \times \{d_\gamma\}$ is the only member of $N(X \times Y)$ containing the point $(x_\gamma, d_\gamma)$. Consequently, $\mathbb{N} = \{n \in \mathbb{N} : (x_\gamma, d_\gamma) \notin St(F_n, N(X \times Y))\} \in \mathscr{I}$, which is absurd as $\mathscr{I}$ is a proper ideal on $\mathbb{N}$. Hence $\mathscr{A} \times Y$ is not SSIH in $X \times Y$.*

**Theorem 7.3.22.** *If $X$ is star-Lindelöf, then every subset of $X$ of cardinality less than $\mathfrak{b}(\mathscr{I})$ is SIH in $X$.*

**Proof:** Consider a subset $Y$ of $X$ with $|Y| < \mathfrak{b}(\mathscr{I})$. Let $(\mathsf{N}_n)$ be a sequence of neighbourhood assignments of $X$. By Lemma 7.3.18, for each $n$ we get a countable subset $A_n = \{x_m^{(n)} : m \in \mathbb{N}\}$ of $X$ such that $X = St(\cup N_n(A_n), N_n(X))$. For each $x \in Y$ pick a $f_x \in \mathbb{N}^{\mathbb{N}}$ such that $St(x, N_n(X)) \cap \mathsf{N}_n(x_{f_x(n)}^{(n)}) \neq \emptyset$ for all $n$. Since the cardinality of $\{f_x : x \in Y\}$ is less than $\mathfrak{b}(\mathscr{I})$, there exists a $g \in \mathbb{N}^{\mathbb{N}}$ such that for each $x \in Y$, $\{n \in \mathbb{N} : f_x(n) > g(n)\} \in \mathscr{I}$. Define a sequence $(F_n)$ as $F_n = \{x_i^{(n)} : i \leq g(n)\}$ for each $n$. We claim that the sequence $(F_n)$ witnesses for $(N_n)$ that $Y$ is *SIH* in $X$. Let $x \in Y$. Choose a $k \in \{n \in \mathbb{N} : x \notin St(\cup N_n(F_n), N_n(X))\}$. Subsequently $St(x, N_k(X)) \cap N_k(x_i^{(k)}) = \emptyset$ for all $i \leq g(n)$. It follows that $f_x(k) > g(k)$ because $St(x, N_n(X)) \cap N_n(x_{f_x(n)}^{(n)}) \neq \emptyset$ for all $n$. Thus $k \in \{n \in \mathbb{N} : f_x(n) > g(n)\}$ and so $\{n \in \mathbb{N} : x \notin St(\cup N_n(F_n), N_n(X)\} \subseteq \{n \in \mathbb{N} : f_x(n) > g(n)\}$. This gives us $\{n \in \mathbb{N} : x \notin St(\cup N_n(F_n), N_n(X))\} \in \mathscr{I}$. Hence $Y$ is *SIH* in $X$.

**Corollary 7.3.23.** *Every star-Lindelöf space of cardinality less than $\mathfrak{b}(\mathscr{I})$ is SIH.*

**Theorem 7.3.24.** *If $X$ is a strongly star-Lindelöf space, then every subset of $X$ of cardinality less than $\mathfrak{b}(\mathscr{I})$ is SSIH in $X$.*

**Proof:** Let $Y$ be a subset of $X$ with $|Y| < \mathfrak{b}(\mathscr{I})$. Let $(N_n)$ be a sequence of neighbourhood assignments of $X$. Since $X$ is strongly star-Lindelöf, by Lemma 7.3.18, for each $n$ there exists a countable subset $A_n = \{x_m^{(n)} : m \in \mathbb{N}\}$ of $X$ such that $X = St(A_n, N_n(X))$. For each $x \in Y$ choose a $f_x \in \mathbb{N}^{\mathbb{N}}$ such that $f_x(n) = \min\{m \in \mathbb{N} : x \in St(x_m^{(n)}, N_n(X))\}$. Since the collection $\{f_x : x \in Y\}$ has cardinality less than $\mathfrak{b}(\mathscr{I})$, we get a $g \in \mathbb{N}^{\mathbb{N}}$ such that for each $x \in Y$, $\{n \in \mathbb{N} : f_x(n) > g(n)\} \in \mathscr{I}$. For each $n$ let $F_n = \{x_i^{(n)} : i \leq g(n)\}$. We now show that the sequence $(F_n)$ witnesses for $(N_n)$

that $Y$ is $SSIH$ in $X$. Pick a $x \in Y$. Let $k \in \{n \in \mathbb{N} : x \notin St(F_n, N_n(X))\}$. Observe that $x \notin St(x_i^{(k)}, N_k(X))$ for all $i \leq g(k)$. Since $x \in St(x_{f_x(k)}^{(k)}, N_k(X))$, we have $f_x(k) > g(k)$. It follows that $k \in \{n \in \mathbb{N} : f_x(n) > g(n)\}$ and hence $\{n \in \mathbb{N} : x \notin St(F_n, N_n(X))\} \subseteq \{n \in \mathbb{N} : f_x(n) > g(n)\}$. Consequently $\{n \in \mathbb{N} : x \notin St(F_n, N_n(X))\} \in \mathscr{I}$. Thus $Y$ is $SSIH$ in $X$.

**Corollary 7.3.25.** *Every strongly star-Lindelöf space of cardinality less than $\mathfrak{b}(\mathscr{I})$ is SSIH.*

**Theorem 7.3.26.** *If a Lindelöf space $X$ is a union of less than $\mathfrak{b}(\mathscr{I})$ SIH spaces, then $X$ is also SIH.*

**Proof:** Let $\kappa$ be a cardinal less than $\mathfrak{b}(\mathscr{I})$ and $X = \cup_{\alpha<\kappa} X_\alpha$, where each $X_\alpha$ is a $SIH$ space. We consider a sequence $(N_n)$ of neighbourhood assignments of $X$ to show that $X$ is S$\mathscr{I}$H. Since $X$ is Lindelöf, by Lemma 7.3.18, for each $n$ we get a countable subset $A_n = \{x_m^{(n)} : m \in \mathbb{N}\}$ of $X$ such that $\cup N_n(A_n) = X$. Clearly for each $n$ the collection $\{N_n(x_m^{(n)}) : m \in \mathbb{N}\}$ is an open cover of $X$. For each $n$ and each $x \in X$ there is a $U_x^{(n)} \in \{N_n(x_m^{(n)}) : m \in \mathbb{N}\}$ such that $x \in U_x^{(n)}$. Define a sequence $(N'_n)$ of neighbourhood of assignments of $X$ as $N'_n(x) = U_x^{(n)}$ for each $n$ and each $x \in X$. For each $\alpha < \kappa$ we can obtain a sequence $(F_n^{(\alpha)})$ of finite subsets of $X$ such that for each $x \in X_\alpha$, $\{n \in \mathbb{N} : x \notin St(\cup N'_n(F_n^{(\alpha)}), N_n(X))\} \in \mathscr{I}$. We define a collection $\{f_\alpha : \alpha < \kappa\} \subseteq \mathbb{N}^{\mathbb{N}}$ as $f_\alpha(n) = \min\{m \in \mathbb{N} : N'_n(F_n^{(\alpha)}) \subseteq \{N_n(x_i^{(n)}) : i \leq m\}\}$ for all $n$. Since the collection $\{f_\alpha : \alpha < \kappa\}$ has cardinality less than $\mathfrak{b}(\mathscr{I})$, there exists a $g \in \mathbb{N}^{\mathbb{N}}$ such that for each $\alpha < \kappa$, $\{n \in \mathbb{N} : f_\alpha(n) > g(n)\} \in \mathscr{I}$. For each $n$ let $F_n = \{x_i^{(n)} : i \leq g(n)\}$. We now show that the sequence $(F_n)$ completes the requirements for $(N_n)$ that $X$ is $SIH$. Let $x \in X$ and choose a $\alpha < \kappa$ such that $x \in X_\alpha$. Pick a $k \in \{n \in \mathbb{N} : x \notin St(\cup N_n(F_n), N_n(X))\}$ and so $x \notin St(\cup \mathsf{N}_k(F_k), \mathsf{N}_k(X))$. If $f_\alpha(k) \leq g(k)$, then $\mathsf{N}'_k(F_k^{(\alpha)}) \subseteq \mathsf{N}_k(F_k)$ because $\mathsf{N}'_k(F_k^{(\alpha)}) \subseteq \{\mathsf{N}_k(x_i^{(k)}) : i \leq f_\alpha(k)\}$. Thus $x \notin St(\cup N'_k(F_k^{(\alpha)}), N_k(X))$ and consequently $k \in \{n \in \mathbb{N} : x \notin St(\cup N'_n(F_n^{(\alpha)}), N_n(X))\}$. If $f_\alpha(k) > g(k)$, then $k \in \{n \in \mathbb{N} : f_\alpha(n) > g(n)\}$. Thus $\{n \in \mathbb{N} : x \notin St(\cup N_n(F_n), N_n(X))\} \subseteq \{n \in \mathbb{N} : x \notin St(\cup N'_n(F_n^{(\alpha)}), N_n(X))\} \cup \{n \in \mathbb{N} : f_\alpha(n) > g(n)\}$ and $\{n \in \mathbb{N} : x \notin St(\cup \mathsf{N}_n(F_n), N_n(X))\} \in \mathscr{I}$ since both $\{n \in \mathbb{N} : x \notin St(\cup N'_n(F_n^{(\alpha)}), N_n(X))\}$ and $\{n \in \mathbb{N} : f_\alpha(n) > g(n)\}$ are members of $\mathscr{I}$. Hence $X$ is $SIH$.

**Theorem 7.3.27.** *If a strongly star-Lindelöf space $X$ is union of less than $\mathfrak{b}(\mathscr{I})$ $\mathscr{I}$H spaces, then $X$ is SSIH.*

**Proof:** Let $\kappa$ be a cardinal less than $\mathfrak{b}(\mathscr{I})$ and $X = \cup_{\alpha<\kappa} X_\alpha$, where each $X_\alpha$ is a $IH$ space. Let $(\mathsf{N}_n)$ be a sequence of neighbourhood assignments of $X$. Since $X$ is strongly star-Lindelöf, by Lemma 7.3.18, for each $n$ there exists a countable subset $A_n = \{x_m^{(n)} : m \in \mathbb{N}\}$ of $X$ such that $X = St(A_n, N_n(X))$. Clearly for each $n$ the collection $\{St(x_m^{(n)}, N_n(X)) : m \in \mathbb{N}\}$ is an open cover of $X$. For each $n$ and each $x \in X$ there is a $U_x^{(n)} \in \{St(x_m^{(n)}, N_n(X)) : m \in \mathbb{N}\}$ such that $x \in U_x^{(n)}$. Define a sequence $(N_n')$ of neighbourhood of assignments of $X$ as $N_n'(x) = U_x^{(n)}$ for each $n$ and each $x \in X$. For each $\alpha < \kappa$ we get a sequence $(F_n^{(\alpha)})$ of finite subsets of $X$ such that for each $x \in X_\alpha$, $\{n \in \mathbb{N} : x \notin \cup N_n'(F_n^{(\alpha)})\} \in \mathcal{I}$. Also define a collection $\{f_\alpha : \alpha < \kappa\} \subseteq \mathbb{N}^{\mathbb{N}}$ as $f_\alpha(n) = \min\{m \in \mathbb{N} : N_n'(F_n^{(\alpha)}) \subseteq \{St(x_i^{(n)}, N_n(X)) : i \leq m\}\}$ for all $n$. Since the collection $\{f_\alpha : \alpha < \kappa\}$ has cardinality less than $\mathfrak{b}(\mathcal{I})$, there exists a $g \in \mathbb{N}^{\mathbb{N}}$ such that for each $\alpha < \kappa$, $\{n \in \mathbb{N} : f_\alpha(n) > g(n)\} \in \mathcal{I}$. For each $n$ let $F_n = \{x_i^{(n)} : i \leq g(n)\}$. We now show that the sequence $(F_n)$ witnesses for $(N_n)$ that $X$ is *SSIH*. Let $x \in X$. Consider a $\alpha < \kappa$ such that $x \in X_\alpha$. We claim that $\{n \in \mathbb{N} : x \notin St(F_n, N_n(X))\} \subseteq \{n \in \mathbb{N} : x \notin \cup N_n'(F_n^{(\alpha)})\} \cup \{n \in \mathbb{N} : f_\alpha(n) > g(n)\}$. Pick a $k \in \{n \in \mathbb{N} : x \notin St(F_n, N_n(X))\}$ and so $x \notin St(F_k, N_k(X))$. If $f_\alpha(k) \leq g(k)$, then $\cup N_k'(F_k^{(\alpha)}) \subseteq St(F_k, N_k(X))$ because $N_k'(F_k^{(\alpha)}) \subseteq \{St(x_i^{(k)}, N_k(X)) : i \leq f_\alpha(k)\}$ and hence $x \notin \cup N_k'(F_k^{(\alpha)})$. This gives us $k \in \{n \in \mathbb{N} : x \notin \cup N_n'(F_n^{(\alpha)})\}$. If $f_\alpha(k) > g(k)$, then $k \in \{n \in \mathbb{N} : f_\alpha(n) > g(n)\}$. Thus $\{n \in \mathbb{N} : x \notin St(F_n, N_n(X))\} \subseteq \{n \in \mathbb{N} : x \notin \cup N_n'(F_n^{(\alpha)})\} \cup \{n \in \mathbb{N} : f_\alpha(n) > g(n)\}$ and consequently $\{n \in \mathbb{N} : x \notin St(F_n, N_n(X))\} \in \mathcal{I}$ as both $\{n \in \mathbb{N} : x \notin \cup N_n'(F_n^{(\alpha)})\}$ and $\{n \in \mathbb{N} : f_\alpha(n) > g(n)\}$ are members of $\mathcal{I}$. Hence $X$ is *SSIH*.

**Theorem 7.3.28.** *If a star-Lindelöf space $X$ is union of less than $\mathfrak{b}(\mathcal{I})$ IH spaces, then $X$ is SIH.*

**Proof:** Let $\kappa$ be a cardinal less than $\mathfrak{b}(\mathcal{I})$ and $X = \cup_{\alpha<\kappa} X_\alpha$, where each $X_\alpha$ is a *IH* space. Let $(N_n)$ be a sequence of neighbourhood assignments of $X$. Since $X$ is star-Lindelöf, by Lemma 7.3.18, for each $n$ there exists a countable subset $A_n = \{x_m^{(n)} : m \in \mathbb{N}\}$ of $X$ such that $X = St(\cup N_n(A_n), N_n(X))$. Clearly for each $n$ the collection $\{St(N_n(x_m^{(n)}), N_n(X)) : m \in \mathbb{N}\}$ is an open cover of $X$. For each $n$ and each $x \in X$ there is a $U_x^{(n)} \in \{St(N_n(x_m^{(n)}), N_n(X)) : m \in \mathbb{N}\}$ such that $x \in U_x^{(n)}$. We now define a sequence $(N_n')$ of neighbourhood of assignments of $X$ by $N_n'(x) = U_x^{(n)}$ for each $n$ and each $x \in X$. For each $\alpha < \kappa$ we get a sequence $(F_n^{(\alpha)})$ of finite subsets of $X$ such that for each $x \in X_\alpha$, $\{n \in \mathbb{N} : x \notin \cup N_n'(F_n^{(\alpha)})\} \in \mathcal{I}$. We also define a collection $\{f_\alpha : \alpha < \kappa\} \subseteq \mathbb{N}^{\mathbb{N}}$ as $f_\alpha(n) = \min\{m \in \mathbb{N} : N_n'(F_n^{(\alpha)}) \subseteq \{St(x_i^{(n)}, N_n(X)) : i \leq m\}\}$ for all $n$. Since the collection $\{f_\alpha : \alpha < \kappa\}$ has cardinality less than $\mathfrak{b}(\mathcal{I})$, there exists a $g \in \mathbb{N}^{\mathbb{N}}$ such that for each $\alpha < \kappa$, $\{n \in \mathbb{N} : f_\alpha(n) > g(n)\} \in \mathcal{I}$. For each $n$ let $F_n = \{x_i^{(n)} : i \leq g(n)\}$. We now show that the sequence $(F_n)$ witnesses for $(N_n)$ that $X$ is *SIH*. Let $x \in X$. Consider a $\alpha < \kappa$ such that $x \in X_\alpha$. We claim that $\{n \in \mathbb{N} : x \notin St(\cup N_n(F_n), N_n(X))\} \subseteq \{n \in \mathbb{N} : x \notin \cup N_n'(F_n^{(\alpha)})\} \cup \{n \in \mathbb{N} : f_\alpha(n) > g(n)\}$. Pick a $k \in \{n \in \mathbb{N} : x \notin St(\cup N_n(F_n), N_n(X))\}$ and so $x \notin St(\cup N_k(F_k), N_k(X))$. If $f_\alpha(k) \leq g(k)$, then $\cup N_k'(F_k^{(\alpha)}) \subseteq St(\cup N_k(F_k), N_k(X))$ because $N_k'(F_k^{(\alpha)}) \subseteq \{St(x_i^{(k)}, N_k(X)) : i \leq f_\alpha(k)\}$ and hence $x \notin \cup N_k'(F_k^{(\alpha)})$. This gives us $k \in \{n \in \mathbb{N} : x \notin \cup N_n'(F_n^{(\alpha)})\}$. If $f_\alpha(k) > g(k)$, then $k \in \{n \in \mathbb{N} : f_\alpha(n) > g(n)\}$.

Thus $\{n \in \mathbb{N} : x \notin St(\cup N_n(F_n), N_n(X))\} \subseteq \{n \in \mathbb{N} : x \notin \cup N'_n(F_n^{(\alpha)})\} \cup \{n \in \mathbb{N} : f_\alpha(n) > g(n)\}$ and consequently $\{n \in \mathbb{N} : x \notin St(\cup N_n(F_n), N_n(X))\} \in \mathscr{I}$ as both $\{n \in \mathbb{N} : x \notin \cup \mathsf{N}'_n(F_n^{(\alpha)})\}$ and $\{n \in \mathbb{N} : f_\alpha(n) > g(n)\}$ are members of $\mathscr{I}$. Hence $X$ is $SIH$.

**Theorem 7.3.29.** *Let $X$ be a regular space of the form $Y \cup Z$ with $Y \cap Z = \emptyset$, where $Y$ is a closed discrete set and $Z$ is a $\sigma$-compact subset of $X$. If $X$ is SSIH, then $|Y| < \mathfrak{b}(\mathscr{I})$.*

**Proof:** If possible suppose that $|Y| \geq \mathfrak{b}(\mathscr{I})$. Clearly $Z$ can be expressed as $Z = \cup_{n \in \mathbb{N}} K_n$, where each $K_n$ is compact and $K_m \subseteq K_n$ when $m \leq n$. Consider a collection $\{f_\alpha : \alpha < \mathfrak{b}(\mathscr{I})\} \subseteq \mathbb{N}^{\mathbb{N}}$ of size $\mathfrak{b}(\mathscr{I})$. For each $\alpha < \mathfrak{b}(\mathscr{I})$ and each $\beta < \omega_1$ choose distinct points $x_{\alpha,\beta} \in Y$. Say $A = \{x_{\alpha,\beta} : \alpha < \mathfrak{b}(\mathscr{I}), \beta < \omega_1\}$. It is easy to observe that for each $\alpha < \mathfrak{b}(\mathscr{I})$, each $\beta < \omega_1$ and each $n$ there exists an open set $U^{(n)}_{x_{\alpha,\beta}}$ containing $x_{\alpha,\beta}$ such that $U^{(n)}_{x_{\alpha,\beta}} \cap Y = \{x_{\alpha,\beta}\}$ and $U^{(n)}_{x_{\alpha,\beta}} \cap K_{f_\alpha(n)} = \emptyset$. Let us define a sequence $(\mathsf{N}_n)$ of neighbourhood assignments of $X$ by

$$N_n(x) = \begin{cases} U^{(n)}_x, & \text{if } x \in A \\ X \setminus A, & \text{otherwise.} \end{cases}$$

We claim that the sequence $(N_n)$ witnesses that $X$ is not $SSIH$. Let $(F_n)$ be any sequence of finite subsets of $X$. Next we define a $g \in \mathbb{N}^{\mathbb{N}}$ as follows. For each $n$

$$g(n) = \begin{cases} \min\{m \in \mathbb{N} : F_n \cap Z \subseteq K_m\} + 1, & \text{if } F_n \cap Z \neq \emptyset \\ 2, & \text{otherwise.} \end{cases}$$

Since the collection $\{f_\alpha : \alpha < \mathfrak{b}(\mathscr{I})\}$ is of size $\mathfrak{b}(\mathscr{I})$, there exists a $\gamma < \mathfrak{b}(\mathscr{I})$ such that $\{n \in \mathbb{N} : f_\gamma(n) \geq g(n)\} \in \mathscr{I}^+$. Also there is a $\delta < \omega_1$ such that $x_{\gamma,\delta} \notin \cup_{n \in \mathbb{N}} F_n$. We now show that $\{n \in \mathbb{N} : x_{\gamma,\delta} \notin St(F_n, N_n(X))\} \in \mathscr{I}^+$. Pick a $k \in \{n \in \mathbb{N} : f_\gamma(n) \geq g(n)\}$ and so $f_\gamma(k) \geq g(k)$. Observe that $x_{\gamma,\delta} \notin St(F_k, N_k(X))$. Indeed, if $x_{\gamma,\delta} \in St(F_k, N_k(X))$, then $N_k(x_{\gamma,\delta}) \cap F_k \neq \emptyset$ because $N_k(x_{\gamma,\delta})$ is the only member of $N_k(X)$ that contains $x_{\gamma,\delta}$. Obviously, there exists a $x \in F_k \cap Z$ such that $x \in N_k(x_{\gamma,\delta})$ since $x_{\gamma,\delta} \notin F_k$. It follows that $F_k \cap Z \subseteq K_{g(k)}$ and hence $F_k \cap Z \subseteq K_{f_\gamma(k)}$ as $f_\gamma(k) \geq g(k)$. This gives us $x \in N_k(x_{\gamma,\delta}) \cap K_{f_\gamma(k)}$, i.e. $N_k(x_{\gamma,\delta}) \cap K_{f_\gamma(k)} \neq \emptyset$. Which leads to a contradiction. Consequently $x_{\gamma,\delta} \notin St(F_k, N_k(X))$. This implies that $k \in \{n \in \mathbb{N} : x_{\gamma,\delta} \notin St(F_n, N_n(X))\}$. Thus $\{n \in \mathbb{N} : f_\gamma(n) \geq g(n)\} \subseteq \{n \in \mathbb{N} : x_{\gamma,\delta} \notin St(F_n, N_n(X))\}$ and so $\{n \in \mathbb{N} : x_{\gamma,\delta} \notin St(F_n, N_n(X))\} \in \mathscr{I}^+$. Thus $X$ is not $SSIH$, which is absurd. Hence $|Y| < \mathfrak{b}(\mathscr{I})$.

**Theorem 7.3.30.** *Let $X$ be a regular space of the form $Y \cup Z$ with $Y \cap Z = \emptyset$, where $Y$ is a closed discrete set and $Z$ is a $\sigma$-compact subset of $X$. If $X$ is strongly star-Lindelöf, then $|Y| < \mathfrak{b}(\mathscr{I})$ if and only if $X$ is SSIH.*

**Proof:** Assume that $|Y| < \mathfrak{b}(\mathscr{I})$. By Theorem 7.3.24, $Y$ is $SSIH$ in $X$ and hence $X$ is $SSIH$. Conversely, if $X$ is $\mathsf{SS}\mathscr{I}\mathsf{H}$, then by Theorem 7.3.29, $|Y| < \mathfrak{b}(\mathscr{I})$.

Let $N(X) = (X \times \{0\}) \cup (\mathbb{R} \times (0,\infty))$ be the Niemytzki plane on a set $X \subseteq \mathbb{R}$. The topology on $N(X)$ is defined as follows. $\mathbb{R} \times (0,\infty)$ has the Euclidean topology and the set $X \times \{0\}$ has the topology generated by all sets of the form $\{(x,0)\} \cup U$, where $x \in X$ and $U$ is an open disc in $\mathbb{R} \times (0,\infty)$ which is tangent to $X \times \{0\}$ at the point $(x,0)$. The topology on $N(X)$ is also called Niemytzki's tangent disc topology. It is to be noted that Niemytzki originally defined $N(\mathbb{R})$ (see [25]).

As a consequence of Theorem 7.3.30, one can obtain the following result.

**Corollary 7.3.31.** *1. The Isbell-Mrówka space $\Psi(\mathscr{A})$ is SSIH if and only if $|\mathscr{A}| < \mathfrak{b}(\mathscr{I})$.*
*2. The Niemytzki plane $N(X)$ is SSIH if and only if $|X| < \mathfrak{b}(\mathscr{I})$.*

### 7.3.3 GAMES

Let $X$ be a space and $\mathscr{I}$ be an ideal. The game $G_{fin}(\Omega, \mathscr{I}\text{-}(\mathscr{I}\text{-}\Lambda)^{gp})$ corresponding to the selection principle $S_{fin}(\Omega, \mathscr{I}\text{-}(\mathscr{I}\text{-}\Lambda)^{gp})$ is played as follows. Players ONE and TWO play an inning for each positive integer $n$. In the $n$th inning ONE chooses an $\omega$-neighbourhood assignment (A neighbourhood assignment $N$ of $X$ is said to be $\omega$-neighbourhood assignment if $N(X)$ is an $\omega$-cover of $X$) $N_n$ of $X$ and TWO responds by selecting a finite subset $F_n$ of $X$. TWO wins the play $N_1, F_1, N_2, F_2, \ldots, N_n, F_n, \ldots$ of this game if $\cup_{n\in\mathbb{N}} N_n(F_n) \in \mathscr{I}\text{-}(\mathscr{I}\text{-}\Lambda)^{gp}$; otherwise ONE wins. The game $G_{ufin}(\mathscr{O}, \mathscr{I}\text{-}\Gamma)$ corresponding to the selection principle $U_{fin}(\mathscr{O}, \mathscr{I}\text{-}\Gamma)$ is played as follows. Players ONE and TWO play an inning for each positive integer $n$. In the $n$th inning ONE chooses a neighbourhood assignment $N_n$ of $X$ and TWO responds by selecting a finite subset $F_n$ of $X$. TWO wins the play $N_1, F_1, N_2, F_2, \ldots, N_n, F_n, \ldots$ of this game if $\{\cup N_n(F_n) : n \in \mathbb{N}\} \in \mathscr{I}\text{-}\Gamma$; otherwise ONE wins. The game $SG^*_{ufin}(\mathscr{O}, \mathscr{I}\text{-}\Gamma)$ (respectively, $SG^*_{fin}(\mathscr{O}, \mathscr{I}\text{-}\Gamma)$) corresponding to the selection principle $U^*_{fin}(\mathscr{O}, \mathscr{I}\text{-}\Gamma)$ (respectively, $SS^*_{fin}(\mathscr{O}, \mathscr{I}\text{-}\Gamma)$) can be similarly defined. Observe that if ONE does not have a winning strategy in $G_{ufin}(\mathscr{O}, \mathscr{I}\text{-}\Gamma)$ (respectively, $SG^*_{ufin}(\mathscr{O}, \mathscr{I}\text{-}\Gamma)$, $SG^*_{fin}(\mathscr{O}, \mathscr{I}\text{-}\Gamma)$, $G_{fin}(\Omega, \mathscr{I}\text{-}(\mathscr{I}\text{-}\Lambda)^{gp})$), then $X$ satisfies $U_{fin}(\mathscr{O}, \mathscr{I}\text{-}\Gamma)$ (respectively, $U^*_{fin}(\mathscr{O}, \mathscr{I}\text{-}\Gamma)$, $SS^*_{fin}(\mathscr{O}, \mathscr{I}\text{-}\Gamma)$, $S_{fin}(\Omega, \mathscr{I}\text{-}(\mathscr{I}\text{-}\Lambda)^{gp})$).

**Lemma 7.3.32.** *[13, Lemma 2.3] If $\mathscr{U}$ is an $\omega$-cover of $X$, then for any finite set $\mathscr{V} \subseteq \mathscr{U}$, $\mathscr{U} \setminus \mathscr{V}$ is also an $\omega$-cover of $X$.*

In [13, Theorem 3.1], authors consider the hypothesis that $X$ is an $\varepsilon$-space. In the following result, we observe that one can drop this hypothesis.

**Theorem 7.3.33.** *If ONE does not have a winning strategy in $G_{ufin}(\mathscr{O}, \mathscr{I}\text{-}\Gamma)$ on $X$, then ONE does not have a winning strategy in $G_{fin}(\Omega, \mathscr{I}\text{-}(\mathscr{I}\text{-}\Lambda)^{gp})$ on $X$.*

**Proof:** Let $\sigma$ be a strategy for ONE in $G_{fin}(\Omega, \mathscr{I}\text{-}(\mathscr{I}\text{-}\Lambda)^{gp})$ on $X$. We define a strategy $\tau$ for ONE in $G^*_{ufin}(\mathscr{O}, \mathscr{I}\text{-}\Gamma)$ on $X$. Let $\sigma(\emptyset) = N_\emptyset$ be the first move of ONE in $G_{fin}(\Omega, \mathscr{I}\text{-}(\mathscr{I}\text{-}\Lambda)^{gp})$. Choose $\tau(\emptyset) = N'_\emptyset = \mathrm{N}_\emptyset$ as the first move of ONE in

$G_{ufin}(\mathscr{O},\mathscr{I}\text{-}\Gamma)$ and then TWO responds with $F_1$. Consider $F_1' = F_1$ as the response of TWO in $G_{fin}(\Omega,\mathscr{I}\text{-}(\mathscr{I}\text{-}\Lambda)^{gp})$. Let $\sigma(F_1') = N_1$ be the second move of ONE in $G_{fin}(\Omega,\mathscr{I}\text{-}(\mathscr{I}\text{-}\Lambda)^{gp})$. Put $A_1 = \{x \in X : N_1(x) \in N_\emptyset'(F_1)\}$. Suppose that $\tau(F_1) = N_1'$ is the second move of ONE in $G_{ufin}(\mathscr{O},\mathscr{I}\text{-}\Gamma)$, where $N_1'$ is a neighbourhood of assignment of $X$ given by $N_1'(x) = N_1(x)$ for all $x \in X\setminus A_1$ and $N_1'(A_1) \subseteq \mathrm{N}_1(X\setminus A_1)$ and then TWO responds with $F_2$. Let $F_2'$ be a finite subset of $X\setminus A_1$ such that $N_1'(F_2) = N_1(F_2')$. Consider $F_2'$ as the response of TWO in $G_{fin}(\Omega,\mathscr{I}\text{-}(\mathscr{I}\text{-}\Lambda)^{gp})$. If $\sigma(F_1',F_2') = N_2$ is the third move of ONE in $G_{fin}(\Omega,\mathscr{I}\text{-}(\mathscr{I}\text{-}\Lambda)^{gp})$, then we take $A_2 = \{x \in X : N_2(x) \in N_\emptyset'(F_1) \cup N_1'(F_2)\}$. Choose $\tau(F_1,F_2) = N_2'$ as ONE's third move in $G_{ufin}(\mathscr{O},\mathscr{I}\text{-}\Gamma)$, where $N_2'$ is a neighbourhood assignment of $X$ given by $N_2'(x) = N_2(x)$ for all $x \in X\setminus A_2$ and $N_2'(A_2) \subseteq N_2(X\setminus A_2)$ and so on. This defines a legitimate strategy $\tau$ for ONE in $G_{ufin}(\mathscr{O},\mathscr{I}\text{-}\Gamma)$ on $X$. Since ONE does not have a winning strategy in $G_{ufin}(\mathscr{O},\mathscr{I}\text{-}\Gamma)$ on $X$, $\tau$ is not a winning strategy for ONE. So there exists a $\tau$-play $\tau(\emptyset), F_1, \tau(F_1), F_2, \tau(F_1,F_2),\ldots$ which is lost by ONE. It follows that for each $x \in X$, the set $\{n \in \mathbb{N} : x \notin \cup\tau(F_1,F_2,\ldots,F_{n-1})(F_n)\} \in \mathscr{I}$. Since the members of the sequence $(\tau(F_1,F_2,\ldots,F_{n-1})(F_n))$ are pairwise disjoint, $\cup_{n\in\mathbb{N}}\tau(F_1,F_2,\ldots,F_{n-1})(F_n) \in \mathscr{I}\text{-}\mathscr{O}^{gp}$. Also observe that $\cup_{n\in\mathbb{N}}\tau(F_1,F_2,\ldots,F_{n-1})(F_n) \in \mathscr{I}\text{-}\Lambda$.
Thus $\cup_{n\in\mathbb{N}}\tau(F_1,F_2,\ldots,F_{n-1})(F_n) \in \mathscr{I}\text{-}(\mathscr{I}\text{-}\Lambda)^{gp}$. From the construction of $\tau$ we can say that the corresponding $\sigma$-play $\sigma(\emptyset), F_1', \sigma(F_1'), F_2', \sigma(F_1',F_2'),\ldots$ in $G_{fin}(\Omega,\mathscr{I}\text{-}(\mathscr{I}\text{-}\Lambda)^{gp})$ is lost by ONE because for each $n$ $\sigma(F_1',F_2',\ldots,F_{n-1}')(F_n') = \tau(F_1,F_2,\ldots,F_{n-1})(F_n)$. This implies that $\sigma$ is not a winning strategy for ONE in $G_{fin}(\Omega,\mathscr{I}\text{-}(\mathscr{I}\text{-}\Lambda)^{gp})$ and hence ONE does not have a winning strategy in $G_{fin}(\Omega,\mathscr{I}\text{-}(\mathscr{I}\text{-}\Lambda)^{gp})$ on $X$.

**Theorem 7.3.34.** *If TWO has a winning strategy in $G_{ufin}(\mathscr{O},\mathscr{I}\text{-}\Gamma)$ on $X$, then TWO has a winning strategy in $G_{fin}(\Omega,\mathscr{I}\text{-}(\mathscr{I}\text{-}\Lambda)^{gp})$ on $X$.*

**Proof:** Let $\sigma$ be a winning strategy for TWO in $G_{ufin}(\mathscr{O},\mathscr{I}\text{-}\Gamma)$ on $X$. We define a strategy $\tau$ for TWO in $G_{fin}(\Omega,\mathscr{I}\text{-}(\mathscr{I}\text{-}\Lambda)^{gp})$ on $X$. Let $N_1$ be the first move of ONE in $G_{fin}(\Omega,\mathscr{I}\text{-}(\mathscr{I}\text{-}\Lambda)^{gp})$. Choose $N_1' = N_1$ as the first move of ONE in $G_{ufin}(\mathscr{O},\mathscr{I}\text{-}\Gamma)$ and then TWO responds with $\sigma(N_1')$. Define $\tau(N_1) = \sigma(N_1')$ as the response's of TWO in $G_{fin}(\Omega,\mathscr{I}\text{-}(\mathscr{I}\text{-}\Lambda)^{gp})$. Let $N_2$ be the second move of ONE in $G_{fin}(\Omega,\mathscr{I}\text{-}(\mathscr{I}\text{-}\Lambda)^{gp})$. Choose $A_1 = \{x \in X : N_2(x) \in N_1'(\sigma(N_1'))\}$. Let $N_2'$ be a neighbourhood assignment of $X$ given by $N_2'(x) = N_2(x)$ for all $x \in X\setminus A_1$ and $N_2'(A_1) \subseteq N_2(X\setminus A_1)$. Take $N_2'$ as ONE's second move in $G_{ufin}(\mathscr{O},\mathscr{I}\text{-}\Gamma)$ and then TWO responds with $\sigma(N_1',N_2')$. Now there exists a finite subset $F_1$ of $X\setminus A_1$ such that $N_2'(\sigma(N_1',N_2')) = N_2(F_1)$. Choose $\tau(N_1,N_2) = F_1$ as the second move of TWO in $G_{fin}(\Omega,\mathscr{I}\text{-}(\mathscr{I}\text{-}\Lambda)^{gp})$. Let $N_3$ be the third move of ONE in $G_{fin}(\Omega,\mathscr{I}\text{-}(\mathscr{I}\text{-}\Lambda)^{gp})$. Choose $A_2 = \{x \in X : N_3(x) \in N_1'(\sigma(N_1')) \cup N_2'(\sigma(N_1',N_2'))\}$. Let $N_3'$ be a neighbourhood assignment of $X$ given by $N_3'(x) = N_3(x)$ for all $x \in X\setminus A_2$ and $N_3'(A_2) \subseteq N_3(X\setminus A_2)$. Consider $N_3'$ as the third move of ONE in $G_{ufin}(\mathscr{O},\mathscr{I}\text{-}\Gamma)$ and so on. Thus we obtain a legitimate strategy $\tau$ for TWO in $G_{fin}(\Omega,\mathscr{I}\text{-}(\mathscr{I}\text{-}\Lambda)^{gp})$. To show that $\tau$ is a winning strategy for TWO in $G_{fin}(\Omega,\mathscr{I}\text{-}(\mathscr{I}\text{-}\Lambda)^{gp})$ we pick a $\tau$-play $N_1, \tau(N_1), N_2, \tau(N_1,N_2), N_3,\ldots$. The corresponding $\sigma$-play is $N'_1, \sigma(N'_1), N'_2, \sigma(N'_1, N'_2), N'_3, \ldots$. Since $\sigma$ is a winning strategy for TWO in $G_{ufin}(\mathcal{O}, \mathcal{I}\text{-}\Gamma)$ on $X$, TWO wins this $\sigma$-play, i.e. for each $x \in X$, $\{n \in \mathbb{N} : x \notin \cup N'_n(\sigma(N'_1, N'_2, \ldots, N'_n))\} \in \mathcal{I}$ and consequently for each $x \in X$, $\{n \in \mathbb{N} : x \notin \cup N_n(\tau(N_1, N_2, \ldots, N_n))\} \in \mathcal{I}$ because for each $n$ $N'_n(\sigma(N'_1, N'_2, \ldots, N'_n)) = N_n(\tau(N_1, N_2, \ldots, N_n))$. It follow that $\cup_{n\in\mathbb{N}} N_n(\tau(N_1, N_2, \ldots, N_n)) \in \mathcal{I}\text{-}\mathcal{O}^{gp}$ since $N_m(\tau(N_1, N_2, \ldots, N_m)) \cap N_n(\tau(N_1, N_2, \ldots, N_n)) = \emptyset$ for $m \neq n$. Also observe that $\cup_{n\in\mathbb{N}} N_n(\tau(N_1, N_2, \ldots, N_n)) \in \mathcal{I}\text{-}\Lambda$ and so $\cup_{n\in\mathbb{N}} N_n(\tau(N_1, N_2, \ldots, N_n)) \in \mathcal{I}\text{-}(\mathcal{I}\text{-}\Lambda)^{gp}$. It follows that TWO wins the $\tau$-play and hence $\tau$ is a winning strategy for TWO in $G_{fin}(\Omega, \mathcal{I}\text{-}(\mathcal{I}\text{-}\Lambda)^{gp})$. This completes the proof.

**Theorem 7.3.35.** *For a paracompact Hausdorff space X, the games $G_{ufin}(\mathcal{O}, \mathcal{I}\text{-}\Gamma)$ and $G^*_{ufin}(\mathcal{O}, \mathcal{I}\text{-}\Gamma)$ are equivalent.*

**Theorem 7.3.36.** *For a metacompact space X, the games $G_{ufin}(\mathcal{O}, \mathcal{I}\text{-}\Gamma)$ and $G^*_{fin}(\mathcal{O}, \mathcal{I}\text{-}\Gamma)$ are equivalent.*

**Proof:** Observe that if ONE has a winning strategy in $\mathrm{SG}^*_{\mathrm{fin}}(\mathcal{O}, \mathcal{I}\text{-}\Gamma)$, then ONE has a winning strategy in $G_{ufin}(\mathcal{O}, \mathcal{I}\text{-}\Gamma)$; moreover if TWO has a winning strategy in $G_{ufin}(\mathcal{O}, \mathcal{I}\text{-}\Gamma)$, then TWO has a winning strategy in $SG^*_{fin}(\mathcal{O}, \mathcal{I}\text{-}\Gamma)$. We now suppose that ONE has a winning strategy $\sigma$ in $G_{ufin}(\mathcal{O}, \mathcal{I}\text{-}\Gamma)$. Let us define a strategy $\tau$ for ONE in $SG^*_{fin}(\mathcal{O}, \mathcal{I}\text{-}\Gamma)$ as follows. Let $\sigma(F_1, F_2, \ldots, F_{n-1}) = N_n$ be the $n$th move of ONE in $G_{ufin}(\mathcal{O}, \mathcal{I}\text{-}\Gamma)$. We choose a point finite open refinement $\mathcal{U}_n$ of $N_n(X)$. Let $N'_n$ be a neighbourhood assignment of $X$ such that $N'_n(X) = \mathcal{U}_n$. Define ONE's move in the $n$th inning of $SG^*_{fin}(\mathcal{O}, \mathcal{I}\text{-}\Gamma)$ to be $\tau(A_1, A_2, \ldots, A_{n-1}) = N'_n$. TWO responds by choosing a finite subset $A_n$ of $X$ in $SG^*_{fin}(\mathcal{O}, \mathcal{I}\text{-}\Gamma)$. Since $N'_n(X)$ is point finite, there exists a finite subset $A'_n$ of $X$ containing $A_n$ such that $St(A_n, N'_n(X)) = \cup N'_n(A'_n)$. Clearly, for each $y \in A'_n$, we get a $x_y \in X$ such that $N'_n(y) \subseteq N_n(x_y)$. Choose $F_n = \{x_y : y \in A'_n\}$ as the response of TWO in $G_{ufin}(\mathcal{O}, \mathcal{I}\text{-}\Gamma)$. Let $\sigma(F_1, F_2, \ldots, F_n) = \mathrm{N}_{n+1}$ be the $(n+1)$th move of ONE in $G_{ufin}(\mathcal{O}, \mathcal{I}\text{-}\Gamma)$ and so on. This defines a winning strategy $\tau$ for ONE in $SG^*_{fin}(\mathcal{O}, \mathcal{I}\text{-}\Gamma)$.

Next we suppose that TWO has a winning strategy in $SG^*_{fin}(\mathcal{O}, \mathcal{I}\text{-}\Gamma)$. Let $\sigma$ be a winning strategy for TWO in $SG^*_{fin}(\mathcal{O}, \mathcal{I}\text{-}\Gamma)$. We define a strategy $\tau$ for TWO in $G_{ufin}(\mathcal{O}, \mathcal{I}\text{-}\Gamma)$ as follows. In the $n$th inning, let $N_n$ be the move of ONE in $G_{ufin}(\mathcal{O}, \mathcal{I}\text{-}\Gamma)$. There exists a point finite open refinement $\mathcal{U}_n$ of $N_n(X)$. Let $N'_n$ be a neighbourhood assignment of $X$ such that $\mathrm{N}'_n(X) = \mathcal{U}_n$. Define ONE's move in the $n$th inning of $SG^*_{fin}(\mathcal{O}, \mathcal{I}\text{-}\Gamma)$ to be $N'_n$. TWO responds by choosing a finite subset $\sigma(N'_1, N'_2, \ldots, N'_n) = A_n$ of $X$ in $SG^*_{fin}(\mathcal{O}, \mathcal{I}\text{-}\Gamma)$. Since $N'_n(X)$ is point finite, there exists a finite subset $A'_n$ of $X$ containing $A_n$ such that $St(A_n, N'_n(X)) = \cup N'_n(A'_n)$. Clearly for each $y \in A'_n$ there exists a $x_y \in X$ such that $N'_n(y) \subseteq N_n(x_y)$. Choose $F_n = \{x_y : y \in A'_n\}$. Define TWO's response in $G^*_{ufin}(\mathcal{O}, \mathcal{I}\text{-}\Gamma)$ to be $\tau(N_1, N_2, \ldots, N_n) = F_n$. Let $\mathrm{N}_{n+1}$ be the $(n+1)$th move of ONE in $G^*_{ufin}(\mathcal{O}, \mathcal{I}\text{-}\Gamma)$ and so on. Thus we obtain a winning strategy $\tau$ for TWO in $G^*_{ufin}(\mathcal{O}, \mathcal{I}\text{-}\Gamma)$.

**Corollary 7.3.37.** *For a paracompact Hausdorff space X, the following games are equivalent.*

1. $G_{ufin}(\mathscr{O}, \mathscr{I}\text{-}\Gamma)$.
2. $SG^*_{fin}(\mathscr{O}, \mathscr{I}\text{-}\Gamma)$.
3. $G^*_{ufin}(\mathscr{O}, \mathscr{I}\text{-}\Gamma)$.

## BIBLIOGRAPHY


1. N. Alam, D. Chandra, Some remarks on games of certain star selection principles, Acta Mathematica Hungarica, **166**(2), (2022), 373–392.
2. K. Alster, On the class of all spaces of weight not greater than $\omega_1$ whose Cartesian product with every Lindelöf space is Lindelöf, Fundamenta Mathematicae, **129**, (1988), 133–140.
3. L. Babinkostova, B.A. Pansera, and M. Scheepers, Weak covering properties and selection principles, Topology and its Applications, **160**, (2013), 2251–2271.
4. M. Bonanzinga, F. Cammaroto, and Lj.D.R. Kocinac, Star-Hurewicz and related properties, Applied General Topology, **5**(1), (2004), 79–89.
5. A. Caserta, and S. Watson, The Alexandroff duplicate and its subspaces, Applied General Topology, **8**(2), (2007), 187–205.
6. D. Chandra, and N. Alam, On localization of the star-Menger selection principle, Hacettepe Journal of Mathematics and Statistics, **50**(4), (2021), 1155–1168.
7. D. Chandra, and N. Alam, Some remarks on star-Menger spaces using box products, Filomat, **36**(5), (2022), 1769–1774.
8. D. Chandra, and N. Alam, Further investigations on certain star selection principles, Topology and its Applications, **328**, (2023), 108446.
9. D. Chandra, and P. Das, Some further investigations of open covers and selection principles using ideals, Topology Proceedings, **39**, (2012), 281–291.
10. P. Das, Certain types of open covers and selection principles using ideals, Houston Journal of Mathematics, **39**(2), (2013), 637–650.
11. P. Das, and D. Chandra, Some further results on $\mathscr{I}$-$\gamma$ and $\mathscr{I}$-$\gamma_k$-covers, Topology and its Applications, **16**, (2013), 2401–2410.
12. P. Das, D. Chandra, and U. Samanta, On certain variations of $\mathscr{I}$-Hurewicz property, Topology and its Applications, **251**, (2018), 363–376.
13. P. Das, Lj.D.R. Kocinac, D. Chandra, Some remarks on open covers and selection principles using ideals, Topology and its Applications, **202**, (2016) 183–193.
14. P. Das, U. Samanta, and D. Chandra, On certain generalized versions of groupability, Topology and its Applications, **258**, (2019), 47–57.
15. P. Das, U. Samanta, and D. Chandra, Some observations on Hurewicz and $\mathscr{I}$-Hurewicz property, Topology and its Applications, **258**, (2019), 202–214.
16. R. Engelking, General Topology, Heldermann Verlag, Berlin, (1989).
17. B. Farkas, and L. Soukup, More on cardinal invariants of analytic $P$-ideals, Commentationes Mathematicae Universitatis Carolinae, **50**(2), (2009), 281–295.
18. J. Gerlits, and Z. S. Nagy, Some properties of $C(X)$, I, Topology and its Applications, **14**, (1982), 151–161.
19. W. Just, A.W. Miller, M. Scheepers, and P.J. Szeptycki, The combinatorics of open covers (II), Topology and its Applications, **73**, (1996), 241–266.
20. Lj.D.R. Kocinac, Star-Menger and related spaces, Publicationes Mathematicae Debrecen, **55**, (1999), 421–431.
21. Lj.D.R. Kocinac, Star selection principles: A survey, Khayyam Journal of Mathematics, **1**(1), (2015), 82–106.

22. Lj.D.R. Kocinac, Variations of classical selection principles: An overview, Quaestiones Mathematicae, **43**(8), (2020), 1121–1153.
23. Lj.D.R. Kocinac, and M. Scheepers, Combinatorics of open covers (VII): Groupability, Fundamenta Mathematicae, **179**, (2003), 131–155.
24. S. Mrówka, On completely regular spaces, Fundamenta Mathematicae, **41**, (1954), 105–106.
25. V. Niemytzki, Über die Axiome des metrischen Raumes, Mathematische Annalen, **104**, (1931), 666–671.
26. M. Scheepers, Combinatorics of open covers I: Ramsey theory, Topology and its Applications, **69**, (1996), 31–62.
27. M. Staniszewski, On ideal equal convergence II, Journal of Mathematical Analysis and Applications, **451**, (2017), 1179–1197.

# 8 Controlled $\nu$-Generalized Metric-type Space and Some Fixed Point Theorems with Application

*Kushal Roy, Mantu Saha, Vahid Parvaneh, Farhan Golkarmanesh, Choonkil Park*

In this chapter, the concept of controlled $\nu$-generalized metric type space has been introduced as a generalization of $b_\nu(s)$-metric spaces and controlled metric type spaces. Some fixed point theorems have been proved for various contractive type mappings over such spaces. Several examples have been given with respect to our new metric-type structure and established theorems. Moreover, our fixed point theorem is applied to find solutions to non-linear integral equations.

## 8.1 INTRODUCTION AND PRELIMINARIES

Nowadays, various generalizations of the usual metric structure had been made by several researchers for their need, particularly in order to study fixed point theory. One of those is the concept of $b$-metric spaces which was introduced by Czerwik [7] in 1993. The concepts of metric and $b$-metric structures have been generalized in many ways (see [3, 5, 9, 10, 11, 18, 19, 20, 21]). Another such generalized metric structure, known as rectangular metric space, was introduced by Branciari [6] in the year 2000, where the triangle inequality had been replaced by a so-called quadrilateral or rectangular inequality. For studying some generalizations of this structure, the reader can refer to [26], [27], [28], [29] and [30].

In the same article, Branciari instigated the concept of $\nu$-generalized metric spaces by using the polygonal inequality. After all such generalizations, Mitrović and Radenović [12] introduced $b_\nu(s)$-metric space by bundling together the concepts of $b$-metric spaces and $\nu$-generalized metric spaces, which generalize all the metric spaces discussed above.

In 2017, Kamran et al. [8] have generalized the concept of $b$-metric space in a new direction. They first show that the coefficient of a $b$-metric space could be considered

DOI: 10.1201/9781003596578-8

as a function instead of just taking a constant. Combining this concept with the concept of $b_\nu(s)$-metric space, Karahan et al. [14] introduced $b_\nu(\theta)$-metric space in the year 2018.

On the other hand, Mlaiki et al. [15] had recently introduced a new extension of $b$-metric spaces, called controlled metric-type spaces, by employing a control function $\alpha(x,y)$ on the right-hand side of the $b$-triangle inequality. Motivated by this new concept in the year 2020, Zubair et al. [17] had established the concept of controlled $b$-Branciari metric-type space which is the combination of rectangular metric spaces and controlled metric-type spaces. Next, we give the definitions of all these metric-type structures below:

**Definition 8.1.1.** *[4, 7] Let $X$ be a nonempty set and $\rho_b : X \times X \to [0,\infty)$ be a mapping. Then $\rho_b$ is called a b-metric if it satisfies the following conditions:*
*(i) $x = y$ if and only if $\rho_b(x,y) = 0$;*
*(ii) $\rho_b(x,y) = \rho_b(y,x)$ for all $x,y \in X$;*
*(iii) $\rho_b(x,y) \leq s[\rho_b(x,a) + \rho_b(a,y)]$ for some $s \geq 1$ and for all $x,y,a \in X$. The pair $(X,\rho_b)$ is called a b-metric space with coefficient s.*

**Definition 8.1.2.** *[8] Let $X$ be a nonempty set and $\rho_\theta$ be a mapping from $X^2$ to $[0,\infty)$. Then $\rho_\theta$ is called an extended b-metric if it satisfies the following conditions:*
*(i) $x = y$ if and only if $\rho_\theta(x,y) = 0$;*
*(ii) $\rho_\theta(x,y) = \rho_\theta(y,x)$ for all $x,y \in X$;*
*(iii) $\rho_\theta(x,y) \leq \theta(x,y)[\rho_\theta(x,a) + \rho_\theta(a,y)]$ for some function $\theta : X \times X \to [1,\infty)$ and for all $x,y,a \in X$. The pair $(X,\rho_\theta)$ is called an extended b-metric space.*

**Definition 8.1.3.** *[6] Let $X$ be a nonempty set and $\rho_r : X \times X \to [0,\infty)$ be a mapping. Then $\rho_r$ is said to be a rectangular metric if it satisfies the following conditions:*
*(i) $\rho_r(x,y) = 0$ if and only if $x = y$;*
*(ii) $\rho_r(x,y) = \rho_r(y,x)$ for all $x,y \in X$;*
*(iii) $\rho_r(x,y) \leq \rho_r(x,a) + \rho_r(a,b) + \rho_r(b,y)$ for all $x,y \in X$ and for all $a,b(a \neq b) \in X \setminus \{x,y\}$. The pair $(X,\rho_r)$ is called a rectangular metric space.*

**Definition 8.1.4.** *[6] Let $X$ be a non-empty set, $\nu \in \mathbb{N}$ and $\rho_{\nu g}$ be a function from $X \times X$ into $[0,\infty)$. Then $(X,\rho_{\nu g})$ is said to be a $\nu$-generalized metric space if for all $x,y \in X$ and for all distinct points $a_1,a_2,...,a_\nu \in X \setminus \{x,y\}$ the following conditions are satisfied:*

*(i) $x = y$ if and only if $\rho_{\nu g}(x,y) = 0$;*
*(ii) $\rho_{\nu g}(x,y) = \rho_{\nu g}(y,x)$;*
*(iii) $\rho_{\nu g}(x,y) \leq \rho_{\nu g}(x,a_1) + \rho_{\nu g}(a_1,a_2) + \cdots + \rho_{\nu g}(a_\nu,y)$.*

**Definition 8.1.5.** *[12] Let $X$ be a non-empty set, $\nu \in \mathbb{N}$ and $\rho_{b\nu}$ be a mapping from $X^2$ to $[0,\infty)$. Then $(X,\rho_{b\nu})$ is called a $b_\nu(s)$-metric space if for all $x,y \in X$ and for all distinct points $a_1,a_2,...,a_\nu \in X \setminus \{x,y\}$ the following conditions hold:*
*(i) $x = y$ if and only if $\rho_{b\nu}(x,y) = 0$;*
*(ii) $\rho_{b\nu}(x,y) = \rho_{b\nu}(y,x)$;*
*(iii) $\rho_{b\nu}(x,y) \leq b[\rho_{b\nu}(x,a_1) + \rho_{b\nu}(a_1,a_2) + \cdots + \rho_{b\nu}(a_\nu,y)]$ for some $b \geq 1$.*

**Example 8.1.6.** *[13] Take $X = X_1 \times X_2$, where $X_1 = \{\frac{1}{n} : n \in \mathbb{N}$ and $n \geq 2\}$ and $X_2 = \{0,1,2\}$. Define $\rho_{bv} : X \times X \to [0,\infty)$ by*

$$\rho_{bv}(x,y) = \begin{cases} |n-m|, & \text{if } x,y \in X_1 \text{ and } x = \frac{1}{n}, y = \frac{1}{m} \text{ and } |n-m| \neq 1,3 \\ \frac{1}{2}, & \text{if } x,y \in X_1 \text{ and } x = \frac{1}{n}, y = \frac{1}{m} \text{ and } |n-m| = 1,3 \\ n, & \text{if } x \in X_1, y \in X_2 \text{ and } x = \frac{1}{n} \\ 5, & \text{if } x,y \in X_2 \text{ and } x \neq y \\ 0, & \text{if } x,y \in X \text{ and } x = y. \end{cases}$$

*Then, it can be easily verified that $(X, \rho_{bv})$ is a $b_3(2)$-metric space.*

**Definition 8.1.7.** *[14] Let X be a non-empty set, $v \in \mathbb{N}$ and $\rho_{bv\theta} : X \times X \to [0,\infty)$ be a mapping. Then $(X, \rho_{bv\theta})$ is called a $b_v(\theta)$-metric space if for all $x,y \in X$ and for all distinct points $a_1, a_2, ..., a_v \in X \setminus \{x,y\}$ the following conditions hold:*
*(i) $x = y$ if and only if $\rho_{bv\theta}(x,y) = 0$;*
*(ii) $\rho_{bv\theta}(x,y) = \rho_{bv\theta}(y,x)$;*
*(iii) $\rho_{bv\theta}(x,y) \leq \theta(x,y)[\rho_{bv\theta}(x,a_1) + \rho_{bv\theta}(a_1,a_2) + \cdots + \rho_{bv\theta}(a_v,y)]$ for some function $\theta : X \times X \to [1,\infty)$.*

**Definition 8.1.8.** *[15] Let X be a nonempty set and $\rho_c : X \times X \to [0,\infty)$ be a mapping. Then $\rho_c$ is said to be a controlled metric type if it satisfies the following conditions:*
*(i) $x = y$ if and only if $\rho_c(x,y) = 0$;*
*(ii) $\rho_c(x,y) = \rho_c(y,x)$ for all $x,y \in X$;*
*(iii) $\rho_c(x,y) \leq \alpha(x,a)\rho_c(x,a) + \alpha(a,y)\rho_c(a,y)$ for some function $\alpha : X \times X \to [1,\infty)$ and for all $x,y,a \in X$. The pair $(X, \rho_c)$ is called a controlled metric-type space.*

**Example 8.1.9.** *[15] Let us consider $X = \mathbb{N}$ and $\rho_c : X \times X \to [0,\infty)$ be defined by*

$$\rho_c(x,y) = \begin{cases} \frac{1}{x}, & \text{if } x \in X \text{ is even and } y \in X \text{ is odd} \\ \frac{1}{y}, & \text{if } y \in X \text{ is even and } x \in X \text{ is odd} \\ 0, & \text{if } x,y \in X \text{ and } x = y \\ 1, & \text{otherwise.} \end{cases}$$

*Then $(X, \rho_c)$ is a controlled metric type space for the function $\alpha : X^2 \to [1,\infty)$ defined as*

$$\alpha(x,y) = \begin{cases} x, & \text{if } x \in X \text{ is even and } y \in X \text{ is odd} \\ y, & \text{if } y \in X \text{ is even and } x \in X \text{ is odd} \\ 1, & \text{otherwise.} \end{cases}$$

**Definition 8.1.10.** *[16, 17] Let X be a nonempty set, $\beta : X \times X \to [1,\infty)$ and $\rho_{cb} : X \times X \to [0,\infty)$ be a function. Then $\rho_{cb}$ is said to be a controlled b-Branciari metric type if it satisfies the following conditions:*
*(i) $x = y$ if and only if $\rho_{cb}(x,y) = 0$;*

*(ii)* $\rho_{cb}(x,y) = \rho_{cb}(y,x)$ *for all* $x,y \in X$;
*(iii)* $\rho_{cb}(x,y) \leq \beta(x,a)\rho_{cb}(x,a) + \beta(a,b)\rho_{cb}(a,b) + \beta(b,y)\rho_{cb}(b,y)$ *for all* $x,y \in X$ *and for all* $a,b(a \neq b) \in X \setminus \{x,y\}$. *The pair* $(X,\rho_{cb})$ *is called a controlled b-Branciari metric-type space.*

**Example 8.1.11.** *[16] Let* $X = [0,\infty)$ *and* $\rho_{cb} : X \times X \to [0,\infty)$ *be given by*

$$\rho_{cb}(x,y) = \begin{cases} \frac{1}{x}, & \text{if } x \geq 1 \text{ and } y \in [0,1) \\ \frac{1}{y}, & \text{if } y \geq 1 \text{ and } x \in [0,1) \\ 0, & \text{if } x,y \in X \text{ and } x = y \\ 1, & \text{otherwise.} \end{cases}$$

*Then* $(X,\rho_{cb})$ *is a controlled b-Branciari metric type space for the function* $\beta : X^2 \to [1,\infty)$ *defined as*

$$\beta(x,y) = \begin{cases} x, & \text{if } x,y \geq 1 \\ 1, & \text{otherwise.} \end{cases}$$

In the next section, we define controlled $\nu$-generalized metric-type space and discuss about some of its important properties.

## 8.2 INTRODUCTION TO CONTROLLED $V$-GENERALIZED METRIC-TYPE SPACE

First here we give the definition of controlled $\nu$-generalized metric type space.

**Definition 8.2.1.** *Let* $\mathscr{D}$ *be a non-empty set,* $\nu \in \mathbb{N}$ *and* $\Upsilon : \mathscr{D} \times \mathscr{D} \to [0,\infty)$ *be a mapping. Then* $(\mathscr{D},\Upsilon)$ *is said to be a controlled* $\nu$*-generalized metric type space if for all* $\theta, \vartheta \in \mathscr{D}$ *and for all distinct points* $\rho_1, \rho_2, \ldots, \rho_\nu \in \mathscr{D} \setminus \{\theta,\vartheta\}$ *the following conditions are satisfied:*

$(c_\nu g1)$ $\theta = \vartheta$ *if and only if* $\Upsilon(\theta,\vartheta) = 0$;

$(c_\nu g2)$ $\Upsilon(\theta,\vartheta) = \Upsilon(\vartheta,\theta)$;

$(c_\nu g2)$ *there exists a function* $\zeta : \mathscr{D} \times \mathscr{D} \to [1,\infty)$ *such that*

$$\begin{aligned} \Upsilon(\theta,\vartheta) \leq \zeta(\theta,\rho_1)\Upsilon(\theta,\rho_1) + \zeta(\rho_1,\rho_2)\Upsilon(\rho_1,\rho_2) + \cdots \\ + \zeta(\rho_{\nu-1},\rho_\nu)\Upsilon(\rho_{\nu-1},\rho_\nu) + \zeta(\rho_\nu,\vartheta)\Upsilon(\rho_\nu,\vartheta). \end{aligned}$$

**Example 8.2.2.** *Let us consider* $\mathscr{D} = \mathbb{R}$ *and* $\Upsilon : \mathscr{D} \times \mathscr{D} \to [0,\infty)$ *is defined by*

$$\Upsilon(\theta,\vartheta) = \frac{1+|\theta-\vartheta|}{3(1+|\theta|+|\vartheta|)}|\theta-\vartheta| \text{ for all } \theta,\vartheta \in \mathscr{D}.$$

*Then* $(\mathscr{D},\Upsilon)$ *is a controlled 2-generalized metric-type space for the function* $\zeta : \mathscr{D} \times \mathscr{D} \to [1,\infty)$ *given by*

$$\zeta(\theta,\vartheta) = \frac{1+|\theta|+|\vartheta|}{1+|\theta-\vartheta|} \text{ for all } \theta,\vartheta \in \mathscr{D}.$$

*But* $\Upsilon$ *is not a rectangular metric space, since*

$$\Upsilon(0.88,1) \approx 0.0156 > \Upsilon(0.88,0.92) + \Upsilon(0.92,0.96) + \Upsilon(0.96,1) \approx 0.0145.$$

*Also* $(\mathscr{D},\Upsilon)$ *is not a* $b_2(\zeta)$*-metric space since,*

$$\Upsilon(0,3) = 1 > \zeta(0,3)[\Upsilon(0,1) + \Upsilon(1,2) + \Upsilon(2,3)] = \frac{11}{18}.$$

**Example 8.2.3.** *Let us take* $\mathscr{D} = (-\infty,\infty)$ *and* $\Upsilon : \mathscr{D}^2 \to [0,\infty)$ *is given by*

$$\Upsilon(\theta,\vartheta) = \Upsilon(\vartheta,\theta) = \begin{cases} 0, & \text{if } \theta = \vartheta \\ \frac{1}{n^2+1}, & \text{if } \theta = 1, \vartheta = \frac{1}{n} \text{ for all } n \geq 2 \\ \frac{1}{n^2-1}, & \text{if } \theta = -1, \vartheta = \frac{1}{n} \text{ for all } n \geq 2 \\ 2, & \text{otherwise.} \end{cases}$$

*Then* $(\mathscr{D},\Upsilon)$ *is a controlled* 2*-generalized metric type space for the function* $\zeta : \mathscr{D} \times \mathscr{D} \to [1,\infty)$ *given by*

$$\zeta(\theta,\vartheta) = \zeta(\vartheta,\theta) = \begin{cases} n+1, & \text{if } \theta = 1, \vartheta = \frac{1}{n} \text{ for all } n \geq 2 \\ n-1, & \text{if } \theta = -1, \vartheta = \frac{1}{n} \text{ for all } n \geq 2 \\ 1, & \text{otherwise.} \end{cases}$$

*Here* $\Upsilon$ *is not a controlled metric-type space since,*

$$\begin{aligned} \Upsilon(1,-1) = 2 &> \zeta(1,\frac{1}{n})\Upsilon(1,\frac{1}{n}) + \zeta(\frac{1}{n},-1)\Upsilon(\frac{1}{n},-1) \\ &= \frac{n+1}{n^2+1} + \frac{1}{n+1} \to 0 \text{ as } n \to \infty. \end{aligned}$$

*Also* $(\mathscr{D},\Upsilon)$ *is not an extended b-metric space since,*

$$\begin{aligned} \Upsilon(1,-1) = 2 &> \zeta(1,-1)\left[\Upsilon(1,\frac{1}{n}) + \Upsilon(\frac{1}{n},-1)\right] \\ &= \frac{1}{n^2+1} + \frac{1}{n^2-1} \to 0 \text{ as } n \to \infty. \end{aligned}$$

**Example 8.2.4.** *Let* $\mathscr{D} = (-\infty,\infty)$ *and* $\Upsilon : \mathscr{D}^2 \to [0,\infty)$ *is defined by*

$$\Upsilon(\theta,\vartheta) = \Upsilon(\vartheta,\theta) = \begin{cases} 0, & \text{if } \theta = \vartheta \\ \frac{1}{n^2}, & \text{if } \theta = 1, \vartheta = 3n \text{ for all } n \geq 2 \\ \frac{1}{(n+1)^2}, & \text{if } \theta = 3n, \vartheta = 3n-1 \text{ for all } n \geq 2 \\ \frac{1}{(n+2)^2}, & \text{if } \theta = 3n-1, \vartheta = 3n-2 \text{ for all } n \geq 2 \\ \frac{1}{(n+3)^2}, & \text{if } \theta = 3n-2, \vartheta = 2 \text{ for all } n \geq 2 \\ 5, & \text{otherwise.} \end{cases}$$

*Then $(\mathscr{D},\Upsilon)$ is a controlled 3-generalized metric-type space for the function $\zeta:\mathscr{D}\times\mathscr{D}\to[1,\infty)$ given by*

$$\zeta(\theta,\vartheta)=\zeta(\vartheta,\theta)=\begin{cases}\frac{5}{4}n^2, & \text{if } \theta=1,\vartheta=3n \text{ for all } n\geq 2\\ \frac{5}{4}(n+1)^2, & \text{if } \theta=3n,\vartheta=3n-1 \text{ for all } n\geq 2\\ \frac{5}{4}(n+2)^2, & \text{if } \theta=3n-1,\vartheta=3n-2 \text{ for all } n\geq 2\\ \frac{5}{4}(n+3)^2, & \text{if } \theta=3n-2,\vartheta=2 \text{ for all } n\geq 2\\ 1, & \text{otherwise.}\end{cases}$$

*But $(\mathscr{D},\Upsilon)$ is not a $b_3(\sigma)$-metric space for any function $\sigma:\mathscr{D}\times\mathscr{D}\to[1,\infty)$ since,*

$$\begin{aligned}\Upsilon(1,2)=5\nleq\sigma(1,2)[&\Upsilon(1,3n)+\Upsilon(3n,3n-1)\\&+\Upsilon(3n-1,3n-2)+\Upsilon(3n-2,2)]\\=\sigma(1,2)&\left[\frac{1}{n^2}+\frac{1}{(n+1)^2}+\frac{1}{(n+2)^2}+\frac{1}{(n+3)^2}\right]\to 0\end{aligned}$$

*as $n\to\infty$.*

**Remark 8.2.5.** *(a) If $\zeta(\theta,\vartheta)=s\geq 1$ for all $\theta,\vartheta\in\mathscr{D}$ then the controlled $\nu$-generalized metric $\Upsilon$ reduces to a $b_\nu(s)$-metric. Now a $b_\nu(s)$-metric structure generalizes several metric-type structures such as metric, b-metric, rectangular metric, rectangular b-metric and $\nu$-generalized metric. Therefore controlled $\nu$-generalized metric also generalizes them.*

*(b) If we consider $\nu=1$ then the controlled $\nu$-generalized metric $\Upsilon$ gives a controlled metric-type structure.*

**Remark 8.2.6.** *1. In a metric space, any convergent sequence converges to some unique point but in a controlled $\nu$-generalized metric space it is not always true. In Example 8.2.4, the sequence $\{\frac{1}{n}\}_{n\geq 2}$ is convergent and converges to two points 1 and $-1$ in $\mathscr{D}$.*

*2. In a metric space, a convergent sequence is always Cauchy but it is not true in case of a controlled $\nu$-generalized metric space. In Example 18.4.8, the sequence $\{\frac{1}{n}\}_{n\geq 2}$ is convergent but $\Upsilon(\frac{1}{n},\frac{1}{m})\nrightarrow 0$ whenever $n,m\to\infty$.*

*3. In a metric space if $\{a_n\}$ and $\{b_n\}$ be two sequences that converge to $a$ and $b$, respectively, then $\Upsilon(a_n,b_n)\to\Upsilon(a,b)$ as $n\to\infty$. In particular, if $a=b$ then $\Upsilon(a_n,b_n)\to 0$ as $n\to\infty$. But this does not always hold in a controlled $\nu$-generalized metric space. In Example 18.4.8, let us consider two sequences $\{\frac{1}{2n}\}_{n\geq 1}$ and $\{\frac{1}{2n+1}\}_{n\geq 1}$ in $\mathscr{D}$. Then both of these two sequences converge to $1\in\mathscr{D}$ but $\Upsilon(\frac{1}{2n},\frac{1}{2n+1})\nrightarrow 0$ as $n\to\infty$.*

**Definition 8.2.7.** *Let $(\mathscr{D},\Upsilon)$ be a controlled $\nu$-generalized metric type space. Then (i) a sequence $\{\xi_n\}$ is said to be $\Upsilon$-Cauchy if for all $p=1,2,\ldots$, $\lim_{n\to\infty}\Upsilon(\xi_n,\xi_{n+p})=0$;*

*(ii) a sequence* $\{\xi_n\}$ *is said to be* $\Upsilon$*-convergent to some* $\vartheta \in \mathscr{D}$ *if* $\lim_{n\to\infty} \Upsilon(\xi_n, \vartheta) = 0$;

*(iii)* $(\mathscr{D}, \Upsilon)$ *is called* $\Upsilon$*-complete if every* $\Upsilon$*-Cauchy sequence* $\{\xi_n\}$ *is* $\Upsilon$*-convergent to some* $\vartheta \in \mathscr{D}$.

**Definition 8.2.8.** *Let* $(\mathscr{D}, \Upsilon)$ *be a controlled v-generalized metric type space and* $\mathscr{P} : \mathscr{D} \to \mathscr{D}$ *be a mapping. Then*

*(i)* $\mathscr{P}$ *is said to be continuous at* $\vartheta \in \mathscr{D}$ *if for any sequence* $\{\xi_n\}$ *converging to* $\vartheta$, $\{\mathscr{P}(\xi_n)\}$ *converges to* $\mathscr{P}(\vartheta)$. *The mapping* $\mathscr{P}$ *is called continuous in* $\mathscr{D}$ *if it is continuous at each point* $\vartheta \in \mathscr{D}$.

*(ii)* $\mathscr{P}$ *is said to be orbitally continuous at* $\theta \in \mathscr{D}$ *if for any sequence* $\{\xi_n\} \subset \mathscr{O}(\mathscr{P}, \kappa)$ *(*$\mathscr{O}(\mathscr{P}, \kappa)$ *is the orbit of* $\mathscr{P}$ *generated by some* $\kappa \in \mathscr{D}$*) converging to* $\theta$, $\{\mathscr{P}(\xi_n)\}$ *converges to* $\mathscr{P}(\theta)$. *The mapping* $\mathscr{P}$ *is called orbitally continuous in* $\mathscr{D}$ *if it is orbitally continuous at each point* $\theta \in \mathscr{D}$.

The following Lemma had been proved in [12] by Mitrović et al. in the context of $b_v(s)$-metric space. We prove the following similar type Lemma over controlled $v$-generalized metric-type space.

**Lemma 8.2.9.** *Let* $(\mathscr{D}, \Upsilon)$ *be a controlled v-generalized metric type space* $(v \geq 2)$, $T : \mathscr{D} \to \mathscr{D}$ *and let* $\theta_n$ *be a sequence in* $\mathscr{D}$ *defined by* $\theta_0 \in X$ *and* $\theta_{n+1} = T\theta_n$ *such that* $\theta_n \neq \theta_{n+1}$, $(n \geq 0)$. *Suppose that* $\lambda \in [0, 1)$ *such that* $\Upsilon(\theta_{n+1}, \theta_n) \leq \lambda \Upsilon(\theta_n, \theta_{n-1})$ *for all* $n \in \mathbb{N}$. *Then* $\theta_n \neq \theta_m$ *for all distinct* $n, m \in \mathbb{N}$.

**Proof:** If $\theta_n = \theta_m$ for some $1 \leq n < m$, then

$$\begin{aligned}
\Upsilon(\theta_{n+1}, \theta_n) = \Upsilon(\theta_{m+1}, \theta_m) &\leq \lambda \Upsilon(\theta_m, \theta_{m-1}) \\
&\leq \lambda^2 \Upsilon(\theta_{m-1}, \theta_{m-2}) \\
&\cdots \\
&\leq \lambda^{m-n} \Upsilon(\theta_{n+1}, \theta_n) < \Upsilon(\theta_{n+1}, \theta_n),
\end{aligned}$$

since $\Upsilon(\theta_{n+1}, \theta_n) > 0$, a contradiction. Therefore $\theta_n \neq \theta_m$ for all $n, m (n \neq m) \in \mathbb{N}$.

**Lemma 8.2.10.** *Let* $(\mathscr{D}, \Upsilon)$ *be a controlled v-generalized metric type space* $(v \geq 2)$ *and* $\{\theta_n\}_{n\geq 1}$ *be a sequence such that* $\theta_n \neq \theta_m$ *for all* $n, m \geq 0$ *with* $n \neq m$. *If* $\{\theta_n\}_{n\geq 1}$ *is* $\Upsilon$*-Cauchy and* $\Upsilon$*-convergent to* $z$ *and* $u$ *in* $\mathscr{D}$ *then* $z = u$ *provided that* $\lim_{n,m\to\infty} \zeta(\theta_n, \theta_m)$, $\lim_{n\to\infty} \zeta(z, \theta_n)$ *and* $\lim_{n\to\infty} \zeta(\theta_n, u)$ *are all finite.*

**Proof:** If possible let the sequence $\{\theta_n\}_{n\geq 1}$ $\Upsilon$-converges to two points $z$ and $u$ in $\mathscr{D}$. Since $\theta_n \neq \theta_m$ for all $n, m \geq 0$ with $n \neq m$ then there exists some $t \geq 1$ such that

$\theta_n \notin \{z,u\}$ for all $n \geq t$. Then

$$\begin{aligned}\Upsilon(z,u) &\leq \zeta(z,\theta_{n+1})\Upsilon(z,\theta_{n+1}) + \cdots \\ &+ \zeta(\theta_{n+\nu-3},\theta_{n+\nu-2})\Upsilon(\theta_{n+\nu-3},\theta_{n+\nu-2}) \\ &+ \zeta(\theta_{n+\nu-2},\theta_{n+\nu-1})\Upsilon(\theta_{n+\nu-2},\theta_{n+\nu-1}) \\ &+ \zeta(\theta_{n+\nu-1},\theta_{n+\nu})\Upsilon(\theta_{n+\nu-1},\theta_{n+\nu}) \\ &+ \zeta(\theta_{n+\nu},u)\Upsilon(\theta_{n+\nu},u) \text{ for all } n \geq t. \end{aligned} \tag{8.1}$$

Taking $n \to \infty$ we see that $\Upsilon(z,u) = 0$, i.e. $z = u$.

Now we are going to prove some fixed point theorems for contractive type mappings over controlled $\nu$-generalized metric type spaces for $\nu \geq 2$ in light of various important works in this area (see [22]–[25]).

## 8.3 FIXED POINT RESULTS OVER CONTROLLED $V$-GENERALIZED METRIC TYPE SPACE

**Theorem 8.3.1.** *(Banach orbital type contraction) Let $(\mathscr{D},\Upsilon)$ be a $\Upsilon$-complete controlled $\nu$-generalized metric type space ($\nu \geq 2$) and $\mathscr{P} : \mathscr{D} \to \mathscr{D}$ be a mapping satisfying*

$$\Upsilon(\mathscr{P}\vartheta,\mathscr{P}^2\vartheta) \leq \beta\Upsilon(\vartheta,\mathscr{P}\vartheta) \text{ for all } \vartheta \in \mathscr{D},\ \beta \in (0,1); \tag{8.2}$$

*where $\lim_{n,m\to\infty} \zeta(\theta_n,\theta_m)$ is finite for the Picard iterating sequence $\{\theta_n\}_{n\geq 1} = \{\mathscr{P}^n\theta_0\}_{n\geq 1}$ generated by some $\theta_0 \in \mathscr{D}$. Then the Picard iterating sequence $\{\theta_n\}_{n\geq 1}$ converges to some $\omega \in \mathscr{D}$. Moreover $\omega$ is a fixed point of $\mathscr{P}$, provided that $\mathscr{P}$ is orbitally continuous at $\omega \in \mathscr{D}$ and also both $\lim_{m\to\infty} \zeta(\omega,\theta_m)$ and $\lim_{m\to\infty} \zeta(\theta_m,\mathscr{P}(\omega))$ exist and finite.*

**Proof:** If $\theta_n = \theta_{n+1}$ for some $n \geq 0$ then $\mathscr{P}$ has a fixed point in $\mathscr{D}$. So let us assume that $\theta_n \neq \theta_{n+1}$ for all $n \geq 0$. Using the contractive condition (8.2)

$$\begin{aligned}\Upsilon(\theta_n,\theta_{n+1}) &= \Upsilon(\mathscr{P}\theta_{n-1},\mathscr{P}^2\theta_{n-1}) \\ &\leq \beta\Upsilon(\theta_{n-1},\mathscr{P}\theta_{n-1}) \\ &= \beta\Upsilon(\theta_{n-1},\theta_n) \text{ for all } n \in \mathbb{N}. \end{aligned} \tag{8.3}$$

Therefore, by Lemma 8.2.9, it follows that $\theta_n \neq \theta_m$ for all $n,m \geq 0$ with $n \neq m$.

Since $\lim_{n,m\to\infty} \zeta(\theta_n,\theta_m)$ is finite say $s \geq 1$; therefore, there exists some $l \in \mathbb{N}$ such that

$$\beta^l < \frac{1}{s}.$$

Now for $1 \leq n < m$,

$$\begin{aligned}
&\Upsilon(\theta_n, \theta_m) \\
&\leq \zeta(\theta_n, \theta_{n+1})\Upsilon(\theta_n, \theta_{n+1}) + \cdots + \zeta(\theta_{n+v-3}, \theta_{n+v-2})\Upsilon(\theta_{n+v-3}, \theta_{n+v-2}) \\
&\quad + \zeta(\theta_{n+v-2}, \theta_{n+l})\Upsilon(\theta_{n+v-2}, \theta_{n+l}) + \zeta(\theta_{n+l}, \theta_{m+l})\Upsilon(\theta_{n+l}, \theta_{m+l}) \\
&\quad + \zeta(\theta_{m+l}, \theta_m)\Upsilon(\theta_{m+l}, \theta_m) \\
&= \sum_{i=n}^{n+v-3} \zeta(\theta_i, \theta_{i+1})\Upsilon(\theta_i, \theta_{i+1}) + \zeta(\theta_{n+v-2}, \theta_{n+l})\Upsilon(\theta_{n+v-2}, \theta_{n+l}) \\
&\quad + \zeta(\theta_{n+l}, \theta_{m+l})\Upsilon(\theta_{n+l}, \theta_{m+l}) + \zeta(\theta_{m+l}, \theta_m)\Upsilon(\theta_{m+l}, \theta_m) \\
&\leq \Upsilon(\theta_0, \theta_1) \sum_{i=n}^{n+v-3} \zeta(\theta_i, \theta_{i+1})\beta^i + \zeta(\theta_{n+v-2}, \theta_{n+l})\beta^n\Upsilon(\theta_{v-2}, \theta_l) + \\
&\quad \zeta(\theta_{n+l}, \theta_{m+l})\beta^l\Upsilon(\theta_n, \theta_m) + \zeta(\theta_{m+l}, \theta_m)\beta^m\Upsilon(\theta_l, \theta_0).
\end{aligned} \tag{8.4}$$

It is given that $\lim_{n,m\to\infty} \zeta(\theta_n, \theta_m)$ exists finitely, therefore

$$\begin{aligned}
b \lim_{n,m\to\infty} \Upsilon(\theta_n, \theta_m) &\leq \lim_{n\to\infty} \Upsilon(\theta_0, \theta_1) \sum_{i=n}^{n+v-3} \zeta(\theta_i, \theta_{i+1})\beta^i \\
&\quad + \lim_{n\to\infty} \zeta(\theta_{n+v-2}, \theta_{n+l})\beta^n\Upsilon(\theta_{v-2}, \theta_l) \\
&\quad + \lim_{n,m\to\infty} \zeta(\theta_{n+l}, \theta_{m+l})\beta^l\Upsilon(\theta_n, \theta_m) \\
&\quad + \lim_{m\to\infty} \zeta(\theta_{m+l}, \theta_m)\beta^m\Upsilon(\theta_l, \theta_0) \\
&\leq \lim_{n\to\infty} \Upsilon(\theta_0, \theta_1) \sum_{i=n}^{n+v-3} \zeta(\theta_i, \theta_{i+1})\beta^i \\
&\quad + \lim_{n\to\infty} \zeta(\theta_{n+v-2}, \theta_{n+l})\beta^n\Upsilon(\theta_{v-2}, \theta_l) \\
&\quad + s\beta^l \lim_{n,m\to\infty} \Upsilon(\theta_n, \theta_m) + \lim_{m\to\infty} \zeta(\theta_{m+l}, \theta_m)\beta^m\Upsilon(\theta_l, \theta_0),
\end{aligned}$$

$$\Rightarrow \lim_{n,m\to\infty} \Upsilon(\theta_n, \theta_m) \leq \frac{\Upsilon(\theta_0, \theta_1)}{1 - s\beta^l} \lim_{n\to\infty} \sum_{i=n}^{n+v-3} \zeta(\theta_i, \theta_{i+1})\beta^i = 0. \tag{8.5}$$

So $\{\theta_n\}$ is $\Upsilon$-Cauchy in $\mathscr{D}$. Since $(\mathscr{D}, \Upsilon)$ is $\Upsilon$-complete, $\{\theta_n\}$ converges to some $\omega \in \mathscr{D}$. Now it is given that $\mathscr{P}$ is orbitally continuous in $\mathscr{D}$ therefore $\{\theta_{n+1}\} = \{\mathscr{P}\theta_n\}$ converges to $\mathscr{P}\omega$. Therefore, due to Lemma 8.2.10, it follows that $\mathscr{P}\omega = \omega$ and thus $\omega$ is a fixed point of $\mathscr{P}$.

The following are some supporting examples of Theorem 8.3.1.

**Example 8.3.2.** *Let us consider the controlled v-generalized metric type space ($v \geq 2$) $(\mathscr{D}, \Upsilon)$ given in Example 8.2.2. First, we show that $(\mathscr{D}, \Upsilon)$ is $\Upsilon$-complete. Let $\{\theta_n\}$*

*be a $\Upsilon$-Cauchy sequence in $\mathscr{D}$. Then for any $p \geq 1$,*

$$\Upsilon(\theta_n, \theta_{n+p}) = \frac{1+|\theta_n - \theta_{n+p}|}{3(1+|\theta_n| + |\theta_{n+p}|)}|\theta_n - \theta_{n+p}| \to 0$$

$$\textit{implies } |\theta_n - \theta_{n+p}| \to 0$$

*as $n \to \infty$. Since $\mathbb{R}$ is complete with respect to the usual metric, then there exists $\varsigma \in \mathscr{D}$ such that $\theta_n \to \varsigma$ as $n \to \infty$. Therefore, it is quite clear that*

$$\Upsilon(\theta_n, \varsigma) = \frac{1+|\theta_n - \varsigma|}{3(1+|\theta_n| + |\varsigma|)}|\theta_n - \varsigma| \to 0 \textit{ as } n \to \infty.$$

*So $(\mathscr{D}, \Upsilon)$ is $\Upsilon$-complete. Let $\mathscr{P} : \mathscr{D} \to \mathscr{D}$ be defined by*

$$\mathscr{P}\vartheta = \begin{cases} 0 & \textit{if } \vartheta \in [-1,1] \\ \vartheta & \textit{if } \vartheta \notin [-1,1]. \end{cases}$$

*If we take $\vartheta \in [-1,1]$, then $\Upsilon(\mathscr{P}\vartheta, \mathscr{P}^2\vartheta) = \Upsilon(0,0) = 0$. Similarly for $\vartheta \notin [-1,1]$, $\Upsilon(\mathscr{P}\vartheta, \mathscr{P}^2\vartheta) = \Upsilon(\vartheta, \vartheta) = 0$. Thus $\mathscr{P}$ is a Banach orbital type contraction on $\mathscr{D}$. Here $\mathscr{P}$ is also orbitally continuous. Hence, all the conditions of Theorem 8.3.1 are satisfied and $\mathscr{P}$ has infinitely many fixed points in $\mathscr{D}$.*

**Example 8.3.3.** *Let us consider the $\Upsilon$-complete controlled $\nu$-generalized metric type space ($\nu \geq 2$) $(\mathscr{D}, \Upsilon)$ given in Example 8.2.2. Let $\mathscr{P} : \mathscr{D} \to \mathscr{D}$ be defined by $\mathscr{P}\vartheta = \begin{cases} 1 & \textit{if } \vartheta \geq 0 \\ -5 & \textit{if } \vartheta < 0. \end{cases}$ If $\vartheta \geq 0$, then $\Upsilon(\mathscr{P}\vartheta, \mathscr{P}^2\vartheta) = \Upsilon(1,1) = 0$. Also for $\vartheta < 0$, $\Upsilon(\mathscr{P}\vartheta, \mathscr{P}^2\vartheta) = \Upsilon(-5,-5) = 0$. Therefore $\mathscr{P}$ is a Banach orbital type contraction on $\mathscr{D}$. $\mathscr{P}$ is also orbitally continuous in $(\mathscr{D}, \Upsilon)$. Then all the conditions of Theorem 8.3.1 are satisfied and $\mathscr{P}$ has two fixed points 1 and $-5$ in $\mathscr{D}$.*

The following corollary comes from our Theorem 8.3.1.

**Corollary 8.3.4.** *(Banach type contraction) Let $(\mathscr{D}, \Upsilon)$ be a $\Upsilon$-complete controlled $\nu$-generalized metric type space ($\nu \geq 2$) and $\mathscr{P} : \mathscr{D} \to \mathscr{D}$ be a mapping which satisfies*

$$\Upsilon(\mathscr{P}\vartheta, \mathscr{P}\nu) \leq \delta\Upsilon(\vartheta, \nu) \textit{ for all } \vartheta, \nu \in \mathscr{D},\ \delta \in (0,1); \tag{8.6}$$

*where $\lim_{n,m\to\infty} \zeta(\theta_n, \theta_m)$ is finite for the Picard iterating sequence $\{\theta_n\}_{n\geq 1} = \{\mathscr{P}^n\theta_0\}_{n\geq 1}$ generated by some $\theta_0 \in \mathscr{D}$. Then the Picard iterating sequence $\{\theta_n\}_{n\geq 1}$ converges to some $\omega \in \mathscr{D}$. Moreover $\omega$ is the unique fixed point of $\mathscr{P}$, provided that $\lim_{m\to\infty} \zeta(\omega, \theta_m)$ and $\lim_{m\to\infty} \zeta(\theta_m, \mathscr{P}(\omega))$ exist and finite.*

**Theorem 8.3.5.** *(Bianchini type contractive mapping) Let $(\mathscr{D}, \Upsilon)$ be a $\Upsilon$-complete controlled $\nu$-generalized metric type space ($\nu \geq 2$) and $\mathscr{P} : \mathscr{D} \to \mathscr{D}$ be a mapping satisfying*

$$\Upsilon(\mathscr{P}\vartheta, \mathscr{P}\varsigma) \leq \eta \max\{\Upsilon(\vartheta, \mathscr{P}\vartheta), \Upsilon(\varsigma, \mathscr{P}\varsigma)\} \textit{ for all } \vartheta, \varsigma \in \mathscr{D},\ \eta \in (0,1); \tag{8.7}$$

*Then the Picard iterating sequence $\{\theta_n\}_{n\geq 1}$ generated by some $\theta_0 \in \mathscr{D}$ converges to some $\rho \in \mathscr{D}$. Moreover $\rho$ is the unique fixed point of $\mathscr{P}$, provided that $\Upsilon(\rho, \mathscr{P}\rho) \leq \limsup_{n\to\infty} \Upsilon(\theta_{n+1}, \mathscr{P}\rho)$.*

**Proof:** Without loss of generality let us assume that $\theta_n \neq \theta_{n+1}$ for all $n \geq 0$. Using the contractive condition (8.7) we get

$$\begin{aligned}\Upsilon(\theta_n, \theta_{n+1}) &= \Upsilon(\mathscr{P}\theta_{n-1}, \mathscr{P}\theta_n)\\ &\leq \eta \max\{\Upsilon(\theta_{n-1}, \mathscr{P}\theta_{n-1}), \Upsilon(\theta_n, \mathscr{P}\theta_n)\}\\ &= \eta \max\{\Upsilon(\theta_{n-1}, \theta_n), \Upsilon(\theta_n, \theta_{n+1})\} \text{ for all } n \in \mathbb{N}. \end{aligned} \tag{8.8}$$

If for some $n \geq 1$, $\Upsilon(\theta_{n-1}, \theta_n) \leq \Upsilon(\theta_n, \theta_{n+1})$ then we see that $\Upsilon(\theta_n, \theta_{n+1}) \leq \eta\Upsilon(\theta_n, \theta_{n+1})$, a contradiction. Therefore, we have $\Upsilon(\theta_n, \theta_{n+1}) \leq \eta\Upsilon(\theta_{n-1}, \theta_n)$ for all $n \in \mathbb{N}$. From the contractive condition (8.7), it follows that

$$\Upsilon(\theta_n, \theta_m) \leq \eta \max\{\Upsilon(\theta_{n-1}, \theta_n), \Upsilon(\theta_{m-1}, \theta_m)\} \text{ for every } n, m \geq 1. \tag{8.9}$$

Using the above inequalities, we can show that $\{\theta_n\}$ is $\Upsilon$-Cauchy in $\mathscr{D}$. By the $\Upsilon$-completeness of $(\mathscr{D}, \Upsilon)$, $\{\theta_n\}$ converges to some $\rho \in \mathscr{D}$. Now for any $n \geq 1$,

$$\Upsilon(\theta_{n+1}, \mathscr{P}\rho) = \Upsilon(\mathscr{P}\theta_n, \mathscr{P}\rho) \leq \eta \max\{\Upsilon(\theta_{n-1}, \mathscr{P}\theta_{n-1}), \Upsilon(\rho, \mathscr{P}\rho)\}$$

implies that $\limsup_{n\to\infty} \Upsilon(\theta_{n+1}, \mathscr{P}\rho) \leq \eta\Upsilon(\rho, \mathscr{P}\rho)$ and by the assumed condition it follows that $\Upsilon(\rho, \mathscr{P}\rho) = 0$. Hence $\rho$ is a fixed point of $\mathscr{P}$. If $\nu$ is another fixed point of $\mathscr{P}$, then

$$\begin{aligned}\Upsilon(\rho, \nu) = \Upsilon(\mathscr{P}\rho, \mathscr{P}\nu) \leq \eta \max\{\Upsilon(\rho, \mathscr{P}\rho), \Upsilon(\nu, \mathscr{P}\nu)\} = 0\\ \Rightarrow \rho = \nu, \text{ leads to a contradiction.}\end{aligned}$$

This shows that $\rho$ is the only fixed point of $\mathscr{P}$.

**Example 8.3.6.** *Let us take $\mathscr{D} = [0,1]$ and the controlled v-generalized metric ($v \geq 2$) $\Upsilon$ defined in Example 8.2.2. Then $(\mathscr{D}, \Upsilon)$ is $\Upsilon$-complete. Let $\mathscr{P} : \mathscr{D} \to \mathscr{D}$ be defined by* $\mathscr{P}\vartheta = \begin{cases} \frac{\vartheta}{4} & \text{if } \vartheta \in [0, \frac{1}{2}) \\ \frac{\vartheta}{5} & \text{if } \vartheta \in [\frac{1}{2}, 1]. \end{cases}$

*Let $\vartheta, \rho \in [0, \frac{1}{2})$, then*

$$\Upsilon(\mathscr{P}\vartheta, \mathscr{P}\rho) = \Upsilon(\frac{\vartheta}{4}, \frac{\rho}{4}) = \frac{1 + \frac{1}{4}|\vartheta - \rho|}{3(1 + \frac{1}{4}|\vartheta| + \frac{1}{4}|\rho|)} \frac{1}{4}|\vartheta - \rho| \leq \frac{1}{12}|\vartheta - \rho|,$$

$$\Upsilon(\vartheta, \mathscr{P}\vartheta) = \Upsilon(\vartheta, \frac{\vartheta}{4}) = \frac{1 + \frac{3}{4}|\vartheta|}{3(1 + |\vartheta| + \frac{1}{4}|\vartheta|)} \frac{3}{4}|\vartheta| \geq \frac{11}{52}|\vartheta|, \text{ and}$$

$$\Upsilon(\rho, \mathscr{P}\rho) = \Upsilon(\rho, \frac{\rho}{4}) = \frac{1 + \frac{3}{4}|\rho|}{3(1 + |\rho| + \frac{1}{4}|\rho|)} \frac{3}{4}|\rho| \geq \frac{11}{52}|\rho|.$$

*Therefore,* $\Upsilon(\mathscr{P}\vartheta, \mathscr{P}\rho) \leq \frac{6}{7} \max\{\Upsilon(\vartheta, \mathscr{P}\vartheta), \Upsilon(\rho, \mathscr{P}\rho)\}$.

*If we take* $\vartheta, \rho \in [\frac{1}{2}, 1]$, *then*

$$\Upsilon(\mathscr{P}\vartheta, \mathscr{P}\rho) = \Upsilon(\frac{\vartheta}{5}, \frac{\rho}{5}) = \frac{1+\frac{1}{5}|\vartheta-\rho|}{3(1+\frac{1}{5}|\vartheta|+\frac{1}{5}|\rho|)}\frac{1}{5}|\vartheta-\rho| \le \frac{1}{15}|\vartheta-\rho|,$$

$$\Upsilon(\vartheta, \mathscr{P}\vartheta) = \Upsilon(\vartheta, \frac{\vartheta}{5}) = \frac{1+\frac{4}{5}|\vartheta|}{3(1+|\vartheta|+\frac{1}{5}|\vartheta|)}\frac{4}{5}|\vartheta| \ge \frac{12}{55}|\vartheta|, \text{ and}$$

$$\Upsilon(\rho, \mathscr{P}\rho) = \Upsilon(\rho, \frac{\rho}{5}) = \frac{1+\frac{4}{5}|\rho|}{3(1+|\rho|+\frac{1}{5}|\rho|)}\frac{4}{5}|\rho| \ge \frac{12}{55}|\rho|.$$

*Therefore,* $\Upsilon(\mathscr{P}\vartheta, \mathscr{P}\rho) \le \frac{6}{7}\max\{\Upsilon(\vartheta, \mathscr{P}\vartheta), \Upsilon(\rho, \mathscr{P}\rho)\}$.

*For* $\vartheta \in [0, \frac{1}{2})$ *and* $\rho \in [\frac{1}{2}, 1]$,

$$\Upsilon(\mathscr{P}\vartheta, \mathscr{P}\rho) = \Upsilon(\frac{\vartheta}{4}, \frac{\rho}{5}) = \frac{1+|\frac{\vartheta}{4}-\frac{\rho}{5}|}{3(1+\frac{1}{4}|\vartheta|+\frac{1}{5}|\rho|)}|\frac{\vartheta}{4}-\frac{\rho}{5}| \le \frac{1}{3}|\frac{\vartheta}{4}-\frac{\rho}{5}|,$$

$$\Upsilon(\vartheta, \mathscr{P}\vartheta) = \Upsilon(\vartheta, \frac{\vartheta}{4}) = \frac{1+\frac{3}{4}|\vartheta|}{3(1+|\vartheta|+\frac{1}{4}|\vartheta|)}\frac{3}{4}|\vartheta| \ge \frac{11}{52}|\vartheta|, \text{ and}$$

$$\Upsilon(\rho, \mathscr{P}\rho) = \Upsilon(\rho, \frac{\rho}{5}) = \frac{1+\frac{4}{5}|\rho|}{3(1+|\rho|+\frac{1}{5}|\rho|)}\frac{4}{5}|\rho| \ge \frac{12}{55}|\rho|.$$

*Therefore* $\Upsilon(\mathscr{P}\vartheta, \mathscr{P}\rho) \le \frac{6}{7}\max\{\Upsilon(\vartheta, \mathscr{P}\vartheta), \Upsilon(\rho, \mathscr{P}\rho)\}$.

*Then* $\mathscr{P}$ *satisfies the contractive condition (8.7) for* $\eta = \frac{6}{7}$, *and all the conditions of Theorem 8.3.5 are also satisfied. Here, we see that* $\mathscr{P}$ *has a unique fixed point* 0 *in* $\mathscr{D}$.

The next corollary follows from our Theorem 8.3.5.

**Corollary 8.3.7.** *(Kannan type contractive mapping) Let* $(\mathscr{D}, \Upsilon)$ *be a* $\Upsilon$*-complete controlled* $\nu$*-generalized metric type space* $(\nu \ge 2)$ *and* $\mathscr{P} : \mathscr{D} \to \mathscr{D}$ *be a mapping satisfying*

$$\Upsilon(\mathscr{P}\vartheta, \mathscr{P}\varsigma) \le \xi[\Upsilon(\vartheta, \mathscr{P}\vartheta) + \Upsilon(\varsigma, \mathscr{P}\varsigma)] \text{ for all } \vartheta, \varsigma \in \mathscr{D}, \ \xi \in \left(0, \frac{1}{2}\right); \quad (8.10)$$

*Then the Picard iterating sequence* $\{\theta_n\}_{n\ge1}$ *generated by some* $\theta_0 \in \mathscr{D}$ *converges to some* $\rho \in \mathscr{D}$. *Moreover,* $\rho$ *is the unique fixed point of* $\mathscr{P}$, *provided that* $\Upsilon(\rho, \mathscr{P}\rho) \le \limsup_{n\to\infty} \Upsilon(\theta_{n+1}, \mathscr{P}\rho)$.

Akram et al. ([1]–[2]) have defined $A$-contractions in the following manner: Let a nonempty set $A$ consisting of all functions $\Omega : \mathbb{R}^3_+ \to \mathbb{R}_+$ satisfying
$(A_1)$ : $\Omega$ is continuous on the set $\mathbb{R}^3_+$ of all triplets of nonnegative reals(with respect to the Euclidean metric on $\mathbb{R}^3$).
$(A_2)$ : $a \le \lambda b$ for some $\lambda \in [0, 1)$ whenever $a \le \Omega(a, b, b)$ or $a \le \Omega(b, a, b)$ or $a \le \Omega(b, b, a)$, for all $a, b$.

**Theorem 8.3.8.** *(A-contraction type mapping) Let* $(\mathscr{D},\Upsilon)$ *be a* $\Upsilon$*-complete controlled v-generalized metric type space* ($v \geq 2$) *and* $\mathscr{P}:\mathscr{D}\to\mathscr{D}$ *be a mapping satisfying*

$$\Upsilon(\mathscr{P}\vartheta,\mathscr{P}\varphi) \leq \Omega(\Upsilon(\vartheta,\varphi),\Upsilon(\vartheta,\mathscr{P}\vartheta),\Upsilon(\varphi,\mathscr{P}\varphi)) \text{ for all } \vartheta,\varphi \in \mathscr{D}; \tag{8.11}$$

*where* $\lim_{n,m\to\infty}\zeta(\theta_n,\theta_m)$ *is finite for the Picard iterating sequence* $\{\theta_n\}_{n\geq 1} = \{\mathscr{P}^n\theta_0\}_{n\geq 1}$ *generated by some* $\theta_0 \in \mathscr{D}$*. Then the Picard iterating sequence* $\{\theta_n\}_{n\geq 1}$ *converges to some* $\omega \in \mathscr{D}$*. Moreover* $\omega$ *is the unique fixed point of* $\mathscr{P}$*, provided that* $\mathscr{P}$ *is orbitally continuous at* $\omega \in \mathscr{D}$ *and also both* $\lim_{m\to\infty}\zeta(\omega,\theta_m)$ *and* $\lim_{m\to\infty}\zeta(\theta_m,\mathscr{P}(\omega))$ *exist and finite.*

**Proof:** For any $\vartheta \in \mathscr{D}$, if we set $\varphi = \mathscr{P}\vartheta$, then from (8.11), we see that

$$\begin{aligned}&\Upsilon(\mathscr{P}\vartheta,\mathscr{P}^2\vartheta) \leq \Omega(\Upsilon(\vartheta,\mathscr{P}\vartheta),\Upsilon(\vartheta,\mathscr{P}\vartheta),\Upsilon(\mathscr{P}\vartheta,\mathscr{P}^2\vartheta))\\ \Rightarrow\ &\Upsilon(\mathscr{P}\vartheta,\mathscr{P}^2\vartheta) \leq \lambda\Upsilon(\vartheta,\mathscr{P}\vartheta).\end{aligned}$$

Therefore, $\mathscr{P}$ satisfies the contractive condition (8.2) and hence due to Theorem 8.3.2 $\omega$ is a fixed point of $\mathscr{P}$. Let $\varepsilon$ and $\zeta$ be two fixed points of $\mathscr{P}$. Then

$$\begin{aligned}\Upsilon(\varepsilon,\zeta) = \Upsilon(\mathscr{P}\varepsilon,\mathscr{P}\zeta) &\leq \Omega(\Upsilon(\varepsilon,\zeta),\Upsilon(\varepsilon,\mathscr{P}\varepsilon),\Upsilon(\zeta,\mathscr{P}\zeta))\\ &= \Omega(\Upsilon(\varepsilon,\zeta),0,0)\\ \Rightarrow \Upsilon(\varepsilon,\zeta) &\leq \lambda.0 = 0\\ &\Rightarrow \varepsilon = \zeta.\end{aligned}$$

**Theorem 8.3.9.** *Let* $(\mathscr{D},\Upsilon)$ *be a* $\Upsilon$*-complete controlled v-generalized metric-type space* ($v \geq 2$) *and* $\mathscr{P}:\mathscr{D}\to\mathscr{D}$ *be a mapping satisfying*

$$\Upsilon(\mathscr{P}\vartheta,\mathscr{P}\varphi) \leq \Omega(\Upsilon(\vartheta,\varphi),\Upsilon(\vartheta,\mathscr{P}\vartheta),\Upsilon(\varphi,\mathscr{P}\varphi)) \text{ for all } \vartheta,\varphi \in \mathscr{D};$$

*where* $\lim_{n,m\to\infty}\zeta(\theta_n,\theta_m)$ *is finite for the Picard iterating sequence* $\{\theta_n\}_{n\geq 1} = \{\mathscr{P}^n\theta_0\}_{n\geq 1}$ *generated by some* $\theta_0 \in \mathscr{D}$*. Then the Picard iterating sequence* $\{\theta_n\}_{n\geq 1}$ *converges to some* $\omega \in \mathscr{D}$*. Moreover,* $\omega$ *is the unique fixed point of* $\mathscr{P}$*, provided that* $\Upsilon(\omega,\mathscr{P}\omega) \leq \limsup_{n\to\infty}\Upsilon(\theta_{n+1},\mathscr{P}\omega)$.

**Proof:** For any $\vartheta \in \mathscr{D}$, if we set $\varphi = \mathscr{P}\vartheta$, then from (8.11), we see that

$$\begin{aligned}&\Upsilon(\mathscr{P}\vartheta,\mathscr{P}^2\vartheta) \leq \Omega(\Upsilon(\vartheta,\mathscr{P}\vartheta),\Upsilon(\vartheta,\mathscr{P}\vartheta),\Upsilon(\mathscr{P}\vartheta,\mathscr{P}^2\vartheta))\\ \Rightarrow\ &\Upsilon(\mathscr{P}\vartheta,\mathscr{P}^2\vartheta) \leq \lambda\Upsilon(\vartheta,\mathscr{P}\vartheta).\end{aligned}$$

Since $\lim_{n,m\to\infty}\zeta(\theta_n,\theta_m)$ is finite for the Picard iterating sequence $\{\theta_n\}_{n\geq 1} = \{\mathscr{P}^n\theta_0\}_{n\geq 1}$ generated by some $\theta_0 \in \mathscr{D}$; therefore, by the similar calculation as in Theorem 8.3.2, we see that $\{\theta_n\}_{n\geq 1}$ converges to some $\omega \in \mathscr{D}$.

By the assumed condition we get,

$$\begin{aligned}\Upsilon(\theta_{n+1},\mathscr{P}\omega)=\Upsilon(\mathscr{P}\theta_n,\mathscr{P}\omega)&\leq\Omega(\Upsilon(\theta_n,\omega),\Upsilon(\theta_n,\mathscr{P}\theta_n),\Upsilon(\omega,\mathscr{P}\omega))\\&=\Omega(\Upsilon(\theta_n,\omega),\Upsilon(\theta_n,\theta_{n+1}),\Upsilon(\omega,\mathscr{P}\omega))\\\Rightarrow\limsup_{n\to\infty}\Upsilon(\theta_{n+1},\mathscr{P}\omega)&\leq\Omega(0,0,\Upsilon(\omega,\mathscr{P}\omega))\\\Rightarrow\Upsilon(\omega,\mathscr{P}\omega)&\leq\Omega(0,0,\Upsilon(\omega,\mathscr{P}\omega))\\\Rightarrow\Upsilon(\omega,\mathscr{P}\omega)&=0\\\Rightarrow\mathscr{P}\omega&=\omega.\end{aligned}\tag{8.12}$$

So $\omega$ is a fixed point of $\mathscr{P}$ and uniqueness of the fixed point is also evident.

**Corollary 8.3.10.** *(Reich-type contractive mapping) Let $(\mathscr{D},\Upsilon)$ be a $\Upsilon$-complete controlled ν-generalized metric-type space ($\nu\geq 2$) and $\mathscr{P}:\mathscr{D}\to\mathscr{D}$ be a mapping satisfying*

$$\Upsilon(\mathscr{P}\vartheta,\mathscr{P}\varphi)\leq r\Upsilon(\vartheta,\varphi)+s\Upsilon(\vartheta,\mathscr{P}\vartheta)+t\Upsilon(\varphi,\mathscr{P}\varphi),\tag{8.13}$$

$$r,s,t\geq 0 \text{ with } r+s+t<1;\tag{8.14}$$

*for all $\vartheta,\varphi\in\mathscr{D}$, where $\lim_{n,m\to\infty}\zeta(\theta_n,\theta_m)$ is finite for the Picard iterating sequence $\{\theta_n\}_{n\geq 1}=\{\mathscr{P}^n\theta_0\}_{n\geq 1}$ generated by some $\theta_0\in\mathscr{D}$. Then the Picard iterating sequence $\{\theta_n\}_{n\geq 1}$ converges to some $\omega\in\mathscr{D}$. Moreover $\omega$ is the unique fixed point of $\mathscr{P}$, provided that $\mathscr{P}$ is orbitally continuous at $\omega\in\mathscr{D}$ and also both $\lim_{m\to\infty}\zeta(\omega,\theta_m)$ and $\lim_{m\to\infty}\zeta(\theta_m,\mathscr{P}(\omega))$ exist and finite.*

**Corollary 8.3.11.** *Let $(\mathscr{D},\Upsilon)$ be a $\Upsilon$-complete controlled ν-generalized metric type space ($\nu\geq 2$) and $\mathscr{P}:\mathscr{D}\to\mathscr{D}$ be a mapping satisfying*

$$\Upsilon(\mathscr{P}\vartheta,\mathscr{P}\varphi)\leq r\Upsilon(\vartheta,\varphi)+s\Upsilon(\vartheta,\mathscr{P}\vartheta)+t\Upsilon(\varphi,\mathscr{P}\varphi),$$
$$r,s,t\geq 0 \text{ with } r+s+t<1;$$

*where $\lim_{n,m\to\infty}\zeta(\theta_n,\theta_m)$ is finite for the Picard iterating sequence $\{\theta_n\}_{n\geq 1}=\{\mathscr{P}^n\theta_0\}_{n\geq 1}$ generated by some $\theta_0\in\mathscr{D}$. Then the Picard iterating sequence $\{\theta_n\}_{n\geq 1}$ converges to some $\omega\in\mathscr{D}$. Moreover $\omega$ is the unique fixed point of $\mathscr{P}$, provided that $\Upsilon(\omega,\mathscr{P}\omega)\leq\limsup_{n\to\infty}\Upsilon(\theta_{n+1},\mathscr{P}\omega)$.*

**Theorem 8.3.12.** *(Berinde weak contractive type mapping) Let $(\mathscr{D},\Upsilon)$ be a $\Upsilon$-complete controlled ν-generalized metric type space ($\nu\geq 2$) and $\mathscr{P}:\mathscr{D}\to\mathscr{D}$ be a mapping satisfying*

$$\Upsilon(\mathscr{P}\vartheta,\mathscr{P}\varphi)\leq r\Upsilon(\vartheta,\varphi)+L\Upsilon(\varphi,\mathscr{P}\vartheta),\ \ 0\leq r<1, L\geq 0;\tag{8.15}$$

*for all $\vartheta,\varphi\in\mathscr{D}$, where $\lim_{n,m\to\infty}\zeta(\theta_n,\theta_m)$ is finite for the Picard iterating sequence $\{\theta_n\}_{n\geq 1}=\{\mathscr{P}^n\theta_0\}_{n\geq 1}$ generated by some $\theta_0\in\mathscr{D}$. Then the Picard iterating* *sequence $\{\theta_n\}_{n\geq 1}$ converges to some $\omega \in \mathscr{D}$. Moreover $\omega$ is a fixed point of $\mathscr{P}$, provided that $\mathscr{P}$ is orbitally continuous at $\omega \in \mathscr{D}$ and also both $\lim_{m\to\infty} \zeta(\omega, \theta_m)$ and $\lim_{m\to\infty} \zeta(\theta_m, \mathscr{P}(\omega))$ exist and finite.*

**Proof:** If we set $\varphi = \mathscr{P}\vartheta$, $\vartheta \in \mathscr{D}$ then from (8.15) we get

$$\Upsilon(\mathscr{P}\vartheta, \mathscr{P}^2\vartheta) \leq r\Upsilon(\vartheta, \mathscr{P}\vartheta) + L\Upsilon(\mathscr{P}\vartheta, \mathscr{P}\vartheta)$$
$$\Rightarrow \Upsilon(\mathscr{P}\vartheta, \mathscr{P}^2\vartheta) \leq r\Upsilon(\vartheta, \mathscr{P}\vartheta).$$

Therefore $\mathscr{P}$ satisfies the contractive condition (8.2) and hence due to Theorem 8.3.2 $\omega$ is a fixed point of $\mathscr{P}$.

Here we furnish our proven result with the following non-trivial example.

**Example 8.3.13.** *Let us take $\mathscr{D} = [0,1]$ and the controlled $v$-generalized metric ($v \geq 2$) $\Upsilon$ defined in Example 8.2.2. Then $\mathscr{D}$ is $\Upsilon$-complete controlled $v$-generalized metric space. Let $\mathscr{P} : \mathscr{D} \to \mathscr{D}$ be defined by*

$$\mathscr{P}\vartheta = \begin{cases} \frac{\vartheta}{3} & \text{if } \vartheta \in [0, \frac{2}{3}] \\ \frac{1}{3} & \text{if } \vartheta \in (\frac{2}{3}, 1]. \end{cases}$$

*Let $\vartheta, \rho \in [0, \frac{2}{3}]$, then $0 \leq |\vartheta - \rho| \leq \frac{2}{3}$, $0 \leq |\vartheta| + |\rho| \leq \frac{4}{3}$ and*

$$\Upsilon(\mathscr{P}\vartheta, \mathscr{P}\rho) = \Upsilon(\frac{\vartheta}{3}, \frac{\rho}{3}) = \frac{1 + \frac{1}{3}|\vartheta - \rho|}{3(1 + \frac{1}{3}|\vartheta| + \frac{1}{3}|\rho|)} \frac{1}{3}|\vartheta - \rho| \leq \frac{1}{9}|\vartheta - \rho|,$$

$$\Upsilon(\vartheta, \rho) = \frac{1 + |\vartheta - \rho|}{3(1 + |\vartheta| + |\rho|)}|\vartheta - \rho| \geq \frac{1 + 0}{3(1 + \frac{4}{3})}|\vartheta - \rho| = \frac{1}{7}|\vartheta - \rho|, \text{ and}$$

$$\Upsilon(\rho, \mathscr{P}\vartheta) = \Upsilon(\rho, \frac{\vartheta}{3}) = \frac{1 + |\rho - \frac{1}{3}\vartheta|}{3(1 + \frac{1}{3}|\vartheta| + |\rho|)}|\rho - \frac{1}{3}\vartheta|.$$

*Therefore, for $\vartheta, \rho \in [0, \frac{2}{3}]$,*

$$\Upsilon(\mathscr{P}\vartheta, \mathscr{P}\rho) \leq \frac{7}{9}\Upsilon(\vartheta, \rho).$$

*Let $\vartheta, \rho \in (\frac{2}{3}, 1]$, then clearly*

$$\Upsilon(\mathscr{P}\vartheta, \mathscr{P}\rho) = 0 \leq \frac{7}{9}\Upsilon(\vartheta, \rho).$$

*Now let $\vartheta \in [0, \frac{2}{3}]$ and $\rho \in (\frac{2}{3}, 1]$, then $\frac{4}{9} < |\rho - \frac{1}{3}\vartheta| \leq 1$, $\frac{2}{3} < \frac{1}{3}|\vartheta| + |\rho| \leq \frac{11}{9}$ and*

$$\Upsilon(\mathscr{P}\vartheta, \mathscr{P}\rho) = \Upsilon(\frac{\vartheta}{3}, \frac{1}{3}) = \frac{1 + \frac{1}{3}|\vartheta - 1|}{3(1 + \frac{1}{3}|\vartheta| + \frac{1}{3})} \frac{1}{3}|\vartheta - \rho| \leq \frac{1}{9}|\vartheta - \rho| \leq \frac{1}{9},$$

$$\Upsilon(\vartheta, \rho) = \frac{1 + |\vartheta - \rho|}{3(1 + |\vartheta| + |\rho|)}|\vartheta - \rho|,$$

$$\Upsilon(\rho, \mathscr{P}\vartheta) = \Upsilon(\rho, \frac{\vartheta}{3}) = \frac{1 + |\rho - \frac{1}{3}\vartheta|}{3(1 + \frac{1}{3}|\vartheta| + |\rho|)}|\rho - \frac{1}{3}\vartheta| \geq \frac{1 + \frac{4}{9}}{3(1 + \frac{11}{9})} \frac{4}{9} = \frac{13}{135}.$$

*Therefore, for $\vartheta \in [0, \frac{2}{3}]$ and $\rho \in (\frac{2}{3}, 1]$,*

$$\Upsilon(\mathscr{P}\vartheta, \mathscr{P}\rho) \leq \frac{7}{9}\Upsilon(\vartheta, \rho) + \frac{15}{13}\Upsilon(\rho, \mathscr{P}\vartheta).$$

*Then $\mathscr{P}$ satisfies the contractive condition (8.15) for $r = \frac{7}{9}$ and $L = \frac{15}{13}$, and all the conditions of Theorem 8.3.12 are also satisfied. $\mathscr{P}$ has only one fixed point 0 in $\mathscr{D}$.*

Let us consider $\mathscr{F}$, the class of all strictly increasing functions $\mathscr{F} : (0, \infty) \to \mathbb{R}$ satisfying the following conditions:

($\mathscr{F}1$) for every sequence $\{t_n\} \subset (0, \infty)$, $\lim_{n\to\infty} t_n = 0$ if and only if $\lim_{n\to\infty} \mathscr{F}(t_n) = -\infty$.

($\mathscr{F}2$) there exists a $k \in (0, 1)$ such that $t^k \mathscr{F}(t) \to 0$ whenever $t \to 0^+$.

There are several examples of such mappings belonging to the family $\mathscr{F}$. We list-up some of them as below:

(a) $\ln(x)$, (b) $x + \ln(x)$, (c) $-\frac{1}{x^n}$ for $0 < n < 1$, (d) $\ln(x + x^2)$.
For more information, one can see [23].

**Theorem 8.3.14.** *(Wardowski F-contraction type mapping) Let $(\mathscr{D}, \Upsilon)$ be a $\Upsilon$-complete controlled $\nu$-generalized metric type space ($\nu \geq 2$) and $\mathscr{P} : \mathscr{D} \to \mathscr{D}$ be a mapping satisfying*

$$\tau + \mathscr{F}(\Upsilon(\mathscr{P}\vartheta, \mathscr{P}\varphi)) \leq \mathscr{F}(\Upsilon(\vartheta, \varphi)) \text{ for all } \mathscr{P}\vartheta \neq \mathscr{P}\varphi, \tag{8.16}$$

*where $\lim_{n,m\to\infty} \zeta(\theta_n, \theta_m) = 1$ for the Picard iterating sequence $\{\theta_n\}_{n\geq 1} = \{\mathscr{P}^n\theta_0\}_{n\geq 1}$ generated by some $\theta_0 \in \mathscr{D}$. Then the Picard iterating sequence $\{\theta_n\}_{n\geq 1}$ converges to some $\omega \in \mathscr{D}$. Moreover, $\omega$ is the unique fixed point of $\mathscr{P}$, provided that both $\lim_{m\to\infty} \zeta(\omega, \theta_m)$ and $\lim_{m\to\infty} \zeta(\theta_m, \mathscr{P}(\omega))$ exist and finite.*

**Proof:** Since $F$ is strictly increasing, from the contractive condition (8.16) it follows that $\Upsilon(\mathscr{P}\vartheta, \mathscr{P}\varphi) \leq \Upsilon(\vartheta, \varphi)$ for all $\vartheta, \varphi \in \mathscr{D}$, showing that the mapping $\mathscr{P}$ is continuous.

For any fixed $n_1, n_2 \in \mathbb{N} \cup \{0\}$,

$$\begin{aligned}\mathscr{F}(\Upsilon(\theta_{n+n_1}, \theta_{n+n_2})) &\leq \mathscr{F}(\Upsilon(\theta_{n-1+n_1}, \theta_{n-1+n_2})) - \tau \\ &\cdots \\ &\leq \mathscr{F}(\Upsilon(\theta_{n_1}, \theta_{n_2})) - n\tau \text{ for all } n \geq 1.\end{aligned} \tag{8.17}$$

So, $\lim_{n\to\infty} \mathscr{F}(\Upsilon(\theta_{n+n_1}, \theta_{n+n_2})) = -\infty$, implying that $\lim_{n\to\infty} \Upsilon(\theta_{n+n_1}, \theta_{n+n_2}) = 0$ by condition (F2). So due to condition (F3)

$$\lim_{n\to\infty} \Upsilon(\theta_{n+n_1}, \theta_{n+n_2})^k \mathscr{F}(\Upsilon(\theta_{n+n_1}, \theta_{n+n_2})) = 0. \tag{8.18}$$

Using (8.17) and (8.18), we get

$$\begin{aligned}
\mathscr{F}(\Upsilon(\theta_{n+n_1},\theta_{n+n_2})) - \mathscr{F}(\Upsilon(\theta_{n_1},\theta_{n_2})) &\le -n\tau \\
\Rightarrow \Upsilon(\theta_{n+n_1},\theta_{n+n_2})^k[\mathscr{F}(\Upsilon(\theta_{n+n_1},\theta_{n+n_2})) - \mathscr{F}(\Upsilon(\theta_{n_1},\theta_{n_2}))] & \qquad (8.19)\\
&\le -\Upsilon(\theta_{n+n_1},\theta_{n+n_2})^k n\tau \\
\Rightarrow \lim_{n\to\infty} \Upsilon(\theta_{n+n_1},\theta_{n+n_2})^k n = 0. & \qquad (8.20)
\end{aligned}$$

Thus, there exists some $N \equiv N(n_1,n_2) \in \mathbb{N}$ such that

$$\begin{aligned}
&\Upsilon(\theta_{n+n_1},\theta_{n+n_2})^k n \le 1 \\
&\Rightarrow \Upsilon(\theta_{n+n_1},\theta_{n+n_2}) \le \frac{1}{n^{\frac{1}{k}}} \le \frac{1}{n} \text{ for all } n \ge N. \qquad (8.21)
\end{aligned}$$

Now we show that $\{\theta_n\}$ is $\Upsilon$-Cauchy in $\mathscr{D}$. For this, let us choose an arbitrary $\varepsilon > 0$. Then we can find some $\Lambda \in \mathbb{N}$ (taken as fixed depending on $\varepsilon$) such that

$$\tau > \mathscr{F}(\varepsilon) - \mathscr{F}\left(\frac{\Lambda}{\Lambda+1}\varepsilon\right).$$

Then, we can find some $N' \in \mathbb{N}$ such that

$$\begin{aligned}
\tau > \mathscr{F}(\varepsilon) - \mathscr{F}\left(\frac{\Lambda}{\Lambda+1}\varepsilon - \frac{v}{n^{\frac{1}{k}}}\right) &> \mathscr{F}(\varepsilon) - \mathscr{F}\left(\frac{\Lambda}{\Lambda+1}\varepsilon\right) \\
\text{for all } n \ge N' (\ge &\left[\left(\frac{(\Lambda+1)v}{\Lambda\varepsilon}\right)^k\right] + 1).
\end{aligned}$$

It is given that $\lim_{n,m\to\infty} \zeta(\theta_n,\theta_m) = 1$, then there exists $N'' \ge 1$ such that

$$\zeta(\theta_n,\theta_m) < 1 + \frac{1}{\Lambda} \text{ for all } n,m \ge N''.$$

Also

$$\begin{aligned}
\Upsilon(\theta_n,\theta_{n+1}) &\le \frac{1}{n^{\frac{1}{k}}} \text{ for all } n \ge N(0,1); \\
\Upsilon(\theta_{n+1},\theta_{n+2}) &\le \frac{1}{n^{\frac{1}{k}}} \text{ for all } n \ge N(1,2); \\
\cdots \quad \cdots & \quad \cdots \quad \cdots \\
\Upsilon(\theta_{n+v-1},\theta_{n+v}) &\le \frac{1}{n^{\frac{1}{k}}} \text{ for all } n \ge N(v-1,v); \\
\Upsilon(\theta_{n+v},\theta_n) &\le \frac{1}{n^{\frac{1}{k}}} \text{ for all } n \ge N(v,0).
\end{aligned}$$

Therefore, these above inequalities hold good for all $n,m \ge \tilde{N}$, where $\tilde{N} = \max\{N',N'',N(0,1),\cdots,N(v-1,v),N(v,0)\}$.

Take an $n \geq \tilde{N}$. We will see that $\Upsilon(\theta_{m+1}, \theta_n) < \varepsilon$, for each $m \geq n+\nu$, if $\Upsilon(\theta_m, \theta_n) < \varepsilon$. Indeed, since $\theta_n, \theta_{n+1}, \cdots, \theta_{n+\nu}$ and $\theta_{m+1}$ are different we get, by the polygonal inequality

$$\begin{aligned}
\Upsilon(\theta_{m+1}, \theta_n) &\leq \zeta(\theta_{m+1}, \theta_{n+1})\Upsilon(\theta_{m+1}, \theta_{n+1}) \\
&\quad + \sum_{i=n+1}^{n+\nu-1} \zeta(\theta_i, \theta_{i+1})\Upsilon(\theta_i, \theta_{i+1}) + \\
&\quad \zeta(\theta_{n+\nu}, \theta_n)\Upsilon(\theta_{n+\nu}, \theta_n) \\
&< \frac{\Lambda+1}{\Lambda}\{\Upsilon(\theta_{m+1}, \theta_{n+1}) + \sum_{i=n+1}^{n+\nu-1} \Upsilon(\theta_i, \theta_{i+1}) + \Upsilon(\theta_{n+\nu}, \theta_n)\} \\
&\leq \frac{\Lambda+1}{\Lambda}\{\Upsilon(\theta_{m+1}, \theta_{n+1}) + \frac{\nu}{n^{\frac{1}{k}}}\}
\end{aligned}$$

$$\Rightarrow \frac{\Lambda}{\Lambda+1}\Upsilon(\theta_{m+1}, \theta_n) - \frac{\nu}{n^{\frac{1}{k}}} < \Upsilon(\theta_{m+1}, \theta_{n+1}). \tag{8.22}$$

If for some $m \geq n+\nu$, $\Upsilon(\theta_{m+1}, \theta_n) \geq \varepsilon$, then from the above inequality we have,

$$\begin{aligned}
\mathscr{F}\left(\frac{\Lambda}{\Lambda+1}\Upsilon(\theta_{m+1}, \theta_n) - \frac{\nu}{n^{\frac{1}{k}}}\right) &\leq \mathscr{F}(\Upsilon(\theta_{m+1}, \theta_{n+1})) \\
&\leq \mathscr{F}(\Upsilon(\theta_m, \theta_n)) - \tau \\
\Rightarrow \mathscr{F}\left(\frac{\Lambda}{\Lambda+1}\varepsilon - \frac{\nu}{n^{\frac{1}{k}}}\right) &\leq F(\varepsilon) - \tau \\
\Rightarrow \tau \leq F(\varepsilon) - \mathscr{F}\left(\frac{\Lambda}{\Lambda+1}\varepsilon - \frac{\nu}{n^{\frac{1}{k}}}\right) &< \tau, \text{ a contradiction.}
\end{aligned} \tag{8.23}$$

This shows that $\Upsilon(\theta_{m+1}, \theta_n) < \varepsilon$, for each $m \geq n+\nu$. Which proves that $\{\theta_n\}$ is $\Upsilon$-Cauchy in $\mathscr{D}$. Since $(\mathscr{D}, \Upsilon)$ is $\Upsilon$-complete, $\{\theta_n\}$ converges to some $\omega \in \mathscr{D}$. As $\mathscr{P}$ is continuous it follows that $\{\theta_{n+1}\}$ is convergent to $\mathscr{P}\omega$. Therefore, due to Lemma 8.2.10, it follows that $\mathscr{P}\omega = \omega$ and hence $\omega$ is a fixed point of $\mathscr{P}$. If there is one more fixed point $\varpi$ of $\mathscr{P}$, then

$$\begin{aligned}
\tau + \mathscr{F}(\Upsilon(\mathscr{P}\omega, \mathscr{P}\varpi)) &\leq \mathscr{F}(\Upsilon(\omega, \varpi)) \\
\Rightarrow \tau + \mathscr{F}(\Upsilon(\omega, \varpi)) &\leq \mathscr{F}(\Upsilon(\omega, \varpi)) \\
&\Rightarrow \tau \leq 0, \text{ which is not possible.}
\end{aligned}$$

So $\omega$ is the unique fixed point of $\mathscr{P}$.

We consider a class $\Phi$ consisting of all those mappings $\phi : [0,\infty) \to [0,\infty)$ such that

($\Phi 1$) $\phi$ is monotonically non-decreasing;

($\Phi 2$) for all $t \geq 0$, $\lim_{n\to\infty} \phi^n(t) = 0$.

If a mapping $\phi \in \Phi$ then $\phi$ also satisfies ($\Phi3$) $\phi(0) = 0$ and ($\Phi4$) $\phi(t) < t$ for all $t > 0$.

One can give several examples of such mappings belonging to the family $\Phi$. Some of them are as the following:

(a) $\mu t$, $\mu \in (0,1)$ (b) $\frac{t}{1+t}$. More details could be found in [25].

**Theorem 8.3.15.** *($\phi$-contraction type mapping) Let $(\mathscr{D}, \Upsilon)$ be a $\Upsilon$-complete controlled v-generalized metric-type space ($v \geq 2$) and $\mathscr{P} : \mathscr{D} \to \mathscr{D}$ be a mapping satisfying*

$$\Upsilon(\mathscr{P}\vartheta, \mathscr{P}\varphi) \leq \phi(\Upsilon(\vartheta, \varphi)), \quad \phi \in \Phi, \text{ for all } \vartheta, \varphi \in \mathscr{D}, \tag{8.24}$$

*where $\lim_{n,m\to\infty} \zeta(\theta_n, \theta_m) = 1$ for the Picard iterating sequence $\{\theta_n\}_{n\geq 1} = \{\mathscr{P}^n \theta_0\}_{n\geq 1}$ generated by some $\theta_0 \in \mathscr{D}$. Then the Picard iterating sequence $\{\theta_n\}_{n\geq 1}$ converges to some $\omega \in \mathscr{D}$. Moreover, $\omega$ is a fixed point of $\mathscr{P}$, provided that both $\lim_{m\to\infty} \zeta(\omega, \theta_m)$ and $\lim_{m\to\infty} \zeta(\theta_m, \mathscr{P}(\omega))$ exist and finite. The fixed point is unique also.*

**Proof:** Due to the condition ($\Phi4$) it is clear that $\mathscr{P}$ is continuous. Without loss of generality, we assume that $\theta_n \neq \theta_{n+1}$ for all $n \in \mathbb{N}$.

For any $n_1, n_2 \in \mathbb{N}$ it follows from the contractive condition (8.24) that

$$\begin{aligned} \Upsilon(\theta_{n+n_1}, \theta_{n+n_2}) &\leq \phi(\Upsilon(\theta_{n-1+n_1}, \theta_{n-1+n_2})) \\ &\cdots \\ &\leq \phi^n(\Upsilon(\theta_{n_1}, \theta_{n_2})) \text{ for all } n \geq 1. \end{aligned} \tag{8.25}$$

Which gives $\lim_{n\to\infty} \Upsilon(\theta_{n+n_1}, \theta_{n+n_2}) = 0$. We now show that $\theta_n \neq \theta_m$ for all $n, m(n \neq m) \in \mathbb{N}$. If not, then there exists some $m > n$ such that $\theta_n = \theta_m$, which implies that $\theta_{n+1} = \theta_{m+1}$ and therefore

$$\begin{aligned} \Upsilon(\theta_n, \theta_{n+1}) = \Upsilon(\theta_m, \theta_{m+1}) &\leq \phi(\Upsilon(\theta_{m-1}, \theta_m)) \\ &\cdots \\ &\leq \phi^{m-n}(\Upsilon(\theta_n, \theta_{n+1})) < \Upsilon(\theta_n, \theta_{n+1}), \end{aligned}$$

a contradiction. Which proves our assertion. Now we prove that $\{\theta_n\}$ is $\Upsilon$-Cauchy in $\mathscr{D}$. For this let us choose an arbitrary $\varepsilon > 0$. Then we can find some $\Gamma \in \mathbb{N}$ (taken as fixed depending on $\varepsilon$) such that

$$\phi(\varepsilon) < \frac{\Gamma}{1+\Gamma}\varepsilon < \varepsilon.$$

From the fact that $\lim_{n,m\to\infty} \zeta(\theta_n, \theta_m) = 1$, there exists $N_1 \geq 1$ such that

$$\zeta(\theta_n, \theta_m) < 1 + \frac{1}{\Gamma} \text{ for all } n, m \geq N_1.$$

From (8.25), we can find $N_2, N_3 \in \mathbb{N}$ such that

$$\Upsilon(\theta_i, \theta_{i+1}) < \frac{\frac{\Gamma}{1+\Gamma}\varepsilon - \phi(\varepsilon)}{\nu} \text{ for all } n+\nu-2 \geq i \geq n \geq N_2;$$

$$\Upsilon(\theta_{n+\nu}, \theta_n) < \frac{\Gamma}{1+\Gamma}\varepsilon - \phi(\varepsilon) - \nu\phi\left(\frac{\frac{\Gamma}{1+\Gamma}\varepsilon - \phi(\varepsilon)}{\nu}\right) \text{ for all } n \geq N_3.$$

Therefore, for any $n \geq \bar{N} = \max\{N_1, N_2, N_3\}$, all the above inequalities hold good.

Take an $n \geq \bar{N}$. We will see that $\Upsilon(\theta_{m+1}, \theta_n) < \varepsilon$, for each $m \geq n+\nu$, if $\Upsilon(\theta_m, \vartheta_n) < \varepsilon$. Indeed, since $\theta_n, \theta_{n+1}, \cdots, \theta_{n+\nu}$ and $\theta_{m+1}$ are different we get, by the polygonal inequality

$$\begin{aligned}
\Upsilon(\theta_{m+1}, \theta_n) &\leq \zeta(\theta_{m+1}, \theta_{n+1})\Upsilon(\theta_{m+1}, \theta_{n+1}) + \sum_{i=n+1}^{n+\nu-1} \zeta(\theta_i, \vartheta_{i+1})\Upsilon(\theta_i, \theta_{i+1}) + \\
&\quad \zeta(\theta_{n+\nu}, \theta_n)\Upsilon(\theta_{n+\nu}, \theta_n) \\
&< \frac{\Gamma+1}{\Gamma}\{\Upsilon(\theta_{m+1}, \theta_{n+1}) + \sum_{i=n+1}^{n+\nu-1} \Upsilon(\theta_i, \theta_{i+1}) + \Upsilon(\theta_{n+\nu}, \theta_n)\} \\
&\leq \frac{\Gamma+1}{\Gamma}\{\phi(\Upsilon(\theta_m, \theta_n)) + \sum_{i=n}^{n+\nu-2} \phi(\Upsilon(\theta_i, \theta_{i+1})) + \Upsilon(\theta_{n+\nu}, \theta_n)\} \\
&\leq \frac{\Gamma+1}{\Gamma}\{\phi(\varepsilon) + (\nu-1)\phi\left(\frac{\frac{\Gamma}{1+\Gamma}\varepsilon - \phi(\varepsilon)}{\nu}\right) + \frac{\Gamma}{1+\Gamma}\varepsilon \\
&\quad - \phi(\varepsilon) - \nu\phi\left(\frac{\frac{\Gamma}{1+\Gamma}\varepsilon - \phi(\varepsilon)}{\nu}\right)\} \\
&< \varepsilon.
\end{aligned} \tag{8.26}$$

This shows that $\Upsilon(\theta_{m+1}, \theta_n) < \varepsilon$, for each $m \geq n+\nu$. Which proves that $\{\theta_n\}$ is $\Upsilon$-Cauchy in $\mathscr{D}$. Since $(\mathscr{D}, \Upsilon)$ is $\Upsilon$-complete, $\{\theta_n\}$ converges to some $\omega \in \mathscr{D}$. As $\mathscr{P}$ is continuous it follows that $\{\theta_{n+1}\}$ is convergent to $\mathscr{P}\omega$. Therefore, Lemma 8.2.10 implies that $\mathscr{P}\omega = \omega$ and hence $\omega$ is a fixed point of $\mathscr{P}$.

If possible let, $\kappa$ and $\mu$ be two distinct fixed points of $\mathscr{P}$, then due to condition ($\Phi$4)

$$\Upsilon(\mathscr{P}\kappa, \mathscr{P}\mu) \leq \phi(\Upsilon(\kappa, \mu)) < \Upsilon(\kappa, \mu), \text{ leads to a contradiction.}$$

Therefore, $\omega$ is the unique fixed point of $\mathscr{P}$.

## 8.4 AN APPLICATION TO NONLINEAR FREDHOLM-TYPE INTEGRAL EQUATIONS

In this section, we discuss about the existence of solutions for nonlinear integral equations as an application of Corollary 8.3.4.

Let $\mathscr{D} = C[a,b]$ be the set of all real-valued continuous functions on $[a,b]$ and $\Upsilon : \mathscr{D}^2 \to [0,\infty)$ be defined by:

$$\Upsilon(u,v) = \max_{a\leq t\leq b} \frac{1+|u(t)-v(t)|}{3(1+|u(t)|+|v(t)|)}|u(t)-v(t)| \text{ for all } u,v \in \mathscr{D}.$$

Then $(\mathscr{D},\Upsilon)$ is a complete controlled $\nu$-generalized metric type space. Now let us consider the integral equation:

$$u(t) = h(t) + \int_a^b F(t,s)\mathscr{K}(t,s,u(s))ds, \tag{8.27}$$

where $h : [a,b] \to \mathbb{R}$, $F : [a,b] \times [a,b] \to \mathbb{R}$ and $\mathscr{K} : [a,b] \times [a,b] \times \mathbb{R} \to \mathbb{R}$ are continuous functions.

**Theorem 8.4.1.** *Assume that the following hypotheses are satisfied:*
*(i) for all $t,s \in [a,b]$ we have*

$$|\mathscr{K}(t,s,u(s)) - \mathscr{K}(t,s,v(s))| \leq L \max_{t\in[a,b]} \frac{1+|u(t)-v(t)|}{1+|u(t)|+|v(t)|}|u(t)-v(t)|, L > 0,$$

*where $u,v \in \mathscr{D}$,*
*(ii) $|F(t,s)| \leq M$ for every $(t,s) \in [a,b] \times [a,b]$, where $M > 0$ is such that $LM(b-a) < 1$.*
*Then, the integral equation* (8.27) *has a unique solution $u \in \mathscr{D}$.*

**Proof:** Let us define $\Omega : \mathscr{D} \to \mathscr{D}$ by

$$\Omega(u)(t) = h(t) + \int_a^b F(t,s)\mathscr{K}(t,s,u(s))ds, \ u \in \mathscr{D}, t,s \in [a,b].$$

Then by conditions $(i)$ and $(ii)$, for all $\varphi, \psi \in \mathscr{D}$, we get

$$\begin{aligned}
&|\Omega(\varphi)(t) - \Omega(\psi)(t)| \\
=&\left|\int_a^b F(t,s)\{\mathscr{K}(t,s,\varphi(s)) - \mathscr{K}(t,s,\psi(s))\}ds\right| \\
\leq&\int_a^b |F(t,s)\{\mathscr{K}(t,s,\varphi(s)) - \mathscr{K}(t,s,\psi(s))\}|ds \\
\leq& M\int_a^b |\mathscr{K}(t,s,\varphi(s)) - \mathscr{K}(t,s,\psi(s))|ds \\
\leq& ML \max_{t\in[a,b]} \frac{1+|\varphi(t)-\psi(t)|}{1+|\varphi(t)|+|\psi(t)|}|\varphi(t)-\psi(t)| \int_a^b ds \\
=& LM(b-a) \max_{t\in[a,b]} \frac{1+|\varphi(t)-\psi(t)|}{1+|\varphi(t)|+|\psi(t)|}|\varphi(t)-\psi(t)| \\
=& \mu \max_{t\in[a,b]} \frac{1+|\varphi(t)-\psi(t)|}{1+|\varphi(t)|+|\psi(t)|}|\varphi(t)-\psi(t)| \text{ for all } t \in [a,b],
\end{aligned} \tag{8.28}$$

where $\mu = LM(b-a) < 1$. Thus

$$\begin{aligned}\Upsilon(\Omega(\varphi),\Omega(\psi)) &= \max_{t\in[a,b]} \frac{1+|\Omega(\varphi)(t)-\Omega(\psi)(t)|}{3(1+|\Omega(\varphi)(t)|+|\Omega(\psi)(t)|)}|\Omega(\varphi)(t)-\Omega(\psi)(t)| \\ &\leq \max_{t\in[a,b]} \frac{1}{3}|\Omega(\varphi)(t)-\Omega(\psi)(t)| \\ &\leq \mu \max_{t\in[a,b]} \frac{1+|\varphi(t)-\psi(t)|}{3(1+|\varphi(t)|+|\psi(t)|)}|\varphi(t)-\psi(t)| = \mu\Upsilon(\varphi,\psi)\end{aligned}$$

for $\mu \in (0,1)$ and for all $\varphi, \psi \in \mathscr{D}$. Hence the contractive condition (8.6) is satisfied and therefore $\Omega$ has a unique fixed point in $\mathscr{D}$, provided that for some $u_0 \in \mathscr{D}$ all other supporting conditions of Corollary 8.3.4 are satisfied, i.e. the nonlinear integral equation (8.27) has a unique solution in $C[a,b]$.

## REFERENCES


1. Akram M., Siddiqui A. A., A fixed point theorem for $A$-contractions on a class of generalised metric spaces, Korean J. Math. Sciences, **10**(2), (2003), 1–5.
2. Akram M., Zafar A. A., Siddiqui A. A., A general class of contractions: $A$-contractions, Novi Sad J. Math., **38**(1), (2008), 25–33.
3. Alolaiyan H., Ali B., Abbas M., Characterization of a $b$-metric space completeness via the existence of a fixed point of Ciric-Suzuki type quasi-contractive multivalued operators and applications. An. Stiint. Univ. Ovidius Constanta Ser. Mat., **27**(1), (2019), 5–33.
4. Bakhtin I. A., The contraction mapping principle in quasi-metric spaces, Functional Analysis, **30**, Ulyanovsk. Gos. Ped. Inst, Ulyanovsk (1989).
5. Basit A., Abbas M., Existence and stability of fixed point set of Suzuki-type contractive multivalued operators in $b$-metric spaces with applications in delay differential equations, J. Fixed Point Theory Appl., **19**, (2017), 2327–2347.
6. Branciari A., A fixed point theorem of Banach-Caccioppoli type on a class of generalized metric spaces, Publ. Math. Debrecen, **57**(2000), 31–37.
7. Czerwik S., Contraction mappings in $b-$metric spaces, Acta Math. Inform. Univ. Ostrav.,**1**, (1993), 5–11.
8. Kamran T., Samreen M., and Ain Q. U., A generalization of $b-$metric space and some fixed point theorems, Mathematics, **5**, 19, (2017).
9. Roy K., Panja S., Saha M., Parvaneh V., An extended $b$-metric type space and related fixed point theorems with an application to nonlinear integral equations, Adv. Math. Phys., **2020**, (2020), 1–7.
10. Roy K., Saha M., Beg I., Fixed point of contractive mappings of integral type over an $S^{JS}$-metric space, Tamkang J. Math., **52**(2) (2021), 267–280.
11. Roy K., Saha M., Generalized contractions and fixed point theorems over bipolar cone$_{tvs}$ $b-$metric spaces with an application to homotopy theory, Mat. Vesnik, **72**(4) (2020), 281–296.
12. Mitrovic Z. D., Radenovic S., The Banach and Reich contractions in $b_\nu(s)$-metric spaces, J. Fixed Point Theory Appl., **19**(4,) (2017), 3087–3095.
13. Garai H., Dey L. K., Mondal P., The contractive principle for mappings in $b_\nu(s)$-metric spaces, (2018), arXiv:1802.03136v1.
14. Karahan I., Isik I., partial $b_\nu(s)$ and $b_\nu(\theta)$ metric spaces and related fixed point theorems,(2018), arXiv:1806.02663v1.

15. Mlaiki N., Aydi H., Souayah N., Abdeljawad T., Controlled metric type spaces and the related contraction principle, Mathematics, **6**(10), (2018), 1–11.
16. N. Alamgir, Q. Kiran, H. Aydi and Y. U. Gaba, On controlled rectangular metric spaces and an application, J. Funct. Spaces, Volume 2021, Article ID 5564324, 9 pages.
17. Zubair S. T., Gopalana K., Abdeljawad T., Controlled $b$-Branciari metric type spaces and related fixed point theorems with applications, Filomat, **34**(13), (2020), 4253–4269.
18. Roy K., Saha M., An abstract metric space and some fixed point theorems with an application to Markov process, J. Anal., (2021), (online published), https://doi.org/10.1007/s41478-021-00325-6.
19. Roy K., Panja S., Saha M., A generalized fixed point theorem in an extended cone $b$-metric space over Banach algebra with its application to dynamical programming, Applied Mathematics E-Notes, **21**, (2021), 209–219.
20. Roy K., Alaeidizaji H., Saha M., Mohammadi B., Parvaneh V., Some fixed-point theorems over a generalized $F$-metric space, Adv. Math. Phys., **2021**, (2021), 1–7.
21. Roy K., Saha M., Parvaneh V., Khorshidi M., A new variant of symmetric distance spaces and an extension of the Banach fixed-point theorem, J. Math., **2022**, (2022), 1–8.
22. Rhoades B. E., A comparison of various definitions of contractive mappings, Trans. Am. Math. Soc., **226**, (1977), 257–290.
23. Wardowski D., Fixed points of a new type of contractive mappings in complete metric spaces, Fixed Point Theory Appl., 2012, **94**, (2012).
24. Berinde V., Approximating fixed points of weak contractions using the Picard iteration, Nonlinear Analysis Forum, **9**(1), (2004), 43–53.
25. Boyd D. W., Wong J. S. W., On nonlinear contractions, Proc. Amer. Math. Soc., **20**, (1969), 458–464.
26. Mustafa Z., Parvaneh V., Jaradat M. M. M., Kadelburg Z., Extended Rectangular b-Metric Spaces and Some Fixed Point Theorems for Contractive Mappings, Symmetry, **11**(4), 594, (2019).
27. Roshan J. R., Parvaneh V., Kadelburg Z., Hussain N., New fixed point results in $b$-rectangular metric spaces, Nonlinear Analysis: Modelling and Control, **21**(5), (2016), 614–634.
28. Kohi M., Hosseinzadeh H., Abazari R., Bagheri A., Parvaneh V., $C^\star$-Algebra-Valued Extended Rectangular $Hb$-Metric Spaces and Presic-Type Fixed Point Theorem with an Application, J. Math., 2022.
29. Hussain N., Parvaneh V., Alamri B. S. A., Kadelburg Z., F-HR-type contractions on $(\alpha,\eta)$-complete rectangular $b$-metric spaces, J. Nonlinear Sci. Appl., **10**(3), (2017), 1030–1043.
30. Parvaneh V., Golkarmanesh F., George R., Fixed points of Wardowski-Ćirić-Presic type contractive mappings in a partial rectangular $b$-metric space, J. Math. Anal., **8**(1), 2017, 183–201.

# 9 Fixed Points of Graphic Contractions in $b-$Metric Spaces with a Directed Graph

*Talat Nazir, Zakaria Ali*

We discuss $b-$metric spaces and their fixed point results utilizing generalized graph set-valued contractions with domain of sets. The main result is demonstrated with the use of an example. We also established the well-posedness of fixed point problem in $b-$metric space. The results derived herein, improve and generalizes many results in the cited references.

## 9.1 INTRODUCTION

Fixed point theory is considered one of the effective appliances for getting clues to solve multiple mathematical problems with multitudinous applications. The research of fixed points of metric spaces furnished with a graph structure is useful in many aspects. As it subsumes the partial ordering structure initially proved by Ran and Reurings [22]. Fixed point results for single and set-valued mappings defined on partially ordered metrics spaces have been proven. (see, for example in [3, 18, 19, 20, 23]). Jachymski and Jozwik [14] initiated the new approach in metric fixed point theory by having the order structure replaced with a graph structure. Abbas et al. [1] established the fixed point of set-valued contraction mappings with this new structure. Other useful results in this direction appear in [4, 6, 7, 13].

Czerwik in [12] introduced the $b-$metric space, a generalization of the metric space. After which, fixed point results with respect to single-valued and set-valued mappings in this space have been proven, see for example in [9, 10, 11, 24, 25].

In this chapter, we obtained two fixed point results of set-valued mapping having domain of sets endowed with a directed graph in a $b-$metric space that satisfies generalized graph contractions. We do not employ any type of continuity of mappings therein. We also present a non-trivial example to support our main result presented therein. Further, we established the well-posedness of fixed point problems

DOI: 10.1201/9781003596578-9

in $b-$metric space. These results improve and generalize many results in the cited references [1, 2, 3, 4, 7, 10, 20, 22, 23].

**Definition 9.1.1.** *[12] Let $Y \neq \emptyset$ equipped with a non-negative real-valued function $\wp : Y \times Y \longrightarrow \mathbb{R}_+$ and a constant $\beta \geq 1$, then $\wp$ is called a b−metric, if it satisfies the following axioms:*

(a) $\wp(x,y) = 0 \Leftrightarrow x = y$;
(b) $\wp(x,y) = \wp(y,x)$;
(c) $\wp(x,z) \leq \beta[\wp(x,y) + \wp(y,z)]$;

*for all $x,y,z \in Y$. The triplet $(Y, \wp, \beta)$ is called a b−metric space.*

We will sometimes denote a $b-$metric as the pair $(Y, \wp)$ or as $Y$ where the constant $\beta$ and (or) metric $\wp$ will be clear from the context.

**Remark 9.1.2.** *Notice that $(Y, \wp, 1)$ is a metric space, that is every metric space is a b−metric space, however, the converse is not true in general. Furthermore axiom 9.1.1 is a* weaker *condition than the usual* triangle inequality, *and hence the b−metric is more general.*

**Example 9.1.3.** *[12] Let $Y = \mathbb{R}_+$ with the metric defined as $\wp(x,y) = (x-y)^2$, then $(Y, \wp)$ is not a metric space. It is a b−metric space with $\beta = 2$.*

Following the initial example in [8], we develop our own shown below:

**Example 9.1.4.** *Let $Y = \{0, 1, \frac{1}{2}\}$, and define $\wp : Y \times Y \longrightarrow \mathbb{R}_+$, as follows:*

$$\wp(c,d) = \begin{cases} 0 \iff c = d, \\ 1 \iff c \neq d, \\ \frac{2}{|c-d|} \iff c \neq d \neq 0. \end{cases}$$

*We can see that $(Y, \wp)$ is not a metric space, as follows:*

$$4 = \wp\left(1, \frac{1}{2}\right) \geq \wp\left(1, 0\right) + \wp\left(0, \frac{1}{2}\right) = 2.$$

*Hence, $(Y, \wp)$ is a b−metric space with $\beta \geq 2$.*

The topology of a $b-$metric space, convergence, Cauchy sequence, continuity and equivalent counterpart definitions.

**Definition 9.1.5.** *[12] Let $(Y, \wp, \beta)$ be a b−metric space, with $y \in Y$ and a positive real number $r > 0$. Then the **open ball** centred at y with radius r is the set defined as follows:*

$$B(y,r) = \{x \in Y : \wp(y,x) < r\}.$$

It is interesting to note that in a $b-$metric space, an open ball is not necessarily an *open set*, as is the case in a usual metric space.

**Definition 9.1.6.** *[12] Let $(Y,\wp,\beta)$ be a $b-$metric space and $S \subseteq Y$. Then $S$ is said to be* **open**, *if for every $s \in S$, there exists $r > 0$ such that the open ball $B(s,r)$ is completely contained in $S$, such an element $s$ is called an* **interior point** *of $S$. Since $S$ is open, it contains all of its interior points.*

We describe the topology $\tau_{\wp}$ on $(Y,\wp,\beta)$ as the family of all open subsets of X, that is,

$$\tau_{\wp} = \Big\{ S \subseteq Y : \forall \;\; s \in S \;\; \exists \;\; r \in \mathbb{R}_+ \text{ such that } B(s,r) \subseteq S \Big\}.$$

A sequence $\{s_n\} \subseteq (Y,\wp,\beta)$ is said to **converge** to a point $s \in Y \iff \wp(s_n,s) \longrightarrow 0$ as $n \longrightarrow \infty$. This is also denoted as $\lim_{n\longrightarrow\infty} s_n = s$.

**Definition 9.1.7.** *[12] Let $\{s_n\} \subseteq (Y,\wp,\beta)$ be a sequence. Then $\{s_n\}$ is said to be* **Cauchy**, *if for any $\varepsilon > 0$ there exists $N(\varepsilon) \in \mathbb{N}$ such that: $\wp(s_m,s_n) < \varepsilon$ whenever $m,n > N(\varepsilon)$.*

The space $(Y,\wp,\beta)$ is said to be **complete**, if every Cauchy sequence $\{s_n\} \subseteq (Y,\wp,\beta)$ converges to a point $s \in Y$.

**Definition 9.1.8.** *[12] Let $(Y,\wp,\beta)$ be a $b-$metric space and $s \in Y$. Then a* **neighbourhood** *$N$ of $s$ is an open ball (not necessarily centred at $s$) containing $s$.*

If $S \subseteq (Y,\wp,\beta)$. Then the set $\overline{S}$ is the **closure** of $S$, that is it contains all *adherent points* of $S$. That is, $s \in \overline{S} \iff N \cap S \neq \emptyset$. Clearly $S \subseteq \overline{S}$.

**Definition 9.1.9.** *[12] Let $S \subseteq (Y,\wp,\beta)$ and $\{s_n\} \subseteq S$. Then $S$ is said to be* **closed** *$\iff \lim_{n\longrightarrow\infty} s_n = s$ such that $s \in S$.*

Another characterization of closedness of a subset $S$ of $Y$ is if $\overline{S} \subseteq S$, by the remark preceding the above definition, $S$ is said to be closed $\iff \overline{S} = S$. Furthermore $S$ is said to be **bounded**, if $\wp(x,y) < \mathfrak{M}$ for all $x,y \in S$, for some real number $\mathfrak{M} > 0$.

**Definition 9.1.10** (Continuity of a $b-$metric). *Let $(Y,\wp,\beta)$ be a $b-$metric space, $s_n \longrightarrow s$ and $t_n \longrightarrow t$ as $n \longrightarrow \infty$. Then $\wp$ is said to be* **continuous** *at $(s,t) \iff \lim_{n\longrightarrow\infty} \wp(s_n,t_n) = \wp(s,t)$. We denote it by $\lim_{n\longrightarrow\infty} \wp(s_n,t_n) = \wp(s,t)$.*

The preceding definition shows that $\wp$ is continuous with respect to both $(s,t)$. We can also say that $\wp$ is continuous to a single variable $\iff$ for any $s \in Y$, where $\lim_{n\longrightarrow\infty} t_n = t$. Then $\lim_{n\longrightarrow\infty} \wp(s,t_n) = \wp(s,t)$. Below, we mention several properties of $(Y,\wp,\beta)$ as follows:

1. $\wp$ is not necessarily continuous in each variable.
2. If $\wp$ is continuous in a single variable, then it is continuous in the second variable.
3. An open ball $B(s,r)$ in $\wp$ is not necessarily open. $B(s,r)$ is open, whenever $\wp$ is continuous.

We now define an important metric induced by the $b-$metric $\wp$. This metric will be important to our main result, which provides a means of evaluating the distance between pairs of elements of $CB^{\wp}(Y)$.

**Definition 9.1.11.** *[17] Let $(Y,\wp)$ be a $b-$metric space. For $U,V \in CB^{\wp}(X)$, the* ***Pompeiu-Hausdorff*** *$PHb-$metric induced by $\wp$, is defined as:*

$$\mathscr{H}_{\wp}(A,B) = \max\left\{\mathscr{D}(A,B), \mathscr{D}(B,A)\right\},$$

*where $\mathscr{D}(A,B) = \sup\{\wp(a,B) : a \in A\}$ and $\wp(a,B) = \inf\{\wp(a,b) : b \in B\}$.*

## 9.2 GRAPH STRUCTURE ON $B-$METRIC SPACE

**Definition 9.2.1.** *Let $Y \neq \emptyset$. We identify $Y$ with a directed graph $G = (V(G), E(G))$, as follows:*

*i) $Y = V(G)$, where $V(G)$ is the vertex set;*
*ii) $\Delta \subseteq E(G)$, where $\Delta \subseteq Y \times Y$, and $E(G)$ is the edge set;*
*iii) Loops at a vertex is allowed, for all $v \in V(G)$;*
*iv) Parallel edges between distinct vertices are prohibited.*

*Since $V(G) = Y$, we shall freely refer to elements of $V(G)$ or $Y$ interchangeably.*

If $u, v \in V(G)$, then a (directed) **path** in $G$ between $u$ and $v$ having length $l \in \mathbb{N}$ is a sequence $\{y_n\}_{n=0}^{l}$ of the vertices so that $u = y_0$, $v = y_l$ also $(y_{i-1}, y_i) \in E(G)$ for $i \in \{1, 2, \ldots, l\}$.

If between every two vertices there is a (directed) path, then $G$ is called connected, and if $\widetilde{G}$ is connected thus it is said to be weakly connected, where $\widetilde{G}$ represents the undirected graph produced by removing the direction of edges of $G$. $G^{-1}$ is the graph formed by reversing the edge direction in $G$, that is $(v,u) \in E(G^{-1})$, if and only if $(u,v) \in E(G)$. We also represent:

$$Y_f = \{y \in Y : (y, fy) \in E(G)\}.$$

Let $(Y, \psi)$ be a metric space. Then we say $G$ is a weighted graph with respect to the metric $\psi$, if a weight $\psi(u,v) \geq 0$ is allocated to every vertex $u, v \in V(G)$. Loops will have weight 0.

**Definition 9.2.2.** *Letting $\emptyset \neq V$ and $W \subseteq Y$. Then,*

*(1) when we declare that there exists an edge among $V$ and $W$, this conveys that there exists an edge among $v \in V$ and $w \in W$ that we may express as $(V,W) \subset E(G)$.*
*(2) when we declare that there is a path $P$ between $V$ and $W$, we are referring to a path between $v \in V$ and $w \in W$, and denote it as $\overrightarrow{UV}$.*

The following is how we define a relation $\mathscr{R}$ on $CB^{\wp}(Y)$: $U\mathscr{R}V$ for $U,V \in CB^{\wp}(Y)$ if and only if there is a path between $U$ and $V$. A relation $\mathscr{R}$ on $CB^{\wp}(Y)$ is called transitive, if $\overrightarrow{UV}$ and $\overrightarrow{VW}$, then $\overrightarrow{UW}$. Furthermore, a subset $\mathscr{C} \subseteq CB^{\wp}(Y)$ is said to be **complete** if for any $U,V \in \mathscr{C}$, there is an edge between $U$ and $V$. The following is the equivalence class of $U$ for $U \in CB^{\wp}(Y)$ determined by $\mathscr{R}$:

$$[U]_G = \{V \in CB^{\wp}(Y) : U\mathscr{R}V\}.$$

Rather than mapping $Y \longrightarrow Y$ or $Y \longrightarrow CB^{\wp}(Y)$, we consider the mapping $T : CB^{\wp}(Y) \longrightarrow CB^{\wp}(Y)$. The set $Y_T$ for a map $T : CB(Y) \longrightarrow CB^{\wp}(Y)$ is interpreted as:

$$Y_T = \{U \in CB^{\chi}(Y) : (U,T(U)) \in E(G)\}.$$

For our main result: $(Y,\wp,\beta)$ is equipped **weighted** directed graph $G$. Where loops have weight $\wp(c,c) = 0$ for all $c \in V(G)$. Moreover, the $PHb-$metric induced by the $b-$metric $\wp$ assigns a non-zero $PHb-$weight to every $U,V \in CB^{\wp}(Y)$. And $\mathscr{H}_{\wp}(U,V) = 0$, whenever $U = V$. Motivated by the work of Abbas et al. [2], we present the following definitions of generalized graph contractive set-valued mappings.

**Definition 9.2.3.** *Let $(Y,\wp,\beta)$ be a $b-$metric space equipped with a directed graph $G$, where $V(G) = Y$ and $\Delta \subseteq E(G)$ and $S : CB^{\wp}(Y) \longrightarrow CB^{\wp}(Y)$ a set-valued mapping. Then $S$ is said to be a* ***generalized rational graph $\lambda^*-$contraction****, if the following conditions are satisfied:*

*i) An edge (path) between $U$ and $V$ implies an edge (path) between $S(U)$ and $S(V)$, for all $U,V \in CB^{\wp}(Y)$.*
*ii) There exists $\lambda^* \in [0,1)$, such that:*

$$\mathscr{H}_{\wp}\Big(S(U),S(V)\Big) \leq \lambda^* N_S(U,V) \text{ for all } U,V \in CB(X),$$

*where*

$$N_S(U,V) = \max\left\{\frac{\mathscr{H}_{\chi}(U,S(V))[1+\mathscr{H}_{\chi}(U,S(U))]}{2[1+\mathscr{H}_{\chi}(U,V)]}, \frac{\mathscr{H}_{\chi}(V,S(V))[1+\mathscr{H}_{\chi}(U,S(U))]}{1+\mathscr{H}_{\chi}(U,V)}, \frac{\mathscr{H}_{\chi}(V,S(U))[1+\mathscr{H}_{\chi}(U,S(U))]}{1+\mathscr{H}_{\chi}(U,V)}\right\}. \tag{9.1}$$

**Definition 9.2.4.** *Let $(Y,\wp,\beta)$ be a $b-$metric space equipped with a directed graph $G$, where $V(G) = Y$ and $\Delta \subseteq E(G)$ and $T : CB^{\wp}(Y) \longrightarrow CB^{\wp}(Y)$ a set-valued mapping. Then $T$ is said to be a* ***generalized graph $\lambda-$contraction****, if the following conditions are satisfied:*

*(a) An edge (path) between $U$ and $V$ implies an edge (path) between $T(U)$ and $T(V)$, for all $U,V \in CB^{\wp}(Y)$.*

*(b) There exists* $\lambda \in [0,1)$*, such that:*

$$\mathscr{H}_{\wp}\Big(T(U),T(V)\Big) \leq \lambda M_T(U,V) \text{ for all } U,V \in CB^{\wp}(Y),$$

*where*

$$\begin{aligned} M_T(U,V) = \max\Big\{&\mathscr{H}_{\chi}(U,V), \mathscr{H}_{\chi}(U,T(U)), \mathscr{H}_{\chi}(V,T(V)), \\ &\mathscr{H}_{\chi}(T(U),T^2(U)), \mathscr{H}_{\chi}(T^2(U),V), \mathscr{H}_{\chi}(T^2(U),T(V)), \\ &\frac{[\mathscr{H}_{\chi}(V,T(U)) + \mathscr{H}_{\chi}(U,T(V))]}{2}\Big\}. \end{aligned} \tag{9.2}$$

**Definition 9.2.5.** *A* $b-$*metric space* $(Y,\wp,\beta)$ *is said to be* $\varepsilon-$*chainable, if for* $\varepsilon > 0$ *and for every* $u,v \in Y$*, there exists an* $\varepsilon-$*chain, a finite set of points* $u = u_0,u_1,u_2,\ldots,u_n = v$*, such that* $\wp(u_{i-1},u_i) < \varepsilon$ *for* $i = 1,2,\ldots,n$*.*

## 9.3 FIXED POINT RESULTS IN $B-$METRIC SPACE WITH DIRECTED GRAPH

**Theorem 9.3.1.** *Let* $(Y,\wp,\beta)$ *be a complete* $b-$*metric space equipped with a directed graph* $G$*, where* $V(G) = Y$ *and* $\Delta \subseteq E(G)$*. If* $S : CB^{\wp}(Y) \longrightarrow CB^{\wp}(Y)$ *is a generalized rational* $\lambda^*-$*contraction mapping with* $\beta\lambda^* < 1$*, then the following statements hold:*

*a)* $F(S) \neq \emptyset$ *is complete* $\Rightarrow$ $PHb-$*weight assigned to* $U,V \in F(S)$ *is zero.*
*b)* $F(S) \neq \emptyset \Longrightarrow Y_S \neq \emptyset$*.*
*c)* $Y_S \neq \emptyset$ *and the weakly connected graph* $\widetilde{G}$ *satisfies* $\overset{*}{\mathscr{P}} \Longrightarrow$ *S has a fixed point.*
*d)* $F(S) \neq \emptyset$ *is complete* $\Longleftrightarrow$ $F(S)$ *is a singleton.*

**Proof:** For $(a)$ : Suppose by way of contradiction that the $PHb-$weight assigned to $U$ and $V$ is $\mathscr{H}_{\wp}(U,V) > 0$. Then by equation (9.1) we have,

$$\begin{aligned} \mathscr{H}_{\wp}\Big(S(U),S(V)\Big) &= \mathscr{H}_{\wp}\Big(S(U),S(V)\Big) \\ &\leq \lambda^* N_S(U,V), \end{aligned} \tag{9.3}$$

where,

$$\begin{aligned} N_S(U,V) = \max\Big\{&\frac{\mathscr{H}_{\wp}(U,S(V))[1+\mathscr{H}_{\wp}(U,S(U))]}{2[1+\mathscr{H}_{\wp}(U,V)]}, \\ &\frac{\mathscr{H}_{\wp}(V,S(V))[1+\mathscr{H}_{\wp}(U,S(U))]}{1+\mathscr{H}_{\wp}(U,V)}, \\ &\frac{\mathscr{H}_{\wp}(V,S(U))[1+\mathscr{H}_{\wp}(U,S(U))]}{1+\mathscr{H}_{\wp}(U,V)}\Big\}. \end{aligned}$$

Note that $\mathscr{H}_\wp\Big(U,U\Big) = \mathscr{H}_\wp\Big(U,S(U)\Big) = 0$ and $\mathscr{H}_\wp\Big(V,V\Big) = \mathscr{H}_\wp\Big(V,S(V)\Big) = 0$, as such, we have

$$N_S(U,V) = \frac{\mathscr{H}_\wp(U,V)}{1+\mathscr{H}_\wp(U,V)}.$$

Now, by inequality (9.3), we have: $[1+\mathscr{H}_\wp(U,V)]\mathscr{H}_\wp(U,V) \leq \lambda^* \mathscr{H}_\wp(U,V)$, a contradiction.

For $(b)$: Letting $V \in F(S)$ and since $\Delta \subseteq E(G)$, clearly we have $V \in Y_S \neq \emptyset$.

For $(c)$: Let $U \in Y_S \neq \emptyset$ and since $U \in CB^\wp(Y)$, then $S(U) \in CB^\wp(Y)$ by the definition of $Y_S$, and by our hypothesis it follows that $CB^\wp(Y) \subseteq [U]_{\widetilde{G}} = \mathrm{P}(Y)$ where $\mathrm{P}(Y)$ is the nonempty power set on $Y$. A similar argument shows that $S(U_i) \in [U]_{\tilde{G}}$ for each $U_i \in [U]_{\tilde{G}}$. Furthermore, there exists an edge between $U$ and $S(U)$. By definition of $S$, we have $\big(S^n(U), S^{n+1}(U)\big) \in E(\widetilde{G})$ for all $n \in \mathbb{N}$.
We now define an iterative sequence, as follows:

$$\begin{aligned}
U &= U_0,\\
S(U_0) &= U_1,\\
S^2(U_0) &= S(U_1) = U_2,\\
&\cdots\\
S^n(U_0) &= S(U_{n-1}) = U_n.
\end{aligned}$$

Assuming that $U_{k+1} \neq U_k$ for all $k \in \{0,1,2,\cdots,n\}$. If $U_{k+1} = U_k$ for some $k$, then $S$ admits a fixed point; $S(U_k) = U_{k+1} = U_k$. There exists a sequence $\{y_i\}_{i=0}^n$ with an edge between consecutive pair of vertices such that $y_i \in U_i$ for $i = 1,2,\cdots,n$, since $\tilde{G}$ is weakly connected. Directly applying the definition of $S$, we have

$$\begin{aligned}
\mathscr{H}_\wp\Big(S^n(U), S^{n+1}(U)\Big) &= \mathscr{H}_\wp(U_n, U_{n+1})\\
&= \mathscr{H}_\wp(S(U_{n-1}), S(U_n))\\
&\leq \lambda^*\Big(N_S(U_{n-1}, U_n)\Big)
\end{aligned} \tag{9.4}$$

for some $\lambda^* \in [0,1)$, where

$$\begin{aligned}
N_S(U_{n-1},U_n) = \max\Bigg\{ & \frac{\mathscr{H}_\wp(U_{n-1},S(U_n))[1+\mathscr{H}_\wp(U_{n-1},S(U_{n-1}))]}{2[1+\mathscr{H}_\wp(U_{n-1},U_n)]},\\
& \frac{\mathscr{H}_\wp(U_n,S(U_n)[1+\mathscr{H}_\wp(U_{n-1},S(U_{n-1}))])}{1+\mathscr{H}_\wp(U_{n-1},U_n)},\\
& \frac{\mathscr{H}_\wp(U_n,S(U_{n-1}))[1+\mathscr{H}_\wp(U_{n-1},S(U_{n-1}))])}{1+\mathscr{H}_\wp(U_{n-1},U_n)}\Bigg\}
\end{aligned}$$

$$
\begin{aligned}
&= \max\left\{\frac{\mathscr{H}_{\wp}(U_{n-1},U_{n+1})[1+\mathscr{H}_{\wp}(U_{n-1},U_n)]}{2[1+\mathscr{H}_{\wp}(U_{n-1},U_n)]},\right.\\
&\qquad \frac{\mathscr{H}_{\wp}(U_n,U_{n+1})[1+\mathscr{H}_{\wp}(U_{n-1},U_n)])}{1+\mathscr{H}_{\wp}(U_{n-1},U_n)},\\
&\qquad \left.\frac{\mathscr{H}_{\wp}(U_n,U_n)[1+\mathscr{H}_{\wp}(U_{n-1},U_n)])}{1+\mathscr{H}_{\wp}(U_{n-1},U_n)}\right\}\\
&= \max\left\{\frac{\mathscr{H}_{\wp}(U_{n-1},U_{n+1})}{2},\mathscr{H}_{\wp}(U_n,U_{n+1})\right\}.
\end{aligned}
$$

Now, if $N_S(U_{n-1},U_n) = \mathscr{H}_{\wp}(U_n,U_{n+1})$, then by inequality (9.4), we have a contradiction. Therefore $N_S(U_{n-1},U_n) = \dfrac{\mathscr{H}_{\wp}(U_{n-1},U_{n+1})}{2}$. It follows that:

$$
\begin{aligned}
\mathscr{H}_{\wp}(U_n,U_{n+1}) &\leq \frac{\lambda^*}{2}\mathscr{H}_{\wp}(U_{n-1},U_{n+1})\\
&\leq \frac{\lambda^*\beta}{2}[\mathscr{H}_{\wp}(U_{n-1},U_n)+\mathscr{H}_{\wp}(U_n,U_{n+1})],
\end{aligned}
$$

that is, for $\eta^* = \dfrac{\lambda^*\beta}{2-\lambda^*\beta} < 1$, we have

$$
\begin{aligned}
\mathscr{H}_{\wp}\Big(S^n(U),S^{n+1}(U)\Big) &= \mathscr{H}_{\wp}(U_n,U_{n+1})\\
&= \mathscr{H}_{\wp}(S(U_{n-1}),S(U_n))\\
&\leq \eta^*\mathscr{H}_{\wp}(U_{n-1},U_n)\\
&= \eta^*\mathscr{H}_{\wp}(S(U_{n-2}),S(U_{n-1}))\\
&\leq \eta^{*2}\mathscr{H}_{\wp}(U_{n-2},U_{n-1})\\
&\leq \cdots \leq \eta^{*n}\mathscr{H}_{\wp}(U_0,U_1) = \eta^{*n}\mathscr{H}_{\wp}(U,S(U)).
\end{aligned}
$$

Now, for $m,n\in\mathbb{N}$ with $m>n$,

$$
\begin{aligned}
&\mathscr{H}_{\wp}\Big(S^n(U),S^m(U)\Big)\\
&\leq \beta\mathscr{H}_{\wp}\Big(S^n(U),S^{n+1}(U)\Big)+\beta^2\mathscr{H}_{\wp}\Big(S^{n+1}(U),S^{n+2}(U)\Big)\\
&+\beta^3\mathscr{H}_{\wp}\Big(S^{n+2}(U),S^{n+3}(U)\Big)+\cdots\\
&+\beta^{m-n-2}\Big[\mathscr{H}_{\wp}\Big(S^{m-2}(U),S^{m-1}(U)\Big)+\mathscr{H}_{\wp}\Big(S^{m-1}(U),S^m(U)\Big)\Big]\\
&\leq \beta\eta^{*n}\mathscr{H}_{\wp}(U,S(U))+\beta^2\eta^{*n+1}\mathscr{H}_{\wp}(U,S(U))+\cdots\\
&+\beta^{m-n-2}\eta^{*m-1}\mathscr{H}_{\wp}(U,S(U))\\
&\leq \beta\lambda^{*n}(1+\beta\eta^*+\cdots)\mathscr{H}_{\wp}(U,S(U))\\
&= \frac{(\beta\eta^*)^n}{1-\beta\eta^*}\mathscr{H}_{\wp}(U,S(U)),
\end{aligned}
$$

since $\beta\eta^* < 1$, then we have

$$\mathscr{H}_{\wp}\Big(S^n(U),S^m(U)\Big) \leq \frac{(\beta\eta^*)^n}{1-\beta\eta^*}\mathscr{H}_{\wp}(U,S(U)) \longrightarrow 0 \text{ as } n \longrightarrow \infty.$$

This proves that $\{S^n(U)\}$ is a Cauchy sequence in $(Y,\wp,\beta)$. Moreover, since the $b-$metric space is complete, there exists $U^*$ such that $S^n(U) \longrightarrow U^*$ as $n \longrightarrow \infty$. It follows that since $\{S^n(U)\}$ is a Cauchy sequence such $\lim_{n\to\infty} S^n(U) = U^*$ and since $\widetilde{G}$ is weakly connected, then by property $\overset{*}{\mathscr{P}}$, there exists a subsequence $\{S^{n_k}(U)\}$ such that $(S^{n_k}(U),U^*) \subseteq E(G)$.

Now, by the triangle inequality 9.1.1 and property 9.2.3 of the generalized rational graph $\lambda^*-$contraction we have the following inequalities, as follows:

$$\begin{aligned}\mathscr{H}_{\wp}S(U^*),U^*) &\leq \beta[\mathscr{H}_{\wp}(S(U^*),S^{n_k}(U^*)) + \mathscr{H}_{\wp}(S^{n_k}(U^*),U^*)]\\ &\leq \beta\lambda^*(N_S(U^*,S^{n_k-1}(U^*))) + \beta\mathscr{H}_{\wp}(S^{n_k}(U^*),U^*) \end{aligned} \tag{9.5}$$

for $\lambda^* \in [0,1)$. Note that, $N_S(U^*,S^{n_k-1}(U^*)) = N_S(S^{n_k-1}(U^*),U^*)$, where

$$\begin{aligned}&N_S(S^{n_k-1}(U^*),U^*)\\ &= \max\left\{\frac{\mathscr{H}_{\wp}(S^{n_k-1}(U^*),S(U^*))[1+\mathscr{H}_{\wp}(S^{n_k-1}(U^*),S(S^{n_k-1}(U^*)))]}{2[1+\mathscr{H}_{\wp}(S^{n_k-1}(U^*),U^*)]},\right.\\ &\quad \frac{\mathscr{H}_{\wp}(U^*,S(U^*))[1+\mathscr{H}_{\wp}(S^{n_k-1}(U^*),S(S^{n_k-1}(U^*)))]}{1+\mathscr{H}_{\wp}(S^{n_k-1}(U^*),U^*)},\\ &\quad \left.\frac{\mathscr{H}_{\wp}(U^*,S(S^{n_k-1}(U^*)))[1+\mathscr{H}_{\wp}(S^{n_k-1}(U^*),S(S^{n_k-1}(U^*)))]}{1+\mathscr{H}_{\wp}(S^{n_k-1}(U^*),U^*)}\right\}.\end{aligned}$$

$$\begin{aligned}&= \max\left\{\frac{\mathscr{H}_{\wp}(S^{n_k-1}(U^*),S(U^*))[1+\mathscr{H}_{\wp}(S^{n_k-1}(U^*),S^{n_k}(U^*)]}{1+\mathscr{H}_{\wp}(S^{n_k-1}(U^*),U^*)},\right.\\ &\quad \frac{\mathscr{H}_{\wp}(U^*,S(U^*))[1+\mathscr{H}_{\wp}(S^{n_k-1}(U^*),S^{n_k}(U^*))]}{1+\mathscr{H}_{\wp}(S^{n_k-1}(U^*),U^*)},\\ &\quad \left.\frac{\mathscr{H}_{\wp}(U^*,S^{n_k}(U^*))[1+\mathscr{H}_{\wp}(S^{n_k-1}(U^*),S^{n_k}(U^*))]}{1+\mathscr{H}_{\wp}(S^{n_k-1}(U^*),U^*)}\right\}.\end{aligned}$$

The following cases follow:

(**Case** 1): If

$$\begin{aligned}N_S(U^*,S^{n_k-1}(U^*)) &= \frac{\mathscr{H}_{\wp}(S^{n_k-1}(U^*),S(U^*))[1+\mathscr{H}_{\wp}(S^{n_k-1}(U^*),S^{n_k}(U^*)]}{1+\mathscr{H}_{\wp}(S^{n_k-1}(U^*),U^*)}\\ &\leq \frac{[\beta\mathscr{H}_{\wp}(S^{n_k-1}(U^*),U^*)+\beta\mathscr{H}_{\wp}(U^*,S(U^*))][1+\mathscr{H}_{\wp}(S^{n_k-1}(U^*),S^{n_k}(U^*)]}{1+\mathscr{H}_{\wp}(S^{n_k-1}(U^*),U^*)}.\end{aligned}$$

It is clear to see that $\lim_{n\to\infty} N_S(U^*, S^{n_k-1}(U^*)) \leq \beta \mathscr{H}_{\wp}(U^*, S(U^*))$. Then by inequality (9.5), we obtain:

$$\mathscr{H}_{\wp}(S(U^*), U^*) \leq \beta^2 \lambda^* \mathscr{H}_{\wp}(U^*, S(U^*)) + \beta \mathscr{H}_{\wp}(S^{n_k}(U^*), U^*)$$
$$\longrightarrow \beta^2 \lambda^* \mathscr{H}_{\wp}(U^*, S(U^*)) \text{ as } n \longrightarrow \infty.$$

Since $\beta^2\lambda^* < 1$, we conclude that in all the above cases, we have that $S(U^*) = U^*$, as required.

(**Case** 2): If

$$N_S(U^*, S^{n_k-1}(U^*)) = \frac{\mathscr{H}_{\wp}(U^*, S(U^*))[1 + \mathscr{H}_{\wp}(S^{n_k-1}(U^*), S^{n_k}(U^*))]}{1 + \mathscr{H}_{\wp}(S^{n_k-1}(U^*), U^*)}$$

This follows a similar argument to *Case* 1.

(**Case** 3): If

$$N_S(U^*, S^{n_k-1}(U^*)) = \frac{\mathscr{H}_{\wp}(U^*, S^{n_k}(U^*))[1 + \mathscr{H}_{\wp}(S^{n_k-1}(U^*), S^{n_k}(U^*))]}{1 + \mathscr{H}_{\wp}(S^{n_k-1}(U^*), U^*)},$$

implies $\lim_{n\to\infty} N_S(U^*, S^{n_k-1}(U^*)) = 0$. Then by inequality (9.5), we obtain:

$$\mathscr{H}_{\wp}(S(U^*), U^*) \leq \beta \mathscr{H}_{\wp}(S^{n_k}(U^*), U^*) \longrightarrow 0 \text{ as } n \longrightarrow \infty.$$

whence, our result follows: $S(U^*) = U^*$.

We have shown that in all resulting cases above we obtain our fixed point $S(U^*) = U^*$.

$(d)$ : Letting $F(S) \neq \emptyset$ be complete, then by 9.3.1 of Theorem 9.3.1, we conclude that $F(S)$ is a singleton. This proves the necessity. Proving the sufficiency, let $F(S)$ be a singleton with $U \in F(S)$, then we have $U \in Y_S$, hence $F(S)$ is complete. □

**Theorem 9.3.2.** *Let $(Y, \wp, \beta)$ be a complete b−metric space equipped with a directed graph G, where $V(G) = Y$ and $\Delta \subseteq E(G)$. If $T : CB^{\wp}(Y) \longrightarrow CB^{\wp}(Y)$ is a generalized graph $\lambda$−contraction mapping with $\beta\lambda < 1$, then the following statements hold:*

1. *$F(T) \neq \emptyset$ is complete $\Longrightarrow$ PHb−weight assigned to $U, W \in F(T)$ is zero.*
2. *$F(T) \neq \emptyset \Longrightarrow Y_T \neq \emptyset$.*
3. *$Y_T \neq \emptyset$ and the weakly connected graph $\widetilde{G}$ satisfies $\overset{*}{\mathscr{P}} \Longrightarrow T$ has a fixed point.*
4. *$F(T) \neq \emptyset$ is complete $\Longleftrightarrow$ $F(T)$ is a singleton.*

**Proof:** For $(a)$ assume $\mathscr{H}_{\wp}(U,W)>0$. Then by equation (9.2), we have:

$$\begin{aligned}M_T(U,W)=\max\Big\{&\mathscr{H}_{\wp}(U,W),\mathscr{H}_{\wp}(U,T(U)),\mathscr{H}_{\wp}(W,T(W)),\\&\mathscr{H}_{\wp}(T(U),T^2(U)),\mathscr{H}_{\wp}(T^2(U),W),\\&\mathscr{H}_{\wp}(T^2(U),T(W)),\frac{\mathscr{H}_{\wp}(W,T(U))+\mathscr{H}_{\wp}(U,T(W))}{2}\Big\}\end{aligned}$$

and after applying the mapping $T$, noting that $T(U)=T^2(U)=U$ we get:

$$\begin{aligned}M_T(U,W)=\max\Big\{&\mathscr{H}_{\wp}(U,W),\mathscr{H}_{\wp}(U,T(U)),\mathscr{H}_{\wp}(W,T(W)),\\&\mathscr{H}_{\wp}(T(U),T^2(U)),\mathscr{H}_{\wp}(T^2(U),W),\mathscr{H}_{\wp}(T^2(U),T(W)),\\&\frac{\mathscr{H}_{\wp}(W,T(U))+\mathscr{H}_{\wp}(U,T(W))}{2}\Big\}\\=\ &\mathscr{H}_{\wp}(U,W),\end{aligned}$$

we obtain a contradiction and our result follows.

The Proof of $(b)$ follows a similar Proof to $(b)$ of Theorem 9.3.1.

(c): Let $U\in Y_T\neq\emptyset$, and since $U\in CB^{\wp}(Y)$, then since $T(U)\in CB^{\wp}(Y)$, by the definition of $Y_T$, and by our hypothesis it follows that $CB^{\wp}(Y)\subseteq[U]_{\widetilde{G}}=\mathrm{P}(Y)$ where $\mathrm{P}(Y)$ is the nonempty power set on $Y$. A similar argument shows that $T(U)\in[U]_{\widetilde{G}}$ for every $U_i\in[U]_{\widetilde{G}}$. Since $U\in Y_T$, there is an edge between $U$ and $T(U)$. By definition of $T$, we have $\big(T^n(U),T^{n+1}(U)\big)\in E(\widetilde{G})$ for all $n\in\mathbb{N}$.

We define an iterative sequence: let $U=U_0$ and $T(U_i)=U_{i+1}$ for $i=1,2,\cdots,n$. Assuming that $U_{i+1}\neq U_i$ for all $i=1,2,\cdots,n$.

Then there exists a sequence $\{y_i\}_{i=0}^{n}$ with an edge between consecutive pair of vertices such that $y_i\in U_i$ for $i=1,2,\cdots,n$. It follows that since $T$ is a graph $\lambda$-contraction, we have

$$\begin{aligned}\mathscr{H}_{\wp}\Big(T^n(U),T^{n+1}(U)\Big)&=\mathscr{H}_{\wp}\big(U_n,U_{n+1}\big)\\&=\mathscr{H}_{\wp}\big(T(U_{n-1}),T(U_n)\big)\\&\leq\lambda\Big(M_T(U_{n-1},U_n)\Big)\end{aligned}\tag{9.6}$$

for some $\lambda \in [0,1)$, where

$$\begin{aligned}
&M_T(U_{n-1},U_n)\\
&= \max\Big\{ \mathcal{H}_{\wp}(U_{n-1},U_n), \mathcal{H}_{\wp}(U_{n-1},T(U_{n-1})), \mathcal{H}_{\wp}(U_n,T(U_n)),\\
&\quad \mathcal{H}_{\wp}(T(U_{n-1}),T^2(U_{n-1})), \mathcal{H}_{\wp}(T^2(U_{n-1}),U_n)),\\
&\quad \mathcal{H}_{\wp}(T^2(U_{n-1}),T(U_n)), \frac{\mathcal{H}_{\wp}(U_n,T(U_{n-1})) + \mathcal{H}_{\wp}(U_{n-1},T(U_n))}{2}\Big\}.
\end{aligned}$$

Applying the mapping $T$, we obtain:

$$\begin{aligned}
&M_T(U_{n-1},U_n)\\
&= \max\Big\{ \mathcal{H}_{\wp}(U_{n-1},U_n), \mathcal{H}_{\wp}(U_{n-1},U_n), \mathcal{H}_{\wp}(U_n,U_{n+1}),\\
&\quad \mathcal{H}_{\wp}(U_n,U_{n+1}), \mathcal{H}_{\wp}(U_{n+1},U_n), \mathcal{H}_{\wp}(U_{n+1},U_{n+1}),\\
&\quad \frac{\mathcal{H}_{\wp}(U_n,U_n) + \mathcal{H}_{\wp}(U_{n-1},U_{n+1})}{2}\Big\}\\
&= \max\Big\{ \mathcal{H}_{\wp}(U_{n-1},U_n), \mathcal{H}_{\wp}(U_n,U_{n+1}), \frac{\mathcal{H}_{\wp}(U_{n-1},U_{n+1})}{2}\Big\}.
\end{aligned}$$

By equation (9.6), if $M_T(U_{n-1},U_n) = \mathcal{H}_{\wp}(U_n,U_{n+1})$ then clearly we get a contradiction. If $M_T(U_{n-1},U_n) = \frac{\mathcal{H}_{\wp}(U_{n-1},U_{n+1})}{2}$, then it follows:

$$\begin{aligned}
\mathcal{H}_{\wp}(U_n,U_{n+1}) &\leq \frac{\lambda\beta}{2-\lambda\beta}\mathcal{H}_{\wp}(U_{n-1},U_n),\\
&\leq \mathcal{H}_{\wp}(U_{n-1},U_n).
\end{aligned}$$

Therefore, $M_T(U_{n-1},U_n) = \mathcal{H}_{\wp}(U_{n-1},U_n)$ and it follows that

$$\begin{aligned}
\mathcal{H}_{\wp}\Big(T^n(U),T^{n+1}(U)\Big) &= \mathcal{H}_{\wp}(U_n,U_{n+1})\\
&= \mathcal{H}_{\wp}(T(U_{n-1}),T(U_n))\\
&\leq \lambda\,\mathcal{H}_{\wp}(U_{n-1},U_n)\\
&= \lambda\,\mathcal{H}_{\wp}(T(U_{n-2}),T(U_{n-1}))\\
&\leq \lambda^2\,\mathcal{H}_{\wp}(U_{n-2},U_{n-1})\\
&\leq \cdots \leq \lambda^n\,\mathcal{H}_{\wp}(U_0,U_1) = \lambda^n\,\mathcal{H}_{\wp}(U,T(U)).
\end{aligned}$$

Now, for $m, n \in \mathbb{N}$, with $m > n$,

$$\begin{aligned}
\mathscr{H}_{\wp}\Big(T^n(U), T^m(U)\Big) &\leq \beta \mathscr{H}_{\wp}\Big(T^n(U), T^{n+1}(U)\Big) + \beta^2 \mathscr{H}_{\wp}\Big(T^{n+1}(U), T^{n+2}(U)\Big) \\
&+ \beta^3 \mathscr{H}_{\wp}\Big(T^{n+2}(U), T^{n+3}(U)\Big) + \cdots \\
&+ \beta^{m-n-2}\Big[\mathscr{H}_{\wp}\Big(T^{m-2}(U), T^{m-1}(U)\Big) + \mathscr{H}_{\wp}\Big(T^{m-1}(U), T^m(U)\Big)\Big] \\
&\leq \beta\lambda^n \mathscr{H}_{\wp}(U, T(U)) + \beta^2\lambda^{n+1} \mathscr{H}_{\wp}(U, T(U)) + \cdots \\
&+ \beta^{m-n-2}\lambda^{m-1} \mathscr{H}_{\wp}(U, T(U)) \\
&\leq \beta\lambda^n (1 + \beta\lambda + \cdots) \mathscr{H}_{\wp}(U, T(U)) \\
&= \frac{\beta\lambda}{1 - \beta\lambda} \mathscr{H}_{\wp}(U, T(U))
\end{aligned}$$

since $\beta\lambda < 1$, then we have

$$\mathscr{H}_{\wp}\Big(T^n(U), T^m(U)\Big) \leq \frac{\beta\lambda}{1 - \beta\lambda} H_{\wp}(U, T(U)) \longrightarrow 0 \text{ as } n \longrightarrow \infty.$$

This proves that $\{T^n(U)\}$ is a Cauchy sequence in $(Y, \wp, \beta)$. Moreover, since the $b-$metric space is complete, there exists $U^*$ such that $T^n(U) \longrightarrow U^*$ as $n \longrightarrow \infty$. It follows that since $\{T^n(U)\}$ is a Cauchy sequence such $\lim_{n \to \infty} T^n(U) = U^*$ and since $\tilde{G}$ is weakly connected, then by property $\overset{*}{\mathscr{P}}$, there exists a subsequence $\{T^{n_k}(U)\}$ such that $(T^{n_k}(U), U^*) \subseteq E(G)$.

Now, by the triangle inequality 9.1.1 and property 9.2.4 of the generalized rational graph $\lambda-$contraction we have the following inequalities, as follows:

$$\begin{aligned}
\mathscr{H}_{\wp}(T(U^*), U^*) &\leq \beta[\mathscr{H}_{\wp}(T(U^*), T^{n_k}(U^*)) + \mathscr{H}_{\wp}(T^{n_k}(U^*), U^*)] \\
&\leq \beta\lambda\big(M_T(U^*, T^{n_k-1}(U^*))\big) + \beta \mathscr{H}_{\wp}(T^{n_k}(U^*), U^*) \qquad (9.7)
\end{aligned}$$

for some $\lambda \in [0,1)$. Note that $M_T(U^*, T^{n_k-1}(U^*)) = M_T(T^{n_k-1}(U^*), U^*)$, where

$$\begin{aligned} M_T(T^{n_k-1}(U^*), U^*) &= \max\Big\{ \mathcal{H}_{\wp}(T^{n_k-1}(U), U^*), \mathcal{H}_{\wp}(T^{n_k-1}(U), T(T^{n_k-1}(U))), \\ &\quad \mathcal{H}_{\wp}(U^*, T(U^*)), \mathcal{H}_{\wp}(T(T^{n_k-1}(U)), T^2(T^{n_k-1}(U))), \\ &\quad \mathcal{H}_{\wp}(T^2(T^{n_k-1}(U)), U^*), \mathcal{H}_{\wp}(T^2(T^{n_k-1}(U)), T(U^*)), \\ &\quad \frac{[\mathcal{H}_{\wp}(U^*, T(T^{n_k-1}(U))) + \mathcal{H}_{\wp}(T^{n_k-1}(U), T(U^*))]}{2} \Big\} \\ &= \max\Big\{ \mathcal{H}_{\wp}(T^{n_k-1}(U^*), U^*), \mathcal{H}_{\wp}(T^{n_k-1}(U^*), T^{n_k}(U^*)), \\ &\quad \mathcal{H}_{\wp}(U^*, T(U^*)), \mathcal{H}_{\wp}(T^{n_k}(U^*), T^{n_k+1}(U^*)), \\ &\quad \mathcal{H}_{\wp}(T^{n_k+1}(U^*), U^*), \mathcal{H}_{\wp}(T^{n_k+1}(U^*), T(U^*)), \\ &\quad \frac{[\mathcal{H}_{\wp}(U^*, T^{n_k}(U^*)) + \mathcal{H}_{\wp}(T^{n_k-1}(U^*), T(U^*))]}{2} \Big\}. \end{aligned}$$

Below are the cases considered:

(**Case** 1): If $M(U^*, T^{n_k-1}(U^*)) = H_{\wp}(T^{n_k-1}(U^*), U^*)$, then by inequality (9.7), we have:

$$\begin{aligned} H_{\wp}(T(U^*), U^*) &\leq \beta\lambda H_{\wp}(T^{n_k-1}(U^*), U^*) \\ &\quad + \beta H_{\wp}(T^{n_k}(U^*), U^*) \longrightarrow 0 \text{ as } n \longrightarrow \infty. \end{aligned}$$

Therefore, a similar argument to the preceding *main result* we have: $T(U^*) = U^*$.

(**Case** 2): If $M(U^*, T^{n_k-1}(U^*)) = \mathcal{H}_{\wp}(T^{n_k-1}(U^*), T^{n_k}(U^*))$, then by inequality (9.7), we have

$$\begin{aligned} \mathcal{H}_{\wp}(T(U^*), U^*) &\leq \beta\lambda \mathcal{H}_{\wp}(T^{n_k-1}(U^*), T^{n_k}(U^*)) \\ &\quad + \beta \mathcal{H}_{\wp}(T^{n_k}(U^*), U^*) \longrightarrow 0 \text{ as } n \longrightarrow \infty. \end{aligned}$$

This is valid since the *first* term on the right-hand side of inequality (9.7) is a Cauchy sequence, and the second term on the right-hand side is identical to that of the above case. Hence, it follows that: $T(U^*) = U^*$.

(**Case** 3): If $M(U^*, T^{n_k-1}(U^*)) = \mathcal{H}_{\wp}(U^*, T(U^*))$, then by a similar argument to that of *Case* 3 of the preceding *main result*, we obtain: $T(U^*) = U^*$.

(**Case** 4): If $M(U^*, T^{n_k-1}(U^*)) = H_b(T^{n_k}(U^*), T^{n_k+1}(U^*))$, then by *Case* 2 above, we have: $T(U^*) = U^*$.

(**Case** 5): If $M(U^*, T^{n_k-1}(U^*)) = \mathscr{H}_{\wp}(T^{n_k+1}(U^*), U^*)$, then similar to *Case* 1 above, we obtain: $T(U^*) = U^*$.

(**Case** 6): If $M(U^*, T^{n_k-1}(U^*)) = \mathscr{H}_{\wp}(T^{n_k+1}(U^*), T(U^*))$, then we obtain

$$\begin{aligned}\mathscr{H}_{\wp}(T(U^*),U^*) &\leq \beta\lambda\mathscr{H}_{\wp}(T^{n_k+1}(U^*),T(U^*)) + \beta\mathscr{H}_{\wp}(T^{n_k}(U^*),U^*)\\ &\leq \beta\lambda\left[\beta\mathscr{H}_{\wp}(T^{n_k+1}(U^*),U^*) + \beta\mathscr{H}_{\wp}(U^*,T(U^*))\right]\\ &\quad + \beta\mathscr{H}_{\wp}(T^{n_k}(U^*),U^*).\end{aligned}$$

It is clear that: $H_{\wp}(T^{n_k+1}(U^*), U^*) \longrightarrow 0$ and $H_{\wp}(T^{n_k}(U^*), U^*) \longrightarrow 0$ as $n \longrightarrow \infty$, and we obtain

$$\mathscr{H}_{\wp}(T(U^*),U^*) \leq \beta^2\lambda\mathscr{H}_{\wp}(U^*,T(U^*))$$

And since $\beta^2\lambda < 1$, then we must have $\mathscr{H}_{\wp}(T(U^*), U^*) = 0$. Whence our result follows: $T(U^*) = U^*$.

(**Case** 7): If

$$M(U^*,T^{n_k-1}(U^*)) = \frac{\mathscr{H}_{\wp}(U^*,T^{n_k}(U^*)) + \mathscr{H}_{\wp}(T^{n_k-1}(U^*),T(U^*))}{2},$$

then it follows:

$$\begin{aligned}\mathscr{H}_{\wp}(T(U^*),U^*) &\leq \beta\lambda\mathscr{H}_{\wp}(U^*,T^{n_k}(U^*)) + \beta\lambda\mathscr{H}_{\wp}(T^{n_k-1}(U^*),T(U^*))\\ &\quad + \beta\mathscr{H}_{\wp}(T^{n_k}(U^*),U^*),\\ &\leq \beta\lambda\mathscr{H}_{\wp}(U^*,T^{n_k}(U^*)),\\ &\quad + \beta\lambda\left[\beta\mathscr{H}_{\wp}(T^{n_k-1}(U^*),U^*) + \beta\mathscr{H}_{\wp}(U^*,T(U^*))\right],\\ &\quad + \beta\mathscr{H}_{\wp}(T^{n_k}(U^*),U^*),\end{aligned}$$

and it is clear to see that by a similar argument to *Case* 6, we then have: $T(U^*) = U^*$.

(d): Proof follows a similar argument to that are in (d) of Theorem 9.3.1. □

**Example 9.3.3.** *Let $Y = \{0,1,2,3\} = V(G)$ and*

$$E(G) = \{(0,0),(1,1),(2,2),(3,3),(0,1),(0,2),(0,3),(1,2),(1,3),(2,3)\}.$$

*Let $V(G)$ be endowed with metric $\wp : Y \times Y \longrightarrow \mathbb{R}_+$ defined by*

| $(x,y)$ | $\wp(x,y)$ |
|---|---|
| $(0,0),(1,1),(2,2),(3,3)$ | 0 |
| $(0,1),(1,0)$ | 1/4 |
| $(0,2),(2,0)$ | 1/2 |
| $(1,2),(2,1),(0,3),(3,0),(1,3),(2,3),(3,2)$ | 4/5 |

Note that, $\wp$ is not a metric on $Y$, since

$$\wp(1,2) = \frac{4}{5} \nleq \frac{1}{4} + \frac{1}{2} = \wp(1,0) + \wp(0,2).$$

But $\wp$ is a $b$-metric on $Y$ with $\beta = 2$, since

$$\wp(1,2) = \frac{4}{5} \leq 2\left(\frac{3}{4}\right) = \beta\left[\wp(1,0) + \wp(0,2)\right].$$

The $PHb-$weight assigned to $V, W \in CB^{\wp}(Y)$, where $\bar{0} = \{0\}, \bar{1} = \{0,1\}, \bar{2} = \{0,2\}$ and $\bar{3} = \{0,3\}$.

The $PHb-$metric on $CB(Y)$ is defined as:

$$\mathscr{H}_{\wp}(V,W) = \begin{cases} \frac{1}{4} & \text{if } V = \bar{0} \text{ and } W = \bar{1}, \\ \frac{1}{2} & \text{if } (V = \bar{0} \text{ and } W = \bar{2}) \text{ or } (V = \bar{1} \text{ and } W = \bar{2}), \\ \frac{4}{5} & \text{if } V \in \{\bar{0}, \bar{1}, \bar{2}\} \text{ and } W = \bar{3}, \\ 0 & \text{if } V = W. \end{cases}$$

Define $T : CB^{\wp}(Y) \longrightarrow CB^{\wp}(Y)$ as follows:

$$T(V) = \begin{cases} \bar{0} & \text{if } V = \bar{0} \text{ or } V = \bar{1}, \\ \bar{1} & \text{if } V = \bar{2} \text{ or } V = \bar{3}. \end{cases}$$

Note that, for all $V, W \in CB^{\wp}(Y)$ with edge (path) between $V$ and $W$, there is an edge (path) between $T(V)$ and $T(W)$. Now for all $V, W \in CB^{\wp}(Y)$, the following cases follow:

(**Case**1): If ($V \neq W \in \{\bar{0}, \bar{1}\}$) or ($V \neq W \in \{\bar{2}, \bar{3}\}$), then

$$\mathscr{H}_{\wp}(T(V), T(W)) = \mathscr{H}_{\wp}(\bar{0}, \bar{0}) = 0$$

or

$$\mathscr{H}_{\wp}(T(V), T(W)) = \mathscr{H}_{\wp}(\bar{1}, \bar{1}) = 0.$$

(**Case** 2): If $V \in \{\bar{0}, \bar{1}\}$ and $W = \bar{2}$, then

$$\begin{aligned} \mathscr{H}_{\wp}(T(V), T(W)) &= \mathscr{H}_{\wp}(\bar{0}, \bar{1}) \\ &= \frac{1}{4} \leq \frac{1}{2}\lambda = \mathscr{H}_{\wp}(V, W). \end{aligned}$$

(**Case** 3): In case $V \in \{\bar{0}, \bar{1}\}$ and $W = \bar{3}$,

$$\begin{aligned} \mathscr{H}_{\wp}(T(V), T(W)) &= \mathscr{H}_{\wp}(\bar{0}, \bar{1}) \\ &= \frac{1}{4} \leq \frac{4}{5}\lambda = \mathscr{H}_{\wp}(V, W). \end{aligned}$$

Hence, for all $V, W \in CB^{\wp}(X)$ having an edge between $V$ and $W$, $T$ is a generalized graph $\lambda-$contraction with $\lambda \in (\frac{1}{2}, 1)$. Thus all the conditions of Theorem 9.3.2 are satisfied. Moreover, $\bar{0}$ is the fixed point of $T$ and $F(T)$ is complete.

**Corollary 9.3.4.** *Let $(Y,\wp,\beta)$ be a $\varepsilon-$chainable complete $b-$metric space for some $\varepsilon>0$ with $\beta\geq 1$, $T: CB^{\wp}(Y)\longrightarrow CB^{\wp}(Y)$ and $0\leq\lambda<1$ with*

$$0<\mathscr{H}_{\wp}(V,W)<\varepsilon.$$

*If*

$$\mathscr{H}_{\wp}(T(V),T(W))\leq\lambda\mathscr{H}_{\wp}(V,W)\text{ for all }V,W\in CB^{\wp}(Y),$$

*then $T$ has a fixed point.*

**Proof:** By Definition 9.2.5 and from $\mathscr{H}_{\wp}(V,W)<\varepsilon$, we have for each $v\in V$, an element $w\in W$ such that $\wp(v,w)<\varepsilon$. Consider the graph $G$ as $V(G)=Y$ and

$$E(G)=\{(v,w)\in Y\times Y: 0<\wp(v,w)<\varepsilon\}.$$

Then the $\varepsilon-$chainability of $(Y,\wp)$ implies that $G$ is connected. For $(V,W)\subset E(G)$, we have from the hypothesis

$$\mathscr{H}_{\wp}(T(V),T(W))\leq\lambda\mathscr{H}_{\wp}(V,W).$$

This implies that $T$ is a generalized graph $\lambda-$contraction mapping. Also, $G$ has property $(\overset{*}{\mathscr{P}})$. So by (c) of Theorem 9.3.1, $T$ has a fixed point. □

## 9.4 WELL-POSEDNESS OF FIXED POINT PROBLEM IN $B-$METRIC SPACE

Well-posedness of fixed point-based problem on $b-$metric spaces is defined and proven.

**Definition 9.4.1.** *Let $(Y,\wp,\beta)$ be a $b-$metric space with $\beta\geq 1$. A fixed point-based problem of the mapping $T: CB^{\wp}(Y)\longrightarrow CB^{\wp}(Y)$ is called* ***well-posed*** *if $T$ has a unique fixed point $V^*\in CB^{\wp}(Y)$ and for any sequence $\{V_n\}\subseteq CB^{\wp}(Y)$ such that $\lim\limits_{n\to\infty}\mathscr{H}_{\wp}(T(V_n),V_n)=0$ implies that $\mathscr{H}_{\wp}(V_n,V^*)=0$.*

**Theorem 9.4.2.** *Let $(Y,\wp,\beta)$ be a complete $b-$metric space with $\beta\geq 1$ and $T: CB^{\wp}(Y)\longrightarrow CB^{\wp}(Y)$ be an operator as defined in Theorem 9.3.2. Then the fixed point-based problem of $T$ is well-posed.*

**Proof:** From Theorem 9.3.2, it follows that $T(V^*)=V^*$. Let $V_n$ be a sequence such that $\lim\limits_{n\to\infty}\mathscr{H}_{\wp}(T(V_n),V_n)=0$. Since $T$ is a generalized graph $\lambda-$contraction and by property 9.1.1, we have:

$$\begin{aligned}\mathscr{H}_{\wp}(V_n,V^*)&\leq\beta\mathscr{H}_{\wp}(V_n,T(V_n))+\beta\mathscr{H}_{\wp}(T(V_n),V^*)\\&=\beta\mathscr{H}_{\wp}(T(V_n),T(V^*))+\beta\mathscr{H}_{\wp}(V_n,T(V_n)),\text{ since }T(V^*)=V^*\\&\leq\beta\lambda M_T(V_n,V^*)+\beta\mathscr{H}_{\wp}(V_n,T(V_n)),\end{aligned}$$

where

$$M_T(V^*,V_n) = \max\Big\{ \mathscr{H}_{\wp}(V^*,V_n), \mathscr{H}_{\wp}(V^*,T(V_n)), \mathscr{H}_{\wp}(V_n,T(V_n)), \mathscr{H}_{\wp}(T(V^*),T^2(V^*)), \mathscr{H}_{\wp}(T^2(V^*),V_n), \mathscr{H}_{\wp}(T^2(V^*),T(V_n)), \frac{\mathscr{H}_{\wp}(V_n,T(V^*)) + \mathscr{H}_{\wp}(V^*,T(V_n))}{2} \Big\}.$$

Now, we consider the following cases:

(**Case** 1): If $M_T(V^*,V_n) = \mathscr{H}_{\wp}(V^*,V_n)$, then

$$\mathscr{H}_{\wp}(V_n,V^*) \leq \beta\lambda\mathscr{H}_{\wp}(V_n,V^*) + \beta\mathscr{H}_{\wp}(V_n,T(V_n)),$$

implies

$$\lim_{n\to\infty}\mathscr{H}_{\wp}(V_n,V^*) \leq \frac{\beta}{1-\beta\lambda}\lim_{n\to\infty}\mathscr{H}_{\wp}(V_n,T(V_n)),$$

since $\beta\lambda < 1$, clearly we have: $\lim_{n\to\infty}\mathscr{H}_{\wp}(V_n,V^*) = 0$.

(**Case** 2): If $M_T(V^*,V_n) = \mathscr{H}_{\wp}(V^*,T(V_n))$, then

$$\begin{aligned}
\mathscr{H}_{\wp}(V_n,V^*) &\leq \beta\lambda\mathscr{H}_{\wp}(V^*,T(V_n)) + \beta\mathscr{H}_{\wp}(V_n,T(V_n)),\\
&\leq \beta\lambda\Big[\beta\mathscr{H}_{\wp}(V^*,V_n) + \beta\mathscr{H}_{\wp}(V_n,T(V_n))\Big] + \beta\mathscr{H}_{\wp}(V_n,T(V_n)),\\
&\leq \beta^2\lambda\mathscr{H}_{\wp}(V^*,V_n) + \beta^2\lambda\mathscr{H}_{\wp}(V_n,T(V_n)) + \beta\mathscr{H}_{\wp}(V_n,T(V_n)),\\
&\leq \frac{\beta^2\lambda}{1-\beta^2\lambda}\mathscr{H}_{\wp}(V_n,T(V_n)) + \frac{\beta}{1-\beta^2\lambda}\mathscr{H}_{\wp}(V_n,T(V_n)),
\end{aligned}$$

taking limits as $n \longrightarrow \infty$ and by a similar argument to *Case* 1, we have our required result: $\lim_{n\to\infty}\mathscr{H}_{\wp}(V_n,V^*) = 0$.

(**Case** 3): If $M_T(V^*,V_n) = \mathscr{H}_{\wp}(V_n,T(V_n))$, then

$$\mathscr{H}_{\wp}(V_n,V^*) \leq \beta\lambda\mathscr{H}_{\wp}(V_n,T(V_n)) + \beta\mathscr{H}_{\wp}(V_n,T(V_n))$$

$$\lim_{n\to\infty}\mathscr{H}_{\wp}(V_n,V^*) \leq \beta\lambda\lim_{n\to\infty}\mathscr{H}_{\wp}(V_n,T(V_n)) + \beta\lim_{n\to\infty}\mathscr{H}_{\wp}(V_n,T(V_n)) \longrightarrow 0$$

hence, $\lim_{n\to\infty}\mathscr{H}_{\wp}(V_n,V^*) = 0$.

(**Case** 4): Since $T(V^*) = V^*$. Then $M_T(V^*,V_n) = \mathscr{H}_{\wp}(T(V^*),T^2(V^*)) = 0$. It follows:

$$\lim_{n\to\infty}\mathscr{H}_{\wp}(V_n,V^*) \leq \beta\lim_{n\to\infty}\mathscr{H}_{\wp}(V_n,T(V_n)) \longrightarrow 0$$

we obtain, $\lim_{n\to\infty}H_{\wp}(V_n,V^*) = 0$.

(**Case** 5) and (**Case** 6): If $M_T(V^*,V_n) = \mathscr{H}_\wp(T^2(V^*),V_n)$ or If $M_T(V^*,V_n) = \mathscr{H}_\wp(T^2(V^*),T(V_n))$, then the result follows similarly as in *Case* 1 and *Case* 2, respectively.

(**Case** 7): If $M_T(V^*,V_n) = \mathscr{H}_\wp(T^2(V^*),V_n)$, then note that:

$$\begin{aligned}\frac{\mathscr{H}_\wp(V_n,T(V^*))+\mathscr{H}_\wp(V^*,T(V_n))}{2} &= \frac{\mathscr{H}_\wp(V_n,V^*)+\mathscr{H}_\wp(V^*,T(V_n))}{2}\\ &\leq \mathscr{H}_\wp(V_n,V^*)+\mathscr{H}_\wp(V^*,T(V_n))\end{aligned}$$

and by a combination of arguments in *Case* 1 and *Case* 2, we obtain: $\lim_{n\to\infty} H_\wp(V_n,V^*) = 0$.

Thus, in all the above cases, we attain: $\lim_{n\to\infty} \mathscr{H}_\wp(V_n,V^*) = 0$, that is, $\lim_{n\to\infty} V_n = V^*$. □

**Theorem 9.4.3.** *Let $(Y,\wp,\beta)$ be a $b-$complete metric space with $\beta \geq 1$ and $S : CB^\wp(Y) \longrightarrow CB^\wp(Y)$ be an operator as defined in Theorem 9.3.1, that is, $S$ is generalized rational contraction. Then the fixed point-based problem of $S$ is well-posed.*

**Proof:** From Theorem 9.3.1, it follows that $S(V^*) = V^*$. Let $V_n$ be a sequence such that $\lim_{n\to\infty} \mathscr{H}_\wp(S(V_n),V_n) = 0$. Since $S$ is a generalized rational graph $\lambda-$contraction and by property 9.1.1, we have

$$\begin{aligned}\mathscr{H}_\wp(V_n,V^*) &\leq \beta\mathscr{H}_\wp(V_n,S(V_n))+\beta\mathscr{H}_\wp(S(V_n),V^*)\\ &= \beta\mathscr{H}_\wp(S(V_n),S(V^*))+\beta\mathscr{H}_\wp(V_n,S(V_n)), \text{ since } S(V^*) = V^*\\ &\leq \beta\lambda N_S(V_n,V^*)+\beta\mathscr{H}_\wp(V_n,S(V_n)),\end{aligned}$$

where

$$\begin{aligned}N_S(V_n,V^*) = \max\Bigg\{&\frac{\mathscr{H}_\wp(V_n,S(V^*))[1+\mathscr{H}_\wp(V_n,S(V_n))]}{2[1+\mathscr{H}_\wp(V_n,V^*)]},\\ &\frac{\mathscr{H}_\wp(V^*,S(V^*))[1+\mathscr{H}_\wp(V_n,S(V_n))]}{1+\mathscr{H}_\wp(V_n,V^*)},\\ &\frac{\mathscr{H}_\wp(V^*,S(V_n))[1+\mathscr{H}_\wp(V_n,S(V_n))]}{1+\mathscr{H}_\wp(V_n,V^*)}\Bigg\}.\end{aligned}$$

We now consider the following cases, taking each component of $N_S(V_n,V^*)$ in sequence:

(**Case** 1): If

$$N_S(V_n,V^*) = \frac{\mathscr{H}_\wp(V_n,S(V^*))[1+\mathscr{H}_\wp(V_n,S(V_n))]}{2[1+\mathscr{H}_\wp(V_n,V^*)]},$$

then

$$\begin{aligned}\mathscr{H}_{\wp}(V_n,V^*) &\leq \beta\lambda\frac{\mathscr{H}_{\wp}(V_n,S(V^*))[1+\mathscr{H}_{\wp}(V_n,S(V_n))]}{2[1+\mathscr{H}_{\wp}(V_n,V^*)]}+\beta\mathscr{H}_{\wp}(V_n,S(V_n))\\ &\leq \beta\lambda\mathscr{H}_{\wp}(V_n,V^*)[1+\mathscr{H}_{\wp}(V_n,S(V_n))]+\beta\mathscr{H}_{\wp}(V_n,S(V_n)),\end{aligned}$$

which implies that

$$\mathscr{H}_{\wp}(V_n,V^*)(1-\beta\lambda)[1+\mathscr{H}_{\wp}(V_n,S(V_n))]\leq\beta\mathscr{H}_{\wp}(V_n,S(V_n)),$$

$$\mathscr{H}_{\wp}(V_n,V^*)\leq\frac{\beta}{1-\beta\lambda}\frac{\mathscr{H}_{\wp}(V_n,S(V_n))}{[1+\mathscr{H}_{\wp}(V_n,S(V_n))]}$$

and it follows that: $\lim_{n\to\infty}\mathscr{H}_{\wp}(V_n,V^*)=0$, since $\beta\lambda<1$.

(**Case** 2): If

$$N_S(V_n,V^*)=\frac{\mathscr{H}_{\wp}(V^*,S(V^*))[1+\mathscr{H}_{\wp}(V_n,S(V_n))]}{1+\mathscr{H}_{\wp}(V_n,V^*)},$$

then note the following:

$$N_S(V_n,V^*)=\frac{\mathscr{H}_{\wp}(V^*,S(V^*))[1+\mathscr{H}_{\wp}(V_n,S(V_n))]}{1+\mathscr{H}_{\wp}(V_n,V^*)}=0,\text{ since }S(V^*)=V^*.$$

Clearly,

$$\mathscr{H}_{\wp}(V_n,V^*)\leq\beta\lim_{n\to\infty}\mathscr{H}_{\wp}(V_n,S(V_n))=0.$$

Then $\lim_{n\to\infty}\mathscr{H}_{\wp}(V_n,V^*)=0$.

(**Case** 3): If

$$N_S(V_n,V^*)=\frac{\mathscr{H}_{\wp}(V^*,S(V_n))[1+\mathscr{H}_{\wp}(V_n,S(V_n))]}{1+\mathscr{H}_{\wp}(V_n,V^*)},$$

then

$$\begin{aligned}&\mathscr{H}_{\wp}(V_n,V^*)\leq\beta\lambda\frac{\mathscr{H}_{\wp}(V^*,S(V_n))[1+\mathscr{H}_{\wp}(V_n,S(V_n))]}{1+\mathscr{H}_{\wp}(V_n,V^*)}+\beta\mathscr{H}_{\wp}(V_n,S(V_n))\\ &\leq[\beta^2\lambda\mathscr{H}_{\wp}(V^*,V_n)+\beta^2\lambda\mathscr{H}_{\wp}(V_n,S(V_n))]\frac{[1+\mathscr{H}_{\wp}(V_n,S(V_n))]}{1+\mathscr{H}_{\wp}(V_n,V^*)}+\beta\mathscr{H}_{\wp}(V_n,S(V_n))\\ &\leq\beta^2\lambda\mathscr{H}_{\wp}(V^*,V_n)[1+\mathscr{H}_{\wp}(V_n,S(V_n))]+\beta^2\lambda\mathscr{H}_{\wp}(V_n,S(V_n))[1+\mathscr{H}_{\wp}(V_n,S(V_n))]\\ &+\beta\mathscr{H}_{\wp}(V_n,S(V_n)),\end{aligned}$$

which implies

$$\begin{aligned}&\mathscr{H}_{\wp}(V_n,V^*)(1-\beta^2\lambda)[1+\mathscr{H}_{\wp}(V_n,S(V_n))])\\ &\leq\beta^2\lambda\mathscr{H}_{\wp}(V_n,S(V_n))[1+\mathscr{H}_{\wp}(V_n,S(V_n))]+\beta\mathscr{H}_{\wp}(V_n,S(V_n)),\end{aligned}$$

that is,

$$\begin{aligned}
&\mathcal{H}_{\wp}(V_n,V^*)\\
&\leq \frac{1}{1-\beta^2\lambda[1+\mathcal{H}_{\wp}(V_n,S(V_n))]}[\beta^2\lambda\mathcal{H}_{\wp}(V_n,S(V_n))[1+\mathcal{H}_{\wp}(V_n,S(V_n))]\\
&+\beta\mathcal{H}_{\wp}(V_n,S(V_n))]
\end{aligned}$$

and we obtain that $\lim_{n\to\infty}\mathcal{H}_{\wp}V_n,V^*)=0$.

Thus, in all the above cases, we attain: $\lim_{n\to\infty}\mathcal{H}_{\wp}(V_n,V^*)=0$, that is, $\lim_{n\to\infty}V_n=V^*$. □

## 9.5 CONCLUSIONS

In this paper, we proved the existence of fixed point for generalized graph set-valued contractions with the domain of sets in $b-$metric space endowed with the directed graph. Moreover, we were able to present an example to illustrate the main result presented therein. Further, the well-posedness of these obtained results is also established. As $b-$metric space is linked with the natural way of having applications of mathematical problems (see [15]), the results in this paper can be investigated in connection with $b-$metric and generalized $b-$metric spaces with some applications.

## BIBLIOGRAPHY


1. M. Abbas, M. R. Alfuraidan, T. Nazir, M. Rashed, Common fixed points of multivalued F-contractions on metric spaces with a directed graph, Carpathian J. Math., **32** (1), (2016), 1–12.
2. M. Abbas, M. R. Alfuraidan, A. R. Khan, T. Nazir, Fixed point results for set-contractions on metric spaces with a directed graph, Fixed Point Theory Appl., **14**, (2015), 1–9.
3. M. Abbas, V. Parvaneh, A. Razani, Periodic points of $T$-Ciric generalized contraction mappings in ordered metric spaces, Georgian Math. J., **19** (4), (2012), 597-610.
4. S. M. A. Aleomraninejad, Sh. Rezapour, N. Shahzad, Some fixed point results on a metric space with a graph, Topology Appl., **159** (3), (2012), 659–663.
5. S. Banach, Sur les opérations dans les ensembles abstraits et leur application aux équations inté grales, Fund. Math., **3** (1), (1922), 133–181.
6. F. Bojor, Fixed point of $\varphi$-contraction in metric spaces endowed with a graph, An. Univ. Craiova Ser. Mat. Inform., **37** (4), (2010), 85–92.
7. C. Chifu, G. Petruşel, Generalized contractions in metric spaces endowed with a graph, Fixed Point Theory Appl., **1**, (2012), 1–9.
8. C. Chifu, G. Petruşel, fixed points for multi-valued contractions in b-metric spaces with applications to fractals, Taiwanese J. Math., **18** (5), (2014), 1365–1375.
9. L. Ciric, M. Abbas, M. Rajovic, B. Ali, Suzuki type fixed point theorems for generalized multi-valued mappings on a set endowed with two b-metrics, Applied Math. Computation, **219** (4), (2012), 1712–1723.
10. S. Czerwik, K. Dlutek, S. Singh, Round-off stability of iteration procedures for operators in b-metric spaces, J. Natural physical sciences, **11**, (1997).
11. S. Czerwik, Nonlinear set-valued contraction mappings in b-metric spaces, Atti Sem. Mat. Fis. Univ. Modena, **46**, (1998), 263–276.

12. S. Czerwik, Contraction Mappings in b-Metric Spaces, itActa Mathematica et Informatica Universitatis Ostraviensis, 1 (1), 5–11 (1993).
13. J. Jachymski, The contraction principle for mappings on a metric space with a graph, Proc. Amer. Math. Soc., **136** (4), (2008), 1359–1373.
14. J. Jachymski, I. Jóźwik, Nonlinear contractive conditions: a comparison and related problems, Banach Center Publ., **1** (77), (2007), 123–146.
15. T. Kamran, M. Samreen, Q. U. Ain, A generalization of b-metric space and some fixed point theorems, Mathematics., **5** (2), (2017), 1–7.
16. A. Latif, T. Nazir, M. A. Kutbi, Common fixed point results for class of set-contraction mappings endowed with a directed graph, RACSAM Rev. R. Acad. Cienc. Exactas Fis. Nat. Ser. A Mat., **113** (4), (2019), 3207–3222.
17. T. Nazir, S. Silvestrov, X. Qi, Fractals of generalized F-Hutchinson operator in b-metric spaces, J. Operator, **2016**, (2016), 1–9.
18. A. Latif, V. Parvaneh, P. Salimi, A. E. Al-Mazrooei, Various Suzuki type theorems in *b*-metric spaces, J. Nonlinear Sci. Appl., **8** (4), (2015), 363–77.
19. Z. Mustafa, J.R. Roshan, V. Parvaneh, Z. Kadelburg, Fixed point theorems for weakly $T$-Chatterjea and weakly $T$-Kannan contractions in *b*-metric spaces, J. Inequal. Appl., **1**, (2014), 1–4.
20. A. Nicolae, D. O'Regan, A. Petruşel, Fixed point theorems for singlevalued and multivalued generalized contractions in metric spaces endowed with a graph, J. Georgian Math. Soc., **18**, (2011), 307–327.
21. J. J. Nieto, R. Rodriguez-López, Contractive mapping theorems in partially ordered sets and applications to ordinary differential equations, Order, **22** (3), (2005), 223–239.
22. A. M. C. Ran, M. C. Reuring, A fixed point theorem in partially ordered sets and some applications to matrix equations, Proc. Am. Math. Soc., **132**, (2004), 1435–1443.
23. J. R. Roshan, N. Shobkolaei, S. Sedghi, V. Parvaneh, S. Radenovic, Common fixed point theorems for three maps in discontinuous $G_b$-metric spaces, Acta Mathematica Scientia, **34** (5), (2014), 1643–1654.
24. J. R. Roshan, N. Hussain, S. Sedghi, N. Shobkolaei, Suzuki type fixed point results in b-metric spaces, Math. Sci., **9** (3), (2015), 153–160.
25. T. Van An, L. Q. Tuyen, and N. Van Dung, Stone-type theorem on b-metric spaces and applications, Top. Application, **185**, (2015), 50–64.

# 10 Common Fixed Point Results via Simulation Functions and Solution of an Anti-Periodic Ordinary Differential Equation in Bipolar Metric Spaces

*Pankaj Kumar, Manoj Kumar*

In this chapter, we define $\alpha$-admissibility of one covariant map with respect to another covariant map and using this prove some common fixed point theorems. We also prove common fixed point theorems for contravariant, one covariant and another contravariant maps. Then, some consequences of our main results are mentioned. In the end, we provide an example to support our result and as an application discuss the solution of an anti-periodic ordinary differential equation.

## 10.1 INTRODUCTION

The groundbreaking research of Banach [4], which was done a century ago, is the place where the concept of fixed point got its start.

In 2015, Abbas *et al.* [1] proved coincident point results for almost $F$-contraction. Then, in 2015 and 2017, they proved fixed point results for generalized Suzuki type contractions and discussed the Ulam-Hyers stability using the proved results in ([2], [3]).

Since Banach's first study, researchers have expanded, enhanced, and generalized this incredibly straightforward but extremely powerful theorem and concept of metric space see ([5]–[17], [19], [21]–[22]). In this respect, the notion of simulation function was given by Khojasteh *et al.* [18] in 2015 as follows:

DOI: 10.1201/9781003596578-10

**Definition 10.1.1.** *[18] A simulation functions is a mapping $\xi : [0,\infty) \times [0,\infty) \to \mathbb{R}$ satisfying the following conditions:*

1. $\xi(0,0) = 0$;
2. $\xi(t,s) < s - t$ *for all* $t,s > 0$;
3. *If* $\{t_n\}, \{s_n\}$ *are sequences in* $(0,\infty)$ *such that* $\lim_{n\to\infty} t_n = \lim_{n\to\infty} s_n > 0$ *then* $\limsup_{n\to\infty} \xi(t_n, s_n) < 0$.

Family of simulation functions will be denoted by $\mathscr{Z}$.
The followings are typical examples of simulation function:

1. $\xi_1(t,s) = \psi(s) - \phi(t)$ for all $t,s \in [0,\infty)$, where $\phi, \psi : [0,\infty) \to [0,\infty)$ are two continuous functions such that $\psi(t) = \phi(t) = 0$ iff $t = 0$ and $\psi(t) < t \leq \phi(t)$ for all $t > 0$.
2. $\xi_2(t,s) = s - \frac{f(t,s)}{g(t,s)}t$ for all $t,s \in [0,\infty)$, where $f,g : [0,\infty) \to [0,\infty)$ are two continuous functions with respect to each variable such that $f(t,s) > g(t,s)$ for all $t,s > 0$.
3. $\xi_3(t,s) = s - \phi(s) - t$ for all $t,s \in [0,\infty)$, where $\phi : [0,\infty) \to [0,\infty)$ is continuous function such that $\psi(t) = \phi(t) = 0$.
4. $\xi_4(t,s) = \frac{s}{s+1} - t$ for all $t,s \in [0,\infty)$.
5. $\xi_5(t,s) = \lambda.s - t$ for all $t,s \in [0,\infty), \lambda \in [0,1)$.
6. $\xi_6(t,s) = s\phi(s) - t$ for all $t,s \in [0,\infty)$, where $\phi : [0,\infty) \to [0,1)$ is a mapping such that $\limsup_{t\to r^+} \phi(t) < 1$ for all $r > 0$.
7. $\xi_7(t,s) = s\eta(s) - t$ for all $t,s \in [0,\infty)$, where $\eta : [0,\infty) \to [0,1)$ is an upper semi continuous mapping such that $\eta(s) < s$ for all $s > 0$ and $\eta(0) = 0$.

The above examples are mentioned in [18]. Many authors proved fixed point results with the help of simulation functions in various metric spaces see ([12], [28]).

In 2016, Karapinar defined the $\alpha$-admissible $\mathscr{Z}$ contraction as explained below:

**Definition 10.1.2.** *[12] Let $T$ be a self map on a metric space $(X,d)$. If there exists $\xi \in \mathscr{Z}$ and $\alpha : X \times X \to [0,\infty)$ such that*

$$\xi(\alpha(x,y)d(Tx,Ty), d(x,y)) \geq 0$$

*for all $x,y \in X$, then $T$ is said to be $\alpha$-admissible $\mathscr{Z}$ contraction.*

Karpinar also proved fixed point theorems with the help of triangular $\alpha$-orbital admissible mappings for the above defined mappings.
In 2016, Mutlu and Gurdal [23] introduced the new notion of bipolar metric space as described below:

**Definition 10.1.3.** *[23] A bipolar metric $d : X \times Y \to [0,\infty)$ is a function such that*

1. $d(x,y) = 0 \Longleftrightarrow x = y$;

2. $d(x,y) = d(y,x) \Longleftrightarrow x,y \in X \cap Y$;

3. $d(x,y') \leq d(x,y) + d(x',y) + d(x',y')$

*for all $x,x' \in X, y,y' \in Y$, where $X,Y$ are two non-void sets.*

The ordered triplet $(X,Y,d)$ is called bipolar metric space. The set $X$ is termed as left pole, and $Y$ is termed as right pole. The elements of left pole are called left elements, and elements of right pole are called right elements. The elements of $X \cap Y$ are called central elements.

**Definition 10.1.4.** *[23] Let $(X_1,Y_1,d_1)$ and $(X_2,Y_2,d_2)$ are two bipolar metric spaces. A map $S : X_1 \cup Y_1 \to X_2 \cup Y_2$ is termed as covariant if $SX_1 \subset SX_2$ and $SY_1 \subset SY_2$ and denoted by $S : (X_1 \cup Y_1) \rightrightarrows (X_2 \cup Y_2)$. Similarly, $S$ is termed as contravariant if $SX_1 \subset SY_2$ and $SY_1 \subset SX_2$ and denoted by $S : (X_1 \cup Y_1) \rightleftarrows (X_2 \cup Y_2)$.*

**Definition 10.1.5.** *[23] Let $(X,Y,d)$ be a bipolar metric space.*

1. *Any sequence $\{\beta_n\} \subseteq X$ is called left sequence and is said to be convergent to right element say $\gamma$ if $d(\beta_n,\gamma) \to 0$ as $n \to \infty$. Similarly, a right sequence $\{\gamma_n\} \subseteq Y$ is said to be convergent to a left element say $\beta$ if $d(\beta,\gamma_n) \to 0$ as $n \to \infty$.*

2. *A sequence $\{(\beta_n,\gamma_n)\}$ on $X \times Y$ is called a bisequence on $(X,Y,d)$.*
3. *If both the sequences $\{\beta_n\}$ and $\{\gamma_n\}$ converge, then the bisequence $\{(\beta_n,\gamma_n\}$ is said to be convergent. If both the sequences $\{\beta_n\}$ and $\{\gamma_n\}$ converge to the same point $u \in X \cap Y$ then the bisequence $\{(\beta_n,\gamma_n)\}$ is called biconvergent.*
4. *A bisequence $\{(\beta_n,\gamma_n)\}$ on $(X,Y,d)$ is said to be Cauchy bisequence, if for each $\varepsilon > 0$ there exists a positive integer $N \in \mathbb{N}$ such that $d(\beta_n,\gamma_m) < \varepsilon$ for all $n,m \geq N$.*
5. *A bipolar metric space is said to be complete if every Cauchy bisequence is convergent in this space.*

**Definition 10.1.6.** *[23] Let $(X_1,Y_1,d_1)$ and $(X_2,Y_2,d_2)$ be two bipolar metric spaces.*

1. *A map $S : (X_1,Y_1,d_1) \rightrightarrows (X_2,Y_2,d_2)$ is called left continuous at a point $\beta_0 \in X$ if for every $\varepsilon > 0$ there exists a $\delta > 0$ such that $d_2(S\beta_0,S\gamma) < \varepsilon$ whenever $d_1(\beta_0,\gamma) < \delta$.*

2. *A map $S : (X_1,Y_1,d_1) \rightrightarrows (X_2,Y_2,d_2)$ is called right continuous at a point $\gamma_0 \in Y$ if for every $\varepsilon > 0$ there exists a $\delta > 0$ such that $d_2(S\beta, S\gamma_0) < \varepsilon$ whenever $d_1(\beta,\gamma_0) < \delta$.*
3. *A covariant map $S : (X_1,Y_1,d_1) \rightrightarrows (X_2,Y_2,d_2)$ is called continuous if it is left continuous at each $\beta_0 \in X$ and right continuous at each $\gamma_0 \in Y$.*
4. *A contravariant map $S : (X_1,Y_1,d_1) \rightleftarrows (X_2,Y_2,d_2)$ is called continuous if and only if it is continuous as a covariant map $S : (X_1,Y_1,d_1) \rightrightarrows (X_2,Y_2,d_2)$.*

**Definition 10.1.7.** *[25] Let $T : (X,Y) \rightrightarrows (X,Y)$ and $\alpha : X \times Y \to [0,\infty)$. Then $T$ is called $\alpha$-admissible if*

$$\alpha(x,y) \geq 1 \text{ implies } \alpha(Tx,Ty) \geq 1$$

*for all $(x,y) \in X \times Y$.*

**Definition 10.1.8.** *[25] Let $T : (X,Y) \rightleftarrows (X,Y)$ and $\alpha : X \times Y \to [0,\infty)$. Then $T$ is called $\alpha$-admissible if*

$$\alpha(x,y) \geq 1 \text{ implies } \alpha(Ty,Tx) \geq 1$$

*for all $(x,y) \in X \times Y$.*

Samet *et al.* [20] introduced the following concept:

**Definition 10.1.9.** *[20] Let $\Psi$ be the family of all the functions $\psi : [0,\infty) \to [0,\infty)$ satisfying the following conditions:*

1. *$\psi$ is non-decreasing;*
2. *$\sum_{n=1}^{\infty} \psi^n(t) < \infty$ for all $t > 0$, where $\psi^n$ is $n^{th}$ iteration of $\psi$.*

It is clear that for each $\psi \in \Psi$ $\psi(t) < t$ for all $t > 0$ and $\psi(0) = 0$.
In literature, these functions are known as (c)-comparison functions or Bianchini-Grandolfi gauge functions.

In 2020, using (c)-comparison functions and Definition 1.7, Gurdal *et al.* [25] defined the following $(\alpha - \psi)$ contractions and proved the fixed point theorem for these contractions in complete bipolar metric space.

**Definition 10.1.10.** *[25] Let $(X,Y,d)$ be a bipolar metric space and $T$ be covariant map on it. Suppose $\alpha : X \times Y \to [0,\infty)$ and $\psi \in \Psi$ such that*

$$\alpha(x,y)d(Tx,Ty) \leq \psi(d(x,y))$$

*for all $(x,y) \in X \times Y$. Then, $T$ is said to be $(\alpha - \psi)$ contractive covariant mapping.*

**Definition 10.1.11.** *[29] Let $(X,Y,d)$ be a bipolar metric space.*

1. *The mappings $S,T$ on bipolar metric space $(X,Y,d)$ are said to be compatible, if for arbitrary bisequences $\{(\beta_n,\gamma_n)\}$, such that $\lim_{n\to\infty} S\beta_n = \lim_{n\to\infty} T\gamma_n = v \in X \cup Y$, then $d(ST\beta_n, TS\gamma_n) \to 0$ as $n \to \infty$.*

2. *If $v = Sv = Tv$ for $v \in X \cup Y$, then $v$ is named as common fixed point of $S$ and $T$.*
3. *If $u = Sv = Tv$ for $u \in X \cup Y$, then $v$ is considered as a coincidence point of $S$ and $T$, then $u$ is termed as the point of coincidence of $S$ and $T$.*
4. *If $S,T$ commutes at all of their coincidence points; that is $STv = TSv$ for all $v \in \{v \in X \cup Y : Sv = Tv\}$, then $S$ and $T$ are called weakly compatible.*

There are number of fixed point results in the setting of bipolar metric space see ([24], [26], [27]).
Motivated from the work of Karapinar [12] and Mutlu *et al.* [27] in this manuscript we shall define $\alpha$-admissible $\mathscr{Z}$ contractions for a pair of covariant maps and proved some common fixed point results for it as well as other pairs like contravariants, one covariant and other contravaraint.

The rest of the paper is organized in the following manner:
Section 2 contains newly defined contractions, common fixed point theorems and an example to support the result. In section 3, we discuss about the consequences of our results. In section 4, we find the solution of an anti-periodic boundary value problem.

## 10.2 MAIN RESULTS

In this section, we prove our fixed point results.

**Definition 10.2.1.** *Let $(X,Y,d)$ be a bipolar metric space and $S,T : (X,Y) \rightrightarrows (X,Y)$ are two covariant maps. Suppose $\alpha : X \times Y \to [0,\infty)$ be a map such that*

$$\alpha(Tx,Ty) \geq 1 \Rightarrow \alpha(Sx,Sy) \geq 1, \tag{10.1}$$

*for all $(x,y) \in X \times Y$. Then, $S$ is said to be covariant $T-\alpha-$admissible mapping.*

**Example 10.2.2.** *Suppose $X = [0,2], Y = [1,4]$ and $d : X \times Y \to [0,\infty)$ such that $d(x,y) = |x-y|$. Then, $(X,Y,d)$ is a bipolar metric space. Define*

$$Tx = \begin{cases} 2x & \text{if } x \in [0,1) \\ x & \text{if } x \in [1,2] \\ \frac{x+1}{2} & \text{if } x \in (2,4] \end{cases} \tag{10.2}$$

*and*

$$Sx = \begin{cases} \frac{3x}{2} & \text{if } x \in [0,1) \\ x & \text{if } x \in [1,2] \\ \frac{2x-1}{3} & \text{if } x \in (2,4]. \end{cases} \tag{10.3}$$

*Taking $\alpha(x,y) = x+y$. Now, we have to evaluate the value of $x,y$ for which $\alpha(Tx,Ty) \geq 1$.*

*Case 1.*
*When $x \in [0,1), y \in [1,2]$*

$$\alpha(Tx,Ty) = Tx + Ty = 2x + y \geq 1,$$

*which is always true for all $x \in [0,1), y \in [1,2]$.*

*Case 2.*
*When* $x \in [0,1), y \in (2,4]$

$$\begin{aligned} \alpha(Tx,Ty) = Tx + Ty = 2x + \frac{y+1}{2} &\geq 1, \\ 2x + \frac{y}{2} &\geq \frac{1}{2}. \end{aligned}$$

*Again, which is true for all* $x \in [0,1), y \in (2,4]$.

*Case 3.*
*When* $x \in [1,2], y \in [1,2]$

$$\alpha(Tx,Ty) = Tx + Ty = x + y \geq 1,$$

*which is always true for all* $x \in [1,2], y \in [1,2]$.

*Case 4.*
*When* $x \in [1,2], y \in (2,4]$

$$\begin{aligned} \alpha(Tx,Ty) = Tx + Ty = x + \frac{y+1}{2} &\geq 1, \\ x + \frac{y}{2} &\geq \frac{1}{2}. \end{aligned}$$

*Again, which is true for all* $x \in [1,2], y \in (2,4]$.
*From the above discussion of cases, we conclude that* $\alpha(Tx,Ty) \geq 1$ *for all* $x \in [0,2], y \in [1,4]$.
*Now, we shall prove that* $\alpha(Sx,Sy) \geq 1$.

*Case 1.*
*When* $x \in [0,1), y \in [1,2]$

$$\alpha(Sx,Sy) = Sx + Sy = \frac{3x}{2} + y \geq 1,$$

*which is always true for all* $x \in [0,1), y \in [1,2]$.

*Case 2.*
*When* $x \in [0,1), y \in (2,4]$

$$\begin{aligned} \alpha(Sx,Sy) = Sx + Sy = \frac{3x}{2} + \frac{2y-1}{3} &\geq 1, \\ \frac{3x}{2} + \frac{2y}{3} &\geq \frac{4}{3}. \end{aligned}$$

*Again, which is true for all* $x \in [0,1), y \in (2,4]$.

*Case 3.*
*When* $x \in [1,2], y \in [1,2]$

$$\alpha(Sx, Sy) = Sx + Sy = x + y \geq 1,$$

*which is always true for all* $x \in [1,2], y \in [1,2]$.

*Case 4.*
*When* $x \in [1,2], y \in (2,4]$

$$\begin{aligned} \alpha(Sx, Sy) = Sx + Sy = x + \frac{2y-1}{3} &\geq 1, \\ x + \frac{2y}{3} &\geq \frac{4}{3}. \end{aligned}$$

*Again, which is true for all* $x \in [1,2], y \in (2,4]$.
*From the above discussion of cases, we conclude that* $\alpha(Sx, Sy) \geq 1$ *for all* $x \in [0,2], y \in [1,4]$.
*So, S is covariant* $T - \alpha-$*admissible mapping.*

**Example 10.2.3.** *In the above example if we take* $\alpha(x,y) = e^{x-y}$, *then for* $x = \frac{1}{2}$ *and* $y = 1$, *we have*
$\alpha(T\frac{1}{2}, T1) = e^{1-1} = 1 \geq 1$, *but* $\alpha(S\frac{1}{2}, S1) = e^{\frac{3}{4}-1} = e^{\frac{-1}{4}} < 1$.
*So, S is not covariant* $T - \alpha-$*admissible mapping.*

**Theorem 10.2.4.** *Let* $(X,Y,d)$ *be a complete bipolar metric space and* $S,T$ *are two commutive continuous covariant maps on* $(X,Y,d)$. *Suppose* $\alpha : X \times Y \to [0,\infty)$ *such that*

$$\xi(\alpha(Tx,Ty)d(Sx,Sy), d(Tx,Ty)) \geq 0, \tag{10.4}$$

*for all* $(x,y) \in X \times Y$.
*Assume*

1. *S is covariant* $T - \alpha-$ *admissible mapping (Definition 2.1);*
2. *There exists* $(x_0, y_0) \in X \times Y$ *such that* $\alpha(Tx_0, Ty_0) \geq 1$ *and* $\alpha(Tx_0, Sy_0) \geq 1$;
3. $\alpha$ *has transitive property that is*
   $\alpha(x,y), \alpha(z,y), \alpha(z,w) \geq 1$ *implies that* $\alpha(x,z) \geq 1$.

*Then, S,T have a common fixed point.*

**Proof:** Let $(x_0, y_0) \in X \times Y$ such that $(Tx_0, Ty_0) \geq 1$ and $(Tx_0, Sy_0) \geq 1$.
Generally, choose $\{(\beta_n, \gamma_n)\}$ in such a way that
$\beta_n = Sx_n = Tx_{n+1}$ and $\gamma_n = Sy_n = Ty_{n+1}$.
Then $\{(\beta_n, \gamma_n)\}$ is a bisequence.
By conditions (1) and (2), we have

$$\begin{aligned} \alpha(Tx_0, Ty_0) \geq 1 \quad &\Rightarrow \quad \alpha(Sx_0, Sy_0) = \alpha(Tx_1, Ty_1) \geq 1; \\ \alpha(Tx_1, Ty_1) \geq 1 \quad &\Rightarrow \quad \alpha(Sx_1, Sy_1) = \alpha(Tx_2, Ty_2) \geq 1; \\ \alpha(Tx_2, Ty_2) \geq 1 \quad &\Rightarrow \quad \alpha(Sx_2, Sy_2) = \alpha(Tx_3, Ty_3) \geq 1. \end{aligned}$$

Continuing like this we get
$\alpha(\beta_n, \gamma_n) = \alpha(Tx_{n+1}, Ty_{n+1}) \geq 1$.
Again, by conditions (1) and (2), we have

$$\begin{aligned} \alpha(Tx_0, Ty_1) = \alpha(Tx_0, Sy_0) \geq 1 &\Rightarrow \alpha(Sx_0, Sy_1) = \alpha(Tx_1, Ty_2) \geq 1; \\ \alpha(Tx_1, Ty_2) \geq 1 &\Rightarrow \alpha(Sx_1, Sy_2) = \alpha(Tx_2, Ty_3) \geq 1; \\ \alpha(Tx_2, Ty_3) \geq 1 &\Rightarrow \alpha(Sx_2, Sy_3) = \alpha(Tx_3, Ty_4) \geq 1. \end{aligned}$$

Continuing like this we get
$\alpha(\beta_n, \gamma_{n+1}) = \alpha(Tx_{n+1}, Ty_{n+2}) \geq 1$.
From the above discussion, we conclude that

$$\alpha(\beta_n, \gamma_n) \geq 1 \text{ and } \alpha(\beta_n, \gamma_{n+1}) \geq 1, \tag{10.5}$$

for all $n \in \mathbb{N}$.
Now, from the transitive property of $\alpha$ and equation (5), it is clear that

$$\alpha(\beta_m, \gamma_n) \geq 1, \tag{10.6}$$

for all $m, n \in \mathbb{N}$.
Using (2) of Definition 1.1 and equation (5) in (4), we have

$$\begin{aligned} 0 &\leq \xi(\alpha(Tx_n, Ty_n)d(Sx_n, Sy_n), d(Tx_n, Ty_n)) \\ 0 &< d(Tx_n, Ty_n) - \alpha(Tx_n, Ty_n)d(Sx_n, Sy_n) \\ \alpha(Tx_n, Ty_n)d(Sx_n, Sy_n) &< d(Tx_n, Ty_n) \\ d(\beta_n, \gamma_n) = d(Sx_n, Sy_n) &< \alpha(Tx_n, Ty_n)d(Sx_n, Sy_n) \\ &< d(Tx_n, Ty_n) = d(\beta_{n-1}, \gamma_{n-1}). \end{aligned} \tag{10.7}$$

So, $\{d(\beta_n, \gamma_n)\}$ is a monotonically decreasing sequence of positive reals, hence convergent. Suppose $d(\beta_n, \gamma_n) \to r$ as $n \to \infty$.
From the inequality (7), we derive that

$$\lim_{n\to\infty} \alpha(Tx_n, Ty_n)d(Sx_n, Sy_n) = \lim_{n\to\infty} \alpha(\beta_{n-1}, \gamma_{n-1})d(\beta_n, \gamma_n) = r. \tag{10.8}$$

Now, setting $s_n = \alpha(\beta_{n-1}, \gamma_{n-1})d(\beta_n, \gamma_n)$ and $t_n = d(\beta_n, \gamma_n)$ in the Definition of simulation function, we obtain

$$0 \leq \limsup_{n\to\infty} \xi(\alpha(\beta_{n-1}, \gamma_{n-1})d(\beta_n, \gamma_n), d(\beta_n, \gamma_n)) < 0$$

a contradiction to $r > 0$.
So,

$$\lim_{n\to\infty} d(\beta_n, \gamma_n) = 0. \tag{10.9}$$

Similarly, using (2) of Definition 1.1 and equation (5) in (4), we have

$$\begin{aligned} 0 &\leq \xi(\alpha(Tx_n, Ty_{n+1})d(Sx_n, Sy_{n+1}), d(Tx_n, Ty_{n+})) \\ 0 &< d(Tx_n, Ty_{n+1}) - \alpha(Tx_n, Ty_{n+1})d(Sx_n, Sy_{n+1}) \end{aligned}$$

$$\begin{aligned} \alpha(Tx_n, Ty_{n+1})d(Sx_n, Sy_{n+1}) &< d(Tx_n, Ty_{n+1}) \\ d(\beta_n, \gamma_{n+1}) = d(Sx_n, Sy_{n+1}) &< \alpha(Tx_n, Ty_{n+1})d(Sx_n, Sy_{n+1}) \\ &< d(Tx_n, Ty_{n+1}) \\ &= d(\beta_{n-1}, \gamma_n). \end{aligned} \tag{10.10}$$

So, $\{d(\beta_n, \gamma_{n+1})\}$ is a monotonically decreasing sequence of positive reals, hence convergent. Suppose $d(\beta_n, \gamma_{n+1}) \to s$ as $n \to \infty$.
From the inequality (10), we derive that

$$\lim_{n\to\infty} \alpha(Tx_n, Ty_{n+1})d(Sx_n, Sy_{n+1}) = \lim_{n\to\infty} \alpha(\beta_n, \gamma_{n+1})d(\beta_n, \gamma_{n+1}) = s. \tag{10.11}$$

Now, setting $s_n = \alpha(\beta_{n-1}, \gamma_n)d(\beta_n, \gamma_{n+1})$ and $t_n = d(\beta_n, \gamma_{n+1})$ in the Definition of simulation function, we obtain

$$0 \leq \limsup_{n\to\infty} \xi(\alpha(\beta_{n-1}, \gamma_n)d(\beta_n, \gamma_{n+1}), d(\beta_n, \gamma_{n+1})) < 0$$

a contradiction to $s > 0$.
So,

$$\lim_{n\to\infty} d(\beta_n, \gamma_{n+1}) = 0. \tag{10.12}$$

Now, we shall prove that$\{(\beta_n, \gamma_n)\}$ is a Cauchy sequence. So, if possible suppose that it is not a Cauchy sequence, then there exists a smallest positive integer $n(k)$ with $n(k) > m(k) > k$ and $\varepsilon > 0$ such that

$$d(\beta_{n(k)}, \gamma_{m(k)}) \geq \varepsilon$$

and

$$d(\beta_{n(k)-1}, \gamma_{m(k)}) < \varepsilon.$$

Now, by using Definition of bipolar metric space, we have

$$\begin{aligned} \varepsilon &\leq d(\beta_{n(k)}, \gamma_{m(k)}) \\ &\leq d(\beta_{n(k)}, \gamma_{n(k)}) + d(\beta_{n(k)-1}, \gamma_{n(k)}) + d(\beta_{n(k)-1}, \gamma_{m(k)}) \end{aligned}$$

Taking $k \to \infty$, we have

$$\lim_{n\to\infty} d(\beta_{n(k)}, \gamma_{m(k)}) = \varepsilon. \tag{10.13}$$

Again

$$\begin{aligned} d(\beta_{n(k)-1}, \gamma_{m(k)-1}) &\leq d(\beta_{n(k)-}, \gamma_{m(k)}) + d(\beta_{m(k)-1}, \gamma_{m(k)}) \\ &+ d(\beta_{m(k)-1}, \gamma_{m(k)-1}) \end{aligned}$$

Taking limit $k \to \infty$, we get

$$\lim_{n\to\infty} d(\beta_{n(k)-1}, \gamma_{m(k)-1}) = \delta, \tag{10.14}$$

where $0 \leq \delta \leq \varepsilon$.

From the equation (4), we have

$$\begin{aligned} 0 &\leq \xi(\alpha(\beta_{n(k)},\gamma_{m(k)})d(\beta_{n(k)},\gamma_{m(k)}),d(\beta_{n(k)-1},\gamma_{m(k)-1})) \\ 0 &< d(\beta_{n(k)},\gamma_{m(k)}) \leq \alpha(\beta_{n(k)},\gamma_{m(k)})d(\beta_{n(k)},\gamma_{m(k)}) \leq d(\beta_{n(k)-1},\gamma_{m(k)-1})). \end{aligned}$$

In the above inequality, taking $k \to \infty$ and using equations (13-14), we conclude that $s_n = \alpha(\beta_{n(k)},\gamma_{m(k)})d(\beta_{n(k)},\gamma_{m(k)}) \to \varepsilon$ and $t_n = d(\beta_{n(k)-1},\gamma_{m(k)-1}) \to \varepsilon$.
Now using condition (3) of Definition 1.1 for the above set up, we get

$$0 \leq \limsup_{k\to\infty} \xi(\alpha(\beta_{n(k)},\gamma_{m(k)})d(\beta_{n(k)},\gamma_{m(k)}),d(\beta_{n(k)-1},\gamma_{m(k)-1})) < 0,$$

By definition of simulation function, this implies that $\varepsilon = 0$. So, a contradiction to $\varepsilon > 0$.
$\{(\beta_n,\gamma_n)\}$ is a Cauchy bisequence. Since, $(X,Y,d)$ is complete, so it is biconvergent that is there exists a $u$ such that $\lim_{n\to\infty}\beta_n = \lim_{n\to\infty}\gamma_n = u \in X \cap Y$.
So,
$\lim_{n\to\infty} Sx_n = \lim_{n\to\infty} Tx_n = u$
and
$\lim_{n\to\infty} Sy_n = \lim_{n\to\infty} Ty_n = u$.
Since $S,T$ are continuous and commutative, so $\lim_{n\to\infty} TSx_n = \lim_{n\to\infty} STx_n = Tu = Su$.
This implies that $u$ is the point of coincidence.
Now, as $S,T$ are commutive.
So, $u$ is the common fixed point of $S$ and $T$.

To prove uniqueness of common fixed point, we add the following condition (U).

**Theorem 10.2.5.** *If $u,v$ are common fixed point of $S$ and $T$, then $\alpha(u,v) \geq 1$. Then $S,T$ have unique common fixed point.*

**Proof:** Proof is trivial using equation (4).

**Theorem 10.2.6.** *Suppose $(X,Y,d)$ be a complete bipolar metric space and $S,T$ are two commutive contravariant maps such that*

$$0 \leq \xi(d(Sy,Sx),d(Ty,Tx)) \tag{10.15}$$

*for all $(x,y) \in X \times Y$.*
*Let $R(S) \subset R(T)$, (where $R$ stands for range of a map) and $T$ is continuous, then $S,T$ have a unique common fixed point.*

**Proof:** Let us choose a bisequence $\{(x_n,y_n)\}$ in such a way that $\gamma_n = Sx_n = Tx_n, \beta_n = Sy_{n-1} = Ty_n$.
Now, for $x = x_n, y = y_n$ equation (15) implies that

$$\begin{aligned} 0 &\leq \xi(d(Sy_n,Sx_n),d(Ty_n,Tx_n)) \\ &= \xi(d(\beta_{n+1},\gamma_n),d(\beta_n,\gamma_n)) \\ d(\beta_{n+1},\gamma_n) &< d(\beta_n,\gamma_n). \end{aligned} \tag{10.16}$$

Again, for $x = x_n, y = y_{n_1}$ equation (15) implies that

$$\begin{aligned} 0 &\leq \xi(d(Sy_{n-1}, Sx_n), d(Ty_{n-1}, Tx_n)) \\ &= \xi(d(\beta_n, \gamma_n), d(\beta_{n-1}, \gamma_n)) \\ d(\beta_n, \gamma_n) &< d(\beta_{n-1}, \gamma_n). \end{aligned} \tag{10.17}$$

Equations (16-17) implies that
The sequences $\{d(\beta_n, \gamma_n)\}$ and $\{d(\beta_n, \gamma_{n-1})\}$ are monotonically decreasing sequences of positive real numbers so convergent. Suppose that these converges to $r_1, r_2 (\geq 0)$, respectively. If possible assume that $r_1, r_2 > 0$.
Then, consider $s_n = d(\beta_n, \gamma_n), t_n = d(\beta_{n-1}, \gamma_{n-1})$ as both are convergent to $r_1$.
Condition (3) of Definition 1.1 implies that

$$0 \leq \limsup_{n\to\infty} \xi(d(\beta_n, \gamma_n), d(\beta_{n-1}, \gamma_{n-1})) < 0$$

holds when $r_1 = 0$.
In the similar way, we can prove $r_2 = 0$.
So,

$$\lim_{n\to\infty} d(\beta_n, \gamma_n) = 0 = \lim_{n\to\infty} d(\beta_{n+1}, \gamma_n) \tag{10.18}$$

for all $n \in \mathbb{N}$.
Now, we shall prove that$\{(\beta_n, \gamma_n)\}$ is a Cauchy sequence. So, if possible suppose that it is not a Cauchy sequence, then there exists a smallest positive integer $n(k)$ with $n(k) > m(k) > k$ and $\varepsilon > 0$ such that

$$d(\beta_{n(k)}, \gamma_{m(k)}) \geq \varepsilon$$

and

$$d(\beta_{n(k)-1}, \gamma_{m(k)}) < \varepsilon.$$

Now, by using Definition of bipolar metric space, we have

$$\begin{aligned} \varepsilon &\leq d(\beta_{n(k)}, \gamma_{m(k)}) \\ &\leq d(\beta_{n(k)}, \gamma_{n(k)}) + d(\beta_{n(k)-1}, \gamma_{n(k)}) + d(\beta_{n(k)-1}, \gamma_{m(k)}) \end{aligned}$$

Taking $k \to \infty$, we have

$$\lim_{n\to\infty} d(\beta_{n(k)}, \gamma_{m(k)}) = \varepsilon. \tag{10.19}$$

Again

$$\begin{aligned} \varepsilon &\leq d(\beta_{n(k)}, \gamma_{m(k)}) \\ &\leq d(\beta_{n(k)}, \gamma_{n(k)}) + d(\beta_{n(k)+1}, \gamma_{n(k)}) + d(\beta_{n(k)+1}, \gamma_{m(k)}) \\ d(\beta_{n(k)+1}, \gamma_{m(k)}) &\leq d(\beta_{n(k)+1}, \gamma_{n(k)}) + d(\beta_{n(k)}, \gamma_{n(k)}) + d(\beta_{n(k)}, \gamma_{m(k)}) \end{aligned}$$

$$\lim_{n\to\infty} d(\beta_{n(k)+1}, \gamma_{m(k)}) = \varepsilon. \tag{10.20}$$

Now, for $x = x_{m(k)}, y = y_{n(k)}$, equation (15) implies that

$$\begin{aligned} 0 &\leq \xi(d(Sy_{n(k)}, Sx_{m(k)}), d(Ty_{n(k)}, Tx_{m(k)})) \\ &= \xi(d(\beta_{n(k)+1}, \gamma_{m(k)}), d(\beta_{n(k)}, \gamma_{m(k)})) \\ &< d(\beta_{n(k)}, \gamma_{m(k)}) - d(\beta_{n(k)+1}, \gamma_{m(k)}) \end{aligned}$$

Taking limit as $k \to \infty$ in the above inequality, we get

$$\lim_{n \to} \xi(d(\beta_{n(k)+1}, \gamma_{m(k)}), d(\beta_{n(k)}, \gamma_{m(k)})) = 0.$$

which is a contradiction to condition (3) of Definition 1.1.
So, $\{(\beta_n, \gamma_n)\}$ is a Cauchy bisequence. Since, $(X, Y, d)$ is complete, so it is biconvergent that is there exists a $u$ such that $\lim_{n\to\infty} \beta_n = \lim_{n\to\infty} \gamma_n = u \in X \cap Y$.
So,
$\lim_{n\to\infty} Sx_n = \lim_{n\to\infty} Tx_n = u$
and
$\lim_{n\to\infty} Sy_n = \lim_{n\to\infty} Ty_n = u$.
Continuity of $T$ and commutive of $S, T$ implies that
$T(\lim_{n\to\infty} Tx_n) = \lim_{n\to\infty} T^2 x_n = Tu$
and
$\lim_{n\to\infty} TSx_n = \lim_{n\to\infty} STx_n = Tu$.
Now, for $x = Tx_n, y = u$ equation (15) implies that

$$\begin{aligned} 0 &\leq \xi(d(STx_n, Su), d(T^2x_n, Tu)) \\ d(STx_n, Su) &\leq d(T^2x_n, Tu) \end{aligned}$$

taking limit as $n \to \infty$, we have

$$d(Su, Tu) = 0.$$

implies that $Su = Tu$.
Again, for $x = Tx_n, y = u$ equation (15) implies that

$$0 \leq \xi(d(Sx_n, Su), d(Tx_n, Tu))$$

taking limit as $n \to \infty$, we have

$$d(Su, u) < d(Tu, u).$$

implies that $Su = Tu = u$.
Uniqueness is obvious.

**Theorem 10.2.7.** *Suppose $(X, Y, d)$ be a complete bipolar metric space and $S, T$ are two commutive maps such that $S$ is covariant and $T$ is contravariant with*

$$0 \leq \xi(d(Sx, Sy), d(Ty, Tx)) \tag{10.21}$$

*for all $(x, y) \in X \times Y$.*
*Let $R(S) \subset R(T)$ and $T$ is continuous, then $S, T$ have a unique common fixed point.*

**Proof:** This theorem can be proved in the same way like Theorem 10.2.6.

**Theorem 10.2.8.** *Suppose $(X,Y,d)$ be a complete bipolar metric space and $S,T$ are two commutative maps such that $S$ is contravariant and $T$ is covariant with*

$$0 \leq \xi(d(Sy,Sx),d(Tx,Ty)) \tag{10.22}$$

*for all $(x,y) \in X \times Y$.*
*Let $R(S) \subset R(T)$ and $T$ is continuous, then $S,T$ have a unique common fixed point.*

**Proof:** Proof is similar like of Theorem 10.2.6.

**Example 10.2.9.** *Consider $X = \{(x,0) : x \in [0,\infty)\}$ and $Y = \{(0,y) : y \in [0,\infty)\}$.*
*Define $d((x,0),(0,y)) = |x-y|^2$.*
*Clearly, $(X,Y,d)$ is a complete bipolar metric space.*
*Let $S,T : X \cup Y \to X \cup Y$ as*

$$S(x,y) = \begin{cases} (0,\frac{3x}{2}) & \text{if } (x,y) \in X \\ (\frac{7y}{2},0) & \text{if } (x,y) \in Y. \end{cases}$$

$$T(x,y) = \begin{cases} (0,x) & \text{if } (x,y) \in X \\ (\frac{7y}{3},0) & \text{if } (x,y) \in Y. \end{cases}$$

*Clearly, $S$ and $T$ are contravarinat maps.*
*It is easy to check that $R(S) \subset R(T)$.*
*Now*

$$\begin{aligned} ST(x,0) &= S(0,x) = (\frac{7x}{2},0) \\ TS(x,0) &= T(0,\frac{3x}{2}) = (\frac{7x}{2},0). \end{aligned}$$

*Similarly, one can check for $(0,y)$.*
*So, $S,T$ are commuting maps.*

$$\begin{aligned} d(S(0,y),S(x,0)) &= \frac{1}{4}(7y-3x)^2 \\ &< \frac{4}{9}(7y-3x)^2 \\ &= \lambda.d(T(0,y),T(x,0)). \end{aligned}$$

*where $\lambda = \frac{4}{9}$. which is a consequence of Theorem 2.9, by taking $\xi(t,s) = \lambda.s - t$.*
*So, $S,T$ have a unique common fixed point.*

Clearly $(0,0)$ is the unique common fixed point.

## 10.3 CONSEQUENCES

**Corollary 10.3.1.** *Let $(X,Y,d)$ be a complete bipolar metric space and $S,T$ are two commutive continuous covariant maps on $(X,Y,d)$. Suppose*

$$\xi(d(Sx,Sy),d(Tx,Ty)) \geq 0, \tag{10.23}$$

*for all* $(x,y) \in X \times Y$.
*Then* $S,T$ *have a unique common fixed point.*

**Proof:** Proof is very standard.

**Corollary 10.3.2.** *(Theorem 3.3 of [25]) Let* $(X,Y,d)$ *be a complete bipolar metric space and S is covariant maps on* $(X,Y,d)$. *Suppose* $\alpha : X \times Y \to [0,\infty)$ *such that*

$$\alpha(x,y)d(Sx,Sy) \leq \psi(d(x,y)), \tag{10.24}$$

*for all* $(x,y) \in X \times Y$.
*Assume*

1. *S is* $\alpha-$ *admissible;*
2. *There exists* $(x_0,y_0) \in X \times Y$ *such that* $\alpha(x_0,y_0) \geq 1$ *and* $\alpha(x_0,Sy_0) \geq 1$;
3. *S is continuous.*

*Then S has a unique fixed point.*

Hence, all the results of ([23]–[26], [27]) can be easily derived from our results. Now, we shall derive Khojasteh *et al.* type results in bipolar metric space from our results:

**Corollary 10.3.3.** *Let* $(X,Y,d)$ *be a complete bipolar metric space and S is covariant maps on* $(X,Y,d)$ *such that*

$$0 \leq \xi(d(Sx,Sy),d(x,y)). \tag{10.25}$$

*Then, S has a unique fixed point.*

**Corollary 10.3.4.** *Let* $(X,Y,d)$ *be a complete bipolar metric space and S is contravariant maps on* $(X,Y,d)$ *such that*

$$0 \leq \xi(d(Sy,Sx),d(x,y)). \tag{10.26}$$

*Then, S has a unique fixed point.*

## 10.4 SOLUTION FOR NONLINEAR FIRST-ORDER CARATHEODORY'S TYPE ORDINARY DIFFERENTIAL EQUATIONS

Consider the following anti-periodic problem for non linear first order Caratheodory's type ordinary differential equations:

$$\mathfrak{Q}'(\mathrm{t}) = \mathfrak{P}(\mathrm{t},\mathfrak{Q}(\mathrm{t})), \ a.e. \ \mathrm{t} \in [0,\mathfrak{L}] \tag{10.27}$$

$\mathfrak{Q}(0) = -\mathfrak{Q}(\mathfrak{L})$.
Here $\mathfrak{P} : [0,\mathfrak{L}] \times \mathbb{R} \to \mathbb{R}$ is a Carathedory type function which satisfies the following conditions:

1. For every $\mathfrak{Q} \in \mathbb{R}, \mathfrak{P}(.,\mathfrak{Q})$ is Lebesgue measurable on $[0,\mathfrak{L}]$ and there exists a point $\mathfrak{w}_0 \in \mathbb{R}$ such that $\int_0^{\mathfrak{L}} \phi(\frac{|\mathfrak{P}(\mathfrak{t},\mathfrak{w}_0)|}{\lambda})dt < \infty$ for some $\lambda = \lambda(\mathfrak{P}(.,\mathfrak{w}_0)) > 0$,
2. for a.e. $\mathfrak{t} \in [0,\mathfrak{L}]$ and all $\mathfrak{Q},\mathfrak{w} \in \mathbb{R}$, there exists a constant $\mathfrak{K} > 0$ such that $\mathfrak{P}(\mathfrak{t},\mathfrak{Q}) - \mathfrak{P}(\mathfrak{t},\mathfrak{w}) \leq \mathfrak{K}|\mathfrak{Q} - \mathfrak{w}|$.
   where $\phi$ is a function satisfying Orlicz condition at infinity and $\lambda > 0$.

for $\mu \in \mathbb{R}$ equation (27) can be rewritten as

$$\mathfrak{Q}'(\mathfrak{t}) + \mu\mathfrak{Q}(\mathfrak{t}) = \mathfrak{P}(\mathfrak{t},\mathfrak{Q}(\mathfrak{t})) + \mu\mathfrak{Q}(\mathfrak{t}), \; a.e. \; \mathfrak{t} \in [0,\mathfrak{L}] \tag{10.28}$$

$\mathfrak{Q}(0) = -\mathfrak{Q}(\mathfrak{L})$,
which is equivalent to the intgral equation

$$\mathfrak{Q}(\mathfrak{t}) = \int_0^L G(\mathfrak{t},\mathfrak{s})[\mathfrak{P}(\mathfrak{s},\mathfrak{Q}(\mathfrak{s})) + \mu\mathfrak{Q}(\mathfrak{s})]ds, \tag{10.29}$$

where $\mathfrak{G}(\mathfrak{t},\mathfrak{s})$ is the Green function defined by

$$\mathfrak{G}(\mathfrak{t},\mathfrak{s}) = \begin{cases} \frac{e^{\mu(\mathfrak{L}+\mathfrak{s}-\mathfrak{t})}}{e^{\mu\mathfrak{L}}+1} & \text{if } 0 \leq \mathfrak{s} \leq \mathfrak{t} \leq \mathfrak{L} \\ \frac{-e^{\mu(\mathfrak{s}-\mathfrak{t})}}{e^{\mu\mathfrak{L}}+1} & \text{if } 0 \leq \mathfrak{t} \leq \mathfrak{s} \leq \mathfrak{L}. \end{cases} \tag{10.30}$$

A function $\mathfrak{Q} : [0,\mathfrak{L}] \to \mathbb{R}$ is a solution of (27) if it satisfies (27) or equivalently (29). Now consider $X = (\mathfrak{C}[0,\mathfrak{L}],[0,\infty)), Y = (\mathfrak{C}[0,\mathfrak{L}],(-\infty,0]$ and $d(\mathfrak{x}(\mathfrak{t}),\mathfrak{y}(\mathfrak{t})) = \sup_{\mathfrak{t}\in[0,\mathfrak{L}]} |\mathfrak{x}(\mathfrak{t}) - \mathfrak{y}(\mathfrak{t})|^2$.
Clearly, $(X,Y,d)$ is a complete bipolar metric space.
Consider a covariant map
$F\mathfrak{Q}(\mathfrak{t}) = \int_0^{\mathfrak{L}} \mathfrak{G}(\mathfrak{t},\mathfrak{s})[\mathfrak{P}(\mathfrak{s},\mathfrak{Q}(\mathfrak{s})) + \mathfrak{Q}(\mathfrak{s})]ds$.
It is clear that if $\mathfrak{Q}$ is the fixed point of $F$ then $\mathfrak{Q}$ is the solution of (27).

**Theorem 10.4.1.** *Suppose that* $\int_0^{\mathfrak{L}} \mathfrak{G}(\mathfrak{t},\mathfrak{s})^2 d\mathfrak{s} \leq \frac{1}{(\mathfrak{K}+2)\mathfrak{L}}$ *for all* $\mathfrak{s},\mathfrak{t} \in [0,\mathfrak{L}]$. *Then equation (27) or (29) has a solution.*

**Proof:** In the above setting consider

$$\begin{aligned} & |F\mathfrak{Q}(\mathfrak{t}) - F\mathfrak{w}(\mathfrak{t})|^2 \\ = & \; |\int_0^{\mathfrak{L}} \mathfrak{G}(\mathfrak{t},\mathfrak{s})[\mathfrak{P}(\mathfrak{s},\mathfrak{Q}(\mathfrak{s})) + \mathfrak{Q}(\mathfrak{s})]d\mathfrak{s} - \int_0^{\mathfrak{L}} \mathfrak{G}(\mathfrak{t},\mathfrak{s})[\mathfrak{P}(\mathfrak{s},\mathfrak{Q}(\mathfrak{s})) + \mathfrak{w}(\mathfrak{s})]d\mathfrak{s}|^2 \\ = & \; |\int_0^{\mathfrak{L}} \mathfrak{G}(\mathfrak{t},\mathfrak{s})[\mathfrak{P}(\mathfrak{s},\mathfrak{Q}(\mathfrak{s})) - \mathfrak{P}(\mathfrak{s},\mathfrak{w}(\mathfrak{s})) + (\mathfrak{Q}(\mathfrak{s}) - \mathfrak{w}(\mathfrak{s}))]d\mathfrak{s}|^2 \\ \leq & \; \int_0^{\mathfrak{L}} \mathfrak{G}(\mathfrak{t},\mathfrak{s})^2 d\mathfrak{s}\mathfrak{L}^2(\mathfrak{K}+1)^2|\mathfrak{Q}(\mathfrak{s}) - \mathfrak{w}(\mathfrak{s})|^2 \\ \leq & \; \frac{(\mathfrak{K}+1)^2}{(\mathfrak{K}+2)^2}|\mathfrak{Q}(\mathfrak{t}) - \mathfrak{w}(\mathfrak{t})|^2 \\ < & \; \lambda|\mathfrak{Q}(\mathfrak{t}) - \mathfrak{w}(\mathfrak{t})|^2. \end{aligned}$$

where $\lambda = \frac{(\mathfrak{K}+1)^2}{(\mathfrak{K}+2)^2} < 1$.

Taking supremum on both sides, we get
$d(F(\mathfrak{Q}(\mathfrak{t}),F(\mathfrak{w}(\mathfrak{t}))) < \lambda d((\mathfrak{Q}(\mathfrak{t}),\mathfrak{w}(\mathfrak{t})))$.
Taking $\xi = \lambda.\mathfrak{s} - \mathfrak{t}$ in Corollary 3.4, we find that $F$ satisfying the above inequality has a fixed point.
So, equation (27) or (29) has a solution.

## 10.5 CONCLUSION

In this chapter, we have introduced a new notion of $\alpha$-admissibility for a pair of mappings and proved fixed point theorems for such mappings as well as mixture of covariant and contravariant mappings using simulation functions. Using our result, we have solved an anti-periodic ordinary differential equation. In future, the researcher can extend our results for maximum as well as interpolative type factors to get new fixed point results.

## BIBLIOGRAPHY


1. Abbas M., Ali M. and Romaguera S., *Coincidence points of generalized multivalued $(f,L)$-almost $F$-contraction with applications*, (2015).
2. Ali B. and Abbas M., *Existence and stability of fixed point set of Suzuki-type contractive multivalued operators in b-metric spaces with applications in delay differential equations*, J. Fixed Point Theory Appl., **19** (2017), 2327-2347.
3. Ali B. and Abbas M., *Existence and Ulam–Hyers stability of fixed point problem of generalized Suzuki type $(\alpha^*, \psi_\phi)$-contractive multivalued operators*, (2017).
4. Banach S., *Sur les operations dans les ensembles abstraits et leur application aux equations integrals*, Funda. Math., **3** (1) (1922), 133-181.
5. Bakhtin I.A., *The contraction mapping principle in quasimetric spaces*, Funct. Anal., **30** (1989), 26-37.
6. Matthews S.G., *Partial metric topology*, In Proc. 8th Summer Conference on General Topology and Applications, Ann. New York Acad. Sci., **728** (1994), 183-197.
7. Mustafa Z. and Sims B., *A new approach to generalized metric spaces*, J. Nonlinear Convex Anal., **7** (2) (2006), 289-297.
8. Chatterjea S.K., *Fixed point theorems*, C.R. Acad. Bulgare Sci., **25** (1972), 727-730.
9. Hardy G.C., Rogers T., *A generalization of fixed point theorem of S. Riech.*, Can. Math. Bull. **16** (1973), 201-206.
10. Jaggi D.S., *Some unique fixed point theorems*, Indian J. Pure Appl. Math., **8** (2) (1977), 223-230.
11. Karapinar E., *A note on common fixed point theorems in partial metric spaces*, Miskolc Math. Notes **12** (2) (2011), 185-191.
12. Karapinar E., *Fixed Points Results via Simulation Functions*, Filomat **30** (8) (2016), 2343-2350.
13. Karapinar E.and Erhan I.M., *Fixed point theorems for operators on partial metric spaces*, Appl. Math. Lett., **24** (2011), 1894-1899.
14. Karapinar E., *Fixed point theorems for cyclic weak $\phi$-contraction*, Appl. Math. Lett., **24** (2011), 822-825.
15. Karapinar E., *Fixed point theorems in cone Banach spaces*, Fixed Point Theory Appl., (2009), Article ID 609281.

16. Karapinar E., *Some non unique fixed point theorems of Ciric type on cone metric spaces*, Abstr. Appl. Anal., (2010), Article ID 123094.
17. Karapinar E., *Weak $\phi$-contractions on partial metric spaces*, J. Comput. Anal. Appl., **14** (2) (2012), 206-210.
18. Khojasteh F., Shukla S. and Radenovic S., *A new approach to the study of fixed point theorems via simulation functions*, Filomat, **29** (2015), 1189-1194.
19. Kirk W.A., Srinavasan P.S. and Veeramani P., *Fixed points for mappings satisfying cyclical contractive conditions*, Fixed Point Appl., **4** (2003), 79-89.
20. Samet B., Vetro C. and Vetro P.,*Fixed ppoint theorems for $(\alpha - \psi)$ contractive types mappings*, Nonlinear Anal., **75** (2012), 2154-2165.
21. Karapinar E., Kumam P. and Salimi P., *On $\alpha - \psi$-Meir-Keeler contractive mappings*, Fixed Point Theory and Applications, **1** (2013), 1-12.
22. Popescu O., *Some new fixed point theorems for $\alpha$-Geraghty contraction type maps in metric spaces*, Fixed Point Theory Appl., **1** (2014), 1-12.
23. Mutlu A. and Gurdal U., *Bipolar metric spaces and some fixed point theorems*, J. Nonlinear Sci. Appl., **9** (9) (2016), 5362-5373.
24. Murthy P.P., Mitrovic Z., Dhuri C.P. and Radenovic S., *The common fixed point theorems in bipolar metric space*, Gulf J. Math., **12** (2) (2022), 31-38.
25. Mutlu A., Gurdal U. and Ozkan K., *Fixed point results for $\alpha$-$\psi$-contractive mappings in bipolar metric spaces*, Journal of inequalities and special functions, **11** (1) (2020), 64-75.
26. Mutlu A., Gurdal U. and Ozkan K., *Fixed point theorems for multivalued mappings on bipolar metric spaces*, Fixed Point Theory, **21** (1) (2020), 271-280.
27. Mutlu A., Ozkan K. and Gurdal U., *Some common fixed point theorems in bipolar metric spaces*, Turk. J. Math. Comput. Sci., **14** (2) (2022), 346-354.
28. Radenovic S., Vetro F. and Vujakovic J., *An alternative and easy approach to fixed point results via simulation functions*, Demonstr. Math., **50** (2017), 223-230.
29. Rao B.S. and Kishore G.N.V., *Common Fixed Point Theorems in Bipolar Metric Spaces with Applications to Integral Equations*, Int. J. Eng. Technol., **7** (4.10) (2018), 1022-1026.

# 11 *wt*-Distance in Extended-Branciari *b*-Distance Metric Spaces and Application

*Reena Jain, Hemant Kumar Nashine, Vahid Parvaneh*

The purpose of this chapter is to extend the notion of *wt*-distance and simulation function in the extended Branciari *b*-distance space and consequently to prove some basic fixed point and periodic point theorems for newly defined simulation function in the underlying space followed by examples. Finally, we present a solution of integral equation which is equivalent form of a differential equation with boundary values as an application of the results.

## 11.1 INTRODUCTION AND AUXILIARY FACTS

Denote $\mathbb{R} :=$ the set of real numbers, $\mathbb{R}_+ := [0, +\infty)$, $\mathbb{N} :=$ the set of natural numbers, and $\mathbb{N}^* := \mathbb{N} \cup \{0\}$.

The distance notion in the metric fixed point theory is introduced and generalized in several different ways by many authors [2]–[4], [7], [11]–[17], [18]. Bakhtin [5] defined the notion of *b*-metric space as follows which is further used by Czerwik in [6, 8].

**Definition 11.1.1.** *[5] Let $S$ be a non-empty set, $s > 1$ be a real number. A function $d : S \times S \to [0, +\infty)$ is called b-metric if it satisfies the following properties for each $x, y, z \in S$.:*

1. $d(x,y) = 0$ *if and only if* $x = y$,
2. $d(x,y) = d(y,x)$
3. $d(x,y) \leq s[d(x,z) + d(z,y)]$

The pair $(S,d)$ is called a b-metric space

Kamran et al. [15] introduced the notion of extended *b*-metric space as follows:

DOI: 10.1201/9781003596578-11

**Definition 11.1.2.** *[15] Let S be a non-empty set and* $\theta : S \times S \to [1, +\infty)$. *A function* $d_\theta : [0, \infty)$ *is called extended b-metric if it satisfies the following properties for all* $x, y, z \in S$.*:*

1. $d_\theta(x,y) = 0$ *if and only if* $x = y$,
2. $d_\theta(x,y) = d_\theta(y,x)$,
3. $d_\theta(x,z) \leq \theta(x,z)[d_\theta(x,y) + d_\theta(y,y)]$ *for all* $x, y, z \in X$.

The pair $(S, d_\theta)$ is called an extended b-metric space.

Later, Branciari [9] extended the metric space and introduced the notion of Branciari distance by changing the property of triangle inequality by rectangular one which is stated as follows:

**Definition 11.1.3.** *[9] Let S be a non-empty set and let* $b : S \times S \to [0, +\infty)$ *such that for all* $x, y \in S$ *and* $u \neq v \in S \backslash \{x, y\}$*:*

1. $b(x,y) = 0$ *if and only if* $x = y$,
2. $b(x,y) = b(y,x)$,
3. $b(x,y) \leq d(x,u) + d(u,v) + d(v,y)$.

In this case, the map b is called a Branciari distance.

Recently Abdeljawad et al. [1] defined the notion of extended Branciari *b*-distance space by combining, extended *b*-metric and Branciari distance.

**Definition 11.1.4.** *[1] Let* $S \neq \emptyset$ *be a set and* $w : S^2 \to \mathbb{R}_+ \backslash (0,1)$. *We say that a function* $d_e : S^2 \to \mathbb{R}_+$ *is called an extended Branciari b-metric (*$d_e$*-metric, in short) if it satisfies:*

(i) $d_e(\vartheta, \upsilon) = 0$ *if and only if* $\vartheta = \upsilon$,
(ii) $d_e(\vartheta, \upsilon) = d_e(\upsilon, \vartheta)$,
(iii) $d_e(\vartheta, \upsilon) \leq w(\vartheta, \upsilon)[d_e(\vartheta, \nu) + d_e(\nu, \rho) + d_e(\rho, \upsilon)]$

*for all* $\vartheta, \upsilon \in S$ *and all distinct* $\nu, \rho \in S \backslash \{\vartheta, \upsilon\}$. *The symbol* $(S, d_e)$ *denotes extended Branciari b-distance space (EBbDS, in short). For* $w(\vartheta, \upsilon) = 1$, $(S, d_e)$ *will be called a Branciari distance space.*

**Example 11.1.5.** *[1] Let* $S = \ell_p$, *where* $1 \leq p < \infty$, *be defined by*

$$\ell_p = \{(\vartheta_n)_{n \geq 1} \subseteq \mathbb{R} : \sum_{n=1}^{\infty} |\vartheta_n|^p < \infty\}.$$

*Define* $d_e : S^2 \to \mathbb{R}_+$ *by*

$$d_e(\vartheta, \upsilon) = (\sum_{n=1}^{\infty} \mid \vartheta_n - \upsilon_n \mid^p)^{\frac{1}{p}})$$

*for all* $\vartheta, \upsilon \in S$. *The function* $w : S^2 \to \mathbb{R}_+ \backslash (0,1)$ *is defined by* $w(\vartheta, \upsilon) = 2^{\frac{1}{p}}$ *for all* $\vartheta, \upsilon \in S$. *Then* $d_e$ *satisfies all the conditions of an EBbDS.*

**Definition 11.1.6.** *[1] Let $S \neq \emptyset$ be a set endowed with extended Branciari b-distance $d_e$.*

*(a) A sequence $\{\vartheta_n\}$ in $S$ converges to $\vartheta$ if for every $\varepsilon > 0$ there exists $N = N(\varepsilon) \in \mathbb{N}$ such that $d_e(\vartheta_n, \vartheta) < \varepsilon$ for all $n \geq N$. For this particular case, we write $\lim_{n\to\infty} \vartheta_n = \vartheta$.*
*(b) A sequence $\{\vartheta_n\}$ in $S$ is called Cauchy if for every $\varepsilon > 0$ there exists $N = N(\varepsilon) \in \mathbb{N}$ such that $d_e(\vartheta_m, \vartheta_n) < \varepsilon$ for all $m, n \geq N$.*
*(c) An EBbDS $(S, d_e)$ is complete if every Cauchy sequence in $S$ is convergent.*

Hussain et al. [10] defined a *wt*-distance in *b*-metric space as extension of a *w*-distance on a metric space given by Kada et al. [14].

**Definition 11.1.7.** *[10] Let $(S, d_b)$ be a b-metric space with constant $\mathfrak{K} \geq 1$. A function $\Lambda : S^2 \to \mathbb{R}_+$ is called the wt-distance on $S$ iff it satisfies the following properties: for all $\vartheta, \upsilon \in S$*

*(1) $\Lambda(\vartheta, \nu) \leq \mathfrak{K}(\Lambda(\vartheta, \upsilon) + \Lambda(\upsilon, \nu))$ for all $\vartheta, \upsilon, \nu \in S$;*
*(2) for any $\vartheta \in S$, $\Lambda(\vartheta, .) : S \to \mathbb{R}_+$ is $\mathfrak{K}$-lower semi-continuous (l.s.c., in short), i.e, if $\vartheta \in S$ and $\upsilon_n \to \upsilon \in S$ then $\Lambda(\vartheta, \upsilon) \leq \liminf_{n\to\infty} \mathfrak{K}\Lambda(\vartheta, \upsilon_n)$;*
*(3) for any $\varepsilon > 0$, there exists $\delta > 0$ such that $\Lambda(\nu, \vartheta) < \delta$ and $\Lambda(\nu, \upsilon) < \delta$ imply $d_b(\vartheta, \upsilon) \leq \delta$.*

A notion of simulation function is defined by Khojasteh et al. which was modified in several ways by many authors. One of the modification was recently introduced by Roldan-Lpez-de-Hierro et al. as:

**Definition 11.1.8.** *A simulation function is a mapping $\zeta : \mathbb{R}^2_+ \to \mathbb{R}$ satisfying the following conditions:*

*(1) $\zeta(0,0) = 0$;*
*(2) $\zeta(\rho, \sigma) < \sigma - \rho$ for all $\rho, \sigma > 0$;*
*(3) if $\{\rho_n\}$ and $\{\sigma_n\}$ are two sequences in $(0, \infty)$ such that $\lim_{n\to\infty} \rho_n = \lim_{n\to\infty} \sigma_n > 0$ and $\rho_n < \sigma_n$ for all $n \in \mathbb{N}$, then, $\limsup_{n\to\infty} \zeta(\rho_n, \sigma_n) < 0$.*

In the present work, we define *wt*-distance in extended Branciari *b*-metric spaces and modified simulation function in the underlying space. Then we define contraction condition on aforesaid notions and based fixed point and periodic point results without exerting the continuity condition on the underlying mapping. Furthermore, we give some examples to furnish our main result followed by an application to the differential equation boundary value problem. With the introduction of *wt*-distance and modified simulation function in the underlying space, our results extend the related existing results proved in the literature.

## 11.2 EXTENSION OF $WT$-DISTANCE AND SIMULATION FUNCTION TO EXTENDED BRANCIARI $B$-METRIC SPACE

We start with modifying the concept of the simulation function and the *wt*-distance to EBbDS.

**Definition 11.2.1.** *Let $S \neq \emptyset$ be a set and $w : S^2 \to \mathbb{R}_+ \backslash (0,1)$ be a mapping. A modified simulation function is a mapping $\zeta_e : \mathbb{R}_+^2 \to \mathbb{R}$ satisfying :*

(1) $\zeta_e(0,0) = 0$;

(2) $\zeta_e(w(\vartheta, \upsilon)\rho, \sigma) < \sigma - w(\vartheta, \upsilon)\rho$ *for all* $\vartheta, \upsilon \in S$;

(3) *if* $\{\rho_n\}$ *and* $\{\sigma_n\}$ *are two sequences in* $(0,\infty)$ *such that* $\lim_{n\to\infty} w\rho_n = \lim_{n\to\infty} \sigma_n > 0$ *and* $\rho_n < \sigma_n$ *for all* $n \in \mathbb{N}$, *then* $\limsup_{n\to\infty} \sigma_e(w(\rho_n, \sigma_n)\rho_n, \sigma_n) < 0$.

*The collection of all modified simulation functions $\zeta_e : \mathbb{R}_+^2 \to \mathbb{R}$ will be denoted by $\mathscr{Z}_e$.*

**Example 11.2.2.** *Let $w : S^2 \to \mathbb{R}_+ \backslash (0,1)$ be a mapping where $S \neq \emptyset$ be a set. Let $\gamma < 1$ and we define the mapping $\zeta_e : \mathbb{R}_+^2 \to \mathbb{R}$ by*

$$\zeta_e(\rho, \sigma) = \begin{cases} \sigma - w(\rho, \sigma), \ \text{if } \sigma < \rho \\ \dfrac{\gamma\sigma - w(\rho, \sigma)\rho}{1 + w(\rho, \sigma)\sigma}, \text{otherwise.} \end{cases}$$

*It is obvious to verify that $\zeta_e(\rho, \sigma)$ satisfies the properties (1) and (2) of definition of $\zeta_e$.*
*Next, we will show that $\zeta_e(\rho, \sigma)$ satisfies the property (3) also. If $\rho_n$ and $\sigma_n$ are sequences in $(0,\infty)$ such that $\limsup_{n\to\infty} w\rho_n = \limsup_{n\to\infty} \sigma_n > 0$ and $\rho_n < \sigma_n$ for all $n \in \mathbb{N}$ then*

$$\begin{aligned} \limsup_{n\to\infty} \zeta_e(w\rho_n, \sigma_n) &= \limsup_{n\to\infty} \frac{\gamma\sigma_n - w.\rho_n}{1 + w.\sigma_n} \\ &< \limsup_{n\to\infty} \frac{\sigma_n - w\rho_n}{w\rho_n} \\ &< \limsup_{n\to\infty} (\frac{\sigma_n}{w\rho_n} - 1) \\ &\leq \limsup_{n\to\infty} \frac{\sigma_n}{w\rho_n} - \liminf_{n\to\infty} 1 \\ &\leq 1 - 1 = 0. \end{aligned}$$

*Thus $\zeta_e$ is a modified simulation function.*

Next we extend the notion of *wt*-distance in the setting of an EBbDS as follows:

**Definition 11.2.3.** *Let $(S, d_e)$ be an EBbDS with $w : S^2 \to \mathbb{R}_+ \backslash (0,1)$. A function $\mathfrak{N} : S^2 \to \mathbb{R}_+$ is called the $wt_e$-distance on $S$ if :*

1 $\mathfrak{N}(\vartheta, \upsilon) \leq w(\vartheta, \upsilon)(\mathfrak{N}(\vartheta, \nu) + \mathfrak{N}(\nu, \rho) + \mathfrak{N}(\rho, \upsilon)$ *for all* $\vartheta, \upsilon \in S$ *and all distinct* $\nu, \upsilon \in S \backslash \{\vartheta, \upsilon\}$.

2 *for any* $\vartheta \in S$, $\mathfrak{N}(x, \cdot) : S \to \mathbb{R}_+$ *is* $wt_e$*-l.s.c. if* $\vartheta \in S$ *and* $\upsilon_n \to \upsilon \in S$, *then* $\mathfrak{N}(\vartheta, \upsilon) \leq \liminf_{n\to\infty} w(\vartheta, \upsilon)\mathfrak{N}(\vartheta, \upsilon_n)$

3 *for any* $\varepsilon > 0$, *there exists* $\rho > 0$ *such that* $\mathfrak{N}(\nu,\vartheta) < \rho, \mathfrak{N}(\nu,\upsilon) < \rho$ *and* $\mathfrak{N}(\upsilon,y) < \rho$ *implies* $d_e(\vartheta,y) \leq \varepsilon$.

The following lemma will have an important role in the establishment of our main results:

**Lemma 11.2.4.** *Let* $(S,d_e)$ *be an EBbDS with* $w : S^2 \to \mathbb{R}_+ \backslash (0,1)$ *and* $\mathfrak{N}$ *be a* $wt_e$*-distance on S. Let* $\{\rho_n\}$, $\{\sigma_n\}$ *be two sequences in S and* $\{\alpha_n\}$ *be a sequence in* $\mathbb{R}_+$ *converging to zero. Then the following conditions hold: for all* $\vartheta, \upsilon \in S$,

*(1) if* $\mathfrak{N}(\rho_n,\rho_m) \leq \alpha_n$ *for all* $n,m \in \mathbb{N}$ *with* $m > n$, *then* $\{\rho_n\}$ *is a Cauchy sequence;*
*(2)* $\mathfrak{N}(\upsilon,\rho_n) \leq \alpha_n$ *for all* $n \in \mathbb{N}$, *then* $\{\rho_n\}$ *is a Cauchy sequence.*

**Proof:** We provide the proof on the lines of the proof given in [14].
Let $\varepsilon > 0$ be given. From the definiton of *wt*-distance, there exists $\rho > 0$ such that $\mathfrak{N}(\nu,\vartheta) < \rho, \mathfrak{N}(\nu,\upsilon) < \rho$ and $\mathfrak{N}(\upsilon,y) < \rho$ implies $d_e(\vartheta,y) \leq \varepsilon$. Choose $n_0 \in N$ such that $\alpha_n \leq \rho$ for every $n \geq n_0$. Then we have, for any $m \geq n \geq n_0$, $\mathfrak{N}(\rho_n,\rho_{n+1}) \leq \alpha_n < \rho$, $\mathfrak{N}(\rho_{n+1},\rho_{n+2}) \leq \alpha_{n+1} < \rho$ and $\mathfrak{N}(\rho_{n+2},\rho_m) \leq \alpha_{n+2} < \rho$ and hence $d_e(\rho_n,\rho_m) \leq \varepsilon$, which concludes that $\{\rho_n\}$ is a Cauchy sequence. (2) is a special case of (1).

## 11.3 MAIN RESULT

Based on the previous section, in this section, we establish some fixed point results for $\mathfrak{Z}\mathfrak{N}_e$-contraction mapping under $wt_e$-distance and $\zeta_e \in \mathscr{Z}_e$.

**Definition 11.3.1.** *Let* $(S,d_e)$ *be an EBbDS with* $w : S^2 \to \mathbb{R}_+ \backslash (0,1)$. *A mapping* $\mathfrak{I} : S \to S$ *is called* $\mathfrak{Z}_e$*-contraction, if there exists* $\zeta_e \in \mathscr{Z}_e$ *such that*

$$\zeta_e(w(\vartheta,\upsilon)d_e(\mathfrak{I}\vartheta,\mathfrak{I}\upsilon),(d_e(\vartheta,\upsilon))^r) \geq 0 \text{ if } d_e(\mathfrak{I}\vartheta,\mathfrak{I}\upsilon) \neq 0 \text{ for } \vartheta,\upsilon \in S$$

*where* $r \in (0,1)$ *is such that* $\sup_{m\geq 1} \lim_{n\to\infty} w(\vartheta_n,\vartheta_m) < \frac{1}{r}$ *and* $\vartheta_n = \mathfrak{I}^n\vartheta_0$ *for* $\vartheta_0 \in S$.

**Definition 11.3.2.** *Let* $(S,d_e)$ *be an EBbDS with* $w : S^2 \to \mathbb{R}_+ \backslash (0,1)$ *and* $\mathfrak{N} : S^2 \to \mathbb{R}_+$ *is a* $wt_e$*-distance on S. A mapping* $\mathfrak{I} : S \to S$ *is called* $\mathfrak{Z}\mathfrak{N}_e$*-contraction, if there exists* $\zeta_e \in \mathscr{Z}_e$ *such that*

$$\zeta_e(w(\vartheta,\upsilon)\mathfrak{N}(\mathfrak{I}\vartheta,\mathfrak{I}\upsilon),(\mathfrak{N}(\vartheta,\upsilon))^r) \geq 0 \text{ if } d_e(\mathfrak{I}\vartheta,\mathfrak{I}\upsilon) \neq 0 \text{ for } \vartheta,\upsilon \in S \quad (11.1)$$

*where* $r \in (0,1)$ *is such that* $\sup_{m\geq 1} \lim_{n\to\infty} w(\vartheta_n,\vartheta_m) < \frac{1}{r}$ *and* $\vartheta_n = \mathfrak{I}^n\vartheta_0$ *for* $\vartheta_0 \in S$.

We denote by $\Upsilon$ the collection of all $\mathfrak{Z}\mathfrak{N}_e$-contraction mappings on an EBbDS *S*.

**Theorem 11.3.3.** *Let* $(S,d_e)$ *be a complete EBbDS with* $w : S^2 \to \mathbb{R}_+ \backslash (0,1)$ *and* $\mathfrak{N} : S^2 \to \mathbb{R}_+$ *is a* $wt_e$*-distance on S. Let* $\mathfrak{I} : S \to S$ *be a given mapping. Suppose that the following conditions hold:*

*(C1)* $\Im \in \Upsilon$;
*(C2)* *there exists* $\vartheta_0 \in S$ *such that* $d_e(\vartheta_0, \Im\vartheta_0) \neq 0$.

*Then* $\Im$ *has a unique fixed point in S, that is,* $\Im\eta = \eta$ *for* $\eta \in \Im$.

**Proof:** Starting with $\vartheta_0 \in S$, we construct the sequence $\{\vartheta_n\}$ as $\vartheta_n = \Im^n\vartheta_0$ for all $n \in \mathbb{N}$. Suppose $\Im^n\vartheta = \Im^{n+1}\vartheta$, then obviously $\Im^n\vartheta$ is a fixed point and the proof is completed. Now, we claim that

$$\lim_{n\to\infty} \mathfrak{N}(\Im^n\vartheta_0, \Im^{n+1}\vartheta_0) = 0. \tag{11.2}$$

Without loss of generality, suppose $\vartheta_n \neq \vartheta_{n+1}$ for all $n \in \mathbb{N}$, i.e., $d_e(\Im\vartheta_{n-1}, \Im\vartheta_n) \neq 0$. Therefore, by equation (11.1)

$$\zeta_e(w(\vartheta_{n-1}, \vartheta_n)\mathfrak{N}(\Im\vartheta_{n-1}, \Im\vartheta_n), (\mathfrak{N}(\vartheta_{n-1}, \vartheta_n))^r) \geq 0.$$

By the property of $\zeta_e$, we have

$$\begin{aligned} 0 &\leq \zeta_e(w(\vartheta_{n-1}, \vartheta_n)\mathfrak{N}(\Im\vartheta_{n-1}, \Im\vartheta_n), (\mathfrak{N}(\vartheta_{n-1}, \vartheta_n))^r) \\ &< (\mathfrak{N}(\vartheta_{n-1}, \vartheta_n))^r - w(\vartheta_{n-1}, \vartheta_n)\mathfrak{N}(\Im\vartheta_{n-1}, \Im\vartheta_n). \end{aligned}$$

As $w(\vartheta, \upsilon) \geq 1$,

$$\mathfrak{N}(\Im\vartheta_{n-1}, \Im\vartheta_n) \leq w(\vartheta_{n-1}, \vartheta_n)\mathfrak{N}(\Im\vartheta_{n-1}, \Im\vartheta_n) < (\mathfrak{N}(\vartheta_{n-1}, \vartheta_n))^r. \tag{11.3}$$

Since $r \in (0,1)$,

$$[\mathfrak{N}(\vartheta_{n-1}, \vartheta_n)]^r < \mathfrak{N}(\vartheta_{n-1}, \vartheta_n).$$

Consequently

$$\mathfrak{N}(\Im\vartheta_{n-1}, \Im\vartheta_n) \leq (\mathfrak{N}(\vartheta_{n-1}, \vartheta_n))^r \leq \mathfrak{N}(\vartheta_{n-1}, \vartheta_n),$$

that is,

$$\mathfrak{N}(\Im\vartheta_{n-1}, \Im\vartheta_n) \leq \mathfrak{N}(\Im\vartheta_{n-2}, \Im\vartheta_{n-1}).$$

Therefore, the sequence $\mathfrak{N}(\Im^n\vartheta_0, \Im^{n+1}\vartheta_0)$ is a decreasing sequence of nonnegative real numbers and convergent to some $t \geq 0$. Suppose that $t > 0$.

**Case I.** If $w(\vartheta_{n-1}, \vartheta_n) > 1$, letting $n \to \infty$ in (11.3), $t < t^r$, a contradiction as $r \in (0,1)$.

**Case II.** If $w(\vartheta_{n-1}, \vartheta_n) = 1$, then putting $\rho_n = \mathfrak{N}(\Im^{n-1}\vartheta_0, \Im^n\vartheta_0)$ and $\sigma_n = \mathfrak{N}(\Im^{n-2}\vartheta_0, \Im^{n-1}\vartheta_0)$, the sequences $\{w\rho_n\}$ and $\{\sigma_n\}$ converges to same limit. In addition, both sequences $\{w\rho_n\}$ and $\{\sigma_n\}$ do have the same positive limit superior but we have $\rho_n < \sigma_n$ for all $n \in \mathbb{N}$. Bye virtue of modified simulation function

$$\begin{aligned} &\lim sup_{n\to\infty}\zeta_e(w(\vartheta_{n-1}, \vartheta_n)\mathfrak{N}(\Im^{n-1}\vartheta_0, \Im^n\vartheta_0), \mathfrak{N}(\Im^{n-2}\vartheta_0, \Im^{n-1}\vartheta_0)) \\ &= \lim sup_{n\to\infty}\zeta_e(w\rho_n, \sigma_n) < 0, \end{aligned}$$

a contradiction, implies that $t = 0$, and we concluded (11.2).

Now we prove that

$$lim_{m,n\to\infty}\mathfrak{N}(\mathfrak{I}^n\vartheta_0,\mathfrak{I}^m\vartheta_0)=0. \tag{11.4}$$

Let the equation (11.4) is not valid. This means we get $\varepsilon_0>0$ with $\{m_k\}$, $\{n_k\}$ such that, for any $m_k>n_k$ say such that

$$\mathfrak{N}(\mathfrak{I}^{n_k}\vartheta_0,\mathfrak{I}^{m_k}\vartheta_0)>\varepsilon_0$$

for $k=0,1,2,3,\ldots$. We can consider that $m_k$ is a least index for which above holds. Therefore,

$$\mathfrak{N}(\mathfrak{I}^{n_k}\vartheta_0,\mathfrak{I}^{m_k-2}\vartheta_0)\le\varepsilon_0.$$

Hence, we have

$$\mathfrak{N}(\mathfrak{I}^{n_k}\vartheta_0,\mathfrak{I}^{m_k}\vartheta_0)\quad\le w(\vartheta_{n_k},\vartheta_{m_k})\left[\begin{array}{c}\mathfrak{N}(\mathfrak{I}^{n_k}\vartheta_0,\mathfrak{I}^{m_k-2}\vartheta_0)\\+\mathfrak{N}(\mathfrak{I}^{m_k-2}\vartheta_0,\mathfrak{I}^{m_k-1}\vartheta_0)\\+\mathfrak{N}(\mathfrak{I}^{m_k-1}\vartheta_0,\mathfrak{I}^{m_k}\vartheta_0)\end{array}\right]. \tag{11.5}$$

Taking $\lim sup_{n\to\infty}$ in (11.5) and using (11.2), we have

$$\varepsilon_0<\lim sup_{n\to\infty}\mathfrak{N}(\mathfrak{I}^{n_k}\vartheta_0,\mathfrak{I}^{m_k}\vartheta_0)\le\varepsilon_0\lim sup_{n\to\infty}w(\vartheta_{n_k},\vartheta_{m_k}) \tag{11.6}$$

or

$$\frac{\varepsilon_0}{\limsup_{n\to\infty}w(\vartheta_{n_k},\vartheta_{m_k})}\le\lim sup_{n\to\infty}\mathfrak{N}(\mathfrak{I}^{n_k+1}\vartheta_0,\mathfrak{I}^{m_k+1}\vartheta_0).$$

Now we prove that

$$\lim sup_{n\to\infty}\mathfrak{N}(\mathfrak{I}^{n_k+1}\vartheta_0,\mathfrak{I}^{m_k+1}\vartheta_0)<\varepsilon_0.$$

On the contrary, we suppose that

$$\lim sup_{n\to\infty}\mathfrak{N}(\mathfrak{I}^{n_k+1}\vartheta_0,\mathfrak{I}^{m_k+1}\vartheta_0)\ge\varepsilon_0,$$

then there exists $k_p$ and $\delta>0$ such that

$$\lim sup_{n\to\infty}\mathfrak{N}(\mathfrak{I}^{n_{k_p+1}}\vartheta_0,\mathfrak{I}^{m_{k_p+1}}\vartheta_0)=\delta\ge\varepsilon_0. \tag{11.7}$$

By the property of $\zeta_e$, we have

$$\begin{array}{rcl}0&\le&\zeta_e(w(\vartheta_{n_{k_p}},\vartheta_{m_{k_p}})\mathfrak{N}(\mathfrak{I}^{n_{k_p}+1}\vartheta_0,\mathfrak{I}^{m_{k_p+1}}\vartheta_0),(\mathfrak{N}(\mathfrak{I}^{n_{k_p}}\vartheta_0,\mathfrak{I}^{m_{k_p}}\vartheta_0))^r)\\&<&(\mathfrak{N}(\mathfrak{I}^{n_{k_p}}\vartheta_0,\mathfrak{I}^{m_{k_p}}\vartheta_0))^r-w(\vartheta_{n_{k_p}},\vartheta_{m_{k_p}})\mathfrak{N}(\mathfrak{I}^{n_{k_p}+1}\vartheta_0,\mathfrak{I}^{m_{k_p+1}}\vartheta_0)\end{array}$$

implies that

$$w(\vartheta_{n_{k_p}},\vartheta_{m_{k_p}})\mathfrak{N}(\mathfrak{I}^{n_{k_p}+1}\vartheta_0,\mathfrak{I}^{m_{k_p+1}}\vartheta_0)\le[\mathfrak{N}(\mathfrak{I}^{n_{k_p}}\vartheta_0,\mathfrak{I}^{m_{k_p}}\vartheta_0)]^r. \tag{11.8}$$

It follows from (11.6), (11.7), (11.9) that

$$\begin{aligned}
& \delta \lim sup_{n\to\infty} w(\vartheta_{n_{kp}}, \vartheta_{m_{kp}}) \\
= \; & \lim sup_{n\to\infty} w(\vartheta_{n_{kp}}, \vartheta_{m_{kp}}). \lim sup_{n\to\infty} \mathfrak{N}(\mathfrak{I}^{n_{kp}+1}\vartheta_0, \mathfrak{I}^{m_{kp}+1}\vartheta_0) \\
\leq \; & [\lim sup_{n\to\infty} \mathfrak{N}(\mathfrak{I}^{n_{kp}}\vartheta_0, \mathfrak{I}^{m_{kp}}\vartheta_0)]^r \\
< \; & \lim sup_{n\to\infty} \mathfrak{N}(\mathfrak{I}^{n_{kp}}\vartheta_0, \mathfrak{I}^{m_{kp}}\vartheta_0) \\
\leq \; & \varepsilon_0 \lim sup_{n\to\infty} w(\vartheta_{n_{kp}}, \vartheta_{m_{kp}}) \\
\leq \; & \delta \lim sup_{n\to\infty} w(\vartheta_{n_{kp}}, \vartheta_{m_{kp}}).
\end{aligned}$$

Therefore, the sequences $\{w(\vartheta_{n_{kp+1}}, \vartheta_{m_{kp+1}})t_{k_p}\} = \{w(\vartheta_{n_{kp+1}}, \vartheta_{m_{kp+1}})\mathfrak{N}(\mathfrak{I}^{n}_{k_p+1}\vartheta_0, \mathfrak{I}^{m_{kp+1}}\vartheta_0)\}$ and $\{s_{k_p}\} = \{\mathfrak{N}(\mathfrak{I}^{n_{kp}}\vartheta_0, \mathfrak{I}^{m_{kp}})\}$ have the same positive limit superior and we have $t_{k_p} < s_{k_p}$ for all $p \in \mathbb{N}$. By the property of $\zeta_e$,

$$\begin{aligned}
0 \; \leq \; & \limsup_{n\to\infty} \zeta_e(w(\vartheta_{n_{kp+1}}, \vartheta_{m_{kp+1}})\mathfrak{N}(\mathfrak{I}^{n_{kp+1}}\vartheta_0, \mathfrak{I}^{m_{kp+1}}\vartheta_0), \mathfrak{N}(\mathfrak{I}^{n_{kp}}\vartheta_0, \mathfrak{I}^{m_{kp}})) \\
\leq \; & \limsup_{n\to\infty} \zeta_e(w(\vartheta_{n_{kp+1}}, \vartheta_{m_{kp+1}})t_{k_p}, s_{k_p}) < 0,
\end{aligned}$$

a contradiction and so (11.4) true. Following Lemma 11.2.4 that $\{\mathfrak{I}^n\vartheta_0\}$ is a Cauchy sequence. Hence, by the completeness of $S$, there exists $\eta \in S$ such that $\{\mathfrak{I}^n\vartheta_0\}$ converges to $\eta$.

Since we have $lim_{m,n\to\infty}\mathfrak{N}(\mathfrak{I}^n\vartheta_0, \mathfrak{I}^m\vartheta_0) = 0$, for each $\varepsilon > 0$, there exists $n_\varepsilon \in \mathbb{N}$ such that $n > n_\varepsilon$ implies

$$\mathfrak{N}(\mathfrak{I}^{n_\varepsilon}\vartheta_0, \mathfrak{I}^n\vartheta_0) < \varepsilon.$$

Since $\mathfrak{N}(x, .)$ is $wt_e$-l.s.c. and the sequence $\{\mathfrak{I}^n\vartheta_0\}$ converges to $\eta$,

$$\mathfrak{N}(\mathfrak{I}^{n_\varepsilon}\vartheta_0, \eta) < \liminf_{n\to\infty} w(\vartheta_{n_\varepsilon - 1}, \vartheta_n)\mathfrak{N}(\mathfrak{I}^{n_\varepsilon}\vartheta_0, \mathfrak{I}^n\vartheta_0) \leq \varepsilon \lim sup_{n\to\infty} w(\vartheta_{n_\varepsilon-1}, \vartheta_n). \tag{11.9}$$

Setting $\varepsilon = \frac{1}{[\lim sup_{n\to\infty} w(\vartheta_{n_\varepsilon-1}, \vartheta_n)]^2}$ and $n_\varepsilon = n_k$, we have from (11.9)

$$\lim_{k\to\infty} \mathfrak{N}(\mathfrak{I}^{n_k}\vartheta_0, \eta) = 0.$$

Next we claim $\eta \in S$ such that $\mathfrak{I}\eta = \eta$. On the contrary, let $\mathfrak{I}\eta \neq \eta$, that is, $d_e(\eta, \mathfrak{I}\eta) > 0$. Using condition (C1), we get

$$\zeta_e(w(\vartheta_{n_k-1}, \vartheta_{n_k})\mathfrak{N}(\mathfrak{I}\vartheta_{n_k-1}, \mathfrak{I}\vartheta_{n_k}), (\mathfrak{N}(\vartheta_{n_k-1}, \vartheta_{n_k}))^r) \geq 0. \tag{11.10}$$

By the property of $\zeta_e$, we have

$$w(\vartheta_{n_k-1}, \vartheta_{n_k})\mathfrak{N}(\mathfrak{I}\vartheta_{n_k-1}, \mathfrak{I}\vartheta_{n_k}) \leq [\mathfrak{N}(\vartheta_{n_k-1}, \vartheta_{n_k})]^r.$$

Also

$$\begin{aligned}
\mathfrak{N}(\eta, \mathfrak{I}\eta) \; = \; & \mathfrak{N}(\lim_{n\to\infty} \mathfrak{I}\vartheta_{n_k-1}, \lim_{n\to\infty} \mathfrak{I}\vartheta_{n_k}) \\
\leq \; & \lim_{n\to\infty} w(\vartheta_{n_k-1}, \vartheta_{n_k})\mathfrak{N}(\lim_{n\to\infty} \mathfrak{I}\vartheta_{n_k-1}, \lim_{n\to\infty} \mathfrak{I}\vartheta_{n_k}) \\
\leq \; & (\mathfrak{N}(\lim_{n\to\infty} \vartheta_{n_k-1}, \lim_{n\to\infty} \vartheta_{n_k}))^r.
\end{aligned}$$

By recursive use of the property of $\zeta_e$, we get

$$\mathfrak{N}(\eta,\Im\eta)) \leq (\mathfrak{N}(\eta,\Im\eta)^{r^n} < \mathfrak{N}(\eta,\Im\eta),$$

a contradiction and so $\Im\eta = \eta$.
Suppose $\nu \in S$ with $\eta \neq \nu$ such that $\Im\nu = \nu$. Then $d_e(\eta,\nu) = d_e(\Im\eta,\Im\nu) \neq 0$.

- If $w(\eta,\nu) = 1$, then from (11.1),

$$0 \leq \zeta_e(\mathfrak{N}(\Im\eta,\Im\nu),(\mathfrak{N}(\eta,\nu))^r) \leq [\mathfrak{N}(\eta,\nu)]^r - \mathfrak{N}(\eta,\nu),$$

that is, $\mathfrak{N}(\eta,\nu) \leq [\mathfrak{N}(\eta,\nu)]^r$ a contradiction. Therefore $\eta = \nu$.

- If $w(\eta,\nu) > 1$, then from (11.1),

$$0 \leq \zeta_e(w(\eta,\nu)\mathfrak{N}(\Im\eta,\Im\nu),(\mathfrak{N}(\eta,\nu))^r) \leq [\mathfrak{N}(\eta,\nu)]^r - w(\eta,\nu)\mathfrak{N}(\eta,\nu),$$

that is, $w(\eta,\nu)\mathfrak{N}(\eta,\nu) \leq \mathfrak{N}(\eta,\nu)^r$. That implies that

$$\mathfrak{N}(\eta,\nu) = \mathfrak{N}(\Im\eta,\Im\nu) \leq w(\eta,\nu)\mathfrak{N}(\Im\eta,\Im\nu) \leq [\mathfrak{N}(\eta,\nu)]^r < \mathfrak{N}(\eta,\nu),$$

a contradiction. Therefore $d_e(\eta,\nu) = 0$. Hence, we have the conclusion.

**Example 11.3.4.** *Let $S = \mathscr{U} \cup \mathscr{V}$, where $\mathscr{U} = \{\frac{1}{2},\frac{1}{3},\frac{1}{4},\frac{1}{5}\}$ and $\mathscr{V} = [\frac{3}{2},2]$. Define $d_e : S^2 \to \mathbb{R}_+$ so that $d_e(\vartheta,\upsilon) = d_e(\upsilon,\vartheta)$ for all $\vartheta,\upsilon \in S$, and*

$$d_e(\frac{1}{2},\frac{1}{3}) = 0.06, d_e(\frac{1}{2},\frac{1}{4}) = 0.02, d_e(\frac{1}{2},\frac{1}{5}) = 0.02,$$

$$d_e(\frac{1}{3},\frac{1}{4}) = 0.02, d_e(\frac{1}{3},\frac{1}{5}) = 0.01, d_e(\frac{1}{5},\frac{1}{4}) = 0.02,$$

*$d_e(\vartheta,\upsilon) = (\vartheta-\upsilon)^2$ otherwise. Then it is clear that $(S,d_e)$ is a EBbDS with $w(\vartheta,\upsilon) = \vartheta+\upsilon+2$ but it is not a Branciari metric space and metric space as well. For instances*

$$d_e(\frac{1}{2},\frac{1}{4}) = 0.06 \not\leq 0.04 = d_e(\frac{1}{2},\frac{1}{3}) + d_e(\frac{1}{3},\frac{1}{4})$$

*and*

$$d_e(\frac{1}{2},\frac{1}{4}) = 0.06 \not\leq 0.05 = d_e(\frac{1}{2},\frac{1}{3}) + d_e(\frac{1}{3},\frac{1}{5}) + d_e(\frac{1}{5},\frac{1}{4})$$

*but*

$$d_e(\frac{1}{2},\frac{1}{4}) = 0.06 \leq 0.1375 = w(\vartheta,\upsilon)\left[d_e(\frac{1}{2},\frac{1}{3}) + d_e(\frac{1}{3},\frac{1}{5}) + d_e(\frac{1}{5},\frac{1}{4})\right].$$

*Similarly, we can take any combination of $\vartheta,\upsilon$ with $\nu,\rho \in S\backslash\{\vartheta,\upsilon\}$ and property (iii) of Definition 11.1.4 will be true.*

*Consider the self-mapping $\Im$ on $S$ given by*

$$\Im(\vartheta) = \begin{cases} \frac{1}{3}, & \text{if } \vartheta \in \mathscr{U} \\ \frac{1}{5}, & \text{if } \vartheta \in \mathscr{V}. \end{cases}$$

*Define functions $wt_e$-distance $\mathfrak{N}$ on $S$ and $\zeta_e : \mathbb{R}_+^2 \to \mathbb{R}$ by $\mathfrak{N}(\vartheta, \upsilon) = |\upsilon|^2$, $\zeta_e(\vartheta, \upsilon) = \frac{\upsilon - w(\vartheta,\nu)\vartheta}{1+w(\vartheta,\upsilon)\upsilon}$ for all $\vartheta, \upsilon \in \mathbb{R}_+$.*
*It's only to show that $\mathfrak{I}$ satisfies the condition (C1) of Theorem 11.3.3, that is,*

$$\zeta_e(w(\vartheta,\upsilon)\mathfrak{N}(\mathfrak{I}\vartheta,\mathfrak{I}\upsilon), [\mathfrak{N}(\vartheta,\upsilon)]^r) \geq 0,$$

*that is,*

$$(\vartheta + \upsilon + 2)^2(\mathfrak{N}(\mathfrak{I}\vartheta, \mathfrak{I}\upsilon)) \leq (\mathfrak{N}(\vartheta,\upsilon))^r,$$

*that is,*

$$(\vartheta + \upsilon + 2)^2(\mathfrak{I}\upsilon)^2 \leq (\upsilon^2)^r \tag{11.11}$$

*for all $\vartheta, \upsilon \in S$ and some $r \in (0,1)$.*

*We will check that $\mathfrak{I}$ satisfy (11.11). There are two non-trivial possible cases when $d_e(\mathfrak{I}\vartheta, \mathfrak{I}\upsilon) > 0$. Here $w(\vartheta,\upsilon) \in [\frac{11}{3}, \frac{9}{2}]$.*

*It is enough to consider when $\vartheta \in \mathscr{U}$ and $\upsilon \in \mathscr{V}$. Then equation (11.11) would be of the form*

$$(\vartheta + \upsilon + 2)^2 \left(\frac{1}{5}\right)^2 \leq \upsilon^{2r}$$

*and it is true for $r \in (0,1)$ and for all $\vartheta \in \mathscr{U}$ and $\upsilon \in \mathscr{V}$. Thus, all the conditions are fulfilled and the $\mathfrak{I}$ have a unique fixed point (which is $\eta = \frac{1}{3}$).*

**Example 11.3.5.** *Consider $S = [0,1]$ and define $d_e : S^2 \to \mathbb{R}_+$ by $d_e(\vartheta, \nu) = |\vartheta - \nu|^2$. Then $(S, d_e)$ is a EBbDS with $w(\vartheta, \nu) = \vartheta + \nu + \frac{5}{2}$ but it is not a Branciari metric space and metric space as well. For instances*

$$d_e(0,1) = 1 \not\leq \frac{1}{2} = d_e(0, \frac{1}{2}) + d_e(\frac{1}{2}, 1)$$

*and*

$$d_e(0,1) = 1 \not\leq 0.4902 = d_e(0, \frac{1}{2}) + d_e(\frac{1}{2}, 0.99) + d_e(0.99, 1)$$

*but*

$$\begin{aligned}
d_e(\vartheta, \nu) &= |\vartheta - \nu|^2 \\
&= |\vartheta - \mu + \mu - \upsilon + \upsilon - \nu|^2 \\
&\leq |\vartheta - \mu|^2 + |\mu - \upsilon|^2 + |\upsilon - \nu|^2 + 2|\vartheta - \mu||\mu - \upsilon| \\
&\quad + 2|\mu - \upsilon||\upsilon - \nu| + 2|\upsilon - \nu||\vartheta - \mu| \\
&\leq \left(\vartheta + \nu + \frac{5}{2}\right) [|\vartheta - \mu|^2 + |\mu - \upsilon|^2 + |\upsilon - \nu|^2] \\
&= w(\vartheta, \nu)[d_e(\vartheta, \mu) + d_e(\mu, \upsilon) + d_e(\upsilon, \nu)]
\end{aligned}$$

*for all $\vartheta, \nu, \mu, \upsilon \in S$.*

*Consider the self-mapping $\mathfrak{I}$ on $S$ given by $\mathfrak{I}(\vartheta) = \frac{\vartheta}{4}$.*

*Define functions $wt_e$-distance $\mathfrak{N}$ on $S$ and $\zeta_e : \mathbb{R}_+^2 \to \mathbb{R}$ by $\mathfrak{N}(\vartheta, \nu) = |\nu|^2$, $\zeta_e(\vartheta, \nu) = \frac{\nu - w(\vartheta,\nu)\vartheta}{1+w(\vartheta,\nu)\nu}$ for all $\vartheta, \nu \in \mathbb{R}_+$.*
*It's only to show that $\mathfrak{I}$ satisfies the condition (C1) of Theorem 11.3.3, that is,*

$$\zeta_e(w(\vartheta,\nu)\mathfrak{N}(\mathfrak{I}\vartheta, \mathfrak{I}\nu), [\mathfrak{N}(\vartheta,\nu)]^r) \geq 0,$$

*that is,*

$$(\vartheta+\nu+2)^2(\mathfrak{N}(\mathfrak{I}\vartheta, \mathfrak{I}\nu)) \leq (\mathfrak{N}(\vartheta,\nu))^r,$$

*that is,*

$$(\vartheta+\nu+2)^2|\mathfrak{I}\nu|^2 \leq |\nu^2|^r \tag{11.12}$$

*for all $\vartheta, \nu \in S$ and some $r \in (0,1)$. That is,*

$$\left(\vartheta+\nu+\frac{5}{2}\right)^2 \frac{|\nu|^2}{16} \leq |\nu|^{2r}$$

*or*

$$\left(\vartheta+\nu+\frac{5}{2}\right)^2 \leq 16|\nu|^{2r-2}$$

*and it is true for $r \in (0,1)$ and for all $\vartheta \in \mathscr{U}$ and $\nu \in \mathscr{V}$. Thus, all the conditions are fulfilled and the $\mathfrak{I}$ have a unique fixed point (which is $\eta = 0$).*

A direct consequence of Theorem 11.3.3 can be given as corollary:

**Corollary 11.3.6.** *Let $(S, d_e)$ be a complete EBbDS with $w : S^2 \to \mathbb{R}_+ \backslash (0,1)$ and $\mathfrak{N} : S^2 \to \mathbb{R}_+$ is a $wt_e$-distance on S. Let $\mathfrak{I} : S \to S$ be a mapping satisfying*

*(i) $w(\vartheta,\nu)\mathfrak{N}(\mathfrak{I}\vartheta, \mathfrak{I}\nu) < [\mathfrak{N}(\vartheta,\nu)]^r$ if $d_e(\mathfrak{I}\vartheta, \mathfrak{I}\nu) \neq 0$ for $\vartheta, \nu \in S$*
*(ii) there exists $\vartheta_0 \in S$ such that $d_e(\vartheta_0, \mathfrak{I}\vartheta_0) \neq 0$.*

*Then $\mathfrak{I}$ has a unique fixed point in S.*

**Proof** If we take $\zeta_e(\vartheta, \nu) = \frac{\nu - w(\vartheta,\nu)\vartheta}{1+w(\vartheta,\nu)\nu}$ for all $\vartheta, \nu \in \mathbb{R}_+$, then result follows from Theorem 11.3.3.

## 11.4 RESULTS ON PERIODIC POINT

If $\mathfrak{I}$ is a self-map which has a fixed point $\vartheta$, then $\vartheta$ is also a fixed point of $\mathfrak{I}^n$ for arbitrary $n \in \mathbb{N}$. However, the converse is obviously false, i.e., a self-map can have a "periodic" point (a point $\vartheta$ satisfying $\mathfrak{I}^n\vartheta = \vartheta$ for some $n \in \mathbb{N}$) which is not its fixed point.

**Definition 11.4.1.** *[13] A mapping $\mathfrak{I} : S \to S$ is said to have the property (P) if it has no periodic points, i.e., if $Fix(\mathfrak{I}^n) = Fix(\mathfrak{I})$ for every $n \in \mathbb{N}$, where $Fix(\mathfrak{I}) := \{\vartheta \in S : \mathfrak{I}\vartheta = \vartheta\}$.*

**Theorem 11.4.2.** *Let $(S,d_e)$ be a complete EBbDS with $w : S^2 \to \mathbb{R}_+\backslash(0,1)$. Let $\Im : S \to S$ be a given mapping satisfying following conditions:*

*(C1) there exists a $wt_e$-distance $\mathfrak{N} : S^2 \to \mathbb{R}_+$ such that*

$$\zeta_e(w(\vartheta,\Im\vartheta)\mathfrak{N}(\Im\vartheta,\Im^2\vartheta),(\mathfrak{N}(\vartheta,\Im\vartheta))^r) \geq 0 \text{ if } d_e(\Im\vartheta,\Im^2\upsilon) \neq 0 \text{ for } \vartheta \in S \tag{11.13}$$

*where $r \in (0,1)$ is such that $\sup_{m\geq 1} \lim_{n\to\infty} w(\vartheta_n,\vartheta_m) < \frac{1}{r}$ and $\vartheta_n = \Im^n\vartheta_0$ for $\vartheta_0 \in S$;*

*(C2) there exists $\vartheta_0 \in S$ such that $d_e(\vartheta_0,\Im\vartheta_0) \neq 0$.*

*Then $\Im$ has the property $(P)$.*

**Proof:** Starting with given $\vartheta_0 \in S$, define the sequence $\{\vartheta_n\}$ by $\vartheta_n = \Im^n\vartheta_0 = \Im\vartheta_{n-1}$, $n \in \mathbb{N}$.

If there exists $n_0 \in \mathbb{N}$ such that $\vartheta_{n_0} = \vartheta_{n_0+1} = \Im\vartheta_{n_0}$, then $\vartheta_{n_0}$ is a fixed point of $\Im$. Hence, we assume $\vartheta_n \neq \vartheta_{n+1}$, i.e., $d_e(\Im\vartheta_{n-1},\Im^2\vartheta_{n-1}) > 0$ for all $n \in \mathbb{N}$.

From (11.13), we have

$$\begin{aligned}0 &\leq \zeta_e(w(\vartheta_{n-1},\Im\vartheta_{n-1})\mathfrak{N}(\Im\vartheta_{n-1},\Im^2\vartheta_{n-1}),(\mathfrak{N}(\vartheta_{n-1},\Im\vartheta_{n-1}))^r)\\ &< (\mathfrak{N}(\vartheta_{n-1},\vartheta_n))^r - w(\vartheta_{n-1},\vartheta_n)\mathfrak{N}(\vartheta_n,\vartheta_{n+1}).\end{aligned}$$

Using property of $w(\vartheta,\nu) \geq 1$ and $r \in (0,1)$, we have

$$\mathfrak{N}(\vartheta_n,\vartheta_{n+1}) \leq \mathfrak{N}(\vartheta_{n-1},\vartheta_n).$$

In similar line of argument as in the proof of Theorem 11.3.3, we get $\Im\eta = \eta$. Thus, $\Im$ has a fixed point and $Fix(\Im^n) = Fix(\Im)$ is true for $n = 1$. It is rest to prove that it is true for $n > 1$. Suppose it is not true, that is, $\rho \in Fix(\Im^n)$ and $\rho \notin Fix(\Im)$; then $d_e(\rho,\Im\rho) > 0$. Applying (11.13), we get

$$\begin{aligned}0 &\leq \zeta_e(w(\Im^{n-1}\rho,\Im(\Im^{n-1}\rho))\mathfrak{N}(\Im(\Im^{n-1}\rho),\Im^2(\Im^{n-1}\rho)),(\mathfrak{N}(\Im^{n-1}\rho,\Im(\Im^{n-1}\rho)))^r)\\ &\leq (\mathfrak{N}(\Im^{n-1}\rho,\rho))^r - w((\Im^{n-1}\rho),\rho)\mathfrak{N}(\rho,\Im\rho),\end{aligned}$$

that is,

$$w((\Im^{n-1}\rho),\rho)\mathfrak{N}(\rho,\Im\rho) \leq (\mathfrak{N}(\Im^{n-1}\rho,\rho))^r = (\mathfrak{N}(\Im^{n-1}\rho,\Im^n\rho))^r.$$

Using property of $w(\vartheta,\nu) \geq 1$ and $r \in (0,1)$, we have

$$\mathfrak{N}(\rho,\Im\rho) \leq (\mathfrak{N}(\Im^{n-1}\rho,\Im^n\rho))^r.$$

Repeating this process, we get $\mathfrak{N}(\rho,\Im\rho) \leq (\mathfrak{N}(\rho,\Im\rho))^{r^{n-1}}$, a contradiction and hence we conclude that $d_e(\rho,\Im\rho) = 0$. Therefore, $Fix(\Im^n) = Fix(\Im)$ for every $n \in \mathbb{N}$.

## 11.5 APPLICATION

Let $S = \mathscr{C}[0,1]$ is the space of all continuous functions defined on $[0,1]$. We define an EBbDS on $S$ by $d_e(\alpha,\beta) = |\alpha-\beta|^2$ where $d_e : S^2 \to \mathbb{R}_+$ and $w : S^2 \to \mathbb{R}_+\backslash(0,1)$ by $w(\alpha,\beta) = |\alpha(\rho)|+|\beta(\rho)|+2$ for $\rho \in [0,1]$.
Consider boundary value problem (BVP) of the form

$$\begin{cases} \alpha^4(\rho) = h(\rho,\alpha(\rho)) \\ \alpha(0) = \alpha'(0) = \alpha''(1) = \alpha'''(0) = 0, \quad \rho \in [0,1]. \end{cases} \tag{11.14}$$

The integral form of above equation is

$$\alpha(\rho) = \int_0^1 \Psi(\rho,\sigma)h(\sigma,\alpha(\sigma))d\sigma, \quad \alpha \in \mathscr{C}[0,1],$$

where

$$\Psi(\rho,\sigma) = \begin{cases} \dfrac{\rho^2(3\sigma-\rho)}{6} & if \ 0 \leq \rho \leq \sigma \leq 1, \\ \dfrac{\sigma^2(3\rho-\sigma)}{6} & if \ 0 \leq \sigma \leq \rho \leq 1. \end{cases}$$

On simple calculation, it is easy to check $\frac{1}{3}\rho^2\sigma^2 \leq \Psi(\rho,\sigma) \leq \frac{1}{2}\sigma^2(or(\frac{1}{2}\rho^2))$.

**Theorem 11.5.1.** *Assume that the following conditions hold:*

*(A1)* $h : [0,1]\times\mathbb{R}\times\mathbb{R}$ *is continuous;*
*(A2) there exists* $r \in (0,1)$ *such that for all* $\alpha,\beta \in S$:

$$|h(\sigma,\alpha(\sigma))| \leq \sqrt{20}\frac{|\alpha(\sigma)|^r}{|\alpha(\sigma)|+|\beta(\sigma)|+2}, \quad \sigma \in [0,1]; \tag{11.15}$$

*(A3) there exists* $\alpha_0 \in S$ *such that for* $\rho \in [0,1]$ *we have*

$$\alpha_0(\rho) \leq \int_0^1 \Psi(\rho,\sigma)h(\sigma,\alpha_0(\sigma))d\sigma. \tag{11.16}$$

*Then the problem BVP (11.14) has a solution in S.*

**Proof:** Let us define the $wt_e$-distance $\mathfrak{N}$ on $S^2$ by $\mathfrak{N}(\vartheta,\upsilon) = |\upsilon|^2$.
Let $f : S \to S$ be the operator defined by

$$f(\alpha)(\rho) = \int_0^1 \Psi(\rho,\sigma)h(\sigma,\alpha(\sigma))d\sigma.$$

Now consider $d_e(f\alpha, f\beta) \neq 0$ and

$$\begin{aligned}
\mathfrak{N}(f\alpha, f\beta) &= |f(\beta)(\rho)|^2 \\
&= \left| \int_0^1 \Psi(\rho,\sigma) h(\sigma,\beta(\sigma)) d\sigma \right|^2 \\
&\leq \int_0^1 (\Psi(\rho,\sigma))^2 |h(\sigma,\beta(\sigma)|^2 d\sigma \\
&\leq \int_0^1 (\Psi(\rho,\sigma))^2 20 \frac{|\beta(\sigma)|^{2r}}{(|\alpha(\sigma)|+|\beta(\sigma)|+2)^2} d\sigma \\
&\leq 20 \times \frac{[\mathfrak{N}(\alpha,\beta)]^r}{(|\alpha|+|\beta|+2)^2} \times \int_0^1 \frac{\sigma^4}{4} d\sigma \\
&= 20 \times \frac{[\mathfrak{N}(\alpha,\beta)]^r}{(|\alpha|+|\beta|+2)^2} \times \frac{1}{20} = \frac{[\mathfrak{N}(\alpha,\beta)]^r}{(|\alpha|+|\beta|+2)^2}
\end{aligned}$$

implies that

$$(|\alpha|+|\beta|+2)^2 \mathfrak{N}(f\alpha, f\beta) \leq [\mathfrak{N}(\alpha,\beta)]^r.$$

Following the Corollary 11.3.6, we conclude that $\eta \in f$ is a fixed point for the operator $f$ and $\eta$ is also a solution to BVP (11.14) in $S$.

## 11.6 OPEN PROBLEM

Generalized orthogonal metric space has been defined by Gordji and Habibi [21] (Also, we can refer to [20] for more results). Is it possible to combine the idea of this paper with the concept of generalized orthogonal metric space to obtain the notion of orthogonal extended Branciari b-distance metric space? Is it possible to combine the ide of wt-distance and the orthogonal extended Branciari b-distance metric space?

## BIBLIOGRAPHY


1. Abdeljawad T., Karapinar E., Panda S. K., Mlaiki N., Solutions of boundary value problems on extended-Branciari b-distance, J. Inequ. Appl., **2020**(103), (2020), 1–16.
2. Aydi H., Karapinar E., Lakzian H., Fixed point results on a class of generalized metric spaces, Math Sci., **6**(46), (2012), 1–6.
3. Azam A., Arshad M., Kannan fixed point theorems on generalized metric spaces, J. Nonlinear Sci. Appl., **1**, (2006), 45–48.
4. Rao G. V. V. J., Nashine H. K., Kadelburg Z., Best proximity point results via simulation functions in metric-like spaces, Kragujevac J. Math., **44**(3), (2020), 401–413.
5. Bakhtin I. A., (1999), The contraction mapping principle in quasi metric spaces, Funkc. Anal. Ulianowsk Gos. Ped. Inst., **30**, (1999), 243–253.
6. Czerwik S., Contraction mappings in b-metric spaces, Acta Math. Inform. Univ. Ostraviensis., **5**, (1993), 5–11.
7. Jleli, M., Samet B,, A new generalization of the Banach contraction principle, J. Inequal. Appl., **2014**(38), (2014), 1–8.
8. Czerwik S., Nonlinear set-valued contraction mappings in b-metric spaces, Atti Sem. Mat. Fis. Univ. Modena., **46**, (1998), 263–276.

9. Branciari A., A fixed point theorem for mappings satisfying a general contractive condition of integral type, Int. J. Math. Math. Sci., **29**(9), (2002), 531–536.
10. Hussain N., Saadati R., Agarwal R. P., On the topology and wt-distance on metric type spaces, Fixed Point Theory Appl., **2014**(88), (2014), 1–14.
11. Jain R., Fixed point theorems for generalized weak contraction mapping in generating space of b-dislocated metric spaces, Asian-European J. Math., **14**(7), (2021), 215–120.
12. Jain R., Fixed point results in partially ordered b-metric spaces, Inter. J. Pure Appl. Mathematics., **120**(8), (2018), 177–185.
13. Jeong G. S., Rhoades B. E., Maps for which $F(T) = F(T^n)$, Fixed Point Theory Appl., **6**, (2005), 87–131.
14. Kada O., Suzuki T., Takahashi W., Nonconvex minimization theorems and fixed point theorems in complete metric spaces, Math. Japon., **44**, (1996), 381–391.
15. Kamran T., Samreen T., Ain Q. UL., A generalization of b-metric space and some fixed point theorems, Mathematics., **5**(2), (2017), 19.
16. Khojasteh F., Shukla S., Radenović S., A new approach to the study of fixed point theorems via simulation functions, Filomat., **96**, (2015), 1189–1194.
17. Mitrović Z., Işik H., Radenović S., The new results in extended b-metric spaces and applications, Int. J. Nonlinear Anal. Appl., **11**(1), (2020), 473–482.
18. Nashine H. K., Aydi H., Generalized altering distances and common fixed points in ordered metric spaces, Inter. j. Math. Math. Sci., **2012**, (2012), 1–23.
19. Roldan-Lopez-de-Hierro A., Karapinar E., Roldán-López-de-Hierro C., Martínez-Moreno J., Coincidence point theorems on metric spaces via simulation function, J. Comput. Appl. Math., **275**, (2015), 345–355.
20. Gordji M. E., Ramezani M., Sen M., Cho Y. J., On orthogonal sets and Banach fixed point theorem, Fixed Point Theory, **18**(2), (2017), 569–578.
21. Gordji M. E., Habibi H., Fixed point theory in generalized orthogonal metric space, J Linear Topolog. Algebra., **6**(3), (2017), 251–260.

# 12 Solution and Stability of New Type of Quadratic Multiplicative Functional Equation

*Jyotsana Jakhar, Shalu Sharma, Jagjeet Jakhar*

In this chapter, the authors find the solution and stability of the following quadratic-multiplicative functional equation in eight variables;

$$\begin{aligned}&\mu(-s_4t_4-s_3t_3-s_2t_2+s_1t_1)+\mu(-s_4t_3+s_3t_4+s_2t_1+s_1t_2)\\&+\mu(s_4t_2+s_3t_1-s_2t_4+s_1t_3)+\mu(-s_4t_1-s_3t_2-s_2t_3+s_1t_4)\\&=(\mu(s_4)+\mu(s_3)+\mu(s_2)+\mu(s_1))(\mu(t_4)+\mu(t_3)+\mu(t_2)+\mu(t_1)).\end{aligned}$$

Our approach involves the utilization of the direct method and fixed point method to derive the desired results.

## 12.1 INTRODUCTION

The emergence of functional equations coincided with the modern formulation of the function concept. The first publications regarding functional equations were authored by D'Alembert [1] during the period between 1747 and 1750. Due to their apparent simplicity and harmonic characteristics, functional equations have captured the interest of numerous renowned mathematicians. In 1940, the first stability problem for functional equations was established by Ulam [2]. The problem is "Let G be a group and $G'$ be a metric group with metric $d(.,.)$. Given $\varepsilon > 0$ does there exists a $\delta > 0$ such that if a function $f : G \to G'$ satisfies the inequality $d(f(xy), f(x)f(y)) < \delta$ for all $x, y \in G$, then there exists homomorphism $H : G \to G'$ with $d(f(x), H(x)) < \varepsilon$ for all $x \in G$?"Hyers [3] provided a partial and positive response to Ulam's question within the framework of Banach spaces. Several mathematicians [4, 5, 6, 7, 8, 9, 10] have been studying several topics related to functional equations since that time. See

DOI: 10.1201/9781003596578-12

[11, 12, 13, 14, 15, 16], for additional information on stability problems with functional equations.
A mapping $\mu:\Omega_1\to\Omega_2$ where, $\Omega_1,\Omega_2$ are real vector spaces, is called the solution of the following functional equation if $\mu$ is the quadratic function;

$$\mu(s-t)+\mu(s+t)=2\mu(s)+2\mu(t). \tag{12.1}$$

Thus, we can say equation (12.1) is the quadratic functional equation. It is widely acknowledged that for every quadratic mapping $\mu:\Omega_1\to\Omega_2$, there is a distinct symmetric biadditive mapping $\mathscr{S}:\Omega_1\times\Omega_1\to\Omega_2$ such that $\mathscr{S}(s,s)=\mu(s)$. Skof [17] was the first mathematician to demonstrate the quadratic functional equation for mapping $\mu:\Omega_1\to\Omega_2$. In this context, $\Omega_1$ refers to a normed space and $\Omega_2$ refers to a Banach space. Cholewa [18] later demonstrated Skof's theorem, stating that the result is still valid even if we take $\Omega_1$ to be an Abelian group $\mathscr{G}$.

**Theorem 12.1.1.** *[17] "Let $\mathscr{G}$ and $\mathscr{R}$ be an abelian group and a real Banach space respectively. If a mapping $\mu:\mathscr{G}\to\mathscr{R}$ satisfies the inequality*

$$||\mu(s+t)+\mu(s-t)-2\mu(s)-2\mu(t)||\leq\delta$$

*for some $\delta>0$ and all $s,t\in\mathscr{G}$, then there exists a unique quadratic mapping $\mathscr{I}:\mathscr{G}\to\mathscr{R}$ such that*

$$||\mu(s)-\mathscr{I}(s)||\leq\frac{1}{2}\delta."$$

**Theorem 12.1.2.** *[18] "Let $\Omega_1$ be a real normed space and $\Omega_2$ be a real Banach space also assume $p\neq 2$ be a positive number. If a mapping $\mu:\Omega_1\to\Omega_2$ satisfies the inequality*

$$||\mu(s+t)+\mu(s-t)-2\mu(s)-2\mu(t)||\leq\varepsilon(||s||^p+||t||^p)$$

*for some $\varepsilon>0$ and all $s,t\in\Omega_1$, then there exists a unique quadratic mapping $\mathscr{I}:\Omega_1\to\Omega_2$ such that*

$$||(\mu(s)-\mathscr{I}(s)||\leq\frac{2\varepsilon}{|4-2^p|}||s||^p."$$

In 1991, Volkmann [19] and Hammer analysed a mapping $\mu:\mathscr{R}\to\mathscr{R}$ which satisfy

$$\mu(s-t)+\mu(s+t)=2\mu(s)+2\mu(t) \tag{12.2}$$

and

$$\mu(ts)=\mu(t)\mu(s)\quad\forall\quad s,t\in\Omega_1, \tag{12.3}$$

then they obtained only continuous solutions for the functional equations (12.2) and (12.3) are $\mu(s)=0$ and $\mu(s)=s^2$.

**Theorem 12.1.3.** *[19] "A function $\mu:\mathscr{R}\to\mathscr{R}$ satisfies the system of the following functional equations*

$$\mu(s+t)+\mu(s-t)=2\mu(s)+2\mu(t)$$

*and*

$$\mu(st) = \mu(s)\mu(t) \quad \forall \quad s,t \in \mathscr{R}$$

*if and only if there exists an additive multiplicative function* $\xi : \mathscr{C} \to \mathscr{C}$ *such that* $f(s) = |\xi(s)|^2$ *for all* $s \in \mathscr{R}$.*"*

**Theorem 12.1.4.** *[20] "A function* $\mu : \mathscr{R} \to \mathscr{R}$ *satisfies the system of the following functional equations:*

$$\mu(s+t)+\mu(s-t) = 2\mu(s)+2\mu(t), \quad \mu(st) = \mu(s)\mu(t) \quad \forall \quad s,t \in \mathscr{R},$$

*if and only if* $\mu \equiv 0$ *or* $\mu(s) = s$ *for all* $s \in \mathscr{R}$.*"*

The stability problems of multiplicative functional equations in several spaces have been broadly investigated by many mathematicians. Motivated by the approach of research by various mathematicians, an effort has been made in this chapter to obtain the solution and stability of the following new type of quadratic multiplicative functional equation in real Banach spaces;

$$\begin{aligned}&\mu(-s_4t_4 - s_3t_3 - s_2t_2 + s_1t_1) + \mu(-s_4t_3 + s_3t_4 + s_2t_1 + s_1t_2)\\&+\mu(s_4t_2 + s_3t_1 - s_2t_4 + s_1t_3) + \mu(-s_4t_1 - s_3t_2 - s_2t_3 + s_1t_4)\\&= (\mu(s_4)+\mu(s_3)+\mu(s_2)+\mu(s_1))(\mu(t_4)+\mu(t_3)+\mu(t_2)+\mu(t_1)).\end{aligned}$$

The stability and solution of this functional equation have broad applications in diverse scientific and engineering fields, ranging from control theory and differential equations to economics, optimization, and physics. This study of stability helps ensure the reliability and predictability of mathematical models and systems in various real-world scenarios.

The main objective of this study is to establish the generalized Hyers-Ulam stability for quadratic multiplicative functional equation within a real Banach space. This chapter is organized into four sections. The section first serves as the introduction. In section second, we derive the general solution of the equation. Moving on to sections third and fourth, our focus shifts to finding the generalized Hyers-Ulam stability of the given functional equation by applying the direct method and fixed point method respectively. Throughout this article, $\mathscr{R}$ and $\mathscr{C}$ represent the set of all real numbers and the set of all complex numbers, respectively.

## 12.2 GENERAL SOLUTION FOR QUADRATIC MULTIPLICATIVE FUNCTIONAL EQUATION

In this segment, the authors find the general solution of the following functional equation:

$$\begin{aligned}&\mu(-s_4t_4 - s_3t_3 - s_2t_2 + s_1t_1) + \mu(-s_4t_3 + s_3t_4 + s_2t_1 + s_1t_2)\\&+\mu(s_4t_2 + s_3t_1 - s_2t_4 + s_1t_3) + \mu(-s_4t_1 - s_3t_2 - s_2t_3 + s_1t_4)\\&= (\mu(s_4)+\mu(s_3)+\mu(s_2)+\mu(s_1))(\mu(t_4)+\mu(t_3)+\mu(t_2)+\mu(t_1)).\end{aligned} \tag{12.4}$$

**Theorem 12.2.1.** *Let us consider a function $\mu : \mathscr{R} \to \mathscr{R}$ which satisfies (12.4) for all $s_1, s_2, s_3, s_4, t_1, t_2, t_3, t_4 \in \mathscr{R}$, then $\mu \equiv 0$, $4\mu \equiv 1$ or $\mu(s) = s^2$ for all $s \in \mathscr{R}$.*

**Proof:** Consider the function $\mu$ to be constant. Then $\mu \equiv 0$ or $4\mu \equiv 1$. Let us proceed with the assumption that $\mu$ is a non-constant function. Take $s_4 = t_4 = s_3 = t_3 = s_2 = t_2 = t_1 = 0$ in (12.4),

$$4\mu(0) = 4\mu(0)[\mu(s_1) + 3\mu(0)], \quad s_1 \in \mathscr{R}.$$

Since $\mu$ is the non-constant function, authors can conclude that $\mu(0) = 0$. Taking $s_2 = t_2 = s_3 = t_3 = s_4 = t_4 = 0$ in (12.4), we acquired

$$\mu(s_1 t_1) = \mu(s_1)\mu(t_1) \quad s_1, t_1 \in \mathscr{R}. \tag{12.5}$$

Taking $t_1 = 1$ in above equation, we have $\mu(1) = 1$, since $\mu$ is non-constant function. Let, $t_1 = t_2 = 1, s_3 = t_3 = s_4 = t_4 = 0$ in (12.1) and using $\mu(1) = 1$, we get

$$\mu(s_1 - s_2) + \mu(s_1 + s_2) = 2\mu(s_1) + 2\mu(s_2).$$

Using Theorem 12.3.1, we can conclude that there is an additive multiplicative function $\xi : \mathscr{C} \to \mathscr{C}$ which satisfies $\mu(t_1) = |\xi(t_1)|^2$ for each $t_1 \in \mathscr{R}$. Since $\mu(1) = 1$, taking $s_2 = t_2 = 1, s_4 = t_4 = s_3 = t_3 = 0$, we acquire

$$\Re[\xi(s_1)\overline{\xi(t_1)}] = \Re[[\xi(s_1)\xi(t_1)], \quad s_1, t_1 \in \mathscr{R},$$

where, $\Re$ represent the real part. Then, $\xi(s_1) \in \mathscr{R}$ or $\xi(t_1) \in \mathscr{R}$. This means that $\xi(\mathscr{R}) \subseteq \mathscr{R}$.
Hence, mapping $\xi : \mathscr{R} \to \mathscr{R}$ is an additive-multiplicative mapping and according to Theorem 12.2.1 we can deduce that $\xi(t_1) = t_1$ for all $t_1 \in \mathscr{R}$. Therefore, $f(t_1) = {t_1}^2$ for all $t_1 \in \mathscr{R}$.

## 12.3 STABILITY OF THE QUADRATIC MULTIPLICATIVE FUNCTIONAL EQUATION USING DIRECT METHOD

With the intention of investigating the stability, the authors embarked on determining the stability of the subsequent functional equation:

$$\begin{aligned}
&\mu(-s_4t_4 - s_3t_3 - s_2t_2 + s_1t_1) + \mu(-s_4t_3 + s_3t_4 + s_2t_1 + s_1t_2) \\
&+\mu(s_4t_2 + s_3t_1 - s_2t_4 + s_1t_3) + \mu(-s_4t_1 - s_3t_2 - s_2t_3 + s_1t_4) \\
&= (\mu(s_4) + \mu(s_3) + \mu(s_2) + \mu(s_1))(\mu(t_4) + \mu(t_3) + \mu(t_2) + \mu(t_1))
\end{aligned} \tag{12.6}$$

by applying the direct method.

**Theorem 12.3.1.** *Let us suppose $\zeta \in \{1,-1\}$ and a function $\nu : \mathscr{R}^8 \to [0,\infty)$ which satisfies*

$$\sum_{l=0}^{\infty} \nu\left(\frac{2^{l\zeta}s_1, 2^{l\zeta}t_1, 2^{l\zeta}s_2, 2^{l\zeta}t_2, 2^{l\zeta}s_3, 2^{l\zeta}t_3, 2^{l\zeta}s_4, 2^{l\zeta}t_4}{4^{l\zeta}}\right) < \infty$$

*for all $s_1, s_2, s_3, s_4, t_1, t_2, t_3, t_4 \in \mathscr{R}$. Also consider a function $\mu : \mathscr{R} \to \mathscr{R}$ which follows the conditions $\mu(0) = 0$ and*

$$|D_\mu(s_1, s_2, s_3, s_4, t_1, t_2, t_3, t_4)| \leq \nu(s_1, t_1, s_2, t_2, s_3, t_3, s_4, t_4) \tag{12.7}$$

*where,*

$$|D_\mu(s_1, t_1, s_2, t_2, s_3, t_3, s_4, t_4)| = \mu(-s_4t_4 - s_3t_3 - s_2t_2 + s_1t_1) +$$
$$\mu(-s_4t_3 + s_3t_4 + s_2t_1 + s_1t_2) + \mu(s_4t_2 + s_3t_1 - s_2t_4 + s_1t_3) +$$
$$\mu(-s_4t_1 - s_3t_2 - s_2t_3 + s_1t_4) - (\mu(s_4) + \mu(s_3) + \mu(s_2) + \mu(s_1))$$
$$(\mu(t_4) + \mu(t_3) + \mu(t_2) + \mu(t_1)).$$

*Then, there is exactly one quadratic function $\mathscr{I} : \mathscr{R} \to \mathscr{R}$ which follows equation (12.6) and*

$$|\mu(s_1) - \mathscr{I}(s_1)| \leq \begin{cases} \frac{1}{4}\sum_{l=0}^{\infty} \frac{g(2^l s_1, 1, 2^l s_1, 1, 2^l s_1, 1, 2^l s_1, 1)}{4^l} \\ +\sum_{l=0}^{\infty} \frac{\nu(2^l s_1, 1, 0, 0, 0, 0, 0, 0)}{4^n}, \zeta = 1, \\ \frac{1}{4}\sum_{l=0}^{\infty} 4^l \nu(\frac{s_1}{2^l}, 1, \frac{s_1}{2^l}, 1, \frac{s_1}{2^l}, 1, \frac{s_1}{2^l}, 1) \\ +\sum_{l=0}^{\infty} 4^l \mu(\frac{s_1}{2^l}, 1, 0, 0, 0, 0, 0, 0), \zeta = -1 \end{cases} \tag{12.8}$$

*for all $s_1 \in \mathscr{R}$, where $\mathscr{I}(s_1) = {s_1}^2$ if $\mathscr{I}(1) = 1$ and $\mathscr{I} \equiv 0$ if $\mathscr{I}(1) \neq 1$.*

**Proof:** We assume that $\zeta = 1$. Let $s_4 = t_4 = s_3 = t_3 = s_2 = t_2 = 0$ in (12.7, we obtain

$$|\mu(s_1t_1) - \mu(s_1)\mu(t_1)| \leq \nu(s_1, t_1, 0, 0, 0, 0, 0, 0) \quad s_1, t_1 \in \mathscr{R}. \tag{12.9}$$

Put $s_1 = 1$ in (12.9), and we obtain

$$|\mu(s_1) - \mu(s_1)\mu(1)| \leq \nu(s_1, 1, 0, 0, 0, 0, 0, 0) \quad s_1 \in \mathscr{R}. \tag{12.10}$$

Letting $t_1 = t_2 = 1, s_4 = s_3 = t_4 = t_3 = 0$ in (12.7), we obtain

$$|\mu(s_1 - s_2) + \mu(s_2 + s_1) - 2\mu(1)[\mu(s_1) + \mu(s_2)]| \leq \nu(s_1, 1, s_2, 1, 0, 0, 0, 0). \tag{12.11}$$

Then, by using the equations (12.10) and (12.11), authors get

$$|\mu(s_1 - s_2) + \mu(s_1 + s_2) - 2[\mu(s_2) + \mu(s_1)]| \leq \Psi(s_1, s_2), \quad s_2, s_1 \in \mathscr{R},$$

where $\Psi(s_1, s_2) = \nu(s_1, 1, s_2, 1, 0, 0, 0, 0) + 2\nu(s_1, 1, 0, 0, 0, 0, 0, 0) + \nu(s_2, 1, 0, 0, 0, 0, 0, 0)$. By [21], there exists one and only one quadratic function $\mathscr{I} : \mathscr{R} \to \mathscr{R}$ in such a way

$$\mathscr{I}(s_1) = lim_{l\to\infty} \frac{f(2^l s_1)}{4^l}, \quad ||\mu(s_1) - \mathscr{I}(s_1)|| \leq \frac{1}{4}\sum_{l=0}^{\infty} \frac{\Psi(2^l s_1, 2^l s_1)}{4^l}.$$

Then from (12.7) and (12.10), we get

$$\mathscr{I}(-s_4t_4-s_3t_3-s_2t_2+s_1t_1)+\mathscr{I}(-s_4t_3+s_3t_4+s_2t_1+s_1t_2)+$$
$$\mathscr{I}(s_4t_2+s_3t_1-s_2t_4+s_1t_3)+\mathscr{I}(-s_4t_1-s_3t_2-s_2t_3+s_1t_4)=$$
$$(\mathscr{I}(s_4)+\mathscr{I}(s_3)+\mathscr{I}(s_2)+\mathscr{I}(s_1))(\mathscr{I}(t_4)+\mathscr{I}(t_3)+\mathscr{I}(t_2)+\mathscr{I}(t_1)).$$

Since $\mathscr{I}(0)=0$, in the end, we find that $\mathscr{I}(1)\in\{0,1\}$. Therefore, according to Theorem 12.3.1), we obtain $\mathscr{I}(s_1)={s_1}^2$ for all $s_1\in\mathscr{R}$, if $\mathscr{I}(1)=1$ and $\mathscr{I}\equiv 0$ if $\mathscr{I}(1)=0$, after that we acquire (3).
In the same way, we can get the result for $\zeta=-1$.
As a result, the theorem was demonstrated.

**Corollary 12.3.2.** *Consider the function $\mu:\mathscr{R}\to\mathscr{R}$ satisfying*

$$|D_f(s_1,s_2,s_3,s_4,t_1,t_2,t_3,t_4)|\le\varepsilon,\quad s_1,s_2,s_3,s_4,t_1,t_2,t_3,t_4\in\mathscr{R} \tag{12.12}$$

*for some $\varepsilon\ge 0$. Then, there are the subsequent statements:*

*$\mu$ is bounded, if $\mu(0)\neq 0$,*
*if $\mu(0)=0$, then there is exactly one quadratic mapping $\mathscr{I}:\mathscr{R}\to\mathscr{R}$, which follows equation (12.6) and*

$$|\mu(s_1)-\mathscr{I}(s_1)|\le\frac{5}{3}\varepsilon,\quad s_1\in\mathscr{R},$$

*where, $\mathscr{I}(s_1)={s_1}^2$ if $\mathscr{I}(1)=1$ and $\mathscr{I}\equiv 0$ if $\mathscr{I}(1)\neq 1$.*

**Proof:** Suppose $\mu(0)\neq 0$ . Let $s_4=s_3=s_2=t_4=t_3=t_2=t_1=0$ in (3.7), we obtain

$$|4\mu(0)-4[\mu(s_1)+\mu(0)]\mu(0)|\le\varepsilon,\quad s_1\in\mathscr{R}.$$

Then, we have

$$|\mu(s_1)|\le|\mu(0)-1|+\frac{\varepsilon}{2|\mu(0)|},\quad s_1\in\mathscr{R}.$$

If $\mu(0)=0$, then the conclusion comes from Theorem 12.3.1.

**Corollary 12.3.3.** *Let us assume $\varepsilon\ge 0$ and a function $\mu:\mathscr{R}^8\to[0,\infty)$. Also consider a function $\nu:\mathscr{R}\to\mathscr{R}$ which satisfying*

$$|D_\mu(s_1,s_2,s_3,s_4,t_1,t_2,t_3,t_4)|\le\varepsilon\nu(s_1,t_1,s_2,t_2,s_3,t_3,s_4,t_4).$$

*Let $\{e_1,e_2,e_3,e_4,e_5,e_6,e_7,e_8\}$ be the standard basis in $\mathscr{R}^8$ and a function $\nu_i:\mathscr{R}\to\mathscr{R}$ defined as $\nu_i(s_1)=\nu(s_1e_i)$ for a given*
*$i\in\{1,2,3,4,5,6,7,8\}$. If $\nu_i$ are bounded for some $i$ and $\mu(0)\neq 0$ then function $\nu$ is bounded.*

**Corollary 12.3.4.** *If $\rho \in (0,2)$, $\varepsilon \geq 0$ and consider a function $\mu : \mathscr{R} \to \mathscr{R}$ which fulfils the following inequality*

$$|D_\mu(s_1,s_2,s_3,s_4,t_1,t_2,t_3,t_4)| \leq \varepsilon(|s_1|^\rho + |t_1|^\rho + |s_2|^\rho + |t_2|^\rho + |s_3|^\rho + |t_3|^\rho + |s_4|^\rho + |t_4|^\rho).$$

*Then there are the following axioms:*

1. *If $\mu(0) \neq 0$, then*
$$|\mu(s_1) - \frac{1}{2}| \leq \varepsilon|s_1|^\rho, \quad s_1 \in \mathscr{R}.$$
2. *If $\mu(0) = 0$ then, there is exactly one quadratic mapping $\mathscr{I} : \mathscr{R} \to \mathscr{R}$ exists which follows equation (3.1) and inequality*
$$|\mu(s) - \mathscr{I}(s_1)| \leq \frac{6\varepsilon}{4-2^\rho}|s_1|^\rho + 2\varepsilon \quad s_1 \in \mathscr{R},$$
*where $\mathscr{I} \equiv 0$ if $\mathscr{I}(1) \neq 1$ and $\mathscr{I}(s_1) = {s_1}^2$ if $\mathscr{I}(1) = 1$.*

**Theorem 12.3.5.** *If $\rho > 4$, $\varepsilon \geq 0$ and consider a function $\mu : \mathscr{R} \to \mathscr{R}$ that fulfills the inequality*

$$|D_\mu(s_1,s_2,s_3,s_4,t_1,t_2,t_3,t_4)| \leq \varepsilon(|s_1|^\rho + |t_1|^\rho + |s_2|^\rho + |t_2|^\rho + |s_3|^\rho + |t_3|^\rho + |s_4|^\rho|t_4|^\rho). \tag{12.13}$$

*Then, there are the subsequent statements:*

1. *If $\mu(0) \neq 0$, then*
$$|\mu(s_1) - \frac{1}{2}| \leq \varepsilon|s_1|^\rho, \quad s_1 \in \mathscr{R}.$$
2. *if $\mu(0) = 0$, then there is exactly one quadratic mapping $\mathscr{I} : \Omega_1 \to \Omega_1$ exists which follows equation (12.6) and the inequality*
$$|\mu(s_1) - \mathscr{I}(s_1)| \leq \frac{6\varepsilon}{\sqrt{2}^\rho - 4}(1 + |s_1|^\rho), \quad s_1 \in \mathscr{R}, \tag{12.14}$$
*where $\mathscr{I} \equiv 0$ if $\mathscr{I}(1) \neq 1$ and $\mathscr{I}(s_1) = {s_1}^2$ if $\mathscr{I}(1) = 1$.*

**Proof:** It is sufficient to show statement (2). Let, $s_4 = s_3 = s_2 = t_4 = t_3 = t_2 = 0$, we obtain

$$|\mu(s_1t_1) - \mu(s_1)\mu(t_1)| \leq \varepsilon(|s_1|^\rho + |t_1|^\rho) \quad s_1,t_1 \in \mathscr{R}. \tag{12.15}$$

Taking, $s_2 = s_1, s_4 = s_3 = 0, t_2 = t_1, t_4 = t_3 = 0$ in (12.13), we get,

$$|\mu(2s_1t_1) - 4\mu(s_1)\mu(t_1)| \leq 2\varepsilon(|s_1|^\rho + |t_1|^\rho) \quad s_1,t_1 \in \mathscr{R}. \tag{12.16}$$

Using (12.15) and (12.16), we get

$$|\mu(2s_1t_1) - 4\mu(s_1t_1)| \leq 6\varepsilon(|s_1|^\rho + |t_1|^\rho) \quad s_1,t_1 \in \mathscr{R}. \tag{12.17}$$

Replacing $s_1$ and $t_1$ by $\frac{s_1}{2^{\frac{(n+1)}{2}}}$ and $\frac{1}{2^{\frac{(n+1)}{2}}}$, respectively, in the (12.17) and multiplying the resulting inequality by $4^n$, we get

$$|4^n\mu(\frac{s_1}{2^n}) - 4^{n+1}\mu\left(\frac{s_1}{2^{n+1}}\right)| \leq \frac{3}{2}\left(\frac{4}{2^{\frac{\rho}{2}}}\right)^{n+1} \varepsilon(1+|s_1|^\rho), \quad s_1 \in \mathscr{R}. \tag{12.18}$$

Then,

$$\begin{aligned}\left|4^n\mu(\frac{s_1}{2^n}) - 4^m\mu(\frac{s_1}{2^m})\right| &= \left|\sum_{i=n}^{m-1}\left[4^i\mu(\frac{s_1}{2^i}) - 4^{i+1}\mu\left(\frac{s_1}{2^{i+1}}\right)\right]\right| \\ &\leq \frac{3}{2}\varepsilon(1+|s_1|^\rho)\sum_{i=n+1}^{m}\left(\frac{4}{2^{\frac{\rho}{2}}}\right)^i \quad s_1 \in \mathscr{R}, m > n \geq 0. \end{aligned} \tag{12.19}$$

As a result, it is implied that the sequence $4^n\{\mu(\frac{s_1}{2^n})\}$ is a Cauchy and is convergent for all $s_1 \in \mathscr{R}$. Now, we set-up a function $\mathscr{I} : \mathscr{R} \to \mathscr{R}$ by

$$\mathscr{I}(s_1) = \lim_{n\to\infty} 4^n\mu\left(\frac{s_1}{2^n}\right).$$

Since $\mu(0) = 0$, we get $\mathscr{I}(0) = 0$. By (12.13) and (12.15), we get $\mathscr{I}$ follows (12.6) and $\mathscr{I}(s_1t_1) = \mathscr{I}(s_1)\mathscr{I}(t_1)$ for all $s_1, t_1 \in \mathscr{R}$. Thus, by Theorem 12.3.1, the authors acquire $\mathscr{I} \equiv 0$ if $\mathscr{I}(1) \neq 1$ and $\mathscr{I}(s_1) = {s_1}^2$ if $\mathscr{I}(1) = 1$. Let $m \to \infty$ and $n = 0$ in (12.3), we get (12.14).

## 12.4 STABILITY OF THE QUADRATIC MULTIPLICATIVE FUNCTIONAL EQUATION USING FIXED POINT METHOD

In this segment, the authors set out to determine the stability of the following functional equation:

$$\begin{aligned}&\mu(-s_4t_4 - s_3t_3 - s_2t_2 + s_1t_1) + \mu(-s_4t_3 + s_3t_4 + s_2t_1 + s_1t_2)\\ &+\mu(s_4t_2 + s_3t_1 - s_2t_4 + s_1t_3) + \mu(-s_4t_1 - s_3t_2 - s_2t_3 + s_1t_4)\\ &= (\mu(s_4) + \mu(s_3) + \mu(s_2) + \mu(s_1))(\mu(t_4) + \mu(t_3) + \mu(t_2) + \mu(t_1))\end{aligned} \tag{12.20}$$

by applying the fixed point method.

**Theorem 12.4.1.** *Let $0 < L < 1$ and $\nu : \mathscr{R}^8 \to [0,\infty)$ be a function such*

$$\begin{aligned}&\lim_{l\to\infty}\frac{\nu(2^ls_1, 2^lt_1, 2^ls_2, 2^lt_2, 2^ls_3, 2^lt_3, 2^ls_4, 2^lt_4)}{4^l} = 0,\\ &\Phi(2s_1, 2t_1) \leq 4L\Phi(s_1, t_1),\end{aligned} \tag{12.21}$$

*where*

$$\begin{aligned}\Phi(s_1, t_1) &= \nu(s_1, 1, s_2, 1, 0, 0, 0, 0) + 2\nu(s_1, 1, 0, 0, 0, 0, 0, 0)\\ &\quad +2\nu(s_2, 1, 0, 0, 0, 0, 0, 0).\end{aligned}$$

*Consider a function $\mu : \mathscr{R} \to \mathscr{R}$ which satisfies $\mu(0) = 0$ and*

$$|D_\mu(s_1,s_2,s_3,s_4,t_1,t_2,t_3,t_4)| \le \nu(s_1,t_1,s_2,t_2,s_3,t_3,s_4,t_4). \tag{12.22}$$

*Then there exists exactly one quadratic function $\mathscr{I} : \mathscr{R} \to \mathscr{R}$, which follows the equation (12.20) and the following inequality*

$$|\mu(s_1) - \mathscr{I}(s_1)| \le \frac{1}{4(1-L)}\Phi(s_1,s_1), \quad s_1 \in \mathscr{R}, \tag{12.23}$$

*where $\mathscr{I} \equiv 0$ if $\mathscr{I}(1) \neq 1$ and $\mathscr{I}(s_1) = {s_1}^2$ if $\mathscr{I}(1) = 1$.*

**Proof:** Take $s_4 = t_4 = s_3 = t_3 = s_2 = t_2 = 0$ in (12.22), we obtain

$$|\mu(s_1t_1) - \mu(s_1)\mu(t_1)| \le \nu(s_1,t_1,0,0,0,0,0,0) \quad s_1,t_1 \in \mathscr{R}. \tag{12.24}$$

Put $t_1 = 1$ in (12.24), and we get

$$|\mu(s_1) - \mu(s_1)\mu(1)| \le \nu(s_1,1,0,0,0,0,0,0) \quad s_1 \in \mathscr{R}. \tag{12.25}$$

Letting $t_2 = t_1 = 1, s_4 = t_4 = s_3 = t_4 = 0$ in (12.22), we obtain

$$|\mu(s_2+s_1) + \mu(s_1s_2) - 2\mu(1)[\mu(s_2) + \mu(s_1)]| \le \nu(s_1,1,t_2,1,0,0,0,0). \tag{12.26}$$

Then, it follows from (12.26) that

$$|\mu(s_1-s_2) + \mu(s_1+s_2) - 2[\mu(s_1) + \mu(s_2)]| \le \Phi(s_1,s_2), \quad s_2,s_1 \in \mathscr{R},$$

where

$$\begin{aligned}\Phi(s_1,s_2) \quad &= \quad \nu(s_1,1,s_2,1,0,0,0,0) + 2\nu(s_1,1,0,0,0,0,0,0) \\ &\quad + \nu(s_2,1,0,0,0,0,0,0).\end{aligned}$$

By [Theorem 4, [22]], there is only one quadratic mapping $\mathscr{I} : \mathscr{R} \to \mathscr{R}$, which ensure that

$$\mathscr{I}(s_1) = \lim_{l\to\infty} \frac{\mu(2^l s_1)}{4^l}, \quad |\mu(s_1) - \mathscr{I}(s_1)| \le \frac{1}{4(1-L)}\Phi(s_1,s_1) \quad s_1 \in \mathscr{R}. \tag{12.27}$$

Then from (12.20), (12.21) and using the definition of $\mathscr{I}$, we get

$$\begin{aligned}&\mathscr{I}(-s_4t_4 - s_3t_3 - s_2t_2 + s_1t_1) + \mathscr{I}(-s_4t_3 + s_3t_4 + s_2t_1 + s_1t_2) + \\ &\mathscr{I}(s_4t_2 + s_3t_1 - s_2t_4 + s_1t_3) + \mathscr{I}(-s_4t_1 - s_3t_2 - s_2t_3 + s_1t_4) = \\ &(\mathscr{I}(s_4) + \mathscr{I}(s_3) + \mathscr{I}(s_2) + \mathscr{I}(s_1))(\mathscr{I}(t_4) + \mathscr{I}(t_3) + \mathscr{I}(t_2) + \mathscr{I}(t_1)).\end{aligned}$$

Since $\mathscr{I}(0) = 0$, by Theorem 12.3.1, the authors obtain $\mathscr{I} \equiv 0$ if $\mathscr{I}(1) \neq 1$ and $\mathscr{I}(s_1) = {s_1}^2$ if $\mathscr{I}(1) = 1$. Then the authors obtain (12.23).

**Theorem 12.4.2.** *Let $0 < L < 1$ and $\nu : \mathscr{R}^8 \to [0,\infty)$ be a function such that*

$$\lim_{l\to\infty} 4^l \nu\left(\frac{s_1}{2^l}, \frac{t_1}{2^l}, \frac{s_2}{2^l}, \frac{t_2}{2^l}, \frac{s_3}{2^l}, \frac{t_3}{2^l}, \frac{s_4}{2^l}, \frac{t_4}{2^l}\right) = 0, \quad \Phi(s_1,t_1) = \frac{L}{4}\Phi(2s_1,2t_1),$$

*where*

$$\Phi(s_1,t_1) = \nu(s_1,1,t_1,1,0,0,0,0) + 2\nu(s_1,1,0,0,0,0,0,0) + 2\nu(t_1,1,0,0,0,0,0,0).$$

*Consider a function $\mu : \mathscr{R} \to \mathscr{R}$ which satisfies $\mu(0) = 0$ and*

$$|D_\mu(s_1,s_2,s_3,s_4,t_1,t_2,t_3,t_4)| \leq \nu(s_1,t_1,s_2,t_2,s_3,t_3,s_4,t_4). \tag{12.28}$$

*Then there is exactly one quadratic function $\mathscr{I} : \mathscr{R} \to \mathscr{R}$, which follows the equation (12.20) and the following inequality*

$$|\mu(s_1) - \mathscr{I}(s_1)| \leq \frac{1}{4(1-L)}\Phi(s_1,s_1), \quad s_1 \in \mathscr{R}, \tag{12.29}$$

*where $\mathscr{I} \equiv 0$ if $\mathscr{I}(1) \neq 1$ and $\mathscr{I}(s_1) = {s_1}^2$ if $\mathscr{I}(1) = 1$.*

**Theorem 12.4.3.** *Consider a mapping $\mu : \mathscr{R} \to \mathscr{R}$ which satisfies $\mu(0) = 0$ and*

$$|D_\mu(s_1,s_2,s_3,s_4,t_1,t_2,t_3,t_4)| \leq \varepsilon \tag{12.30}$$

*where, $\varepsilon \geq 0$ and*

$$\max\{|s_1,s_2,s_3,s_4,t_1,t_2,t_3,t_4|\} \geq d.$$

*Then there is exactly one quadratic function $\mathscr{I} : \mathscr{R} \to \mathscr{R}$,which follows the equation (12.20) and the following inequality*

$$|\mu(s_1) - \mathscr{I}(s_1)| \leq \varepsilon, \tag{12.31}$$

*where $\mathscr{I} \equiv 0$ if $\mathscr{I}(1) \neq 1$ and $\mathscr{I}(s_1) = {s_1}^2$ if $\mathscr{I}(1) = 1$.*

**Proof:** Let $s_2 = t_2 = s_3 = t_3 = s_4 = t_4 = 0$ in (12.30), we get

$$|\mu(s_1t_1) - \mu(s_1)\mu(t_1)| \leq \varepsilon. \tag{12.32}$$

Taking $t_1 = t_2$ and $t_3 = t_4 = 0 = s_3 = s_4$ in (12.30)

$$|\mu(s_1t_1 - s_1t_2) + \mu(s_1t_1 + s_1t_2) - 2\mu(t_1)[s_1 + s_2]| \leq \varepsilon, \tag{12.33}$$

$s_1, s_2, t_1 \in \mathscr{R}, |s_1| \geq d$. From equations (12.32) and (12.4), we get

$$|\mu(s_1t_1 - t_1s_2) + \mu(s_1t_1 + t_1s_2) - 2\mu(s_1t_1) - 2\mu(t_1s_1)| \leq 3\varepsilon. \tag{12.34}$$

Taking $t_1 = d$ and substituting $s_1$ and $s_2$ with $\frac{s_1}{d}$ and $\frac{s_2}{d}$ in (12.4), respectively, we obtain

$$|\mu(s_1 - s_2) + \mu(s_1 + s_2) - 2\mu(s_1) - 2(s_2)| \leq 3\varepsilon.$$

Therefore, from [Theorem 1, [23]], there is exactly one quadratic function $\mathscr{I}:\mathscr{R}\to\mathscr{R}$ such that

$$\mathscr{I}(s_1)=\lim_{l\to\infty}\frac{\mu(2^l s_1)}{4^l},\quad |\mu(s_1)-\mathscr{I}(s_1)|\le\varepsilon. \tag{12.35}$$

Using equations (12.30) and (12.35), we get

$$\begin{aligned}&|D_{\mathscr{I}}s_1,s_2,s_3,s_4,t_1,t_2,t_3,t_4)|\\&=\lim_{l\to\infty}\frac{1}{128^l}|D_\mu(2^l s_1,2^l t_1,2^l s_2,2^l t_2,2^l s_3,2^l t_3,2^l s_4,2^l t_4)|\\&\le\lim_{l\to\infty}\frac{1}{128^l}\varepsilon=0.\end{aligned}$$

Since $\mathscr{I}(0)=0$, from Theorem 12.3.1, we obtain $\mathscr{I}\equiv 0$ if $\mathscr{I}(1)\neq 1$ and $\mathscr{I}(s_1)={s_1}^2$ if $\mathscr{I}(1)=1$.

**Corollary 12.4.4.** *Let $\mu:\mathscr{R}\to\mathscr{R}$ be a non-zero function which satisfies $\mu(0)=0$ and*

$$\lim_{\max\{|s_1|,|t_1|,|s_2|,|t_2|,|s_3|,|t_3|,|s_4|,|t_4|\}\to+\infty}\sup|D_\mu(s_1,s_2,s_3,s_4,t_1,t_2,t_3,t_4)|=0. \tag{12.36}$$

*Then, $\mu(s_1)={s_1}^2$ for all $s_1\in\mathscr{R}$.*

**Proof:** Suppose a sequence $\zeta_n$ of positive real numbers decreases monotonically and converges to zero. According to equation (12.36), there exists a sequence $k_n$ comprising positive real numbers such that
$\lim_{n\to\infty}k_n=+\infty$ and

$$\begin{aligned}|D_Q(s_1,s_2,s_3,s_4,t_1,t_2,t_3,t_4)|&\le\zeta_n,\\ \max\{|s_1|,|t_1|,|s_2|,|t_2|,|s_3|,|t_3|,|s_4|,|t_4|\}&\ge k_n.\end{aligned}$$

From Theorem 12.4.3, there is exactly one quadratic function
$Q_n:\mathscr{R}\to\mathscr{R}$ satisfying equation (12.20) and

$$|\mu(s_1)-Q_n(s_1)|\le\varepsilon,\quad s_1\in\mathscr{R}.$$

Then, the quadratic function $Q:\mathscr{R}\to\mathscr{R}$ satisfying

$$|\mu(s_1)-Q(s_1)|\le\varepsilon.$$

Consequently, the uniqueness implies $Q_n=Q_1$ for all n. Therefore,

$$|\mu(s_1)-Q_1(s_1)|\le\varepsilon_n.$$

Taking $n\to+\infty$, we obtain $f(s_1)=Q_1(s_1)$ for all $s_1\in\mathscr{R}$. So, $\mu$ satisfies equation (12.20) and $\mu(0)=0$. According to Theorem 12.3.1, we obtain, $\mu(s_1)={s_1}^2$ for all $s_1\in\mathscr{R}$.

**Corollary 12.4.5.** *Let us consider a non-zero function* $\mu : \mathscr{R} \to \mathscr{R}$ *with* $\mu(0) = 0$ *that fulfil one of the conditions below:*

$$(1) \limsup_{s_1 \to +\infty} |D_\mu(s_1, s_2, s_3, s_4, t_1, t_2, t_3, t_4)| = 0;$$

$$(2) \limsup_{t_1 \to +\infty} |D_\mu(s_1, s_2, s_3, s_4, t_1, t_2, t_3, t_4)| = 0;$$

$$(3) \limsup_{s_2 \to +\infty} |D_\mu(s_1, s_2, s_3, s_4, t_1, t_2, t_3, t_4)| = 0;$$

$$(4) \limsup_{t_2 \to +\infty} |D_\mu(s_1, s_2, s_3, s_4, t_1, t_2, t_3, t_4)| = 0;$$

$$(5) \limsup_{s_3 \to +\infty} |D_\mu(s_1, s_2, s_3, s_4, t_1, t_2, t_3, t_4)| = 0;$$

$$(6) \limsup_{t_3 \to +\infty} |D_\mu(s_1, s_2, s_3, s_4, t_1, t_2, t_3, t_4)| = 0;$$

$$(7) \limsup_{s_4 \to +\infty} |D_\mu(s_1, s_2, s_3, s_4, t_1, t_2, t_3, t_4)| = 0;$$

$$(8) \limsup_{t_4 \to +\infty} |D_\mu(s_1, s_2, s_3, s_4, t_1, t_2, t_3, t_4)| = 0;$$

$$(9) \qquad \limsup_{|s_1|+|t_1|+|s_2|+|t_2|+|s_3|+|t_3|+|s_4|+|t_4| \to +\infty} |D_\mu(s_1, s_2, s_3, s_4, t_1, t_2, t_3, t_4)| = 0.$$

*Then,* $\mu(s_1) = {s_1}^2$ *for all* $s_1 \in \mathscr{R}$.

## 12.5 CONCLUSION

In this chapter, the authors have formulated a novel quadratic multiplicative functional equation, determined its solution, and investigated stability in a real Banach space. They employed fundamental methods, including the direct method and fixed point method. Our findings deepen the understanding of the behavior of the equation and provide insights into potential applications in various mathematical contexts, contributing to the field of functional analysis.

In the future, it is possible to extend the stability findings of this functional equation to diverse spaces, such as non-Archimedean space, orthogonal space, fuzzy space, random normed space, and others.

## BIBLIOGRAPHY

1. Alembert J. D., Addition au Mémoire sur la courbe que forme une corde tendue mise en vibration, Hist. Acad. Berlin, (1750), 355–360.
2. Ulam S. M., A Collection of Mathematical Problems, Interscience Publishers, **8**, (1960).
3. Hyers D. H., Isac G., Rassias T., Stability of Functional Equations in Several Variables, Springer Sci. Busin. Med., **34**, (2012).
4. Hyers D. H., On the Stability of the Linear Functional Equation, Proc. Nat. Acad. Sci. U.S.A. **27** (4) (1941) 222–224.

5. Jakhar J., Chugh R., Jakhar J., Solution and Intuitionistic Fuzzy Stability of 3-Dimensional Cubic Functional Equation: Using two Different Methods, J. Math. Comput. Sci., **25**, (2022), 103–114.
6. Jakhar J., Chugh R., Dubey R., Mishra V. N., Non-Linear Fuzzy Stability of Functional Equations in Various Spaces, AIP Conf. Proc. Vol. 2364
7. Jakhar J., Chugh R., Jaiswal S., Dubey R., Stability of various functional equations in non-Archimedean $(n,\beta)$-normed spaces, J. Anal., **30**(4), (2022), 1653–1669.
8. Tamilvanan K., Alkhaldi A. H., Jakhar J., Chugh R., Jakhar J., Rassias J. M., Ulam Stability Results of Functional Equations in Modular Spaces and 2-Banach Spaces, Math., **11**(2), (2023), 371–394.
9. Rassias T. M., On the Stability of Functional Equations Originated by a Problem of Ulam, Mathematica., **44** (67), (2002).
10. Ashish, Chugh R., Hyers-Ulam-Rassias Stability Of Orthogonally Cubic and Quartic Functional Equations. Int. J. Pure Appl. Math., **81**(1), (2012), 9–20.
11. Kim H. M., Shin H. Y., Refined Stability of Additive and Quadratic Functional Equations in Modular Spaces. J. Inequali. Appli. **2017**(1), (2017), 1–13.
12. Kim S. O., Rassias J. M., Stability of the Apollonius Type Additive Functional Equation in Modular Spaces and Fuzzy Banach Spaces, Math., **7** (11), (2019) 11–25.
13. Park C., Bodaghi A., Two multi-Cubic Functional Equations and Some Results On The Stability in Modular Spaces, J. Inequal. Applic., **2020**(1), (2020), 1–16.
14. Jakhar J., Chugh R., Dubey R., Mishra V. N., Non Linear –Fuzzy Stability of Functional Equations in Various Spaces, AIP Conf. Proc. Volume 2364, AIP Publishing LLC, (2021), 020030.
15. Jakhar J., Chugh R., Jaiswal S., Dubey R., Stability of Various functional Equations in Non-Archimedean$(n,\beta)$-Normed Spaces, The J. Anal., (2022), 1–17.
16. Kim H. M., Chang I. S., Son E., Stability of Cauchy Additive Functional Equation in Fuzzy Banach Spaces, Math. Inequal. Applica., **16** (4), (2013), 1123–1136.
17. Skof F., Proprieta localie approssimazione di operatori, Rend. Sem. Mth. Fis. Milano., **53**, (1983), 113–129.
18. Cholewa P. W., Remarks on the stability of functional equations, Aequationes Math., **27**, (1984) 76–86.
19. Hammer C., Volkmann P., Die upsilonltiplikativen Losungen der Parallelogrammgleichung, Abh. Math. Sem. Univ. Hamburg., **61**, (1991) 197–201.
20. Sahoo P. K., Kannappan P., Introduction to functional Equations, CRC Press, Boca Raton, Florida (2011).
21. Borelli C., Forti G. L., On a general Hyers-Ulam stability result, Int. J. Math. Math. Sci., **18**, (1995), 229–236.
22. Jung S., Kim T., Lee K., A fixed point approach to the stability of quadratic functional equation, Bull. Korean Math. Soc., **43**, (2006), 1–12.
23. Czerwik S., On the stability of the quadratic mapping in normed spaces, Abh. Math. Sem. Hamburg., **62**, (1992) 59–64.

# 13 Existence and Uniqueness of Solution of Nonlinear Fractional Dynamic Equation Involving Initial Condition on Time Scales

*Bikash Gogoi, Walickta Singkai, Satyabrata Gogoi*

The chapter is devoted to the existence and uniqueness of the solution of an implicit type fractional dynamic equation on a time scale using the Caputo nabla derivative (Caputo derivative in the nabla sense) based on the fixed point theorem. First, we discussed the existence of the solution by using Schauder's fixed point theorem, then uniqueness has been addressed by employing Banach fixed point theorem. Additionally, a practical example is provided to illustrate the application of the theoretical findings.

## 13.1 INTRODUCTION

The manuscript presents the concept of dynamic equation, which combines the differential and difference equation in a single domain called time scale, denoted by $\mathbb{T}$. Time scale calculus was first developed by Stefan Hilger (see [11, 12]) in the year of 1988 in his Ph.D work supervised by Bernd Aulbach. Certain important phenomena do not possess only continuous or discrete data but rather hybrid data. An example of this can be seen in the species "Periodical cicada" which lives as a moth for 13 years or 17 years and then as mature for 28 days. In this particular case, the problem should be considered in the domain which forms time scale

$$\mathbb{T} = \cup_{J=0}^{\infty}\left[J(13y+28d), J(13y+28d)+28d\right] \text{ or}$$
$$\mathbb{T} = \cup_{J=0}^{\infty}\left[J(17y+28d), J(17y+28d)+28d\right],$$

where $d$ = days, $y$ = years and $J \in \mathbb{N} \cup \{0\}$. In such instances, the solution of the dynamic equation might provide the necessary data for the dynamic model under

DOI: 10.1201/9781003596578-13

discussion. This new and assertive field is more general and versatile than the traditional theories of differential and difference equations, and hence it is an optimal way for accurate and malleable mathematical modelling. Recently the topic has attained too much attention due to its application in Engineering and Biological sciences.

The combination of fractional calculus and time scale calculus was begun in the PhD dissertation of the N. R. O. Bastos, where he introduced the derivative and integral of the time scale calculus in the sense of fractional derivative and integral operator. The study of fractional order calculus gives more accurate results than ordinary calculus in real-world problems, so we claim that the combination of fractional calculus in time scale calculus extends the concept of ordinary calculus and gives a bright direction in the branch of mathematical science. In the literature, an ample amount of work can be found on fractional linear and non-linear dynamic equations involving different operators such as Caputo, Riemann-Liouville, Caputo-Hadamard, and many more. For convenience, we refer the readers to go through [3, 5, 6, 15, 16, 17, 18, 20, 21, 23, 24, 25] and the references cited therein. Since real situations often deal with nonlinear differential equations, it is worthwhile to study the varieties of nonlinear dynamic equations. Motivated by the above work, we study the implicit type fractional dynamic equation of the type

$$\begin{cases} {}^{C}D_0^{\psi}x(t) = \mathscr{H}\left(t, x(t), {}^{C}D_0^{\psi}x(t)\right) \\ x(t)|_{t=0} = x_0, \ \forall\, x_0 \in \mathbb{R}. \end{cases} \tag{13.1}$$

For $t \in \mathscr{J}$, $\mathscr{J} = [0,T] \cap \mathbb{T}_{\kappa}$, $T \in \mathbb{T}^{+}$ and $\mathscr{H} : \mathscr{J} \times \mathbb{R} \times \mathbb{R}$ be a ld continuous function which is discussed in. Moreover, for the detailed study of the time scale calculus, we refer the readers to go through the book [4] and the manuscripts [1, 2, 4, 7, 8, 9, 22].

The rest of the paper is organized as follows: In section 2, we have highlighted some auxiliary results related to fractional dynamic equations on time scales. In Section 3, we have presented the existence and uniqueness of the dynamic equation (13.1). In Section 4, we have given an example related to our main findings which make the manuscript easier to understand, and the conclusion part of the paper is presented in Section 5.

## 13.2 PRELIMINARIES

**Definition 13.2.1.** *[7] Time scale $\mathbb{T}$ is a non empty subset of $\mathbb{R}$ which is closed in $\mathbb{R}$. For connectedness of $\mathbb{T}$, there are two jump operators called backward and forward jump operators. Let $\rho : \mathbb{T} \to \mathbb{R}$, defined by*

$$\rho(t) = \sup\{s \in \mathbb{T} : s < t\}, \mathit{for}\ t \in \mathbb{T},$$

*then $\rho(t)$ is a backward jump operator and*

$$\sigma(t) = \inf\{s \in \mathbb{T} : s > t\}$$

*is a forward jump operator. Later, if* $t > \inf \mathbb{T}$ *then* $\rho(t)$ *is said to be left scattered and left dense if* $\rho(t) < t$ *and* $\rho(t) = t$*, respectively. If d is the possible minimum right scattered, then we write* $\mathbb{T}_\kappa = \mathbb{T} \setminus d$*, else* $\mathbb{T}_\kappa = \mathbb{T}$*.*

*Since our manuscript is concerned with the nabla derivative, the results related to the delta derivative are omitted and so is also forward jump operator.*

**Definition 13.2.2.** *[8] Let* $v : \mathscr{J} \to \mathbb{R}$*. If at the left dense point in* $\mathscr{J}$*, v is continuous and a right-sided limit exists at right dense points of* $\mathscr{J}$*, then v is said to be ld continuous.*

*The set* $\xi(\mathscr{J}, \mathbb{R})$ *is used to denote all ld continuous functions from* $\mathscr{J}$ *to* $\mathbb{R}$*.*

**Definition 13.2.3.** *[9] Let* $x \in \xi(\mathbb{T}, \mathbb{R})$*, if* $X^{\nabla}(t) = x(t)$*, the nabla integral of x is presented by*

$$\int_{t_0}^{t} x(v) \nabla v = X(t) - X(t_0),$$

*where* $t_0 \in \mathbb{T}$*.*

**Definition 13.2.4.** *[8] Consider a ld continuous function* $v : \mathbb{T}_\kappa \to \mathbb{R}$*, the nabla derivative of v(t) of order* $\psi \in (0,1)$ *is*

$$v^{(\psi)}(t) = \begin{cases} \frac{v(t)-v(\rho(t))}{\mu(t)^{\psi}}, & if\ \rho(t) < t \\ \lim_{s \to t} \frac{v(t)-v(s)}{(t-s)^{\psi}}, & if\ \rho(t) = t, \end{cases}$$

*where* $\mu$ *denotes the backward graininess function. Also for* $t \in \mathbb{T}$ *value of* $\mu(t) = t - \rho(t)$*.*

**Definition 13.2.5.** *[7, Riemann-Liouville integro differential equation] Let* $x : \mathbb{T}_\kappa \to \mathbb{R}$ *be a ld continuous function, then the Riemann-Liouville integro-differential equation of order* $\psi \in (0,1)$ *is given by*

$${}^{RL}D_a^{-\psi} x(t) = \mathbb{I}_a^{\psi} x(t) = \frac{1}{\Gamma(\psi)} \int_a^t \left(t - \rho(v)\right)^{\psi - 1} x(v) \nabla v.$$

*For any two ld continuous function x(t) and y(t), the operator* $\mathbb{I}^{\psi}$ *is linear, that is,*

$$\mathbb{I}^{\psi} x(t) - \mathbb{I}^{\psi} y(t) = \mathbb{I}^{\psi}\left(x(t) - y(t)\right).$$

**Definition 13.2.6.** *[8, Caputo nabla derivative] Let* $\mathscr{X}$ *be a ld continuous function, where* $\mathscr{X}_{\nabla}^{(n)}$ *exists on* $\mathbb{T}_\kappa$*. Then the Caputo nabla derivative of order* $\psi > 0$ *is defined by*

$${}^{C}D_a^{\psi} \mathscr{X}(t) = \frac{1}{\Gamma(n-\psi)} \int_a^t (t - \mu(u))^{n-\psi-1} \mathscr{X}_{\nabla}^{(n)}(u) \nabla(u),$$

*where* $n = [\psi] + 1$*. In general if* $\psi \in (0,1)$ *then,*

$${}^{C}D_a^{\psi} \mathscr{X}(t) = \frac{1}{\Gamma(1-\psi)} \int_a^t \left(t - \mu(u)\right)^{-\psi} \mathscr{X}_{\nabla}(u) \nabla u.$$

**Definition 13.2.7.** *[8, Schauder's fixed point theorem] Let $\mathscr{G}$ be a nonempty convex and closed subset of a Banach space F and let $x : G \rightarrow F$ be a continuous mapping such that $x(G)$ is relatively compact in F. Then the function x contains a fixed point in $\mathscr{G}$.*

**Theorem 13.2.8.** *[14] A function $u : \mathscr{X} \rightarrow \mathscr{Y}$ is a completely continuous mapping if $B \subseteq \mathscr{X}$, such that B is bounded then $u(\mathscr{X})$ is relatively compact in $\mathscr{X}$.*

**Proposition 13.2.9.** *[8] Let $\xi(\mathscr{J}, \mathbb{R})$ is a set of ld-continuous functions from the time scale domain $\mathscr{J}$ to $\mathbb{R}$.*

*The set $\xi(\mathscr{J}, \mathbb{R})$ formed a Banach space equipped by the norm*

$$||x||_\xi = \sup_{t \in \mathscr{J}} |x(t)|.$$

*The set $\xi^1(\mathscr{J}, \mathbb{R})$ is used to denote all ld-continuously nabla derivable function from $\mathscr{J}$ to $\mathbb{R}$.*

**Definition 13.2.10.** *[8, Arzelà-Ascoli theorem]Let $\mathscr{D} \subseteq \xi(\mathscr{J}, \mathbb{R})$. Then $\mathscr{D}$ is said to be relatively compact if and only if at the same time, $\mathscr{D}$ is bounded and equicontinuous.*

## 13.3 MAIN FINDINGS

The dynamic equations (13.1) can be related to a population dynamics problem involving a start-stop phenomenon, such as the population of insects in an area that thrives during the incubation season and declines in winter. By considering the toxic effect on the population, we can model the change in population in that area. Here, x(t) represents the insect population at time $t$, ${}^C D_0^\psi x(t)$ denotes the rate of change of the population with respect to $t$ and $x(0) = x_0$ signifies the initial insect population at the starting time in that area.

For the existence of the solution of the equation (13.1), we reduce the equation into a fixed point problem and we claim that the solution of (13.1) satisfies the Volterra integral equation

$$x(t) = x_0 + \frac{1}{\Gamma(\psi)} \int_0^t (t - \rho(v))^{\psi - 1} \mathscr{H}(t, x(v), {}^C D^\psi x(v)) \nabla v. \tag{13.2}$$

To show our findings, we need to present the following results.

**Proposition 13.3.1.** *[8] Let $r, v \in \mathbb{T}$ such that $r < v$. Consider a ld continuous function g on $\mathscr{J}$, then*

$$\begin{aligned}\int_r^v g(t)\nabla t &= \int_{\rho(r)}^r g(t)\nabla t + \int_r^{\rho(v)} g(t)\nabla t + \int_{\rho(v)}^v g(t)\nabla t \\ &= [r - \rho(r)]g(r) + [v - \rho(v)]g(v) + \int_r^{\rho(v)} g(t)\nabla t.\end{aligned}$$

**Theorem 13.3.2.** *Let a ld continuous function g on a time scale interval* $[r,v]_T$, *and if the function* $\mathscr{G}$ *is extension of g in real interval* $[r,v]$, *then*

$$\mathscr{G}(s) = \begin{cases} g(s), & if s \in \mathbb{T} \\ g(t), & if t \in (\rho(t),t) \notin \mathbb{T}, \end{cases}$$

*hence*

$$\int_r^v g(t)\nabla t \leq \int_r^v \mathscr{G}(t)dt.$$

**Proof:** Let $a \in [r,v]$, such that $\rho(a) < a$. From the Proposition 13.3.1, we obtain

$$\int_{\rho(a)}^a g(t)\nabla t = [a-\rho(a)]g(a).$$

Since the function $g(t)$ is increasing, so $\mathscr{G}(t)$, its extension is also increasing. So using the mean value theorem of integration, we get

$$[a-\rho(a)]\mathscr{G}(\rho(a)) \leq \int_{\rho(a)}^a \mathscr{G}(t)dt \leq [a-\rho(a)]\mathscr{G}(a)$$

$$[a-\rho(a)]g(\rho(a)) \leq \int_{\rho(a)}^a \mathscr{G}(t)dt \leq [a-\rho(a)]g(a)$$

and hence,

$$\int_{\rho(a)}^a g(t)\nabla t \leq \int_{\rho(a)}^a \mathscr{G}(t)dt.$$

Now if the interval $[r,v]$ has only one left scattered point say a, then we acquire the following by using Proposition 13.3.1 and the above results:

$$\begin{aligned} \int_r^v g(t)\nabla t &= \int_r^{\rho(a)} g(t)\nabla t + \int_{\rho(a)}^a g(t)\nabla t + \int_a^v g(t)\nabla t \\ &\leq \int_r^{\rho(a)} \mathscr{G}(t)dt + \int_{\rho(a)}^a \mathscr{G}(t)dt + \int_a^v \mathscr{G}(t)dt. \end{aligned}$$

Similarly we can extend the above result for $n$ left scattered point on $[r,v]$, and thus obtain

$$\int_r^v g(t)\nabla t \leq \int_r^v \mathscr{G}(t)dt.$$

**Lemma 13.3.3.** *Any function* $h \in \xi^1(\mathscr{J},\mathbb{R})$ *is a solution of* (13.1), *if* $h \in \xi(\mathscr{J},\mathbb{R})$ *satisfy the following Voltera integral equation*

$$h(t) = x_0 + \frac{1}{\Gamma(\psi)}\int_0^t (t-\rho(v))^{\psi-1}x(v)\nabla v, \tag{13.3}$$

*where x(t) is a solution of the integral equation* (13.2).

**Proof:** Let ${}^{C}D^{\Psi}x(t) = h(t)$ with $x(t)|_{t=0} = x_0$, then from the equation (13.2), we obtain

$$x(t) = x_0 + \frac{1}{\Gamma(\psi)} \int_0^t \left(t - \rho(v)\right)^{\psi-1} h(v) \nabla v$$
$$= x_0 + \mathbb{I}^{\Psi} h(t).$$

Moreover, the dynamic equation (13.1) is equivalent to

$$h(t) = \mathscr{H}\left(t, x_0 + \mathbb{I}^{\Psi} h(t), h(t)\right). \tag{13.4}$$

Later, we create some hypotheses that will be used to support our principal result.

(L1) The mapping $\mathscr{H} : \mathscr{J} \times \mathbb{R} \times \mathbb{R} \to \mathbb{R}$ is always ld continuous in its first variable and continuous in its second and third variables separately.

(L2) There exist $E_1 \in \xi(\mathscr{J}, \mathbb{R})$ and two constants $E_2$ and $E_3$ such that $E_2 > 0$ and $0 < E_3 < 1$ satisfying

$$|\mathscr{H}(t, h_1, h_2)| \leq |E_1(t)| + E_2|h_1| + E_3|h_2|,$$

for all $t \in \mathscr{J}$ and $(h_1, h_2) \in \mathbb{R} \times \mathbb{R}$.

(L3) There exist two constants $F_1 > 0$ and $0 < F_2 < 1$ such that

$$|\mathscr{H}(t, h_1, h_2) - \mathscr{H}(t, r_1, r_2)| \leq F_1|h_1 - r_1| + F_2|h_2 - r_2|,$$

for all $t \in \mathscr{J}$ and $h_i, r_i \in \mathbb{R} \times \mathbb{R}$ for i = 1, 2.

To prove the existence and uniqueness of the solution of the IVP (13.1). For this, first, we give the following essentials:

Consider a set $\mathscr{M} \subset \xi(\mathscr{J}_T, \mathbb{R})$ defined as $\mathscr{M} = \{h \in \xi(\mathscr{J}_{\mathscr{R}}, \mathbb{R}) : \|h\|_{\xi} \leq \mathscr{L}\} \subseteq \xi(\mathscr{J}, \mathbb{R})$ and an operator $\mathscr{Z} : \mathscr{M}_{\mathscr{L}} \to \mathscr{M}_{\mathscr{L}}$ such that

$$\mathscr{Z}(h) = \mathscr{H}\left(t, x_0 + \mathbb{I}^{\Psi} h(t), h(t)\right). \tag{13.5}$$

The following proposition and Lemmas will help to prove the existence and uniqueness of the solution of the equation (13.1).

**Proposition 13.3.4.** *For $t \in \mathscr{J}$ and $x \in \xi(\mathscr{J}, \mathbb{R})$, the operator $\mathbb{I}^{\Psi}x(t)$ is bounded concerning the norm defined in the Proposition 13.2.9.*

**Proof:** Since, $(t - \rho(v))^{\psi-1}$ is an increasing monotonic function, so by using Theorem 13.3.2 we get

$$\int_0^t (t - \rho(v))^{\psi-1} \nabla v \leq \int_0^t (t - v)^{\psi-1} dv. \tag{13.6}$$

Now from Definition 13.2.5, we obtain

$$\begin{aligned}
||\mathbb{I}^{\psi}x||_{\xi} &= \sup_{t\in\mathscr{J}} |\mathbb{I}^{\psi}x(t)| \\
&\leq \left|\frac{1}{\Gamma(\psi)}\int_0^t (t-\rho(v))^{\psi-1}x(v)\nabla v\right| \\
&< \frac{1}{\Gamma(\psi)}\int_0^t (t-s)^{\psi-1}dv \sup_{v\in\mathscr{J}} |x(v)| \\
&\leq \frac{T^{\psi}}{\Gamma(\psi+1)}||x||_{\xi},
\end{aligned}$$

that is

$$||\mathbb{I}^{\psi}x||_{\xi} \leq \frac{T^{\psi}}{\Gamma(\psi+1)}||x||_{\xi}. \tag{13.7}$$

**Proposition 13.3.5.** *For any $h(t), r(t) \in \xi(\mathscr{J}, \mathbb{R})$, $t \in \mathscr{J}$, then by using the Definition 13.2.5, we get*

$$\begin{aligned}
||\mathbb{I}^{\psi}h - \mathbb{I}^{\psi}r||_{\xi} &= \sup_{t\in\mathscr{J}} |\mathbb{I}^{\psi}h(t) - \mathbb{I}^{\psi}r(t)| \\
&\leq |\mathbb{I}^{\psi}(h(t) - r(t))| \\
&\leq \frac{T^{\psi}}{\Gamma(\psi+1)}||h-r||_{\xi},
\end{aligned}$$

*that is*

$$||\mathbb{I}^{\psi}h - \mathbb{I}^{\psi}r||_{\xi} \leq \frac{T^{\psi}}{\Gamma(\psi+1)}||h-r||_{\xi}. \tag{13.8}$$

*Thus the operator $\mathbb{I}^{\psi}$ is linear concerning the norm indicated by the Proposition 13.2.9.*

The next theorem is based on Schauder's fixed theorem which shows the existence of the solution of the IVP (13.1).

**Theorem 13.3.6.** *Consider that (L1) $\rightarrow$ (L3) hold true, then for $t \in \mathscr{J}$ the equation (13.1) contain a unique solution.*

**Proof:** The proof of the theorem is presented in the following steps:

Step 1: Let $\mathscr{Z}:\mathscr{M}_{\mathscr{L}}\to\mathscr{M}_{\mathscr{L}}$ and a sequence $(h_n)$ such that $h_n\to h$, then for $t\in\mathscr{J}$, we obtain

$$\begin{aligned}
||\mathscr{Z}h_n-\mathscr{Z}h||_\xi &= \sup_{t\in\mathscr{J}}|\mathscr{Z}(h_n(t))-\mathscr{Z}(h(t))|\\
&\le\left|\mathscr{H}\big(t,x_0+\mathbb{I}^\Psi(h_n(t)),h_n(t)\big)-\mathscr{H}\big(t,x_0+\mathscr{I}^\Psi(h(t)),h(t)\big)\right|\\
&\le F_1\big|\mathbb{I}^\Psi h_n(t)-\mathbb{I}^\Psi h(t)\big|+F_2|h_n(t)-h(t)|\\
&\le\frac{F_1T^\Psi}{\Gamma(\psi+1)}|h_n(t)-h(t)|+F_2|h_n(t)-h(t)|\\
&\le\frac{F_1T^\Psi}{\Gamma(\psi+1)}||h_n-h||_\xi+F_2||h_n-h||_\xi.
\end{aligned}$$

Since $||h_n-h||\to 0$ as $n\to\infty$, we get $||\mathscr{Z}(h_n)-\mathscr{Z}(h)||_\xi\to 0$, as $n\to\infty$. Thus the mapping $\mathscr{Z}:\mathscr{M}_{\mathscr{L}}\to\mathscr{M}_{\mathscr{L}}$ is continuous.
Step 2: For each $h\in\mathscr{M}_{\mathscr{L}}$, the mapping $\mathscr{Z}$ conveyed bounded sets of $\xi(\mathscr{J},\mathbb{R})$ into bounded set of $\xi(\mathscr{J},\mathbb{R})$.

$$\begin{aligned}
||\mathscr{Z}h||_\xi &= \sup_{t\in\mathscr{J}}|\mathscr{Z}(h(t))|\\
&\le|\mathscr{H}(t,x_0+\mathbb{I}^\Psi h(t),h(t)|\\
&\le|E_1(t)|+E_2|x_0+\mathbb{I}^\Psi h(t)|+E_3|h(t)|\\
&\le|E_1(t)|+E_2|x_0|+E_2|\mathbb{I}^\Psi h(t)|+E_3|h(t)|\\
&<||E_1||_\xi+E_2|x_0|+E_2\Big|\frac{1}{\Gamma(\psi)}\int_0^t(t-\rho(s))^{\psi-1}h(s)\nabla s\Big|+E_3||h||_\xi\\
&\le||E_1||_\xi+E_2||x_0||_\xi+\frac{E_2T^\Psi}{\Gamma(\psi+1)}||h||_\xi+E_3||h||_\xi\\
&\le||E_1||_\xi+E_2||x_0||_\xi+\Big(\frac{E_2T^\Psi}{\Gamma(\psi+1)}+E_3\Big)||h||_\xi\\
&\le\mathscr{L},
\end{aligned}$$

that is

$$||\mathscr{Z}h||_\xi\le\mathscr{L}. \tag{13.9}$$

Thus from the equation (13.13), we obtain the operator $\mathscr{Z}$ is bounded.
Step 3: Let $\mathscr{Z}:\mathscr{M}_{\mathscr{L}}\to\mathscr{M}_{\mathscr{L}}$, and cosider $t_1,t_2\in\mathscr{J}$ such that $t_1<t_2$, then for any $h\in\mathscr{M}_{\mathscr{L}}$, the map $\mathscr{Z}$ conveyed bounded set into equicontinuous set.

$$\begin{aligned}
||(\mathscr{Z}h)(t_2)-(\mathscr{Z}h)(t_1)||_\xi &= \sup_{t\in\mathscr{J}}|\mathscr{Z}(h(t_2))-\mathscr{Z}(h(t_1))|\\
&\le\left|\mathscr{H}\big(t_2,x_0+\mathbb{I}^\Psi h(t_2)),h(t_2)\big)-\mathscr{H}\big(t_1,x_0+\mathbb{I}^\Psi h(t_1)),h(t_1)\big)\right|\\
&\le\frac{F_1T^\Psi}{\Gamma(\psi+1)}|h(t_2)-h(t_1)|+F_2|h(t_2)-h(t_1)|,
\end{aligned}$$

that is

$$\|(\mathscr{L}h)(t_2)-(\mathscr{L}h)(t_1)\|_\xi \le \frac{F_1T^\psi}{\Gamma(\psi+1)}|h(t_2)-h(t_1)|+F_2|h(t_2)-h(t_1)|. \quad (13.10)$$

When $t_2 \to t_1$, then $|h(t_2)-h(t_1)| \to 0$. So $||(\mathscr{L}h)(t_2)-(\mathscr{L}h)(t_1)||_\xi \to 0$.

So as a consequences of Step 1 $\to$ Step 3 and using the Definition 13.2.10, we conclude that the mapping $\mathscr{L}$ is continuous completely.

Step 4 (Priori bound): It has remained to show that the set given by

$$\mathscr{N}=\{h\in\mathscr{M}_{\mathscr{L}}:h=\sigma\mathscr{L}(h),0<\sigma<1\}$$

is a bounded.

Let $h(t)\in\mathscr{N},t\in\mathscr{J}$ and for $0<\sigma<1$ we have from the equations (13.4) and (13.11), that $\mathscr{N}$ is a bounded set.

Thus in view of the Definition 13.2.7, we must say that the operator $\mathscr{L}$ consists of a fixed point, which is a solution of the dynamic equation (13.1).

**Theorem 13.3.7.** *Consider that (L1), (L2) and (L3) hold true, and if*

$$\frac{F_1T^\psi}{\Gamma(\psi+1)}+F_2<1,$$

*then the equation* (13.1) *consists of a solution.*

**Proof:** Let ${}^CD^\psi x(t)=h(t)$ for $t\in\mathscr{J}$. Using the equation (13.4), we obtain

$$\begin{aligned}|h(t)| =&\left|\mathscr{H}\left(t,x_0+\mathbb{I}^\psi h(t),h(t)\right)\right|\\ \le&|E_1(t)|+E_2|x_0|+E_2|\mathbb{I}^\psi h(t)|+E_3|h(t)|\\ \le&|E_1(t)|+E_2|x_0|+\frac{E_2T^\psi}{\Gamma(\psi+1)}|h(t)|+E_3|h(t)|\\ \le&\frac{|E_1(t)|+E_2|x_0|}{1-\left(\frac{E_2T^\psi}{\Gamma(\psi+1)}+E_3\right)}\\ \le&\gamma,\end{aligned} \quad (13.11)$$

where

$$\gamma=\frac{|E_1(t)|+E_2|x_0|}{1-\left(\frac{E_2T^\psi}{\Gamma(\psi+1)}+E_3\right)}.$$

Taking the norm of $\xi\left(\mathscr{J},\mathbb{R}\right)$ on both sides of the equation (13.11), we get

$$\begin{aligned}||h||_\xi \le&\frac{||E_1||_\xi+E_2||x_0||_\xi}{1-\left(\frac{E_2T^\psi}{\Gamma(\psi+1)}+E_3\right)}\\ \le&\mathscr{L},\end{aligned} \quad (13.12)$$

where

$$\mathscr{L} = \frac{||E_1||_\xi + E_2||x_0||_\xi}{1-\left(\frac{E_2 T^\psi}{\Gamma(\psi+1)} + E_3\right)}. \tag{13.13}$$

Now consider a set $\mathscr{M}_\mathscr{L} = \{h \in \xi(\mathscr{J}_\mathscr{R}, \mathbb{R}) : ||h||_\xi \leq \mathscr{L}\} \subseteq \xi(\mathscr{J}, \mathbb{R})$ and an operator $\mathscr{Z} : \mathscr{M}_\mathscr{L} \to \mathscr{M}_\mathscr{L}$ such that

$$\mathscr{Z}(h) = \mathscr{H}\big(t, x_0 + \mathbb{I}^\Psi h(t), h(t)\big). \tag{13.14}$$

Moreover, for $h, r \in \mathscr{M}_\mathscr{L}$ and using the Proposition 13.3.4 and 13.3.5, we get

$$\begin{aligned}
||\mathscr{Z}(h) - \mathscr{Z}(r)||_\xi &= \sup_{t\in\mathscr{J}} |\mathscr{Z}(h(t)) - \mathscr{Z}(r(t))| \\
&\leq F_1|\mathscr{H}(t, x_0 + \mathbb{I}^\Psi h(t), h(t)) - \mathscr{H}(t, x_0 + \mathbb{I}^\Psi r(t), r(t))| \\
&\leq F_1\left|\mathbb{I}^\Psi h(t) - \mathbb{I}^\Psi r(t)\right| + F_2|h(t) - r(t)| \\
&\leq \frac{F_1 T^\psi}{\Gamma(\psi+1)}|h(t) - r(t)| + F_2|h(t) - r(t)| \\
&\leq \frac{F_1 T^\psi}{\Gamma(\psi+1)}||h - r||_\xi + F_2||h - r||_\xi \\
&\leq \left(\frac{F_1 T^\psi}{\Gamma(\psi+1)} + F_2\right)||h - r||_\xi,
\end{aligned}$$

that is,

$$||\mathscr{Z}(h) - \mathscr{Z}(r)||_\xi \leq \left(\frac{F_1 T^\psi}{\Gamma(\psi+1)} + F_2\right)||h - r||_\xi. \tag{13.15}$$

Since $\frac{F_1 T^\psi}{\Gamma(\psi+1)} + F_2 < 1$, which is a contraction mapping, this hinted the solution to the equation (13.1) is unique.

For sufficient conditions of the existence of the solution of (13.1), we will use the conditions of Definition 13.2.7.

## 13.4 EXAMPLE

**Example 13.4.1.** Consider an initial value problem on $\mathbb{T} = [0,1] \cup [2,3]$

$$\begin{cases} {}^C D^\Psi x(t) = \dfrac{e^{-2t}}{(e^{3t}+7)\left[1+|x(t)|+|{}^C D^\Psi x(t)|\right]} \\ x(0) = 1. \end{cases}$$

For $t \in [0,3] \cap \mathbb{T}_\kappa$ and $\psi = \frac{1}{2}$, we set

$$\mathscr{Z}(t, u, v) = \frac{e^{-2t}}{(e^{3t}+7)[1+u+v]}.$$

The right-hand side of the problem satisfies all the conditions of the Definition 13.2.7. To solve the Example 13.4.1 completely, the condition of the Theorem 13.3.7 is yet to be solved. Now,

$$\begin{aligned}|\mathscr{Z}(t,u_2,v_2)-\mathscr{Z}(t,u_1,v_1)| &= \frac{e^{-2t}}{(e^{3t}+7)}\left[\frac{1}{1+u_2,v_2}-\frac{1}{1+u_1,v_1}\right]\\ &= \frac{e^{-2t}}{(e^{3t}+7)}\left[\frac{(u_2-u_1)+(v_2-v_1)}{(1+u_2+v_2)(1+u_1+v_1)}\right]\\ &\leq \frac{e^{-2t}}{(e^{3t}+7)}\big(|u_2-u_1|+|v_2-v_1|\big)\\ &\leq \frac{1}{8}|u_2-u_1|+\frac{1}{8}|v_2-v_1|,\end{aligned}$$

that is,

$$|\mathscr{Z}(t,u_2,v_2)-\mathscr{Z}(t,u_1,v_1)| \leq \frac{1}{8}|u_2-u_1|+\frac{1}{8}|v_2-v_1|. \tag{13.16}$$

From the conditions (L3) of Lemma 13.3.3, we get $F_1=\frac{1}{8}$ and $F_2=\frac{1}{8}$. Now

$$\begin{aligned}\frac{F_1T^{\psi}}{\Gamma(\psi+1)}+F_2 &= \frac{\frac{1}{8}T^{\frac{1}{2}}}{\Gamma(\frac{1}{2}+1)}+\frac{1}{8}\\ &\leq \frac{3^{\frac{1}{2}}}{8\times 0.887}+\frac{1}{8}\\ &< 1,\end{aligned}$$

that is,

$$\frac{F_1T^{\psi}}{\Gamma(\psi+1)}+F_2<1.$$

Thus, the Banach contraction principle is satisfied. Hence, the equation (13.1) has a unique solution.

**Example 13.4.2.** Consider a time scale $\mathbb{T}=[0,2]\cup[3,5]$ and let $T=3$. Now $\mathscr{J}=[0,T]\cap\mathbb{T}=[0,3]\cap\mathbb{T}=[0,2]\cup\{3\}$. Consider a dynamic equation with the initial condition as follows:

$$\begin{cases}{}^{C}D^{0.5}x(t)=\frac{1}{8e^{4t}}[t+sin|x(t)|+5|{}^{C}D^{0.5}x(t)|]\\ x(0)=1,\ t\in\mathscr{J}.\end{cases} \tag{13.17}$$

Now we setting $\mathscr{Z}(t,x(t),{}^{C}D^{0.5}x(t))=\frac{1}{8e^{4t}}\left[t+sin|x(t)|+5|{}^{C}D^{0.5}x(t)|\right]$, which satisfies (L1). Again let ${}^{C}D^{0.5}x(t)=h(t)$, consider that

$$\mathscr{Z}(t,x(t),h(t))\leq\frac{3}{8}+\frac{1}{8}|x(t)|+\frac{1}{8}|h(t)|.$$

Thus, (L2) satisfied with $E_1=\frac{3}{8}, E_2=\frac{1}{8}, E_3=\frac{1}{8}$. Again for $i=1,2, {}^C D^{0.5} x_i(t)=h_i(t)$ consider that

$$|\mathscr{X}(t,x_1(t),h_1(t))-\mathscr{X}(t,x_2(t),h_2(t))| \leq \frac{1}{8}|x_1(t)-h_1(t)|+\frac{1}{8}|x_2(t)-h_2(t)|.$$

Hence, (L3) is satisfied with $F_1=\frac{1}{8}, F_2=\frac{1}{8}$. Putting the above data in

$$\frac{F_1 T^{0.5}}{\Gamma(1.5)}+F_2 \leq (\equiv 0.365) < 1.$$

Thus from Theorems 13.3.6 and 13.3.7, the equation 13.4.2 has a solution which is unique.

## 13.5 CONCLUSION

In this manuscript, we explore an implicit type nonlinear fractional dynamic equation incorporating the Caputo nabla fractional derivative, utilizing Schauder's fixed point theorem to present a novel result within the context of time scales. Given the practical implications of the Caputo derivative, the findings of this study can be readily applied to real-world problems within the time scale framework. Additionally, the future scope of this research includes analyzing the dependency and stability characteristics of the solutions. The topic has recently gained significant attention in engineering due to its applications in fractional two-dimensional time scales [19].

## BIBLIOGRAPHY

1. Abbas S., Dynamic equation on time scale with almost periodic coefficients. Nonautonomous Dynamical Systems, **7**(1), (2020), 151–162.
2. Agarwal R. P., Bohner M., ORegan D., Peterson A., Dynamic equations on time scales: a survey. Journal of Computational and Applied Mathematics, **141**(1-2), (2002), 1–26.
3. Benkhettou N., Hammoudi A., Torres D. F., Existence and uniqueness of solution for a fractional Riemann–Liouville initial value problem on time scales. Journal of King Saud University-Science, **28**(1), (2016), 87–92.
4. Bohner M., Peterson A., *Dynamic equations on time scales: An introduction with applications*, (2001), Springer Science & Business Media.
5. Samko S. G., *Fractional integrals and derivatives: Theory and applications*, (1993).
6. Du N. H., Liem N. C., Chyan C. J., Lin S. W., Lyapunov stability of quasilinear implicit dynamic equations on time scales. Journal of Inequalities and Applications, (2011), 1–27.
7. Gogoi B., Saha U. K., Hazarika B., Torres D. F., Ahmad H., Nabla fractional derivative and fractional integral on time scales. Axioms, **10**(4), (2021), 3–17.
8. Gogoi, B., Saha, U. K., Hazarika, B. (2024). Existence of solution of a nonlinear fractional dynamic equation with initial and boundary conditions on time scales. The Journal of Analysis, **32**(1), (2022), 85–102.
9. Gogoi B., Hazarika B., Saha U. K., Tikare S., Periodic Boundary Value Problems for Fractional Dynamic Equations on Time Scales. Results in Mathematics, **78**(6), (2023), 228.

10. Gogoi B., Hazarika B., Saha U. K., Impulsive Fractional Dynamic Equation with Non-local Initial Condition on Time Scales. arXiv preprint arXiv:2207.01517 (2022).
11. Hilger S., Analysis on measure chains—a unified approach to continuous and discrete calculus. Results in mathematics, **18**(1), (1990), 18–56.
12. Hilger S., Differential and difference calculus—unified!. Nonlinear Analysis: Theory, Methods & Applications, **30**(5), (1997), 2683–2694.
13. Tikare S., Nonlocal initial value problems for first-order dynamic equations on time scales. Appl. Math. E-Notes, **21**, (2021), 410–420.
14. Tikare S., Bohner M., Hazarika B., Agarwal R. P., Dynamic local and nonlocal initial value problems in Banach spaces. Rendiconti del Circolo Matematico di Palermo Series, **2**, (2021), 1–16.
15. Telli B., Souid M. S., $L^1$-solutions of the initial value problems for implicit differential equations with Hadamard fractional derivative. Journal of Applied Analysis, **28**(1), (2022), 1–9.
16. Kou C., Liu J., Ye Y., Existence and Uniqueness of Solutions for theCauchy-TypeProblems of Fractional Differential Equations. Discrete Dynamics in Nature and Society, **2010**(1), (2010), 142–175.
17. Letnikov A. V., Theory of differentiation with an arbtraly indicator. Matem Sbornik, **3**, (1968), 1–68.
18. K. S. Miller, B. Ross, An introduction to the fractional calculus and fractional differential equations. (No Title), (1993).
19. Ozturk O., On oscillatory behavior of two-dimensional time scale systems. Advances in Difference Equations, **1**(18), (2018), 1–12.
20. Podlubny I., *Fractional Differential Equation*, Academic Press, New York, (1999).
21. Rynne B. P., $L^2$ spaces and boundary value problems on time-scales. Journal of Mathematical Analysis and Applications, **328**(2), (2007), 1217–1236.
22. Torres D. F., Cauchy's formula on nonempty closed sets and a new notion of Riemann–Liouville fractional integral on time scales. Applied Mathematics Letters, **121**, (2021), 107407.
23. Zhao D. F., You X. X., Cheng J., Remarks on conformable fractional derivative on time scales. Adv Theor Appl Math, **11**(1), (2016), 61–68.
24. Zhu J., Wu L., Fractional Cauchy problem with Caputo nabla derivative on time scales. In Abstract and Applied Analysis, **2015**(1), (2015), 486054.
25. Wu L., Zhu J., Fractional Cauchy Problem with Riemann-Liouville Derivative on Time Scales, In Abstract and Applied Analysis, **1**, (2013), 795701.

# 14 Remarks on Absolute and Non-absolute Integrability of Different Integrals for Multifunctions

*Cristina Stamate, Anca Croitoru*

In this chapter, we present some properties concerning the absolute and nonabsolute integrability for some real and vector integrals for functions and multifunctions by respect to a submeasure or a multisubmeasure of the Sugeno, Choquet, Aumann, Brooks and Henstock-Kurzweil type.

## 14.1 INTRODUCTION

In the last decades, different problems in game theory, decision making, economy, fuzzy logic, computer science were the basis for the consideration of non-additive or set-valued measures and integrals ([1], [5], [7], [8], [9], [10], [19], [20], [21], [39], [41], [44], [45], [48], [49], [53]). Starting with the Lebesgue integral, a long way has been completed in order to find an integral that makes differentiation and integration inverse processes. Some quantum mechanics problems in which the Lebesgue integration theory cannot be used ([22]) also led to the emergence of a non-absolute integration theory. This goal was fulfilled when the Henstock-Kurzweil integral was defined, apparently with a minor change in the definition of the Riemann integral, but with profound consequences in the study of functions with applications in various fields, such as differential and integral equations, operator theory etc.

As is known, the Henstock-Kurzweil integral ([23], [31]) is a non-absolute convergent integral and this motivates its use in the study of physical phenomena whose models involve singular functions or with wide oscillations, as well as in the study of associated solutions of differential equations with minimal regularity conditions which cannot be approached with the Lebesgue integration theory.

The transition to the case of vector functions was achieved, as usual, by scalarization using the Pettis-type integral ([20], [24], [37]) and for real functions defined on vector spaces, partitions were used, as a generalization of the Riemann integral.

DOI: 10.1201/9781003596578-14

Different non-absolute integrals have been defined in [22], [35], [36]. A non-absolute multiple integral was defined in [38] constructively on the Euclidean space, as an extension of the special Denjoy integral to higher dimensions. Also, a Henstock-Kurzweil type non-absolute integral is defined in [33] on a measure space endowed with a metrizable topology.

Some interesting surveys and motivation papers concerning the non-absolute integrals can be found in [6], [27], and their properties and applications in [2], [3], [4], [12], [22], [26], [28], [29], [30], [32], [34], [43], [51].

Our interest in what follows is to study the absolute and non-absolute integrability for some real or vector integrals for functions and multifunctions by respect to a submeasure or a multisubmeasure.

In this chapter, we will address Henstock-Kurzweil integrability for (multi) functions with values in Euclidean spaces, in a unitary framework regarding absolute integrability for various types of linear and non-linear integrals and different notions of absolute integrability that will be introduced. Obviously, for the vector case, the absolute value will be replaced by a real function, sometimes a set norm ([14]).

The structure of this chapter is as follows. Section 2 is for some preliminaries. In Section 3, a survey on real integrals for functions and multifunctions is presented. We define in Section 4 some types of absolute integrability for multifunctions and multisubmeasures with values in Euclidean spaces and we present several properties and relationships among them.

## 14.2 PRELIMINARIES

Let's consider $(X, \|\cdot\|)$ be a Banach algebra (with identity $e \neq 0$) ordered by $X_+ \subset X$, a nonempty, closed, convex, pointed ($X_+ \bigcap -X_+ = \emptyset$) cone. Denote $X^*$ the dual space of $X$ and $X_+^* = \{x^* \in X^* \mid x^*(x) \geq 0\}$, the dual cone of $X_+$. As usually, we may consider a biggest element denoted $+\infty$ and a smallest element denoted $-\infty$. $p_0(X)$ is the family of nonempty sets of $X$, $p_b(X)$ is the family of all nonempty bounded subsets of $X$, $p_k(X)$ is the family of all nonempty compact subsets of $X$ and $p_{kc}(X)$ is the family of all nonempty compact convex subsets of $X$. Denote by $D$ the Hausdorff metric on $p_{kc}(X)$. For a bounded set $A \subset X$, we denote $\|A\| = \sup\{\|a\|, a \in A\}$. Let $\mathfrak{K} \subset p_{kc}(X)$ and for $A, B \in \mathfrak{K}$ we consider two operations "+" and "·" defined by:

$$A + B = \{x + y \mid x \in A, y \in B\};$$

$$A \cdot B = \{x \cdot y \mid x \in A, y \in B\}.$$

We say that $(\mathfrak{K}, +, \cdot)$ is a semialgebra denoted $\mathfrak{K}$ if the following properties hold:

1. $A, B \in \mathfrak{K}$ then $A + B \in \mathfrak{K}$ and $A \cdot B \in \mathfrak{K}$;
2. $\exists\, \theta \in \mathfrak{K}$ such that $A + \theta = \theta + A = A$;
3. $\{e\} \in \mathfrak{K}$;
4. $A, B, C \in \mathfrak{K}$ then $A \cdot (B + C) = A \cdot B + A \cdot C$;
5. $A \in \mathfrak{K}$, $\alpha \in [0, +\infty) \to \alpha A \in \mathfrak{K}$.

For a nonempty set $T$, we denote by $\mathscr{C}$ a $\sigma-$algebra of subsets of $T$. A multifunction (also called a set-valued function) denoted $F :\rightrightarrows X$ is a function $F : T \to \mathscr{P}_0(X)$. We denote by $\mathscr{F}(T,X)$ the set of functions $f : T \to X$ and by $\mathscr{S}(F) = \{f \mid f \in \mathscr{F}(T,X),\ f(t) \in F(t), \forall t \in T\}$, the selections set of $F$.

**Definition 14.2.1.** *[11] A multifunction $F : T \rightrightarrows X$ is called measurable if $F^{-1}(U) = \{t \in T \mid F(t) \cap U \neq \emptyset\} \in \mathscr{C}$, for every open set $U \subseteq X$.*
*A function $f : T \to X$ is measurable if and only if the multifunction $F$ given by $F(t) = \{f(t)\}$ for all $t \in T$ is measurable.*

We denote by $\|F\|$ the function $\|F\| : T \to \mathbb{R}_+$ given by $\|F\|(t) = \|F(t)\|$. If $F,\ G : T \to p_{kc}(X)$ are multifunctions, denote by $D(F,G) : T \to \mathbb{R}_+$ the function defined by $D(F,G)(t) = D(F(t),G(t)),\ \forall t \in T$.

**Definition 14.2.2.** *A function $\lambda : \mathscr{C} \to X_+ \cup \{+\infty\}$ is called*

1. *fuzzy vector measure (or a vector submeasure) if $\lambda(\emptyset) = 0$ and $\lambda(A) \leq \lambda(B)$, $\forall A, B \in \mathscr{C}$;*
2. *vector measure if $\lambda(\emptyset) = 0$ and $\lambda(A \cup B) = \lambda(A) + \lambda(B)$ for all $A, B \subset X$, $A \cap B = \emptyset$;*
3. *subadditive if for every $A, B \in c$ there holds $\lambda(A \cup B) \leq \lambda(A) + \lambda(B)$;*
4. *submodular if for every $A, B \in c$, there holds $\lambda(A \cup B) + \lambda(A \cap B) \leq \lambda(A) + \lambda(B)$;*
5. *supermodular if for every $A, B \in c$, there holds $\lambda(A \cup B) + \lambda(A \cap B) \geq \lambda(A) + \lambda(B)$;*
6. *lower semi-continuous if for every increasing sequence $\{A_n\}_{n\geq 1}$ of measurable sets from c, we have $\lambda(\bigcup_{n\in\mathbb{N}} A_n) = \lim_{n\to+\infty} \lambda(A_n)$;*
7. *upper semi-continuous if for every decreasing sequence $\{A_n\}_{n\geq 1}$ of measurable sets from c and $\lambda(A_1) < +\infty$, we have $\lambda(\bigcap_{n\in\mathbb{N}} A_n) = \lim_{n\to+\infty} \lambda(A_n)$;*
8. *continuous if it is lower and upper semi-continuous.*

**Remark 14.2.3.** *If $X = \mathbb{R}$ then $\lambda$ from (1) (respectively (2)) will be called real submeasure (respectively real measure).*

**Definition 14.2.4.** *1. A multifunction $\mu : \mathscr{C} \rightrightarrows X_+ \cup \{+\infty\}$ with $\mu(\emptyset) = \{0\}$ is called a fuzzy vector multimeasure (or a vector multisubmeasure) if $A \subset B$ implies $\mu(A) \subset \mu(B) - X_+$ and $\mu(B) \subset \mu(A) + X_+$, $\forall A, B \in \mathscr{C}$;*
*2. [5] A multifunction $\mu : c \to p_0(X)$ is said to be a vector multimeasure if $\mu(\emptyset) = \{0\}$ and $\mu(A \cup B) = \mu(A) + \mu(B)$ if $A, B \in c$, $A \cap B = \emptyset$.*

**Remark 14.2.5.** *If $X = \mathbb{R}$ then $\mu$ from (1) (respectively (2)) will be called real multisubmeasure (respectively real multimeasure).*

**Remark 14.2.6.** *A function $\lambda : \mathscr{C} \to X_+ \cup \{+\infty\}$ is a vector submeasure (respectively vector measure) if and only if the multifunction $\mu : \mathscr{C} \rightrightarrows X_+ \cup \{+\infty\}$ given by $\mu(A) = \{\lambda(A)\}$ for all $A \in c$ is a vector multisubmeasure (respectively vector multimeasure).*

If $\mu : \mathscr{C} \rightrightarrows X_+$ is a multisubmeasure we consider a real function denoted $\|\mu\|$ given by $\|\mu\| : \mathscr{C} \to \mathbb{R}_+$, $\|\mu\|(A) = \|\mu(A)\|$ for all $A \in c$ and let remark that if the norm of $X$ is a monotone norm then $\|\mu\|$ is a real submeasure. Let $\widehat{\mathscr{G}} \subset \mathscr{F}(\mathscr{P}_0(X), \mathscr{P}_0(\mathbb{R}))$ such that for all $widehatg \in \widehat{\mathscr{G}}$ and $x \in X$ we have $\widehat{g}(\{x\}) = \{y\}$, $y \in \mathbb{R}$. Let's suppose that the element $\widehat{g}$ of $\widehat{g}$ satisfies the following monotonicity condition: $\widehat{g}(A) \subseteq \widehat{g}(B) - \mathbb{R}_+$ if $A \subseteq B - X_+$.

We consider $\mathscr{G} = \{g \mid g : X \to \mathbb{R},\ \exists \widehat{g} \in \widehat{g},\ \widehat{g}(\{x\}) = \{g(x)\},\ \forall x \in X\}$ and for a cone $\emptyset \neq Y_+ \subseteq X_+$ and $A \in p_0(X)$ we say that $\mathscr{G}$ has $(H_0^A)$ properties if $x, y \in A, \&g(x) \leq g(y), \&\forall g \in g \setminus \{0\} \Longleftrightarrow x - y \in Y_+$;
$x, y \in A, \&g(x) < g(y), \&\forall g \in g \setminus \{0\} \Longleftrightarrow x - y \in Y_+ \setminus \{0\}$; $g(0) = 0$, '$\&\forall g \in g.(\mathrm{H}_0^{\mathrm{A}})$
If $A = X$, $Y_+ = X_+$ we denote simply $(\mathrm{H}_0)$. For more details see [46], [47].

In what follows in this chapter, we write that a real function is $\mathscr{T}$-integrable if it fulfills one of the following types of integrability: Sugeno, Choquet, Dunford, Henstock-Kurzweil, Brooks, Lebesgue. The corresponding integral for $f$ by respect to a submeasure (or a measure) $\lambda$ on $T$ will be denoted by the symbol $(\mathscr{T}) \int\limits_T f d\lambda$.

For a multifunction $F$ and a submeasure (or multisubmeasure) $\mu$, we write that $F$ is Aumann $\mathscr{T}$-integrable if $F$ has selections which are $\mathscr{T}$-integrable and the Aumann integral of $F$ corresponding to the $\mathscr{T}$-integrability of the selections will be denoted by the symbol $(A\mathscr{T}) \int\limits_T F d\mu$.

For vector functions $f : T \to X$ (respectively multifunctions $F : T \rightrightarrows X$), we introduced in [46] the $g$ integral of $f$ (respectively $F$) by respect to a vector submeasure $\lambda : c \to X_+ \cup \{+\infty\}$ (respectively a vector multisubmeasure $\mu : c \rightrightarrows X_+ \cup \{+\infty\}$) as follows.

**Definition 14.2.7.** *Let be $\widehat{\mathscr{G}} \subset \mathscr{F}(\mathscr{P}_0(X), \mathscr{P}_0(\mathbb{R}))$ with $\widehat{g}(\{x\}) \in \mathbb{R}$ and suppose that $g = \{g_0 \mid g_0 : X \to \mathbb{R},\ \exists \widehat{g_0} \in widehatg,\ \{g_0(x)\} = \widehat{g_0}(\{x\}),\ \forall x \in X\}$ has $(H_0)$ properties. By $\widehat{g_0}oF$ (respectively $g_0oF$) we denote the multifunction from $X$ to $p_0(\mathbb{R})$ given by $\widehat{g_0}oF(x) = \widehat{g_0}(F(x))$ (respectively $g_0oF(x) = \{g_0(y) \mid y \in F(x)\}$) for each $x \in X$.*

1. *Let $\lambda : c \to X_+ \cup \{+\infty\}$ be a vector submeasure, $f : T \to X$ be a function such that $g_0of$ is $\mathscr{T}$-integrable on $T$ by respect to $g_0o\lambda$ for each $g_0 \in g$. We say that $f$ is $g_0$ $\mathscr{T}$-integrable on $T$ by respect to $\lambda$ if there exists $I \in X$ such that for each $g_0 \in g$ we have*
$$g_0(I) = (\mathscr{T}) \int\limits_T g_0ofdg_0o\lambda.$$
*We denote $(g)(\mathscr{T}) \int\limits_T f d\lambda = I$.*
*The integral from the right side may be the real integral of Sugeno, Choquet, Dunford, Henstock-Kurzweil, Brooks or even Lebesgue (if $g_0o\lambda$ is the Lebesgue measure) type.*
2. *Let $\mu : c \rightrightarrows X_+ \cup \{+\infty\}$ be a vector multisubmeasure and $F : T \rightrightarrows X$ be a multifunction such that for each $\widehat{g_0} \in \widehat{g}$, $\widehat{g_0}oF$ is Aumann $\mathscr{T}$-integrable on $T$ by respect*

*to the multisubmeasure $\widehat{g_0}o\mu$. We say that $F$ is $\widehat{g}$ $\mathscr{T}$-integrable on $T$ by respect to $\mu$ if there exists $I \in p_0(X)$ such that for each $\widehat{g_0} \in \widehat{g}$ we have*

$$\widehat{g_0}(I) = (A\mathscr{T}) \int_T \widehat{g_0}oFd\widehat{g_0}o\mu$$

*and for each $I' \in p_0(X)$ with $\widehat{g_0}(I') = \widehat{g_0}(I)$, $\forall \widehat{g_0} \in g$ we have $I' \subseteq I$.*
*We denote $(\widehat{g})(A\mathscr{T}) \int_T Fd\mu = I$. The integral from the right side is the Aumann type integral for real multifunctions corresponding for each type of integrals (Sugeno, Choquet, Dunford, Henstock-Kurzweil, Lebesgue) for the integrable selections of $\widehat{g_0}oF$ by respect to the selections of $g_0o\mu$ which are submeasures.*

3. *Let $\mu : c \rightrightarrows X_+ \cup \{+\infty\}$ be a vector multisubmeasure and $F : T \rightrightarrows X$ be a multifunction such that for each $g_0 \in g$, $g_0oF$ is Aumann $\mathscr{T}$-integrable on $T$ by respect to the multisubmeasure $g_0o\mu$. We say that $F$ is $g$ $\mathscr{T}$-integrable on $T$ by respect to $\mu$ if there exists $I \in p_0(X)$ such that for each $g_0 \in g$ we have*

$$g_0(I) = (A\mathscr{T}) \int_T g_0oFdgo\mu$$

*and for each $I' \in p_0(X)$ with $g_0(I') = g_0(I)$, $\forall g_0 \in g$ we have $I' \subseteq I$.*
*We denote $(g)(A\mathscr{T}) \int_T Fd\mu = I$. The integral from the right side is the Aumann type integral for real multifunctions corresponding for each type of integrals (Sugeno, Choquet, Dunford, Henstock-Kurzweil, Lebesgue) for the integrable selections of $g_0oF$ by respect to the selections of $g_0o\mu$ which are submeasures.*

Let remark that we may define the $g$ (respectively $\widehat{g}$ $\mathscr{T}$-integrability for a function $f : T \to X$ by respect to a vector multisubmeasure $\mu : c \rightrightarrows X_+ \cup \{+\infty\}$ if we consider Definition 14.2.7 (2), (3) for $F(t) = \{f(t)\}$, $\forall t \in T$ and $\mu$. Similarly, we may define the $g$(respectively $\widehat{g}$) $\mathscr{T}$-integrability for a multifunction $F : T \rightrightarrows X$ by respect to a vector submeasure $\lambda : c \to X_+ \cup \{+\infty\}$ if we consider Definition 14.2.7 (2), (3) for $\mu(A) = \{\lambda(A)\}$, $\forall A \in c$ and $F$.

We observe that for $g \subseteq X_+^*$ s.t. the linear hull of $g$ is $X^*$ if the $\mathscr{T}$-integral from Definition 14.2.7 (i) is linear and $\lambda$ is a vector measure $\lambda : c \to X_+ \cup \{+\infty\}$, then the $g\mathscr{T}$-integral is the Pettis type integral and we have

$$x^*((g)(\mathscr{T}) \int_T fd\lambda) = (\mathscr{T}) \int_T x^*ofdx^*o\lambda$$

for each $x^* \in X_+^* \setminus \{0\}$.

If the $\mathscr{T}$-integral isn't linear (for example the Sugeno or Choquet integral), the relation above does hold under more restrictive conditions concerning $f$ and $\lambda$ as we can see in [46].

**Theorem 14.2.8.** *Let $X = \mathbb{R}^n$ ordered by the cone $\mathbb{R}_+^n$ and $\{e_i,\, i = \overline{1,n}\}$ the canonical base of $\mathbb{R}^n$. Let denote*

1. $\widehat{g}_e = \{\widehat{g}_{e_i} \mid \widehat{g}_{e_i}(C) = < e_i, c >,\ c \in C\},\ \forall C \in p_{kc}(\mathbb{R}^n)\}$.
2. $\widehat{g}_s = \{\widehat{g0}_{e_i} \mid \widehat{g0}_{e_i}(C) = \sup\{| < e_i, c > |,\ c \in C\},\ \forall C \in p_{kc}(\mathbb{R}^n)\}$;
3. $\widehat{g}_* = \widehat{g}'_* \bigcup \widehat{g}''_*$ *where*
   $\widehat{g}'_* = \{\widehat{g0}'_{e_i} \mid \widehat{g0}'_{e_i}(C) = \sup\{< e_i, c >,\ c \in C\},\ \forall C \in p_{kc}(\mathbb{R}^n)\}$ *and*
   $\widehat{g}''_* = \{\widehat{g0}''_{e_i} \mid \widehat{g0}''_{e_i}(C) = \inf\{< e_i, c >,\ c \in C\},\ \forall C \in p_{kc}(\mathbb{R}^n)\}$.

*and the corresponding $g_e$, $g_s$ and $g_*$ as we have seen in the presentation of $(H_0^A)$ property.*

1. *For $i = \overline{1,n}$ let $f_i : T \to \mathbb{R}$, $f = (f_i)_{i=\overline{1,n}}$, and $\lambda = (\lambda_i)_{i=\overline{1,n}}$ where $\lambda_i : c \to \mathbb{R}_+ \cup \{+\infty\}$ are real submeasures. If $f_i$ is $\mathcal{T}$-integrable by respect to $\lambda_i$ for each $i = \overline{1,n}$, the $g_e$ $\mathcal{T}$-integral of $f$ by respect to $\lambda$ is*

$$(g_e)(\mathcal{T}) \int_T f d\lambda = ((\mathcal{T}) \int_T f_i d\lambda_i)_{i=\overline{1,n}}$$

*(it coincides with the integral introduced in [42] for the Choquet case).*

2. *For $i = \overline{1,n}$, let $F_i : T \rightrightarrows \mathbb{R}$ and $F : T \rightrightarrows \mathbb{R}^n$ given by $F(t) = \prod_{i=\overline{1,n}} F_i(t)$ for each $t \in T$ and let $\mu_i : c \rightrightarrows \mathbb{R}_+ \cup \{+\infty\}$ be real multisubmeasures, $\mu : c \rightrightarrows \mathbb{R}^n_+ \cup \{+\infty\}$ given by $\mu(A) = \prod_{i=\overline{1,n}} \mu_i(A)$ for each $A \in c$.*

   *2(1). If for each $i = \overline{1,n}$, $F_i$ are Aumann $\mathcal{T}$-integrable by respect to $\mu_i$ then we have*

$$(\widehat{g}_e)(\mathcal{T}) \int_T F d\mu = (g_e)(A\mathcal{T}) \int_T F d\mu = \prod_{i=\overline{1,n}} ((A\mathcal{T}) \int_T F_i d\mu_i).$$

   *2(2). If $F_i : T \to p_{kc}(\mathbb{R})$ and $\mu_i : c \rightrightarrows p_{kc}(\mathbb{R})$ then we have*

$$(\widehat{g}_s)(\mathcal{T}) \int_T F d\mu = \prod_{i=\overline{1,n}} [-(\mathcal{T}) \int_T \|F_i\| d\|\mu_i\|, (\mathcal{T}) \int_T \|F_i\| d\|\mu_i\|],$$

   *if the functions $\|F_i\|$ are $\mathcal{T}$-integrable by respect to the real submeasure $\|\mu_i\|$. If $\|F_i\|$ is Choquet integrable by respect to $\mu_i = \lambda$, $\forall i = \overline{1,n}$ where $\lambda : c \to \mathbb{R}_+$ is a submeasure, we get the integral introduced by Park in [40].*
   *Also, we have*

$$(g_s)(A\mathcal{T}) \int_T F d\mu = \prod_{i=\overline{1,n}} (A\mathcal{T}) \int_T G_i d\mu_i,$$

   *where $G_i(t) = \{|y|,\ y \in F_i(t)\}$.*

3. *If $F_i : T \to p_{kc}(\mathbb{R})$ and $\mu(A) = \lambda(A)e$ for each $A \in c$ where $\lambda : c \to \mathbb{R}$ is a real submeasure and $e = (1, 1, ..., 1)$ we have*

$$(\widehat{g}_*)(\mathcal{T}) \int_T F d\mu = \prod_{i=\overline{1,n}} [(\mathcal{T}) \int_T f_{i*} d\lambda, (\mathcal{T}) \int_T f_i^* d\lambda]$$

*if the functions* $f_{i*} : T \to \mathbb{R}$ *given by* $f_{i*}(t) = \inf\{y,\ y \in F_i(t)\}$ *and* $f_i^* : T \to \mathbb{R}$ *given by* $f_i^*(t) = \sup\{y,\ y \in F_i(t)\}$ *are* $\mathscr{T}$*-integrable by respect to* $\lambda$.
*In this case*

$$(g_*)(A\mathscr{T})\int_T F d\mu = \prod_{i=\overline{1,n}} (A\mathscr{T})\int_T F_i d\mu_i.$$

Let remark that for each $\tilde{g}_{e_i} \in \widehat{g}_e$ (respectively $\widehat{g0}_{e_i} \in \widehat{g}_s$, $\widehat{g0}'_{e_i} \in \widehat{g}_*$ and $\widehat{g0}''_{e_i} \in \widehat{g}_*$) we have $\widehat{g0}_{e_i} o f = f_i$ (respectively $\widehat{g0}_{e_i} oF = \|F_i\|$, $\widehat{g0}'_{e_i} oF = f_{i*}$ and $\widehat{g0}''_{e_i} oF = f_i^*$). Following Definition 14.2.7, the corresponding integral for each case is the maximal set by respect to the sets inclusion and thus we get the conclusions.

## 14.3 A SURVEY ON REAL INTEGRALS FOR FUNCTIONS AND MULTI-FUNCTIONS

In this section, we present some definitions for the real integrals which will be used in what follows.

**Definition 14.3.1.** *(The Sugeno type integrals)*

1. *[48], [50] Let* $\lambda : \mathscr{C} \to \mathbb{R}_+ \cup \{+\infty\}$ *be a fuzzy measure and let* $f : T \to \mathbb{R}_+$ *be a measurable function. The Sugeno integral of* $f$ *with respect to* $\lambda$ *on* $A \in \mathscr{C}$ *is defined by*

$$(S)\int_A f d\lambda = \sup_{\alpha \geq 0} \min\{\alpha, \lambda(A \cap f^{-1}([\alpha, +\infty)))\}.$$

   $f$ *is called Sugeno integrable if* $(S)\int_A f d\lambda < +\infty$.

2. *[16] Let* $\mu : c \rightrightarrows \mathbb{R}_+ \cup \{+\infty\}$ *be a multisubmeasure and let* $F : T \rightrightarrows \mathbb{R}_+$ *be a measurable multifunction. The strong integral of F with respect to* $\mu$ *on* $A \in \mathscr{C}$ *is defined by*

$$(s)\int_A F d\mu = \sup_{\alpha \in [0,+\infty]} \inf(\alpha \cup \mu(A \cap F^{-1}([\alpha, +\infty)))).$$

   $F$ *is called strong integrable if* $(s)\int_A F d\mu < +\infty$.

3. *[15] Let* $\lambda : c \to \mathbb{R}_+ \cup \{+\infty\}$ *be a submeasure and let* $F : T \rightrightarrows \mathbb{R}_+$ *be a measurable multifunction. The weak integral of F with respect to* $\lambda$ *on* $A \in \mathscr{C}$ *is defined by*

$$(w)\int_A F d\lambda = \sup_{\alpha \in [0,+\infty]} \min\{\alpha, \lambda(A \cap \|F\|^{-1}([\alpha, +\infty)))\}.$$

   $F$ *is called weak integrable if* $(w)\int_A F d\lambda < +\infty$.

4. *[52] Suppose* $F : T \rightrightarrows \mathbb{R}$ *is a multifunction and* $\lambda : \mathscr{C} \to \mathbb{R}_+ \cup \{+\infty\}$ *is a submeasure. The Guo integral of F on* $A \in \mathscr{C}$ *with respect to* $\lambda$ *is defined as*

$$(G)\int_A F d\lambda = \{(S)\int_A f d\lambda \mid f \in S_S(F, \lambda)\},$$

*where $S_S(F,\lambda)$ is the set of all measurable selections Sugeno integrable with respect to $\lambda$.*
*We may extend this integral for the case of a multisubmeasure $\mu : \mathscr{C} \rightrightarrows \mathbb{R}_+ \cup \{+\infty\}$ as*

$$(AS)\int_A F d\mu = \{(S)\int_A f d\mu_s \mid \mu_s \in S(\mu),\ f \in S_S(F,\mu_s)\},$$

*where $S(\mu)$ is the set of selections of $\mu$ which are submeasures and $S_S(F,\mu_s)$ is the set of all measurable selections Sugeno integrable with respect to the selection $\mu_s$. We say that $F$ is Aumann Sugeno integrable with respect to $\mu$ on $A$ if $(AS)\int_A F d\mu \neq \emptyset$.*

5. *Let $\mu : c \to \mathbb{R}_+ \cup \{+\infty\}$ be a multisubmeasure and let $F : T \rightrightarrows \mathbb{R}_+$ be a measurable multifunction. The Aumann weak integral of $F$ with respect to $\mu$ on $A \in \mathscr{C}$ is*

$$(Aw)\int_A F d\mu = \{(w)\int_A F d\mu_s \mid \mu_s \in S(\mu)\},$$

*where $S(\mu)$ is the set of selections of $\mu$ which are submeasures. $F$ is called Aumann w-integrable if $(Aw)\int_A F d\mu \neq \emptyset$.*

**Definition 14.3.2.** *(The Choquet type integrals)*

1. *[13] Let $\lambda : c \to \mathbb{R}_+ \cup \{+\infty\}$ be a fuzzy measure and let $f : T \to \mathbb{R}_+$ be a measurable function.*
*The Choquet integral of $f$ with respect to $\lambda$ on $A \in \mathscr{C}$ is defined by*

$$(C)\int_A f d\lambda = \int_0^{+\infty} \lambda(f^{-1}((\alpha,+\infty)) \cap A) d\alpha,$$

*where the integral from the right side is the Lebesgue integral. We say that $f$ is Choquet integrable with respect to $\lambda$ on $A$ if $(C)\int_A f d\lambda < +\infty$.*

2. *[19] Let $\lambda : \mathscr{C} \to \mathbb{R}_+ \cup \{+\infty\}$ be a fuzzy measure and let $f : T \to \mathbb{R}$ be a measurable function.*
*The Choquet (asymmetric integral) of $f$ with respect to $\lambda$ on $A$ is defined by*

$$(C_a)\int_A f d\lambda = (C)\int_A f^+ d\lambda - (C)\int_A f^- d\overline{\lambda},$$

*where $f^+ = f \vee 0$, $f^- = -f \vee 0$ and $\overline{\lambda}$ is the dual (conjugate) of $\lambda$, given by $\overline{\lambda}(A) = \lambda(T) - \lambda(T \setminus A)$, $\forall A \in \mathscr{C}$. If the right-hand is $\infty - \infty$, then the integral is not defined. We say that $f$ is Choquet integrable with respect to $\lambda$ on $A$ if $(C_a)\int_A f d\lambda \in \mathbb{R}$.*

3. *[49] Let $\lambda : \mathscr{C} \to \mathbb{R}_+ \cup \{+\infty\}$ be a fuzzy measure and let $f : T \to \mathbb{R}$ be a measurable function. The Choquet symmetric integral (Sipos, 1979) of $f$ with respect to $\lambda$ on $A \in \mathscr{C}$ is defined as*

$$(C_s)\int_A f d\lambda = (C)\int_A f^+ d\lambda - (C)\int_A f^- d\lambda,$$

*where $f^+ = f \vee 0$, $f^- = -f \vee 0$. If the right-hand is $\infty - \infty$, then the integral is not defined. We say that $f$ is Choquet integrable with respect to $\lambda$ on $A$ if $(C_s)\int_A f d\lambda \in mathbbR$.*

4. *[25] Suppose $F : T \rightrightarrows \mathbb{R}_+$ is a multifunction and $\lambda : \mathscr{C} \to \mathbb{R}_+ \cup \{+\infty\}$ is a submeasure. The Aumann-Choquet integral of F by respect to $\lambda$ on $A \in \mathscr{C}$ is defined by*

$$(AC)\int_A F d\lambda = \{(C)\int_A f d\lambda \mid f \in \mathscr{S}_C(F,\lambda)\},$$

*where $\mathscr{S}_C(F,\lambda) = \{f \mid f : T \to \mathbb{R}_+,\ f$ is Choquet integrable with respect to $\lambda$, $f(t) \in F(t),\ t \in T\ \mu - a.e.\}$. $F$ is called Aumann-Choquet integrable on $A$ if $(AC)\int_A F d\mu \neq \emptyset$.*

5. *Suppose $F : T \rightrightarrows \mathbb{R}_+$ is a multifunction and $\mu : \mathscr{C} \rightrightarrows \mathbb{R}_+$ is a multisubmeasure. The Aumann-Choquet integral of F with respect to $\mu$ on $A \in \mathscr{C}$ is defined by*

$$(AC)\int_A F d\mu = \{(C)\int_A f d\mu_s \mid \mu_s \in S(\mu),\ f \in \mathscr{S}_C(F,\mu_s)\},$$

*where $\mathscr{S}_C(F,\mu_s) = \{f \mid f : T \to \mathbb{R}_+,\ f(t) \in F(t),\ t \in T\ \mu -$ a.e., $f$ is Choquet integrable by respect to $\mu_s\}$, $S(\mu) = \{\mu_s \mid \mu_s(C) \in \mu(C),$ for all $C \in c$, $\mu_s$ is a fuzzy measure$\}$. $F$ is called Aumann-Choquet integrable on $A$ by respect to $\mu$ if $(AC)\int_A F d\mu \neq \emptyset$.*

In what follows we present the definition for the Henstock-Kurzweil integral of a real valued function which is a linear integral in contrast to the Choquet and Sugeno type integrals and which is an important non-absolute integral.

In this case we will consider $T$ endowed with a metric $d$ and we suppose that the topology $\tau_d$ induced by $d$ is included in $\mathscr{C}$. We denote $\tau_1$ the set of all open balls $B(t,r) = \{y \in T : d(t,y) < r\}$ where $t \in T$ and $r \in (0,+\infty)$. Following [34], let consider

$$\mathscr{H}_\infty = \{\overline{B_1} \setminus \overline{B_2} :\ B_1, B_2 \in \tau_1,\ B_1 \not\subseteq B_2,\ B_2 \not\subseteq B_1\},$$

$$\mathscr{H}_\in = \{\bigcap_{i \in I} H_i \neq \emptyset :\ H_i \in \mathscr{H}_\infty,\ I \in \mathscr{P}_f(\mathbb{N})\},$$

where $\mathscr{P}_f(\mathbb{N})$ denote the family of finite subsets of $\mathbb{N}$.

The elements of $\mathscr{H}_\in$ are called intervals and a finite union of mutually disjoint intervals are called elementary set.

A set $D = \{(I_i, x_i) : i = \overline{1,n}\}$ is called a partial division of the elementary set $E$ if $I_i$, $i = \overline{1,n}$ are mutually disjoint subintervals of $E$ such that $E \setminus \bigcup_{i=\overline{1,n}} I_i$ is either empty or an elementary subset of $E$ and for each $i$ we have $x_i \in I_i$. If $E = \bigcup_{i=\overline{1,n}} I_i$, $D$ is a division of $E$. Let denote $\overline{E}$ the closure of $E$.

A positive function $\delta : \overline{E} \to \mathbb{R}_+$ is called gauge on $E$. A division $D$ of $E$ is called $\delta$-fine if for each $(I_i, x_i) \in D$, we have $I_i \subset B(x_i, \delta(x_i))$.

In the sequel we consider an elementary set $E$, a countably additive measure $\lambda : \mathscr{C} \to [0, +\infty]$ such that $\lambda(E) < +\infty$ and we suppose that for every measurable set $C \in \mathscr{C}$ and every $\varepsilon > 0$ there exist an open set $U$ and a closed set $Y$ such that $\lambda(U \setminus Y) < \varepsilon$.

**Definition 14.3.3.** *[34] (The Henstock-Kurzweil type integral)*

*A function $f : T \to \mathbb{R}$ is called Henstock-Kurzweil integrable on E if there exists a real number $\alpha$ such that: for each $\varepsilon > 0$, there exists a gauge $\delta$ on E and for every $\delta$-fine division $D = \{(I_i, x_i),\ i \in I,\ I \in \mathscr{P}_f(\mathbb{N})\}$ of E, we have*

$$\Big|\sum_{i\in I} f(x_i)\lambda(I_i) - \alpha\Big| < \varepsilon.$$

*We write $(H)\int_E f d\lambda = \alpha$.*

*The Aumann-Henstock-Kurzweil integral for a real multifunction $F : T \rightrightarrows \mathbb{R}$ by respect to the real measure $\lambda$ is defined as the set*

$$(AH)\int_E F d\lambda = \{(H)\int_E f d\lambda \mid f \in S_H(F, \lambda)\}$$

*where $S_H(F, \lambda)$ denotes the set of all selections of F which are Henstock-Kurzweil integrable by respect to $\lambda$.*

## 14.4 PROPERTIES FOR INTEGRALS OF $\mathbb{R}^N$-VALUED FUNCTIONS AND MULTIFUNCTIONS $(N \geq 1)$

In this section, we present some properties concerning the absolute and non-absolute integrability for the integrals presented in the precedent sections.

Let be a multifunction $F : T \to \mathscr{P}_0(\mathbb{R}^n)$, $F = (F_i)_{i=\overline{1,n}}$ and a vector multisubmeasure $\mu : \mathscr{C} \rightrightarrows \mathbb{R}^n_+ \cup \{+\infty\}$, $\mu = (\mu_i)_{i=\overline{1,n}}$. For a function $f : T \to \mathbb{R}^n$ and a submeasure $\lambda : \mathscr{C} \to \mathbb{R}_+ \cup \{+\infty\}$ we denote by $\tilde{F} : T \rightrightarrows \mathbb{R}^n$ a multifunction with $\tilde{F}_i(t) = \{f(t)\}$, for all $t \in T$, $i = \overline{1,n}$ and by $\tilde{\mu} : \mathscr{C} \rightrightarrows \mathbb{R}^n_+ \cup \{+\infty\}$ the vector multisubmeasure with $\tilde{\mu}_i(A) = \{\lambda(A)\}$, for all $A \in \mathscr{C}$ and $i = \overline{1,n}$.

**Definition 14.4.1.** *We say that a multifunction $F : T \to \mathscr{P}_b(\mathbb{R}^n)$ is strongly absolute $\mathscr{T}$-integrable by respect to the multisubmeasure $\mu$ if F and $\|F\|$ are Aumann $\mathscr{T}$-integrable by respect to $\mu$.*

*A function $f : T \to \mathbb{R}^n$ is called strongly absolute $\mathscr{T}$-integrable by respect to the submeasure $\lambda$ if $\tilde{F}$ is strongly absolute $\mathscr{T}$-integrable by respect to the vector multisubmeasure $\tilde{\mu}$.*

*Similarly, a multifunction F (respectively a function $f$) is called strongly absolute integrable by respect to a submeasure $\lambda$ (respectively to a multisubmeasure $\mu$) if F is strongly absolute $\mathscr{T}$-integrable by respect to $\tilde{\mu}$ (respectively $\tilde{F}$ is strongly absolute $\mathscr{T}$-integrable by respect to $\mu$).*

*We say that a multifunction $F : T \to \mathscr{P}_0(\mathbb{R}^n)$ is Aumann absolute $\mathscr{T}$-integrable by respect to the multisubmeasure $\mu$ if F and G are Aumann $\mathscr{T}$-integrable by respect to $\mu$ where $G : T \to \mathscr{P}_0(\mathbb{R}_+)$, $G(t) = \{\|y\|, y \in F(t)\}$, $\forall t \in T$.*

*We say that a multifunction $F : T \to \mathscr{P}_0(\mathbb{R}^n)$ is $\tilde{\mathscr{G}}$ absolute $\mathscr{T}$-integrable by respect to the multisubmeasure $\mu$ if for each $\tilde{g} \in \tilde{\mathscr{G}}$ we have $\tilde{g} \circ F$ is Aumann absolute $\mathscr{T}$-integrable by respect to $\tilde{g} \circ \mu$.*

*A function $f : T \to \mathbb{R}^n$ is called $\tilde{\mathscr{G}}$ absolute $\mathscr{T}$-integrable by respect to the submeasure $\lambda$ if $\tilde{F}$ is $\tilde{\mathscr{G}}$ absolute $\mathscr{T}$-integrable by respect to the multisubmeasure $\tilde{\mu}$.*

*Similarly, a multifunction F (respectively a function $f$) is called $\tilde{\mathscr{G}}$ absolute $\mathscr{T}$-integrable by respect to the submeasure $\lambda$ (respectively to the multisubmeasure $\mu$) if F is $\tilde{\mathscr{G}}$ absolute $\mathscr{T}$-integrable by respect to $\tilde{\mu}$ (respectively $\tilde{F}$ is $\tilde{\mathscr{G}}$ absolute $\mathscr{T}$-integrable by respect to $\mu$).*

**Remark 14.4.2.** *If $n = 1$ and $\tilde{\mathscr{G}}_0 = \{\tilde{g} \in \mathscr{F}(\mathscr{C}, \mathscr{P}_0(\mathbb{R})) \mid \tilde{g}(C) = C, \forall C \in \mathscr{C}\}$, the $\tilde{\mathscr{G}}_0$ absolute $\mathscr{T}$-integrability and the strong absolute $\mathscr{T}$-integrability for a function $f$ by respect to a submeasure $\lambda$ are the usual notion of absolute $\mathscr{T}$ integrability i.e. $f$ and $|f|$ are $\mathscr{T}$-integrable by respect to $\lambda$.*

*If $\tilde{\mathscr{G}}_{\|\cdot\|} = \{\tilde{g} \in \mathscr{F}(\mathscr{C}, \mathscr{P}_0(\mathbb{R}^n)) \mid \tilde{g}(C) = \{\|C\|\}, \forall C \in \mathscr{C}\}$ then the $\tilde{\mathscr{G}}_{\|\cdot\|}$ absolute $\mathscr{T}$-integrability of F by respect to $\mu$ is equivalent to the $\mathscr{T}$-integrability for $\|F\|$ by respect to $\|\mu\|$.*

*The $\tilde{\mathscr{G}}_e$ absolute $\mathscr{T}$-integrability of F by respect to $\mu$ is equivalent with the Aumann absolute $\mathscr{T}$-integrability for each $F_i$ by respect to $\mu_i$, $i = \overline{1,n}$.*

*The $\tilde{\mathscr{G}}_*$ absolute $\mathscr{T}$-integrability of F by respect to $\mu = \lambda e$ is equivalent with the absolute $\mathscr{T}$-integrability by respect to $\lambda$ for each real functions $f_{i*}$ and $f_i^*$ (defined in Theorem 14.2.8 (iii)), $i = \overline{1,n}$.*

### 14.4.1 SOME PROPERTIES FOR THE SUGENO INTEGRAL

Firstly, let remark that for a real function $f : T \to \mathbb{R}_+$, the Sugeno integrability by respect to a fuzzy measure $\lambda : T \to \mathbb{R}_+$ is equivalent with the Sugeno absolute integrability of $f$ by respect to $\lambda$ and it is equivalent with the condition $(S)\int_T f d\lambda < +\infty$.

**Proposition 14.4.3.** *Every measurable bounded function $f : T \to \mathbb{R}_+$ is Sugeno and Choquet integrable by respect to each finite submeasure $\lambda$ (i.e. $\lambda(T) < +\infty$).*

Following the properties concerning the calculus rules for the Sugeno and Choquet integral we have $(C)\int_T f d\lambda \leq (\sup f)\mu(T) < +\infty$ respectively $(S)\int_T f d\lambda \leq \max\{(\sup f), \mu(T)\} < +\infty$.

**Theorem 14.4.4.** *Let $(T,d)$ be a metric space such that $\tau_d \subset c$, $\lambda : T \to \mathbb{R}_+$ be a fuzzy measure and $f : T \to \mathbb{R}_+$ be a function bounded on sets $A \in c$ with $\lambda(A) < +\infty$. Then $f$ is Sugeno integrable if and only if $f$ is bounded on $T$.*

**Proof:** Let suppose that $(S)\int_T f d\lambda \le M$. In this case for each $\alpha \ge M$, $\min\{\alpha, \lambda(\{t: f(t) \ge \alpha\})\} = \lambda(\{t: f(t) \ge \alpha\}) \le M$. Thus, there exist $m \in \mathbb{R}$ such that $f(t) \le m$ for each $t \in \{t: f(t) \ge \alpha\}$. For $M' = \max\{M, m\}$ we get $f(t) \le M'$ for each $t \in T$, so $f$ is bounded on $T$.

The following results follows from the definitions and Theorem 14.4.4.

**Theorem 14.4.5.** *Let $F : T \to p_b(\mathbb{R}_+)$ and $\lambda : c \to \mathbb{R}_+$ be a submeasure.*

1. *For each selection $f$ of $F$, we have $(S)\int_T f d\lambda \le (w)\int_T F d\lambda \le (S)\int_T \|F\| d\lambda$;*
2. *If $\|F\|$ is Sugeno integrable by respect to $\lambda$ then $F$ is weak integrable, strong integrable and each selection $f$ of $F$ is Sugeno integrable by respect to $\lambda$.*
3. *If $F : T \to p_{kc}(\mathbb{R}_+)$ and $\|F\|$ is a bounded function on $T$, then $\|F\|$ is Sugeno integrable by respect to $\lambda \Longleftrightarrow F$ is weak integrable by respect to $\lambda \Longleftrightarrow S_S(F,\lambda) = S(F) \Longleftrightarrow (AS)\int_T F d\lambda \neq \emptyset$.*

   *Also, in this case, each selection $f : T \to \mathbb{R}_+$ of $F$ which is bounded on sets $A \in c$ with $\lambda(A) < +\infty$ is bounded on $T$.*

**Theorem 14.4.6.** *Let $F : T \to p_k(\mathbb{R}_+)$ and $\mu : c \rightrightarrows \mathbb{R}_+$ be a multisubmeasure. We have*

1. $\|(AS)\int_T F d\mu\| \le \|(Aw)\int_T F d\mu\| \le \|(AS)\int_T \|F\| d\mu\|$;
2. $\|(AS)\int_T F d\mu\| \le \|(AS)\int_T F d\|\mu\|\|$;
3. $\|(Aw)\int_T F d\mu\| \le (w)\int_T F d\|\mu\|$;
4. $(AS)\int_T \|F\| d\mu \neq \emptyset \Rightarrow (Aw)\int_T F d\mu \neq \emptyset \Rightarrow (AS)\int_T F d\mu \neq \emptyset$;
5. *If $F$ has compact values then we have equality in (i) and equivalence in (iv).*

**Corollary 14.4.7.** *If $F : T \to p_k(\mathbb{R}_+)$ and $\|F\|$ is a bounded function on $T$, then*

1. *$F$ is strongly absolute Sugeno integrable by respect to $\lambda$ if and only if $F$ is Aumann-Sugeno integrable by respect to $\lambda$ if and only if $F$ is weak integrable.*
2. *$F$ is $\widehat{g}_{\|\cdot\|}$ absolute Sugeno integrable by respect to $\lambda$ if and only if $F$ is Aumann-Sugeno integrable by respect to $\lambda$ if and only if $F$ is weak integrable by respect to $\lambda$.*

Let consider now the case $n \ge 2$.
As we have seen in Theorem 14.2.8, the $\mathcal{G}(\mathcal{T})$ integral and the $\widehat{g}(\mathcal{T})$ integral have the following forms:

1. If $f = (f_i)_{i=\overline{1,n}}$, $\lambda = (\lambda_i)_{i=\overline{1,n}}$ and for each $i = \overline{1,n}$, $f_i$ are Sugeno integrable by respect to the submeasure $\lambda_i$, the $g_e$ Sugeno integral of $f$ by respect to $\lambda$ has the following form
$$(g_e)(S)\int_T f d\lambda = ((S)\int_T f_i d\lambda_i)_{i=\overline{1,n}}.$$
Thus $f$ is $g_e$ Sugeno integrable (or Sugeno vector integrable) by respect to $\lambda$ if and only if $(g_e)(S)\int_T f d\lambda \in \mathbb{R}^n$.
2. Let $F = (F_i)_{i=\overline{1,n}}$ and $\mu = (\mu_i)_{i=\overline{1,n}}$ where $\mu_i : c \rightrightarrows \mathbb{R}_+$ are multisubmeasures.
(2(a).) If $F_i$ are Aumann-Sugeno integrable by respect to $\mu_i$, then the $\widehat{g}_e$ Sugeno integral is
$$(g_e)(S)\int_T F d\mu = \prod_{i=\overline{1,n}} (AS)\int_T F_i d\mu_i.$$
Thus $F$ is $\widehat{g}_e$ Sugeno integrable (or Aumann-Sugeno vector integrable) by respect to $\mu$ if and only if $(g_e)(S)\int_T F d\mu \neq \emptyset$.
(2(b).) If $F_i$ are Aumann weak integrable by respect to $\mu_i$, then the $\widehat{g}_e$ weak integral is
$$(g_e)(w)\int_T F d\mu = \prod_{i=\overline{1,n}} (Aw)\int_T F_i d\mu_i.$$
Thus $F$ is $\widehat{g}_e$ weak integrable (or Aumann weak vector integrable) by respect to $\mu$ if and only if $(g_e)(w)\int_T F d\mu \neq \emptyset$.
3. If $F = (F_i)_{i=\overline{1,n}}$, where $F_i : T \rightrightarrows p_{kc}(\mathbb{R})$ and $\mu(C) = \lambda(C)e$ for each $C \in c$ with $\lambda : c \to \mathbb{R}$ is a real submeasure and $e = (1,1,...,1)$, the $\widehat{g}_s$ Sugeno integral is
$$(g_s)(S)\int_T F d\mu = \prod_{i=\overline{1,n}} [-(S)\int_T f_i d\lambda, (S)\int_T f_i d\lambda]$$
if the functions $f_i : T \to \mathbb{R}$ given by $f_i(t) = \sup\{|y|,\ y \in F_i(t)\}$ are Sugeno integrable by respect to $\lambda$ for each $i = \overline{1,n}$.
Thus $F$ is $\widehat{g}_s$ Sugeno integrable (or strictly Sugeno vector integrable) if $(g_s)(S)\int_T F d\mu \subseteq \mathbb{R}^n$.

**Remark 14.4.8.** *If F is strictly Sugeno vector integrable, then F is Aumann weak vector integrable.*

By using the properties from the case $n = 1$, we obtain the following properties for the vector case.

**Theorem 14.4.9.** *Let $f : T \to \mathbb{R}^n$, $f = (f_i)_{i=\overline{1,n}}$, $\lambda : c \to \mathbb{R}^n$, $\lambda = (\lambda_i)_{i=\overline{1,n}}$, $\lambda_i$ are submeasures.*

*1. If $\|f\|$ is a real Sugeno integrable function by respect to the real submeasure $\lambda_m(C) = \max\{\lambda_i(C),\ i = \overline{1,n}\}$ for each $C \in c$, then $f$ is Sugeno vector integrable* *by respect to* $\lambda$. *Equivalently, this means that if* $f$ *is* $g_{\|\cdot\|}$ *absolute Sugeno integrable by respect to* $\lambda$ *then* $f$ *is Sugeno vector integrable.*

2. *Suppose for every* $i = \overline{1,n}$ $f_i$ *are bounded on each set with a finite submeasure* $\lambda_i$. *Then the following conditions are equivalent:*
   a) $f$ *is Sugeno vector integrable by respect to* $\lambda$;
   b) $\|f\|$ *is Sugeno integrable by respect to* $\lambda_m$;
   c) $\|f\|$ *is Sugeno integrable by respect to each* $\lambda_i$;
   d) $f$ *is* $g_{\|\cdot\|}$ *absolute Sugeno integrable by respect to* $\lambda$.
3. *Suppose for all* $i = \overline{1,n}$, $f_i$ *are bounded on each set with a finite submeasure* $\lambda$. *Then* $f$ *is* $g_e$ *absolute Sugeno integrable by respect to* $\lambda$ *if and only if* $f$ *is bounded on* $T$.

**Theorem 14.4.10.** *1. Let* $f : T \to \mathbb{R}^n_+$, $f = (f_i)_{i=\overline{1,n}}$ *and* $\lambda_i : c \to \mathbb{R}_+$ *be fuzzy measures for each* $i = \overline{1,n}$. *For* $\lambda = (\lambda_i)_{i=\overline{1,n}}$ *and* $\mu : c \to \mathbb{R}^n$, $\mu_i = \|\lambda\|$ *for each* $i = \overline{1,n}$, *if* $f$ *is Sugeno vector integrable by respect to* $\mu$ *then* $f_i$ *are Sugeno integrable by respect to each* $\lambda_i$ *i.e.* $f$ *is Sugeno vector integrable by respect to* $\lambda$.

2. *Let* $F : T \rightrightarrows \mathbb{R}$ *be a real multifunction and* $\mu : T \rightrightarrows \mathbb{R}_+$ *be a multisubmeasure with compact values. If* $F$ *is Aumann Sugeno integrable by respect to* $\|\mu\|$ *then* $F$ *is Aumann Sugeno integrable by respect to* $\mu$.

**Proposition 14.4.11.** *Let* $F_i : T \rightrightarrows \mathbb{R}$, $F(t) = \prod_{i=\overline{1,n}} F_i(t)$ *for each* $t \in T$ *and* $\mu(C) = \lambda(C)e$ *for each* $C \in c$ *where* $\lambda : c \to \mathbb{R}_+$ *is a real submeasure and* $e = (1,1,...,1)$. *If* $F_i$ *has compact values then* $F$ *is* $\widehat{g}_s$ *absolute Sugeno integrable by respect to* $\mu$ *if and only if each selection of* $F$ *is strongly absolute Sugeno integrable by respect to* $\lambda$.

### 14.4.2 SOME PROPERTIES FOR THE CHOQUET INTEGRAL

**Theorem 14.4.12.** *Let* $f : T \to \mathbb{R}$ *be a real function and* $\lambda : c \to \mathbb{R}_+$ *be a fuzzy measure. If* $f^+$ *and* $f^-$ *are measurable and Choquet integrable by respect to a submodular submeasure* $\lambda$, *then*

1. $|f|$ *is Choquet integrable by respect to* $\lambda$;
2. $f^-$ *is Choquet integrable by respect to* $\overline{\lambda}$;
3. $f$ *is absolute Choquet (symmetric and asymmetric) integrable by respect to* $\lambda$.

**Proof:** Since $|f| = f^+ + f^-$ and $\lambda$ is submodular we get (i) from the inequality

$$(C)\int_T |f|d\lambda \leq (C)\int_T f^+d\lambda + (C)\int_T f^-d\lambda.$$

Also, the submodularity of $\lambda$ implies $\bar{\lambda}(A) \leq \lambda(A)$ and thus $(C)\int_T f^-d\bar{\lambda} \leq (C)\int_T f^-d\lambda < +\infty$. Hence, $(C_a)\int_T fd\lambda \in \mathbb{R}$, $(C_s)\int_T fd\lambda \in \mathbb{R}$ and (2), (3) follows.

Similarly we get the following propositions.

**Proposition 14.4.13.** *Let $f : T \to \mathbb{R}$ be a real function and $\lambda : c \to \mathbb{R}_+$ be a fuzzy measure. If $f$ is symmetric Choquet absolute integrable and $\lambda$ is supermodular then*

1. *$f^+$, $f^-$ are Choquet integrable by respect to $\lambda$.*
2. *$f$ is asymmetric Choquet absolute integrable by respect to $\lambda$.*

**Proposition 14.4.14.** *Let $f : T \to \mathbb{R}$ be a real function and $\lambda : c \to \mathbb{R}_+$ be a fuzzy measure. If $f$ is asymmetric Choquet absolute integrable and $\lambda$ is submodular then*

1. *$f^+$, $f^-$ are Choquet integrable by respect to $\lambda$.*
2. *$f$ is symmetric Choquet absolute integrable by respect to $\lambda$.*

**Theorem 14.4.15.** *Let $f : T \to \mathbb{R}^n$ be a vector function, $f = (f_i)_{i=\overline{1,n}}$, such that each $f_i$ are measurable, $\lambda : T \to \mathbb{R}_+$ is a fuzzy measure and $\mu : c \to \mathbb{R}^n_+$, $\mu = (\mu_i)_{i=\overline{1,n}}$ such that $\mu_i = \lambda e$. If $\lambda$ is submodular, then $f$ is $\widehat{g}_e$ absolute Choquet integrable (symmetric or asymmetric) by respect to $\mu$ if and only if $f$ is $\widehat{g}_{\|\cdot\|}$ absolute Choquet integrable by respect to $\lambda$.*

Following the definition, if $f$ is $\widehat{g}_e$ absolute Choquet integrable by respect to $\mu$ then $f_i$ are absolute Choquet integrable by respect to $\lambda$ for each $i = \overline{1,n}$. Since $\lambda$ is submodular and $\|\mu\| = \lambda$ we get

$$(C)\int_T \|f\|d\lambda \leq \sum_{i=\overline{1,n}} (C)\int_T |f_i|d\lambda < +\infty.$$

Remark 14.4.2 leads to fact that $f$ is $\widehat{g}_{\|\cdot\|}$ absolute Choquet integrable by respect to $\lambda$. For the converse, let observe that if $\|f\|$ is Choquet integrable by respect to $\lambda$ we have that $|f_i|$, $f_i^+$ and $f_i^-$ are Choquet integrable by respect to $\lambda$ for each $i = \overline{1,n}$. Theorem 14.4.12 implies the fact that $f_i$ are absolute Choquet integrable (symmetric and asymmetric) and the conclusion follows.

**Theorem 14.4.16.** *Let $F : T \rightrightarrows \mathbb{R}^n$ be a measurable vector multifunction, $F = (F_i)_{i=\overline{1,n}}$ and $\mu : c rightrightarrows \mathbb{R}^n_+$ be a multisubmeasure $\mu = (\mu_i)_{i=\overline{1,n}}$. The following properties hold:*

1. *IF $F$ is $\widehat{g}_{\|\cdot\|}$ absolute Choquet integrable by respect to $\mu$ then $F_i$ are $\widehat{g}_{\|\cdot\|}$ absolute Choquet integrable by respect to $\mu_i$ for each $i = \overline{1,n}$. If $\mu$ is submodular we have equivalence.*
2. *If $F_i$ has compact values for all $= \overline{1,n}$, then $F$ is $\widehat{g}_{\|\cdot\|}$ absolute Choquet integrable by respect to $\mu$ if and only if each measurable selection of $F$ is $g_{\|\cdot\|}$ absolute Choquet integrable by respect to $\mu$.*
   *Obviously, in this case $F$ is $\widehat{g}_{\|\cdot\|}$ absolute Choquet integrable by respect to $\mu$.*

**Proof:** (1) Let observe that $(C)\int_T \|F\|d\|\mu\| \geq (C)\int_T \|F_i\|d\mu_i$ and $\|F\| = \sum_{i=\overline{1,n}} \|F_i|$.

Hence (i) follows obviously.

(2) If $F$ has compact values, then $\|F_i\|$ are measurable selections of $F_i$ and (i) implies that $F_i$ are $\widehat{g}_{\|\cdot\|}$ absolute Choquet integrable by respect to $\mu_i$. If $f \in S_C(F, \mu_s)$, $\mu_s \in S(\mu)$ then $\|f\| \leq \|F\|$, $\|\mu_s\| \leq \|\mu\|$ and thus $f$ is $g_{\|\cdot\|}$ absolute Choquet integrable by respect to $\mu$. Since $(\|F_i\|)_{i=\overline{1,n}}$ is a selection of $F$, this selection is $\widehat{g}_{\|\cdot\|}$ absolute Choquet integrable by respect to $\mu$ i.e. $\|(\|F_i\|)_{i=\overline{1,n}}\| = \|F\|$ is Choquet integrable by respect to $\|\mu\|$ and the conclusion follows.

Let remark that if $F$ has compact values then $\|F_i\|(t) \in G_i(t)$ for each $i = \overline{1,n}$ and if $\|F_i\|$ are Choquet integrable by respect to $\lambda_i$ for each $i = \overline{1,n}$ then $F$ is $\widehat{g}_s$ Choquet integrable by respect to a submeasure $\mu$ (see Theorem 14.2.8) and this integral is

$$(\widehat{g}_s)(C)\int_T F d\lambda = \prod_{i=\overline{1,n}} [-(C)\int_T \|F_i\| d\lambda_i, (C)\int_T \|F_i\| d\lambda_i].$$

In this case the following results hold.

**Proposition 14.4.17.** *A vector multifunction $F : T \rightrightarrows \mathbb{R}^n$ is $\widehat{g}_s$ absolute Choquet integrable by respect to $\mu : c \rightrightarrows \mathbb{R}^n_+$ if and only if $F$ is $\widehat{g}_{\|\cdot\|}$ absolute Choquet integrable by respect to $\mu$.*

**Theorem 14.4.18.** *Let $\lambda_i : c \to \mathbb{R}_+$ be continuous fuzzy measures and $\lambda = (\lambda_i)_{i=\overline{1,n}}$. Let $F : T \rightrightarrows \mathbb{R}^n$, $F = (F_i)_{i=\overline{1,n}}$, be a measurable multifunction such that $F_i(t) = [f_i(t), g_i(t)]$, $f_i(t) < g_i(t)$ for $t \in T$ and the functions $T_i : \mathrm{f}(T, \mathbb{R}^n) \to \mathbf{f}(T, \mathbb{R}_+)$ given by $T_i(f) = |f_i|$ $(|f_i|(t) = |f_i(t)|$, $\forall t \in T)$ are comonotonic for $i = \overline{1,n}$.*
*Thus $f_i$ and $g_i$ are absolute Choquet integrable by respect to $\lambda_i$ for each $i = \overline{1,n}$ if and only if $F$ is $g_e$ Choquet absolute integrable by respect to $\lambda$.*

The conditions from the hypothesis imposed to $F$ and $\lambda$ ensure that $f_i$ and $g_i$ are Choquet integrable by respect to $\lambda_i$ for each $i$ and

$$(AC)\int_T F_i d\lambda_i = [(C)\int_T f_i d\lambda_i, (C)\int_T g_i d\lambda_i].$$

The condition imposed for $T_i$ implies that we have one of the following two cases:
a)For $t \in T$, $\|F_i\|(t) = \sup(|f_i|(t), |g_i|(t)) = |g_i|(t)$ for each $i$ or
b)For $t \in T$, $\|F_i\|(t) = \sup(|f_i|(t), |g_i|(t)) = |f_i|(t)$ for each $i$.
In the case a) for each $i = \overline{1,n}$ we must have $g_i(t) > 0$ and $|f_i(t)| < g_i(t)$ which implies that every Choquet integrable selection $h_i$ of $F_i$ satisfy $|h_i|(t) < g_i(t)$ and thus it is Choquet absolute integrable and $F_i$ is Aumann Choquet absolute integrable. Thus, $F$ is $g_e$ absolute Choquet integrable.

Similarly, for the case b) every selection $h_i(t)$ of $F_i(t)$ satisfies $|h_i|(t) < |f_i(t)|$ and the conclusions follow by the same arguments. The converse is obvious.

### 14.4.3 SOME PROPERTIES FOR THE HENSTOCK-KURZWEIL INTEGRAL

Let $f : T \to \mathbb{R}^n$, $f = (f_i)_{i=\overline{1,n}}$, and $\lambda : c \to \mathbb{R}_+$ a countably additive measure. We remark that Definition 14.3.3 may be extended for this vector-valued functions by replacing the absolute value from the real case with the space norm.

**Definition 14.4.19.** *1. A vector measurable function $f : T \to \mathbb{R}^n$ is called Henstock-Kurzweil integrable on an elementary set E by respect to a countably additive measure $\lambda : c \to \mathbb{R}_+$ if there exists $\alpha \in \mathbb{R}^n$ such that: for each $\varepsilon > 0$, there exists a gauge $\delta$ on E and for every $\delta$-fine division $D = \{(I_i, x_i)_{i\in I}, I \in p_f(\mathbb{N})\}$ of E, we have*

$$\|\sum_{i\in I} f(x_i)\lambda(I_i) - \alpha\| < \varepsilon.$$

*We write $(H)\int_E f d\lambda = \alpha$.*

*2. The Aumann-Henstock-Kurzweil integral for a multifunction $F : T \rightrightarrows \mathbb{R}^n$ by respect to $\lambda$ is the set*

$$(AH)\int_E F d\lambda = \{(H)\int_E f d\lambda : f \in S_H(F,\lambda)\}$$

*where $S_H(F,\lambda)$ denotes the set of measurable selections of F which are Henstock-Kurzweil integrable with respect to $\lambda$.*

Let observe that if the function $f : T \to \mathbb{R}^n$, $f = (f_i)_{i=\overline{1,n}}$ is Henstock-Kurzweil integrable with respect to the measure $\lambda : c \to \mathbb{R}_+$ then for each $i$, $f_i$ are Henstock-Kurzweil integrable with respect to the measure $\lambda : c \to \mathbb{R}_+$ and

$$(H)\int_E f d\lambda = \prod_{i=\overline{1,n}} (H)\int_E f_i d\lambda.$$

Also, let remark that for a vector function $f : T \to \mathbb{R}^n$, $f = (f_i)_{i=\overline{1,n}}$, if $f_i$ are real functions Henstock-Kurzweil integrable with respect to a countably additive measure $\lambda_i : c \to \mathbb{R}_+$ then from Theorem 14.2.8, the $\gamma_e$ Henstock-Kurzweil integral for $f$ with respect to $\lambda : c \to \mathbb{R}^n_+$, $\lambda = (\lambda_i)_{i=\overline{1,n}}$ is

$$(g_e)(H)\int_E f d\lambda = \prod_{i=\overline{1,n}} (H)\int_E f_i d\lambda_i$$

and thus $(g_e)(H)\int_E f d\lambda = (H)\int_E f d\lambda$.

It is an well-known fact that the Henstock-Kurzweil integral is not an absolute one and a real function is Henstock-Kurzweil absolute integrable on an elementary set $E$ by respect to the Lebesgue measure $\lambda$ (i.e. $f$ and $|f|$ are Henstock-Kurzweil integrable on $E$ by respect to $\lambda$) if and only if $f$ is Lebesgue integrable on $E$. Thus, since $\|f\|$ is a positive real function, it is Henstock-Kurzweil integrable if and only if it is Lebesgue integrable. As a consequence, if $f$ is Henstock integrable and $\|f\|$ is not Henstock-Kurzweil integrable, then at least one from the components $f_i$ of $f$ is not Lebesgue integrable.

It is more adequate to consider in this case a notion of weakly $\mathcal{T}$-integrability and weakly absolute $\mathcal{T}$-integrability as it was presented in [20].

**Definition 14.4.20.** *Let $f : T \to X$ and $\lambda : c \to X$ be a countably additive measure. We say that $f$ is weakly $\mathcal{T}$-integrable (respectively, weakly absolute $\mathcal{T}$-integrable) if for each $x^* \in X^*$, $x^* o f$ is a real function $\mathcal{T}$-integrable (respectively absolute $\mathcal{T}$-integrable) by respect to $x^* o \lambda$.*

In our case, for $X = \mathbb{R}^n$, $f$ is weakly $\mathcal{T}$-integrable (respectively, weakly absolute $\mathcal{T}$-integrable ) if and only if $f_i$ are $\mathcal{T}$-integrable (respectively, absolute $\mathcal{T}$-integrable). For the Henstock-Kurzweil integrability we have the following result.

**Proposition 14.4.21.** *Let $f : T \to \mathbb{R}^n$, $f = (f_i)_{i=\overline{1,n}}$ be a measurable function and let $\lambda_i : c \to \mathbb{R}_+$ be countable additive measures for each $i = \overline{1,n}$. Then $f$ is weakly absolute Henstock-Kurzweil integrable on an elementary set $E$ with respect to the measure $\lambda : c \to \mathbb{R}^n$, $\lambda = (\lambda_i)_{i=\overline{1,n}}$ if and only if for each $i = \overline{1,n}$, $f_i$ are absolute Henstock-Kurzweil integrable on $E$ by respect to $\lambda_i$. If $\lambda_i$ is the Lebesgue measure for each $i$, then $f$ is weakly absolute Henstock-Kurzweil integrable on an elementary set $E$ by respect to the measure $\lambda : c \to \mathbb{R}^n$, $\lambda = (\lambda_i)_{i=\overline{1,n}}$ if and only if for each $i = \overline{1,n}$, $f_i$ are Lebesgue integrable on $E$.*

**Proof:** For a set-valued function $F : T \rightrightarrows \mathbb{R}^n$ and a measure $\lambda : c \to \mathbb{R}^n$, $\lambda = (\lambda_i)_{i=\overline{1,n}}$ where $\lambda_i$ is a countably additive measure for each $i$, the $\widehat{g}_e$ Henstock-Kurzweil integral with respect to $\lambda$ (as it was presented in Theorem 14.2.8) exists if the Aumann-Henstock-Kurzweil integral for each $F_i$ by respect to $\lambda_i$ is nonempty and

$$(g_e)(AH)\int_E Fd\lambda = (\widehat{g}_e)(H)\int_E Fd\lambda = \prod_{i=\overline{1,n}} (AH)\int_E F_i d\lambda_i.$$

Corresponding to family $\widehat{g}_s$ and $\mu(C) = \lambda(C)e$ for each $C \in c$ where $\lambda$ is a real measure countable additive and $e = (1, 1, ..., 1)$, the $\widehat{g}_s$ Henstock-Kurzweil integral of $F : T \to p_{kc}(\mathbb{R}^n)$ with respect to $\mu$ is

$$(g_s)(H)\int_T Fd\mu = \prod_{i=\overline{1,n}} [-(H)\int_T f_i d\lambda, (H)\int_T f_i d\lambda]$$

if the functions $f_i : T \to \mathbb{R}$ given by $f_i(t) = \sup\{|y|, y \in F_i(t)\}$ are Henstock-Kurzweil integrable by respect to $\lambda$.

If we consider a multifunction $F : T \to p_{kc}(\mathbb{R}^n)$ and a measure $\mu : c \to \mathbb{R}^n_+$, $\mu(C) = \lambda(C)e$ for each $C \in c$ where $\lambda : c \to \mathbb{R}$ is a real measure countable additive and $e = (1, 1, ..., 1)$, then the $\widehat{g}_*$ Henstock-Kurzweil integral of $F$ by respect to $\mu$ will be denoted $(\widehat{g}_*)(H)\int_T Fd\mu$ and is

$$(\widehat{g}_*)(H)\int_T Fd\mu = \prod_{i=\overline{1,n}} [(H)\int_T f_{i*} d\lambda, (H)\int_T f_i^* d\lambda]$$

where $f_{i*}(t) = \inf\{y \in F_i(t)\}$ and $f_i^*(t) = \sup\{y \in F_i(t)\}$ if these functions are Henstock-Kurzweil integrable by respect to $\lambda$.

Let remark that $(\widehat{g}_*)(H)\int_T F d\mu = (KHP)\int_T F d\mu$ where the integral from the right hand is the $(KHP)$-integral from [20].

In particular if $F$ is an interval valued multifunction, i.e. for each $i = \overline{1,n}$, $F_i(t) = [f_i(t), g_i(t)]$, $\forall t \in T$ with $f_i \leq g_i$, then we have the following properties.

**Theorem 14.4.22.** *Let $F : T \rightrightarrows \mathbb{R}^n$ $F = (F_i)_i$ be a measurable multifunction which is interval valued, i.e. for each i, $F_i(t) = [f_i(t), g_i(t)]$, with $f_i(t) \leq g_i(t)$ $\forall t \in T$ and $\mu : c \to \mathbb{R}^n_+$, $\mu(C) = \lambda(C)e$ for each $C \in c$ where $\lambda : c \to \mathbb{R}$ is a real measure countable additive and $e = (1, 1, ..., 1)$. The following assertions hold:*

1. *If $F$ is $g_*$ Henstock-Kurzweil integrable with respect to $\mu$, then the selections $h_i$ of $F_i$, $h_i = \gamma f_i + (1-\gamma)g_i$, $\gamma \in [0,1]$ are Henstock-Kurzweil integrable by respect to $\lambda$.*
2. *If $F$ is $g_*$ Henstock-Kurzweil integrable with respect to $\mu$, then $F$ is $\widehat{g}_e$ Henstock-Kurzweil integrable with respect to $\mu$ and*

$$(\widehat{g}_*)(H)\int_T F d\mu = (\widehat{g}_e)(H)\int_T F d\mu.$$

The theorem's hypothesis ensure that $f_i$ and $g_i$ are measurable and Henstock-Kurzweil integrable. Since the Henstock-Kurzweil integral is linear, each combination $h_i = \gamma f_i + (1-\gamma)g_i$, $\gamma \in [0,1]$ is Henstock-Kurzweil integrable and we have the equality $(\widetilde{\mathscr{G}}_e)(H)\int_T F_i d\mu = [(H)\int_T f_i d\mu, (H)\int_T g_i d\mu]$.

**Definition 14.4.23.** *A measurable multifunction $F : T \to p_{kc}(\mathbb{R}^n)$ is called weakly Henstock-Kurzweil integrable (respectively, weakly absolute Henstock-Kurzweil integrable) by respect to a real countable measure $\lambda$ if for each $x^* \in X^*$, the functions $\sup x^* oF$ are Henstock integrable (respectively, absolute Henstock-Kurzweil integrable) by respect to $\lambda$.*

The following property follows obviously.

**Proposition 14.4.24.** *A measurable multifunction $F : T \to p_{kc}(\mathbb{R}^n)$ is weakly absolute Henstock-Kurzweil integrable on an elementary set $E$ by respect to a real countable measure $\lambda$ if and only if for each $i = \overline{1,n}$, the functions $f_{i*}$ and $f_i^*$ (given by $f_{i*}(t) = \inf F_i(t)$ and $f_i^*(t) = \sup F_i(t)$) are absolute Henstock-Kurzweil integrable on $E$ by respect to $\lambda$. If $\lambda$ is the Lebesgue measure, $F$ is weakly absolute Henstock-Kurzweil integrable on a measurable set $E$ by respect to $\lambda$ if and only if $f_{i*}$ and $f_i^*$ are Lebesgue integrable on $E$.*

### 14.4.4 SOME PROPERTIES FOR THE BROOKS INTEGRAL

Let recall from [18] the general definition of the $\phi$-integral for a multifunction $F : T \to \mathscr{K} \subset p_{kc}(\mathbb{R}^n)$ with respect to a multimeasure $\phi : c \to p_{kc}(\mathbb{R}^n)$.

**Definition 14.4.25.** *1. A multifunction $F : T \to \mathfrak{K}$ is said to be simple if $F = \sum_{i}^{n} C_i \chi_{A_i}$, where $C_i \in \mathfrak{K}$, $A_i \in c$, $A_i \cap A_j = \emptyset$, $i \neq j$, $\bigcup_{i=\overline{1,n}} A_i = T$ and $\chi_{A_i}$ is the characteristic function of $A_i$.*
*The integral of $F$ over $E \in c$ with respect to $\phi$ is defined by:*

$$(\phi)\int_E F d\phi = \sum_{i=1}^{n} C_i \phi(A_i \cap E) \in \mathfrak{K}.$$

2. *A multifunction $F : T \to \mathfrak{K}$ is said to be $\phi$-totally measurable if there exists a sequence $(F_n)_n$ of simple multifunctions $F_n : T \to \mathfrak{K}$ such that:*
   *(1) $D(F_n, F)$ is $\nu$-measurable where $\nu$ is the variation of $\phi$, $\forall n \in \mathbb{N}$,*
   *(2) $D(F_n, F) \xrightarrow{\nu} 0$.*
3. *If $F$ is a $\phi$-totally measurable multifunction then $F$ is called $\phi$-integrable on $T$ if there exists a sequence $(F_n)_n$ of simple multifunctions $F_n : T \to \mathfrak{K}$ satisfying the conditions (1) and (2) of (ii) and*

$$\lim_{n,m\to+\infty} \int_T D(F_n, F_m) d\nu = 0.$$

*The sequence $(F_n)$ is called a defining sequence of $F$. The integral of $F$ on $T$ relative to $\phi$ is defined as $(\phi)\int_T F d\phi = \lim_{n\to+\infty} (\phi)\int_T F_n d\phi$. The integrability on every $E \in c$ is defined in the usual way.*

We recall from [17], [18] the following theorem which will be useful for studying the $\phi$-integrability for the norm of vector functions and multifunctions.

**Theorem 14.4.26.** *Let $F : T \to \mathfrak{K}$ be a $\phi$-integrable multifunction. Then the function $\|F\|$ is Brooks integrable with respect to $\phi$. Moreover, if $X = \mathbb{R}$ and $\mathfrak{K} = \{A \subset \mathbb{R}_+ : A \in p_{kc}(\mathbb{R})\}$, then we have:*

$$(\phi)\int_E \|F\| d\phi \subseteq (\phi)\int_E F d\phi, \ \forall E \in c.$$

**Remark 14.4.27.** *It is obvious that the converse does not hold i.e. if the function $\|F\|$ is Brooks integrable with respect to $\phi$ does not imply that the multifunction $F$ is $\phi$ integrable. It is easy to see this if we consider an interval valued multifunction $F = [f, g]$, $0 < f(t) < g(t)$, $\forall t \in T$ and $\mathfrak{K} = p_{kc}(\mathbb{R})$. In this case, the $\phi$-integrability of $\|F\|$ is equivalent with the Brooks integrability of $g$ by respect to $\phi$, while the $\phi$-integrability of $F$ involves the Brooks integrability of $f$ and $g$ by respect to $\phi$.*

Following Theorem 14.2.8, a function $f : T \to \mathbb{R}^n_+$, $f = (f_i)_i$ is $g_e$ $\phi$-integrable if and only if $f_i$ are $\phi$-integrable and $(g_e)(\phi)\int_T f d\phi = ((\phi)\int_T f_i d\phi)_{i=\overline{1,n}}$. Using the properties of the Brooks integral we get the following propositions.

**Proposition 14.4.28.** *Let $f : T \to \mathbb{R}^n_+$, $f = (f_i)_i$ be a $g_e$ $\phi$-integrable function. Then $\|f\|$ is Brooks integrable by respect to $\phi$ and $(\phi)\int_T \|f\| d\mu = \|(g_e)(\phi)\int_T f d\mu\|$ if $\|f\|$ is $\phi$-totally measurable.*

Also, let remark that a multifunction $F : T \to k \subset p(\mathbb{R}^n)$ is $\widehat{g}_e$ $\phi$-integrable if and only if $F_i$ are $\phi$-integrable for each $i$ and $(\widehat{g}_e)(\phi)\int_T F d\phi = ((\phi)\int_T F_i d\phi)_{i=\overline{1,n}}$. The following proposition does hold.

**Proposition 14.4.29.** *Let $F : T \to k \subset p(\mathbb{R}^n)$ be a $\widehat{g}_e$ $\phi$-integrable multifunction. Then the function $\|F\|$ is Brooks integrable with respect to $\phi$ . Moreover, $(\phi)\int_T \|F\| d\phi \leq \|(\widehat{g}_e)(\phi)\int_T F d\phi\|$ and $(B)\int_T \|F_i\| d\phi = \|(\widehat{g}_e)\int_T F_i d\phi\|$ if $\|F\|$ is $\phi$-totally measurable.*

**Proposition 14.4.30.** *Let $F : T \to k \subset p(\mathbb{R}^n)$ be an interval valued multifunction, i.e. for each i, $F_i = [f_i, g_i]$, $0 < f_i < g_i$. In this case F is $\phi$-integrable if and only if F is $\widehat{g}_s$ $\phi$-integrable and*

$$(\widehat{g}_s)(\phi)\int_T F d\phi = \prod_{i=\overline{1,n}} [(\phi)\int_T f_i d\phi, (\phi)\int_T g_i d\phi].$$

*In this case, $(\phi)\int_T \|F\| d\phi = \|(\widehat{g}_s)(\phi)\int_T F d\phi\|$.*

## 14.5 CONCLUSION

In this chapter we study the absolute and non-absolute integrability for some real or vector integrals for functions and multifunctions by respect to a submeasure or a multisubmeasure. We will address Henstock-Kurzweil integrability for (multi)functions with values in Euclidean spaces, in a unitary framework regarding absolute integrability for various types of linear and non-linear integrals and different notions of absolute integrability that will be introduced. We define some types of absolute integrability for multifunctions and multisubmeasures with values in Euclidean spaces and we present several properties and relationships among them.

## BIBLIOGRAPHY


1. Aumann R. J., Integrals of set-valued maps, J. Math. Anal. Appl., **12**, (1965), 1–12.
2. Boccuto A., Candeloro D., Sambucini A. R., A Fubini Theorem in Riesz Spaces for the Kurzweil-Henstock Integral, J. Funct. Spaces Appl., **9**, (2011), 283–304.
3. Boccuto A., Minotti A. M., Sambucini, A. R., Set-valued Kurzweil-Henstock integral in Riesz space setting, PanAmerican Math. J., **23** (1), (2013), 57–74.
4. Bongiorno B., Giertz M., Pfeffer W. F., Some nonabsolutely convergent integrals in the real line, Boll. Un. Mat. Ital., **7** (6-B), (1992), 371–402.
5. Brooks J. K., An integration theory for set-valued measures, I/II, Bull. Soc. Roy. Sci. Liege, **37**, (1968), 312–319/ 375–380.

6. Bullen P. S., Non-absolute integrals-A survey, Real Anal. Exchange, **5**, (1979-1980), 195–259.
7. Candeloro D., Piazza L. D., Musial K., Sambucini A. R., Relations among gauge and Pettis integrals for multifunctions with weakly compact convex values, Annali di Matematica, **197**(1), (2018), 171–183.
8. Candeloro D., Croitoru A., Gavrilut A., Iosif A., Sambucini A. R., Properties of the Riemann-Lebesgue integrability in the non-additive case, Rend. Circ. Mat. Palermo, Serie **2**(69), (2020), 577–589.
9. Candeloro D., Croitoru A., Gavrilut A., Sambucini A. R., An extension of the Birkhoff integrability for multifunctions, Mediterranean J. Math., **13**(5), (2016), 2551–2575.
10. Cascales B., Kadels V., Rodriguez J. , The Pettis integral for multivalued functions via single-valued ones, J. Math. Anal. Appl., **332**, (2007), 1–10.
11. Castaing C., Valadier M., *Convex Analysis and Measurable Multifunctions*, Springer-Verlag, Lecture Notes in Math., **580**, (1977).
12. Cichon K., Cichon M., *Some Applications of Nonabsolute Integrals in the Theory of Differential Inclusions in Banach Spaces*, Operator Theory: Advances and Applications, Birkhauser Verlag Basel/Switzerland, **201**, (2009).
13. Choquet G., Theory of capacities, Ann. Inst. Fourier, **5**, (1953), 131–295.
14. Croitoru A., Set-norm continuity of set multifunctions, ROMAI J. **6**, (2010), 47–56.
15. Croitoru A., Fuzzy integral of measurable multifunctions, Iran. J. Fuzzy Syst., **9**, (2012), 133–140.
16. Croitoru A., Strong integral of multifunctions relative to a fuzzy measure, Fuzzy Sets Syst., **244**, (2014), 20–33.
17. Croitoru A., An integral for multifunctions with respect to a multimeasure, An. St. Univ. "Al. I. Cuza" Iasi, **49**, (2003), 95–106.
18. Croitoru A., On set-valued integral, Carpathian J. Math., **19** (1), (2003), 41–50.
19. Denneberg D., *Non additive measures and integrals*, Kluwer Acad. Publ. Ser. B, **27**, (1994).
20. Piazza L. D., Musial K., Set-valued Kurzweil-Henstock-Pettis integral, Set-Valued Analysis, **13**, (2005), 167–179.
21. Piazza L. D., Marraffa V., Satco B., Set valued integrability and measurability in non separable Frechet spaces and applications, Math. Slovaca, **66**(5), (2016), 1119–1138.
22. Feynman R. P., Space-time approach to non-relativistic quantum mechanics, Rev. Modern Phys., **20**(2), (1948), 367–387.
23. Henstock R., *Theory of integration*, Butterwoths, London, (1963).
24. Hess C., On the Pettis integral of closed valued multifunctions, Set-Valued Analysis, **8**, (2000), 329–360.
25. Jang L. C., Kil B. M., Kim Y. K., Kwan J. S., Some properties of Choquet integrals of set-valued function, Fuzzy Sets Syst., **91**, (1997), 95–98.
26. Jarnik J., Kurzweil J., A nonabsolutely convergent integral which admits transformation and can be used for integration on manifolds, Czech. Math. J., **35**, (1986), 116–139.
27. Kalita H., Hazarika B., Why we need non absolute integral in place of Lebesgue integral?, Dera Natung Gov. College Res. j., **6**(1), (2022), 86–91.
28. Kalita H., Hazarika H., *The Theory of Nonabsolute Integrable Function Spaces over Riesz and Its Various Applications*, Ch. 15 in Advances in Number Theory and Applied Analysis, (2023), 343–363.
29. Kalita H., Agarwal R. P., Hazarika B., Convergence of ap-Henstock-Kurzweil integral on locally compact spaces, Czech Math J (2023). https://doi.org/10.21136/CMJ.2023.0450-22.

30. Kalita H., Hazarika B., Becerra T. P., On AP-Henstock-Kurzweil Integrals and Non-Atomic Radon Measure, Mathematics, **11**, (2023), 11, 1552.
31. Kurzweil J., Generalized Ordinary Differential Equations and Continuous Dependence on a Parameter, Czech. Math. Journal, **7**(3), (1957), 418–449.
32. Lee T. Y., Multipliers for some non-absolute integrals in Euclidean spaces, Real Analysis Exchange, **24**(1), (1998/99), 149–160.
33. Len N. W., *Nonabsolute Integration on Measure Spaces*, Series in Real Analysis, World Scientific, **14**, (2017).
34. Leng N. W., A nonabsolute integral on measure spaces that include the Davies-Mcshane integral, New Zealand J. Math., **30**, (2001), 147–155.
35. Muldowney P., Feynman's path integrals and Henstock's non-absolute integration, J. Appl. Anal., **6**(1), (2000), 1–24.
36. Muldowney P., Introduction to Feynman integration, J. Math. Study, **27**(1), (1994), 127–132.
37. Musial K., *Topics in the theory of Pettis integrations*, In: School of Measure Theory and Real Analysis, Grado, Italy, May (1992).
38. Nakanishi S., Non-absolute multiple integral defined constructively on the Euclidean space and iterated integral, Scientiae Mathematicae Japonicae Online, **8**(2003), 249–282.
39. Pap E., Iosif A., Gavrilut A., Integrability of an Interval-valued Multifunction with respect to an Interval-valued Set Multifunction, Iran. J. Fuzzy Syst., **15**(3), (2018), 47–63.
40. Park C. K., Set-valued Choquet-Pettis integrals, Korean J. Math., **20**(4), (2012), 381–393.
41. Precupanu A., Gavrilut A., Croitoru A., A fuzzy Gould type integral, Fuzzy Sets and Syst., **161**, (2010), 661–680.
42. Sambucini A. R., The Choquet Integral with Respect to Fuzzy Measures and Applications, Math. Slovaca, **67**(6), (2017), 1427–1450.
43. Schwabik S., On non-absolutely convergent integrals, Mathematica Bohemica, **121**(4), (1996), 369–383.
44. Stamate C., About fuzzy integrals for vector valued multifunctions, Recent Researchers in Neural Networks, Fuzzy Systems, Evolutionary Computing and Automation (2011), 110–113.
45. Stamate C., Croitoru A., Non-linear integrals, properties and relationships, Recent Advances in Telecommunications, Signals and Systems (Proceedings of NOLASC **13**). WSEAS Press (2013), 118–123.
46. Stamate C., Croitoru A., The General Pettis-Sugeno integral of vector multifunctions relative to a fuzzy multimeasure, Fuzzy Sets Syst., **327**, (2017), 123–136.
47. Stamate C., Croitoru A., Aumann-Pettis-Sugeno integrals for vector multifunctions relative to a fuzzy vector measure, Fuzzy Sets Syst., **444**, (2022), 172–181. x
48. Sugeno M., *Theory of fuzzy integrals and its applications,* Ph.D. Thesis, Tokyo Institute of Technology, (1971).
49. Sipos J., Non-linear integrals, Math Slovaca, **29**, (1979), 257–270.
50. Wu C., Mamadou T., An extension of Sugeno integral, Fuzzy Sets Syst., **138**, (2003), 537–550.
51. Yee L. P., Seng C. T., On convergence theorems for nonabsolute integrals, Bull. Austral. Math. Soc., **34**, (1986), 133–140.
52. Zhang D., Guo C., Generalized fuzzy integrals of set-valued functions, Fuzzy Sets Syst., **76**, (1995), 365–373.
53. Zhang D., Guo C., Liu D., Set-valued Choquet integrals revisited, Fuzzy Sets Syst., **147**, (2004), 475–485.

# 15 Statistically Riemann Integrable Sequence of Functions on Time Scales

*Hemen Bharali, Vikuozonuo Sekhose, Hemanta Kalita*

We explore the idea of Riemann integrable sequence of functions which converges statistically on time scales and establish some interesting results pertaining to this. Efficient definitions of statistically Riemann $\Delta$-integrable and statistically Riemann $\nabla$-integrable sequence of functions are put forward. A relation between Riemann integrable sequence of functions and its corresponding statistical convergence is obtained. Notion of statistically Cauchy-Riemann sequence of functions is defined and its equivalence with statistically Riemann integrable sequence of functions is formulated.

Uniform statistical Riemann integrable sequence of functions and uniform statistically Cauchy-Riemann sequence of functions are defined and their relation with statistically Riemann integrable sequence of functions are established. Equivalence of uniform statistically Riemann integrable sequence of functions and uniform statistically Cauchy-Reimann sequence of functions is also formulated.

## 15.1 LITERARY PREFACE AND MOTIVATION

Statistical convergence, as a theoretical extension of ordinary convergence, was introduced by Fast in the year 1951 [5].
Suppose $D \subseteq \mathbb{N}$, the density (or asymptotic density) denoted by 's' is defined as-

$$s(D) = \lim_{x \to \infty} \frac{|\{y : y \leq x \text{ and } y \in D\}|}{x}$$

if the limit exists, where $|\{y : y \leq x \text{ and } y \in D\}|$ denotes the cardinality of the set $\{y : y \leq x \text{ and } y \in D\}$. Theoretical definition of statistical convergence, according to Fast [5], is formulated as- a given sequence $\{h_y\}_{y \in \mathbb{N}}$ is said to statistically converge

DOI: 10.1201/9781003596578-15

to h provided that for every $\varepsilon > 0$ the set $D_\varepsilon = \{y : y \leq x \text{ and } |h_y - h| \geq \varepsilon\}$ has density 0, i.e.

$$s(D_\varepsilon) = \lim_{x \to \infty} \frac{|\{y : y \leq x \ \text{ and } \ |h_y - h| \geq \varepsilon\}|}{x} = 0.$$

Schoenberg, 1959, also discussed statistical convergence calling it D-convergence [16] and examined the concept as a summability method. Another noteworthy paper leading to the initial development of the theory of statistical convergence was that of Šalát [15] published in 1980. As observed by Šalát [15], the condition $s(D_\varepsilon) = 0$ is equivalent to the condition $s(D'_\varepsilon) = 1$, where $D'_\varepsilon = \{y : y \leq x \ \text{ and } \ |h_y - h| < \varepsilon\}$.

The theory of statistically Cauchy sequence was introduced in 1985 by J. A. Fridy [6], formulated as- a given sequence $\{h_y\}_{y \in \mathbb{N}}$ is said to be statistically Cauchy sequence if for every $\varepsilon > 0$ there exists $z \in \mathbb{N}$ depending on $\varepsilon$ such that the set $D_\varepsilon = \{y : y \leq x \ \text{ and } \ |h_y - h_z| \geq \varepsilon\}$ has density 0, i.e.

$$s(D_\varepsilon) = \lim_{x \to \infty} \frac{|\{y : y \leq x \ \text{ and } \ |h_y - h_z| \geq \varepsilon\}|}{x} = 0.$$

Fridy [6] also proved the equivalence of the statistically Cauchy sequence with the concept of statistical convergence.

The idea of statistical convergence of series was introduced by Tripathy [21] in 1998, formulated as- a given series $\sum h_y$ is said to statistically converge to h, if the sequence of partial sums $\{m_y\}$ of $\sum h_y$ (here $m_y = h_1 + h_2 + \ldots + h_y$, $y \in \mathbb{N}$) converge statistically to h provided that for every $\varepsilon > 0$ the set $S_\varepsilon = \{y : y \leq x \text{ and } |m_y - h| \geq \varepsilon\}$ has density 0, i.e.

$$s(S_\varepsilon) = \lim_{x \to \infty} \frac{|\{y : y \leq x \ \text{ and } \ |m_y - h| \geq \varepsilon\}|}{x} = 0.$$

Gökhan and Güngör in 2002 [9] introduced the notion of pointwise statistical convergence, the same authors also introduced uniform statistical convergence and uniform statistical Cauchy sequence in 2005 [10].
Pointwise statistical convergence is formulated as- a given sequence of functions $\{g_y\}_{y \in \mathbb{N}}$ is said to pointwise statistically converge to function g on set, say A, provided that for every $\varepsilon > 0$ the set $D_\varepsilon = \{y : y \leq x \text{ and } |g_y(a) - g(a)| \geq \varepsilon$ for each $a \in A\}$ has density 0, i.e.

$$s(D_\varepsilon) = \lim_{x \to \infty} \frac{|\{y : y \leq x \ \text{ and } \ |g_y(a) - y(a)| \geq \varepsilon \ \text{ for each } \ a \in A\}|}{x} = 0.$$

Uniform statistical convergence is formulated as- a given sequence of functions $\{g_y\}_{y \in \mathbb{N}}$ is said to uniform statistically converge to function g on set, say A, provided that for every $\varepsilon > 0$ the set $D_\varepsilon = \{y : y \leq x \text{ and } |g_y(a) - g(a)| \geq \varepsilon$ for all $a \in A\}$ has density 0, i.e.

$$s(D_\varepsilon) = \lim_{x \to \infty} \frac{|\{y : y \leq x \ \text{ and } \ |g_y(a) - g(a)| \geq \varepsilon \ \text{ for all } \ a \in A\}|}{x} = 0.$$

Uniform statistically Cauchy sequence is formulated as- a given sequence of functions $\{g_y\}_{y\in\mathbb{N}}$ on set, say A, is said to be uniform statistically Cauchy sequence if for every $\varepsilon > 0$ there exists $z \in \mathbb{N}$ depending on $\varepsilon$ such that the set $D_\varepsilon = \{y : y \leq x \text{ and } |g_y(a) - g_z(a)| \geq \varepsilon \text{ for all } a \in A\}$ has density 0, i.e.

$$s(D_\varepsilon) = \lim_{x\to\infty} \frac{|\{y : y \leq x \text{ and } |g_y(a) - g_z(a)| \geq \varepsilon \text{ for all } a \in A\}|}{x} = 0,$$

and proved its equivalence to uniform statistical convergence.

Srivastava et al. in 2021 [19] and 2022 [20] introduced the idea of statistically Riemann integrable and statistically Lebesgue integrable sequence of functions; and the concept of statistical Riemann summability/integrability and statistical Lebesgue summability/integrability via deferred weighted mean. Another paper by Raj and Sharma [14], 2023, discussed applications of statistical Riemann and Lebesgue integrability of sequence of functions. With the same motivation as presented by Srivastava et al. [19, 20] and Raj et al. [14] we wish to explore statistically Riemann integrable sequence of functions on time scales, and establish some interesting results pertaining to this.

The theory of time scale calculus, initially known as measure chain calculus, was introduced by Hilger in 1988 as part of his doctoral thesis, transcripts of which were published in 1990 [11]. The theoretical and foundational framework is presented in his paper [11] (also view [12]) as a system of three axioms. Any set, say T, that satisfies these axioms are called time scales. By nature any closed subset T of $\mathbb{R}$ is a time scale, an excerpt " *...any closed subset of $\mathbb{R}$ bears the structure of a measure chain in a natural manner.*" [11] concludes this.

Hilger introduced two operators [11]. The forward jump operator denoted by $\sigma$ is defined as a mapping $\sigma : T \to T$ such that $\sigma(t) = \inf\{b \in T : b > t\}$. And the backward jump operator is denoted by $\rho$ and is defined as a mapping $\rho : T \to T$ such that $\rho(t) = \sup\{b \in T : b < t\}$.

Hilger introduced the delta derivative ($\Delta$-derivative) formulated using the notion of forward jump operator [11]. A decade later in 2000, Ahlbrandt et al. [1] introduced a notion of derivative, which they called the alpha derivative consisting both the $\Delta$-derivative and another derivative called the nabla derivative ($\nabla$-derivative) as special cases. This $\nabla$-derivative was formulated using the notion of backward jump operator, official named so in 2002 by Atici and Guseinov [3].

The integration of the above derivatives is extensively discussed in literature including the Riemann integration.

The Riemann $\Delta$-integral for real-valued functions using the concept of Darboux sum definition was first introduced in 1992 by Sailer [4, 7]; and re-investigated in 2002 by Guseinov and Kaymakçalan [7] using both Darboux sum and Riemann sum approach, they also presented the nabla definition of the integral (also view [8]). The intervals on which these integrals are defined, i.e. intervals on time scale T are defined as-

Assuming $e \leq f$ (refer [4])
$[e,f]_T = \{t \in T : e \leq t \leq f\}$; $(e,f)_T = \{t \in T : e < t < f\}$;
$[e,f)_T = \{t \in T : e \leq t < f\}$; $(e,f]_T = \{t \in T : e < t \leq f\}$.

A little over a decade ago, statistical convergence on time scales picked up interest. The concept was first explored in 2012 by M. S. Seyyidoglu and N. Ö. Tan [17] discussing the notions: Δ-density, Δ-convergence and Δ-Cauchy of Δ-measurable sets. C. Turan and O. Duman [22] in 2013 introduced the concept of density and statistical convergence of delta measurable real-valued functions; in the same year in [23] they studied the notion of lacunary statistical convergence on time scales. Uniform lacunary statistical convergence was explored in [24]. The concept of uniform statistical convergence was discussed in [2]. Various statistical convergence on time scales like that of $\lambda$-statistical convergence [25]; $\alpha\beta$-statistical convergence [18] and deferred statistical convergence [13] were also studied.

As mentioned earlier, we wish to explore statistically Riemann integrable sequence of functions on time scales, and establish some interesting results.

## 15.2 STATISTICAL RIEMANN INTEGRABILITY OF SEQUENCE OF FUNCTIONS

The concept of Riemann summability and integrability of a sequence of functions is formulated followed by the concept of its statistical convergence.

We let $[e,f]_T$ be a closed interval on T. We partition the interval as $\{e = t_0 < t_1 < \ldots < t_y = f\}$, $y \in \mathbb{N}$; P will denote the collection of all possible partitions of $[e,f]_T$.

For the sake of clarity $G \in P$ will represent the partition for the Δ-integral and $H \in P$ will represent the partition for the ∇-integral.

### 15.2.1 RIEMANN Δ-INTEGRABILITY OF SEQUENCE OF FUNCTIONS

Let $G = \{e = t_0 < \ldots < t_y = f\} \in P$. Subintervals for the Δ-integral are to be taken of the form $[t_{j-1}, t_j)_T$, $1 \leq j \leq y$; we will refer to them as Δ-subintervals. From each Δ-subinterval we choose $\vartheta_j \in [t_{j-1}, t_j)_T$ arbitrarily; we refer to them as Δ-tags. The point-interval collection is defined as $\breve{G} = \{(\vartheta_j, [t_{j-1}, t_j)_T)\}_{j=1}^{y}$; which we call the Δ-tagged partition.

Mesh of partition refers to the size of partition; here mesh-$(G)$ is defined as mesh-$(G) = \max_{1 \leq j \leq y} [t_j - t_{j-1}] > 0$. For some $\delta > 0$, $G_\delta$ will represent a partition of $[e,f]_T$ with mesh $\delta$ satisfying the property: for each $j = 1, 2, \ldots, y$ we have either $t_j - t_{j-1} \leq \delta$ or $t_j - t_{j-1} > \delta \wedge \rho(t_j) = t_{j-1}$ (here $\wedge$ stands for "and"). Henceforth $\breve{G}_\delta$ will mean a Δ-tagged partition with mesh $\delta$.

Let $\{g_y\}_{y \in \mathbb{N}}$ be a sequence of functions defined as $g_y : [e,f]_T \to \mathbb{R}$ for every $y \in \mathbb{N}$. The Riemann Δ-sum of the sequence of functions $\{g_y\}_{y \in \mathbb{N}}$, denoted by $(R_{sq})_\Delta(g_y; \breve{G}_\delta)$, evaluated at the Δ-tags is formulated as-

$$(R_{sq})_\Delta(g_y; \breve{G}_\delta) := \sum_{j=1}^{y} g_j(\vartheta_j)[t_j - t_{j-1}].$$

**Definition 15.2.1.** *A sequence of functions* $\{g_y\}_{y\in\mathbb{N}}$ *is* $\Delta$*-integrable to function g in the Riemann sense over* $[e,f]_T$ *if, for each* $\varepsilon > 0$ *there exists* $\delta > 0$ *hence for any* $\Delta$*-tagged partition* $\breve{G}_\delta$ *we have* $\left|(\mathrm{R}_{sq})_\Delta(g_y;\breve{G}_\delta) - g\right| < \varepsilon$.

The set of all Riemann $\Delta$-integrable sequence of functions on $[\mathrm{e},\mathrm{f}]_\mathrm{T}$ will be denoted by $(\mathtt{Seq}_\mathrm{R})_\Delta$.

### 15.2.2 RIEMANN $\nabla$-INTEGRABILITY OF SEQUENCE OF FUNCTIONS

Let $H = \{\mathrm{e} = \mathrm{t}_0 < \ldots \mathrm{t}_i = \mathrm{f}\} \in \mathrm{P}$. Subintervals for the $\nabla$-integral will be taken of the form $[\mathrm{t}_{j-1},\mathrm{t}_j)_\mathrm{T}$, $1 \leq j \leq \mathrm{y}$; we will refer to them as $\nabla$-subintervals. From each $\nabla$-subinterval, we choose $\xi_j \in [\mathrm{t}_{j-1},\mathrm{t}_j)_\mathrm{T}$ arbitrarily; we refer to them as $\nabla$-tags. The point-interval collection is defined as $\breve{H} = \left\{\left(\xi_j,(\mathrm{t}_{j-1},\mathrm{t}_j]_\mathrm{T}\right)\right\}_{j=1}^{\mathrm{y}}$; which we call the $\nabla$-tagged partition.

Mesh-$(H)$ is defined as mesh-$(H) = \max_{1\leq j\leq \mathrm{y}}\left[\mathrm{t}_j - \mathrm{t}_{j-1}\right] > 0$. For some $\delta > 0$, $H_\delta$ will represent a partition of $[\mathrm{e},\mathrm{f}]_\mathrm{T}$ with mesh $\delta$ satisfying the property: for each $j = 1,2,\ldots,\mathrm{y}$ we have either $\mathrm{t}_j - \mathrm{t}_{j-1} \leq \delta$ or $\mathrm{t}_j - \mathrm{t}_{j-1} > \delta \wedge \mathrm{t}_j = \sigma(\mathrm{t}_{j-1})$. Henceforth $\breve{H}_\delta$ will mean a $\nabla$-tagged partition with mesh $\delta$.

Let $\{\mathrm{g_y}\}_{\mathrm{y}\in\mathbb{N}}$ be a sequence of functions defined as $\mathrm{g_y} : [\mathrm{e},\mathrm{f}]_\mathrm{T} \to \mathbb{R}$ for every $\mathrm{y} \in \mathbb{N}$. The Riemann $\nabla$-sum of the sequence of functions $\{\mathrm{g_y}\}_{\mathrm{y}\in\mathbb{N}}$, denoted by $(\mathrm{R}_{sq})_\nabla(\mathrm{g_y};\breve{H}_\delta)$, evaluated at the $\nabla$-tags is formulated as-

$$(\mathrm{R}_{sq})_\nabla(\mathrm{g_y};\breve{H}_\delta) := \sum_{j=1}^{\mathrm{y}} \mathrm{g}_j(\xi_j)\left[\mathrm{t}_j - \mathrm{t}_{j-1}\right].$$

**Definition 15.2.2.** *A sequence of functions* $\{g_y\}_{y\in\mathbb{N}}$ *is* $\nabla$*-integrable to function g in the Riemann sense over* $[e,f]_T$ *if, for each* $\varepsilon > 0$ *there exists* $\delta > 0$ *hence for any* $\nabla$*-tagged partition* $\breve{H}_\delta$ *we have* $\left|(\mathrm{R}_{sq})_\nabla(g_y;\breve{H}_\delta) - g\right| < \varepsilon$.

The set of all Riemann $\nabla$-integrable sequence of functions on $[\mathrm{e},\mathrm{f}]_\mathrm{T}$ will be denoted by $(\mathtt{Seq}_\mathrm{R})_\nabla$.

The function 'g' (as given in Definition 15.2.1 and Definition 15.2.2) is unique to each corresponding sequence of functions $\mathrm{g_y}$, given it is Riemann $\Delta$-integrable and Riemann $\nabla$-integrable sequence of functions respectively.

We now define statistically Riemann $\Delta$-integrability for sequence of functions $\{\mathrm{g_y}\}_{\mathrm{y}\in\mathbb{N}}$.

### 15.2.3 STATISTICALLY RIEMANN $\Delta$-INTEGRABLE SEQUENCE OF FUNCTIONS

**Definition 15.2.3.** *Statistically Riemann* $\Delta$*-integrable sequence of functions: A sequence of functions* $\{g_y\}_{y\in\mathbb{N}}$ *is said to be statistically Riemann* $\Delta$*-integrable to function g on* $[e,f]_T$ *if, for each* $\varepsilon > 0$ *and every* $t \in [e,f]_T$ *there exists* $\delta > 0$ *hence for any* $\Delta$*-tagged partition* $\breve{G}_\delta$ *the set*

$$D_\varepsilon = \left\{y : y \leq x \ \text{ and } \ \left|(\mathrm{R}_{sq})_\Delta(g_y;\breve{G}_\delta) - g\right| \geq \varepsilon\right\}$$

*has zero density. That is, for every* $\varepsilon > 0$

$$\mathrm{s}(D_\varepsilon) = \lim_{x \to \infty} \frac{\left|\left\{y : y \leq x \ \text{ and } \ |(\mathrm{R}_{sq})_\Delta(g_y; \breve{G}_\delta) - g| \geq \varepsilon\right\}\right|}{x} = 0.$$

*We write,*

$$\overline{stat_{Rie}} \int_e^f g_y(t)\Delta t = g.$$

The set of all statistically Riemann $\Delta$-integrable sequence of functions on $[\mathrm{e}, \mathrm{f}]_\mathrm{T}$ will be denoted by $(\mathtt{Stat}_\mathrm{R})_\Delta$.

In short we write, a sequence of functions $\{\mathrm{g_y}\}$ is statistically Riemann $\Delta$-integrable if and only if for every $\varepsilon > 0$ we have $\mathrm{s}\Big(\{\mathrm{y} : \mathrm{y} \leq \mathrm{x} \ \text{ and } \ |(\mathrm{R}_{sq})_\Delta(\mathrm{g_y}; \breve{G}_\delta) - \mathrm{g}| \geq \varepsilon\}\Big) = 0.$

Definition 15.2.3 can also be alternatively defined as follows.

**Definition 15.2.4.** *A sequence of functions* $\{g_y\}_{y \in \mathrm{N}}$ *is said to be statistically Riemann* $\Delta$*-integrable to function g on* $[e,f]_T$ *if, for each* $\varepsilon > 0$ *and every* $t \in [e,f]_T$ *there exists* $\delta > 0$ *hence for any* $\Delta$*-tagged partition* $\breve{G}_\delta$ *the set*

$$D'_\varepsilon = \left\{y : y \leq x \ \text{ and } \ |(\mathrm{R}_{sq})_\Delta(g_y; \breve{G}_\delta) - g| < \varepsilon\right\}$$

*has unit density. That is, for every* $\varepsilon > 0$

$$\mathrm{s}(D'_\varepsilon) = \lim_{x \to \infty} \frac{\left|\left\{y : y \leq x \ \text{ and } \ |(\mathrm{R}_{sq})_\Delta(g_y; \breve{G}_\delta) - g| < \varepsilon\right\}\right|}{x} = 1.$$

In short we write, a sequence of functions $\{\mathrm{g_y}\}$ is statistically Riemann $\Delta$-integrable if and only if for every $\varepsilon > 0$ we have $\mathrm{s}\Big(\{\mathrm{y} : \mathrm{y} \leq \mathrm{x} \ \text{ and } \ |(\mathrm{R}_{sq})_\Delta(\mathrm{g_y}; \breve{G}_\varepsilon) - \mathrm{g}| < \varepsilon\}\Big) = 1.$

Following we define the $\nabla$-version of the integral, theorems and statements are however limited to the $\Delta$-integral due to their similarity.

### 15.2.4 STATISTICALLY RIEMANN $\nabla$-INTEGRABLE SEQUENCE OF FUNCTIONS

**Definition 15.2.5.** *Statistically Riemann* $\nabla$*-integrable sequence of functions: A sequence of functions* $\{g_y\}_{y \in \mathrm{N}}$ *is said to be statistically Riemann* $\nabla$*-integrable to function g on* $[e,f]_T$ *if, for each* $\varepsilon > 0$ *and every* $t \in [e,f]_T$ *there exists* $\delta > 0$ *hence for any* $\nabla$*-tagged partition* $\breve{H}_\delta$ *we have*

$$\lim_{x \to \infty} \frac{\left|\left\{y : y \leq x \ \text{ and } \ |(\mathrm{R}_{sq})_\nabla(g_y; \breve{H}_\delta) - g| \geq \varepsilon\right\}\right|}{x} = 0,$$

*or equivalently,*

$$\lim_{x\to\infty} \frac{\left|\left\{y : y \le x \text{ and } \left|(R_{sq})_\nabla(g_y;\breve{H}_\delta) - g\right| < \varepsilon\right\}\right|}{x} = 1.$$

*We write,*

$$\underline{stat_{Rie}} \int_e^f g_y(t)\nabla t = g.$$

The set of all statistically Riemann $\nabla$-integrable sequence of functions on $[e,f]_T$ will be denoted by $(\mathtt{Stat}_R)_\nabla$.

**Theorem 15.2.6.** *If a sequence of functions $\{g_y\}_{y\in\mathbb{N}}$ is statistically Riemann $\Delta$-integrable to function g, then g is unique.*

**Proof:** Let us assume that $\{g_y\}_{y\in\mathbb{N}}$ is statistically Riemann $\Delta$-integrable to functions $g_1$ and $g_2$. Given $\varepsilon > 0$, there exists $\delta^1_{\frac{\varepsilon}{2}} > 0$ such that for $\Delta$-tagged partition $\breve{G}_{\delta^1_{\frac{\varepsilon}{2}}}$, we have

$$s\left(\left\{y : y \le x \text{ and } \left|(R_{sq})_\Delta(g_y;\breve{G}_{\delta^1_{\frac{\varepsilon}{2}}}) - g_1\right| < \frac{\varepsilon}{2}\right\}\right) = 1.$$

Also let there exists $\delta^2_{\frac{\varepsilon}{2}} > 0$ such that for $\Delta$-tagged partition $\breve{G}_{\delta^2_{\frac{\varepsilon}{2}}}$, we have

$$s\left(\left\{y : y \le x \text{ and } \left|(R_{sq})_\Delta(g_y;\breve{G}_{\delta^2_{\frac{\varepsilon}{2}}}) - g_2\right| < \frac{\varepsilon}{2}\right\}\right) = 1.$$

Hence letting $\delta_\varepsilon = \min\left\{\delta^1_{\frac{\varepsilon}{2}}, \delta^2_{\frac{\varepsilon}{2}}\right\} > 0$ with corresponding $\Delta$-tagged partition $\breve{G}_{\delta_\varepsilon}$, we have

$$s\left(\left\{y : y \le x \text{ and } \left|(R_{sq})_\Delta(g_y;\breve{G}_{\delta_\varepsilon}) - g_1\right| < \frac{\varepsilon}{2}\right\}\right) = 1,$$

and

$$s\left(\left\{y : y \le x \text{ and } \left|(R_{sq})_\Delta(g_y;\breve{G}_{\delta_\varepsilon}) - g_2\right| < \frac{\varepsilon}{2}\right\}\right) = 1,$$

whence it follows from triangle inequality that,

$$s\left(\left\{y : y \le x \text{ and } \left|g_1 - g_2\right| < \frac{\varepsilon}{2}\right\}\right) = 1.$$

Since $\varepsilon > 0$ is arbitrary, thus we conclude that $g_1 = g_2$.

Below we establish a relation which depicts that if a sequence of functions is Riemann integrable (Definition 15.2.1) to a function than it must also be statistically Riemann integrable (Definition 15.2.3) to the same function.

**Theorem 15.2.7.** *If a sequence of functions $\{g_y\}_{y\in\mathbb{N}}$ is Riemann $\Delta$-integrable to function g on $[e,f]_T$, then $\{g_y\}_{y\in\mathbb{N}}$ is statistically Riemann $\Delta$-integrable to the same function g on $[e,f]_T$.*

**Proof:** Given $\varepsilon > 0$ there exists $\delta > 0$ such that for $\Delta$-tagged partition $\breve{G}_\delta$, we get

$$\left|(\mathrm{R}_{sq})_\Delta(\mathrm{g_y};\breve{G}_\delta) - \mathrm{g}\right| < \varepsilon.$$

Since for each $\varepsilon > 0$, $\breve{G}_\delta$ is a $\Delta$-tagged partition with mesh $\delta$, so we have

$$\lim_{\mathrm{x}\to\infty} \frac{\left|\left\{\mathrm{y} : \mathrm{y} \leq \mathrm{x} \ \text{ and } \ |(\mathrm{R}_{sq})_\Delta(\mathrm{g_y};\breve{G}_\delta) - \mathrm{g}| \geq \varepsilon\right\}\right|}{\mathrm{x}}$$

$$\leq \lim_{\mathrm{x}\to\infty} \left|(\mathrm{R}_{sq})_\Delta(\mathrm{g_x};\breve{G}_\delta) - \mathrm{g}\right| < \varepsilon,$$

since $\varepsilon > 0$ is arbitrary hence we conclude that

$$\overline{stat_{Rie}} \int_{\mathrm{e}}^{\mathrm{f}} \mathrm{g_y}(\mathrm{t})\Delta \mathrm{t} = \mathrm{g}.$$

The converse of Theorem 15.2.7 is not necessarily true.

Linearity property holds given a sequence of functions is statistically Riemann $\Delta$-integrable-

**Theorem 15.2.8.** *If* $\{g_y\}_{y\in\mathbb{N}}, \{k_y\}_{y\in\mathbb{N}} \in (\mathtt{Stat}_\mathrm{R})_\Delta$ *such that* $\overline{stat_{Rie}} \int_e^f g_y(t)\Delta t = g$ *and* $\overline{stat_{Rie}} \int_e^f k_y(t)\Delta t = k$*; and* $r \in \mathbb{R}$*. Then,*

1. $\overline{stat_{Rie}} \int_e^f \Big(rg_y(t)\Big)\Delta t = rh.$
2. $\overline{stat_{Rie}} \int_e^f \Big(g_y(t) + k_y(t)\Big)\Delta t = h + k.$

The notion of statistically Cauchy-Riemann $\Delta$-sequence of functions is presented here and theorem establishing its equivalence with Definition 15.2.3 is formulated (as Theorem 15.2.11); the $\nabla$-definition of the integral (limited to definition only) is also presented.

**Definition 15.2.9.** *Statistically Cauchy-Riemann $\Delta$-sequence of functions: A sequence of functions $\{g_y\}_{y\in\mathbb{N}}$ is said to be statistically Cauchy-Riemann $\Delta$-sequence on $[e,f]_T$ if, for each $\varepsilon > 0$ and every $t \in [e,f]_T$ there exists $\delta > 0$, hence for any $\Delta$-tagged partition $\breve{G}_\delta$ and $l \in \mathbb{N}$ depending on $\varepsilon$, we have*

$$\lim_{x\to\infty} \frac{\left|\left\{y : y \leq x \ \text{ and } \ |(\mathrm{R}_{sq})_\Delta(g_y;\breve{G}_\delta) - (\mathrm{R}_{sq})_\Delta(g_l;\breve{G}_\delta)| \geq \varepsilon\right\}\right|}{x} = 0,$$

*or equivalently,*

$$\lim_{x\to\infty} \frac{\left|\left\{y : y \leq x \ \text{ and } \ |(\mathrm{R}_{sq})_\Delta(g_y;\breve{G}_\delta) - (\mathrm{R}_{sq})_\Delta(g_l;\breve{G}_\delta)| < \varepsilon\right\}\right|}{x} = 1.$$

**Definition 15.2.10.** *Statistically Cauchy-Riemann $\nabla$-sequence of functions: A sequence of functions $\{g_y\}_{y\in\mathbb{N}}$ is said to be statistically Cauchy-Riemann $\nabla$-sequence on $[e,f]_T$ if, for each $\varepsilon > 0$ and every $t \in [e,f]_T$ there exists $\delta > 0$, hence for any $\Delta$-tagged partition $\breve{H}_\delta$ and $l \in \mathbb{N}$ depending on $\varepsilon$, we have*

$$\lim_{x\to\infty} \frac{\left|\left\{y : y \leq x \text{ and } \left|(R_{sq})_\nabla(g_y;\breve{H}_\delta) - (R_{sq})_\nabla(g_l;\breve{H}_\delta)\right| \geq \varepsilon\right\}\right|}{x} = 0,$$

*or equivalently,*

$$\lim_{x\to\infty} \frac{\left|\left\{y : y \leq x \text{ and } \left|(R_{sq})_\nabla(g_y;\breve{H}_\delta) - (R_{sq})_\nabla(g_l;\breve{H}_\delta)\right| < \varepsilon\right\}\right|}{x} = 1.$$

**Theorem 15.2.11.** *A sequence of functions $\{g_y\}_{y\in\mathbb{N}}$ is statistically Cauchy-Riemann $\Delta$-sequence on $[e,f]_T$ if and only if $\{g_y\}_{y\in\mathbb{N}}$ is statistically Riemann $\Delta$-integrable on $[e,f]_T$.*

**Proof:** Suppose $\{g_y\}_{y\in\mathbb{N}}$ is statistically Cauchy-Riemann $\Delta$-sequence but not statistically Riemann $\Delta$-integrable. There exists $l \in \mathbb{N}$ depending on $\varepsilon$ such that density of set $D_\varepsilon = \left\{y : y \leq x \text{ and } \left|(R_{sq})_\Delta(g_y;\breve{G}_\delta) - (R_{sq})_\Delta(g_l;\breve{G}_\delta)\right| \geq \varepsilon\right\}$ is zero. Consequently, $s\left(\left\{y : y \leq x \text{ and } \left|(R_{sq})_\Delta(g_y;\breve{G}_\delta) - (R_{sq})_\Delta(g_l;\breve{G}_\delta)\right| < \varepsilon\right\}\right) = 1$. In particular, we write

$$\left|(R_{sq})_\Delta(g_y;\breve{G}_\delta) - (R_{sq})_\Delta(g_l;\breve{G}_\delta)\right| \leq 2\left|(R_{sq})_\Delta(g_y;\breve{G}_\delta) - g\right| < \varepsilon \tag{15.1}$$

if $\left|(\overline{R}_{sq})(g_y;\breve{G}_\delta) - g\right| < \frac{\varepsilon}{2}$.

Letting $A_\varepsilon = \left\{y : y \leq x \text{ and } \left|(R_{sq})_\Delta(g_y;\breve{G}_\delta) - h\right| \geq \varepsilon\right\}$ and $B_\varepsilon = \left\{y : y \leq x \text{ and } \left|(R_{sq})_\Delta(g_l;\breve{G}_\delta) - g\right| \geq \varepsilon\right\}$.

Since $\{g_y\}_{y\in\mathbb{N}}$ is not statistically convergent hence $s(A_\varepsilon) = 1$ i.e. $s\left(\left\{y : y \leq x \text{ and } \left|(R_{sq})_\Delta(g_y;\breve{G}_\delta) - g\right| < \varepsilon\right\}\right) = 0$. Therefore, by Eq. 15.1, the density of set $\left\{y : y \leq x \text{ and } \left|(R_{sq})_\Delta(g_y;\breve{G}_\delta) - (R_{sq})_\Delta(g_l;\breve{G}_\delta)\right| < \varepsilon\right\}$ is zero, i.e. $s(D_\varepsilon) = 1$ which is a contradiction. Hence, $\{g_y\}_{y\in\mathbb{N}}$ is statistically Riemann $\Delta$-integrable.

Conversely, let $\{g_y\}_{y\in\mathbb{N}}$ be statistically Riemann $\Delta$-integrable to g. Then for every $\varepsilon > 0$, the density of set $\left\{y : y \leq x \text{ and } \left|(R_{sq})_\Delta(g_y;\breve{G}_\delta) - g\right| \geq \varepsilon\right\}$ is zero. Now choosing $l$ such that $\left|(R_{sq})_\Delta(g_l;\breve{G}_\delta) - g\right| \geq \varepsilon$. Then $D_\varepsilon \subseteq A_\varepsilon \cup B_\varepsilon$ and therefore $s(D_\varepsilon) \leq s(A_\varepsilon) + s(B_\varepsilon) = 0$. Hence $\{g_y\}_{y\in\mathbb{N}}$ is statistically Cauchy-Riemann $\Delta$-sequence.

This concludes the equivalence of statistically Riemann $\Delta$-integrable sequence of functions and statistically Cauchy-Riemann $\Delta$-sequence of functions.

## 15.3 UNIFORM STATISTICAL RIEMANN INTEGRABILITY

**Definition 15.3.1.** *Uniform statistically Riemann Δ-integrable sequence of functions: A sequence of functions $\{g_y\}_{y\in\mathbb{N}}$ is said to be uniformly statistically Riemann Δ-integrable to function g on $[e,f]_T$ if, for every $\varepsilon > 0$ and all $t \in [e,f]_T$ there exists $\delta > 0$; hence, for any Δ-tagged partition $\breve{G}_\delta$, the set*

$$D_\varepsilon = \left\{y : y \le x \text{ and } \left|(\mathrm{R}_{sq})_\Delta(g_y(t);\breve{G}_\delta) - g(t)\right| \ge \varepsilon \text{ for all } t \in [e,f]_T\right\}$$

*has zero density. That is, for every $\varepsilon > 0$*

$$\mathrm{s}(D_\varepsilon) = \lim_{x\to\infty} \frac{\left|\left\{y : y \le x \text{ and } \left|(\mathrm{R}_{sq})_\Delta(g_y(t);\breve{G}_\delta) - g(t)\right| \ge \varepsilon\ \forall\, t \in [e,f]_T\right\}\right|}{x} = 0.$$

*Or equivalently,*

$$\lim_{x\to\infty} \frac{\left|\left\{y : y \le x \text{ and } \left|(\mathrm{R}_{sq})_\Delta(g_y(t);\breve{G}_\delta) - g(t)\right| < \varepsilon \text{ for all } t \in [e,f]_T\right\}\right|}{x} = 1.$$

*We write,*

$$\overline{stat_{Rie}} \int_e^f g_y(t)\Delta t = g \text{ uniformly on } [e,f]_T.$$

Uniform statistically Riemann Δ-integrable sequence of functions are also statistically Riemann Δ-integrable.

**Theorem 15.3.2.** *If a sequence of functions $\{g_y\}_{y\in\mathbb{N}}$ is uniform statistically Riemann Δ-integrable on $[e,f]_T$ then, $\{g_y\}_{y\in\mathbb{N}}$ is statistically Riemann Δ-integrable on $[e,f]_T$.*

**Proof:** Proof follows directly from Definition 15.3.1 and Definition 15.2.3.

The converse of Theorem 15.3.2 is not necessarily true; below we provide a theorem with additional conditions which make the converse of Theorem 15.3.2 true

**Theorem 15.3.3.** *If a sequence of functions $\{g_y\}_{y\in\mathbb{N}}$ is statistically Riemann Δ-integrable on $[e,f]_T$ and $\left\{y : y \le x \text{ and } \sup_{t\in[e,f]_T}\left|(\mathrm{R}_{sq})_\Delta(g_y(t);\breve{G}_\delta) - g(t)\right| = 0\right\}$, then $\{g_y\}_{y\in\mathbb{N}}$ is uniform statistically Riemann Δ-integrable on $[e,f]_T$.*

**Proof:** Suppose $\{\mathrm{g_y}\}_{\mathrm{y}\in\mathbb{N}}$ is statistically Riemann Δ-integrable to g on $[\mathrm{e},\mathrm{f}]_\mathrm{T}$ thus

$$\lim_{\mathrm{x}\to\infty} \frac{\left|\left\{\mathrm{y} : \mathrm{y} \le \mathrm{x} \text{ and } \left|(\mathrm{R}_{sq})_\Delta(\mathrm{g_y};\breve{G}_\delta) - \mathrm{g}\right| < \varepsilon\right\}\right|}{\mathrm{x}} = 1,$$

and $\left\{\mathrm{y} : \mathrm{y} \le \mathrm{x} \text{ and } \sup_{\mathrm{t}\in[\mathrm{e},\mathrm{f}]_\mathrm{T}}\left|(\mathrm{R}_{sq})_\Delta(\mathrm{g_y}(\mathrm{t});\breve{G}_\delta) - \mathrm{g}(\mathrm{t})\right| = 0\right\}$.

For every $\varepsilon > 0$ we have

$$\left\{\mathrm{y} : \mathrm{y} \le \mathrm{x} \text{ and } \sup_{\mathrm{t}\in[\mathrm{e},\mathrm{f}]_\mathrm{T}}\left|(\mathrm{R}_{sq})_\Delta(\mathrm{g_y}(\mathrm{t});\breve{G}_\delta) - \mathrm{g}(\mathrm{t})\right| < \varepsilon\right\}.$$

Hence,

$$\left\{\mathrm{y} : \mathrm{y} \leq \mathrm{x} \text{ and } \left|(\mathrm{R}_{sq})_{\Delta}(\mathrm{g}_{\mathrm{y}}(\mathrm{t}); \breve{G}_{\delta}) - \mathrm{g}(\mathrm{t})\right| < \varepsilon \text{ for all } \mathrm{t} \in [\mathrm{e}, \mathrm{f}]_{\mathrm{T}}\right\},$$

which completes the proof.

Linearity property holds given a sequence of functions is uniform statistically Riemann Δ-integrable-

**Theorem 15.3.4.** *If* $\{g_y\}_{y\in\mathbb{N}}, \{k_y\}_{y\in\mathbb{N}} \in (\mathtt{Stat}_{\mathrm{R}})_{\Delta}$ *such that* $\overline{stat_{Rie}} \int_e^f g_y(t)\Delta t = g$ *uniformly on* $[e,f]_T$ *and* $\overline{stat_{Rie}} \int_e^f k_y(t)\Delta t = k$ *uniformly on* $[e,f]_T$*; and* $r \in \mathbb{R}$*. Then,*

1. $\overline{stat_{Rie}} \int_e^f \left(rg_y(t)\right)\Delta t = rh$ *uniformly on* $[e,f]_T$.
2. $\overline{stat_{Rie}} \int_e^f \left(g_y(t) + k_y(t)\right)\Delta t = h + k$ *uniformly on* $[e,f]_T$.

**Definition 15.3.5.** *Uniform statistically Cauchy-Riemann Δ-sequence of functions: A sequence of functions* $\{g_y\}_{y\in\mathbb{N}}$ *is said to be uniformly statistically Cauchy-Riemann Δ-sequence on* $[e,f]_T$ *if, for every* $\varepsilon > 0$ *and all* $t \in [e,f]_T$ *there exists* $\delta > 0$*, hence for any Δ-tagged partition* $\breve{G}_{\delta}$*, we have*

$$\lim_{x\to\infty} \frac{\left|\left\{y : y \leq x \text{ and } \left|(\mathrm{R}_{sq})_{\Delta}(g_y(t); \breve{G}_{\delta}) - (\mathrm{R}_{sq})_{\Delta}(g_l(t); \breve{G}_{\delta})\right| \geq \varepsilon\ \forall\, t \in [e,f]_T\right\}\right|}{x} = 0.$$

*Or equivalently,*

$$\lim_{x\to\infty} \frac{\left|\left\{y : y \leq x \text{ and } \left|(\mathrm{R}_{sq})_{\Delta}(g_y(t); \breve{G}_{\delta}) - (\mathrm{R}_{sq})_{\Delta}(g_l(t); \breve{G}_{\delta})\right| < \varepsilon\ \forall\, t \in [e,f]_T\right\}\right|}{x} = 1.$$

**Theorem 15.3.6.** *A sequence of functions* $\{g_y\}_{y\in\mathbb{N}}$ *is uniform statistically Cauchy-Riemann Δ-sequence on* $[e,f]_T$ *if and only if* $\{g_y\}_{y\in\mathbb{N}}$ *is uniform statistically Riemann Δ-integrable on* $[e,f]_T$.

**Proof:** Suppose $\{\mathrm{g_y}\}_{\mathrm{y}\in\mathbb{N}}$ is uniform statistically Cauchy-Riemann Δ-sequence but not uniform statistically Riemann Δ-integrable. There exists $l \in \mathbb{N}$ depending on $\varepsilon$ such that density of set

$$\mathrm{D}_{\varepsilon} = \left\{\mathrm{y} : \mathrm{y} \leq \mathrm{x} \text{ and } \left|(\mathrm{R}_{sq})_{\Delta}(\mathrm{g}_{\mathrm{y}}(\mathrm{t}); \breve{G}_{\delta}) - (\mathrm{R}_{sq})_{\Delta}(\mathrm{g}_{l}(\mathrm{t}); \breve{G}_{\delta})\right| \geq \varepsilon\ \forall\, \mathrm{t} \in [\mathrm{e}, \mathrm{f}]_{\mathrm{T}}\right\}$$

is zero. Consequently,

$$\mathrm{s}\left(\left\{\mathrm{y} : \mathrm{y} \leq \mathrm{x} \text{ and } \left|(\mathrm{R}_{sq})_{\Delta}(\mathrm{g}_{\mathrm{y}}(\mathrm{t}); \breve{G}_{\delta}) - (\mathrm{R}_{sq})_{\Delta}(\mathrm{g}_{l}(\mathrm{t}); \breve{G}_{\delta})\right| < \varepsilon\ \forall\, \mathrm{t} \in [\mathrm{e}, \mathrm{f}]_{\mathrm{T}}\right\}\right) = 1.$$

In particular, we write

$$\left|(\mathrm{R}_{sq})_{\Delta}(\mathrm{g}_{\mathrm{y}}(\mathrm{t}); \breve{G}_{\delta}) - (\mathrm{R}_{sq})_{\Delta}(\mathrm{g}_{l}(\mathrm{t}); \breve{G}_{\delta})\right| \leq 2\left|(\mathrm{R}_{sq})_{\Delta}(\mathrm{g}_{\mathrm{y}}(\mathrm{t}); \breve{G}_{\delta}) - \mathrm{g}(\mathrm{t})\right| < \varepsilon \quad (15.2)$$

for all $\mathrm{t} \in [\mathrm{e}, \mathrm{f}]_{\mathrm{T}}$, if $\left|(\mathrm{R}_{sq})_{\Delta}(\mathrm{g}_{\mathrm{y}}(\mathrm{t}); \breve{G}_{\delta}) - \mathrm{g}(\mathrm{t})\right| < \frac{\varepsilon}{2}$.

Letting $A_\varepsilon = \left\{y : y \leq x \text{ and } \left|(R_{sq})_\Delta(g_y(t); \breve{G}_\delta) - g(t)\right| \geq \varepsilon \ \forall\, t \in [e,f]_T\right\}$ and $B_\varepsilon = \left\{y : y \leq x \text{ and } \left|(R_{sq})_\Delta(g_l(t); \breve{G}_\delta) - g(t)\right| \geq \varepsilon \ \forall\, t \in [e,f]_T\right\}$.

Since $\{g_y\}_{y\in N}$ is not uniform statistically Riemann $\Delta$-integrable hence $s(A_\varepsilon) = 1$ i.e. $s\left(\left\{y : y \leq x \text{ and } \left|(R_{sq})_\Delta(g_y(t); \breve{G}_\delta) - g(t)\right| < \varepsilon \text{ for all } t \in [e,f]_T\right\}\right) = 0$. Therefore, by Eq. 15.2, the density of set $\left\{y : y \leq x \text{ and } \left|(R_{sq})_\Delta(g_y(t); \breve{G}_\delta) - (R_{sq})_\Delta(g_l(t); \breve{G}_\delta)\right| < \varepsilon \text{ for all } t \in [e,f]_T\right\}$ is zero, i.e. $s(D_\varepsilon) = 1$ which is a contradiction. Hence, $\{g_y\}_{y\in N}$ is uniform statistically Riemann $\Delta$-integrable.

Conversely, let $\{g_y\}_{y\in N}$ be uniform statistically Riemann $\Delta$-integrable to g. Then for every $\varepsilon > 0$, the density of set $\left\{y : y \leq x \text{ and } \left|(\overline{R}_{sq})(g_y(t); \breve{G}_\delta) - g(t)\right| \geq \varepsilon \text{ for all } t \in [e,f]_T\right\}$ is zero. Now choosing $l$ such that $\left|(R_{sq})_\Delta(g_l(t); \breve{G}_\delta) - g(t)\right| \geq \varepsilon$ for all $t \in [e,f]_T$. Then $D_\varepsilon \subseteq A_\varepsilon \cup B_\varepsilon$ and therefore $s(D_\varepsilon) \leq s(A_\varepsilon) + s(B_\varepsilon) = 0$. Hence, $\{g_y\}_{y\in N}$ is uniform statistically Cauchy-Riemann $\Delta$-sequence.

The above theorem proves the equivalence of uniform statistically Cauchy-Riemann $\Delta$-sequence and uniform statistically Riemann $\Delta$-integrable sequence of functions.

**Theorem 15.3.7.** *If a sequence of functions $\{g_y\}_{y\in N}$ is uniform statistically Cauchy-Riemann $\Delta$-sequence on $[e,f]_T$, then $\{g_y\}_{y\in N}$ is statistically Riemann $\Delta$-integrable on $[e,f]_T$.*

**Proof:** Proof follows directly from Theorem 15.3.2 and Theorem 15.3.6.

The $\nabla$-integral definitions of Definition 15.3.1 and Definition 15.3.5 are provided below; theorems are, however, omitted due to similarity.

**Definition 15.3.8.** *Uniform statistically Riemann $\nabla$-integrable sequence of functions: A sequence of functions $\{g_y\}_{y\in N}$ is said to be uniformly statistically Riemann $\nabla$-integrable to function g on $[e,f]_T$ if, for every $\varepsilon > 0$ and all $t \in [e,f]_T$ there exists $\delta > 0$; hence, for any $\nabla$-tagged partition $\check{H}_\delta$, we have*

$$\lim_{x\to\infty} \frac{\left|\left\{y : y \leq x \text{ and } \left|(R_{sq})_\nabla(g_y(t); \check{H}_\delta) - g(t)\right| \geq \varepsilon \text{ for all } t \in [e,f]_T\right\}\right|}{x} = 0,$$

*or equivalently,*

$$\lim_{x\to\infty} \frac{\left|\left\{y : y \leq x \text{ and } \left|(R_{sq})_\nabla(g_y(t); \check{H}_\delta) - g(t)\right| < \varepsilon \text{ for all } t \in [e,f]_T\right\}\right|}{x} = 1.$$

**Definition 15.3.9.** *Uniform statistically Cauchy-Riemann $\nabla$-sequence of functions: A sequence of functions $\{g_y\}_{y\in N}$ is said to be uniformly statistically Cauchy-Riemann*

*$\nabla$-sequence on $[e,f]_T$ if, for every $\varepsilon > 0$ and all $t \in [e,f]_T$ there exists $\delta > 0$, hence, for any $\nabla$-tagged partition $\check{H}_\delta$, we have*

$$\lim_{x\to\infty} \frac{\left|\left\{y : y \leq x \text{ and } \left|(\mathrm{R}_{sq})_\nabla(g_y(t);\check{H}_\delta) - (\mathrm{R}_{sq})_\nabla(g_l(t);\check{H}_\delta)\right| \geq \varepsilon \ \forall\, t \in [e,f]_T\right\}\right|}{x} = 0.$$

*Or equivalently,*

$$\lim_{x\to\infty} \frac{\left|\left\{y : y \leq x \text{ and } \left|(\mathrm{R}_{sq})_\nabla(g_y(t);\check{H}_\delta) - (\mathrm{R}_{sq})_\nabla(g_l(t);\check{H}_\delta)\right| < \varepsilon \ \forall\, t \in [e,f]_T\right\}\right|}{x} = 1.$$

## 15.4 CONCLUSION

Statistically Riemann integrable sequence of functions on time scales is explored, and some interesting results pertaining to it is established. Definitions of statistically Riemann $\Delta$-integrable and statistically Riemann $\nabla$-integrable sequence of functions are put forward. A relation between Riemann integrable sequence of functions and its corresponding statistical convergence is obtained. Notion of statistically Cauchy-Riemann sequence of functions is defined, and its equivalence with statistically Riemann integrable sequence of functions is formulated.

Uniform statistically Riemann integrable sequence of functions and uniform statistically Cauchy-Riemann sequence of functions are defined and their relation with statistically Riemann integrable sequence of functions are established. Equivalence of uniform statistically Riemann integrable sequence of functions and uniform statistically Cauchy-Reimann sequence of functions is also formulated.

## REFERENCES


1. C. D. Ahlbrandt, M. Bohner and J. Ridenhour, Hamiltonian Systems on Time Scales, J. Math. Anal. Appl., 250 (2), (2000), 561–578.
2. Y. Altin, H. Koyunbakan and E. Yilmaz, Uniform Statistical Convergence on Time Scales, J. Appl. Math. (2014), 1–6.
3. F. M. Atici and G. Sh. Guseinov, On Green's functions and positive solutions for boundary value problems on time scales, J. Comput. Appl. Math. **141** (1), (2002), 75–99.
4. M. Bohner and S. G. Georgiev, *Multivariable Dynamic Calculus on Time Scales*, Springer International Publishing Switzerland, (2016).
5. H. Fast, *Sur la convergence statistique*, Colloq. Math., **2**, (1951), 241–244.
6. J. A. Fridy, *On statistical convergence*, Analysis, **5**, (1985), 301–313.
7. Guseinov G. S., Kaymakcalan B., *Basics of Riemann Delta and Nabla Integration on Time Scales*, J. Difference Equ. Appl. **8**(11), (2002), 1001–1017.
8. Guseinov G. S., Integration on time scales, J. Math. Anal. Appl., **285**, (2003), 107–127.
9. Gökhan A., Güngör M., On pointwise statistical convergence, Indian J. Pure Appl. Math. **33** (9), (2002), 1379–1384.
10. Güngör M., Gökhan A., On uniform statistical convergence, Int. J. Pure Appl. Math., **19** (1), (2005), 17–24.

11. Hilger S., Analysis on Measure Chains-A unified approach to continuous and discrete calculus, Results Math. **18**, (1990), 18–56.
12. Hilger S., Differential and Difference Calculus- Unified, Nonlinear Anal., Theory, Methods and Appl. **30** (5), (1997), 2683–2694.
13. Cinar M., Yilmaz E., Et M., Deferred statistical convergence on time scales, Proceedings of the Romanian Academy, Series A., **22** (4), (2021), 301–306.
14. Raj K., Sharma S., Applications of statistical Riemann and Lebesgue integrability of sequence of functions, J. Nonlinear Sci. Appl., **16**, (2023), 30–40.
15. Salát T., On statistically convergent sequences of real numbers, Math. Slovaca **30** (2), (1980), 139–150.
16. Schoenberg I. J., The Integrability of Certain Functions and Related Summability Methods, The American Mathematical Monthly **66** (5), (1959), 361–375.
17. Seyyidoğlu M. S., Tan N. Ö., A note on statistical convergence on time scale, J. Inequal. Appl., **219** (1), (2012), 1–8.
18. Sozbir B., Altundag S., $\alpha\beta$-statistical convergence on time scales, Facta universitatis (Nis) Ser. Math. Inform., **35** (1), (2020), 141–150.
19. Srivastava H. M., Jena B. B., Paikray S. K., Statistical Riemann and Lebesgue Integrable Sequence of Functions with Korovkin-Type Approximation Theorems, Axioms, **229** (10),(2021), 1–17.
20. Srivastava H. M., Jena B. B., Paikray S. K., Some Korovkin-Type Approximation Theorems Associated with a Certain Deferred Weighted Statistical Riemann-Integrable Sequence of Functions, Axioms, **128** (11), (2022), 1–18.
21. Tripathy B. C., On statistical convergence, Proc. Estonian Acad. Sci. Phys. Math., **47** (4), (1998), 299–303.
22. Turan C., Duman O., Statistical convergence on timescales and its characterizations, Springer Proc. Math. Stat. **41** (2013), 57–71.
23. Turan C., Duman O., Convergence methods on time scales, AIP Conf. Proc. **1558**, (2013), 1120–1123.
24. Yilmaz E., Mohiuddine S. A., Altin Y., Koyunbakan H., Uniform Lacunary statistical convergence on time scales, International Journal of Analysis and Applications, **14** (1), (2017), 99–106.
25. Yilmaz E., Altin Y., Koyunbakan H., $\lambda$-statistical convergence on time scales, Dynamics of Continuous, Discrete and Impulsive Systems; Series A: Mathematical Analysis **23** (1), (2016), 69–78.

# 16 On McShane-type Integrals for Functions with Values in Topological Vector Spaces

*Hemanta Kalita*

In this chapter, we introduce an ap-McShane integrable function on a topological vector spaces. Basic results of ap-McShane integrable functions are discussed here. Various convergence theorems of ap-McShane integrable function via equi-integrability of ap-McShane integrals are discussed. For ap-McShane integrable sequences of functions a convergence theorem based on the concept of equi-integrability is proved and it is shown that this theorem is equivalent to the Vitali convergence theorem.

## 16.1 INTRODUCTION AND PRELIMINARIES

In 1902, Henri Lebesgue, introduced Lebesgue integral theory. The Lebesgue integral has been considered by a lot of mathematicians as the official or standard integral in mathematical research. A further approach to the problem of the primitives was introduced in 1957 by Kurzweil and in 1963 by Henstock, independently. They defined a generalized version of the Riemann integral that is known as the Henstock-Kurzweil integral, also abbreviated as the HK-integral. Later on,the approximate Henstock integral (ap-Henstock-Kurzweil integral) further generalizes the Henstock integral by using the concept of the approximate derivatives and the gauge function in the Henstock integral is generalized to the choice in the ap-Henstock-Kurzweil integral. Every Henstock integrable function is the ap-Henstock integrable and the integral are the same Furthermore, the ap-Henstock integral recovers an approximate continuous function from its approximate derivative. Researchere can see [7] and their references therein for details of ap-Henstock-Kurzweil integrals and related work. In the late 1960's, McShane defined a Riemann-type integral and proved

DOI: 10.1201/9781003596578-16 

that this integral is equivalent to the Lebesgue integral. As a Riemann-type integral, it is simpler to handle than the Lebesgue integral. See [3] and their references therein for McShane integrals. The approximate McShane integral (ap-McShane integral) futher generalizes the McShane integral by using the approximate derivatives. In this study, we introduce the concept of ap- McShane integral on a topological vector space. See [4, 5, 6] and their references within for details of ap-Henstock-Kurzweil, and ap-McShane integrals in different settings. The work of [1] motivated us to construct ap-McShane integrals on topological vector spaces. Let $\mathscr{X}$ be a Hausdorff topological space. We say that $\mathscr{X}$ is a topological vector space if $\mathscr{X}$ is a real vector space and the operations, vector addition and scalar multiplication, are continuous. The next lemma is a property of a topological vector space that we will use in this paper.

**Lemma 16.1.1.** *Let $\mathscr{X}$ be a topological vector space. Then there is a local base $B$ of $\theta$ (the zero vector), satisfying the following:*

1. *If $U,V \in B$, then there is a $W \in B$ with $W \subseteq U \cap V$.*
2. *If $U \in B$ and $x \in U$, there is a $V \in B$ such that $x+V \subseteq U$ .*
3. *If $U \in B$, there is a $V \in B$ such that $V+V \subseteq U$.*
4. *If $U \in B$ and $x \in X$ , then there is $k \in R$ such that $x \in kU$.*
5. *If $U \in B$ and $0 < |k| \leq 1$, then $kU \subseteq U$ and $kU \in B$.*
6. *$T\{U : U \in B\} = \{\theta\}$.*

Conversely, given a collection B of subsets containing $\theta$ and satisfying the above conditions, there is a topology for $\mathscr{X}$ making $\mathscr{X}$ a topological vector space and having $B$ as a local base at $\theta$.

Let $\mathscr{X}$ be a non empty set. A family $F = \{A_\nu : \nu \in N\}$ of subsets of $\mathscr{X}$ is a filter in $\mathscr{X}$ if the following are satisfied:

1. For every $\nu \in N$, $A_\nu$, $\phi$.
2. For $A, B \in F$ then $A \cap B \in F$.
3. If $A \in F$, $B \subseteq X$ and $A \subseteq B$ then $B \in F$.

The filter F converges to $x \in \mathscr{X}$ if for every $\theta$-nbd $U$ ($\theta$ is zero vector of $\mathscr{X}$) there exists $A \in F$ such that $A - x \subseteq U$. We say $F$ is Cauchy if for every $\theta$-nbd U there exists $-A \in F$ such that $A \subseteq U$. Let $\mathscr{X}$ be a topological vector space. We say that $\mathscr{X}$ is complete if every Cauchy filter in $\mathscr{X}$ converges. We say that $\mathscr{X}$ is locally convex if there is a local base at $\theta$ whose members are convex.

**Definition 16.1.2.** *A function $F : [a,b] \to \mathscr{X}$ is an approximate Lusin function on $[a,b]$, if for every measurable set $E \subseteq [a,b]$ of measurable zero, and for every ap$\theta$-nbd $\mathfrak{U}$, there exists an $\varepsilon > 0$ such that for any S-partition $P = \{(t_i, [c_i, d_i]) : 1 \leq i \leq n\}$ of $[a,b]$ that is subordinate to $\nu$ with $\sum_{i=1}^{n} d_i - c_i < \varepsilon$, there exist ap $\theta$-nbd $\mathfrak{U}_1, \mathfrak{U}_2, \ldots, \mathfrak{U}_n$ such that $\sum_{i=1}^{n} \mathfrak{U}_i \subseteq \mathfrak{U}$ and $\sum_{P} F([c_i, d_i]) \in \mathfrak{U}_i$ for $i = 1, 2, \ldots$.*

## 16.2 *AP*-MCSHANE INTEGRALS ON TOPOLOGICAL VECTOR SPACES

An approximate $\theta-$nbd (in short ap $\theta$-nbd) of $x \in [a,b]$ is a measurable set $S_x \subset [a,b]$ containing $x$ as a point of density 1. Let $E \subset [a,b]$. For every $x \in E \subset [a,b]$, choose an approximate $\theta-$nbd $S_x \subset [a,b]$ of $x$. Then $S = \{S_x : x \in E\}$ is a choice on $E$. We assume each point of $S_x$ is a point of density 1.
A free tagged interval $(x, [c,d])$ is said to be subordinate to the choice $S = \{S_x\}$ if $c,d \in S_x$. Let $P$ be a finite collection of non-overlapping free tagged intervals. If $(x_i, [c_i,d_i])$ is subordinate to $S$ for each $i$, then we call $P$ is subordinate to $S$. If $P$ is subordinate to $S$ and $[a,b] = \bigcup\limits_{i=1}^{\infty} [c_i,d_i]$. Then we call $P$ is free tagged partition of $[a,b]$ that is $S$-fine and subordinate to $\nu$.

**Definition 16.2.1.** *[4]*

*a.* *$F : [a,b] \to \mathbb{R}$ is approximately continuous at $c \in [a,b]$ if there exists a measurable ap $\theta-$nbd $\mathfrak{U} \subset [a,b]$ with density 1 at $c$ such that $f(x) - f(c) \in U$ whenever $|x-c| < \delta$.*

*b.* *We say $F$ is approximately differentiable at $c$ if there exists a real number $A$ and a measurable ap $\theta$-nbd $\mathfrak{U} \subset [a,b]$ such that the density of $\mathfrak{U}$ at $c$ is 1 and $\frac{f(x)-f(c)}{x-c} - A \in \mathfrak{U}$. We denote the approximate differentiation at $c$ by $F'_\theta(c)$.*

For a tagged partition $P = \{(t_i, [u_i,v_i]) : 1 \leq i \leq n\}$ of $[a,b]$ we define the Riemann sum as

$$S(f,P) = \sum_{i=1}^{n} f(t_i)(v_i - u_i) \text{ if it exists.}$$

**Definition 16.2.2.** *Let $f : [a,b] \to \mathscr{X}$ is said to be ap-McShane integrable on $[a,b]$ if $\exists$ an $A \in \mathscr{X}$ such that for a choice $S$ on $[a,b]$ whenever $P$ is a $S$-fine free tagged partition i.e.*

$$P = \{(t_i, [x_{i-1},x_i]) : 1 \leq i \leq n\}$$

*is sub coordinate to $\nu$ and we have,*

$$S(f,P) - A \in U$$

*(for $\theta$ nbd). And it is denoted by $A = (AM)\int_a^b f$.*

Let us consider $M([a,b],\mathscr{X})$ be the set of all $ap-$McShane integrable $\mathscr{X}$-valued functions on $[a,b]$. The function $f$ is $ap$-McShane integrable on a measurable set $E \subseteq [a,b]$ if $f\chi_E$ is $ap$-McShane integrable on $[a,b]$, where $\chi_E$ is the characteristic functions on $E$. In this setting, the McShane integrable functions are certainly the $ap$-McShane integrable. It is not hard to prove $ap$-McShane integral $A = (AM)\int_a^b f$ is unique. Also every McShane integrable function is also an ap-McShane integrable function, as shown by the definitions of the two integrals, and every ap-McShane integrable function is also an ap-Henstock-Kurzweil integrable function.

The following proposition can be proved in a standard way.

**Proposition 16.2.3.** *Let $f : [a,b] \to \mathscr{X}$ and $g : [a,b] \to \mathscr{X}$ be two ap-McShane integrable functions, then:*

*a. The function $f+g$ is McShane integrable;*
*b. for each $\alpha \in \mathbb{R}$ the function $\alpha f$ is ap-McShane integrable;*

**Theorem 16.2.4.** *Let $f : [a,b] \to \mathscr{X}$. If $f = \theta$ a.e. in $[a,b]$ then $f$ is ap-McShane integrable on $[a,b]$ and $(AP)\int_a^b f = \theta$.*

**Proof:** Let $F : [a,b] \to \mathscr{X}$ by $F(t) = \theta$ is a primitive of $f$. Let $\mathfrak{V}$ be an ap $\theta$-nbd. Let $\mathfrak{V}$ be an absorbing, balanced, and convex ap $\theta$-nbd with $\mathfrak{U} \subseteq \mathfrak{V}$. Consider $\varepsilon > 0$ be a given. Consider the set $\mathscr{T}$ as

$$\mathscr{T} = \{t \in [a,b];\ f(t) \neq 0\} \text{ and for each } n \in \mathbb{N}$$

$$\mathscr{T}_n = \{t \in \mathscr{T},\ n-1 \leq |f(t)| < n\}.$$

Clearly $\{\mathscr{T}_n\}_{i=1}^{\infty}$ is pairwise disjoint. Also since $\mathfrak{U}$ is absorbing, there is a positive integer $r$ such that $f(t) \in r\mathfrak{U}$. Consider $r$ to be the smallest positive integer with above property, then $f(t) \in \mathscr{T}_r$. Since $\operatorname{diam}(\mathscr{T}) = 0$, we also $\operatorname{diam}(\mathscr{T}_n) = 0$ for each $n \in \mathbb{N}$ and therefore there are open sets $\mathscr{G}_n$ such that $\mathscr{T}_n \subset \mathscr{G}_n$ and $\operatorname{diam}(\mathscr{G}_n) < \frac{\varepsilon}{n.2^n}$. Consider a gauge $\nu : [a,b] \to (0,\infty)$ in such a way that

$$\nu(t) = \begin{cases} 1 \text{ if } t \in [a,b] \setminus \mathscr{T} \\ B(t,\nu(t)) \text{ if } t \in \mathscr{T}_n, \end{cases}$$

where $B(t,\nu(t))$ is a ball centred at $t$. Consider $P = \{(t_i,[c_i,d_i]) : i = 1,2,...,p\}$ is a $S$-fine partition that is subordinate to $\nu$ on $[a,b]$. Then

$$\begin{aligned}
\sum_{i=1}^{p} f(t_i)(d_i - c_i) &\leq \sum_{n=1}^{\infty} \sum_{i=1,t_i\in\mathscr{T}_n}^{p} f(t_i)(d_i - c_i) \\
&\leq \sum_{n=1}^{\infty} n \sum_{i=1,t_i\in\mathscr{T}_n}^{p} (d_i - c_i) \\
&< \sum_{n=1}^{\infty} n\operatorname{diam}(\mathscr{G}_n) \\
&< \sum_{n=1}^{\infty} n.\frac{\varepsilon}{n.2^n} = \varepsilon.
\end{aligned}$$

Now,

$$\begin{aligned}
F(d_i) - F(c_i) - f(t_i)(d_i - c_i) &= \theta - \theta - f(t_i)(d_i - c_i) \\
&= -f(t_i)(d_i - c_i) \in r\mathfrak{U}.
\end{aligned}$$

Convexity of $\mathfrak{U}$ implies $\sum_{i=1}^{n}(r_i\mathfrak{U} \subseteq \mathfrak{U}$. Hence, $S(f,P)-\theta \in \mathfrak{U}$. Consequently, $f$ is $ap$-McShane integrable on $[a,b]$ and $(AP)\int_a^b f = F(b)-F(a) = \theta-\theta = \theta$.

**Theorem 16.2.5.** (Cauchy's criterion) Let $\mathscr{X}$ be a complete topological vector space. Then, $f \in AM([a,b],\mathscr{X})$ if and only if for every ap $\theta$ -nbd $\mathfrak{U}$ there exists a positive function $\nu$ on $[a,b]$ such that for any $\nu$-fine free tagged partitions $P_1$ and $P_2$ of $[a,b]$ that are $S$-fine, we have

$$S(f,P_1)-S(f,P_2) \in \mathfrak{U} \tag{16.1}$$

**Proof:** Let $(AM)\int_a^b f = \mathfrak{l}$. Consider $\mathfrak{U}$ is an ap $\theta$-nbd. Then by definition of ap $\theta$-nbd there exists an ap $\theta$-nbd $\mathfrak{V}$ such that

$$\mathfrak{V}-\mathfrak{V} \subseteq \mathfrak{U}.$$

Thus, there exists a gauge $\nu$ on $[a,b]$ such that if $P$ is $\nu$-fine free tagged partition of $[a,b]$,

$$S(f,P)-\mathfrak{l} \in \mathfrak{V}.$$

Hence, for any two $S$-fine partitions $P_1 = \{(t_i,I_i): \ i = 1,2,...,p\}$ and $P_2 = \{(s_j,I_j): \ j = 1,2,...,r\}$ on $[a,b]$, we have

$$S(f,P_1)-S(f,P_2) = S(f,P_1)-\mathfrak{l}+\mathfrak{l}-S(f,P_2) \in \mathfrak{V}$$

Given $\varepsilon > 0$ assume that (18.1) holds for any $S$-fine partitions $P_1$, and $P_2$ on $[a,b]$. Let

$$\begin{aligned} S(\varepsilon) &= \left\{S(f,P) = \sum_{i=1}^{s} f(t_i)(k_i-k_{i-1});\ P = \{(t_i,[k_{i-1},k_i],\ i = 1,2,...,s\}\right\} \\ &\subset \mathfrak{U}, \end{aligned}$$

where $P$ is an arbitrary $S$-fine partition of $[a,b]$ that is subordinate to $\nu$. Clearly $S(\varepsilon) \subset \mathfrak{U}$ is nonempty. Then from (18.1),

$$S(f,P_1)-S(f,P_2) \in \mathfrak{U}$$

for all $S$-fine partitions $P_1$ and $P_2$ of $[a,b]$ with $\mathrm{diam} S(\varepsilon) < \varepsilon$. That is $S(\varepsilon_1) \subset S(\varepsilon_2)$ provided $\varepsilon_1 < \varepsilon_2$ whenever $\nu_1(t) \leq \nu_2(t)$ $t \in [a,b]$. Since $\mathscr{X}$ is complete. So, $\bigcap_{\varepsilon>0} \overline{S(\varepsilon)} = S \in \mathscr{X}$, where $\overline{S(\varepsilon)}$ is closure of $S(\varepsilon)$ consisting of a single point. Hence $S(f,P)-S \in \mathfrak{U}$ whenever $P$ is an arbitrary $S$-fine partition of $[a,b]$ that is subordinate to $\nu$. Consequently, $f \in AM([a,b],\mathscr{X})$. We now use the Cauchy's criterion to demonstrate an integrability on subinterval.

**Theorem 16.2.6.** *On a complete topological vector space $\mathscr{X}$, if $f \in AM([a,b],\mathscr{X})$. Then for every $[c,d] \subseteq [a,b]$, $f \in AM([c,d],\mathscr{X})$.*

**Proof:** By Theorem 16.2.5 for any given $\varepsilon > 0$ there exists a gauge $\nu : [a,b] \to (0,\infty)$ such that for every $S$-fine partitions $P = \{(t_i,[c_i,d_i]) : i = 1,2,...,p\}$ and $Q = \{(t_j,[c_j,d_j] : j = 1,2,...,r\}$ of $[a,b]$ that are subordinates to $\nu$, the inequality (18.1) is satisfies. Let $R = \{\sigma_i,[c'_i,d'_i]) : i = 1,2,...,q\}$ and $S = \{\sigma_j,[c'_j,d'_j]) : j = 1,2,...,s\}$ are $S$-fine partitions of $[c,d] \subseteq [a,b]$ that are subordinates to $\nu$. The complement $[a,b] \setminus [c,d]$ can be expressed as a finite union of intervals contained in $[a,b]$. Now, consider an arbitrary $S$-fine partition $P$ of each of those intervals $[c'_i,d'_i]$ and $[c'_j,d'_j]$ of $[c,d]$. Clearly

$$S(f,R) - S(f,S) \in \mathfrak{U}$$

since the remaining $\sum_{i=1}^{t} f(\rho_i)(d'_i - c'_i)$ will either $S(f,R)$ or $S(f,S)$ on $[c,d]$. By Theorem 16.2.4, $f \in AM([a,b],\mathscr{X})$. Then for every $[c,d] \subseteq [a,b]$, $f \in AM([c,d],\mathscr{X})$. We shall now demonstrate some of the $ap$-McShane integral's properties using its primitive. We define a primitive of an $ap$-McShane integrable function $f : [a,b] \to \mathscr{X}$ as follows:

**Definition 16.2.7.** *Let $f : [a,b] \to \mathscr{X}$ be an ap-McShane integrable on $a,b]$ and let $F(t) = (AM)\int_a^t f$ for each $t \in [a,b]$. We call $F$ is a primitive of $f$.*

We will provide a necessary and sufficient condition of an $ap$-McShane integral via a primitive of the $ap$-McShane integrable function.

**Theorem 16.2.8.** *Let $\mathscr{X}$ be a complete topological vector space. A function $f : [a,b] \to \mathscr{X}$ is an ap-McShane integral on $[a,b]$ if and only if the following condition holds: For each an ap $\theta$-nbd $\mathfrak{U}$ there exists a S-fine gauge $\nu$ on $[a,b]$, and a S-fine free tagged partition $P = \{(t_i,[x_{i-1},x_i]) : i = 1,2,...,n\}$ of $[a,b]$, also there exist open sets $\mathfrak{U}_i : i = 1,2,...$ with $\sum_{i=1}^{n} \mathfrak{U}_i \subseteq \mathfrak{U}$. Then $F : [a,b] \to \mathscr{X}$ is a primitive of $f : [a,b] \to \mathscr{X}$ so that*

$$F(x_i) - F(x_{i-1}) - (x_i - x_{i-1})f(t_i) \in \mathfrak{U}_i \;\forall\, i.$$

**Theorem 16.2.9.** *Let $f : [a,b] \to \mathscr{X}$ be an ap-McShane integrable on $[a,b]$ and $F(t) = \int_a^t f$ for each $t \in [a,b]$. Then*

*a. The function $F$ is measurable.*
*b. The function $F$ is approximately continuous on $[a,b]$.*
*c. The function $F$ is approximately differentiable a.e. on $[a,b]$ and $F'_\theta = f$ a.e. on $[a,b]$.*
*d. The function $f$ is measurable.*

**Theorem 16.2.10.** *Consider $F : [a,b] \to \mathscr{X}$ is a primitive of $f : [a,b] \to \mathscr{X}$. If $F$ is an approximate Lusin function and approximately differentiable almost everywhere on $[a,b]$, and $F'_\theta = F$ a.e. then $f$ is an ap-McShane integral on $[a,b]$ and*

$$(AM)\int_a^b f = F(b) - F(a).$$

**Proof:** Suppose $F$ is an approximate Lusin function and approximately differentiable almost everywhere on $[a,b]$ and $F'_\theta = F$ a.e. on $[a,b]$. Let $\mathscr{E} = \{t \in [a,b] : F'_\theta(t) \neq f(t)\}$. Then $\mu(\mathscr{E}) = 0$. Let $\mathscr{D} = [a,b] \setminus \mathscr{E}$ and let $\varepsilon > 0$. Then for each $t \in \mathscr{D}$ there exists a measurable set $\mathscr{D}_t \subseteq [a,b]$ such that $t \in \mathscr{D}_t$. Let $\varepsilon > 0$. Then for each $x \in \mathscr{D}_t$, there exists a measurable set $\mathscr{D}_x \subseteq [a,b]$ such that $x \in \mathscr{D}_x$ and

$$\frac{F(x) - F(t)}{x - t} - F'_\theta(t) \in \mathfrak{U},$$

for $x - t \in \mathfrak{U}$, $x \in \mathscr{D}_x$. Hence, there exists a $S$-fine gauge $\nu_x > 0$ such that for every $x \in \mathscr{D}_x \cap (x - \nu_x, x + \nu_x) = S_x$. So, $F(x) - F(t) - F'_\theta(t)(x-t) \in \mathfrak{U}$. That is, if $[t, [c,d])$ is a tagged interval that is subordinate to $\{S_x\}$, then

$$F(d) - F(c) - F'_\theta(t)(d-c) \in \mathfrak{U}.$$

Also, there exists a choice $S'$ on $\mathscr{D}_t$ such that

$$\sum_P f(t_i)(d_i - c_i) - \sum_P F([c,d]) \in \mathfrak{U}$$

whenever $P$ is a free tagged partition that is subordinate to $S'$.

Again from the fact that $F$ is approximate Lusin on $[a,b]$, there exists a choice $S''$ on $\mathscr{E}$ such that

$$\sum_P f(t)(d_i - c_i) \in \mathfrak{U}$$

and

$$\sum_P F([c,d]) \in \mathfrak{U}$$

whenever $P$ is subordinate to $S''$. If $S = S' \cup S''$. Then $S$ is a choice on $[a,b]$. If $P$ is a free tagged partition of $[a,b]$ that is subordinate to $S$, $P_{\mathscr{E}} \subset P$ that has free tags in $\mathscr{E}$, $P_{\mathscr{D}_t} = P \setminus P_{\mathscr{E}}$ then

$$\begin{aligned}\sum_P f(t)(d_i - c_i) - \sum_P F([c,d]) &= \sum_{P_{\mathscr{D}_t}} f(t)(d_i - c_i) - \sum_P F([c,d]) \\ &\quad + \sum_{P_{\mathscr{E}}} f(t)(d_i - c_i) + \sum_{P_{\mathscr{E}}} F([a,b]) \in \mathfrak{U}.\end{aligned}$$

Hence $f$ is an ap-McShane integrable on $[a,b]$ and

$$(AM)\int_a^b f = \sum_P F([a,b]) = F(b) - F(a).$$

The following example shows that an *ap*-McShane integrable function need not a McShane integral.

**Example 16.2.11.** *Consider $\{(a_n,b_n)\}$ is a sequence of disjoint open intervals in $(a,b)$ with the following conditions:*

a. $b_1 < b$ and $b_{n+1} < b_n$ for all $n$.
b. $a_n \to a$;
c. $a$ is a point of dispersion of $\bigcup_{n=1}^{\infty}(a_n,b_n) = \mathfrak{O}$.

*Consider $F : [a,b] \to \mathscr{X}$ by*

$$F(t) = \begin{cases} 0 & if\, t \in [a,b]\setminus \mathscr{O}, \\ sin^2\left(\frac{t-a_n}{b_n-a_n}\right)\pi & if\, t \in (a_n,b_n). \end{cases}$$

*Then $F$ is differentiable on $(a,b]$ and approximately differentiable at $a$ but $F$ is not continuous at $a$. Hence $F' = F'_\theta$ a.e. on $[a,b]$. Clearly $f$ is an approximate Lusin function. See [7, Example 2.5]. By Theorem 16.2.10, $f$ is an ap-McShane integrable on $[a,b]$. By [1, Theorem 2.8], $f$ is not a McShane integrable on $[a,b]$.*

## 16.3 CONVERGENCE THEOREM ON AP-MCSHANE INTEGRALS VIA EQUI-INTEGRABILITY

We describe several convergence theorems in this section for ap-McShane integrals of functions taking values in topological vector space. Convergence theorems based on the idea of equi-integrability for ap-McShane integrable sequences of functions are established.

Let us start with the following.

**Definition 16.3.1.** *Let $f : [a,b] \to \mathscr{X}$ be measurable function. Let $\{f_k\}$ be a sequence of integrable function defined on $[a,b]$. The sequence $\{f_k\}$ is said to be an equi-integrable on $[a,b]$ if $\{f_k\}$ is ap-McShane integrable on $[a,b]$ if for each $\varepsilon > 0$ there exists a choice $S$ that is subordinate to $\nu$ such that*

$$S(f_k,P) - (AM)\int_a^b f_k \in \mathfrak{U} \tag{16.2}$$

*hold for each $S$-fine partition $P = (t_i,[x_{i-1},x_i])_{i=1}^n$ of $[a,b]$, an ap $\theta$-nbd of $\mathfrak{U}$ and $n \in \mathbb{N}$.*

It is observe that if $(u_n)$ be a pointwise bounded sequence of function $f_n : [a,b] \to \mathscr{X}$ and let $E$ be subset of $[a,b]$ such that $\mu(S\setminus E) = 0$ . Then the sequence $(f_n)$ is ap-McShane integrable if and only if $f_n.\mathscr{X}_E$ is ap-McShane integrable.

The following theorems are related to the concept of equi-integrability of ap-McShane integral.

**Theorem 16.3.2.** *A family $M_e$-of functions $f_n : [a,b] \to \mathscr{X}$ is equi-integrable if and only if for every $\varepsilon > 0$ there exists a gauge $\nu$ such that*

$$S(f_n,P_1) - S(f_n,P_2) \in \mathfrak{U}$$

*for every S-fine M-partitions $P_1 = \{(t_i,[c_i,d_i])\}$ and $P_2 = \{(s_j,[c_j,d_j])\}$ of $[a,b]$, an ap $\theta$-nbd $\mathfrak{U}$ for every $n \in \mathbb{N}$ and any $f_n \in M_e$.*

**Proof:** Let $M_e = \left\{ f_n : n \in \mathbb{N} \right\}$ is equi-integrable for any gauge $\nu$ which corresponds to $\varepsilon > 0$ in the Definition 16.3.1 of equi-integrability. If the condition 16.2 fulfilled, then by Theorem 16.2.5, every individual function $f$ is ap-McShane integrable with the same gauge $\nu$ for a given $\varepsilon > 0$ independently of the choice of $f$ with same ap $\theta$-nbd $\mathfrak{U}$. Remaining part of the proof is very standard.

**Theorem 16.3.3.** *Assume $M_e = \{f_n : [a,b] \to \mathscr{X} : n \in \mathbb{N}\}$ is an equi-integrable sequence such that*

$$\lim_{n\to\infty} f_n(t) = f(t),\ t \in [a,b].$$

*Then $f : [a,b] \to \mathscr{X}$ is ap-McShane integrable and*

$$\lim_{n\to\infty} (AM) \int_a^b f_n(t) = (AM) \int_a^b f.$$

**Proof:** Let $M_e = \{f_n : [a,b] \to \mathscr{X},\ n \in \mathbb{N}\}$ be a family of equi-integrable function. Then by Definition 16.3.1, there exists a gauge $\nu : [a,b] \to (0,\infty)$ such that for every $\nu$-fine partition $P = \{(t_i,[c_i,d_i]) : i = 1,2,...,k\}$ of $a,b]$,

$$S(f_n,P) - \int_a^b f_n \in \mathfrak{U} \tag{16.3}$$

hold for each ap $\theta$-nbd $\mathfrak{U}$. If $P$ is fixed then $\lim_{n\to\infty} f_n(t) = f(t),\ t \in [a,b]$ yields,

$$\lim_{n\to\infty} S(f_n,P) - S(f,P) \in \mathfrak{U}.$$

Choose $k_0 \in \mathbb{N}$ such that for $k > k_0$ such that

$$\sum_i f_n(t_i)(d_i - c_i) - \sum_i f(t_i)(d_i - c_i) \in \mathfrak{U} \text{ holds.}$$

Then we have

$$\sum_i f(t_i)(d_i - c_i) - (AM) \int_a^b f_n \in \mathfrak{U}$$

for $k > k_0$. This gives for $k,l > k_0$,

$$(AM) \int_a^b f_n - \int_a^b f \in \mathfrak{U}.$$

This shows that the sequence $(AM)\int_a^b f_n$, $n \in \mathbb{N}$ is a Cauchy. Therefore

$$\lim_{n\to\infty}(AM)\int_a^b f_n = A \in \mathbb{R} \text{ exists.} \tag{16.4}$$

Consider for $\varepsilon > 0$, there exists a gauge $\nu$ such that (16.3) holds for all $n$ whenever $P$ is a $S$-fine partition of $[a,b]$. By (16.4) choose an $N \in \mathbb{N}$ with an ap $\theta$-nbd $\mathfrak{U}$ such that

$$(AM)\int_a^b f_n - A \in \mathfrak{U} \text{ for all } k \geq N. \tag{16.5}$$

Since $f_n \to f$ pointwise, there exists $k_1 \geq N$ such that

$$\sum_i f_{n_1}(t_i)(d_i - c_i) - \sum_i f(t_i)(d_i - c_i) \in \mathfrak{U}.$$

Hence, $\sum_i f(t_i)(d_i - c_i) - A \in \mathfrak{U}$. This shows that $f$ is ap-McShane integrable and $A = (AM)\int_a^b f$. Consequently, $\lim_{n\to\infty}(AM)\int_a^b f_n(t) = (AM)\int_a^b f$.

**Lemma 16.3.4.** (Saks-Henstock Lemma) Assume that a family $M_e$ of functions $f : [a,b] \to \mathscr{X}$ is equi-integrable. Given $\varepsilon > 0$ assume that the gauge $\nu$ on $[a,b]$ such that

$$\sum_i f(t_i)(d_i - c_i) - (AM)\int_a^b f \in \mathfrak{U}$$

for every $S$-fine partition $\{([t_i,[c_i,d_i]) : i = 1,2,...,n\}$ of $[a,b]$ that is subordinate to $\nu$, and $f \in M_e$. Then if $\{([r_j,[c_j,d_j]) : i = 1,2,...,m\}$ is an arbitrary $S$-fine partition that is subordinate to $\nu$, we have

$$\sum_j \left( f(r_j)(d_j - c_j) - (AM)\int_{c_j}^{d_j} f \right) \in \mathfrak{U}$$

for every $f \in M_e$.

**Proof:** Since $\{([r_j,[c_j,d_j]) : j = 1,2,...,m\}$ is a $S$-fine that is subordinate to a gauge $\nu$, so $[a,b] \setminus \left(\bigcup_j [c_j,d_j]\right)^o$ is expressible as finite point-interval pairs with $[c_i,d_i] \cap [c_j,d_j] = \emptyset$ for $i \neq j$. The function $f \in M_e$ are equi-integrable, they are equi-integrable with each $S$-fine partitions $P_l$, $l = 1,2,...,r$ and we have

$$\sum_i f(s_i^l)(d_i^l - c_i^l) - (AM)\int_{c_i^l}^{d_i^l} f \in \mathfrak{U}$$

provided $\left\{(s_i^l,[c_i^l,d_i^l])\right\}$ is a $S$-fine partition that is subordinate to $\nu$ on $[c_i^l,d_i^l]$ and $f \in M_e$. Hence,

$$\sum_j f(r_j)(d_j - c_j) + \sum_l \sum_i f(s_i^l)(d_i^l - c_i^l) - (AM)\int_a^b f \in \mathfrak{U}.$$

Consequently,

$$\sum_j \left( f(r_j)(d_j - c_j) - (AM)\int_{c_j}^{d_j} f \right) \in \mathfrak{U}$$

for every $f \in M_e$. Next theorem is as follows:

**Theorem 16.3.5.** *Let $f_n : [a,b] \to \mathscr{X}$, $n \in \mathbb{N}$ are ap-McShane integrable on $[a,b]$ such that*

*i) $f_n(t) \to f(t)$ for $t \in [a,b]$;*
*ii) The set $\{f_n : n \in \mathbb{N}\}$ is uniformly absolutely continuous;*

*Then the set $\{f_n : n \in \mathbb{N}\}$ is equi-integrable on $[a,b]$.*

**Proposition 16.3.6.** *Let $f_n : [a,b] \to \mathscr{X}$, $n \in \mathbb{N}$ are ap-McShane integrable on $[a,b]$ such that*

*i) $f_n(t) \to f(t)$ for $t \in [a,b]$;*
*ii) The set $\{f_n : n \in \mathbb{N}\}$ is equi-integrable on $[a,b]$;*

*Then $f_n \cdot \chi_{\mathscr{E}}$, $n \in \mathbb{N}$ is an equi-integrable sequence for every measurable set $\mathscr{E} \subset [a,b]$.*

The following theorem shows relaxed Vitali convergence.

**Theorem 16.3.7.** *Let $f_n : [a,b] \to \mathscr{X}$, $n \in \mathbb{N}$, are ap-McShane integrable functions such that $f_n(t) \to f(t)$ for $t \in [a,b]$. Then the set $\{f_n : n \in \mathbb{N}\}$ is an equi-integrable sequences if and only if $\{f_n : n \in \mathbb{N}\}$ is uniformly absolutely continuous.*

**Proof:** The proof is similar to the techniques of [2, Theorem 9].

## 16.4 CONCLUSION

In this chapter, we defined ap-McShane integral on topological vector space. Several properties of our newly introduce integral have been discussed. Finally, we prove relaxed Vitali convergence-type results for ap-McShane integrals in our settings.

## BIBLIOGRAPHY

1. Paluga R. N., Canoy S. R., On the McShane Integral in Topological Vector Spaces, Chamchuri Journal of Mathematics, **2**, (2010), 45–56.
2. Kurzweil J., Schwabik S., McShane equi-integrability and Vitali's convergence theorem, Mathematica Bohemica, **129**(2), (2004), 141–157.
3. Pal S. K. Surinder, Rana I. K., Some Alternative approaches to the McShane integral, Real Analysis Exchange, **35** (1), (2010), 229–233.
4. Kalita H., Hazarika B., A convergence theorem for *ap*-Henstock-Kurzweil integral and its relation to topology, Filomat, **36** (20), (2022), 6831–6839

5. Kalita H., Agarwal R. P., Hazarika B., Convergence of ap-Henstock-Kurzweil integral on locally compact spaces, Czec. Math. J., (2023), 1–19.
6. Kalita H., Bharali H., Kalita S., A McShane type integral on complete metric measure space, Filomat, **38** (15), (2024), 1–9.
7. Park J. M., Oh J. J., Park C. G., Daejeon D., Lee H., Kongju, The ap-Denjoy and ap-Henstock integrals, Czechoslovak Mathematical Journal, **57** (132), (2007), 689–696.
8. Yoon J. H., The ap-Henstock integral of vector-valued functions, J. Chung. Math. Soc., **30** (1), (2017), 151–157.

# 17 On Existence and Uniqueness of Solutions of Volterra Integrals in Sense of McShane and Strong McShane

*Hemanta Kalita, Talat Nazir*

In this chapter, McShane and strong McShane integrals of a vector-valued function endowed with the Alexiewicz norm are discussed. Several properties of vector-valued McShane and strong McShane integrals including fundamental theorem of calculus are enclosed in this manuscript. We extend this theory to study the existence and uniqueness of solutions of

$$x(t) - \int_a^b \mathscr{A}(t,s)x(s)d(s) = f(t),\ t \in [a,b]$$

where $x$ and $f$ are McShane integrable functions defined on $[a,b]$ with values in a Banach space.

## 17.1 INTRODUCTION AND PRELIMINARIES

The Riemann, Lebesgue, or Dushnik integrals are the primary topics of study for integral equations in the literature (see [9]). The Lebesgue integral's classical definition may be challenging to understand, the Riemann integral is weak, and while the Dushnik vector integral is more general than the Riemann-Stieltjes integral, it might not coincide with the Kurzweil-Henstock vector integral, which is more general than both the Riemann-Stieltjes and Lebesgue-Stieltjes integrals. The main topics of study for integral equations in the literature are the Rieann, Lebesgue, or Dushnik integrals (see [9]). While the Dushnik vector integral is more general than the Riemann-Stieltjes integral, it might not coincide with the Kurzweil-Henstock vector integral,

DOI: 10.1201/9781003596578-17

which is more general than both the Riemann-Stieltjes and Lebesgue-Stieltjes integrals. The classical definition of the Lebesgue integral may be difficult to understand. The Riemann integral is weak. Riemannian definition is found to be easier to manage when integral equations are considered in the context of the Kurzweil-Henstock integral. Federson et al. explored numerous conditions for the existence of a solution to a Volterra linear integral equation, given by the variational integral of Henstock, which coincides with the Kurzweil integral in real-valued case (see [4, 5, 6]).

McShane is a kind of non-absolute integral. For real-valued functions the McShane integral is equivalent to the Lebesgue integral and the Henstock-Kurzweil integral is equivalent to the Perron integral. It is not hard to see from the corresponding definitions that for Banach space-valued functions the McShane integrability implies Henstock-Kurzweil integrability (see [2]), but not conversely. The main topics of study for integral equations in the literature are the Rieann, Lebesgue, or Dushnik integrals (see [9]). When integral equations are taken into account in relation to the Kurzweil-Henstock integral, it is discovered that managing Riemannian definition is simpler. In the real-valued case, the variational integral of Henstock coincides with the Kurzweil integral. Federson et al. investigated multiple conditions for the existence of a solution to a Volterra linear integral equation (see [4, 5, 6]).

One type of non-absolute integral is McShane. The McShane integral and the Henstock-Kurzweil integral are equivalent to the Lebesgue and Perron integrals, respectively, for real-valued functions. From the corresponding definitions, it can be easily observed that McShane integrability for Banach space-valued functions implies Henstock-Kurzweil integrability (see [2]), but not the other way around. While the Dushnik vector integral is more general than the Riemann-Stieltjes integral, it might not coincide with the Kurzweil-Henstock vector integral, which is more general than both the Riemann-Stieltjes and Lebesgue-Stieltjes integrals. The classical definition of the Lebesgue integral may be difficult to understand. The Riemann integral is weak. Pérez-Becerra et al. [1] applied the Henstock and Kurzweil integrals to find an integration by parts theorem and some Riesz representation theorems. The fundamental research of Pérez-Becerra et al. [1] motivated us to construct the spaces of McShane and strong McShane integrable functions with the Alexiewicz norm, which is equivalent to the $L^1$-norm for the strong McShane integral. Using the Henstock and Kurzweil integrals, Pérez-Becerra et al. [1] discovered an integration by parts theorem and several Riesz representation theorems. We constructed the spaces of McShane and strong McShane integrable functions with the Alexiewicz norm, which is equivalent to the $L^1$-norm for the strong McShane integral, inspired by the fundamental research of Pérez-Becerra et al. [1]. The definitions that for Banach space-valued McShane integrable functions implies Henstock-Kurzweil integrability of [2] motivated us to consider the existence and uniqueness of solution of

$$x(t) - \int_a^b \mathscr{A}(t,s)x(s)d(s) = f(t),\ t \in [a,b]$$

where $x$ and $f$ are McShane integrable functions defined on $[a,b]$ with values in a Banach space implies work of [4].

Our work is organized as follows: We go over some key definitions and findings in Section 1 that are pertinent to the remainder of the work. Several facts concerning McShane and strong McShane integrable Banach-space valued functions are covered in Section 2. We also go over the connections between the McShane and Bochner integrable functions. This section covers a number of McShane-Stieltjes integral existence criteria. In Section 3, we provide a necessary and sufficient condition for the existence of a solution to the linear Volterra integral equation in the McShane integral sense.

Throughout this chapter $\mathscr{X}$ and $\mathscr{Y}$ symbolized for Banach spaces. Let us denote their respective norms as $\|\cdot\|_{\mathscr{X}}$ and $\|\cdot\|_{\mathscr{Y}}$. Dual spaces are $\mathscr{X}'$ and $\mathscr{Y}'$, respectively. The space of bounded operators on $\mathscr{X}$ with values in $\mathscr{Y}$ is denoted by $L(\mathscr{X},\mathscr{Y})$. Any finite set of closed non-overlapping sub-intervals $[t_{i-1},t_i]$ of $[a,b]$ such that the union of all intervals $[t_{i-1},t_i]$ equals $[a,b]$ is called a division of $[a,b]$. Let $\mathscr{D}$ denotes the set of all divisions of $[a,b]$. Let $d=(t_i)\in\mathscr{D}$. Let $n(d)$ be the number of subintervals in which $[a,b]$ is divided through a given $d\in\mathscr{D}$.

**Definition 17.1.1.** *For a given function $f:[a,b]\to\mathscr{X}$ and $d=(t_i)\in\mathscr{D}$, $V_d(f)=\sum_{i=1}^{n(d)}\Big\|f(t_i)-f(t_{i-1})\Big\|$ and the variation of $f$ is $V(f)=\sup\Big\{V_d(f):\ d\in\mathscr{D}\Big\}$. If $V(f)<\infty$, then $f$ is a function of bounded variation and we denote it by $f\in BV_a^b(\mathscr{X})$.*

Recall the bounded semi-variation of a function $\mathscr{A}:[a,b]\to L(\mathscr{X},\mathscr{Y})$ as follows:

**Definition 17.1.2.** *Let $\mathscr{X}$ and $\mathscr{Y}$ be Banach spaces and $\mathscr{A}:[a,b]\to L(\mathscr{X},\mathscr{Y})$ is given. Consider $d=(t_i)\in\mathscr{D}$. The semi-variation of $\mathscr{A}$ as $SV(\mathscr{A})=\sup\Big\{SV_d(\mathscr{A}):\ d\in\mathscr{D}\Big\}$ where*

$$SV_d(\mathscr{A})=\sup\Bigg\{\Bigg\|\sum_{i=1}^{n(d)}\Big(\mathscr{A}(t_i)-\mathscr{A}(t_{i-1})\Big)x_i\Bigg\|:\ x_i\in\mathscr{X},\ \|x_i\|\le 1\Bigg\}.$$

*If $SV(\mathscr{A})<\infty$, then $\mathscr{A}$ is a function of bounded semi-variation.*

**Definition 17.1.3.** *1. A function $\mathscr{A}:[a,b]\to\mathscr{X}$ is absolute continuous ($^*AC$) on $[a,b]$ if for each $\varepsilon>0$ there exists $\nu>0$ such that $\Bigg\|\sum_{i=1}^{n(d)}\Big[\mathscr{A}(t_i)-\mathscr{A}(t_{i-1})\Big]\Bigg\|<\varepsilon$ whenever $\{[t_{i-1},t_i]:\ 1\le i\le n\}$ is a collection of non-overlapping intervals of $[a,b]$ and satisfies $\sum_{i=1}^{n(d)}(t_i-t_{i-1})<\nu$.*

2. *A function $\mathscr{A}:[a,b]\to\mathscr{X}$ is strongly absolute continuous (sAC) on $[a,b]$ if for each $\varepsilon>0$ there exists $\nu>0$ such that $\sum_{i=1}^{n(d)}\Bigg\|\Big[\mathscr{A}(t_i)-\mathscr{A}(t_{i-1})\Big]\Bigg\|<\varepsilon$ whenever $\{[t_{i-1},t_i]:\ 1\le i\le n(d)\}$ is a collection of non-overlapping intervals of $[a,b]$ and satisfy $\sum_{i=1}^{n(d)}(t_i-t_{i-1})<\nu$.*

## 17.2 MCSHANE AND STRONG MCSHANE INTEGRATION OF VECTOR-VALUED FUNCTIONS

In this section, we describe two extensions of the Lebesgue integral for vector-valued context: McShane and strong McShane integrals. We recommend Stefan Schwabik and Ye Guoju's book *"Topics in Banach space integration"* ([12]) to know a series of interesting properties and examples of these Riemann-type integrals.

The McShane and strong McShane integrals are two extensions of the Lebesgue integral for vector-valued context that are discussed in this section. The book Topics in Banach space integration" by Stefan Schwabik and Ye Guoju ([12]) contains a number of fascinating characteristics and examples of these Riemann-type integrals.

A pair $d = (\xi_i, t_i)$ is a tagged division of $[a,b]$, if $(t_i) \in \mathscr{D}$ and $\xi_i \in [a,b] = \bigcup[t_{i-1}, t_i]$, for every $i$. Let $\delta : E \subset [a,b] \to (0,\infty)$ be a gauge function. Let us consider a free tagged division

$$P = \{([t_{i-1}, t_i], \xi_i) : i = 1, \ldots, n\}$$

with $t_i, \xi_i \in [a,b]$ and $\cup_i [t_{i-1}, t_i] = [a,b]$. The free tagged division $P$ is called a $M$-division subordinated to $\delta$ if for all $0 \le i \le n$, $(t_{i-1}, t_i) \subset (\xi_i - \delta(\xi_i), \xi_i + \delta(\xi_i))$. Note that unlike other integrals based on the gauge function $\delta$ (such as Henstock or Kurzweil), the labels $\xi_i$ do not necessarily belong to their respective interval $[t_{i-1}, t_i]$. If $\xi_i \in [t_{i-1}, t_i]$, for each $i = 1, \cdots, n$, the division is called an $H$-division.

We now specify our primary definition as follows:

**Definition 17.2.1.** *Let $\mathscr{A} : [a,b] \to L(\mathscr{X}, \mathscr{Y})$ and $f : [a,b] \to \mathscr{X}$ be given. We say that $\mathscr{A}$ is McShane $f$-integrable on $[a,b]$ if there is an element $\Delta \in \mathscr{Y}$ with the property: for each $\varepsilon > 0$, there exists a gauge $\delta$ such that*

$$\left\| \sum_{i=1}^{n} \mathscr{A}(\xi_i)(f(t_i) - f(t_{i-1})) - \Delta \right\| < \varepsilon$$

*whenever $P$ is a $M$-partition subordinated to $\delta$.*

In this occasion, $\Delta = (M)\int_a^b \mathscr{A}(t)df(t)$ and $\mathscr{A} \in M_f([a,b], L(\mathscr{X}, \mathscr{Y}))$ which is the set of all McShane integrable functions on $[a,b]$ with values in $\mathscr{X}$. If we change in definition 17.2.1 the $M$-division by $H$-division, we obtain the definition of the Kurzweil integral.

**Definition 17.2.2.** *We say that $f$ is McShane $\mathscr{A}$-integrable if there exists $\Delta \in \mathscr{Y}$ such that for a given $\varepsilon > 0$, there is a gauge $\delta$ on $[a,b]$ such that*

$$\left\| \sum_{i=1}^{n(d)} \Big[ \mathscr{A}(t_i) - \mathscr{A}(t_{i-1}) \Big] f(\xi_i) - \Delta \right\| < \varepsilon$$

*whenever $d$ is subordinated to $\delta$. In this occasion, $\Delta = (M)\int_a^b d\mathscr{A}(t)f(t)$ and $f \in M^{\mathscr{A}}([a,b], \mathscr{X})$.*

Analogous to the Definition 17.2.1:

**Definition 17.2.3.** *Let $f : [a,b] \to \mathscr{X}$ be given. A function $\mathscr{A} : [a,b] \to L(\mathscr{X},\mathscr{Y})$ is said to be strong McShane $f$-integrable on $[a,b]$ if there exists $\mathscr{F}_f : [a,b] \to \mathscr{Y}$ with the property: for each $\varepsilon > 0$, there exists a gauge $\delta$ such that*

$$\sum_{i=1}^{n(d)} \left\| \mathscr{A}(\xi_i)(f(t_i) - f(t_{i-1})) - \left[\mathscr{F}_f(t_i) - \mathscr{F}_f(t_{i-1})\right] \right\| < \varepsilon$$

*whenever $P$ is a $M$-division subordinate to $\delta$.*

In this case, $(SM)\int_a^b \mathscr{A}(t)df(t) = \mathscr{F}_f(b) - \mathscr{F}_f(a)$. We can find linearity and integrability over subintervals. If we consider $H$-divisions instead $M$-divisions, we obtain the definition of the Henstock integral and write $(H)\int_a^b \mathscr{A}(t)df(t)$ for the value of the integral. We denote by $SM_f([a,b],L(\mathscr{X},\mathscr{Y}))$ the space of all strong McShane integrable functions on $[a,b]$. Direct from the definitions, the McShane and the strong McShane integral of a function $\mathscr{A}$ is unique on $[a,b]$ and

$$SM_f([a,b],L(\mathscr{X},\mathscr{Y})) \subset M_f([a,b],L(\mathscr{X},\mathscr{Y})).$$

and the contention is proper as we prove it with the next example (adapted of example 2.1 in [3]).

**Example 17.2.4.** *A family $(x_i)_{i\in[a,b]}$ of elements in $\mathscr{X}$ is summable with sum $x \in \mathscr{X}$ which is written as $\sum_{i\in[a,b]} x_i = x$, if for every $\varepsilon > 0$, there exists a finite subset $E_\varepsilon \subset [a,b]$ such that for every finite subset $F \subset [a,b]$ which contains to $E_\varepsilon$, we have*

$$\left\| x - \sum_{i\in F} x_i \right\|_{\mathscr{X}} < \varepsilon.$$

*Define the set of all $(x_i)_{i\in[a,b]}$, with $x_i \in \mathbb{R}$, such that $\left(x_i^2\right)_{i\in[a,b]}$ is summable, as*

$$\mathscr{L}_2([a,b]) := \left\{ x = (x_i)_{i\in[a,b]},\ x_i \in \mathbb{R} \mid \sum_{i\in[a,b]} x_i^2 < \infty \right\}.$$

*The space $\mathscr{L}_2([a,b])$ is complete with the norm*

$$\|x\|_2 = \left( \sum_{i\in[a,b]} x_i^2 \right)^{\frac{1}{2}}.$$

*Consider the function $\mathscr{A} : [a,b] \to \mathscr{L}_2([a,b])$ as $\mathscr{A}(t) = e_t$, $t \in [a,b]$, where*

$$e_t(i) = \begin{cases} 1, & if \quad i = t, \\ 0, & if \quad i \neq t. \end{cases}$$

*The function $\mathscr{A}$ is McShane integrable since for given $\varepsilon > 0$, we can take $\delta > 0$ such that $\delta^{\frac{1}{2}} < \frac{\varepsilon}{(b-a)^{\frac{1}{2}}}$ and an arbitrary M-division $P = \{([t_{i-1}, t_i], \xi_i) : i = 1, \ldots, n\}$ subordinated to $\frac{\delta}{2}$, then*

$$\begin{aligned} \left\| \sum_{i=1}^{n} \mathscr{A}(\xi_i)(f(t_i) - f(t_{i-1})) - 0 \right\|_2 &= \left\| \sum_{i=1}^{n} e_{\xi_i}(f(t_i) - f(t_{i-1})) \right\|_2 \\ &= \left[ \sum_{i=1}^{n} |f(t_i) - f(t_{i-1})|^2 \right]^{\frac{1}{2}} \\ &= \delta^{\frac{1}{2}} \left[ \sum_{i=1}^{n} (f(t_i) - f(t_{i-1})) \right]^{\frac{1}{2}} \\ &< \varepsilon. \end{aligned}$$

*But $\mathscr{A}$ is not strong McShane integrable since*

$$\sum_{i=1}^{n} \|\mathscr{A}(\xi_i)(f(t_i) - f(t_{i-1})) - 0\|_2 = b - a.$$

On the other hand, $SM([a,b], \mathscr{X}) = M([a,b], \mathscr{X})$ if $\mathscr{X}$ is of finite dimension ([12, Proposition 3.6.6]).

**Definition 17.2.5.** *We say that $f$ is strongly McShane $\mathscr{A}$-integrable if there exists $\mathscr{F}_f : [a,b] \to \mathscr{Y}$ with the property: for each $\varepsilon > 0$ there exists a gauge $\delta$ such that*

$$\sum_{i=1}^{n(d)} \left\| \left[ \mathscr{A}(t_i) - \mathscr{A}(t_{i-1}) \right] f(\xi_i) - \left[ \mathscr{F}_f(t_i) - \mathscr{F}_f(t_{i-1}) \right] \right\| < \varepsilon$$

*whenever $P$ is a M-division subordinated to $\delta$.*

In this occasion, $\Delta = (SM) \int_a^b d\mathscr{A}(t) f(t)$ and $f \in SM^{\mathscr{A}}([a,b], \mathscr{X})$. Clearly, $M^{\mathscr{A}}([a,b], \mathscr{X}) \subset SM^{\mathscr{A}}([a,b], \mathscr{X})$. Consider $f : [a,b] \to \mathscr{X}$ and $\mathscr{A} \in M_f([a,b], L(\mathscr{X}, \mathscr{Y})$. If we define $\widehat{\mathscr{A}_f}(t) = (M) \int_a^t \mathscr{A}(s) df(s)$, $t \in [a,b]$, then for $\mathscr{F} \in M_f([a,b], L(\mathscr{X}, \mathscr{Y})$, $\widehat{\mathscr{A}_f}(t) = \mathscr{F}_f(t) - \mathscr{F}_f(a)$ for every $t \in [a,b]$. In similar way, for $\mathscr{A} : [a,b] \to L(\mathscr{X}, \mathscr{Y})$ we can define the following:

$$\widehat{f^{\mathscr{A}}}(t) = (M) \int_a^t d\mathscr{A}(s) f(s),\ t \in [a,b] \text{ for each } f \in M^{\mathscr{A}}([a,b], \mathscr{X}).$$

Let $A \subset [a,b]$ and $\chi_A$ be the characteristic function on $A$ and $f_{|_A}$ denotes the restriction of $f$ to $A$. Let $\mathscr{A} : [a,b] \to L(\mathscr{X}, \mathscr{Y})$ be given and $[c,d] \subset [a,b]$. It is not hard to find the following facts:

**Theorem 17.2.6.** *1.* If $f \in SM^{\mathscr{A}}([a,b].X)$ (respectively $f \in M^{\mathscr{A}}([a,b],X)$), then $f \in SM^{\mathscr{A}}([c,d].X)$ (respectively $f \in M^{\mathscr{A}}([c,d],X)$).

2. If $f \in SM^{\mathscr{A}}([c,d],X)$, then $(M)\int_a^b d\mathscr{A}(t)\chi_{[c,d]}(t)f(t) = (M)\int_c^d d\mathscr{A}(t)f(t)$.
3. Let $f \in SM^{\mathscr{A}}([c,d],X)$ and let $\widehat{f} : [a,b] \to X$ be such that $\widehat{f}_{|_{[c,d]}} = f$ and $\widehat{f}(t) = 0$ otherwise. Then $(M)\int_a^b d\mathscr{A}(t)\widehat{f}(t) = (M)\int_c^d d\mathscr{A}(t)f(t)$.

Let $C([a,b],\mathscr{X})$ and $R_g([a,b],\mathscr{X})$ be the Banach spaces of continuous and regulated functions from $[a,b]$ to $\mathscr{X}$ endowed with the supremum norm $||.||_\infty$, respectively. Let $f(\xi+)$ and $f(\xi-)$ be the right and left limits of $f : [a,b] \to \mathscr{X}$ at $\xi$ of $[a,b]$ if exists. Let $C^\beta([a,b],L(\mathscr{X},\mathscr{Y}))$ be the set of all functions $\mathscr{A} : [a,b] \to L(\mathscr{X},\mathscr{Y})$ that are weakly continuous and $R_g^\beta([a,b],L(\mathscr{X},\mathscr{Y}))$ be the set of all weakly regulated functions. Let $SV([a,b],L(X,Y))$ be the space of all functions $\mathscr{A} : [a,b] \to L(\mathscr{X},\mathscr{Y})$ of bounded semivariation with semivariation $SV(\mathscr{A})$ and let $VB([a,b],\mathscr{X})$ be the space of all functions $f : [a,b] \to \mathscr{X}$ of bounded variation with variation denoted by $V(\mathscr{A})$. Recall that $BV([a,b],L(\mathscr{X},\mathscr{Y})) \subset SV([a,b],L(\mathscr{X},\mathscr{Y}))$. If $\mathscr{X}$ is of finite dimensional, $SV([a,b],L(\mathscr{X})) = BV([a,b],L(\mathscr{X}))$. Moreover, the space $BV_a([a,b],\mathscr{X}) = \left\{ f \in BV([a,b],\mathscr{X}) : \ f(a) = 0 \right\}$ is complete.

**Theorem 17.2.7.** *Let $f \in AC([a,b],L(\mathscr{X},\mathscr{Y})$ and $\mathscr{A} \in M_f([a,b],L(\mathscr{X},\mathscr{Y})$. Let $\mathscr{B} \in M_f([a,b]$ and $L(\mathscr{X},\mathscr{Y})$ such that $\mathscr{B} = \mathscr{A}$ $\mu$-a.e.. Then $\mathscr{B} \in M_f([a,b],L(\mathscr{X},\mathscr{Y})$ and $\widehat{\mathscr{B}}_f(t) = \widehat{\mathscr{A}_f}(t)$, for every $t \in [a,b]$.*

We state a crucial definition as below:

**Definition 17.2.8.** *Let $f \in AC([a,b],\mathscr{X})$. Two functions $\mathscr{A},\mathscr{B} \in M_f([a,b],L(\mathscr{X},\mathscr{Y})$ are equivalent if and only if $\widehat{\mathscr{A}_f} = \widehat{\mathscr{B}}_f$. Analogously, let $f \in AC([a,b],\mathscr{X})$. Two functions $\mathscr{A},\mathscr{B} \in SM_f([a,b],L(\mathscr{X},\mathscr{Y})$ are equivalent if and only if $\widehat{\mathscr{A}_f} = \widehat{\mathscr{B}}_f$.*

Throughout our work, we denote by $\mathbb{M}_f([a,b],L(\mathscr{X},\mathscr{Y})$ and $\mathbb{SM}_f([a,b],L(\mathscr{X},\mathscr{Y})$ the space of all equivalent classes of functions of $M_f([a,b],L(\mathscr{X},\mathscr{Y})$ and $SM_f([a,b],L(\mathscr{X},\mathscr{Y})$ respectively endowed with Alexiewicz norm $||\mathscr{A}||_{A,f} = \sup\left\{ \left|\left| (M)\int_a^t \mathscr{A}(s)df(s) \right|\right| : \ t \in [a,b] \right\} = \left|\left| \widehat{\mathscr{A}_f} \right|\right|_\infty$.

Let's observe that the integral on the left is of McShane-type. This implies the equivalence of the Alexiewicz norm and the $L^1$-norm on the space $SM([a,b],\mathscr{X})$.

The following theorems give a relevant characteristic of the primitive of a McShane and strong McShane integrable function.

**Theorem 17.2.9.** *[12, Theorem 7.4.10] If a function $\mathscr{A} : [a,b] \to L(\mathscr{X},\mathscr{Y})$ is McShane integrable and $\mathscr{F}_f$ is an associate function of $\mathscr{A}$, then $\mathscr{F}_f$ is absolutely continuous (*AC) on $[a,b]$.*

The Fundamental Theorem of Calculus is generally invalid for the McShane integral, as we see in the following example.

**Example 17.2.10.** *Consider the function $\mathscr{A}$ in example 17.2.4. Its associate function $\mathscr{F}_f(t) = (M)\int_a^t \mathscr{A}(s)ds = 0$ for every $t \in [a,b]$. The function $\mathscr{F}_f$ is differentiable on $[a,b]$, but $\mathscr{F}_f^{'}(t) = 0 \neq \mathscr{A}(t)$ on $[a,b]$.*

But we have the following part of the Fundamental Theorem of Calculus.

**Theorem 17.2.11.** *Let $f : [a,b] \to \mathscr{X}$ and $\mathscr{A} : [a,b] \to L(\mathscr{X},\mathscr{Y})$ are given. If the associate function $\mathscr{F}_f$ of $\mathscr{A}$ is $^*AC$ on $[a,b]$ and $\mathscr{F}_f^{'}(t) = f(t)$ a.e. on $[a,b]$, then $\mathscr{A} : [a,b] \to L(\mathscr{X},\mathscr{Y})$ is McShane integrable on $[a,b]$ with the associate function $\mathscr{F}_f$.*

**Proof:** This is an adaptation of the proof of Theorem 7.4.13 in [12].

Let $S$ be the set of points at which either $\mathscr{F}_f(t)$ does not exist or is not equal to $\mathscr{A}(t)$. By hypothesis, $S$ is of measure zero.

For $\xi \in [a,b]\setminus S$, let $\varepsilon > 0$ such that if $\xi \in [u,v] \subset (\xi - \delta(\xi), \xi + \delta(\xi))$ we obtain

$$\|\mathscr{F}_f(v) - \mathscr{F}_f(u) - \mathscr{A}(\xi)(f(v) - f(u))\| \leq \varepsilon(f(v) - f(u)).$$

Let $S_j$ be the set

$$S_j = \{f(t) : t \in S,\ j-1 \leq \|\mathscr{A}(t)\| < j,\ j \in \mathbb{N}\}.$$

Since $\mathscr{F}_f$ is $^*AC$ on $[a,b]$, then $\mathscr{F}_f$ is also $^*AC$ on $S_j$. So, there is a $\eta_j < \frac{\varepsilon 2^{-j}}{j}$ such that for any collection of non-overlapping intervals $\{[c_k,d_k]\}$ with at least one endpoint of $[c_k,d_k]$ belonging to $S_j$ and satisfying $\sum_k (d_k - c_k) < \eta_j$ we have

$$\left\|\sum_k [\mathscr{F}(d_k) - \mathscr{F}(c_k)]\right\| < \varepsilon 2^{-j}.$$

Define $G_j$ as the union of a sequence of open intervals such that

$$\mu(G_j) < \eta_j \text{ and } S_j \subset G_j.$$

For $\xi = f(x_j) \in S_j$, $j = 1,2,\ldots$, take $\delta(\xi) > 0$ such that $(\xi - \delta(\xi), \xi + \delta(\xi)) \subset G_j$. The function $\delta$ is a gauge on $[a,b]$.

Take any M-division $\{([t_{i-1},t_i],\xi_i) : i = 1,\ldots,n\}$ of $[a,b]$ subordinated to $\delta$. Then we obtain

$$\begin{aligned}
&\left\|\sum_{i=1}^{n(d)} \mathscr{A}(\xi_i)(f(t_i)-f(t_{i-1})) - [\mathscr{F}_f(b)-\mathscr{F}_f(a)]\right\| \\
&\leq \left\|\sum_{i,\xi\notin S} \{\mathscr{A}(\xi_i)(f(t_i)-f(t_{i-1})) - [\mathscr{F}_f(t_i)-\mathscr{F}_f(t_{i-1})]\}\right\| \\
&+ \left\|\sum_{i,\xi\in S} \{\mathscr{A}(\xi_i)(f(t_i)-f(t_{i-1})) - [\mathscr{F}_f(t_i)-\mathscr{F}_f(t_{i-1})]\}\right\| \\
&\leq \sum_{i,\xi\notin S} \|\mathscr{A}(\xi_i)(f(t_i)-f(t_{i-1})) - [\mathscr{F}_f(t_i)-\mathscr{F}_f(t_{i-1})]\| \\
&+ \sum_{i,\xi\in S} \|\mathscr{A}(\xi_i)\|(f(t_i)-f(t_{i-1})) + \left\|\sum_{i,\xi\in S} [\mathscr{F}_f(t_i)-\mathscr{F}_f(t_{i-1})]\right\| \\
&< \varepsilon(f(b)-f(a)) + \sum_j j\eta_j + \sum_j \varepsilon 2^{-j} \\
&< \varepsilon(f(b)-f(a)) + 2\varepsilon.
\end{aligned}$$

Then $\mathscr{A}$ is McShane integrable and $\mathscr{F}_f$ is its associate function.
And for the strong McShane integral, we have the following result.

**Theorem 17.2.12.** *[12, Theorems 7.4.9, 7.4.11 and 7.4.13] Let $f : [a,b] \to \mathscr{X}$ be given. If a function $\mathscr{A} : [a,b] \to L(\mathscr{X},\mathscr{Y})$ is strong McShane integrable with the associate function $\mathscr{F}_f$, then $\mathscr{F}_f$ is differentiable a.e., $\mathscr{F}_f'(t) = \mathscr{A}(t)$ a.e. on $[a,b]$, and $\mathscr{F}_f$ is strongly absolutely continuous (AC) on $[a,b]$. Moreover, If $\mathscr{F}_f$ is AC and $\mathscr{F}_f'(t) = \mathscr{A}(t)$ a.e. on $[a,b]$, then $\mathscr{A} : [a,b] \to L(\mathscr{X},\mathscr{Y})$ is strongly McShane integrable on $[a,b]$ with the associate function $\mathscr{F}_f$.*

**Theorem 17.2.13.** *(Saks-Henstock Lemma, [12, Lemma 3.4.1]) Let $f : [a,b] \to \mathscr{X}$ be given. Assume that $\mathscr{A} : [a,b] \to L(\mathscr{X},\mathscr{Y})$ is McShane integrable. Let for given $\varepsilon > 0$ there is a gauge $\delta$ on $[a,b]$ such that*

$$\left\|\sum_{i=1}^{n(d)} \mathscr{A}(\xi_i)(f(t_i)-f(t_{i-1})) - (M)\int_a^b \mathscr{A}(t)df\right\| < \varepsilon$$

*for every M-division $P = \{([t_{i-1},t_i],\xi_i) : i = 1,\ldots,n\}$ subordinated to $\delta$. Then if $\{([s_{i-1},s_i],\eta_i) : i = 1,\ldots,m\}$ is an arbitrary M-subdivision subordinated to $\delta$, then we have*

$$\left\|\sum_{i=1}^{m(d)} \left(\mathscr{A}(\eta_i)(f(s_i)-f(s_{i-1})) - (M)\int_{s_{i-1}}^{s_i} \mathscr{A}(t)df\right)\right\| \leq \varepsilon.$$

**Theorem 17.2.14.** *(Strong Saks-Henstock Lemma, [12, Lemma 3.6.15]) Let* $f : [a,b] \to \mathscr{X}$ *be given. Assume that* $\mathscr{A} : [a,b] \to L(\mathscr{X}, \mathscr{Y})$ *is strong McShane integrable then given* $\varepsilon > 0$ *there exists a gauge* $\delta$ *on* $[a,b]$ *such that if* $P = \{([t_{i-1}, t_i], \xi_i) : i = 1, \ldots, n\}$ *is an arbitrary M-subdivision subordinate to* $\delta$ *we have*

$$\sum_{i=1}^{n(d)} \|\mathscr{A}(\xi_i)(f(t_i) - f(t_{i-1})) - [\mathscr{F}_f(t_i) - \mathscr{F}_f(t_{i-1})]\| < \varepsilon.$$

Now, we give an alternative proof of the following theorem which appears in [12, Theorem 5.1.5].

**Theorem 17.2.15.** *Let* $f : [a,b] \to \mathscr{X}$ *be given. Suppose* $\mathscr{A} : [a,b] \to L(\mathscr{X}, \mathscr{Y})$ *is Bochner integrable on* $[a,b]$*, then* $\mathscr{A}(t)$ *has a McShane integral and*

$$(B)\int_a^b \mathscr{A}(t)df = (M)\int_a^b \mathscr{A}(t)df.$$

**Proof:** Consider $E \subset [a,b]$ be a measurable set. Let us fix $\mathscr{A}(t) = x \cdot \chi_E(t)$, $x \in \mathscr{X}$ where $\chi_E(t)$ is the characteristic function of $E$. Let $F$ be a compact subset of $E$ and $\varepsilon > 0$ be chosen. Let $D \subset [a,b]$ be an open set containing $E$ in a way that $\mu(D \setminus F) < \frac{\varepsilon}{\|x\|_{\mathscr{X}}}$. Consider a gauge function $\delta : [a,b] \to (0, \infty)$ which is defined as

$$\delta(t) = \begin{cases} d(t, [a,b] \setminus D), & if\ t \in E, \\ d(t, F), & if\ t \in [a,b] \setminus E, \end{cases}$$

where $d(t, [a,b] \setminus D) = \inf\{|t-s| : s \in [a,b] \setminus D\}$ and $d(t,F) = \inf\{|t-s| : s \in F\}$. Suppose $P = \{t_0, \xi_1, t_1, \xi_2, t_2, \xi_3, t_3, \ldots, \xi_n, t_n\}$ is a $M$-partition subordinated to $\delta$. If $\xi_i \in E$, $1 \le i \le n$, then $(t_{i-1}, t_i) \subset D$ and so

$$\left\| \sum_{i=1}^{n} \mathscr{A}(\xi_i)(f(t_i) - f(t_{i-1})) - x\mu(D) \right\| = \|x\|_{\mathscr{X}} \left[ \mu(D) - \sum_{\xi \in E} (t_i - t_{i-1}) \right]. \quad (17.1)$$

If $\xi_i \notin E$, then $(t_{i-1}, t_i)$ and $F$ are disjoint and

$$\left\| \sum_{i=1}^{n} \mathscr{A}(\xi_i)(f(t_i) - f(t_{i-1})) - x\mu(F) \right\| = \|x\|_{\mathscr{X}} \left[ \sum_{\xi \notin E} (f(t_i) - f(t_{i-1})) - \mu(F) \right]. \quad (17.2)$$

Combining (17.1) and (17.2), we have

$$\begin{aligned} \left\| \sum_{i=1}^{n} \mathscr{A}(\xi_i)(f(t_i) - f(t_{i-1})) - x\mu(F) \right\| &= \|x\|_{\mathscr{X}} \left[ \sum_{\xi_i \in E} (f(t_i) - f(t_{i-1})) - \mu(E) \right] \\ &\le \|x\|_{\mathscr{X}} \left[ \mu(D) - \mu(E) \right] \\ &\le \|x\|_{\mathscr{X}} \mu(D \setminus F) < \varepsilon. \end{aligned}$$

Now, Bochner integrability of $\mathscr{A}$ with $\mathscr{A}(t)df = \sum_{j=1}^{\infty} x_j \chi_{E_j}(t)$ implies

$$(B)\int_a^b \mathscr{A}(t)df = \sum_{j=1}^{\infty} x_j \mu(E_j)$$

and

$$(L)\int_a^b \|\mathscr{A}(t)\|df = \sum_{j=1}^{\infty} \|x_j\|_{\mathscr{X}} \mu(E_j).$$

Also, since $\mathscr{A}(t)$ is a Bochner integrable vector-valued function, then it is uniformly measurable on $[a,b]$. Then from the definition, there exists a sequence $\{\mathscr{A}_j(t)f_j(t)\}$ of countably vector-valued functions in $L(\mathscr{X},\mathscr{Y})$ which converges to $\mathscr{A}(t)f(t)$ in uniform topology such that

$$\lim_{j\to\infty}(L)\int_a^b \|\mathscr{A}_j(t)f_j(t) - \mathscr{A}(t)f(t)\|df = 0$$

and

$$(B)\int_a^b \mathscr{A}(t)df = \lim_{j\to\infty}(B)\int_a^b \mathscr{A}_j(t)df.$$

Since $\mathscr{A}_j(t)$ are countably valued,

$$(M)\int_a^b \mathscr{A}_j(t)df = (B)\int_a^b \mathscr{A}_j(t)df.$$

Hence

$$(B)\int_a^b \mathscr{A}(t)df = \lim_{j\to\infty}(M)\int_a^b \mathscr{A}_j(t)df.$$

Finally, since every Lebesgue integral is McShane integral (see [7]), $f_j(t) = ||\mathscr{A}_j(t) - \mathscr{A}(t)||$ has McShane integral and hence

$$\lim_{j\to\infty}(M)\int_a^b \mathscr{A}_j(t)df = 0.$$

Hence,

$$\left\|(B)\int_a^b \mathscr{A}(t)df - (M)\int_a^b \mathscr{A}_m(t)df\right\| < \varepsilon$$

whenever $m$ is large. Consequently, $(B)\int_a^b \mathscr{A}(t)df = (M)\int_a^b \mathscr{A}(t)df$. In fact, the strong McShane integral is equivalent to the Bochner integral:

**Theorem 17.2.16.** *[12, Theorem 5.1.4] A function $\mathscr{A} : [a,b] \to L(\mathscr{X},\mathscr{Y})$ is Bochner integrable if and only if $\mathscr{A}$ is strong McShane integrable.*

### 17.2.1 MCSHANE-STIELTJES INTEGRAL

Let $\phi : [a,b] \to \mathbb{R}$ be an increasing function and $f : [a,b] \to \mathscr{X}$ be a vector-valued function. Consider

$$\mathbb{P} = \{t_0, \xi_1, t_1, \xi_2, t_2, ..., \xi_n, t_n\}$$

be a $M$-partition. We denote the Riemann-Stieltjes sum as

$$S(f,\phi,\mathbb{P}) = \sum_{i=1}^{n} f(\xi_i)[\phi(t_i) - \phi(t_{i-1})].$$

**Definition 17.2.17.** *The function $f$ is McShane-Stieltjes integrable with respect to $\phi$ on $[a,b]$ with McShane-Stieltjes integral $I_{MS}$ if for each chosen $\varepsilon > 0$ there exists a gauge $\delta$ such that*

$$\|S(f,\phi,\mathbb{P}) - I_{MS}\|_{\mathscr{X}} < \varepsilon,$$

*whenever $\mathbb{P}$ is a $M$-partition which is subordinated to $\delta$.*

**Definition 17.2.18.** *Let $f : [a,b] \to \mathscr{X}$ and $g : [a,b] \to \mathscr{Y}$ be two vector-valued functions. Consider*

$$P = \{t_0, \xi_1, t_1, \xi_2, t_2, ..., \xi_n, t_n\}$$

*be a $M$-tagged division. We denote the Riemann-Stieltjes sum as*

$$S(f,g,P) = \sum_{i=1}^{n} f(\xi_i)[g(t_i) - g(t_{i-1})].$$

*The function $f$ is McShane-Stieltjes integrable with respect to $g$ on $[a,b]$ with McShane-Stieltjes integral $I_{MS}$ if for each chosen $\varepsilon > 0$ there exists a gauge $\delta$ such that*

$$\left\| \sum_{n=1}^{n} f(\xi_i)[g(t_i) - g(t_{i-1})] - I_{MS} \right\| < \varepsilon,$$

*whenever $P = \{t_0, \xi_1, t_1, \xi_2, t_2, ..., \xi_n, t_n\}$ is a $M$-tagged division which is subordinated to $\delta$. We denote it by $(MS)\int_a^b (f,dg) = I_{MS}$. If the division is an $H$-tagged division, we have the Kurzweil-Stieltjes integral and write $(KS)\int_a^b (f,dg)$ the value of the integral. If the gauge $\delta$ is constant, then we call the function $f$ to be Riemann-McShane-Stieltjes integrable on $[a,b]$ and denote the integral as $(RMS)\int_a^b (f,dg)$.*

*The function $f$ is strong McShane-Stieltjes integrable with respect to $g$ on $[a,b]$ if there exists a function $F : [a,b] \to L(X,Y)$ with the property: for each chosen $\varepsilon > 0$ there exists a gauge $\delta$ such that*

$$\sum_{i=1}^{n} \|f(\xi_i)[g(t_i) - g(t_{i-1})] - [F(t_i) - F(t_{i-1})]\| < \varepsilon,$$

*whenever $P$ is a $M$-tagged division which is subordinated to $\delta$. We denote it by $(SMS)\int_a^b (f,dg) = F(b) - F(a)$. If the division is an $H$-tagged division, we have defined the Henstock-Stieltjes integral and write it by $(HS)\int_a^b (f,dg)$ for the value of the integral.*

Let $\kappa = \max\{t_i - t_{i-1}\}$. Then the McShane-Stieltjes integrals are given by

$$\begin{aligned}(MS)\int_a^b d\mathscr{A} f(t) &= (MS)\int_a^b d\mathscr{A}(t) f(t)\\ &= \lim_{\kappa\to 0}\sum_i [\mathscr{A}(t_i) - \mathscr{A}(t_{i-1})] f(\xi_i)\end{aligned}$$

and $(MS)\int_a^b \mathscr{A} df = (MS)\int_a^b \mathscr{A}(t) df(t) = \lim_{\kappa\to 0}\sum_i \mathscr{A}(\xi_i)[f(t_i) - f(t_{I-1})]$.

**Theorem 17.2.19.** *1. If $\mathscr{A} \in SV([a,b], L(X,Y))$ and $f \in C([a,b],X)$, then $(MS)\int_a^b d\mathscr{A} f$ exists and $||(MS)\int_a^b d\mathscr{A} f|| \leq SV(\mathscr{A}||f||_\infty$.*

*2. If $\mathscr{A} \in C([a,b], L(X,Y))$ and $f \in BV([a,b],X)$, then $(MS)\int_a^b \mathscr{A} df$ exists and $||(MS)\int_a^b \mathscr{A} df|| \leq ||\mathscr{A}||_\infty V(f)$.*

**Theorem 17.2.20.** *$(MS)\int_a^b d\mathscr{A} df$ exists if and only if $(MS)\int_a^b \mathscr{A} df$ exists and $(MS)\int_a^b d\mathscr{A} f = \mathscr{A}(b)f(b) - \mathscr{A}(a)f(a) - (MS)\int_a^b \mathscr{A} df$ holds.*

**Proof:** Let the McShane-Stieltjes integral $(MS)\int_a^b d\mathscr{A} f$ exists. By definition of McShane-Stieltjes integral: for every $\varepsilon > 0$, there is a $\delta > 0$ such that $||\sum_i [\mathscr{A}(t_i) - \mathscr{A}(t_{i-1})] f(\xi_i) - (MS)\int_a^b d\mathscr{A} f|| < \varepsilon$ whenever $t_i$ are in free tagged division of free tagged partition $P$, and $\max_i\{t_i - t_{i-1}\} < \delta$. Next,

$$\begin{aligned}&\left\|\left\{\mathscr{A}(b)f(b) - \mathscr{A}(a)f(a) - (MS)\int_a^b d\mathscr{A} f\right\} - \sum_i \mathscr{A}(\xi_i)[f(t_i) - f(t_{i-1})\right\|\\ &= \left\|\sum_i [\mathscr{A}(t_i)f(t_i) - \mathscr{A}(t_{i-1})f(t_{i-1})] - (MS)\int_a^b d\mathscr{A} f - \sum_i \mathscr{A}(\xi_i)[f(t_i) - f(t_{i-1})]\right\|\\ &= \left\|\sum_i [\mathscr{A}(t_i) - \mathscr{A}(\xi_i)] f(t_i) + \sum_i [\mathscr{A}(\xi_i) - \mathscr{A}(t_{i-1})] f(t_{i-1})\right.\\ &\left. - (MS)\int_a^b d\mathscr{A} f - \sum_i \mathscr{A}(\xi_i)[f(t_i) - f(t_{i-1})]\right\|\\ &< \varepsilon.\end{aligned}$$

So, $(MS)\int_a^b d\mathscr{A} f$ exists if $(MS)\int_a^b \mathscr{A} df$ exists and $(MS)\int_a^b d\mathscr{A} f = \mathscr{A}(b)f(b) - \mathscr{A}(a)f(a) - (MS)\int_a^b \mathscr{A} df$.

**Corollary 17.2.21.** *The McShane-Stieltjes integrals $(MS)\int_a^b d\mathscr{A} f$ and $(MS)\int_a^b \mathscr{A} df$ exist and the integration by parts formula holds if one of the following conditions is satisfied:*

1. *$\mathscr{A} \in SV([a,b], L(X,Y))$ and $f \in C([a,b],X)$;*
2. *$\mathscr{A} \in C([a,b], L(X,Y))$ and $f \in BV([a,b],X)$.*

**Lemma 17.2.22.** *Let $\mathscr{A} : [a,b] \to L(X,Y)$ and $f \in SM^{\mathscr{A}}([a,b],X)$. If for every $\varepsilon > 0$, the gauge $\delta$ of $[a,b]$ is such that for every $\delta$-fine free tagged partition $P$,*

$$\left\| \sum_{i=1}^{n} [\mathscr{A}(t_i) - \mathscr{A}(t_{i-1})] f(\xi_i) - (M) \int_a^b d\mathscr{A}(t) f(t) \right\| < \varepsilon,$$

*then for $a \leq c_1 \leq v_1 \leq d_1 \leq c_2 \leq v_2 \leq d_2 \leq .. \leq c_k \leq v_k \leq d_k \leq b$ with $[c_j, d_j] \subset \Big( v_j - \delta(n_j), v_j + \delta(v_j) \Big)$ for every $j$,*

$$\left\| \sum_{j=1}^{k} \left\{ [\mathscr{A}(d_j) - \mathscr{A}(c_j)] f(v_j) - (M) \int_{c_j}^{d_j} d\mathscr{A}(t) f(t) \right\} \right\| \leq \varepsilon.$$

**Proof:** Proof is analogous to [11, Lemma 16], so we have omitted the proof.

**Theorem 17.2.23.** *If $\mathscr{A} \in R_g^{\beta}([a,b], L(X,Y))$* (resp $\mathscr{A} \in C^{\beta}([a,b], L(X,Y))$) and $f \in SM^{\mathscr{A}}([a,b],X)$, then $\widehat{f}^{\mathscr{A}} \in R_g([a,b],X)$ (resp $\widehat{f}^{\mathscr{A}} \in C([a,b],X)$ ).

*Proof.* Let $f \in SM^{\mathscr{A}}([a,b],X)$. By definition, for given $\varepsilon > 0$, there is a gauge $\delta$ of $[a,b]$ such that for every $\delta$-fine $M$-tagged division $P$,

$$\sum_i ||[\mathscr{A}(t_i) - \mathscr{A}(t_{i-1})] f(\xi_i) - (M) \int_a^b d\mathscr{A} f(t)|| < \frac{\varepsilon}{2}.$$

Let $\mathscr{A} \in R_g^{\beta}([a,b], L(X,Y))$. Then there exists $(\mathscr{A}x)(\xi+)$, for every $x \in X$. In particular, there exists $\mu > 0$, such that for every $0 < \rho < \mu$, $||[\mathscr{A}(\xi+\rho) - \mathscr{A}(\xi+)] f(\xi)|| < \frac{\varepsilon}{2}$. If $\delta(\xi) < \mu$ and $0 < \rho < \delta(\xi)$, by Lemma 17.2.22, $||\mathscr{A}(\xi+\rho) - \mathscr{A}(\xi)] f(\xi) - (M) \int_{\xi}^{\xi+\rho} d\mathscr{A} f(t)|| \leq \frac{\varepsilon}{2}$. Thus

$$\begin{aligned}
&||\widehat{f}^{\mathscr{A}}(\xi+) - \widehat{f}^{\mathscr{A}}(\xi) - [\mathscr{A}(\xi+) - \mathscr{A}(\xi)] f(\xi)|| \\
&= ||(MS) \int_{\xi}^{\xi+\rho} d\mathscr{A} f(t) - [\mathscr{A}(\xi+) - \mathscr{A}(\xi)] f(\xi)|| \\
&\leq ||(MS) \int_{\xi}^{\xi+\rho} d\mathscr{A} f(t) - [\mathscr{A}(\xi+\rho) - \mathscr{A}(\xi)] f(\xi)|| \\
&+ ||[\mathscr{A}(\xi+\rho) - \mathscr{A}(\xi)] f(\xi) - [\mathscr{A}(\xi+) - \mathscr{A}(\xi)] f(\xi)|| < \varepsilon.
\end{aligned}$$

□

## 17.3 VOLTERRA-MCSHANE LINEAR INTEGRAL EQUATIONS

In this section, we give a necessary condition for the existence of solution of a linear Volterra integral equation in the sense of McShane integral. Let us define the following condition for Bochner and McShane integrable functions. We provide a necessary condition in this section for the existence of a linear Volterra integral equation solution in the context of the McShane integral. First, let us define the following conditions for integrable functions, Bochner and McShane.

Let $\{X_n\}$ be a sequence of closed subsets of $[a,b]$ with $X_n \subset X_{n+1}$ for all $n$ and $\bigcup_{n=1}^{\infty} X_n = [a,b]$. A function $f$ defined on $[a,b]$ is said to fulfil the condition "B" on $\{X_n\}$ if $f$ is Bochner integrable on each $X_n$ and $(B)\int_{X_n\cap[a,x]} f(t)dt$ converge uniformly on $[a,b]$. Also, a McShane integrable function $f$ is said to fulfil the condition "M" on $\{X_n\}$ if for each $n$ there exists $\delta_n(t) > 0$ satisfying $\Big(t-\delta_n(t), t+\delta_n(t)\Big) \subset (a,b)\setminus X_n$ when $t \in (a,b)\setminus X_n$ such that $\lim_{n\to\infty} H_n = 0$ where $H_n(x) = \sup_P \left|\sum_{t\notin X_n} f(t)(v-u)\right|$ whenever the supremum is taken over all $\delta_n$-fine divisions $P = \Big\{([u,v],t)\Big\}$ of $[a,x]$ and the sum is taken over $([u,v],t)$ in $P$ with $t \notin X_n$ and $H_n = \sup_{a\le x\le b} H_n(x)$.

**Lemma 17.3.1.** *If $f \in M_f([a,b],\mathscr{X})$, then there exists a sequence of closed sets $\{X_n\}_{n\in\mathbb{N}}$ such that $X_n \subset [a,b]$, for every $n \in \mathbb{N}$ and $\bigcup X_n = [a,b]$ and $f \in Boc(X_n,\mathscr{X})$ for every $n \in \mathbb{N}$ and $\lim_{n\to\infty}(B)\int_{X_n\cap[a,t]} f = (M)\int_a^t f$ uniformly for every $t \in [a,b]$.*

**Proof:** If $f \in M_f([a,b],\mathscr{X})$, then $f$ is Bochner integrable on $\mathscr{X}$. Let $\delta_n(t) > 0$ satisfying $\Big(t-\delta_n(t), t+\delta_n(t)\Big) \subset (a,b)\setminus X_n$ when $t \in (a,b)\setminus X_n$ such that for any $\delta_n$-fine division $P = \Big\{([u,v],t)\Big\}$ of $[a,x]$ we have,

$$\left|\sum\Big(f(t)-f_n(t)\Big)(v-u) - (M)\int_a^x (f(t)-f_n(t))dt\right| < \frac{1}{2k}.$$

Next, for any $\delta_n$-fine division $P$ of $[a,x]$, $\Big(n(k) \le n < n(k+1)\Big)$ we have $\left|\sum_{t\notin X_n} f(t)(v-u)\right| < \frac{1}{k}$ where $H_n(x) = \left|\sum_{t\notin X_n} f(t)(v-u)\right|$ whenever $x \in [a,b]$ and $n = n(k), n(k)+1, \ldots, n(k+1)-1$. Let us assume $H_n = \sup_{a\le x\le b} H_n(x)$. Then clearly $\lim_{n\to\infty} H_n = 0$. Hence, the above-mentioned condition where $H_n = \sup_{a\le x\le b} H_n(x)$.

**Lemma 17.3.2.** *[10, Theorem 7.1.2] Let $\{T_n\}_{n\in\mathbb{N}}$ be a sequence of functions from $\mathscr{X}$ to $\mathscr{X}$ such that each function has a fixed point $x_n = T(x_n)$. Let $T : \mathscr{X} \to \mathscr{X}$ be such that for some integer m, $T^m$ is a contraction, where $T^m$ is the composition of $T$ m times, and suppose that $T_n \to T$ uniformly. Then $x_n \to x_0 = T(x_0)$.*

**Theorem 17.3.3.** *Let $\mathscr{A} \in SM_f([a,b],L(\mathscr{X},\mathscr{Y}))$ and $f : [a,b] \to \mathscr{X}$ be given. Then $\mathscr{A}f \in SM_f([a,b],\mathscr{Y})$ if $f \in C([a,b],\mathscr{X})$ and $\widehat{\mathscr{A}} \in BV([a,b],L(\mathscr{X},\mathscr{Y}))$.*

**Theorem 17.3.4.** *Consider linear Volterra-McShane integral equation*

$$x(t)+(M)\int_a^t \mathscr{A}(s)x(s)ds = f(t),\ t\in[a,b] \tag{17.3}$$

*whenever* $\mathscr{A}\in SM_f\Big([a,b],L(\mathscr{X})\Big)$ *is bounded. Consider linear Volterra-Bochner integral equation*

$$x(t)+(B)\int_a^t \chi_{X_n}(s)\mathscr{A}(s)x(s)ds = f(t),\ t\in[a,b] \text{ and } n\in\mathbb{N}. \tag{17.4}$$

*Assume that the function* $T: BV_a([a,b],\mathscr{X})\to BV_a([a,b],\mathscr{X})$ *by*

$$(Tx)(t) = f(t)-(M)\int_a^t \mathscr{A}(s)x(s)ds,\ t\in[a,b]$$

*where* $x,f\in BV_a([a,b],\mathscr{X})$. *If* $\left\|\chi_{X_n}(.)\mathscr{A}(.)\right\|<1$ *for each* $n\in\mathbb{N}$, *then (17.4) admits one and only one solution* $x_n\in BV_a([a,b],\mathscr{X})$. *Let us consider the following:*

1. $(x_n)_{n\in\mathbb{N}}$ *has a convergent subsequence* $x_{n_k}\to x_0$;
2. $T^m$ *is a contraction for some* $m>1$.

*Then* $x_0\in BV_a([a,b],\mathscr{X})$ *is a solution of (17.3). Consequently, there exists* $x=\lim_n x_n$ *&* $x\in BV_a([a,b],\mathscr{X})$ *satisfies (17.3).*

**Proof:** For each $n\in\mathbb{N}$, $||\chi_{X_n}(.)\mathscr{A}(.)||<1$. So, for each $T_n$ it follows that:

$$(T_nx)(t) = f(t)+(B)\int_a^t \chi_{X_n}(s)\mathscr{A}(s)x(s)ds,\ t\in[a,b]$$

is a contraction. By contraction principle it has a unique fixed point $x_n$. Therefore, (17.4) admits one and only one solution $x_n\in BV_a([a,b],\mathscr{X})$. Next, if (1) holds, we can see $||x_0-Tx_0||\to 0$. On the other hand, if (2) holds, by Lemma 17.3.1, $T_n\to T$. Next, as $T_n\to T$ uniformly, using Lemma 17.3.2, $x_n\to x=Tx$. Hence, the proof is now completed.

**Theorem 17.3.5.** *Consider a mapping* $T: C([a,b],\mathscr{X})\to C([a,b],\mathscr{X})$ *defined by*

$$(Tx)(t) = f(t)-(M)\int_a^t \mathscr{A}(s)x(s)ds,\ t\in[a,b]$$

*where* $x,\ f\in C([a,b],\mathscr{X})$. *If* $\left\|\chi_{X_n}(.)\mathscr{A}(.)\right\|<1$ *for each* $n\in\mathbb{N}$, *then (17.4) admits one and only one solution* $x_n\in C([a,b],\mathscr{X})$. *Let we consider the followings:*

1. $(x_n)_{n\in\mathbb{N}}$ *has a convergent sub sequence* $x_{n_k}\to x_0$;
2. $\mathscr{A}$ *is bounded and* $T^m$ *is a contraction for some* $m>1$.

*Then* $x_0\in C([a,b],\mathscr{X})$ *is a solution of (17.3). Consequently, there exists* $x=\lim_n x_n$ *&* $x\in C([a,b],\mathscr{X})$ *satisfies (17.3).*

**Proof:** Theorem 17.2.16 gives: a function is Bochner integrable if and only if it is strong McShane integrable. Since every strong McShane integrable function is McShane integrable, the remaining proof is an analogous to the Theorem 17.3.4.

Next, we formulate a uniqueness solution. Let $SM_f^1([a,b],\mathscr{X})_A$ denote the space of all equivalence classes of functions of $SM_f([a,b],\mathscr{X})$ endowed with the Alexiewicz norm which is equivalent to $L^1$-norm for the strong McShane integral. The function $f \in SM_f^1([a,b],X)_A$ means we consider a function $f = f_\Phi \in \Phi$ where $\Phi \in SM_f^1([a,b],X)_A$.

**Lemma 17.3.6.** *Given $\mathscr{A} \in SV([a,b],L(\mathscr{X}))$ with $\mathscr{A}(b) = 0$. Consider the Volterra-strong McShane (resp Volterra-McShane) linear integral equation*

$$x(t) + (SM)\int_a^t \mathscr{A}(s)x(s)ds = f(t),\ t \in [a,b]$$

*where $x, f \in SM_f^1([a,b],\mathscr{X})_A$. Then, for every $f \in SM_f^1([a,b],\mathscr{X})_A$, there exists one and only one solution $x \in SM_f^1([a,b],\mathscr{X})_A$ with*

$$x(t) = f(t) - (SM)\int_a^t \rho(t,s)x(s)ds,\ t \in [a,b]$$

*where the kernel $\rho : [a,b] \times [a,b] \to L(\mathscr{X})$ is bounded and can be given by the Neumann series which converges in $SM_f(SM_f^1([a,b],\mathscr{X})_A)$.*

**Proof:** Let $g(t) = (SM)\int_a^t \rho(t,s)f(s)ds$ and $y = x - f$. Then $x(t) = f(t) - (SM)\int_a^t \rho(t,s)x(s)ds$ is equivalent to $y(t) - (SM)\int_a^t \rho(t,s)y(s)ds = g(t),\ a \le t \le b$, with $g, y \in C([a,b],\mathscr{X})$. By Theorem (17.2.16), it is equivalent to $y(t) - (SM)\int_a^t \rho(t,s)y(s)ds = g(t),\ a \le t \le b$, with $g, y \in C([a,b],\mathscr{X})$. Rest portion of the proof is similar to [9, Theorem 3.5].

**Theorem 17.3.7.** *Given $\mathscr{A} \in SM_f([a,b],L(\mathscr{X}))$ and finite $I \subset [a,b]$. Suppose that $\mathscr{A}$ is of bounded variation on $[a,b]\setminus I$ and consider the Volterra-McShane linear integral equation*

$$x(t) + (SM)\int_a^t \mathscr{A}(s)x(s)ds = f(t),\ t \in [a,b] \tag{17.5}$$

*and*

$$x(t) + (SM)\int_a^t \chi_{x_n}(s)\mathscr{A}(s)x(s)ds = f(t),\ t \in [a,b],\ and\ n \in \mathbb{N} \tag{17.6}$$

*where $x, f \in BV_a([a,b],\mathscr{X})$ and the sequence $\{X_n\}_{n\in\mathbb{N}}$ satisfies the conditions of Lemma 17.3.1 for $\mathscr{A} \in SM_f^1([a,b],L(\mathscr{X}))$. Suppose that $f \in I([a,b],\mathscr{X})$, and that $||\chi_{x_n}()\mathscr{A}()||_A < 1$ for every $n \in \mathbb{N}$. Then (17.6) has one and only one solution $x_n \in SM([a,b],\mathscr{X})$.*

**Proof:** By Lemma 17.3.6, each Eqn. (17.6) has a unique solution $x_n \in SM([a,b],\mathscr{X})$ and $x(t) = f(t) - (SM)\int_a^t \rho(t,s)x(s)ds,\ t \in [a,b]$ with $\rho : [a,b] \times [a,b] \to L(\mathscr{X})$ is bounded and can be expressed by Neumann series. Since $f \in BV_a([a,b],\mathscr{X})$ and the indefinite integral function is of bounded variation, so, each $T_n$ has a fixed point whenever $T_n : BV_a([a,b],\mathscr{X}) \to BV_a([a,b],\mathscr{X})$ defined by $(T_n x)(t) = f(t) - (SM)\int_a^t \chi x_n(s)\mathscr{A}(s)x(s)ds,\ t \in [a,b]$. Remaining steps are alike as Theorem 17.3.4.

## BIBLIOGRAPHY


1. Becerra T. P., Perales S. S., Escamilla-Reyna J. A., Henstock-Kurzweil Vector Distributions, Mediterr. J. Math., **17**(195), (2020), 1–19.
2. Cao S. S., The Henstock integral for Banach-valued functions, SEA Bull. Math. **16**, (1992), 35–40.
3. Federson M., Some peculiarities of the Henstock and Kurzweil integrals of Banach space-valued functions. Real analysis Exchange, **29**(1), (2003), 439–460.
4. Federson M., Bianconi R., Linear Volterra-Stieltjes integral equations in the sense of the Kurzweil-Henstock integral, Arc. Math., **37**(4), (2001), 307–328.
5. Federson M., Bianconi R., Linear Fredholm integral equations and the integral of Kurzweil, J. Appl. Anal., **8**(1), (2002), 83–110.
6. Federson M., Bianconi R., A fredholm-type theorem for linear integral equations of Stieltjes type, Integration: Mathematical Theory and Applications, **1**(2),(2008), 25–57.
7. Gordon R. A., The McShane integral of Banach-valued functions, Illinois J.Math., **34**, (1990), 556–567.
8. Gordon R. A., *The integrals of Lebesgue, Denjoy, Perron and Henstock*, volume 4 of Graduate Studies in Mathematics. American Math. Soc., Providence, RI, (1994).
9. Hönig C. S., Volterra-Stieltjes integral equations, Math. Studies, North-Holland Publ. Comp., **16**, (1975), 1–18.
10. Istratescu V. I., *Fixed Point Theory: An introduction*, D. Reidel Pub. Comp., (1981).
11. Schwabik S., Abstract Perron-Stieltjes integral, Math. Bohem., **121**(4), (1996), 425–447.
12. Schwabik S., G. Ye., *Topics in Banach Space Integration*, volume 10 of Real Analysis. World Scientific, Singapore, (2005).
13. Yoon Ju H., Eun G. S., Lee Y. C., On McShane-Stieltjes integral, Kangweon-Kyungki Math. J., **5**(2), (1997), 217–225.

# 18 Weighted H-Orlicz Spaces and their Inclusion Properties

*Hemanta Kalita, Anca Croitoru*

In this chapter we introduce the weak H-Orlicz spaces generated by convex functions. We present sufficient and necessary conditions for the inclusion relation between two weighted Orlicz spaces. In particular, we obtain some sufficient and necessary conditions for the inclusion properties of H-Orlicz spaces.

## 18.1 INTRODUCTION

Weighted spaces are studied in most cases as a generalization of a special case. Orlicz spaces are generalization of Lebesgue space were introduced by Birnbaum and Orlicz in 1931 (see [5]). L. Maligranda discussed about inclusion properties of Orlicz space. Masta et al. [3] discussed about inclusion properties between Orlicz spaces and weak Orlicz spaces. Osancliol [4] have prove a necessary and sufficient condition for inclusion relation betwwen two weighted Orlicz spaces.

The theory of Henstock-Kurzweil integral was developed by Henstock and Kurzweil independently during 1955–1957 (see [7]). Henstock-Kurzweil integrals can be viewed as a generalized Riemann integral, contain Lebesgue intrgral also. In general, a function is Henstock-Kurzweil integral if and only if is Lebesgue integrable [7] also (see [1]).

In 2020, B. Hazarika et al. [2] introduced an Orlicz type spaces so called H-Orlicz spaces containing Henstock-Kurzweil integrals. Orlicz spaces are subspace of H-Orlicz spaces. We motivate the approach of [3] and various technique of [4] to introduce *a weighted H-Orlicz* spaces to find a necessary and sufficient conditions of inclusions between two weighted H-Orlicz spaces with the condition $\prec$ and $\prec\prec$ of Young functions and weight functions that are associated in H-Orlicz spaces.

The chapter is organized as follows: in Section 2, the basic concepts and terminology are in- troduced together with some definitions and results. In Section 3, we introduce *weighted H-Orlicz* spaces generated by convex functions. Fundamental properties of weighted H-Orlicz spaces are discussed in this section. In Section 4, necessary and sufficient conditions for the inclusion relation between two weighted

DOI: 10.1201/9781003596578-18 

Orlicz spaces are discussed. In particular, we obtain some sufficient and necessary conditions for the inclusion properties of H-Orlicz spaces.

## 18.2 PRELIMINARIES

In this section, we recall definitions, results that will be useful in our main section.

**Definition 18.2.1.** *[5] A non zero function $\theta : [0,+\infty) \to [0,+\infty]$ is called a Young function if $\theta$ is convex and satisfies $\lim\limits_{x\to 0+} \theta(x) = \theta(0) = 0$ and $\lim\limits_{x\to +\infty} \theta(x) = +\infty$.*

Recall that

1. $\theta_1 \prec \theta_2$ if there exist $\varepsilon > 0$ and $x_0 \geq 0$ such that $\theta_1(x) \leq \theta_2(\varepsilon x)\ \forall\, x \geq x_0$.
2. $\theta_1 \prec\prec \theta_2$ if for each $\varepsilon > 0$, there exists $x_\varepsilon \geq 0$ such that $\theta_1(x) \leq \theta_2(\varepsilon x)\ \forall\, x \geq x_\varepsilon$.

Throughout the article, $w$ is a weight on $X$. Recall $w : X \to (0,+\infty)$ is a measurable function. Recall that $w_1 \prec\prec w_2$ if there exists a $C > 0$ such that $w_1(x) \leq C w_2(x)$ for all $x \in X$. If $w_1 \prec\prec w_2$ and $w_2 \prec\prec w_1$ then wo weights $w_1$ and $w_2$ on $X$ is with $w_1 \approx w_2$.

Let $(X,\Sigma,\mu)$ be a measure space and $X$ is a Banach space. Recalling the Orlicz space $\mathrm{L}^\theta(X,\Sigma,\mu)$ or $\mathrm{L}^\theta(X)$ as

$$\mathrm{L}^\theta(X) = \left\{ f : X \to \mathbb{C} :\ (L)\int_X \theta(\alpha|f(x)|)\,d\mu(x) < +\infty \ \textit{for some}\ \alpha > 0 \right\},$$

where $f$ is $\mu$-equivalence classes of measurable functions. Orlicz space is a Banach space with the norm

$$||f||_{L^\theta} = \inf\left\{ \alpha > 0 :\ (L)\int_X \theta\left(\frac{|f(x)|}{\alpha}\right) d\mu(x) \leq 1 \right\}.$$

The weighted Orlicz space $\mathrm{L}^\theta_w(X)$ is the set of all measurable functions $f : X \to \mathbb{R}$ such that $wf \in \mathrm{L}^\theta(X)$. This space is a Banach space equipped with the norm

$$||f||_{L^\theta_w} = ||wf||_{L^\theta} = \inf\left\{ \alpha > 0 :\ (L)\int_X \theta\left(\frac{|w(x).f(x)|}{\alpha}\right) d\mu(x) \leq 1 \right\}.$$

**Definition 18.2.2.** *[1] A function $f : X \to \mathbb{R}$ is said to be Henstock-Kurzweil integrable on $X$ if there exists $A \in \mathbb{R}$ with the following property: given $\varepsilon > 0$ there exists a gauge $\delta : X \to \mathbb{R}^+$ such that*

$$||S(P,f) - A||_X < \varepsilon$$

*for each $\delta$-fine partition $P$ of $X$.*

We denote $HK(X)$, the class of Henstock-Kurzweil integrable functions on $X$. For details of Henstock-Kurzweil integrals (see [1, 7]). Recalling the relation between Young functions and Henstock-Kurzweil integral as follows:

**Proposition 18.2.3.** *[2] Let $\theta$ be a Young function, then $\theta(f): X \to \mathbb{R}$ is a Henstock-Kurzweil integrable function.*

**Definition 18.2.4.** *[2] Let $(X, \Sigma, \mu)$ is a measurable space and $\theta$ is a Young function. The H-Orlicz spaces is the set of measurable functions $f : X \to \mathbb{R}$ such that $(HK) \int_X \theta(\alpha f) d\mu(x) < \infty$ for some $\alpha > 0$.*

H-Orlicz space is a Banach space with the norm

$$||f||_\theta = \inf\left\{\alpha > 0 : \ (HK) \int_X \theta\left(\frac{|f(x)|}{\alpha}\right) d\mu(x) \leq 1\right\}.$$

For detailed of H-Orlicz (see [2]).

**Theorem 18.2.5.** *[2] For any Young function $\theta$, $\mathrm{H}^\theta(X) \subset HK(X)$.*

## 18.3 WEIGHTED H-ORLICZ SPACES

Let $(X, \Sigma, \mu)$ be a measure space and $\theta$ be a $N$-functions. If $w$ is a weight on $X$. That is $w : X \to \mathbb{R}^+$ is a measurable function and positive a.e.. Now, we define the *weighted H-Orlicz* space as follows:

**Definition 18.3.1.** *The* weighted H-Orlicz *space $\mathrm{H}^\theta_w(X)$ is the set of all measurable function $f : X \to \mathbb{R}$ such that $wf \in \mathrm{H}^\theta(X)$. That is*

$$\mathrm{H}^\theta_w(X) = \left\{f : \ wf \in \mathrm{H}^\theta(X)\right\}.$$

In condition,

$$||f||_{\theta,w} = ||wf||_\theta \ \forall \ f \in \mathrm{H}^\theta_w(X). \tag{18.1}$$

Hence, $||f||_{\theta,w} = \inf\left\{\alpha > 0 : \ (HK) \int_X \theta\left(\frac{|w(x).f(x)|}{\alpha}\right) d\mu(x) \leq 1\right\}$. Then $||.||_{\theta}, w$ defines a norm on $\mathrm{H}^w_\theta(X)$. We called this norm as *weighted H-Orlicz* norm.

**Remark 18.3.2.** *If $w = 1$, then* (18.1) reduces to H-Orlicz norm and we can come back to $(\mathrm{H}^\theta(X), ||.||_\theta)$.

We will find fundamental properties of the *weighted H-Orlicz* spaces below.

**Proposition 18.3.3.** *Norm convergence in* weighted H-Orlicz *spaces are norm convergence in H-Orlicz spaces.*

**Proof:** Let $\{f_n\} \subset \mathrm{H}^\theta_w(X)$, such that $||f_n - f||_{\theta,w} \to 0$. Let us consider $\varepsilon > 0$ and $\theta$ be an $N$-function such that $\varepsilon\theta(t) < 1$. Then there is an $n_0 \in \mathbb{N}$ such that $||f_n - f||_{\theta,w} < \varepsilon\theta(t) < 1$ whenever $n \geq n_0$. So, $||(f_n - f)w||_{\theta,w} \to 0$. Consequently, $||f_n - f||_\theta \to 0$. It is very easy to see the following results:

**Lemma 18.3.4.** *1. $f \in \mathtt{H}^{\theta}_{w}(X)$ if and only if $-f \in \mathtt{H}^{\theta}_{w}(X)$.*
*2. Let $f, \mathrm{g}$ be measurable and $|f| < |\mathrm{g}|$ a.e., then $f \in \mathtt{H}^{\theta}_{w}(X)$ if $\mathrm{g} \in \mathtt{H}^{\theta}_{w}(X)$.*
*3. If $||f_n - f||_{\theta,w} \to 0$ then $f_n \to f$ in measure.*

**Proposition 18.3.5.** *The norm convergence of $\{f_n\}$ in a* weighted H-Orlicz space *$\mathtt{H}^{\theta}_{w}(X)$ implies the convergence a.e. of some some sequence $\{f_{n_k}\}$ of $X$.*

**Proposition 18.3.6.** $\left(\mathtt{H}^{\theta}_{w}(X), ||.||_{\theta,w}\right)$ *is a Banach space.*

**Proof:** Since $\mathtt{H}^{\theta}(X)$ is a Banach space (see [2, Theorem 3.7]), there exists a function $\mathrm{g} \in \mathtt{H}^{\theta}(X)$ such that

$$\sum_{n=1}^{N} ||f_n w - \mathrm{g}||_{\theta,w} \to 0 \ \text{ as } N \to \infty.$$

If $f = \frac{\mathrm{g}}{w}$ then $f \in \mathtt{H}^{\theta}_{w}(X)$ and $|| \sum_{n=1}^{N} f_n - f||_{\theta,w} = \left\| \left( \sum_{n=1}^{N} f_n \right) w - f w \right\|_{\theta,w} \to 0$ as $N \to \infty$. So, $\mathtt{H}^{\theta}_{w}(X)$ is a Banach space.

We will establish a relationship of classical *weighted Orlicz space* and a *weighted H-Orlicz space*.

**Theorem 18.3.7.** *The classical weighted-Orlicz space $\mathtt{L}^{\theta}_{w}(X)$ is a dense subspace of $\mathtt{H}^{\theta}_{w}(X)$ as continuous dense embeddings. That is, $\mathtt{L}^{\theta}_{w}(X) \hookrightarrow \mathtt{H}^{\theta}_{w}(X)$ is continuous dense embeddings.*

**Proof:** Let $f \in \mathtt{L}^{\theta}_{w}(X)$. Then $f \in \mathtt{L}^{1}_{w}(X)$ with $||f||_{L,\theta} < \infty$. Then for some $\alpha > 0$, we have

$$\inf\left\{ (HK) \int_X \theta\left( \frac{|w(x).f(x)|}{\alpha} \right) d\mu(x) \right\} \leq \inf\left\{ (L) \int_X \theta\left( \frac{|w(x).f(x)|}{\alpha} \right) d\mu(x) \right\} \leq 1.$$

So, for some $\alpha > 0, \inf\left\{ \alpha > 0 : (L) \int_X \theta\left( \frac{|w(x).f(x)|}{\alpha} \right) d\mu(x) \leq 1 \right\}$, we get the following

$$\inf\left\{ \alpha > 0 : (HK) \int_X \theta\left( \frac{|w(x).f(x)|}{\alpha} \right) d\mu(x) \leq 1 \right\}.$$

Hence $f \in \mathtt{H}^{\theta}_{w}(X)$ with $||f||_{\theta,w} \leq ||f||_{L^{\theta}_{w}}$. Hence the proof.

**Remark 18.3.8.** *In compact support, every Henstock-Kurzweil integrable functions are Lebesgue integrable and these two integrals are equal. Hence $\mathtt{H}^{\theta_1}_{w_1}(X)$ and $\mathtt{H}^{\theta_2}_{w_2}(X)$ becomes weighted Lebesgue spaces $\mathtt{L}^{\theta_1}_{w_1}(X)$ and $\mathtt{L}^{\theta_2}_{w_2}(X)$ respectively.*

## 18.4 INCLUSION RELATION BETWEEN TWO WEIGHTED H-ORLICZ SPACES

In this section, we will discuss various inclusion relations between two weighted H-Orlicz spaces. Let us consider $w_1$ and $w_2$ are two weights. We consider $\theta_1$, $\theta_2$ are two different Young functions. We start inclusion results of a *weak H-Orlicz* spaces when the Young functions $\theta_1$, $\theta_2$ satisfies $\prec$ and $\prec\prec$ conditions.

**Proposition 18.4.1.** *Let $\theta_1, \theta_2$ be two Young functions. If there exist $C > 0$ such that $\theta_1 \prec \theta_2$ then $\mathrm{H}_w^{\theta_2}(X) \subseteq \mathrm{H}_w^{\theta_1}(X)$ with $||f||_{\mathrm{H}_w^{\theta_1}(X)} \leq C||f||_{\mathrm{H}_w^{\theta_2}(X)}$ for every $f \in \mathrm{H}_w^{\theta_2}(X)$.*

**Proof:** Let $f \in \mathrm{H}_w^{\theta_2}(X)$, we will prove that $f \in \mathrm{H}_w^{\theta_1}(X)$. Since,

$$\begin{aligned}
(HK)\int_X \theta_1\left(\frac{|w(x).f(x)|}{C.||f||_{\mathrm{H}_w^{\theta_2}(X)}}\right)d\mu(x) &\leq (HK)\int_X \theta_2\left(\frac{C|w(x).f(x)|}{C.||f||_{\mathrm{H}_w^{\theta_2}(X)}}\right)d\mu(x)\\
&= (HK)\int_X (HK)\int_X \theta_1\left(\frac{|w(x).f(x)|}{||f||_{\mathrm{H}_w^{\theta_2}(X)}}\right)d\mu(x)\\
&\leq 1.
\end{aligned}$$

Hence, $f \in \mathrm{H}_w^{\theta_1}(X)$. Consequently, $||f||_{\mathrm{H}_w^{\theta_1}(X)} \leq C||f||_{\mathrm{H}_w^{\theta_2}(X)}$.

The following inclusion result of $\mathrm{H}_{w_1}^{\theta_1}$ and $\mathrm{H}_{w_2}^{\theta_2}$ will be with $\theta_1 \prec \theta_2$ and $w_1 \preceq w_2$ as follows:

**Theorem 18.4.2.** *If $\theta_1$, $\theta_2$ are Young functions such that $\theta_1 \prec \theta_2$ and $w_1 \preceq w_2$ then $\mathrm{H}_{w_2}^{\theta_2}(X) \subseteq \mathrm{H}_{w_1}^{\theta_1}(X)$.*

**Proof:** Suppose $\theta_1 \prec \theta_2$ and $w_1 \preceq w_2$. Consider $f \in \mathrm{H}_{w_2}^{\theta_2}(X)$ then there exists $\alpha > 0$ such that

$$(HK)\int_X \theta_2(\alpha.w_2(x)|f(x)|)d\mu(x) < +\infty.$$

In the other side, since $w_1 \preceq w_2$ and $\theta_1 \prec \theta_2$, there exists a number $C > 0$ and $C' > 0$ such that

$$w_1(x) \leq C'w_2(x) \ \forall\, x \in X \tag{18.2}$$

and

$$\theta_1(x) \leq \theta_2(Cy) \ \forall\, y \geq 0. \tag{18.3}$$

If $\frac{\alpha}{C.C'} = \beta > 0$ then from (18.2)

$$\begin{aligned}
\theta_1\Big(\beta w_1(x).|f(x)|\Big) &\leq \theta_1\Big(\beta.C'w_1(x).|f(x)|\Big)\\
&= \theta_1\left(\frac{\alpha}{C}.w_2(x).|f(x)|\right) \ \forall\, x \in X.
\end{aligned}$$

Using the increasing nature of Young's function, we get for all $x \in X$ as follows:

$$\theta_1\left(\frac{\alpha}{C}.w_2(x).|f(x)|\right) \le \theta_2\left(C.\frac{\alpha}{C}.w_2(x).|f(x)|\right) = \theta_2\left(\alpha.w_2(x).|f(x)|\right).$$

Now from (18.3),

$$(HK)\int_X \theta_1\left(\beta w_1(x).|f(x)|\right)d\mu(x) \le (HK)\int_X \theta_2\left(\alpha.w_2(x).|f(x)|\right)d\mu(x) < +\infty.$$

Thus $f \in \mathtt{H}_{w_1}^{\theta_1}(X)$. Hence $\mathtt{H}_{w_2}^{\theta_2}(X) \subseteq \mathtt{H}_{w_1}^{\theta_1}(X)$. %endproof The converse of the above Proposition does not holds. To prove this, we can consider

$$\theta_1(x) = \frac{x^{n_1}}{n_1},\ \theta_2(x) = \frac{x^{n_2}}{n_2}$$

with $x \ge 0,\ 1 < n_1 < n_2 < +\infty$.

In compact support, from the Remark 18.3.8, we can find $\mathtt{L}_w^{n_2}(X) \subseteq \mathtt{L}_w^{n_1}(X)$. But, $\theta_2$ is stronger then $\theta_1$. So, $\theta_1 \not\prec \theta_2$ for all $x \ge 0$. Hence $\mathtt{H}_{w_2}^{\theta_2}(X) \not\subseteq \mathtt{H}_{w_1}^{\theta_1}(X)$.

**Proposition 18.4.3.** *If $\theta_1 \approx \theta_2$ and $\mu(X) < \infty$ with $w_1 \approx w_2$ then $\mathtt{H}_{w_1}^{\theta_1} = \mathtt{H}_{w_2}^{\theta_2}$.*

**Proof:** The proof is an analogous of the Theorem 18.4.2.

**Lemma 18.4.4.** *Let $w_1, w_2$ are two weight functions and $\theta_1, \theta_2$ are two Young functions. If $\mathtt{H}_{w_2}^{\theta_2}(X) \subset \mathtt{H}_{w_1}^{\theta_1}(X)$ with $||f||_{\theta_1,w_1} < +\infty$ and $||f||_{\theta_2,w_2} < +\infty$. The expression*

$$|||f||| = ||f||_{\theta_1,w_1} + ||f||_{\theta_2,w_2} \tag{18.4}$$

*is a norm of $\mathtt{H}_{w_2}^{\theta_2}(X)$.*

**Proof:** The following conditions must be met in order to establish the expression (18.4) as a norm:
(i) If $f = 0$ a.e. then $||f||_{\theta_1,w_1} = ||f||_{\theta_2,w_2} = 0$ gives $|||f||| = 0$.
Conversely, suppose $|||f||| = 0$. Then, $||f||_{\theta_1,w_1} + ||f||_{\theta_2,w_2} = 0$. This implies $||f||_{\theta_1,w_1} = 0$ and $||f||_{\theta_2,w_2} = 0$. Suppose $|f| > 0$, on a set of positive measure. If possible then there exists a weight function $w_1$, $w_1(x) > 0$ such that

$$A = \{p:\ |f(p)| \ge w_1(x)\}$$

satisfies $\mu(A) > 0$. Now,

$$\begin{aligned}\theta_1\left(|w_1(x).f(x)|\right)\mu(A) &= \int_A \theta_1(|w_1(x).f(x)|)d\mu(x) \\ &\le \int_A \theta_1(|w_1(x).f(x)|)d\mu(x) \\ &\le \int_X \theta_1(|w_1(x).f(x)|)d\mu(x).\end{aligned}$$

Since $\mu(A) > 0$, this gives $\theta_1(|w_1(x).f(x)|)d\mu(x) \to \infty$. But $||f||_{\theta_1,w_1} = 0$, gives

$$\inf\left\{\alpha > 0:\ \theta\left(\frac{|w_1(x).f(x)|}{\alpha}\right)d\mu(x) \leq 1\right\} = 0,$$

only if $\theta_1(|w_1(x).f(x)|)d\mu(x) \to 0$. This is a contradiction on $\mu(A) > 0$, gives $\theta_1(|w_1(x).f(x)|)d\mu(x) \to \infty$. Hence $f = 0$ a.e..
(ii) For a scalar $K \in \mathbb{R}$ or $\mathbb{C}$ we have,

$$\begin{aligned}
|||kf||| &= ||kf||_{\theta_1,w_1} + ||kf||_{\theta_2,w_2} \\
&= |k|\left[\inf\left\{\frac{\alpha}{|k|} > 0:\ (HK)\int_X \theta_1\left(\frac{|w_1(x).f(x)|}{\frac{\alpha}{|k|}}\right)d\mu(x) \leq 1\right\}\right. \\
&\left.+\inf\left\{\frac{\alpha}{|k|} > 0:\ (HK)\int_X \theta_2\left(\frac{|w_2(x).f(x)|}{\frac{\alpha}{|k|}}\right)d\mu(x) \leq 1\right\}\right] \\
&= |k|\left[||f||_{\theta_1,w_1} + ||f||_{\theta_2,w_2}\right] \\
&= |k||||f|||.
\end{aligned}$$

(iii) Lastly, we need to prove $|||f+g||| \leq |||f||| + |||g|||$ for every $f,g \in \mathrm{H}_{w_2}^{\theta_2}(X)$. If $f,g \in \mathrm{H}_{w_2}^{\theta_2}(X)$, then $||f||_{\theta_2,w_2} < +\infty$. Now,

$$|||f+g||| = ||f+g||_{\theta_1,w_1} + ||f+g||_{\theta_2,w_2}$$

Since, $||.||_{\theta_1,w_1}$, $||.||_{\theta_2,w_2}$ are norm. Hence,

$$|||f+g||| \leq ||f||_{\theta_1,w_1} + ||g||_{\theta_1,w_1} + ||f||_{\theta_2,w_2} + ||g||_{\theta_2,w_2}.$$

Consequently, $|||f+g||| \leq |||f||| + |||g|||$. Hence the expression (18.4) form a norm.
Moreover, it is easy to see $\left(\mathrm{H}_{w_2}^{\theta_2}(X), |||.|||\right)$ form a Banach space.

**Proposition 18.4.5.** *Let $w_1, w_2$ are weight functions. $\theta_1$ and $\theta_2$ are two Young functions. If $\mathrm{H}_{w_2}^{\theta_2}(X) \subseteq \mathrm{H}_{w_1}^{\theta_1}(X)$, then the inclusion map $i : \left(\mathrm{H}_{w_2}^{\theta_2}(X), ||.||_{\theta_2,w_2}\right) \to \left(\mathrm{H}_{w_1}^{\theta_1}(X), ||.||_{\theta_1,w_1}\right)$ is continuous.*

Next, results will be a necessary and sufficient condition of inclusions between $\mathrm{H}_w^{\theta_1}$ and $\mathrm{H}_w^{\theta_2}$.

**Theorem 18.4.6.** *Let $\theta_1$, $\theta_2$ be Young functions and $w : X \to (0,\infty)$ be a measurable and positive a.e.. Then $\mathrm{H}_w^{\theta_2}(X) \subseteq \mathrm{H}_w^{\theta_1}(X)$ if and only if there exists a constant $C > 0$*

*such that*

$$||f||_{\mathtt{H}_w^{\theta_1}(X)} \leq C||f||_{\mathtt{H}_w^{\theta_2}(X)},$$

*for every* $f \in \mathtt{H}_w^{\theta_2}(X)$.

**Proof:** Let $\mathtt{H}_{w_2}^{\theta_2}(X) \subseteq \mathtt{H}_{w_1}^{\theta_1}(X)$. Let $f \in \mathtt{H}_{w_2}^{\theta_2}(X)$. Then $||f||_{\theta_2,w_2} < +\infty$ and $||f||_{\theta_1,w_1} < +\infty$. Let $\mathtt{T} : \left(\mathtt{H}_{w_2}^{\theta_2}(X), |||.|||\right) \to \left(\mathtt{H}_{w_2}^{\theta_2}(X), ||.||_{\theta_2,w_2}\right)$ by $Tf = f$ for all $f \in \mathtt{H}_{w_2}^{\theta_2}(X)$. Then from the Proposition 18.4.5 and the closed graph theorem, there exists a $C > 0$ such that

$$||f||_{\theta_1,w_1} \leq |||f||| \leq C.||f||_{\theta_2,w_2}$$

for all $f \in \mathtt{H}_{w_2}^{\theta_2}(X)$.
Conversely, suppose

$$||f||_{\theta_1,w_1} \leq |||f||| \leq C.||f||_{\theta_2,w_2}$$

for all $f \in \mathtt{H}_{w_2}^{\theta_2}(X)$. It is clear from the definition of the norm of $||.||_{\theta-1,w_1}$ and $||.|_{\theta_2,w_2}$ that $\mathtt{H}_w^{\theta_2}(X) \subseteq \mathtt{H}_w^{\theta_1}(X)$.

### 18.4.1 INCLUSION RELATION BETWEEN H-ORLICZ SPACES

In the sequel from the Remark 18.3.2, we can come back to H-Orlicz space with $w(x) = 1$. We discuss a necessary and sufficient condition of inclusion properties on H-Orlicz space. We start with the following Proposition.

**Proposition 18.4.7.** *Let $\theta_1, \theta_2$ be two Young functions. If there exist $C > 0$ such that $\theta_1(t) \leq \theta_2(Ct)$ for every $t > 0$. Then $\mathtt{H}_2^{\theta}(X) \subseteq \mathtt{H}_1^{\theta}(X)$ with $||f||_{\mathtt{H}_1^{\theta}} \leq C||f||_{\mathtt{H}_2^{\theta}}$ for each $f \in \mathtt{H}_2^{\theta}(X)$.*

**Proof:** The proof is of a similar technique to Proposition 18.4.1, so we have ommitted. We need the following lemma to reach our necessary and sufficient condition of inclusion of H-Orlicz space.

**Lemma 18.4.8.** *Let $\theta$ be a Young function, $a \in X$, and $r > 0$. Then $||\chi_{B(a,r)}||_{\mathtt{H}^{\theta}(X)} = \frac{1}{\theta^{-1}(\frac{1}{\mu(B(a,r))})}$ where $\mu(B(a,r))$ denote the volume of $B(a,r)$.*

**Proof:** We have,

$$
\begin{aligned}
&(HK)\int_X \theta\left(\frac{\chi_{B(a,r)}}{\theta^{-1}(\frac{1}{\mu(B(a,r))})}\right)d\mu(x)\\
&=(HK)\int_X \theta\left(\chi_{B(a,r)}\theta^{-1}\left(\frac{1}{\mu(B(a,r))}\right)\right)d\mu(x)\\
&=(HK)\int_{B(a,r)} \theta\left(\chi_{B(a,r)}\theta^{-1}\left(\frac{1}{\mu(B(a,r))}\right)\right)d\mu(x)\\
&=(HK)\int_{B(a,r)} \theta\left(\theta^{-1}\left(\frac{1}{\mu(B(a,r))}\right)\right)d\mu(x)\\
&\leq (HK)\int_{B(a,r)} \left(\frac{1}{\mu(B(a,r))}\right)d\mu(x)\\
&=\frac{1}{\mu(B(a,r))}\left\{(HK)\int_{B(a,r)} d\mu(x)\right\}\\
&=1.
\end{aligned}
$$

It is easy to see $||\chi_{B(a,r)}||_{\mathtt{H}^\theta(X)} \leq \frac{1}{\theta^{-1}\left(\frac{1}{\mu(B(a,r)}\right)}$. Now, suppose, $||\chi_{B(a,r)}||_{\mathtt{H}^\theta(X)} < \frac{1}{\theta^{-1}(\frac{1}{\mu(B(a,r))})}$, then $(HK)\int_X \theta\left(\frac{\chi_{B(a,r)}}{\theta^{-1}(\frac{1}{\mu(B(a,r))})}\right)d\mu(x) \leq 1$.
Therefore, $||\chi_{B(a,r)}||_{\mathtt{H}^\theta(X)} = \frac{1}{\theta^{-1}(\frac{1}{\mu(B(a,r))})}$.

**Proposition 18.4.9.** *Let $\theta_1$, $\theta_2$ be two Young functions, and $C > 0$. Then the following statements are equivalent:*

1. $\theta_1 \prec \theta_2$,
2. $\mathtt{H}^{\theta_2}(X) \subseteq \mathtt{H}^{\theta_1}(X)$,
3. $||f||_{\theta_1} \leq C||f||_{\theta_2}$.

**Proof:** For $(1) \implies (2)$: Suppose that $f \in \mathtt{H}^{\theta_2}(X)$, then we have $(HK)\int_X \theta_2\left(\frac{f}{||f||_{\theta_2}}\right) \leq 1$. Now, $\theta_1 \prec \theta_2$ gives

$$
\begin{aligned}
(HK)\int_X \theta_1\left(\frac{f}{C||f||_{\theta_2}}\right) &\leq (HK)\int_X \theta_2\left(\frac{Cf}{C||f||_{\theta_2}}\right)\\
&=(HK)\int_X \theta_2\left(\frac{f}{||f||_{\theta_2}}\right)\\
&\leq 1.
\end{aligned}
$$

Hence, $f \in \mathtt{H}^{\theta_1}(X)$.
From the definition, it is easy to see $(2) \implies (3)$.

For (3) $\implies$ (1) : Using the Lemma 18.4.8, we can have

$$\frac{1}{\theta^{-1}(\frac{1}{\mu(B(a,r))})} = ||\chi_{B(a,r)}||_{\theta_1}$$
$$\leq C||f||_{\theta_2}$$
$$= \frac{C}{\frac{1}{\theta^{-2}(\frac{1}{\mu(B(a,r))})}}.$$

Hence, for any arbitrary $a \in X$ and $r > 0$, we have $C\theta_1^{-1}\left(\frac{1}{\mu(B(a,r))}\right) \geq \theta_2^{-1}\left(\frac{1}{\mu(B(a,r))}\right)$. Consequently, $\theta_1(x) \leq \theta_2(Cx)$ for all $C > 0$. Hence, $\theta_1 \prec \theta_2$.

Finally, the following theorem gives a necessary and sufficient condition of inclusion of H-Orlicz space.

**Theorem 18.4.10.** *Let $\theta_1$, $\theta_2$ be Young functions. Then $\mathtt{H}^{\theta_2}(X) \subseteq \mathtt{H}^{\theta_1}(X)$ if and only if there exists a constant $C > 0$ such that*

$$||f||_{\theta_1} \leq C||f||_{\theta_2},$$

*for every $f \in \mathtt{H}^{\theta_2}(X)$.*

**Proof:** The proof is of a similar technique to the Theorem 18.4.6, so we have ommitted.

## 18.5 CONCLUSIONS

In this chapter, we introduce *a-weighted H-Orlicz* spaces. Fundamental properties like the completeness of the *weighted H-Orlicz* spaces are discussed here. A relationship of weighted Orlicz spaces and *H-Orlicz* space is also discussed in Section 2. Various inclusion properties of *H-Orlicz* spaces are discussed in the respective section. A necessary and sufficient condition of inclusion of a *weighted H-Orlicz space* and H-Orlicz spaces are discussed in Theorem 18.4.6 and Theorem 18.4.10, respectively.

## BIBLIOGRAPHY


1. Gordon R. A., *The Integrals of Lebesgue, Denjoy, Perron, and Henstock*, Ameri. Math. Soc., (1994).
2. Hazarika B., Kalita H., Henstock-Orlicz space and its dense space, Asian-Europian Journal of Math, **14**(7), 2021, 1–18.
3. Masta A., Ifronika L., Taqiyuddin M., A note on inclusion properties of weighted Orlicz spaces, J. Indones. Math. Soc, **26** (1), (2020), 128–136.
4. Osanclol A., Inclusions between weighted Orlicz spaces, J. Inequal. Appl., **390**, 2014, 1–8.

5. Rao M. M., Ren Z. D., *Theory of Orlicz Spaces,* 1st edn. CRC Press, New York (1991).
6. Zhang X., Zhang C., Weak Orlicz spaces generated by concave functions, International Conference on Information Science and Technology (ICIST) (2011), 42–44
7. Swartz C., *Introduction to gauge integral,* World Scientific, (2008).
8. Singer I., *Bases in Banach spaces*, **1**, Springer, (1970).
9. Singer I., *Bases in Banach spaces*, **2**, Springer, (1981).

# Index


$(\alpha - \psi)$ contractive covariant mapping, 176
$H$-division, 283
$M$-division subordinated, 283
$T-$summable, 39
$\Delta$-integrable sequence, 258
$\Upsilon$-Cauchy, 133
$\Upsilon$-complete, 134
$\Upsilon$-convergent, 134
$\alpha$-admissible $\mathscr{Z}$ contraction, 174
$\varepsilon-$chainable, 156
$\mathfrak{Z}_e$-contraction, 194
$\nabla$-subinterval, 258
$b$-metric, 129
$b_v(\theta)$-metric space, 130
$f$-orbitally complete metric space, 9
$q-$Bell matrix, 21
$q-$Bell numbers, 22
$q-$combination, 22
$q-$factorial, 22
$\nu$-generalized metric space, 129
$wt_e$-distance, 193
Ćirić generalized contraction, 8
Ćirić quasi-contraction, 10

$\mathscr{I}-$statistical convergent, 53

Abel's double partial summation, 82
admissible ideal, 53
almost convergent, 40
approximate $\theta-$nbd, 270
approximate Lusin function, 269
approximately differentiation, 270
asymptotic density, 254

backward and forward jump operators, 219
Banach contraction principle, 6
Banach limit, 92
Bell core, 44
bipolar metric, 175
bisequence, 175
bounded semi-variation, 282
Branciari distance, 191
Branciari distance space, 191

Cauchy bisequence, 175
coincidence point, 177
common fixed point, 177
compatible, 176
complex-valued double sequences, 71
continuity of a $b-$metric, 153
contraction, 6
controlled $b$-Branciari metric type, 130
controlled $\nu$-generalized metric type space, 131
controlled metric type, 130
covariant $T - \alpha-$admissible mapping, 177

density of $E$, 51
direct path, 154
double sequence spaces, 75

extended $b$-metric, 129
extended Branciari $b$-distance space, 191

filter of $\mathbb{N}$, 53
free tagged interval, 270
free tagged partition, 270

generalized graph $\lambda-$contraction, 155
generalized rational graph $\lambda^*-$contraction, 155
gradual normed linear space, 54
gradually bounded, 54
gradually convergent, 55
gradually lacunary statistical convergent, 55

Henstock-Kurzweil integrable, 15
Henstock-Kurzweil integrable, 299

jump operators, 219

Kuelbs-Steadman space, 16

lacunary statistical convergent, 52
lacunary statistical limit, 52
ld continuous, 220

matrix summability method, 2
McShane $f$-integrable, 283
modified simulation function, 193
modulus, 94
multiplier space, 4

nabla integral, 220

paranorm on $X$, 95
periodic point, 200
point of coincidence, 177
Pompeiu-Hausdorff $PHb-$metric, 154
property (P), 200

rectangular metric, 129
regularly convergent, 72

Riemann-Liouville integro differential equation, 220

simulation function, 173, 192
statistically Cauchy sequence, 255
statistically convergent of order $\alpha$, 52
strongly McShane $\mathscr{A}$-integrable, 285
subordinate, 270

tagged division, 283

uniform statistically Riemann $\Delta$-integrable, 263

weakly compatible, 177
weighted H-Orlicz norm, 300
weighted H-Orlicz space, 300
well-posed, 167

Young function, 299

DFG

Occupational Toxicants

WILEY-VCH

DFG Deutsche Forschungsgemeinschaft

# Occupational Toxicants

## Critical Data Evaluation for MAK Values and Classification of Carcinogens

## Volume 16

Edited by Helmut Greim

Commission for the Investigation of Health Hazards of Chemical Compounds in the Work Area

WILEY-VCH

Weinheim · New York · Chichester · Brisbane · Singapore · Toronto

Prof. Dr. Helmut Greim
Senatskommission
zur Prüfung gesundheitsschädlicher Arbeitsstoffe
der Deutschen Forschungsgemeinschaft
Technische Universität München
Hohenbachernstraße 15-17
85354 Freising



Translation and layout: Dr. A. E. Wild and J. Handwerker-Sharman

Library of Congress Card No. applied for.

A catalogue record for this book is available from the British Library.

Deutsche Bibliothek Cataloguing-in-Publication Data:
Occupational toxicants : critical data evaluation for MAK values and classification of carcinogens / DFG, Deutsche Forschungsgemeinschaft ; Commission for the Investigation of Health Hazards of Chemical Compounds in the Work Area.
- Weinheim ; New York ; Chichester ; Brisbane ; Singapore ; Toronto : Wiley-VCH
Einheitssacht : Gesundheitsschädliche Arbeitsstoffe <engl.>
Erscheint unregelmäßig. - Aufnahme nach Vol. 1 (1991)
Vol. 1 (1991) -


Printing and Bookbinding: Franz Spiegel Buch GmbH, D-89026 Ulm.
Printed in the Federal Republic of Germany.

# Note

In the present volume reference is frequently made to the various sections of the *List of MAK and BAT Values** (MAK List). The MAK List was introduced and its sections described in detail in the first volume of this series.

* Available from the publisher, WILEY-VCH, D-69451 Weinheim, both in the original German edition and as an English translation

# Contents of Volume 16

Acrolein . . . 1
$n$-Butyl acrylate . . . 35
Ethyl acrylate . . . 41
2-Ethylhexyl acrylate . . . 47
Ethyl methacrylate . . . 51
Glutaraldehyde . . . 59
Hexachlorobenzene . . . 65
2-Hydroxyethyl acrylate . . . 89
Hydroxypropyl acrylate (all isomers) . . . 95
2-Hydroxypropyl methacrylate . . . 105
Lindane . . . 113
Methanol . . . 143
Methyl acrylate . . . 177
Methyl methacrylate . . . 181
Pentaerythritol triacrylate . . . 193
Trimethylolpropane triacrylate . . . 201

**Metal-working fluid components** . . . 207
Citric acid and its alkali metal salts . . . 209
5-Ethyl-3,7-dioxa-1-azabicyclo[3.3.0]octane (EDAO) . . . 231
Hexylene glycol . . . 233
3-Iodo-2-propynyl butylcarbamate . . . 247
Isodecyl oleate . . . 257
2-Octyl-4-isothiazolin-3-one . . . 263
Oleoyl sarcosine . . . 281

**Dusts** . . . 287
Aerosols – Ultrafine aerosol particles, their aggregates and agglomerates . 289
Tantalum . . . 293
Wollastonite . . . 297

Authors of Documents in Volume 16 . . . 317
**Index** . . . 319

# Acrolein

| | |
|---|---|
| **MAK value** | – |
| **Peak limitation** | – |
| **Absorption through the skin** | – |
| **Sensitization** | – |
| **Carcinogenicity (1997)** | **Category 3B** |
| **Prenatal toxicity** | – |
| **Germ cell mutagenicity** | – |
| **BAT value** | – |
| Synonyms | acrylic aldehyde<br>acrylaldehyde<br>acraldehyde |
| Chemical name (CAS) | 2-propenal |
| CAS number | 107-02-8 |
| Structural formula | $H_2C{=}CH{-}CHO$ |
| Molecular formula | $C_3H_4O$ |
| Molecular weight | 56.06 |
| Melting point | –88 to –87°C |
| Boiling point | 52°C |
| Density at 20°C | 0.84 g/cm$^3$ |
| Vapour pressure at 20°C | 290 hPa |
| log $P_{ow}$* | –1.10 to 0.9 |
| **1 ml/m$^3$ (ppm) ≙ 2.33 mg/m$^3$** | **1 mg/m$^3$ ≙ 0.43 ml/m$^3$ (ppm)** |

This document also discusses studies of cyclophosphamide which are relevant to the question of potential carcinogenic effects of acrolein, as metabolic activation of cyclophosphamide releases acrolein.

---

* *n*-octanol/water distribution coefficient

# 1 Toxic Effects and Mode of Action

Because acrolein is highly reactive, its main effects are local toxic effects. The toxicity of acrolein is high after oral and inhalation exposure, and moderate after dermal exposure. Even in highly diluted solution, acrolein causes severe irritation of the skin and mucous membranes, so that undiluted acrolein can be expected to have corrosive effects. In studies with volunteers, acrolein caused eye irritation from concentrations of about 0.15 $ml/m^3$. The substance causes severe sensory irritation in the mouse with $RD_{50}$ values around 2 $ml/m^3$. Cell proliferation and slight histological changes in the nasal epithelium of the rat are observed after exposing the animals three times to 0.25 $ml/m^3$.

In medium and long-term inhalation studies with various species, irritation was observed in the respiratory tract, mainly in the nose, and even hyperplastic and metaplastic changes in the nasal epithelium, depending on the concentration. In a 90-day study with rats, the most sensitive species, histologically detectable low-grade irritation of the nasal epithelium was observed after exposure of the animals to concentrations as low as 0.4 $ml/m^3$. A NOEL (no observed effect level) could not be determined for rats; for hamsters it was 0.4 $ml/m^3$. Only unspecific systemic-toxic effects are seen after oral administration, and the NOEL was found to be 0.5 mg/kg body weight for rats.

Acrolein was not found to have sensitizing effects in the maximization test with guinea pigs.

With oral doses up to the parentally toxic range, in rats acrolein neither has adverse effects on fertility nor does it cause embryotoxicity; in rabbits it causes neither teratogenicity nor embryotoxicity or foetotoxicity. Only at parentally toxic doses are reversible delays in the growth of the offspring seen in rats.

Acrolein reacts *in vitro* with DNA bases to form cyclic adducts. Cyclophosphamide, from which acrolein and other alkylating metabolites are formed, causes *in vivo* the formation of DNA adducts with both acrolein and the other alkylating metabolites. *In vitro*, acrolein was directly genotoxic in various test systems. In *Drosophila*, mutations were induced in both germ cells and somatic cells. Two *in vivo* studies of the mutagenicity and cytogenetics in rats yielded negative results. Carcinogenicity studies with dermal, inhalation and oral exposure of hamsters, rats and mice yielded no evidence of carcinogenic effects. In experiments with animals and after long-term therapeutic use, cyclophosphamide caused haemorrhagic cystitis, cystic toxicity and bladder tumours. The cystic toxicity is caused by acrolein formed during the metabolism of cyclophosphamide. Acrolein is probably also involved in the formation of the bladder tumours.

# 2 Mechanism of Action

As a result of the conjugation of the C–C double bond with the carbonyl group, acrolein is a highly reactive substance which can react with numerous nucleophilic cell components. Acrolein and to a greater extent its glutathione adduct 3-oxopropylglutathione stimulate the release of oxygen radicals (Adams and Klaidman 1993). These findings may explain the severe local irritation and the genotoxic effects. The DNA single strand breaks probably occur, on the one hand, as a result of the known formation of alkali-labile DNA adducts via alkali-labile sites and, on the other hand, as a result of incomplete repair. In V79–HPRT mutants, the exocyclic acrolein adducts found in deoxyguanosine and deoxyadenosine and C–C intrastrand crosslinks may be involved in the production of mutations (Smith *et al.* 1996). In tests with *Drosophila melanogaster* it was shown that acrolein is not only directly mutagenic, but is also activated by CYP450 and glutathione conjugation, and inactivated by enzymes induced with phenobarbital (Barros *et al.* 1994b). For other 2-alkenals it has been demonstrated *in vitro*, however, that the induction of DNA strand breaks correlates inversely with the glutathione level in the cells (Eisenbrand *et al.* 1995). The cytotoxicity of acrolein can be reduced by glutathione, so that conjugation with glutathione is to be regarded primarily as detoxification. Complete detoxification is, however, not achieved by glutathione conjugation, as the conjugates or the resulting mercapturic acid gradually release the alkenal in the absence of glutathione. After systemic distribution, acrolein conjugates can reach tissues with low levels of glutathione and after cleavage produce genotoxic effects there (Eisenbrand *et al.* 1995).

Haemorrhagic cystitis after cyclophosphamide treatment is caused by the acrolein released in the bladder and not by the other alkylating metabolites, which do, however, also reach the bladder (see Section 3.2). This was clearly demonstrated by the toxic effects on the bladder of a cyclophosphamide derivative whose alkyl residue was not chlorinated and thus could not have an alkylating effect; metabolism of the cyclophosphamide derivative did, however, yield acrolein. Conversely, a derivative with chloroalkyl residues which could not release acrolein, had only minimal effects on the bladder (Brock *et al.* 1981a, Cox 1979). 2-Mercaptoethanesulfonate (MESNA) reduces the irritative effects of cyclophosphamide on the bladder (Brock *et al.* 1982, Kunze *et al.* 1984) and the proliferation of the urothelium induced by cyclophosphamide (Kunze *et al.* 1984), and is therefore used together with cyclophosphamide in therapy. MESNA reacts both with the 4-hydroxy-cyclophosphamide initially formed during metabolism and with acrolein, hardly, however, with the alkylating cyclophosphamide metabolites. As a result of the pharmacokinetic properties of MESNA, the reaction takes place only after the substance has passed through the kidneys. The reaction with 4-hydroxy-cyclophosphamide leads to a relatively stable condensation product, so that release of the alkylating metabolites and acrolein is reduced. The reaction with acrolein yields a non-toxic thioether. The MESNA adducts with 4-hydroxy-cyclophosphamide and acrolein have been detected in urine (Brock *et al.* 1979, 1981b). It can therefore be assumed that not only free acrolein but also the concentration of alkylating metabolites in the bladder is reduced by MESNA.

Another mechanism suggested as being responsible for the toxic effects of cyclophosphamide on the bladder is the β-elimination of acrolein from acrolein-mercapturic acids or their directly cytotoxic effects on the bladder epithelium (Ramu *et al.* 1995).

Enzyme polymorphism in man was held responsible for individual differences in metabolism of cyclophosphamide and its toxic effects on the bladder (Fleming *et al.*

1996). In animals of different strains, differences in sensitivity to the toxic effects of cyclophosphamide and acrolein have been observed. After administration of cyclophosphamide, differences between strains were also observed in excretion of acrolein and of substances which release acrolein after acidic hydrolysis (Fraiser and Kehrer 1992).

In experiments with animals, the formation of bladder tumours after doses of cyclophosphamide is prevented by the administration of MESNA (Habs and Schmähl 1983, Schmähl and Habs 1983). It was not described in the studies whether the animals developed cystitis. It is unclear whether the protective effects of MESNA on the bladder are mainly a result of its trapping the acrolein released or its stabilisation of 4-hydroxy-cyclophosphamide. The reaction with acrolein seems, however, to be more important (Brock *et al.* 1979). The tumours are thus either to be attributed to a direct genotoxic effect of acrolein in the bladder or to the strongly irritative effects of acrolein on the bladder epithelium in combination with genotoxic effects. These can be induced by acrolein or the other alkylating metabolites, which also reach the bladder.

# 3 Toxicokinetics and Metabolism

## 3.1 Absorption, distribution, elimination

In the isolated upper respiratory tract of F344 rats, at a flow rate of 200 ml/min and concentrations of 0.86, 4.3 or 8.6 ml/m$^3$, during the last 20 minutes of the 40-minute exposure period 62 %, 38 % and 28 % of an administered acrolein dose was absorbed. Independent of the flow rate and concentration, absorption never reached equilibrium, but decreased slowly in the course of the experiment. During simultaneous exposure to the two substances, acrolein increased the absorption of acetone (Morris 1990).

Seven days after fasting rats had been given single oral doses of 2.5 mg radioactively labelled acrolein per kg body weight, 52 % to 63 % of the radioactivity was eliminated with the urine, 12 % to 15 % with the faeces and about 30 % as $CO_2$ in exhaled air. About 90 % of the dose had been eliminated after 24 hours. Also after repeated doses, the excretion pattern was similar. Of a single dose of 15 mg/kg body weight about 40 % was detected in urine, about 30 % in the faeces and about 30 % as $CO_2$. About 90 % of the dose had been eliminated after 48 hours. Intravenous administration of 2.5 mg/kg body weight led to renal excretion of about 70 % of the radioactivity and exhalation as $CO_2$ of 27 %. Small amounts of radioactivity were found in the organism (Parent *et al.* 1996b; see also the study of Sharp *et al.* 1994, Section 3.2).

## 3.2 Metabolism

### 3.2.1 Acrolein

The main pathway of acrolein metabolism involves enzymatic or non-enzymatic conjugation with glutathione, which, during inhalation by rats, takes place in the first tissues exposed, in the nasal mucosa: after exposure of F344 rats to 2.5 ml/m$^3$ for 3 hours, the glutathione concentration in the respiratory mucosa decreased by about 60 % and at 0.1 ml/m$^3$ by about 20 % (Lam *et al.* 1985, McNulty *et al.* 1984). In rats given acrolein doses of about 5 mg/kg body weight by intraperitoneal injection, the glutathione concentration in the liver was decreased (Cooper *et al.* 1992). In mammalian cells incubated with acrolein, complete glutathione depletion was observed from 10 µM (Dypbukt *et al.* 1993, Eisenbrand *et al.* 1995, Grafström 1990, Grafström *et al.* 1988).

After oral or intravenous administration of acrolein (for conditions see Parent *et al.* 1996b, Section 3.1) the main urinary metabolite was found to be *S*-(hydroxypropyl)-mercapturic acid. Other urinary metabolites were malonic acid, *S*-(2-carboxy-2-hydroxyethyl)-mercapturic acid, *S*-(carboxyethyl)-mercapturic acid and 3-hydroxypropionic acid. Oxalic acid was another important metabolite, but was only detected after oral administration. The metabolite pattern was comparable for all modes of administration. In the urine, 25 % to 77 % of the administered acrolein was recovered. In the faeces an inert homopolymer of acrolein (15 % of the total dose after acrolein doses of 2.5 mg/kg body weight and 30 % after 15 mg/kg body weight) was detected only after oral administration. The authors assumed that oxalic acid and the polymer were formed in the intestinal tract. From this study which was published only as an abstract, quantitative data on the metabolite pattern are not available (Sharp *et al.* 1994). The extent to which the substance was exhaled as $CO_2$ was not given, but is stated in another publication by the same research group (Parent *et al.* 1996b, see Section 3.1).

Further metabolites found *in vitro* were acrylic acid, glycidaldehyde and glyceraldehyde (BUA 1995).

After inhalation of acrolein in concentrations of 10, 18, 33 or 54 ml/m$^3$ for one hour, rats excreted 10.9 %, 13.3 %, 16.7 % and 21.5 % of the absorbed amount as the sum of the two mercapturic acids *S*-(carboxyethyl)-mercapturic acid and *S*-(hydroxypropyl)-mercapturic acid. About 30 % of intraperitoneal doses of 8.9 to 35.7 µmol/kg body weight (0.5–2 mg/kg body weight) was excreted as the mercapturic acids, independently of the dose (Linhart *et al.* 1996).

78.5 % of an oral acrolein dose of 13 mg/kg body weight was excreted by rats as *S*-(hydroxypropyl)-mercapturic acid. Other metabolites were not investigated (Sanduja *et al.* 1989).

The oral or intravenous administration of radioactively labelled acrolein to fasting rats led to the renal excretion of 34 % of the administered radioactivity as *S*-(carboxyethyl)-mercapturic acid and 7 % as *S*-(hydroxypropyl)-mercapturic acid. The other 28 % of the administered radioactivity in the urine was not identified (no further details; Parent *et al.* 1993b; c.f. the study of Sharp *et al.* 1994).

The known pathways of acrolein metabolism are shown in Figure 1.

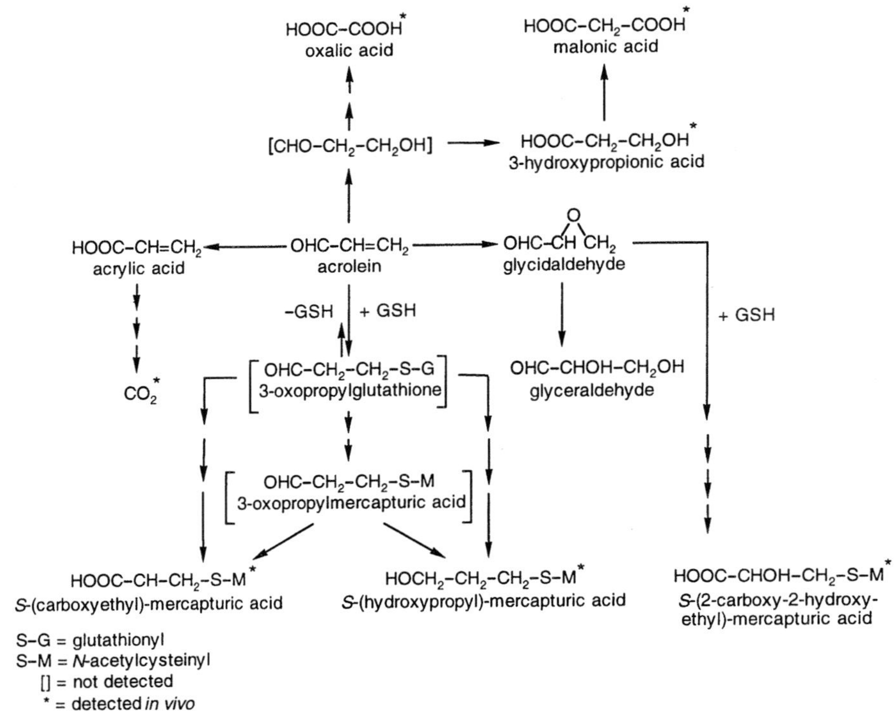


**Figure 1.** Pathways of acrolein metabolism

The following section describes studies of the stability and toxicity of the acrolein-mercapturic acids and the glutathione adduct.

*In vitro*, with increasing pH, small amounts of acrolein were released from 3-oxopropylmercapturic acid *S*-oxide, 3-oxopropylmercapturic acid and 3-oxopropylglutathione, conceivable intermediates of acrolein metabolism which have not yet been detected *in vivo*. From 3-oxopropylmercapturic acid *S*-oxide at pH 8 a maximum of 16 % acrolein was released and at pH 7.4 a maximum of 8 %, from 3-oxopropylmercapturic acid at pH 8 a maximum of 7 %, at pH 7.4 a maximum of 4 %. From 3-oxopropylglutathione and the detected metabolite *S*-(hydroxypropyl)-mercapturic acid at pH 8, a maximum of 2 % was released. The values for pH 7.4 were not given. The amount of acrolein released was evaluated by determining absorption at 210 nm over 120 minutes; after 30–60 minutes there was no further increase in absorption. Very similar values were obtained with the above substances at a pH of 8.0 also when the released acrolein was determined by means of the reaction with semicarbazide. However, when the reaction with *m*-aminophenol was used, 23 % to 75 % acrolein was shown to be released at a pH of 8.0. It was checked that the reaction product came mainly from acrolein and was not the result of the reaction with the aldehyde group of the initial substances. Also here the amount of acrolein released decreased in the order 3-oxopropylmercapturic acid *S*-oxide > 3-oxopropylmercapturic acid > 3-oxopropylglutathione > *S*-(hydroxypropyl)-mercapturic acid. The authors did not explain the quantitative differences between the methods of detection. To investigate the toxic effects on the bladder, the above substances and acrolein were instilled in concentrations of 5 to 20 mM in buffer solutions at pH 7.4 into the bladders of rats. This led to concentration-dependent damage to the bladder, detected as increased levels of haemoglobin in the bladder. The potency of the individual substances correlated with the amount of acrolein released as described above. However, *S*-(hydroxypropyl)-mercapturic acid, the main metabolite found *in vivo*, did not have toxic effects on the bladder (Ramu *et al.* 1995). The release of acrolein from the other metabolites detected *in vivo* was not investigated.

The release of acrolein from 3-oxopropylmercapturic acid was detected indirectly, deduced from its cytotoxicity *in vitro*, which was prevented by glutathione or *N*-acetylcysteine (Perry *et al.* 1995).

3-Oxopropylmercapturic acid *S*-oxide, 3-oxopropylmercapturic acid, and to a lesser extent 3-oxopropylglutathione, were cytotoxic for lung carcinoma cells. The toxicity was decreased by glutathione and increased by glutathione depletion. The structurally similar 4-oxobutylmercapturic acid, which cannot release acrolein, was not toxic; the authors therefore concluded that the toxicity is caused by acrolein and not by the aldehyde group of the mercapturic acids. The corresponding acids, e.g. *S*-(carboxyethyl)-mercapturic acid which was found *in vivo*, were not toxic (Ramu *et al.* 1996).

Under alkaline conditions (pH 8) acrolein was released from 3-oxopropylmercapturic acid *S*-oxide, but not from 3-oxopropylmercapturic acid itself, which suggests that for acrolein to be released the oxidation of 3-oxopropylmercapturic acid to form the sulfoxide is necessary. At lower pH values less acrolein was released. 3-Oxopropylmercapturic acid *S*-oxide and 3-oxopropylmercapturic acid were found to be cytotoxic for LLC-PK1 cells and isolated proximal tubular kidney cells of the rat. With the isolated kidney cells, incubation with the monooxygenase inhibitor methimazole, which inhibits oxidation to form the sulfoxide and thus the release of acrolein, resulted in a decrease in the cytotoxicity of 3-oxopropylmercapturic acid but not of the *S*-oxide. With LLC-PK1 cells, however, the toxicity of 3-oxopropylmercapturic acid could not be reduced by incubation with methimazole. The authors tried to explain these discrepancies in terms of possible differences in the inactivation of methimazole in the two types of cell. They are of the opinion that the release of acrolein from the mercapturic acids is decisive for the toxicity (Hashmi *et al.* 1992).

To summarize, in all studies the release of acrolein was held responsible for the toxicity of the aldehydic mercapturic acids. The results of Hashmi *et al.* (1992) with LLC-PK1 cells show, however, that also the mercapturic acids themselves can be cytotoxic.

To date, the release of significant amounts of acrolein has only been demonstrated from mercapturic acids which are not found *in vivo*. It is, however, unclear whether this could have been detected under the analytical conditions in the other studies. It is unclear whether acrolein is also released under physiological conditions, as the pH of urine is lower than that used in these studies. The available results indicate that the *S*-(hydroxypropyl)-mercapturic acid found in urine is relatively stable and does not have toxic effects on the bladder. The other main metabolite, *S*-(carboxyethyl)-mercapturic acid, is not cytotoxic.

### 3.2.2 Cyclophosphamide

The main metabolite of cyclophosphamide is 4-hydroxy-cyclophosphamide, which is formed in the liver and is in equilibrium with aldophosphamide. Aldophosphamide decomposes spontaneously to form acrolein and phosphoramide mustard, which can be metabolized to nornitrogen mustard. The alkylating agent is phosphoramide mustard and also nornitrogen mustard. In addition dechloroethyl-cyclophosphamide, 4-keto-cyclophosphamide, alcophosphamide and carboxyphosphamide are formed; these substances do not yield alkylating species or are not as strongly alkylating or mutagenic (Chan *et al.* 1994, Ellenberger and Mohn 1977, Pool *et al.* 1988). In addition, chloroethyloxazolidin-2-one (Chan *et al.* 1994), and acrolein, and substances which release acrolein after acid hydrolysis (Al-Rawithi *et al.* 1993, Fraiser and Kehrer 1992) have been detected in urine.

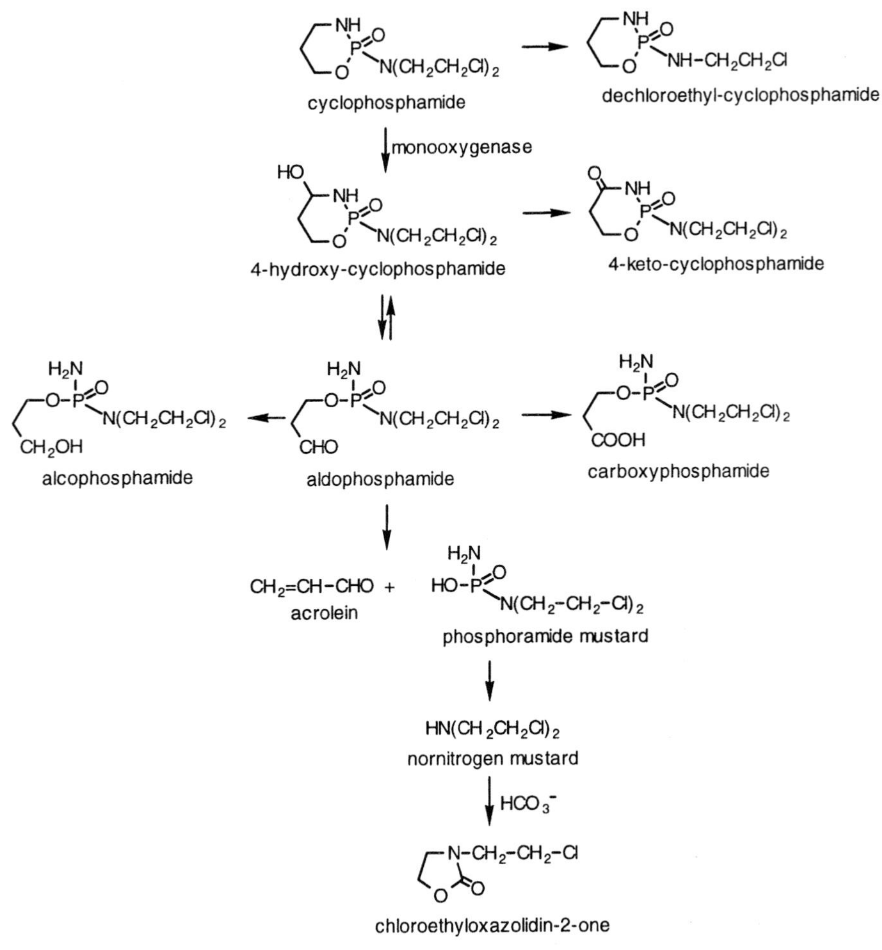


**Figure 2.** Metabolic pathways of cyclophosphamide

The main metabolic pathways of cyclophosphamide are shown in Figure 2 and have been described in detail elsewhere (Chan *et al.* 1994, Sladek 1988).

Children and adolescents given therapeutic doses of cyclophosphamide excreted between 0.7 % and 23.4 % as phosphoramide mustard (Tasso *et al.* 1992). In 6 patients 10.8 % was found unchanged in the urine, 0.5 % as hydroxy-cyclophosphamide and 39.0 % as phosphoramide mustard. Most of it was excreted within the first 24 hours (Chan *et al.* 1994). These patients were not given MESNA. In another publication the excretion of 18.5 % phosphoramide mustard in 24-hour urine was reported (Hadidi *et al.* 1986). These and other studies (Fleming *et al.* 1996, Newman *et al.* 1990) show that high levels of alkylating and mutagenic metabolites are excreted with the urine.

2.6 % of an oral cyclophosphamide dose of 160 mg/kg body weight was excreted by rats as *S*-(hydroxypropyl)-mercapturic acid (Sanduja *et al.* 1989). Rats given intraperitoneal cyclophosphamide doses of 50 mg/kg body weight excreted 9 % of the dose as *S*-(hydroxypropyl)-mercapturic acid. The mercapturic acid was also detected in urine from patients given cyclophosphamide (Alarcon 1976).

# 4 Effects in Man

There are no data available for toxic effects on human reproduction.

## 4.1 Single exposures

Accidental or non-accidental dermal absorption or ingestion of acrolein led to severe local irritation or corrosion at the site of contact. Intoxication after accidental inhalation of acrolein, which in some cases was lethal, has been reported (BUA 1995).

## 4.2 Repeated exposure

Vapour containing formaldehyde and probably acrolein produced during the processing of polyethylene bags caused irritative or allergic reactions in workers (Høvding 1969). As the level of exposure to acrolein is unclear, these results cannot be evaluated.

## 4.3 Local effects on skin and mucous membranes

Accidental or non-accidental exposure of the skin or mucous membranes to acrolein led to severe local irritation or corrosion (BUA 1995).

The irritative effects on the eyes of 9 to 14 test persons were estimated after exposure for 30 minutes via a mask to acrolein concentrations of 0.15, 0.27 or 0.54 ml/m$^3$ by determining the increase in the frequency of eye blinking. From the concentration–effect curve the authors concluded that irritation can be observed from 0.15 ml/m$^3$. In addition it was shown that during the 30-minute exposure to acrolein in concentrations of 0.54 ml/m$^3$, the blinking frequency increased with time to a maximum of 4 times the control value (van Eick 1977).

In a review of toxicological studies with acrolein, it is reported that exposure for 5 minutes to 0.25 ml/m$^3$ led to moderate irritation of the sensory organs. Tolerance to the irritative effects is not thought to develop (Shell 1958).

Another study cannot be evaluated as the acrolein concentration was increased during the experiment (Weber-Tschopp *et al.* 1977).

## 4.4 Allergenic effects

In the literature only one case of an allergic reaction to acrolein is described, in a smoker (Rappaport and Hoffman 1941).

## 4.5 Genotoxicity

Acrolein–DNA adducts were detected immunologically in leukocytes of patients treated with cyclophosphamide (McDiarmid *et al.* 1991). The absence of dose-dependency, the fact that adducts were still found even 4 years after the last cyclophosphamide treatment and the inadequate characterization of the monoclonal antibody call the results into question.

With $^{32}$P-postlabelling, cyclic acrolein–DNA adducts, probably from endogenous sources, were found in control persons (see also Section 5.6).

## 4.6 Carcinogenicity

For acrolein itself there are no studies with man available. Cyclophosphamide was classified by the IARC as carcinogenic in man because its effects include an increase in the incidence of bladder tumours (IARC 1981). It is known from animal experiments (see Section 2) that acrolein is responsible for the toxic effects of cyclophosphamide on the bladder and that the formation of bladder tumours after administration of cyclophosphamide is prevented by trapping the acrolein with MESNA. To clarify the question of whether acrolein may be responsible for the bladder tumours seen after therapy with cyclophosphamide, the epidemiological studies with cyclophosphamide in which the relationship between toxic effects on the bladder and cancer of the bladder was investigated are reviewed below.

In a study with 145 patients treated with cyclophosphamide, non-glomerular haematuria was found to be an indicator of haemorrhagic cystitis and significantly ($p < 0.01$) associated with the development of cancer of the bladder (maximum total observation period 27 years, median 8.5 years). The patients underwent urological examination at 1 to 3 month intervals. Non-glomerular haematuria occurred in 73 of the patients. In 18 patients this was diagnosed only after the end of the cyclophosphamide therapy. 70 % of the patients with haematuria also had cystitis. In all, 7 patients had cancer of the bladder and also microscopic or macroscopic haematuria. In one of the patients, cancer of the bladder was diagnosed as little as 0.6 years after the beginning of the cyclophosphamide therapy, so that a relationship with the therapy is improbable. On the basis of data from an American cancer register, the risk was said to be significantly increased by a factor of 31. Five of the six patients were smokers, while in the total group about 50 % were smokers. None of the 72 patients without haematuria developed cancer of the bladder. The study showed that the risk of developing non-glomerular haematuria increases with the duration of therapy and the total amount of cyclophosphamide administered. In addition, smokers developed non-glomerular haematuria sooner than non-smokers. The smokers developed cancer of the bladder after lower total amounts of cyclophosphamide and shorter therapy periods. The 6 patients with cancer of the bladder each received more than 100 g cyclophosphamide, while in the total group 65 % of all patients were given less than 100 g (Talar-Williams *et al.* 1996).

Of 119 patients who suffered from rheumatoid arthritis and were given cyclophosphamide, 6 of the treated patients developed bladder tumours. Two of them did not have haemorrhagic cystitis. In the similarly sized control group (patients with rheumatoid arthritis and a different treatment) no bladder tumours occurred (Baker *et al.* 1987). In the follow-up study of the same 119 treated patients, 9 bladder tumours were diagnosed (maximum total observation period 25 years). Three of these patients had haemorrhagic cystitis and 2 others microscopic haematuria. It is not described how and how often the patients underwent urological examination. Six of the patients with bladder tumours were heavy smokers. The smoking habits of the remaining treated patients were not described. The patients who developed cancer of the bladder were treated for longer periods and therefore received more cyclophosphamide than the others. In the control group no bladder tumours were found (Radis *et al.* 1995).

Of 471 patients treated with cyclophosphamide because of non-Hodgkin's lymphomas, 33 patients developed haemorrhagic cystitis and 9 a bladder tumour (maximum total observation period 17 years). The patients did not undergo urological examination until clinically visible haematuria developed. Seven of the nine patients with bladder tumours did not have haemorrhagic cystitis. The risk of developing cancer of the bladder relative to that in the general population was higher in patients with haemorrhagic cystitis (with a value of 11.3) than in patients without haemorrhagic cystitis (6.4); the difference could not, however, be shown to be statistically significant. Six of the nine patients were moderate smokers. The number of smokers in the whole group was not stated. The risk of developing haemorrhagic cystitis increased during the treatment period. Eight of the nine patients received more than 80 g cyclophosphamide, including the two patients with haemorrhagic cystitis (Pedersen-Bjergaard *et al.* 1988).

The study of Talar-Williams *et al.* (1996) revealed a clear relationship between non-glomerular microscopic haematuria and bladder tumours after cyclophosphamide therapy. The authors emphasize that the occurrence of non-glomerular microscopic haematuria allowed a subgroup of patients with a high risk of developing cancer of the bladder to be identified. Of particular importance is the fact that none of the patients without haematuria developed cancer of the bladder. The reason why the relationship between cancer of the bladder and haemorrhagic cystitis was not so clear in the other studies may be that the patients were not subjected to such intensive urological examination and microscopic haematuria is a more sensitive indicator of toxic effects on the bladder. It must also be borne in mind that the number of tumour cases is small, which limits the statistical power of the studies. In the study of Radis *et al.* (1995), the average amount of cyclophosphamide administered was higher in the cases of cancer of the bladder than in the other patients. Also in the two other studies the patients with cancer of the bladder had received high amounts of cyclophosphamide. Smoking could also have contributed towards the toxic effects on the bladder and cancer of the bladder. Overall the results indicate that there is a relationship between damage to the bladder such as results from acrolein entering the bladder during cyclophosphamide therapy, and the formation of bladder tumours.

# 5 Animal Experiments and *in vitro* Studies

## 5.1 Acute toxicity

The 4-hour $LC_{50}$ for rats was 21 mg/m$^3$ and for hamsters 58 mg/m$^3$. 30–minute $LC_{50}$ values between 60 (Dow Chemical 1976) and 130 ml/m$^3$ (BUA 1995) have been determined. Oral $LD_{50}$ values of 42 and 46 mg/kg body weight were given for rats (BUA 1995). The dermal $LD_{50}$ for rabbits was found to be between 164 and 1022 mg/kg body weight, depending on the dilution and vehicle. The dermal $LD_{50}$ of undiluted acrolein was 562 mg/kg body weight. For 10 % and 20 % aqueous acrolein solutions, dermal $LD_{50}$ values of 335 and 1022 mg/kg body weight were obtained. The duration of exposure was not given (Shell 1958). At these concentrations skin irritation probably occurred (see Section 5.3), which accelerated the absorption of acrolein. The $RD_{50}$ values were found to be 13.7 ml/m$^3$ for rats and between 1 and 2.9 ml/m$^3$ for mice (BUA 1995).

Focal hyperplasia of the bladder mucosa of rats was observed after intraperitoneal administration of acrolein doses of 25 mg/kg body weight, but not after oral administration of the same dose; with both modes of administration this dose was lethal for many of the animals. It is conceivable that a direct toxic effect of acrolein after diffusion from the peritoneal cavity into the bladder could be responsible for the hyperplasia. Both modes of administration resulted in multifocal eosinophilic degeneration in the liver. No adverse effects in the kidneys were found. After intraperitoneal administration of acrolein doses of 0.5, 1, 2, 4 and 6 mg/kg body weight in a concentration of 0.002 %, the incorporation of $^3$H-methylthymidine into the DNA of the bladder mucosa, an indicator of cell proliferation, was increased 5 days after the last injection only after doses of 6 mg/kg body weight. This may also be a direct toxic effect of acrolein after diffusion from the peritoneal cavity into the bladder (Sakata *et al.* 1989).

The instillation of acrolein solutions in concentrations up to 6 mM into the mouse bladder led to haemorrhage like that seen after systemic administration of cyclophosphamide. The two strains of mouse used differed in sensitivity to both the acrolein and cyclophosphamide treatment (Fraiser and Kehrer 1992).

The acrolein–glutathione adduct 3-oxopropylglutathione was nephrotoxic in rats after intravenous administration of 0.5 or 1 mmol/kg body weight (corresponding to acrolein doses of about 28 or 56 mg/kg body weight). The toxicity was prevented by inhibition of γ-glutamyltranspeptidase (Horvath *et al.* 1992). As glutathione adducts are not transported efficiently into liver cells, intravenous administration reduces the level of oxidation or reduction in the liver whereas when the glutathione adduct is formed intracellularly in the liver it is subject to oxidative or reductive detoxification (Hashmi *et al.* 1992). Therefore the above experiment does not reflect the conditions *in vivo*.

## 5.2 Subacute, subchronic and chronic toxicity

### 5.2.1 Inhalation

Groups of 5–6 rats were exposed via the nose to acrolein concentrations of 0.25, 0.67 or 1.40 ml/m$^3$, 6 hours/day for 3 days, and the nasal epithelium was then examined. From concentrations of 0.25 ml/m$^3$ slight disarrangement, basal cell hyperplasia and an increased number of cells in mitosis were observed in the respiratory and transitional epithelia. The findings were moderately severe in half of the animals exposed to 0.67 ml/m$^3$. The changes in this concentration group were similar in severity to those found under the same conditions in animals exposed to a formaldehyde concentration of 3.2 ml/m$^3$. A concentration-dependent increase in cell proliferation, measured as the expression of the proliferating cell nuclear antigen and the incorporation of BrdU into DNA, was found in the nose in standard plane of section II. In plane of section III both parameters were significantly increased, but no longer clearly concentration-dependent. No effects were found after single exposures (Cassee *et al.* 1996).

A similar study of cell proliferation in the nasal epithelium of rats (Roemer *et al.* 1993) cannot be included in the evaluation because of inadequate methods.

Male rats were exposed to analytically determined acrolein concentrations of 0.17, 1.07 or 2.98 ml/m$^3$ for 6 hours/day on 5 days/week for 3 weeks. No changes were found in the lungs of the animals. Only in the highest concentration group were body weight gains delayed. In the nasal cavity, metaplastic, hyperplastic and dysplastic changes in the respiratory and olfactory epithelia were found mainly in the septum and in the frontal and ventral areas. In the dorsal area, there was degeneration and, in some cases, even complete disintegration of the respiratory and olfactory epithelia and inflammation with leukocyte infiltration. The NOEL was found to be 1.07 ml/m$^3$ (Leach *et al.* 1987).

Groups of 10 male rats were exposed to acrolein in concentrations of 0.15, 0.51 or 1.52 mg/m$^3$, 24 hours/day for 61 days. From 0.51 mg/m$^3$ (0.22 ml/m$^3$) weight loss, epithelial proliferation and migration of eosinophils into the bronchi were observed.

The highest concentration resulted in bronchiolitis, bronchopneumonia, and myocardial and hepatic dystrophy. The NOEL was found to be 0.15 mg/m$^3$ (0.065 ml/m$^3$). The extent of the histopathological examination is unclear; the nasal mucosa of the animals was probably not examined (Gusev *et al.* 1966).

Groups of 12 Wistar rats, 20 Syrian hamsters and 4 rabbits were exposed to 0.4, 1.4 or 4.9 ml/m$^3$, 6 hours/day on 5 days/week for 13 weeks. In the rats exposed to concentrations from 0.4 ml/m$^3$, metaplasia of the nasal epithelium and leukocyte infiltration were found. Histopathological examination did not reveal any changes in the lung. For hamsters and rabbits the NOEL was 0.4 ml/m$^3$. At concentrations of 1.4 ml/m$^3$, metaplasia and leukocyte infiltration in the nasal epithelium were detected in hamsters, while in rabbits there was only a reduction in body weight gains. After exposure to concentrations of 4.9 ml/m$^3$, 6 of the 12 rats died and in all species metaplasia was observed in the nose, larynx and trachea. In addition, in rats and rabbits the bronchi, bronchioles and alveoli were inflamed. In no species was occult blood found in the urine, or lesions in the bladder, kidneys or adrenal glands (Feron *et al.* 1978).

In three other studies, rats were exposed to concentrations of 0.4, 1.4 or 4.0 ml/m$^3$, 6 hours/day on 5 days/week for 62 weeks. These studies are described below.

With F344 rats, concentrations of 0.4 ml/m$^3$ were not found to have adverse effects; the nose, however, was not investigated. The higher concentrations led to inflammation of the lower respiratory tract and increased levels of collagen in the lungs (Kutzman *et al.* 1985). The diffusing capacity for CO in the lung was determined in groups of 24 F344 rats per concentration; from 0.4 ml/m$^3$ it was increased. At 0.4 ml/m$^3$ the flow–volume dynamics were enhanced without any histological findings in the lung. At 1.4 ml/m$^3$ there were bronchial lesions in 10 % of the animals. At 4 ml/m$^3$ in the 9 surviving animals the flow–volume curve was depressed, the lung volumes increased and severe bronchial lesions were found. Only the lungs were examined histologically (Costa *et al.* 1986). Groups of 5 female Dahl rats resistant to salt-induced hypertension (DR) and 5 sensitive to salt-induced hypertension (DS) underwent histological examination 7 days after the end of exposure. In the lungs of 8/10 DS rats and 5/10 DR rats exposed to concentrations of 0.4 and 1.4 ml/m$^3$ slight epithelial hyperplasia, in some cases with squamous metaplasia, was found. Aggregations of intra-alveolar macrophages with foamy cytoplasm in the vicinity of damaged terminal bronchioles were observed in 7/10 DS and 5/10 DR rats. The incidences were, however, not given separately for the two concentration groups. At concentrations of 4 ml/m$^3$ the incidence of such findings was increased, and mortality, delayed body weight gains, oedema of the lung, and increased levels of protein, hydroxyproline and elastin in the lung were observed. There was a trend towards increase in relative lung weights in all concentration groups, but significant levels of about twice those found in the controls were reached only in the 4 ml/m$^3$ group. The nasal turbinates of the surviving animals were not found to be damaged; the extent of the examination is, however, not known. The nasal turbinates of the animals that died were not available for examination (Kutzman *et al.* 1984).

Groups of rats, guinea pigs, dogs and monkeys were exposed for eight hours to concentrations of 0.7 or 3.7 ml/m$^3$ on 5 days a week for 6 weeks; 3.7 ml/m$^3$ led to visible irritation and reduced body weight gains. Inflammatory changes and occasionally emphysema were found in the lungs of the animals from concentrations of 0.7 ml/m$^3$; the effects were more severe in dogs and monkeys. In the animals exposed to 3.7 ml/m$^3$ treatment-related squamous metaplasia and basal cell hyperplasia were found in the trachea of dogs and monkeys, bronchopneumonia in dogs and obliterating bronchiolitis in monkeys. In all species unspecific inflammation was observed in the lungs, liver and kidneys. The nasal mucosa of the animals was not examined (Lyon *et al.* 1970).

Summary: in a 90-day study with intermittent exposure, no NOEL was determined in the most sensitive species, the rat, for histopathological changes in the nose. The LOEL (lowest observed effect level) was found to be 0.4 ml/m$^3$. For hamsters and rabbits this concentration was the NOEL. The evidence of bronchial damage in rats found in one study from concentrations of 0.4 ml/m$^3$ could not be confirmed in two other studies with larger numbers of animals.

### 5.2.2 Ingestion

The carcinogenicity studies with oral doses yielded for rats and mice NOELs in the range of 0.5–2 mg/kg body weight and day (see Section 5.7).

## 5.3 Local effects on skin and mucous membranes

Even 1 % acrolein solutions cause severe irritation of the rabbit skin and eye, so that corrosion is to be expected with the undiluted substance. Eye irritation was observed in various inhalation studies, but not quantified. Habituation to the irritative effects on the respiratory tract was reported (BUA 1995).

## 5.4 Allergenic effects

A Magnusson and Kligman maximization test yielded negative results (BUA 1995).

## 5.5 Reproductive and developmental toxicity

Groups of 30 male and female SD rats were given gavage doses of acrolein of 1, 3 or 6 mg/kg body weight in 5 ml water/kg body weight for 70 days before mating and during the 14-day to 21-day cohabitation period. The female animals received acrolein also during gestation and lactation. Groups of 40 $F_1$ animals were given the same acrolein doses as the parent animals from day 22 of life. The $F_2$ animals were not treated and were examined on day 21 of life. Mortality was significantly increased in the high dose group for male and female $F_0$ and $F_1$ animals; aspiration of the test substance could not be excluded as the cause of death. In many animals of the 6 mg/kg group erosion, hyperplasia and hyperkeratosis was found in the glandular mucosa of the stomach or in the forestomach. In 4 female $F_0$ animals of the 3 mg/kg group hyperplasia and hyperkeratosis were observed in the forestomach. The reproductive organs of the $F_0$ and $F_1$ animals were unchanged. The following parameters did not differ in the treated animals and controls: mating behaviour, fertility index, incidence of abortion, relative number of live births, viability, growth, morphology of the young animals, number of implantations, duration of pregnancy, number of dams with abortions or stillbirths, lactation index, litter size and sex ratio. The only effect was reduced body weights after 6 mg/kg body weight in the male $F_1$ animals from day 1 to day 85 of life and in the female $F_1$ animals from day 1 to day 22 of life (Parent *et al.* 1992a). Even at the parentally toxic dose of 3 mg/kg, no impairment of fertility occurred and after 6 mg/kg body weight only delayed growth, sometimes reversible, was observed in the offspring.

Groups of 14 to 17 rabbits were given acrolein doses of 0.1, 0.75 or 2 mg/kg body weight from days 7 to 19 of gestation. Except for temporarily reduced food consumption and body weight gains during the first three days of treatment, no effects were observed. The numbers of abortions and implantations, the *corpus luteum* count, the relative number of live foetuses, the litter size, sex ratio, the number of resorptions, the number of skeletal and visceral malformations, and the ossification disturbances did not differ from those found in the control animals. In a dose-finding study, maternally toxic effects and mortality were caused by 4 and 6 mg/kg body weight (Parent *et al.* 1993a). Doses up to just below the maternally toxic range were neither teratogenic nor embryotoxic or foetotoxic.

As cyclophosphamide can have teratogenic effects under certain conditions, the following studies of cyclophosphamide are relevant in a review of the reproductive toxicity of acrolein. Teratogenic effects of cyclophosphamide have been described for the mouse, rat, rabbit and man (Beauchamp *et al.* 1985, Greenberg and Tanaka 1964). In *in vitro* studies it was shown, however, that metabolism via the microsomal monooxygenases of the liver is necessary before cyclophosphamide can have teratogenic effects.

Studies of the teratogenic effects of the two cyclophosphamide metabolites, phosphoramide mustard and acrolein, have been carried out. After intra-amniotic injection (day 13 after mating), teratogenic effects were described in rats for both substances. In these studies acrolein induced a range of malformations which were qualitatively like those induced by cyclophosphamide, but acrolein was found to have more pronounced effects than cyclophosphamide or phosphoramide mustard. After acrolein doses of 1 µg/foetus, malformations were detected in 85 % of the living foetuses, after 0.1 µg, no embryotoxic or teratogenic effects were observed. After treatment with phosphoramide mustard only part of the range of malformations described for cyclophosphamide was observed. After phosphoramide mustard doses of 1 µg/foetus no changes were detected (Hales 1982).

In an *in vitro* study with 10-day-old rat embryos, however, phosphoramide mustard was described as the "ultimate" teratogenic metabolite of cyclophosphamide. The authors observed the same effects on the growth and development of the embryos as with activated cyclophosphamide. No teratogenic effects were induced by treatment with acrolein (5 µg/ml) (Mirkes *et al.* 1981).

In a study of the toxic effects after inhalation, 3 male rats and 21 females were exposed to average acrolein concentrations of 0.56 ml/m$^3$ from four days before mating until the end of the gestation period. The number of pregnancies and the weights of the

foetuses were unaffected. As the exact data are not given in the study it cannot be included in the present evaluation (Bouley *et al.* 1976).

Acrolein doses of 3, 4.5 and 6 mg/kg body weight administered by intravenous injection on day 9 of gestation were lethal for rabbit embryos; the numbers of deaths increased with the dose. All the administered doses were also toxic (in some cases lethal) for the dams (Claussen *et al.* 1980).

In addition to these *in vivo* studies with acrolein, results are described from various *in vitro* studies.

Rat embryos (10.5 days old) were exposed to acrolein concentrations of 50 to 250 µM for 48 hours. The authors found the substance to be lethal to the embryos in a narrow dose range, and observed dose-dependent effects on the growth of the embryos, but no malformations (Schmid *et al.* 1981).

In a similar study, 10-day-old rat embryos were treated in a serum-free medium with acrolein concentrations of 0.005 to 5 µg/ml. From 0.5 µg/ml, dose-dependent effects on the development and survival of the embryos were observed. The malformations characteristic of cyclophosphamide were not observed (Mirkes *et al.* 1984).

The treatment of limb bud cultures of the mouse with acrolein concentrations of 3 to 10 µg/ml produced a significant, dose-dependent impairment of the differentiation of the limb buds, evidence that acrolein has embryotoxic potential. In the opinion of the authors the results could, however, also be interpreted as cytotoxic effects (Stahlmann *et al.* 1985).

## 5.6 Genotoxicity

In addition to studies with acrolein, also studies with cyclophosphamide are discussed below.

The genotoxicity of cyclophosphamide after metabolic activation *in vitro* and *in vivo* has been demonstrated in practically all the systems investigated. Only relevant studies are therefore discussed here, such as investigations of DNA adducts, studies on the influence of MESNA on the mutagenicity of cyclophosphamide and studies in which the genotoxicity of acrolein was compared with that of cyclophosphamide and its other metabolites.

### 5.6.1 *In vitro*

#### DNA adducts

Under physiological conditions *in vitro*, acrolein forms cyclic adducts with isolated DNA or nucleotides (BUA 1995). In *Salmonella typhimurium* TA100 and TA104 incubated with acrolein, DNA adducts were found and characterized as premutagenic lesions (Foiles *et al.* 1989). DNA adducts were found also in CHO cells (a cell line derived from Chinese hamster ovary) after incubation with acrolein (Foiles *et al.* 1990).

*In vitro*, cyclophosphamide caused monoalkylation of the N7 of guanine, and formation of DNA–DNA crosslinks and of phosphoester adducts resulting from reactions of DNA with metabolites such as phosphoramide mustard and nornitrogen mustard. N7 alkylation was twice as frequent as the formation of phosphoester adducts (Maccubbin *et al.* 1991).

Comparative *in vitro* studies of the frequency of monoalkyl adducts and other conceivable cyclic adducts produced by the acrolein released from cyclophosphamide are not available.

### Strand breaks and crosslinks

Acrolein in concentrations from 10 μM induces DNA single strand breaks and DNA–protein crosslinks in mammalian cells *in vitro*; the concentrations effective in the various cell types are inversely correlated with the glutathione level in the cells (Dypbukt *et al.* 1993, Eisenbrand *et al.* 1995, Grafström *et al.* 1988, Grafström 1990). Glutathione is essential for DNA repair (Lai *et al.* 1989). As a result of rapid non-enzymatic glutathione depletion at acrolein concentrations from about 10 μM, even at this concentration DNA repair is almost completely inhibited (Dypbukt *et al.* 1993). In V79 cells, abasic sites (Smith *et al.* 1994) and C–C intrastrand crosslinks (Smith *et al.* 1995) are produced by acrolein.

Acrolein is more effective than formaldehyde in the induction of strand breaks, cytotoxicity, the depletion of glutathione, the increasing of intracellular calcium levels and the induction of terminal differentiation. If the effects are standardized to the same cytotoxicity, formaldehyde, with the exception of the effects on terminal differentiation, is more potent and of greater genotoxicity than acrolein (Grafström 1990).

### Tests in microbial systems

Acrolein was found to be mutagenic in the *Salmonella* mutagenicity test (Table 1) with strain TA100 after preincubation (Foiles *et al.* 1989, Lutz *et al.* 1982, Parent *et al.* 1996b, Waegemaekers and Bensink 1984). The addition of S9 reduces or prevents the mutagenic effects (Lutz *et al.* 1982, Parent *et al.* 1996b). In the standard plate test (Bartsch *et al.* 1980, Basu and Marnett 1984, Lijinsky and Andrews 1980, Loquet *et al.* 1981, Rosen *et al.* 1980) and the spot test (Florin *et al.* 1980) no mutagenic effects could be found in strain TA100. Because of their unusual concentration dependency, the mutagenic effects observed in strain TA98 are questionable (Lijinsky and Andrews 1980) or marginal (Parent *et al.* 1996b).

In the *Salmonella* mutagenicity test with TA1535 and S9, the mutagenicity of cyclophosphamide was not reduced by the simultaneous presence of MESNA. In a second test, rats were given cyclophosphamide doses of 30 mg/kg body weight (0.128 mmol/kg body weight) without and with MESNA (0.09 mmol/kg body weight), and the urine of the animals was tested in strain TA1535. Neither with nor without S9 was the mutagenicity of the urine of the animals also treated with MESNA lower than that after cyclophosphamide treatment alone (about 30–50 times the spontaneous number of revertants) (Lähdetie *et al.* 1990).

In the *Salmonella* mutagenicity test with TA1535 without S9, 4-hydroperoxy-cyclophosphamide, nornitrogen mustard and phosphoramide mustard were found to be clearly mutagenic, while acrolein at the same concentrations (7.8–62.5 nmol/25 ml) was inactive. In TA1535 and TA100, the urine of rats given cyclophosphamide with MESNA was of lower mutagenicity than that from rats without additional MESNA treatment (Pool *et al.* 1988).

Negative results were obtained in tests for forwards mutation (Ellenberger and Mohn 1977) and reversion in *Escherichia coli* (Ellenberger and Mohn 1977, Parent *et al.* 1996b). Other results for the mutagenicity in bacteria and yeasts could not be included in the evaluation because of inadequate documentation (Hemminki *et al.* 1980, Jung *et al.* 1992, Khudoley *et al.* 1987, Marnett *et al.* 1985), test systems that have not been validated (Izard 1973) or low purity of the acrolein used (Haworth *et al.* 1983).

**Table 1.** Mutagenicity tests with acrolein in microbial test systems

| Strain; test system | Lowest effective or highest tested concentration; toxicity | Result –S9/+S9 | References |
|---|---|---|---|
| TA100, TA104; preincubation (20 mins, then the addition of GSH) | 4 mM; at 13 mM reduced by GSH | +/n.d. | Foiles *et al.* 1989 |
| TA100; preincubation (20 mins); | 50 µg/plate; > 50 µg/plate | –/+ | Haworth *et al.* 1983 |
| TA98, TA1535, TA1537; preincubation (20 mins) | 100 µg/plate; > 50 µg/plate | –/– | |
| TA100; preincubation (90 mins) | about 20 µM in the incubation mixture; ≥ about 20 µM; at 75 µM 20 % survival | +/– | Lutz *et al.* 1982 |
| TA 104; preincubation (20 mins, then the addition of GSH) | 0.35 µmol/plate; > 0.35 µmol/plate | +/n.d. | Marnett *et al.* 1985 |
| TA100, TA98; preincubation (20 mins); | 6.7 µg/plate; ≥ 20 µg/plate | +/+ | Parent *et al.* 1996a |
| TA102, TA104, TA1535, TA1537, TA1538; preincubation (20 mins) | 100 µg/plate; ≥ 33 µg/plate | –/– | |
| TA 100; preincubation (90 mins) | 90 µM; >130 µM | +/n.d. | Waegemaekers and Bensink 1984 |
| TA98, TA100, TA1535, TA1538; standard plate test | 250 µM; not specified | n.d./– | Bartsch *et al.* 1980 |
| hisD3052; standard plate test | not specified; not specified | n.d./– | Basu and Marnett 1984 |

**Table 1.** continued

| Strain; test system | Lowest effective or highest tested concentration; toxicity | Result −S9/+S9 | References |
|---|---|---|---|
| TA1535; standard plate test | not specified; 10 µg/plate | –/+(?) | Hales 1982 |
| TA102; standard plate test | 5000 µg/plate; not specified | –/– | Jung *et al.* 1992 |
| TA98, TA100; standard plate test | not specified; not specified | –/+ | Khudoley *et al.* 1987 |
| TA98; standard plate test; | 20 µg/plate; >100 µg/plate | +/– | Lijinsky and Andrews 1980 |
| TA100, TA1535, TA1537, TA1538; standard plate test | not specified; not specified | –/– | |
| TA98, TA100, TA1535; standard plate test | 50 µg/plate; not specified | –/– | Loquet *et al.* 1981 |
| TA100; standard plate test | not specified; not specified | –/– | Lutz *et al.* 1982 |
| TA100; standard plate test | not specified; ≥ 1.1 µg/plate | –/– | Rosen *et al.* 1980 |
| TA98, TA100, TA1535, TA1537; spot test | 150 µg/plate; 150 µg/plate | –/– | Florin *et al.* 1980 |

n.d. not determined, (?) questionably positive

### Tests in mammalian cells

Acrolein can cause HPRT mutations in mammalian cells if suitable experimental parameters are selected. Mutagenicity has been detected to date only after treatment of the cells in a serum-free medium and, as a result of its extreme cytotoxicity, with acrolein concentrations of 2 µM or less (Curren *et al.* 1988, Smith *et al.* 1990). Incubation with acrolein in concentrations up to 120 µM with and without added S9, in the presence of 5 % foetal calf serum did not, however, produce HPRT mutations (Parent *et al.* 1991a). The potentially mutagenic damage to the DNA can apparently be repaired well by normal cells (Curren *et al.* 1988).

In tests for SCE (sister chromatid exchange) in CHO cells, weak concentration-dependent effects were seen with acrolein concentrations in the range from 0.1 to 1 µg/ml (1.8–18 µM) only without S9 (Au *et al.* 1980) and more clearly positive results, but with a reduced mitotic index, in the range from 10 to 100 µM (Galloway *et al.* 1987). Also in human leukocyte cultures, acrolein yielded positive results in the concentration range from 0.0014 to 40 µM. Fourteen times more SCE is induced by 3.39 µM phosphoramide mustard than by 20 µM acrolein. 40 µM acrolein was found to be cytotoxic, so that mitosis could not be evaluated. MESNA reduced the incidence of SCE and the cytotoxicity induced by acrolein (Wilmer *et al.* 1986).

Weak clastogenic effects were detected in a study with S9 and 40 µM acrolein (Au *et al.* 1980). In three other studies acrolein was not found to be clastogenic (Galloway *et al.* 1987, Wilmer *et al.* 1986, 1990). Phosphoramide mustard, however, was clearly clastogenic (Wilmer *et al.* 1986). Thus the metabolite acrolein plays only a minor role in the clastogenicity of cyclophosphamide (Wilmer *et al.* 1986, 1990).

In summary, acrolein can be shown to be mutagenic but the mutagenicity is slight compared with that of other cyclophosphamide metabolites.

**Other effects**

Increased terminal differentiation was observed in cultured human bronchial epithelial cells from 1 µM acrolein, an effect which is induced in these cells also by the skin tumour promoters 12-*O*-tetradecanoylphorbol-13-acetate (TPA), aplysiatoxin and teleocidin B (Grafström *et al.* 1988).

DNA cytosin-5 methyltransferase, which plays a role in the regulation of gene expression, is inhibited by acrolein ($K_i$: 6.7 µM) (Cox *et al.* 1988).

### 5.6.2 *In vivo*

DNA–protein crosslinks were induced by the incubation of a homogenate of rat nasal mucosa with acrolein. DNA-protein crosslinks were, however, not detected in a homogenate of rat nasal mucosa after the animals had inhaled acrolein concentrations of 2 ml/m$^3$ for 6 hours. The authors assume that acrolein is inactivated *in vivo* by reaction with SH groups (Lam *et al.* 1985).

**Acrolein-DNA adducts**

Studies of DNA adduct formation after administration of acrolein *in vivo* are not available.

Cyclic acrolein-DNA adducts (mainly propanodeoxyguanosine and, to a lesser extent, ethenodeoxyguanosine) were detected using $^{32}$P-postlabelling in DNA from liver, mamma and leukocytes in man and in organs from untreated laboratory animals (mouse: liver, skin; rat: liver, lung, kidneys, brain, mamma, prostate, intestine and leukocytes) (Nath and Chung 1994, Nath *et al.* 1994, 1996, Swenberg *et al.* 1995). These adducts, like those with crotonaldehyde, are possibly the result of the lipid peroxidation of cell membranes. The level of propano adducts with acrolein and crotonaldehyde (only given as the sum) and of etheno adducts, which are formed with both acrolein and crotonaldehyde, was 1 adduct per $10^6$–$10^7$ guanines (Chung *et al.* 1996). The level of the most important acrolein propano adduct in the liver DNA in man was 0.03–0.74, in rats 0.01–0.04 and in mice 0.23–0.67 µmol/mol guanine (Nath and Chung 1994).

By means of $^{32}$P-postlabelling, hydroxypropanodeoxyguanosine–DNA adducts were detected in the lymphocytes of a dog treated with cyclophosphamide in doses of 6.6 mg/kg body weight, but not in untreated animals. The adducts were not quantified (Wilson *et al.* 1991).

In mice treated with cyclophosphamide, two adducts with deoxyguanosine were found in the hepatic DNA at levels of 1 adduct per 2.7–4.1 × $10^7$ normal nucleotides (37.5 and 24.1 nmol/mol). Control animals were found to have adduct levels of 0.4 and 0.6 nmol/mol. The adducts had the same $R_f$ values in two-dimensional chromatography as the adducts obtained *in vitro* with acrolein and deoxyguanosine monophosphate (Maccubbin *et al.* 1990).

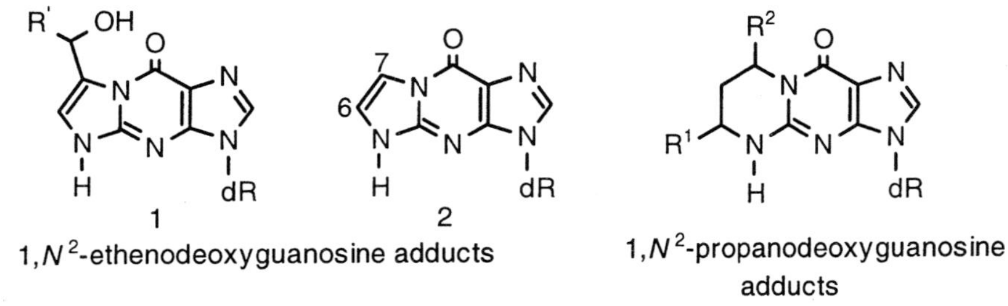

**Figure 3.** Cyclic adducts of acrolein or crotonaldehyde with deoxyguanosine.

Acrolein: R' = H; $R^1$ = OH and $R^2$ = H; $R^1$ = H and $R^2$ = OH
Crotonaldehyde: R' = $CH_3$; $R^1$ = OH and $R^2$ = $CH_3$; $R^1$ = $CH_3$ and $R^2$ = OH
dR = deoxyribose
Adduct 2 is formed by base-catalysed cleavage from adduct 1 (Chung *et al.* 1996).

### DNA adducts found after administration of cyclophosphamide

After administration of cyclophosphamide, 0.2 to 0.4 guanine–N7 adducts per $10^6$ nucleotides were found in the DNA of mice and rats by means of $^{32}$P-postlabeling (Maccubbin *et al.* 1991). In the DNA of rats given cyclophosphamide which was radioactively labelled in the chloroethyl side chain, *N*-(2-hydroxyethyl)-*N*-[2-(7-guaninyl)-ethyl]amine was found by HPLC to be the main adduct.

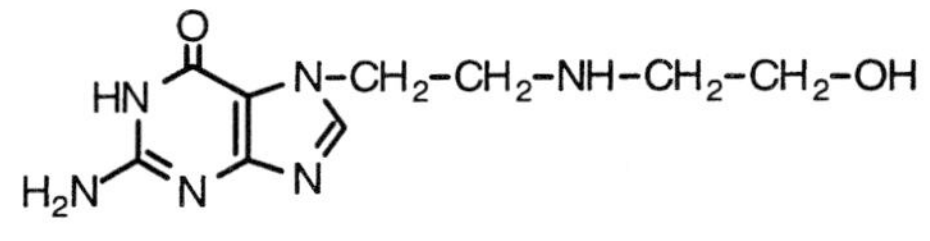

The highest levels of radioactivity in the DNA were found in the bladders of the animals; 0.2 alkyl adducts were detected per $10^6$ nucleotides. Adducts with acrolein were not investigated (Benson *et al.* 1988).

### *Drosophila* tests

Acrolein (concentration not specified, yet lethal for at least 75 % of the larvae and eggs) induced in *Drosophila melanogaster* X-chromosomal recessive lethal mutations (SLRL, sex-linked recessive lethal) in 2.23 % of the male offspring compared with an incidence of 0.19 % in the controls (Rapoport 1948). The "eye spot" test with acrolein concentrations of 500 to 2000 mg/kg diet, yielded a positive result at 500 mg/kg. The higher concentrations were toxic (Vogel and Nivard 1993). Two tests ("wing spot" and "eye spot") for somatic mutations with acrolein in concentrations of 5 to 20 mM (280 to 1120 mg/kg) in the diet yielded reproducibly positive, but not concentration-dependent results. An SLRL test with injection of 0.2 µl per animal of an acrolein solution in concentrations of 2 to 7 mM yielded concentration-dependent positive results. Another germ cell mutation test (SCL, sex chromosome loss) yielded negative results (Sierra *et al.* 1991). Adult *Drosophila* males were injected with 0.2 µl of an acrolein solution in concentrations of 2 to 5 mM and were mated with normal or repair-deficient females. Mating with normal females revealed the concentration-dependent induction of SLRL mutations. Mating with excision repair-deficient females revealed an even higher increase in mutations. After mating with females with mitochondrial DNase deficiency, the incidence of mutations was less than that seen after mating with normal females. The authors conclude from their results that acrolein lesions are repaired by the excision repair mechanism (Barros *et al.* 1994a). Adult *Drosophila* males were fed for 19 hours with phenobarbital (to stimulate xenobiotic metabolism), with phenylimidazole or iproniazid (CYP450 inhibitors) or with diethyl maleate (for glutathione depletion), before they were fed or injected with acrolein solutions. The results of the SLRL tests show that acrolein is not only a directly acting mutagen, but is also activated by CYP450 and glutathione conjugation. On the other hand, acrolein was also inactivated by the enzymes induced with phenobarbital (Barros *et al.* 1994b). Also negative results in the germ cell tests with feeding or injection have been published (Zimmering *et al.* 1985, 1989).

**Chromosomal aberration and mutation tests**

Groups of 5 to 7 male mice were given doses of acrolein of 1.5 or 2.2 mg/kg body weight by intraperitoneal injection and mated for 8 weeks in a mating ratio of 1:3 with untreated virgin females. Acrolein did not induce dominant lethal mutations in the germ cells. The study does not, however, meet present day standards (Epstein and Shafner 1968, Epstein *et al.* 1972).

An *in vivo* chromosomal aberration study with rats yielded negative results after intraperitoneal administration of acrolein doses of 1, 2 or 4.1 mg/kg body weight. Doses of 8.2 mg/kg body weight were lethal (BUA 1995).

## 5.7 Carcinogenicity

### 5.7.1 Initiation promotion studies

A 0.5 % solution of acrolein in acetone was applied to the skin of 15 mice for 10 weeks in doses of 0.3 ml until a total dose of 12.6 mg/animal had been applied; 25 days after the first application, croton oil was applied 18 times at weekly intervals. Three tumours were found in 2/15 animals from the acrolein group, while treatment with croton oil alone caused 4 tumours in 4/19 animals (Salaman and Roe 1956). According to the test protocol, application was not occlusive (Roe and Salaman 1955).

To investigate the initiating potential of acrolein on the bladder, groups of 30 rats were given intraperitoneal doses of acrolein of 2 mg/kg body weight twice a week for 6 weeks. *N*-[4-(5-Nitro-2-furyl)-2-thiazolyl]formamide (FANFT) was used as a positive control, and water as a negative control. During the promotion phase 3 % uracil was administered to each group for 20 weeks (formation of bladder stones); after this period the animals were given standard diet for 6 weeks. A fourth group was given only acrolein during the initiation period, followed by standard diet for 26 weeks. In all, the animals treated with acrolein received 3.3 mg acrolein. Toxic effects developed during the administration of acrolein so that the treatment had to be interrupted. The animals developed peritonitis. In the positive control group, 9 papillomas and 21 carcinomas of the bladder were diagnosed. In the group given acrolein and uracil, 18 animals developed papillomas while 8 of the animals given only uracil developed papillomas. In both groups one carcinoma was found. From this significant increase in the incidence of papillomas the authors concluded that acrolein has initiating potential (Cohen *et al.* 1992).

In the same study the tumour-promoting potential of acrolein was investigated. In the initiation phase, acrolein or FANFT was administered as described above. For promotion the animals were given 24 intraperitoneal injections of acrolein during the next 15 weeks. According to the protocol a promotion phase of 100 weeks was planned for this test. As a result of the general toxicity, the dose and the frequency of administration had to be reduced even during the initiation phase. The animals remained untreated for a further 32 weeks. In all, the doses were between 8.3 mg acrolein per animal (initiation with FANFT) and 11.9 mg acrolein per animal (initiation with acrolein). In half of the animals hyperplasia of the bladder epithelium was observed. Also the administration of FANFT without a promoter led to a comparable incidence of hyperplasia of the bladder epithelium. Neither with FANFT nor with acrolein as initiator did bladder papillomas or carcinomas develop (Cohen *et al.* 1992). It is questionable whether acrolein has been found to have initiating effects, as the irritant effects of acrolein during the 6-week initiation phase could also have contributed to the increased incidence of papillomas. It is unclear whether acrolein or other metabolites enter the bladder during repeated intraperitoneal administration or whether acrolein diffuses directly from the peritoneal cavity into the bladder and there produces the observed practically irreversible hyperplasia of the bladder mucosa. It has been demonstrated that (mechanical) irritation of the bladder can result in papillomas.

### 5.7.2 Long-term studies

Groups of 20 female rats were exposed to acrolein concentrations of 8 ml/m$^3$, one hour a day on 5 days a week for 10 or 18 months. During the first acrolein inhalation the animals developed severe symptoms of respiratory irritation, later habituation developed. The extensive histological examination which included the bladder, adrenal gland and nose did not reveal an increased tumour incidence. The body weights of the animals were not changed relative to the control value (Le Bouffant *et al.* 1980). There are no data available for survival, the number of animals investigated or the incidence of non-neoplastic histological changes in the nasal mucosa.

Groups of 36 male and female hamsters were exposed to acrolein concentrations of 4 ml/m$^3$, 7 hours a day for 52 weeks. Half of the animals received physiological saline by tracheal instillation once a week. Six animals were examined after 52 weeks, the rest 81 weeks after the end of the experiment. During the second week of the experiment the animals became adapted to the irritative effects of acrolein. After 52 weeks increased lung weights, irritative effects and metaplasia of the nasal mucosa were observed. Body weight gains were delayed, but survival was unchanged. After the last exposure all of the male animals and 28 of the females were still alive. After week 80 of the experiment, 23 males and 17 females were still alive. Thickening of the submucosa of the nasal cavity and exudate in the lumen were still detected in 20 % of the animals in the follow-up period. Lesions were not found in the bladder or adrenal glands of the animals. Treatment-related tumours were not found (Feron and Kruysse 1977). The treatment period was too short to be able to exclude carcinogenic effects. A clear local carcinogenic potential would probably, however, have been detected.

In addition, the influence of acrolein on the respiratory tract tumours caused by benzo[*a*]pyrene or diethylnitrosamine was investigated. Only the number of respiratory tract tumours caused by benzo[*a*]pyrene was increased, but not significantly, after simultaneous treatment with acrolein. The authors did not conclude that acrolein is cocarcinogenic (Feron and Kruysse 1977).

In a 2-year study, groups of 20 male F344 rats were given drinking water containing acrolein in concentrations of 100, 250 or 625 mg/l on 5 days/week. 20 female rats were given 625 mg/l. The acrolein dose taken up at this concentration was estimated to be 50 mg/kg body weight and day. The volatility and instability of acrolein in aqueous solution was not taken into account; the value is therefore too high. In the females, 5 adenomas and 2 hyperplastic nodules in the adrenal cortex were found. In the control group this tumour occurred in only one of 20 animals. In the males no adrenal tumours developed. Increased incidences of other tumours were not found (Lijinsky and Reuber 1987). The concentration dependency was not investigated and makes interpretation of the tumour findings in the female animals difficult. Tumour incidences of up to 4.8 % in the historical controls were recorded. With other substances investigated which are probably also metabolized to form acrolein—acrolein diethylacetal, acrolein oxime and allyl alcohol—no adrenal tumours developed. In a re-evaluation of the study by Parent *et al.* (1992b) no adrenal tumours were found in the treated group, but 3 phaeochromocytomas. According to the authors, however, their incidence was still within the range in the historical controls. Therefore the adrenal tumours described cannot be considered as evidence of a carcinogenic effect. Also in the other carcinogenicity studies the incidence of adrenal tumours was not increased. Other points of criticism are that the control group was introduced into the study 10 months later than the exposed group, no original sections were available for re-evaluating the female animals, the MTD (maximum tolerated dose) was not determined, and no data for body weight gains or toxicity were given. The instability of acrolein in aqueous solution was not taken into consideration.

Acrolein diethylacetal, acrolein oxime and allyl alcohol, but not acrolein, were administered by gavage to groups of 20 male hamsters for 53 to 64 weeks, depending on the substance. In each group 13 to 17 animals lived for longer than 48 weeks. After a total of 90 to 92 weeks at most, the animals were examined. In the control group the incidence of adrenal tumours was high (8/20). Tumours of the forestomach and the efferent duct of the pancreas were detected with the diethylacetal in 1/13 animals and with the oxime in 2/17 animals (Lijinsky and Reuber 1987). The data are not sufficient for evaluation but speak against the induction of adrenal tumours in hamsters by substances related to acrolein.

Groups of 60 SD rats per sex and dose were given gavage doses of acrolein of 0.05, 0.5 or 2.5 mg/kg body weight in 10 ml water per kg body weight for 102 weeks. The initial increase in mortality in the high dose group indicates that the maximum tolerated dose was reached. Necropsy yielded no evidence of a possible cause of death. In the extensive histopathological examination no substance-related increase in the incidence of tumours was found. Male animals of the high dose group which died prematurely were found to have focal reddening of the stomach. At the end of the study, however, the incidence of microscopically detectable stomach lesions was not increased (Parent *et al.* 1992b).

CD-1 mice were given acrolein doses of 0.5, 2 or 4.5 mg/kg body weight in 10 ml water/kg body weight on 5 days/week for 18 months. In the highest dose group 75 animals of each sex were used, in the other groups 70 animals of each sex per dose. In the male animals an initial increase in mortality was observed, so that the highest dose was considered to be in the range of the maximum tolerated dose. Necropsy yielded no evidence of a possible cause of death. In the extensive histopathological examination no substance-related increase in the incidence of tumours was found. At the end of the study the incidence of microscopically detectable stomach lesions was not increased. The authors discuss as a possible reason for the absence of specific systemic effects the degradation and reaction of acrolein in the stomach, e.g. with mercapto, hydroxy and amino groups, so that potential target tissues are not reached (Parent *et al.* 1991b).

In the two studies of Parent *et al.* (1991b, 1992b), no occult blood was found in the urine. The histological examination revealed no changes in the bladder or adrenal glands. Despite the initial increase in mortality, the extent of internal exposure to acrolein is unclear. The lack of systemic effects could be due to a reaction between acrolein and proteins in the feed or degradation in the stomach, as is discussed by the authors in the study of the toxic effects on reproduction with rats (Parent *et al.* 1992a) and the study with mice (Parent *et al.* 1991b). The absence of local effects is probably the result of the low concentration used (0.02 %). At a higher concentration (0.12 %), and to a lesser extent with 0.06 %, marked irritation of the stomach was found in the study of toxic effects on reproduction with rats. The relatively high renal excretion of acrolein metabolites in a metabolism study (Parent *et al.* 1996b) does not represent evidence of a generally good bioavailability of acrolein after oral administration, as the animals were fasted overnight and thus the possibility of a reaction of acrolein with the contents of the stomach is reduced. In addition, the reaction with glutathione can take place actually during absorption, as a result of the high level of glutathione in the mucosa of stomach and intestine (Siegers *et al.* 1988, 1989), so that only the less reactive glutathione conjugate is systemically available.

# 6 Manifesto

Acrolein is genotoxic in *in vitro* tests and in tests in *Drosophila melanogaster*. Because of the great reactivity of acrolein and the uncertain internal exposure to the substance, the negative results of the oral carcinogenicity studies carried out with acrolein to date cannot refute the suspicion that the substance has carcinogenic effects. A valid experiment with sufficiently long inhalation exposure has not been carried out. Therefore acrolein is classified in Carcinogen category 3B of the *List of MAK and BAT Values*.

The MAK value valid until 1997 was withdrawn because acrolein is genotoxic and at present no concentration can be given at which carcinogenic effects in man can be excluded. The previous classification in Pregnancy risk group D has also been withdrawn.

As damage to the barrier layers of the skin resulting from the irritative or corrosive effects of the substance is of importance for the percutaneous absorption of acrolein and as no *in vivo* data with non-irritative concentrations are available, the substance is at present not designated with an "H".

There is no evidence from animal studies that the substance causes sensitization. The one reported case of sensitization in man does not provide justification for designating acrolein with an "Sh".

# 7 References

Adams JD, Klaidman LK (1993) Acrolein-induced oxygen radical formation. *Free Radical Biol Med 15*: 187–193

Al-Rawithi S, El-Yazigi A, Nicholls PJ (1993) Determination of acrolein in urine by liquid chromatography and fluorescence detection of its quinoline derivative. *Pharm Res 10*: 1587–1590

Alarcon RA (1976) Studies on the *in vivo* formation of acrolein: 3-hydroxypropylmercapturic acid as an index of cyclophosphamide (NSC-26271) activation. *Cancer Treat Rep 60*: 327–335

Au W, Sokova OI, Kopnin B, Arrighi FE (1980) Cytogenetic toxicity of cyclophosphamide and its metabolites *in vitro*. *Cytogenet Cell Genet 26*: 108–116

Baker GL, Kahl LE, Zee BC, Stolzer BL, Agarwal AK, Medsger TA (1987) Malignancy following treatment of rheumatoid arthritis with cyclophosphamide. *Am J Med 83*: 1–9

Barros AR, Comendador MA, Sierra LM (1994a) Acrolein genotoxicity in *Drosophila melanogaster*. II. Influence of mus201 and mus308 mutations. *Mutat Res 306*: 1–8

Barros AR, Sierra LM, Comendador MA (1994b) Acrolein genotoxicity in *Drosophila melanogaster*. III. Effects of metabolic modification. *Mutat Res 321*: 119–126

Bartsch H, Malaveille C, Camus A-M, Martel-Planche G, Brun G, Hautefeuille A, Sabadie N, Barbin A, Kuroki T, Drevon C, Piccoli C, Montesano R (1980) Validation and comparative studies on 180 chemicals with *S. typhimurium* strains and V79 Chinese hamster cells in the presence of various metabolizing systems. *Mutat Res 76*: 1–50

Basu AK, Marnett LJ (1984) Molecular requirements for the mutagenicity of malondialdehyde and related acroleins. *Cancer Res 44*: 2848–2854

Beauchamp RO, Andjelkovich DA, Kligerman AD, Morgan KT, Heck H d'A, Feron VJ (1985) A critical review of the literature on acrolein toxicity. *Crit Rev Toxicol 14*: 309–380

Benson AJ, Martin CN, Gardner RC (1988) *N*-(2-Hydroxyethyl)-*N*-[2-(7-guaninyl)ethyl]amine, the putative major DNA adduct of cyclophosphamide *in vitro* and *in vivo* in the rat. *Biochem Pharmacol 37*: 2979–2985

Bouley G, Dubreuil A, Godin J, Boisset M, Boudáene Cl (1976) Phenomena of adaptation in rats continuously exposed to low concentrations of acrolein. *Ann occup Hyg 19*: 27–32

Brock N, Stekar J, Pohl J, Niemeyer U, Scheffler G (1979) Acrolein, the causative factor of urotoxic side-effects of cyclophosphamide, ifosfamide, trofosfamide and sufosfamide. *Arzneim-Forsch 29*: 659–661

Brock N, Pohl J, Stekar J (1981a) Studies on the urotoxicity of oxazaphosphorine cytostatics and its prevention – I. Experimental studies on the urotoxicity of alkylating compounds. *Eur J Cancer 17*: 595–607

Brock N, Pohl J, Stekar J (1981b) Detoxification of urotoxic oxazaphosphorines by sulfhydryl compounds. *J Cancer Res Clin Oncol 100*: 311–320

Brock N, Pohl J, Stekar J, Scheef W (1982) Studies on the urotoxicity of oxazaphosphorine cytostatics and its prevention – III. Profile of action of sodium 2-mercaptoethane sulfonate (Mesna). *Eur J Cancer Clin Oncol 18*: 1377–1387

BUA (Beratergremium für umweltrelevante Altstoffe) (1995) *Acrolein, Stoffbericht 157*, Hirzel, Stuttgart

Cassee FR, Groten JP, Feron VJ (1996) Changes in the nasal epithelium of rats exposed by inhalation to mixtures of formaldehyde, acetaldehyde, and acrolein. *Fundam Appl Toxicol 29*: 208–218

Chan KK, Hong PS, Tutsch K, Trump DL (1994) Clinical pharmacokinetics of cyclophosphamide and metabolites with and without SR-2508. *Cancer Res 54*: 6421–6429

Chung F-L, Chen H-JC, Nath RG (1996) Lipid peroxidation as a potential endogenous source for the formation of exocyclic DNA adducts. *Carcinogenesis 17*: 2105–2111

Claussen U, Hellmann W, Pache G (1980) Embryotoxicity of the cyclophosphamide metabolite acrolein in rabbits, tested *in vivo* by i.v. injection and by the yolk-sac method. *Arzneimittel-Forsch 30*: 2080–2083

Cohen SM, Garland EM, St John M, Okamura T, Smith RA (1992) Acrolein initiates rat urinary bladder carcinogenesis. *Cancer Res 52*: 3577–3581

Cooper KO, Witz G, Witmer C (1992) The effects of α,β-unsaturated aldehydes on hepatic thiols and thiol-containing enzymes. *Fundam Appl Toxicol 19*: 343–349

Costa DL, Kutzman RS, Lehmann JR, Drew RT (1986) Altered lung function and structure in the rat after subchronic exposure to acrolein. *Am Rev Respir Dis 133*: 286–291

Cox PJ (1979) Cyclophosphamide cystitis – identification of acrolein as the causative agent. *Biochem Pharmacol 28*: 2045–2049

Cox R, Goorha S, Irving CC (1988) Inhibition of DNA methylase activity by acrolein. *Carcinogenesis 9*: 463–465

Curren RD, Yang LL, Conklin PM, Grafström RC, Harris CC (1988) Mutagenesis of xeroderma pigmentosum fibroblasts by acrolein. *Mutat Res 209*: 17–22

Dow Chemical (1976) Letter submitting multiple studies on multiple chemicals required for docket OPTS-82036 with attachments. NTIS/OTS 0535413, EPA/OTS Doc ID 86-920000855S

Dypbukt JM, Atzori L, Edman CC, Grafström RC (1993) Thiol status and cytopathological effects of acrolein in normal and xeroderma pigmentosum skin fibroblasts. *Carcinogenesis 14*: 975–980

van Eick AJ (1977) *Het effect van acroleine in lucht op de frequentie van de ooglidslag van mens en cavia.* Projekt MBL 1977-1, Medisch Biologisch Laboratorium TNO, Niederlande

Eisenbrand G, Schuhmacher J, Gölzer P (1995) The influence of glutathione and detoxifying enzymes on DNA damage induced by 2-alkenals in primary rat hepatocytes and human lymphoblastoid cells. *Chem Res Toxicol 8*: 40–46

Ellenberger J, Mohn GR (1977) Mutagenic activity of major mammalian metabolites of cyclophosphamide toward several genes of *Escherichia coli. J Toxicol Environ Health 3*: 637–650

Epstein SS, Arnold E, Andrea J, Bass W, Bishop Y (1972) Detection of chemical mutagens by the dominant lethal assay in the mouse. *Toxicol Appl Pharmacol 23*: 288–325

Epstein SS, Shafner H (1968) Chemical mutagens in the human environment. *Nature (London) 219*: 385–387

Feron VJ, Kruysse A, Til HP, Immel HR (1978) Repeated exposure to acrolein vapour: subacute studies in hamsters, rats and rabbits. *Toxicology 9*: 47–57

Feron VJ, Kruysse A (1977) Effects of exposure to acrolein vapor in hamsters simultaneously treated with benzo[*a*]pyrene or diethylnitrosamine. *J Toxicol Environ Health 3*: 379–394

Fleming RA, Cruz JM, Webb CD, Kucera GL, Perry JJ, Hurd DD (1996) Urinary elimination of cyclophosphamide alkylating metabolites and free thiols following two administration schedules of high-dose cyclophosphamide and mesna. *Bone Marrow Transplant 17*: 497–501

Florin I, Rutberg L, Curvall M, Enzell CR (1980) Screening of tobacco smoke constituents for mutagenicity using the Ames test. *Toxicology 18*: 219–232

Foiles PG, Akerkar SA, Chung F-L (1989) Application of an immunoassay for cyclic acrolein deoxyguanosine adducts to assess their formation in DNA of *Salmonella typhimurium* under conditions of mutation induction by acrolein. *Carcinogenesis 10*: 87–90

Foiles PG, Akerkar SA, Miglietta LM, Chung F-L (1990) Formation of cyclic deoxyguanosine adducts in Chinese hamster ovary cells by acrolein and crotonaldehyde. *Carcinogenesis 11*: 2059–2061

Fraiser L, Kehrer JP (1992) Murine strain differences in metabolism and bladder toxicity of cyclophosphamide. *Toxicology 75*: 257–272

Galloway SM, Armstrong MJ, Reuben C, Colman S, Brown B, Cannon C, Bloom AD, Nakamura F, Ahmed M, Duk S, Rimpo J, Margolin BH, Resnick MA, Anderson B, Zeiger E (1987) Chromosome aberrations and sister chromatide exchanges in Chinese hamster ovary cells. *Environ Mol Mutagen 10, Suppl 10*: 1–175

Grafström RC, Dybpukt JM, Willey JC, Sundqvist K, Edman CC, Atzori L, Harris CC (1988) Pathobiological effects of acrolein in cultured human bronchial epithelial cells. *Cancer Res 40*: 1717–1721

Grafström RC (1990) *In vitro* studies of aldehyde effects related to human respiratory carcinogenesis. *Mutat Res 238*: 175–184

Greenberg LH, Tanaka KR (1964) Congenital anomalies probably induced by cyclophosphamide. *J Am med Ass 188(5)*: 423–426

Gusev MI, Svechnikova AI, Dronov IS, Grebenskova MD, Golovina AI (1966) Determination of the daily average maximum permissible concentration of acrolein in the atmosphere. *Hyg Sanit (GUS) 31*: 8–13

Habs MR, Schmähl D (1983) Prevention of urinary bladder tumors in cyclophosphamide treated rats by additional medication with the uroprotectors sodium 2-mercaptoethane sulfonate (Mesna) and disodium 2,2'-dithio-bis-ethane sulfonate (Dimesna). *Cancer 51*: 606–609

Hadidi A-HFA, Coulter CEA, Idle JR (1986) Phenotypically deficient urinary elimination of carboxyphosphamide after cyclophosphamide administration to cancer patients. *Cancer Res 48*: 5167–5171

Hales BF (1982) Comparison of the mutagenicity and teratogenicity of cyclophosphamide and its active metabolites, 4-hydroxycyclophosphamide, phosphoramide mustard, and acrolein. *Cancer Res 42*: 3016–3021

Hashmi M, Vamvakas S, Anders MW (1992) Bioactivation of *S*-(3-oxopropyl)-*N*-acetyl-L-cysteine, the mercapturic acid of acrolein. *Chem Res Toxicol 5*: 360–365

Haworth S, Lawlor T, Mortelmans K, Speck W, Zeiger E (1983) *Salmonella* mutagenicity test results for 250 chemicals. *Environ Mutagen 1, Suppl 3*: 1–142

Hemminki K, Falck K, Vainio H (1980) Comparison of alkylation rates and mutagenicity of directly acting industrial and laboratory chemicals. *Arch Toxicol 46*: 277–285

Horvath JJ, Witmer CM, Witz G (1992) Nephrotoxicity of the 1:1 acrolein-glutathione adduct in the rat. *Toxicol Appl Pharmacol 117*: 200–207

Høvding G (1969) Occupational dermatitis from pyrolysis products of polyethene. *Acta Derm Venereol 49*: 147–149

IARC (International Agency for Research on Cancer) (1981) *IARC monographs on the evaluation of carcinogenic risks to humans*, Vol 26, IARC, Lyon

Izard C (1973) Recherches sur les effets mutagenes de l'acroleine et de ses deux epoxydes: le glycidol et le glycidal, sur *Saccharomyces cerevisiae. C R Acad Sci, Ser D 276*: 3037–3040

Jung R, Engelhart G, Herbolt B, Jäckh R, Müller W (1992) Collaborative study of mutagenicity with *Salmonella typhimurium* TA102. *Mutat Res 278*: 265–270

Khudoley VV, Mizgireuv I, Pliss GB (1987) The study of mutagenic activity of carcinogens and other chemical agents with *Salmonella typhimurium* assays: testing of 126 compounds. *Arch Geschwulstforsch 6*: 453–462

Kunze E, Köhnecke B, Engelhardt W, Steinröder H, Brock N, Pohl J (1984) Effect of the uroprotector Mesna on the proliferation of the bladder urothelium in the rat after administration of cyclophosphamide. *Urol Int 39*: 61–67

Kutzman RS, Wehner RW, Haber SB (1984) Selected responses of hypertension-sensitive and resistant rats to inhaled acrolein. *Toxicology 31*: 53–65

Kutzman RS, Popenoe EA, Schmaeler M, Drew RT (1985) Changes in rat lung structure and composition as a result of subchronic exposure to acrolein. *Toxicology 34*: 139–151

Lähdetie J, Räty R, Sorsa M (1990) Interaction of Mesna with the mutagenicity of cyclophosphamide *in vitro* and *in vivo. Mutat Res 245*: 27–32

Lai G-M, Ozols RF, Young RC, Hamilton TC (1989) Effect of glutathione on DNA repair in cisplatin-resistant human ovarian cancer cells lines. *J Nat Cancer Inst 81*: 535–539

Lam C-W, Casanova M, Heck H d'A (1985) Depletion of nasal mucosal glutathione by acrolein and enhancement of formaldehyde-induced DNA-protein cross-linking by simultaneous exposure to acrolein. *Arch Toxicol 58*: 67–71

Leach CL, Hatoum NS, Ratajczak HV, Gerhart JM (1987) The pathologic and immunologic effects of inhaled acrolein in rats. *Toxicol Lett 39*: 189–198

Le Bouffant L, Martin JC, Daniel H, Henin JP, Normand C (1980) Action of intensive cigarette smoke inhalations on the rat lung. *J Nat Cancer Inst 64*: 273–284

Lijinsky W, Andrews AW (1980) Mutagenicity of vinyl compounds in *Salmonella typhimurium. Teratogen Carcinogen Mutagen 1*: 259–267

Lijinsky W, Reuber MD (1987) Chronic carcinogenesis studies of acrolein and related compounds. *Toxicol Ind Health 3*: 337–345

Linhart I, Frantík E, Vodicková L', Vosmanská M, Šmejkal J, Mitera J (1996) Biotransformation of acrolein in the rat: excretion of mercapturic acids after inhalation and intraperitoneal injection. *Toxicol Appl Pharmacol 136*: 155–160

Loquet C, Toussaint G, LeTalaer JY (1981) Studies on mutagenic constituents of apple brandy and various alcoholic beverages collected in western France, a high incidence area for oesophageal cancer. *Mutat Res 88*: 155–164

Lutz D, Eder E, Neudecker T, Henschler D (1982) Structure-mutagenicity relationship in α,β-unsaturated carbonylic compounds and their corresponding allylic alcohols. *Mutat Res 93*: 305–315

Lyon JP, Jenkins LJ, Jones RA, Coon RA, Siegel J (1970) Repeated and continuous exposure of laboratory animals to acrolein. *Toxicol Appl Pharmacol 17*: 726–732

Maccubbin AE, Caballes L, Scappaticci F, Struck RF, Gurtoo HL (1990) $^{32}$P-Postlabeling analysis of binding of the cyclophosphamide metabolite, acrolein, to DNA. *Cancer Commun 2*: 207–211

Maccubbin AE, Caballes L, Riordan JM, Huang DH, Gurtoo HL (1991) A cyclophosphamide/DNA phosphoester adduct formed *in vitro* and *in vivo*. *Cancer Res 51*: 886–892

Marnett LJ, Hurd HK, Hollstein MC, Levin DE, Esterbauer H, Ames BN (1985) Naturally occurring carbonyl compounds are mutagens in *Salmonella* tester strain TA104. *Mutat Res 148*: 25–34

McDiarmid MA, Iype PT, Kolodner K, Jacobson-Kram D, Strickland PT (1991) Evidence for acrolein-modified DNA in peripheral blood leukocytes of cancer patients treated with cyclophosphamide. *Mutat Res 248*: 93–99

McNulty MJ, Heck H d'A, Casanova-Schmitz M (1984) Depletion of glutathione in rat respiratory mucosa by inhaled acrolein. *Fed Proc 43*: 575

Mirkes PE, Fantel AG, Greenaway JC, Shepard TH (1981) Teratogenicity of cyclophosphamide metabolites: phosphoramide mustard, acrolein, and 4-ketocyclophosphamide in rat embryos cultured *in vitro*. *Toxicol appl Pharmacol 58*: 322–330

Mirkes PE, Greenaway JC, Rogers JG, Brundrett RB (1984) Role of acrolein in cyclophosphamide teratogenicity in rat embryos *in vitro*. *Toxicol appl Pharmacol 72*: 281–291

Morris JB (1990) Uptake of acrolein in the upper respiratory tract of the F344 rat. *Inhalat Toxicol 8*: 387–403

Nath RG, Chung F-L (1994) Detection of exocyclic 1,N$^2$-propanodeoxyguanosine adducts as common DNA lesions in rodents and humans. *Proc Nat Acad Sci USA 91*: 7491–7495

Nath RG, Chen H-JC, Nishikawa A, Young-Sciame R, Chung F-L (1994) A $^{32}$P-postlabeling method for simultaneous detection and quantification of exocyclic etheno and propano adducts in DNA. *Carcinogenesis 15*: 979–984

Nath RG, Ocando JE, Chung F-L (1996) Detection of 1,N$^2$-propanodeoxyguanosine adducts as potential endogenous DNA lesions in rodent and human tissues. *Cancer Res 56*: 452–456

Newman MA, Que Hee SS, Schoeny RS (1990) Mutagenesis assays on urines produced by patients administered adriamycin and cyclophosphamide. *Environ Mol Mutagen 16*: 189–203

Parent RA, Caravello HE, Harbell JW (1991a) Gene mutation assay of acrolein in the CHO/HGPRT test system. *J Appl Toxicol 11*: 91–95

Parent RA, Caravello HE, Long JE (1991b) Oncogenicity study of acrolein in mice. *J Am Coll Toxicol 10*: 647–659

Parent RA, Caravello HE, Hoberman AM (1992a) Reproductive study of acrolein on two generations of rats. *Fundam Appl Toxicol 19*: 228–237

Parent RA, Caravello HE, Long JE (1992b) Two-year toxicity and carcinogenicity study of acrolein in rats. *J Appl Toxicol 12*: 131–139

Parent RA, Caravello HE, Christian MS, Hoberman AM (1993a) Developmental toxicity of acrolein in New Zealand White rabbits. *Fundam Appl Toxicol 20*: 248–256

Parent RA, Caravello HE, Schrimpf MK, Sharp DE (1993b) Acrolein metabolite identification in rats. *Toxicologist 13*: 401

Parent RA, Caravello HE, San RH (1996a) Mutagenic activity of acrolein in *S. typhimurium* and *E. coli*. *J Appl Toxicol 16*: 103–108

Parent RA, Caravello HE, Sharp DE (1996b) Metabolism and distribution of [2,3-$^{14}$C]acrolein in Sprague-Dawley rats. *J Appl Toxicol 16*: 449–457

Pedersen-Bjergaard J, Ersbøll J, Hansen VL, Sørensen BL, Christoffersen K, Hou-Jensen K, Nissen NI, Knudsen JB, Hansen MM (1988) Carcinoma of the urinary bladder after treatment with cyclophosphamide for non-Hodgkin's lymphoma. *N Engl J Med 318*: 1028–1032

Perry CS, Liu X, Lund LG, Whitman CP, Kehrer JP (1995) Differential toxicities of cyclophosphamide and its glutathione metabolites to A549 cells. *Toxicol In Vitro 9*: 21–26

Pool BL, Bos RP, Niemeyer U, Theuws JLG, Schmähl D (1988) *In vitro/in vivo* effects of Mesna on the genotoxicity and toxicity of cyclophosphamide – a study aimed at clarifying the mechanism of Mesna's anticarcinogenic activity. *Toxicol Lett 41*: 49–56

Radis CD, Kahl LE, Baker GL, Wasko MC, Cash JM, Gallatin A, Stolzer BL, Agarwal AK, Medsger TA, Kwoh CK (1995) Effects of cyclophosphamide on the development of malignancy and on long-term survival of patients with rheumatoid arthritis. *Arthritis Rheum 38*: 1120–1127

Ramu K, Fraiser LH, Mamiya B, Ahmed T, Kehrer JP (1995) Acrolein mercapturates: synthesis, characterization, and assessment of their role in the bladder toxicity of cyclophosphamide. *Chem Res Toxicol 8*: 515–524

Ramu K, Perry CS, Ahmed T, Pakenham G, Kehrer JP (1996) Studies on the basis for the toxicity of acrolein mercapturates. *Toxicol Appl Pharmacol 140*: 487–498

Rapoport IA (1948) Mutations under the influence of unsaturated aldehydes (Russian). *Dok Akad Nauk SSSR 61*: 713–715

Rappaport BZ, Hoffman MM (1941) Urticaria due to aliphatic aldehydes. A clinical and experimental study. *J Am Med Assoc 116*: 2656–2659

Roe FJC, Salaman MH (1955) Further studies on incomplete carcinogenesis: triethylene melamine (T.E.M.), 1,2-benzanthracene and β-propiolactone, as initiators of skin tumour formation in the mouse. *Br J Cancer 9*: 177–203

Roemer E, Anton HJ, Kindt R (1993) Cell proliferation in the respiratory tract after acute inhalation of formaldehyde or acrolein. *J Appl Toxicol 13*: 103–107

Rosen JD, Segall Y, Casida JE (1980) Mutagenic potency of haloacroleins and related compounds. *Mutat Res 78*: 113–119

Sakata T, Smith RA, Garland EM, Cohen SM (1989) Rat urinary bladder epithelial lesions induced by acrolein. *J Environ Pathol Toxicol 9*: 159–170

Salaman MH, Roe FJC (1956) Further tests for tumor-initiating activity: *N,N*-di-(2-chloroethyl)-*p*-aminophenylbutyric acid (CB 1348) as an initiator of skin tumour formation in the mouse. *Br J Cancer 10*: 363–378

Sanduja R, Ansari GAS, Boor PJ (1989) 3-Hydroxypropylmercapturic acid: a biologic marker of exposure to allylic and related compounds. *J Appl Toxicol 9*: 235–238

Schmähl D, Habs MR (1983) Prevention of cyclophosphamide-induced carcinogenesis in the urinary bladder of rats by administration of mesna. *Cancer Treat Rev 10, Suppl A*: 57–61

Schmid BP, Goulding E, Kitchin K, Sanyal MK (1981) Assessment of the teratogenic potential of acrolein and cyclophosphamide in a rat embryo culture system. *Toxicology 22*: 235–243

Sharp DE, Paust DE, Schrimpf MK, Talaat RE, Doane RA, Parent RA (1994) Metabolism of 2,3-$^{14}$C-acrolein in Sprague-Dawley rats: identification of urinary and fecal metabolites. *ISSX Proc 6*: 233

Shell (1958) *Toxicity data sheet acrolein*, Shell Chemical Corporation, April 1958

Siegers C-P, Riemann D, Thies E, Younes M (1988) Glutathione and GSH-dependent enzymes in the gastrointestinal mucosa of the rat. *Cancer Lett 40*: 71–76

Siegers C-P, Bartels L, Riemann D (1989) Effects of fasting and glutathione depletors on the GSH-dependent enzyme system in the gastrointestinal mucosa of the rat. *Pharmacology 38*: 121–128

Sierra LM, Barros AR, Garcia M, Ferreiro JA, Comendador MA (1991) Acrolein genotoxicity in *Drosophila melanogaster*. I. Somatic and germinal mutagenesis under proficient repair conditions. *Mutat Res 260*: 247–256

Sladek NE (1988) Metabolism of oxazaphosphorines. *Pharmacol Ther 37*: 301–355

Smith RA, Cohen SM, Lawson T (1990) Acrolein mutagenicity in the V79 assay. *Carcinogenesis 11*: 497–498

Smith RA, Smith TE, Cohen SM, Lawson T (1994) $^{32}$P-Postlabeling of abasic sites in acrolein modified DNA. *Proc Am Assoc Cancer Res 35*: 111

Smith RA, Smith TE, Cohen SM, Lawson T (1995) DNA crosslinks in V79 cells exposed to acrolein. *Proc Am Assoc Cancer Res 36*: 135

Smith RA, Kubik GS, Cohen SM, Lawson T (1996) Acrolein-induced mutations in V79 cells. *Proc Am Assoc Cancer Res 37*: 141

Stahlmann R, Bluth U, Neubert D (1985) Effects of the cyclophosphamide metabolite acrolein in mammalian limb bud cultures. *Arch Toxicol 57*: 163–167

Swenberg JA, La DK, Scheller NA, Wu KY (1995) Dose-response relationships for carcinogens. *Toxicol Lett 82/83*: 751–756

Talar-Williams C, Hijazi YM, Walter MCM, Linehan WM, Hallahan CW, Lubensky I, Kerr GS, Hoffman GS, Fauci AS, Sneller MC (1996) Cyclophosphamide-induced cystitis and bladder cancer in patients with Wegener granulomatosis. *Ann Intern Med 124*: 477–484

Tasso MJ, Boddy AV, Price L, Wyllie RA, Pearson ADJ, Idle JR (1992) Pharmacokinetics and metabolism of cyclophosphamide in paediatric patients. *Cancer Chemother Pharmacol 30*: 207–211

Vogel EW, Nivard MJ (1993) Performance of 181 chemicals in a *Drosophila* assay predominantly monitoring interchromosomal mitotic recombination. *Mutagenesis 8*: 57–81

Waegemaekers THJM, Bensink MPM (1984) Non-mutagenicity of 27 aliphatic acrylate esters in the *Salmonella*-microsome test. *Mutat Res 137*: 95–102

Weber-Tschopp A, Fischer T, Gierer R, Grandjean E (1977) Experimentelle Reizwirkungen von Akrolein auf den Menschen. *Int Arch Occup Environ Health 40*: 117–130

Wilmer JL, Erexson GL, Kligerman AD (1986) Attenuation of cytogenetic damage by 2-mercaptoethanesulfonate in cultured human lymphocytes exposed to cyclophosphamide and its reactive metabolites. *Cancer Res 46*: 203–210

Wilmer JL, Erexson GL, Kligerman AD (1990) Effect of acrolein on phosphoramide-induced sister chromatid exchanges in cultured human lymphocytes. *Cancer Res 50*: 4635–4638

Wilson VL, Foiles PG, Chung F-L, Povey AC, Frank AA, Harris CC (1991) Detection of acrolein and crotonaldehyde DNA adducts in cultured human cells and canine peripheral blood lymphocytes by $^{32}P$-postlabeling and nucleotide chromatography. *Carcinogenesis 12*: 1483–1490

Zimmering S, Mason JM, Valencia R, Woodruff RC (1985) Chemical mutagenesis testing in *Drosophila*. II. Results of 20 coded compounds tested for the National Toxicology Program. *Environ Mutagen 7*: 87–100

Zimmering S, Mason JM, Valencia R (1989) Chemical mutagenesis testing in Drosophila. VII. Results of 22 coded compounds tested in larval feeding experiments. *Environ Mutagen 14*: 245–251

completed 18.06.1997

# *n*-Butyl acrylate

| | |
|---|---|
| **MAK value (1996)** | **2 ml/m$^3$ (ppm) ≙ 11 mg/m$^3$** |
| **Peak limitation (1985, 2000)** | **Category 1 (excursion factor 2)** |
| **Absorption through the skin** | **–** |
| **Sensitization (1985)** | **Sh** |
| **Carcinogenicity** | **–** |
| **Prenatal toxicity (1985)** | **Pregnancy risk group D** |
| **Germ cell mutagenicity** | **–** |
| **BAT value** | **–** |
| Synonyms | acrylic acid *n*-butyl ester |
| Chemical name (CAS) | 2-propenoic acid butyl ester |
| CAS number | 141-32-2 |

*n*-Butyl acrylate is an acrylate monomer used in the production of polymers for paints and lacquers, for glues in the production of textiles and plastics and in the production of elastomers. The potential for sensitizing effects on the skin was already known in 1985 (see the 1986 MAK documentation for *n*-butyl acrylate, Volume 6 of the present series), and the substance was designated accordingly with an "Sh". This supplement to the previous documentation presents the subsequently published data on the sensitizing effects of *n*-butyl acrylate.

## 1 Allergenic Effects

### 1.1 Effects in man

That *n*-butyl acrylate has sensitizing effects on the skin is apparent from numerous, more or less well-documented reports of reactions to the substance in patch tests (see Table 1). Only in a few cases can the sensitization be causally attributed to *n*-butyl acrylate. The products used almost always contain acrylate mixtures, so that a clear demonstration that the person was exposed to *n*-butyl acrylate is only rarely possible, e.g. as a component of a sticking-plaster glue. Some reactions are therefore likely to be cross-reactions after sensitization by other acrylates or methacrylates, or seldom unspecific concomitant reactions. With a test concentration of not more than 0.1 %, *n*-butyl acrylate is not likely to cause toxic reactions.

**Table 1.** Results with *n*-butyl acrylate in patch tests on patients with suspected acrylate allergy

| Persons tested | Concentration (vehicle) | Results | Contact / comments | References |
|---|---|---|---|---|
| 1 patient | 0.1 % (petrolatum) | 2+ (read after 72 hours) | patient with contact dermatitis from artificial finger nails / reaction to 8 acrylates and methacrylates | Casse *et al.* 1998 |
| 2 patients | 2 % (petrolatum) | positive (no other details) | plaster containing *n*-butyl acrylate / no reactions in controls (no further details) | Daecke *et al.* 1994 |
| 5 patients | 0.5 % (petrolatum) | 2/5 (each 2+) | artificial finger nails / also (strong) reaction to several methacrylates and urethane diacrylate | Hemmer *et al.* 1996 |
| 769 patients | 0.1 % (petrolatum) | 5/769 (3 × 1+, 2 × 2+) | only results of the 72-hour reading; test period 1989–1998 | IVDK 1999 |
| 24 patients | 0.5 % (petrolatum) | 1/24 (no other details) | anamnesis: exposure to acrylates / a dentist reacted to 14 acrylates and methacrylates | Kanerva *et al.* 1988 |
| 1 dentist and 4 dental assistants | 0.5 % (petrolatum) | 2/5 (2+ and 3+) | patients with suspected sensitization to acrylates in synthetic resins / reactions to 15 acrylates and methacrylates | Kanerva *et al.* 1989 |
| 1 dentist with pharyngitis | 0.1 % (petrolatum) | 2+ (read on day 6) | patient without skin symptoms / reactions to 16 acrylates and methacrylates | Kanerva *et al.* 1992 |
| 1 dental technician | 0.1 % (petrolatum) | 2+ | reactions to several acrylates and methacrylates | Kanerva *et al.* 1993a |
| 1 petro-chemical company employee | 0.17 % (petrolatum) | 2+ and 3+ (read on day 2, and days 3 and 4) | accidental sensitization with undiluted methyl acrylate | Kanerva *et al.* 1993b |
| 1 automobile industry employee | 0.1 % (petrolatum) | 1+ (read on days 2, 3 and 4) | glue with 24.6 % 2-hydroxyethyl methacrylate / reactions to 10 acrylates and methacrylates | Kanerva *et al.* 1995 |
| 1 dentist with derma-titis of the finger tips | 0.1 % (petrolatum) | 1+ (read on days 2 and 4), 2+ (on day 3) | dental plastics containing acrylate / marked reactions to 5 acrylates and methacrylates | Kanerva *et al.* 1996 |
| 242 patients | 0.1–0.5 % (petrolatum) | 12/242 (no other details) | anamnesis: exposure to acrylates / summary of the test results 1985–1995 | Kanerva *et al.* 1997 |
| 23 patients | 0.1 % (petrolatum) | 1/23 (2+) | anamnesis: exposure to acrylates / reaction in 1 dental technician | Kiec-Swierczynska 1996 |
| 15 patients | 0.1 % (petrolatum) | 2/15 (no other details) | both patients with artificial finger nails | Koppula *et al.* 1995 |

**Table 1.** continued

| Persons tested | Concentration (vehicle) | Results | Contact / comments | References |
|---|---|---|---|---|
| 1 patient | 0.1 % (petrolatum) | 2+ (read after 96 hours) | glue of an electrosurgical earthing plate containing hydroxyethyl acrylate and hydroxyethyl methacrylate / reaction also to ethyl acrylate and 4 methacrylates | Miranda-Romero *et al.* 1998 |
| 1 patient | 0.1 % (petrolatum) | 2+ (after 48 and 96 hours) | acrylate lacquer of a leg prosthesis / reaction to 13 acrylates and methacrylates | Romaguera *et al.* 1990 |
| 35 dental technicians with allergic contact dermatitis | 0.1 % (petrolatum) | 1/35 (no other details) | | Rustemeyer and Frosch 1996 |
| 4 patients with contact dermatitis from acrylates | 0.1 % and 1 % (petrolatum) | 2/4 (no other details) | | Taylor 1989 |
| 244 patients | 0.5 % (petrolatum) | 9/244 (no other details) | anamnesis: exposure to acrylates | Tucker and Beck 1999 |

One female patient was sensitized by a patch test with *n*-butyl acrylate. She reacted to 1 % *n*-butyl acrylate and 1 % ethyl acrylate in a patch test with irritation after 24 and 48 hours, and with positive results at a reading after 21 days. She also produced a very strong reaction to the two acrylates when she was tested again 2 months later. Another patient, who was later sensitized by a patch test with 0.5 % ethyl acrylate, 2-hydroxyethyl acrylate and 2-hydroxypropyl acrylate, reacted to 0.5 % *n*-butyl acrylate when she was retested (no further details) (Kanerva *et al.* 1988). In a third, incompletely documented case, a patch test with 0.1 % *n*-butyl acrylate was also regarded as the cause of sensitization. The patient, who had not produced a reaction at the 48-hour and 96-hour readings, developed a delayed reaction 3 weeks later to some of the tested acrylates and methacrylates. Retesting produced strong reactions at the 96-hour reading to 0.1 % ethyl acrylate and *n*-butyl acrylate and to 2 % 2-hydroxyethyl methacrylate and methyl methacrylate (Vozmediano and Manrique 1998).

There are no data available for allergenic effects on the respiratory passages.

## 1.2 Animal studies

With a modified Draize test and with various test protocols using adjuvant (modified maximization test, FCA test, Polak test, split adjuvant test) sensitizing effects were detected in animals (Parker and Turk 1983, van der Walle *et al.* 1982) (Table 2).

**Table 2.** Sensitizing effects of *n*-butyl acrylate in the guinea pig

| Method, strain, sex | **Intradermal induction** injection volume, concentration (vehicle); amount/animal | **Epicutaneous induction** day, application, concentration (vehicle) | **Provocation** day, application, concentration (vehicle) | Number of animals which produced reactions | Comments | References |
|---|---|---|---|---|---|---|
| maximization test, Himalayan white spotted, ♀ | injections in duplicate 100 µl of substance, substance in FCA and FCA, 6.4 % (arachis oil); 25.6 mg | day 7, 48 hours occlusive, 12.8 % (80 % ethanol) | day 21, 24 hours occlusive; day 35, non-occlusive, 1.3 % (2-butanone:arachis oil, 2:1) | in each case **7/10** | purity > 99 % | van der Walle *et al.* 1982 |
| FCA test, Dunkin-Hartley, ♀ | 100 µl on days 0, 2, 4, 7, 9, 6.4 % (FCA); 32 mg | – | days 21 and 35, non-occlusive, 38.4 % (2-butanone:arachis oil 2:1) | **8/8** and **7/8** | purity > 99 % | van der Walle *et al.* 1982 |
| Polak test, Hartley, ♂♀ | subcutaneous doses of 100 µl in each paw and in the neck area, 0.2 % (physiological NaCl: ethanol 4:1 in FCA); 1 mg | – | day 7, non-occlusive 20 µl, 0.5 %, 1 % or 5 % (acetone:olive oil 4:1) | not specified; mean reaction score: 0.3, 0.5 and 1.7 | n = 11, number of animals with reactions not specified | Parker and Turk 1983 |
| split adjuvant test, Hartley ♂♀ | 5 doses of 50 µl FCA on day 0 and 100 µl on day 1, 0.2 % (physiological NaCl:ethanol 100:1) in the dorsal flank; 0.2 mg | – | from day 14, one non-occlusive dose per week of 20 µl, non-irritative concentrations, max. 5 % (acetone:olive oil 4:1) | **6/6** | incompletely documented | Parker and Turk 1983 |
| modified Draize test, Hartley, ♂♀ | – | 10 non-occlusive doses: 100 µl on each of days 0–4 and 7–11, 10 % (acetone:olive oil 1:1) | from day 21, one non-occlusive dose per week of 20 µl, non-irritative concentrations, max. 5 % | **3/6** | incompletely documented | Parker and Turk 1983 |

Repeated non-occlusive epicutaneous application revealed no evidence of sensitizing effects: with a cotton wool swab, repeated doses of about 0.1 ml 20 % or 50 % *n*-butyl acrylate in alcohol and undiluted *n*-butyl acrylate were applied to the depilated flank skin of guinea pigs until irritation was observed in the area of the treated skin. After a 9-day treatment-free interval the skin lesions had healed, and the treated animals and 3 control animals were treated with a single dose of a solution of 5 % *n*-butyl acrylate in chloroform. No skin reactions were observed after 8, 12 or 24 hours (BASF 1958).

After induction treatment consisting of the subcutaneous injection of $3 \times 10^7$ haptenized macrophages into the ears, paws and flanks, six male and six female inbred guinea pigs (no further details) reacted to non-occlusive provocation 12–14 days later with 8 mg *n*-butyl acrylate in 2-butanone:arachis oil 2:1 (no further details). To prepare the macrophages, 4 days after intraperitoneal injection of 12 ml paraffin oil, cells were taken from the peritoneal exudate, washed twice and incubated at 37°C for 30 minutes with 3.9 mM *n*-butyl acrylate, centrifuged and washed again twice (von Blomberg-van der Flier *et al.* 1984).

## 2 Manifesto (sensitization)

The sensitizing effects of *n*-butyl acrylate in man have been documented in numerous patch tests on patients with occupational contact dermatitis or contact dermatitis acquired away from the workplace. These findings are supported by positive results in several experiments with animals both with and without the use of adjuvant. Designation with "Sh" has therefore been retained. There are no data available for sensitizing effects on the respiratory passages.

## 3 References


BASF (1958) *Bericht über die Prüfung der Hautwirkung verschiedener Acrylsäureester. Gewerbehygienisch-pharmakologisches Institut*, Report No. VII/308–310, unpublished

von Blomberg-van der Flier M, Scheper RJ, Boerrightre GH, Polak L (1984) Induction of contact sensitivity to a broad variety of allergens with haptenized macrophages. *J Inv Dermatol 83*: 91–95

Casse V, Salmon-Ehr V, Mohn C, Kalis B (1998) Dépigmentation durable secondaire à des tests positifs aux dérivés des méthacrylates. *Ann Dermatol Venereol 125*: 56–57

Daecke C, Schaller J, Goos M (1994) Acrylates as potent allergens in occupational and domestic exposures. *Contact Dermatitis 30*: 190–191

Hemmer W, Focke M, Wantke F, Götz M, Jarisch R (1996) Allergic contact dermatitis to artificial fingernails prepared from UV light-cured acrylates. *J Am Acad Dermatol 35*: 377–380

IVDK (Informationsverbund Dermatologischer Kliniken) (1999) *Auswertung der zwischen 1989 und 1998 vom IVDK erfassten Daten zu n-Butylacrylat, 15.06.1999*

Kanerva L, Estlander T, Jolanki R (1988) Sensitization to patch test acrylates. *Contact Dermatitis 18*: 10–15

Kanerva L, Estlander T, Jolanki R (1989) Allergic contact dermatitis from dental composite resins due to aromatic epoxy acrylates and aliphatic acrylates. *Contact Dermatitis 20*: 201–211

Kanerva L, Estlander T, Jolanki R, Pekkarinen E (1992) Occupational pharyngitis associated with allergic patch test reactions from acrylics. *Allergy 47*: 571–573

Kanerva L, Estlander T, Jolanki R, Tarvainen K (1993a) Occupational allergic contact dermatitis caused by exposure to acrylates during work with dental prostheses. *Contact Dermatitis 28*: 268–275

Kanerva L, Jolanki R, Estlander T (1993b) Accidental occupational sensitization caused by methyl acrylate. *Eur J Dermatol 3*: 195–198

Kanerva L, Jolanki R, Leino T, Estlander T (1995) Occupational allergic contact dermatitis from 2-hydroxyethyl methacrylate and ethylene glycol dimethacrylate in a modified acrylic structural adhesive. *Contact Dermatitis 33*: 84–89

Kanerva L, Estlander T, Jolanki R (1996) False negative patch test reaction caused by testing with dental composite acrylic resin. *Int J Dermatol 35*: 189–192

Kanerva L, Estlander T, Jolanki R (1997) 10 years of patch testing with the (meth)acrylate series. *Contact Dermatitis 37*: 255–258

Kiec-Swierczynska M (1996) Occupational allergic contact dermatitis due to acrylates in Lodz. *Contact Dermatitis 34*: 419–422

Koppula SV, Fellman JH, Storrs FJ (1995) Screening allergens for acrylate dermatitis associated with artificial nails. *Am J Contact Dermatitis 6*: 78–85

Miranda-Romero A, Martínez M, Sanchez-Sambucety P, Aragoneses H, Garcia Munoz CM (1998) Allergic contact dermatitis from the acrylic adhesive of a surgical earthing plate. *Contact Dermatitis 38*: 279–280

Parker D, Turk JL (1983) Contact sensitivity to acrylate compounds in guinea pigs. *Contact Dermatitis 9*: 55–60

Romaguera C, Vilaplana J, Grimalt F, Ferrando J (1990) Contact sensitivity to meth(acrylates) in a limb prosthesis. *Am J Contact Dermatitis 1*: 183–185

Rustemeyer T, Frosch PJ (1996) Occupational skin diseases in dental laboratory technicians. (I). Clinical picture and causative factors. *Contact Dermatitis 34*: 125–133

Taylor JS (1989) Acrylic reactions – ten years' experience. in: Frosch PJ, Dooms-Goossens A, Lachapelle JM (Eds) *Current topics in contact dermatitis*, Springer, Berlin, 346–351

Tucker SC, Beck MH (1999) A 15-year study of patch testing to (meth)acrylates. *Contact Dermatitis 40*: 278–279

Vozmediano J, Manrique A (1998) Active sensitization to (meth)acrylates. *Contact Dermatitis 39*: 314

van der Walle HB, Klecak G, Geleick H, Bensink T (1982) Sensitizing potential of 14 mono (meth) acrylates in the guinea pig. *Contact Dermatitis 8*: 223–235

completed 02.03.1999

# Ethyl acrylate

| | |
|---|---|
| **MAK value (1986)** | **5 ml/m$^3$ (ppm) ≙ 20 mg/m$^3$** |
| **Peak limitation (1987, 2000)** | **Category 1 (excursion factor 2)** |
| **Absorption through the skin** | **–** |
| **Sensitization (1986)** | **Sh** |
| **Carcinogenicity** | **–** |
| **Prenatal toxicity (1990)** | **Pregnancy risk group D** |
| **Germ cell mutagenicity** | **–** |
| **BAT value** | **–** |
| Synonyms | acrylic acid ethyl ester |
| Chemical name (CAS) | 2-propenoic acid ethyl ester |
| CAS number | 140-88-5 |

Ethyl acrylate is an acrylate monomer used in the production of water-based polymer emulsions or polymer solutions in paints and lacquers, and in the paper, leather and textile industries. The potential for sensitizing effects on the skin was already known in 1986 (see the 1986 MAK documentation for ethyl acrylate in Volume 6 of the present series), and the substance was designated accordingly with an "Sh". This supplement to the previous documentation presents more recent data on the sensitizing effects of ethyl acrylate.

## 1 Allergenic Effects

### 1.1 Effects in man

That ethyl acrylate has sensitizing effects on the skin is apparent from numerous, more or less well-documented reports of reactions to the substance in patch tests. Only in few cases was previous exposure to ethyl acrylate demonstrated specifically, as the products probably responsible for the skin reactions often contained several acrylates in a mixture and its composition was unknown or incompletely known. It is often also not stated whether the positive results obtained in tests may be possible cross-reactions after sensitization to other acrylates or even unspecific concomitant reactions (see Table 1).

**Table 1.** Results with ethyl acrylate in patch tests on patients with suspected acrylate allergy

| Persons tested | Concentration (vehicle) | Results | Contact / comments | References |
|---|---|---|---|---|
| 1 patient with contact dermatitis | 0.1 % (petrolatum) | 2+ (read after 72 hours) | artificial finger nails / depigmentation of the test area; reaction to 8 acrylates and methacrylates | Casse *et al.* 1998 |
| 6 patients | 1 % (petrolatum) | 2/6 (both 2+) | anaerobic sealant | Condé-Salazar *et al.* 1988 |
| 1 hearing aid manufacturer with conjunctivitis | 0.1 % (petrolatum) | 2+ | plastics and UV-cured lacquers containing methyl methacrylate / reaction to 15 acrylates and methacrylates | Estlander *et al.* 1996 |
| 3899 patients | 0.1 % (petrolatum) | 33/3899 (25 × 1+, 8 × 2+) | only results of the 72-h reading; test period 1989–1998; reactions in 11/215 dental technicians | IVDK 1999 |
| 4 patients | 0.1 % (petrolatum) | 4/4 (3 × 2+, 1 × 3+ after 72 hours) | glue containing hydroxyethyl acrylate for an electrosurgical earthing plate / also strong reaction to hydroxyethyl acrylate | Jagtman 1998 |
| 5 patients | 5 % (olive oil) | 3/5 (each 2+) | reactions only in repeat tests; test substance moderately irritative in most control persons | Jordan 1975 |
| 24 patients | 0.5 % (petrolatum) | 1/24 (no other details) | anamnesis: exposure to acrylates / a dentist reacted to 14 acrylates and methacrylates | Kanerva *et al.* 1988 |
| 1 dentist and 4 dental assistants | 0.17 % (petrolatum) | 2/5 (1+ and 3+) | patients with suspected sensitization to acrylates in synthetic resins / reactions to 15 acrylates and methacrylates | Kanerva *et al.* 1989 |
| 1 dentist with pharyngitis | 0.1 % (petrolatum) | 3+ (read on day 6) | patient without skin symptoms; reactions to 16 acrylates and methacrylates | Kanerva *et al.* 1992b |
| 1 employee of a petrochemical company | 0.17 % (petrolatum) | 1+ (read on days 2, 3 and 4) | accidental sensitization with undiluted methyl acrylate | Kanerva *et al.* 1993 |
| 1 dentist with dermatitis of the finger tips | 0.1 % (petrolatum) | 2+ (read on days 2, 3 and 4) | dental plastics containing acrylate / marked reactions to 5 acrylates and methacrylates | Kanerva *et al.* 1996a |

**Table 1.** continued

| Persons tested | Concentration (vehicle) | Results | Contact / comments | References |
|---|---|---|---|---|
| 1 beautician | 0.1 % (petrolatum) | 2+ | UV-cured plastics for artificial finger nails / marked reactions to 10 acrylates and meth-acrylates | Kanerva *et al* 1996b |
| 192 patients | 0.1–0.5 % (petrolatum) | 16/192 (no other details) | anamnesis: exposure to acrylates / summary of the test results 1985–1995 | Kanerva *et al* 1997 |
| 1 dental assistant with dermatitis on the finger tips | 0.1 % (petrolatum) | 2+ | plastic containing hydroxyethyl meth-acrylate / reactions to 11 acrylates and meth-acrylates | Kanerva *et al* 1998 |
| 23 patients | 0.1 % (petrolatum) | 2/23 (1+ and 2+ after 72 hours) | anamnesis: exposure to acrylates / reactions in 2 dentists | Kiec-Swierczynska 1996 |
| 18 patients with contact dermatitis from acrylates | 0.1 % (petrolatum) | 13/18 (no other details) | reactions in 10/11 patients with artificial finger nails | Koppula *et al.* 1995 |
| 1 patient | 0.1 % (petrolatum) | 3+ (read after 96 hours) | glue of an electrosurgical earthing plate containing hydroxyethyl acrylate and hydroxyethyl methacryl-ate / reaction also to butyl acrylate and 4 methacryl-ates | Miranda-Romero *et al.* 1998 |
| 35 dental technicians with allergic contact dermatitis | 0.1 % (petrolatum) | 3/35 (no other details) | | Rustemeyer and Frosch 1996 |
| 109 dental technicians | 0.1 % (petrolatum) | 4/109 (no other details) | | Schnuch *et al.* 1998 |
| 255 patients | 0.5 % (petrolatum) | 22/255 (no other details) | anamnesis: exposure to acrylates | Tucker and Beck 1999 |

With a test concentration of not more than 0.1 %, ethyl acrylate is not likely to cause irritative effects. Higher test concentrations, however, can lead to false positive reactions. In addition, in individual cases sensitization was induced by patch tests with ethyl acrylate. In 8/24 patients skin irritation was observed in tests with 0.5 % ethyl acrylate in petrolatum and in 3 of these patients a second test carried out 18–19 days after the first yielded positive results; this was interpreted as sensitization resulting from the first patch test. Renewed testing of 2 of the patients confirmed this assumption (no further details). A fourth patient reacted to 1 % ethyl acrylate and 1 % butyl acrylate in a patch test with irritation after 24 and 48 hours, and a positive reaction after 21 days. She produced a very strong reaction when she was tested again 2 months later (Kanerva *et al.* 1988).

In another case a patch test with 0.1 % ethyl acrylate (and 2-hydroxyethyl acrylate and 2-hydroxypropyl acrylate) in petrolatum also resulted in sensitization. When the patch test was repeated, the patient produced a marked crescendo reaction: even after 21 days a clear reaction could be observed. As a result of this further case, if allergy to acrylates is suspected, the authors no longer carry out routine testing with the three acrylates (Kanerva *et al.* 1992a).

There are no data available for allergenic effects on the respiratory passages.

## 1.2 Animal studies

In an FCA test and in a Buehler test, positive results were obtained with ethyl acrylate in guinea pigs (Parsons and Baldwin 1981, van der Walle *et al.* 1982). A non-occlusive patch test and a maximization test, however, yielded negative results (Klecak 1985, van der Walle *et al.* 1982) (see Table 2). With repeated non-occlusive epicutaneous application there was also no evidence of sensitizing effects: with a cotton wool swab repeated doses of about 0.1 ml 20 % or 50 % ethyl acrylate in alcohol and undiluted ethyl acrylate were applied to the depilated flank skin of guinea pigs, until irritation was observed in the area of the treated skin. After a 9-day treatment-free interval the effects on the skin had healed, and the treated animals and 3 control animals were treated with a single dose of a solution of 5 % ethyl acrylate in chloroform. No skin reactions were observed after either 8, 12 or 24 hours (BASF 1958).

# 2 Manifesto (sensitization)

The sensitizing effect of ethyl acrylate in man has been demonstrated in several reports of positive results in patch tests and in some cases of sensitization induced by patch tests. These findings are supported by positive results in two experiments with animals both with and without the use of adjuvant. Designation with “Sh” has therefore been retained. There are no data available for sensitizing effects on the respiratory passages.

# 3 References

BASF (1958) *Bericht über die Prüfung der Hautwirkung verschiedener Acrylsäureester.* Gewerbehygienisch-pharmakologisches Institut, Report No. VII/308–310, unpublished

Casse V, Salmon-Ehr V, Mohn C, Kalis B (1998) Dépigmentation durable secondaire à des tests positifs aux dérivés des méthacrylates. *Ann Dermatol Venereol 125*: 56–57

**Table 2.** Sensitizing effects of ethyl acrylate in the guinea pig

| Method, strain, sex | **Intradermal induction** injection volume, concentration (vehicle); amount/animal | **Epicutaneous induction** day, application, concentration (vehicle) | **Provocation** day, application, concentration (vehicle) | Number of animals with reactions | Comments | References |
|---|---|---|---|---|---|---|
| maximization test, Himalayan white spotted, ♀ | injections in duplicate: 100 µl of substance, substance in FCA and FCA, 5 % (arachis oil); 20 mg | day 7, 48 hours occlusive, 20 % (80 % ethanol) | day 21, 24 hours occlusive; day 35, non-occlusive, 3 % (2-butanone:arachis oil, 2:1) | in each case **0/8** | purity > 99 % | van der Walle *et al.* 1982 |
| FCA test, Dunkin-Hartley, ♀ | 100 µl on days 0, 2, 4, 7, 9, 5 % (FCA); 25 mg | – | days 21 and 35, non-occlusive, 30 % (2-butanone: arachis oil, 2:1) | in each case **7/8** | purity > 99 % | van der Walle *et al.* 1982 |
| non-occlusive patch test, Himalayan white spotted, ♂♀ | – | days 0–20, daily 100 µl non-occlusive on 8 $cm^2$, (not specified) | days 21 and 35, 25 µl non-occlusive on 2 $cm^2$, minimum irritative and maximum non-irritative concentration (no further details) | in each case **0/6** | incompletely documented | Klecak 1985 |
| Buehler test, Hartley, ♂♀ | – | 9 × within 3 weeks, 6 hours occlusive, 0.4 % (80 % ethanol) | day 35, 24 hours occlusive, 0.8 % (acetone);<br><br>day 42, 24 hours, occlusive, 0.4 % (acetone) | **17/20** (after 24 hours) **12/20** (after 48 hours)<br><br>**8/20** (after 24 hours) **7/20** (after 48 hours) | 2/10 and 0/10, and 0/20 and 1/20 control animals reacted after 24 and 48 hours to 0.8 % and 0.4 % | Parsons and Baldwin 1981 |

Conde-Salazar L, Guimaraens D, Romero LV (1988) Occupational allergic contact dermatitis from anaerobic acrylic sealants. *Contact Dermatitis 18*: 129–132

Estlander T, Kanerva L, Kari O, Jolanki R, Mölsä K (1996) Occupational conjunctivitis associated with type IV allergy to methacrylates. *Allergy 51*: 56–59

IVDK (Informationsverbund Dermatologischer Kliniken) (1999) *Auswertung der zwischen 1989 und 1998 vom IVDK erfaßten Daten zu Ethylacrylat*, 05.05.1999

Jagtman BA (1998) Contact dermatitis from acrylates in an electrosurgical earthing plate. *Contact Dermatitis 38*: 280–281

Jordan Jr WP (1975) Cross-sensitization patterns in acrylate allergies. *Contact Dermatitis 1*: 13–15

Kanerva L, Estlander T, Jolanki R (1988) Sensitization to patch test acrylates. *Contact Dermatitis 18*: 10–15

Kanerva L, Estlander T, Jolanki R (1989) Allergic contact dermatitis from dental composite resins due to aromatic epoxy acrylates and aliphatic acrylates. *Contact Dermatitis 20*: 201–211

Kanerva L, Estlander T, Jolanki R (1992a) Double active sensitization caused by acrylics. *Am J Contact Dermatitis 3*: 23–26

Kanerva L, Estlander T, Jolanki R, Pekkarinen E (1992b) Occupational pharyngitis associated with allergic patch test reactions from acrylics. *Allergy 47*: 571–573

Kanerva L, Jolanki R, Estlander T (1993) Accidental occupational sensitization caused by methyl acrylate. *Eur J Dermatol 3*: 195–198

Kanerva L, Estlander T, Jolanki R (1996a) False negative patch test reaction caused by testing with dental composite acrylic resin. *Int J Dermatol 35*: 189–192

Kanerva L, Lauerma A, Estlander T, Alanko K, Henriks-Eckerman M-L, Jolanki R (1996b) Occupational allergic contact dermatitis caused by photobonded sculptured nails and a review of (meth)acrylates in nail cosmetics. *Am J Contact Dermatitis 7*: 109–115

Kanerva L, Estlander T, Jolanki R (1997) 10 years of patch testing with the (meth)acrylate series. *Contact Dermatitis 37*: 255–258

Kanerva L, Mikola H, Henriks-Eckerman M-L, Jolanki R, Estlander T (1998) Fingertip paresthesia and occupational allergic contact dermatitis caused by acrylics in a dental nurse. *Contact Dermatitis 38*: 114–116

Kiec-Swierczynska M (1996) Occupational allergic contact dermatitis due to acrylates in Lodz. *Contact Dermatitis 34*: 419–422

Klecak G (1985) The Freund's complete adjuvant test and the open epicutaneous test. *Curr Probl Dermatol 14*: 152–171

Koppula SV, Fellman JH, Storrs FJ (1995) Screening allergens for acrylate dermatitis associated with artificial nails. *Am J Contact Dermatitis 6*: 78–85

Miranda-Romero A, Martínez M, Sanchez-Sambucety P, Aragoneses H, Munoz CM Garcia (1998) Allergic contact dermatitis from the acrylic adhesive of a surgical earthing plate. *Contact Dermatitis 38*: 279–280

Parsons RD, Baldwin RC (1981) Delayed contact hypersensitivity of guinea pigs to ethyl acrylate. *Toxicologist 1*: 17

Rustemeyer T, Frosch PJ (1996) Occupational skin diseases in dental laboratory technicians. (I). Clinical picture and causative factors. *Contact Dermatitis 34*: 125–133

Schnuch A, Uter W, Geier J, Frosch PJ, Rustemeyer T (1998) Contact allergies in healthcare workers. Results from the IVDK. *Acta Derm Venereol (Stockh) 78*: 358–363

Tucker SC, Beck MH (1999) A 15-year study of patch testing to (meth)acrylates. *Contact Dermatitis 40*: 278–279

van der Walle HB, Klecak G, Geleick H, Bensink T (1982) Sensitizing potential of 14 mono (meth) acrylates in the guinea pig. *Contact Dermatitis 8*: 223–235

completed 02.03.1999

# 2-Ethylhexyl acrylate

| | |
|---|---|
| **MAK value** | – |
| **Peak limitation** | – |
| **Absorption through the skin** | – |
| **Sensitization (1997)** | **Sh** |
| **Carcinogenicity** | – |
| **Prenatal toxicity** | – |
| **Germ cell mutagenicity** | – |
| **BAT value** | – |
| Synonyms | acrylic acid 2-ethylhexyl ester<br>2-ethylhexyl propenoate |
| Chemical name (CAS) | 2-propenoic acid 2-ethylhexyl ester |
| CAS number | 103-11-7 |

2-Ethylhexyl acrylate is used, mainly together with vinyl acetate, in the production of vinyl acrylate plastics and in the production of polymers for glues and other plastics.

## 1 Allergenic Effects

### 1.1 Effects in man

According to several case reports, low 2-ethylhexyl acrylate concentrations produced reactions in patch tests. Data for the clinical relevance of the test results are, however, lacking.

In 2/4 patients with dermatitis from UV-cured lacquers, reactions were observed to e.g. 1 % 2-ethylhexyl acrylate in acetone (no other details) (Björkner and Dahlquist 1979). A patient with a work-related allergy to polyethylene glycol dimethacrylate produced in the patch test 3+ reactions to 0.5 % 2-ethylhexyl acrylate and ethyl acrylate at the 72-hour reading and 2+ reactions to several other acrylates (Senff *et al.* 1992). One patient developed dermatitis on the skin area covered with a plaster containing acrylates used to protect the wound after an operation. Patch tests with the plaster, with 2 % 2-ethylhexyl acrylate or with 2 % butyl acrylate in petrolatum yielded positive results (Daecke *et al.* 1994). After his leg prosthesis was repainted with a resinous lacquer containing acrylates, a patient developed skin reactions which were initially restricted to the amputation stump and then spread to other areas. At the 48-hour and 72-hour readings the patch test produced 2+ reactions to 14 (meth)acrylates, including 0.1 % 2-ethylhexyl acrylate in petrolatum (Romaguera *et al.* 1989, 1990).

The sticking plaster used for test patches was regarded in 5 cases as the cause of sensitization to 2-ethylhexyl acrylate. The adhesive on the sticking plaster contained three other acrylates in addition to 2-ethylhexyl acrylate. Although the patients produced strong reactions—which were visible for up to 6 weeks—to 5 % 2-ethylhexyl acrylate in olive oil after 48-hour occlusive application (Jordan 1975), an allergic genesis is not certain because the provocation concentration used was very high. Of 124 patients with anamnestic data for exposure to acrylates, 23 reacted to at least one substance in the patch test with 30 different acrylates, but none of the persons to 2-ethylhexyl acrylate in concentrations of 0.1 % to 0.5 % in petrolatum (Kanerva *et al.* 1995).

## 1.2 Animal studies

In three experimental studies in which FCA was used (maximization test, FCA test and Polak test, see Table 1), 2-ethylhexyl acrylate was found to have sensitizing effects in guinea pigs.

In an earlier study in which FCA was not used, the mode of application and amount of substance applied are not stated exactly. Induction was carried out three times a week with epicutaneous or intradermal application of 0.1 % 2-ethylhexyl acrylate in water for 3 weeks. Ten days after the last induction treatment 10 animals reacted to topical provocation with 0.1 % 2-ethylhexyl acrylate. On intradermal provocation with the same concentration 5/10 animals produced reactions. The reading was carried out 48 hours after provocation; controls were not included in the study (Hunter *et al.* 1966).

# 2 Manifesto (sensitization)

There are some reports of positive results in patch tests for the allergenic effects of 2-ethylhexyl acrylate on the skin. Also in experiments with animals, 2-ethylhexyl acrylate was found to have allergenic effects. 2-Ethylhexyl acrylate is therefore designated with "Sh". There are no data available for allergenic effects on the respiratory passages.

**Table 1.** Sensitizing effects of 2-ethylhexyl acrylate in the guinea pig

| Method, strain, sex | **Intradermal induction** injection volume, concentration (vehicle); amount/animal | **Epicutaneous induction** day, application, concentration (vehicle); amount/animal | **Provocation** day, application, concentration (vehicle); amount/animal | Number of animals which produced reactions | Comments | References |
|---|---|---|---|---|---|---|
| maximization test, SSc:Al, ♀ | injections in duplicate: 50 µl of substance, substance in FCA and FCA; 1 % (soybean oil); about 1.8 mg | day 8, 48 hours occlusive, 25 % (petrolatum); about 88 mg | day 21, 24 hours occlusive with Finn chamber, 0.3 % or 3 % in petrolatum; about 0.07 or 0.7 mg | 0.3 %: **5/19**<br>3 %: **11/19** | 24 h before epicutaneous induction non-occlusive application of 250 mg 10 % sodium dodecylsulfate in petrolatum; control animals included (no other details); no data on purity | Clemmensen 1984 |
| Polak test, Hartley, ♀♂ | 4 × 100 µl substance in paw, 1 × 100 µl in neck; 2 mg/ml (ethanol:NaCl 4:1, FCA); 1 mg | – | from day 7 every 7 days, non-occlusive, 1 % or 5 % in acetone:olive oil 4:1; 0.18 or 0.9 mg | **not specified**; 1 %: mean reaction score 0.2 (day 7, n = 6)<br>5 %: mean reaction score 0.8 (day 7, n = 6) | no data on purity or controls | Parker and Turk 1983 |
| FCA test, Dunkin-Hartley, ♀ | 100 µl substance on days 0, 5 and 9; 0.17 or 0.5 M (FCA); about 9 or 27 mg | – | day 21, non-occlusive, 3 M (about 60 %) (irritative) in arachis oil:2-butanone 1:2; further provocations every 14 days, non-occlusive, 1 M; about 4.6 mg (on day 21 13.8 mg) | 0.17 M:<br>**13/16** (day 21),<br>**13/16** (day 35),<br>**11/16** (day 49)<br>0.5 M:<br>**6/16** (day 21),<br>**11/16** (day 35),<br>**9/16** (day 49) | purity 94 %, addition of 120 mg hydroquinone/l; 4 animals sensitized to ethylhexyl acrylate reacted also to 1 % hydroquinone; in controls 0/10 (day 21), 3/10 (day 35) and 1/10 (day 49) positive reactions to ethylhexyl acrylate | Wagemaekers and van der Walle 1983 |

# 3 References

Björkner B, Dahlquist I (1979) Contact allergy caused by UV-cured acrylates. *Contact Dermatitis 5*: 403–404

Clemmensen S (1984) Cross-reaction patterns in guinea pigs sensitized to acrylic monomers. *Drug Chem Toxicol 7*: 527–540

Daecke C, Schaller J, Goos M (1994) Acrylates as potent allergens in occupational and domestic exposures. *Contact Dermatitis 30*: 190–191

Hunter CG, Brown VK, Ferrigan LW (1966) Experimental studies on skin hazard with 'Versatic' 9–11 acid and its monoglycidyl and vinyl esters. *Br J Ind Med 23*: 137–141

Jordan WP jr (1975) Cross-sensitization patterns in acrylate allergies. *Contact Dermatitis 1*: 13–15

Kanerva L, Estlander T, Jolanki R, Tarvainen K (1995) Statistics on allergic patch test reactions caused by acrylate compounds, including data on ethyl methacrylate. *Am J Contact Dermatitis 6*: 75–77

Parker D, Turk JL (1983) Contact sensitivity to acrylate compounds in guinea pigs. *Contact Dermatitis 9*: 55–60

Romaguera C, Vilaplana J, Grimalt F, Ferrando J (1989) Contact sensitivity to meth(acrylates) in a limb prosthesis. *Contact Dermatitis 21*: 125

Romaguera C, Vilaplana J, Grimalt F, Ferrando J (1990) Contact sensitivity to meth(acrylates) in a limb prosthesis. *Am J Contact Dermatitis 1*: 183–185

Senff H, Köllner A, Kunze J (1992) Allergisches Kontaktekzem auf Polyethylenglycoldimethacrylat. *Dermatosen 40*: 24–26

Waegemaekers TH, van der Walle HB (1983) The sensitizing potential of 2-ethylhexyl acrylate in the guinea pig. *Contact Dermatitis 9*: 372–376

completed 05.05.1997

# Ethyl methacrylate

| | |
|---|---|
| **MAK value** | – |
| **Peak limitation** | – |
| **Absorption through the skin** | – |
| **Sensitization (1999)** | **Sh** |
| **Carcinogenicity** | – |
| **Prenatal toxicity** | – |
| **Germ cell mutagenicity** | – |
| **BAT value** | – |
| Synonyms | ethyl α-methacrylate<br>ethyl 2-methyl-2-propenoate<br>methacrylic acid ethyl ester |
| Chemical name (CAS) | 2-methyl-2-propenoic acid ethyl ester |
| CAS number | 97-63-2 |

Ethyl methacrylate is a methacrylate monomer used in the production of polymers in dental plastics, bone cement and glues for artificial finger nails, in plastics for hearing aids, and in the printing, lacquer, leather, paper and textile industries.

## 1 Allergenic Effects

### 1.1 Effects in man

There are numerous reports of positive results obtained in patch tests with ethyl methacrylate. In almost all cases glues, lacquers or other plastics containing several acrylates and methacrylates are responsible for the allergic skin reactions described. Only in few cases, therefore, was previous exposure to ethyl methacrylate in these plastics demonstrated specifically, e.g. as a component of glues for artificial finger nails. It is not stated to what extent the other positive results obtained in tests might be cross-reactions after sensitization to other acrylates or methacrylates, or even unspecific concomitant reactions. At least the frequent positive test results with ethyl methacrylate observed in dental technicians and patients with artificial finger nails suggest, however, a causal relationship between previous exposure to this substance and the contact dermatitis or positive test results.

6/6 and 15/15 patients from the clinics of the IVDK (Information Network of Departments of Dermatology) with positive results in the ethyl methacrylate test, reacted also to hydroxyethyl methacrylate and ethylene glycol dimethacrylate, respectively. Particularly in the case of the methacrylates used in dental plastics, however, co-exposure is very likely, so that probably only some of these simultaneous reactions can be interpreted as cross-reactions. Of the patients who reacted to ethyl methacrylate, 14/19 reacted also to methyl methacrylate, 9/19 to butanediol dimethacrylate and 8/18 to triethylene glycol dimethacrylate (IVDK 1999).

There are no data available for allergenic effects on the respiratory passages.

**Table 1.** Positive results with ethyl methacrylate in patch tests on patients with suspected acrylate allergy

| Persons tested | Concentration (vehicle) | Results | Contact / comments | References |
|---|---|---|---|---|
| 1 patient with contact dermatitis | 2 % (petrolatum) | 2+ (read after 72 hours) | artificial finger nails / reactions to 8 acrylates and methacrylates; depigmentation of the test area | Casse *et al.* 1998 |
| 1 patient | 1 % (petrolatum) | 2+ | anaerobic sealant | Dempsey 1982 |
| 1 manufacturer of hearing aids, 1 dental technician with conjunctivitis | 2 % (petrolatum) | 2/2 (1 × 1+ and 1 × 2+) | lacquers and plastics containing acrylates / reactions to numerous acrylates and methacrylates | Estlander *et al.* 1996 |
| 5 patients | 1 % (petrolatum) | 5/5 (4 × strong positive, 1 × weak positive) | artificial finger nails / in 2 patients perhaps primary sensitization to methyl methacrylate; reactions to several methacrylates in all persons | Fisher 1980 |
| 1 patient | 2 % (petrolatum) | 2+ | UV-curing glue / reactions to 11 acrylates and methacrylates | van den Hove *et al.* 1996 |
| 4201 patients | 2 % (petrolatum) | 22/4201 (18 × 1+, 4 × 2+) | only results of the 72-hour reading, test period 1989–1998 | IVDK 1999 |

**Table 1.** continued

| Persons tested | Concentration (vehicle) | Results | Contact / comments | References |
|---|---|---|---|---|
| 4 patients | 2 % (petrolatum) | 2/4 (1+ and 2+ after 72 hours) | glue containing hydroxy-ethyl acrylate for an electrosurgical earthing plate; strong reaction to hydroxyethyl acrylate | Jagtman 1998 |
| 1 dentist | 2 % (petrolatum) | positive (no other details) | reactions to 14 acrylates and methacrylates | Kanerva *et al* 1988 |
| 1 dentist and 4 dental assistants | 2 % (petrolatum) | 1/5 (2+) | patients with suspected sensitization to acrylates in synthetic resins / reactions to 15 acrylates and methacrylates | Kanerva *et al* 1989 |
| 1 patient | 2 % (petrolatum) | 2+ (read on day 13) | possible cross-reaction after sensitization to ethyl, hydroxyethyl and hydroxypropyl acrylate | Kanerva *et al* 1992a |
| 1 dentist with pharyngitis | 2 % (petrolatum) | 3+ (read on day 6) | patient without skin symptoms; reactions to 16 acrylates and meth-acrylates | Kanerva *et al* 1992b |
| 1 dental technician | 2 % (petrolatum) | 2+ | reactions to 7 acrylates and methacrylates | Kanerva *et al* 1993 |
| 1 employee in the automobile industry | 2 % (petrolatum) | 3+ (read on days 2 and 3; on day 4: 2+) | glue with 24.6 % hydroxyethyl meth-acrylate / reactions to 10 acrylates and meth-acrylates | Kanerva *et al* 1995 |
| 1 dentist with dermatitis of the finger tips | 2 % (petrolatum) | 2+ (read after 72 and 96 hours) | dental plastics containing acrylates / marked reactions to 5 acrylates and methacrylates | Kanerva *et al* 1996a |
| 1 beautician | 2 % (petrolatum) | 2+ | UV-curing plastics for artificial finger nails / marked reactions to 10 acrylates and meth-acrylates | Kanerva *et al* 1996b |
| 243 patients | 2 % (petrolatum) | 18/243 (no other details) | anamnesis: exposure to acrylates / summary of the test results 1985–1995 | Kanerva *et al* 1997 |
| 1 dental assistant with dermatitis on the finger tips | 2 % (petrolatum) | 2+ | patient with leukodermia on test areas treated earlier with acrylates and methacrylates / reactions to 8 methacrylates and *n*-butyl acrylate | Kanerva and Estlander 1998 |

**Table 1.** continued

| Persons tested | Concentration (vehicle) | Results | Contact / comments | References |
|---|---|---|---|---|
| 1 dental assistant with dermatitis on the finger tips | 2 % (petrolatum) | 2+ | plastic containing hydroxyethyl methacrylate / reactions to 11 acrylates and methacrylates | Kanerva *et al.* 1998 |
| 23 patients | 2 % (petrolatum) | 2/23 (1 × 2+, 1 × 3+ read after 72 hours) | anamnesis: exposure to acrylates / reactions in 2 dentists | Kiec-Swierczynska 1996 |
| 22 patients with contact dermatitis from acrylates | 5 % (petrolatum) | 14/22 (no other details) | reactions in 7/11 patients with artificial finger nails | Koppula *et al.* 1995 |
| 542 consecutively tested patients | 1 % (petrolatum) | 2/542 (no other details) | paints containing acrylates / no cross-reactions with methyl methacrylate | Maibach *et al.* 1978 |
| 1 patient | 5 % (petrolatum)<br>1 % (ethanol) | in both cases 2+ (read after 48 and 96 hours) | artificial finger nails / also reactions to methyl acrylate and *n*-butyl acrylate | Marks *et al.* 1979 |
| 93 dental technicians | not specified | 10/93 (no other details) | patients with manifest skin diseases tested | Peiler *et al.* 1996 |
| 1 patient | 2 % (petrolatum) | 2+ (read after 48 and 96 hours) | acrylate lacquer of a leg prosthesis / reactions to 13 acrylates and methacrylates | Romaguera *et al.* 1990 |
| 35 dental technicians with allergic contact dermatitis | 2 % (petrolatum) | 6/35 (no other details) | | Rustemeyer and Frosch 1996 |
| 110 dental technicians | 2 % (petrolatum) | 10/110 (no other details) | | Schnuch *et al.* 1998 |
| 1 beautician | 1 % (petrolatum) | 2+ (read after 48 and 72 hours) | plastic for artificial finger nails / no reaction to 5 % methyl methacrylate | Schubert *et al.* 1992 |
| 11 patients with contact dermatitis from acrylates | 5 % (petrolatum) | 2/11 (no other details) | | Taylor 1989 |
| 246 patients | 2 % (petrolatum) | 11/246 (no other details) | anamnesis: exposure to acrylates; test period 1983–1998 | Tucker and Beck 1999 |

**Table 2.** Sensitizing effects of ethyl methacrylate in the guinea pig

| Method, strain, sex | **Intradermal induction** injection volume, concentration (vehicle); amount/animal | **Epicutaneous induction** day, application, concentration (vehicle) | **Provocation** day, application, concentration (vehicle) | Number of animals which produced reactions | Comments | References |
|---|---|---|---|---|---|---|
| maximization test, Himalayan white spotted, ♀ | injections in duplicate: 100 µl of substance, substance in FCA and FCA, 0.5 M (5.7 %) (arachis oil); 22.8 mg | day 7, 48 hours occlusive, 1 M (11.4 %) (80 % ethanol) | day 21, 24 hours occlusive; day 35, non-occlusive, 3 M (34.2 %) (2-butanone:arachis oil 2:1) | in each case **1/10** | purity > 98 % | van der Walle *et al.* 1982 |
| maximization test, Hartley, ♀ | injections in duplicate: 100 µl of substance, substance in FCA and FCA, 1 M (11.4 %); 0.1 M (1.14 %) or 0.01 M (0.114 %) (olive oil); 45.6 mg; 4.6 mg or 0.46 mg | day 7, 48 hours occlusive, 1 M (11.4 %); 0.1 M (1.14 %); or 0.01 M (0.114 %) (olive oil) | day 21, 24 hours occlusive, 1 M (11.4 %) (acetone) | 11.4 %: **3/5**<br>1.14 %: **1/5**<br>0.11 %: **0/5** | | Kanazawa *et al.* 1999 |
| FCA test, Dunkin-Hartley, ♀ | 100 µl on days 0, 2, 4, 7, 9, 0.5 M (5.7 %) (FCA); 22.8 mg | – | days 21 and 35, non-occlusive, 3 M (34.2 %) (2-butanone:arachis oil 2:1) | in each case **2/6** | purity > 98 % | van der Walle *et al.* 1982 |

## 1.2 Animal studies

Earlier experimental studies in which tests with adjuvant were used yielded contradictory results (van der Walle *et al.* 1982). Positive results were obtained in a more recent maximization test on provocation with 11.4 % ethyl methacrylate in acetone of animals previously treated intradermally and epicutaneously for induction with 11.4 % ethyl methacrylate in olive oil. Tests with lower induction concentrations yielded one borderline positive and one negative result (Kanazawa *et al.* 1999) (see Table 2).

One study cannot be included in the present evaluation. After the injection of FCA doses of 100 µl into the paws, 9 animals were subjected once to non-occlusive topical induction treatment with ethyl methacrylate in olive oil and 25 animals three times with ethyl methacrylate in ethanol. On provocation with 2 % and 5 % ethyl methacrylate in olive oil or ethanol 60 days later, reactions were recorded in 9/9 animals treated with olive oil as vehicle, but in none of the 25 animals treated with an ethanolic solution. The authors describe reactions of similar severity after a second provocation with either a 2 % or a 0.4 % preparation. In a test for cross-reactions, both the animals pretreated with olive oil as vehicle and those treated with ethanolic solutions produced reactions on provocation with preparations of 1 % methyl methacrylate, ethyl methacrylate and *n*-butyl methacrylate in olive oil (Chung and Giles 1977).

# 2 Manifesto (sensitization)

Ethyl methacrylate produced positive patch test results in numerous persons with contact dermatitis, which was work-related in some cases. In several cases relevant exposure to ethyl methacrylate was demonstrated or can be regarded as probable. These findings are supported by positive results in two animal studies in which adjuvant was used. There are no data available for sensitizing effects on the respiratory passages.

Ethyl methacrylate is therefore designated with “Sh”.

# 3 References

Casse V, Salmon-Ehr V, Mohn C, Kalis B (1998) Dépigmentation durable secondaire à des tests positifs aux dérivés des méthacrylates. *Ann Dermatol Venereol 125*: 56–57

Chung CW, Giles AL (1977) Sensitization potential of methyl, ethyl, and *n*-butyl methacrylates and mutual cross-sensitivity in guinea pigs. *J Invest Dermatol 68*: 187–190

Dempsey KJ (1982) Hypersensitivity to Sta-Lok and Loctite anaerobic sealants. *J Am Acad Dermatol 7*: 779–784

Estlander T, Kanerva L, Kari O, Jolanki R, Mölsä K (1996) Occupational conjunctivitis associated with type IV allergy to methacrylates. *Allergy 51*: 56–59

Fisher AA (1980) Cross reactions between methyl methacrylate monomer and acrylic monomers presently used in acrylic nail preparations. *Contact Dermatitis 6*: 345–347

van den Hove J, Jacobs M-C, Tennstedt D, Lachapelle J-M (1996) Allergic contact dermatitis from acrylates in insulin pump infusion sets. *Contact Dermatitis 35*: 108

IVDK (Informationsverbund Dermatologischer Kliniken) (1999) *Auswertung der zwischen 1989 und 1998 vom IVDK erfaßten Daten zu Ethylmethacrylat, 5.5.1999*

Jagtman BA (1998) Contact dermatitis from acrylates in an electrosurgical earthing plate. *Contact Dermatitis 38*: 280–281

Kanazawa Y, Yoshida T, Kojima K (1999) Structure-activity relationships in allergic contact dermatitis induced by methacrylates. *Contact Dermatitis 40*: 19–23

Kanerva L, Estlander T, Jolanki R (1988) Sensitization to patch test acrylates. *Contact Dermatitis 18*: 10–15

Kanerva L, Estlander T, Jolanki R (1989) Allergic contact dermatitis from dental composite resins due to aromatic epoxy acrylates and aliphatic acrylates. *Contact Dermatitis 20*: 201–211

Kanerva L, Estlander T, Jolanki R (1992a) Double active sensitization caused by acrylics. *Am J Contact Dermatitis 3*: 23–26

Kanerva L, Estlander T, Jolanki R, Pekkarinen E (1992b) Occupational pharyngitis associated with allergic patch test reactions from acrylics. *Allergy 47*: 571–573

Kanerva L, Estlander T, Jolanki R, Tarvainen K (1993) Occupational allergic contact dermatitis caused by exposure to acrylates during work with dental prostheses. *Contact Dermatitis 28*: 268–275

Kanerva L, Jolanki R, Leino T, Estlander T (1995) Occupational allergic contact dermatitis from 2-hydroxyethyl methacrylate and ethylene glycol dimethacrylate in a modified acrylic structural adhesive. *Contact Dermatitis 33*: 84–89

Kanerva L, Estlander T, Jolanki R (1996a) False negative patch test reaction caused by testing with dental composite acrylic resin. *Int J Dermatol 35*: 189–192

Kanerva L, Lauerma A, Estlander T, Alanko K, Henriks-Eckerman M-L, Jolanki R (1996b) Occupational allergic contact dermatitis caused by photobonded sculptured nails and a review of (meth)acrylates in nail cosmetics. *Am J Contact Dermatitis 7*: 109–115

Kanerva L, Estlander T, Jolanki R (1997) 10 years of patch testing with the (meth)acrylate series. *Contact Dermatitis 37*: 255–258

Kanerva L, Estlander T (1998) Contact leukoderma caused by patch testing with dental acrylics. *Am J Contact Dermatitis 9*: 196–198

Kanerva L, Mikola H, Henriks-Eckerman M-L, Jolanki R, Estlander T (1998) Fingertip paresthesia and occupational allergic contact dermatitis caused by acrylics in a dental nurse. *Contact Dermatitis 38*: 114–116

Kiec-Swierczynska M (1996) Occupational allergic contact dermatitis due to acrylates in Lodz. *Contact Dermatitis 34*: 419–422

Koppula SV, Fellman JH, Storrs FJ (1995) Screening allergens for acrylate dermatitis associated with artificial nails. *Am J Contact Dermatitis 6*: 78–85

Maibach HI, Hjorth N, Menghini C, Bandman HJ, Malten K, Pirillä V, Magnusson B, Cronin E, Calnan C, Wilkinson D, Marzulli F (1978) Butyl methacrylate monomer and ethyl methacrylate monomer – frequency of reactions. *Contact Dermatitis 4*: 60

Marks JG, Bishop ME, Willis WF (1979) Allergic contact dermatitis to sculptured nails. *Arch Dermatol 115*: 100

Peiler D, Rustemeyer T, Frosch PJ (1996) Dermatosen bei Zahntechnikern – Irritantien und Allergene. *Allergologie 19*: 93–94

Romaguera C, Vilaplana J, Grimalt F, Ferrando J (1990) Contact sensitivity to meth(acrylates) in a limb prosthesis. *Am J Contact Dermatitis 1*: 183–185

Rustemeyer T, Frosch PJ (1996) Occupational skin diseases in dental laboratory technicians. (I). Clinical picture and causative factors. *Contact Dermatitis 34*: 125–133

Schnuch A, Uter W, Geier J, Frosch PJ, Rustemeyer T (1998) Contact allergies in healthcare workers. Results from the IVDK. *Acta Derm Venereol (Stockh) 78*: 358–363

Schubert HJ, Lindner K, Prater E (1992) Kontaktallergie im Nagelstudio. *Z Hautkr 67*: 1067–1069

Taylor JS (1989) Acrylic reactions – 10 years experience. in: Frosch PJ, Dooms-Goossens A, Lachapelle JM *et al.* (Eds), *Current topics in contact dermatitis*, Springer, Berlin, 346–351

Tucker SC, Beck MH (1999) A 15-year study of patch testing to (meth)acrylates. *Contact Dermatitis 40*: 278–279

van der Walle HB, Klecak G, Geleick H, Bensink T (1982) Sensitizing potential of 14 mono (meth) acrylates in the guinea pig. *Contact Dermatitis 8*: 223–235

completed 02.03.1999

# Glutaraldehyde

| | |
|---|---|
| **MAK value (1994)** | **0.1 ml/m$^3$ (ppm) ≙ 0.42 mg/m$^3$** |
| **Peak limitation (1983, 2000)** | **Category 1 (excursion factor 1)*** |
| **Absorption through the skin** | **–** |
| **Sensitization (1981)** | **Sh** |
| **Carcinogenicity** | **–** |
| **Prenatal toxicity (1994)** | **Pregnancy risk group C** |
| **Germ cell mutagenicity** | **–** |
| **BAT value** | **–** |
| Synonyms | 1,3-diformylpropane<br>glutaral<br>glutaric dialdehyde<br>1,5-pentanedial |
| Chemical name (CAS) | pentanedial |
| CAS number | 111-30-8 |

The documentation of the MAK value for glutaraldehyde from 1994 (Volume 8 of the present series) describes the irritative effects of the substance and cases of contact allergy. Glutaraldehyde vapour can have irritative effects on the respiratory passages, but there was no evidence of sensitization after inhalation. In the meantime studies have become available which report cases of occupational asthma said to be caused by exposure to glutaraldehyde. In particular there are 2 extensive epidemiolcgical studies from the hospital sector and a study carried out in a factory which allow re-evaluation of the question of glutaraldehyde-induced allergic asthma.

## 1 Sensitization of the Respiratory Passages

In Great Britain, as a result of cases reported in the Surveillance of Work-related and Occupational Respiratory Disease (SWORD) project, glutaraldehyde is thought to have asthma-inducing effects. SWORD is a voluntary register which includes 12 categories of

* A momentary value of 0.2 ml/m$^3$ (0.83 mg/m$^3$) should not be exceeded.

disease and is based on a prescribed list of hazardous substances (Meredith *et al.* 1991). Glutaraldehyde was included in 1990 in the list of substances which induce occupational asthma. The justification for this ordinance is, however, not specifically documented (DSS 1990), and scientific evidence for sensitizing effects on the respiratory passages cannot be derived from it. While in 1989 only 2 of the total 554 registered cases of occupational asthma were ascribed to glutaraldehyde exposure (Meredith *et al.* 1991), the number of asthma cases attributed to glutaraldehyde and registered under the SWORD project from 1989–1990 was 20 out of 1085 (Meredith 1993), and in the whole period from 1989–1991 30 cases out of 1528 (Meredith and McDonald 1994). In 1993, with an estimated total number of cases of work-related asthma of 1047, 13 cases of disease after exposure to glutaraldehyde were registered, mainly from the health services (Sallie *et al.* 1994).

The scientific content of this information is in general limited: the diagnostic criteria which led to registration are unknown, and there are no results available for the exposure levels, so that a causal relationship between disease and exposure cannot be derived.

## 1.1 Case reports

As an example, a case is reported of a nurse from an endoscopy department who suffered asthma attacks at the workplace after 7 years exposure to glutaraldehyde. In a first challenge test with 0.032 ml/m$^3$ (0.133 mg/m$^3$) a dual asthmatic reaction was registered, and after provocation with methacholine bronchial hyperreactivity was observed. Similar reactions also occurred, however, after exposure to NaCl solution, and incorrect spirometric readings, probably in connection with an existing infection of the respiratory passages, could not be excluded. Provocation was therefore repeated with glutaraldehyde in concentrations of 0.01–0.32 ml/m$^3$ (0.042–1.331 mg/m$^3$). This yielded negative results even with the highest concentration. The test was carried out as a double blind test with intermittent NaCl controls, the samples were administered and dosed with a respirator, and the woman was prevented from detecting any odours with a nose clip. Under controlled conditions it could therefore not be confirmed that her asthma was caused by glutaraldehyde. The authors draw attention to the necessity for stringent blind conditions, to exclude wrong diagnoses in challenge tests. In this case psychological factors were regarded as decisive for the induction of the asthmoid reactions (Stenton *et al.* 1994).

A study describes seven women employed in various hospitals (four endoscopy nurses, one operating theatre nurse, one x-ray assistant and one x-ray secretary), who were exposed to glutaraldehyde and referred to a special clinic because of asthmatic symptoms. The latency period between the first glutaraldehyde exposure and the occurrence of the symptoms was 6 months to 23 years, in five of the seven persons more than 4 years. One of the employees already suffered from asthma, and two others produced atopic reactions in the prick test with standard allergens. To confirm the asthma diagnosis, serial analyses of the peak expiratory flow (PEF) were taken over 4 weeks; the data are, however, not available. In three of four persons investigated, bronchial hyperreactivity was diagnosed in a test with inhalation of methacholine. The other three persons were tested with inhaled histamine and their reactivity was found to be normal.

Determinations of the glutaraldehyde concentrations in 13 different endoscopy rooms (which, however, were carried out at the workplace of the persons investigated in only three cases), yielded values of 0.003 to 0.94 mg/m$^3$, in six x-ray dark rooms less than 0.009 mg/m$^3$. Bronchial challenge tests were carried out in exposure chambers; the test persons were exposed for 10 minutes to a 2 m$^2$ piece of cardboard which was coated with aqueous 2 % glutaraldehyde, 1 % formaldehyde or 0.9 %r NaCl solutions. The forced expiratory volume of all 7 persons was decreased by 15 % to 44 % in the first second ($FEV_1$) after the exposure to glutaraldehyde (positive from values equal to 15 % of the NaCl control). A $FEV_1$ decrease of 27 % to 33 % was found in 3 of 6 of the persons exposed to formaldehyde. Two of the women produced dual reactions after exposure to glutaraldehyde, the other 5 only delayed reactions, also after formaldehyde. Glutaraldehyde concentrations simulated in the exposure chamber (i.e. determinations carried out outside the test period) yielded values of 0.064 to 0.081 mg/m$^3$ (Gannon *et al.* 1995). The authors base their supposition that the asthma was caused by glutaraldehyde mainly on the positive results in bronchial challenge tests. The relevance of these results is, however, questionable, as testing with this highly odorous substance was carried out without masks. The long latency period between the beginning of exposure and the occurrence of symptoms, and the delayed reactions and dual reactions to relatively low concentrations in the bronchial challenge test in this study are indicative of sensitizing effects of glutaraldehyde on the respiratory passages. The concentrations of the substance which induced the symptoms were, however, not determined directly. There is no mention of a control group, and the criteria for the selection of the test persons from the total collective of persons exposed in the hospitals are unknown. In addition, direct analysis of the exposure and workplace concentrations was not carried out. The concentrations determined at the workplaces cannot be matched to the individual cases. There is also no mention of immunological reactions, and specific skin tests were not carried out (Gannon *et al.* 1995).

In a subsequent study, glutaraldehyde-specific IgE was determined in the blood of 20 persons exposed to glutaraldehyde and in the stored blood samples of 21 persons said not to have been exposed to the substance. In 13 of the exposed persons, work-related asthma had been clinically diagnosed. Seven persons of this group are the same cases described by Gannon *et al.* (1995) and only for these cases is there a description of the diagnostic criteria and exposure values. The other 7 of the 20 exposed persons claimed to have respiratory symptoms which, however, were not described in any more detail. Only in one person was a challenge test with glutaraldehyde carried out and it yielded a negative result. With the criteria given by the authors (RAST $> 0.88$ % with a total IgE $\leq 150$ kU/l), the radio-allergo-sorbent test (RAST) showed 2 persons of the group described by Gannon *et al.* (1995) and 2 exposed persons without asthmatic symptoms to have glutaraldehyde-specific IgE. The authors concluded from their results that there is no correlation between respiratory symptoms or occupational asthma and glutaraldehyde-specific IgE (Curran *et al.* 1996).

## 1.2 Cohort studies

218 workers from a chemical company in the USA, who between 1959–1992 were employed in the production and packaging of glutaraldehyde, were investigated in a study. Up to 1978 the workers were also exposed to a series of other substances which have sensitizing effects on the skin and respiratory passages. The 8-hour average values for glutaraldehyde determined in the period 1977 to 1988 during routine air analyses were found to be 0.05 ml/m$^3$ (0.208 mg/m$^3$). After 1989 15-minute peak values between 0.01 and 0.34 ml/m$^3$ (0.042–1.414 mg/m$^3$) were determined. Contact allergy, sensitization of the respiratory passages and allergic blepharoconjunctivitis which could be attributed to glutaraldehyde did not occur (Teta *et al.* 1995).

In an Australian cross-sectional study in 26 hospitals, 135 employees (121 nurses, 14 male nurses), who regularly sterilized instruments with glutaraldehyde (105 in the operating area, 18 in endoscopy rooms), were asked about their state of health and symptoms during the previous 12 months. In addition, workplace analyses were carried out and the exposure levels determined. A similar group of 132 persons employed in the nursing professions (115 nurses, 17 male nurses), without direct exposure to glutaraldehyde, served as controls. The average concentration of glutaraldehyde in the ambient air in the endoscopy area was found to be 0.053 ml/m$^3$ (0.22 mg/m$^3$) and in the operating area 0.028 ml/m$^3$ (0.116 mg/m$^3$). For about 20 % of the analyses in the vicinity of the employees, exposure values of more than 0.1 ml/m$^3$ (0.416 mg/m$^3$) were registered. Irritation of the skin, eyes and throat was reported significantly more often by the persons exposed to glutaraldehyde than by the control group. Surprisingly, the prevalence of the non-dermal symptoms correlated inversely with the concentration of glutaraldehyde in the workplace air. A relationship between the symptoms and the glutaraldehyde exposure is therefore unlikely. The authors could find no explanation for the lack of dependency of the symptoms on the inhaled concentration of glutaraldehyde, but suspect that glutaraldehyde-tolerant persons were over-represented in the population investigated. The presence of other irritants or allergens in the ambient air, e.g. latex, was not investigated in this study. According to more recent studies, in the medical sector latex allergens from protective gloves play an important role in the occurrence of respiratory diseases, including bronchial asthma (Heilman *et al.* 1996, Kujala and Reijula 1996). There was, therefore, no evidence of an increase in the occurrence of respiratory symptoms, in particular of asthmatic complaints, in employees exposed to glutaraldehyde in hospitals (Pisaniello *et al.* 1994).

An even more extensive study was carried out in 59 endoscopy departments of 58 hospitals in Great Britain. This cross-sectional study included 348 employees from endoscopy departments (319 nurses), of whom 318 had been exposed to glutaraldehyde during the sterilization of instruments for one month to 19 years (on average 2.18 years), and 16 also to peroxyacetic acid and another 16 to a succinaldehyde/formaldehyde mixture. The other 30 persons used only succinaldehyde/formaldehyde. There was no control group as it was difficult to identify hospital personnel who worked under the ambient air conditions of the endoscopy departments, but were never exposed to glutaraldehyde. The detailed questionnaires included questions on the state of health and workplace, and the investigations comprised prick tests with standard allergens and latex, blood sampling to determine the total IgE, and glutaraldehyde and latex-specific IgE, lung function tests (forced vital capacity FVC and forced expiratory volume in one second $FEV_1$), and PEF (peak expiratory flow) tests during 4 weeks for persons with work-related respiratory symptoms. The glutaraldehyde concentration was determined both in the breathing zone of the persons exposed, to evaluate the peak values, and in the ambient air of the endoscopy departments during the work period. Average peak values of 0.057 mg/m$^3$ (less than 0.001–1.08 mg/m$^3$) and average ambient air values of 0.0143 mg/m$^3$ (0.002–0.15 mg/m$^3$) were found. Of the 318 persons primarily exposed to glutaraldehyde, 121 (38.1 %) were found to be asymptomatic. Work-related nasal irritation was stated by 63 persons (19.8 %), eye irritation by 43 (13.5 %) and pulmonary symptoms by 27 (8.5 %). These symptoms were, however, more often regarded by the persons questioned as not work-related (23.3 %, 15.1 % and 23.3 %), and nasal irritation was the only symptom which was of low significance ($p = 0.041$) and associated with peak exposures to glutaraldehyde (after spilling solutions). Lung function tests revealed no significant deviations from the normal values, and neither did evaluation of the PEF diagrams. In the study population there was, therefore, no evidence of asthma. The immunological tests with 321 blood samples yielded only one positive result for glutaraldehyde-specific IgE (RAST $\geq$ 0.88 % with a total IgE $\leq$ 120 kU/l), and 13 positive results for latex IgE. There was no relationship between the immunoglobulin status and the symptoms. The prick tests produced 20 reactions to latex in 336 persons.

As part of this investigation 18 former employees were questioned, all of whom had been exposed to glutaraldehyde, and of whom 12 stated respiratory complaints to be the reason for their leaving the job. Ten of these nurses still had asthmatic complaints, although 9 of them no longer had direct contact with glutaraldehyde. In one of these cases latex-specific IgE was detected, and another nurse with persistent asthma reacted in the prick test to latex allergens (Pickering and Vyas 1997).

# 2 Manifesto (sensitization)

Compared to the high number of persons exposed to glutaraldehyde in health care institutions there are only a few, regional case descriptions of asthma apparently caused by glutaraldehyde. The causal relationships have not been explained satisfactorily and are not sufficient evidence of sensitizing effects of glutaraldehyde on the respiratory passages. A relationship between respiratory symptoms and glutaraldehyde-specific IgE could not be found to date.

The extensive epidemiological studies which include the risk populations in the hospital sector also did not reveal asthma-inducing effects of glutaraldehyde. The exposure levels found in these studies are in the range of the currently valid MAK value, in some cases even above it. No evidence was found of allergenic effects on the respiratory passages, and the results of spirometry did not reveal the presence of irritative bronchial hyperreactivity. Glutaraldehyde has, therefore, not been designated with an "Sa", the existing designation "Sh" has been retained.

# 3 References

Curran AD, Burge PS, Wiley K (1996) Clinical and immunological evaluation of workers exposed to glutaraldehyde. *Allergy 51*: 826–832

DSS (Department of Social Security) (1990). *Occupational asthma.* Report by the Industrial Injuries Advisory Council (Command 1244) London: HMSO

Gannon PF, Bright P, Campbell M, O'Hickey SP, Burge PS (1995) Occupational asthma due to glutaraldehyde and formaldehyde in endoscopy and *x* ray departments. *Thorax 50*: 156–159

Heilman DK, Jones RT, Swanson MC, Yunginger JW (1996) A prospective, controlled study showing that rubber gloves are the major contributor to latex aeroallergen levels in the operating room. *J Allergy Clin Immunol 98*: 325–330

Kujala VM, Reijula KE (1996) Glove-related rhinopathy among hospital personnel. *Am J Ind Med 30*: 164–170

Meredith SK (1993) Reported incidence of occupational asthma in the United Kingdom, 1989–1990. *J Epidemiol Comm Health 47*: 459–463

Meredith SK, McDonald JC (1994) Work-related respiratory disease in the United Kingdom, 1989–1992: report on the SWORD project. *Occup Med 44*: 183–189

Meredith SK, Taylor VM, McDonald JM (1991) Occupational respiratory disease in the United Kingdom, 1989: a report to the British Thoracic Society and the Society of Occupational Medicine by the SWORD project group. *Br J Ind Med 48*: 292–298

Pickering CAC, Vyas A (1997) *Survey of hospital endoscopy nursing staff with particular reference to symptoms, respiratory function, immunology and occupational exposures.* Department of Occupational and Environmental Medicine, North West Lung Centre, Wythenshawe Hospital, Manchester, England, August 1997

Pisaniello D, Gun R, Tkaczuk M, Nitschke M, Crea J (1994) *Glutaraldehyde exposure among endoscopy nurses.* Final Report for Worksafe Australia. University of Adelaide and Occupational Health Division, South Australian Department for Industrial Affairs, May 1994

Sallie BA, Ross DJ, Meredith SK, McDonald JC (1994) SWORD '93. Surveillance of work-related and occupational respiratory disease in the UK. *Occup Med 44*: 177–182

Stenton SC, Beach JR, Dennis JH, Keaney NP, Hendrick DJ (1994) Glutaraldehyde, asthma and work – a cautionary tale. *Occup Med 44*: 95–98

Teta MJ, Avashia BH, Cawley TJ, Yamin AT (1995) Absence of sensitizations and cancer increases among glutaraldehyde workers. *Toxic Subst Mech 14*: 293–305

completed 15.12.1997

# Hexachlorobenzene

| | |
|---|---|
| **MAK value** | – |
| **Absorption through the skin (1998)** | **H** |
| **Sensitization** | – |
| **Carcinogenicity (1998)** | **Category 4** |
| **Prenatal toxicity** | – |
| **Germ cell mutagenicity** | – |
| **BAT value (1984)** | **150 µg/l serum**<br>**Sampling time: not fixed** |
| Synonyms | HCB<br>perchlorobenzene |
| Chemical name (CAS) | hexachlorobenzene |
| CAS number | 118-74-1 |
| Structural formula | 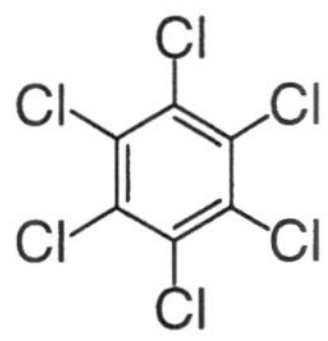 |
| Molecular formula | $C_6Cl_6$ |
| Molecular weight | 284.78 g/mol |
| Melting point | 227–231°C |
| Boiling point | 322–326°C |
| Density at 20°C | 2.075 g/cm$^3$ |
| Vapour pressure at 20°C | 1.1–1.68 x $10^{-3}$ Pa |
| log $P_{ow}$* | 5.5 |
| **1 ml/m$^3$ (ppm) ≙ 11.82 mg/m$^3$** | **1 mg/m$^3$ ≙ 0.085 ml/m$^3$ (ppm)** |

The present document is based on a report on hexachlorobenzene by the "Beratergremium für umweltrelevante Altstoffe" (BUA 1994).

---

* *n*-octanol/water distribution coefficient

Hexachlorobenzene was introduced in the 1940s as a fungicide for cereal seed. It was also used for pyrotechnics. Its use has been forbidden since 1975 in western Germany and since 1984 in eastern Germany. It is, however, formed in large quantities during the production of chlorinated solvents, occurs as a contaminant in pesticides and is a metabolite of lindane (BUA 1994).

# 1 Toxic Effects and Mode of Action

Hexachlorobenzene is a lipophilic substance which accumulates in adipose tissues and is released again only very slowly. The biological half-life in man is about 2 years and is much longer than the half-life in the rat (up to about 5 months).

Ingested hexachlorobenzene is absorbed readily from oily solutions but poorly from aqueous suspensions or when administered in crystalline form. It has been shown to be absorbed through the skin.

Mortality was high among the persons affected during the 4 to 5 years of a mass poisoning with fungicide containing hexachlorobenzene. The symptoms described were general weakness, skin lesions, porphyria, hyposomia, osteoporosis, arthritis and neuritis. Enlargement of the liver, thyroid and lymph nodes was also reported.

In animal studies the acute toxicity is low. The $LD_{50}$ for the rat is of the order of 3500–10000 mg/kg body weight. The symptoms of intoxication include tremor, convulsions and ataxia.

In animals given the substance for longer periods by oral administration, dermal lesions such as hair loss, blister and scab formation and hyperpigmentation were seen. In experimental animals the main target organ is the liver. Porphyria is observed especially in female rats and pigs (in both cases after doses of 0.5 mg/kg body weight and day administered for periods of 15 and 13 weeks, respectively). After long-term administration of hexachlorobenzene to rats in the diet, degenerative effects are seen in the livers of animals treated with concentrations of 8 ppm (doses of about 0.8 mg/kg body weight and day), preneoplastic foci in those treated with 40 ppm (about 4 mg/kg body weight and day) and liver tumours at 75 ppm (about 7.5 mg/kg body weight and day). Liver tumours also develop in mice given 100 ppm hexachlorobenzene in the diet (about 12 mg/kg body weight and day) and in hamsters given 50 ppm (about 4 mg/kg body weight and day). In rats, bile duct tumours, renal adenomas and adrenal pheochromocytomas are also seen. The liver, bile duct and kidney tumours may be attributed to the induction of cytochrome P450 (CYP) isozymes and the consequent formation of oxygen radicals which results in cell damage and compensatory hyperplasia. Tumour-promoting effects of the substance in the liver have been demonstrated in rats.

The thyroid gland is also a target organ. Hexachlorobenzene causes increased excretion of thyroxine (T4) which can result in stimulation of the thyroid and thyroid hyperplasia. In hamsters, thyroid adenomas develop (200 ppm, about 20 mg/kg body weight and day).

Hexachlorobenzene causes immunomodulation. In rats, hexachlorobenzene causes an increase in humoral and cell-mediated immune responses and autoimmune reactions. In mice, on the other hand, a suppression of immune responses was generally observed.

Hexachlorobenzene doses of 0.01 mg/kg body weight and day, administered to female monkeys for 13 weeks, caused ultrastructural changes in the ovaries. However, effects on fertility were not seen even after administration of doses as high as 10 mg/kg body weight and day for 90 days.

Hexachlorobenzene crosses the placenta and accumulates in the foetus. In a study with rats an increased incidence of 14th ribs was observed, and renal anomalies were seen in the offspring of mice given maternally toxic doses. These findings are conceivably a result of the general toxicity of hexachlorobenzene. During lactation, hexachlorobenzene is mobilized with the fat reserves and excreted in the milk. In breast-fed babies and suckling animal offspring whose mothers are exposed to hexachlorobenzene, severe toxic effects and increased mortality is observed.

Mutagenic effects have not been detected but there is evidence of weak clastogenicity.

## 2 Mechanism of Action

Hexachlorobenzene induces CYP1A1 in the liver, lungs and kidneys of the rat. CYP1A2 is induced mainly in the liver and lungs and CYP2B1 in the liver (Gulyaeva *et al.* 1994). The induction of cytochrome P450 isozymes and the redox reactions involving hexachlorobenzene metabolites such as tetrachlorohydroquinone result in the formation of oxygen radicals and in lipid peroxidation (Almeida *et al.* 1997, Billi de Catabbi *et al.* 1997). The consequent cell damage and compensatory hyperplasia have been suggested as the mechanism of induction of the liver tumours seen in experimental animals (Carthew and Smith 1994). In monkeys, ultrastructural alterations of the ovarian follicle membrane could be considered to be a result of the induction of cytochrome P450 (Bourque *et al.* 1995).

The porphyria induced by hexachlorobenzene is a result of changes in activities of the enzymes of haem biosynthesis, especially the inhibition of uroporphyrinogen decarboxylase. This results in an accumulation of intermediates of haem biosynthesis, especially in the liver, and in increased porphyrin excretion (Billi de Catabbi *et al.* 1997, BUA 1994, Mylchreest and Charbonneau 1997). The oxidative biotransformation and thus the formation of the very reactive tetrachloro-1,4-benzoquinone is directly related to the porphyrinogenic action of hexachlorobenzene (van Ommen *et al.* 1989). Later studies demonstrated, however, that a reactive intermediate formed during the cytochrome P450-mediated oxidation of hexachlorobenzene to pentachlorophenol could be responsible for the porphyrinogenic effects (Rietjens *et al.* 1995). There is also evidence that thyroid hormones play an important role in hexachlorobenzene-induced porphyria (Soprena de Kracoff *et al.* 1994).

Reduced T4 concentrations in serum were observed in man and rats exposed to hexachlorobenzene. This may be due in part to the induction of various UDP glucuronosyl transferases in the rat liver and the consequent increase in glucuronidization and excretion of T4 in the bile (van Raaij *et al.* 1993a). In addition, the hexachlorobenzene metabolite pentachlorophenol was shown to compete with T4 for the binding site on its transport protein (transthyretin) and to have an affinity for the transport protein which was twice that of T4 (van Raaij *et al.* 1993b). The increased excretion of T4 resulted in TSH-mediated stimulation of the thyroid and thus in hypertrophy and hyperplasia.

There is evidence that immunomodulation reactions are involved in the aetiology of the hexachlorobenzene induced skin lesions (Michielsen *et al.* 1997).

The mechanisms of the neurological and behavioural effects of the substance have not yet been clarified. Prenatal and postnatal exposure to hexachlorobenzene impairs scheduled-controlled behaviour in rats (Lilienthal *et al.* 1996). It has been suggested that hexachlorobenzene interferes with myelination during development or that it modifies the activity of neurotransmitters. Effects of altered thyroid hormone concentrations on the developing nervous system are also conceivable (see Lilienthal *et al.* 1996).

Studies of the mechanism of action of hexachlorobenzene on the endocrine and immune systems have not been published. In analogy to 2,3,7,8-tetrachlorodibenzo-*p*-dioxin, hexachlorobenzene could act by Ah receptor-mediated mechanisms.

# 3 Toxicokinetics and Metabolism

Two days after oral administration of a hexachlorobenzene dose of 20 mg/kg body weight in olive oil to female Wistar rats, 83.2 % of the dose had been absorbed. When the substance was administered in water (16 mg/kg body weight) only 19.7 % of the dose had been absorbed after 3 days (BUA 1994).

Hexachlorobenzene was administered by occlusive application of a dermal dose of 10 mg/kg body weight in tetrachloroethylene to male Fischer rats. After 6 hours 1.05 % of the dose had been absorbed, after 24 hours 2.67 % and after 72 hours 9.71 % (BUA 1994).

Hexachlorobenzene is distributed into all organs. The time required to reach equilibrium depends markedly on the administration route; after oral administration of hexachlorobenzene in oil, it is 2 to 5 days. The highest concentrations are found in adipose tissue (BUA 1994). In the rat, high concentrations are also found in endocrine tissues such as the thyroid gland, adrenal glands and ovaries (Foster *et al.* 1993).

Hexachlorobenzene is oxidized by CYP3A to pentachlorophenol and then to tetrachlorohydroquinone. In addition, the conjugation products *N*-acetyl-*S*-(pentachlorophenyl)cysteine and mercaptotetrachlorothioanisole are found (den Besten *et al.* 1994). The rate of metabolism is, however, low (BUA 1994).

Hexachlorobenzene is excreted only very slowly. Fourteen days after intraperitoneal administration of a hexachlorobenzene dose of 4 mg/kg body weight in oil to Wistar rats,

34 % of the dose had been excreted with the faeces (80 % as unchanged hexachlorobenzene) and 5 % with the urine (96 % as metabolites) (Koss and Koransky 1975). The excretion kinetics are biphasic or multiphasic. The elimination half-time in the rat was up to 5 months, in the Rhesus monkey 2.5 to 3 years (BUA 1994). The biological half-life in workers exposed to hexachlorobenzene was found to be about 2 years (Henschler and Lehnert 1984).

# 4 Effects in Man

There are no data available for the effects of single exposures of persons to hexachlorobenzene nor for its effects on skin or mucous membranes or allergenic effects.

## 4.1 Repeated exposure

### 4.1.1 Workers

In a group of 50 workers in the production of chlorinated hydrocarbons who were monitored in the years 1974 to 1977, the average plasma hexachlorobenzene level varied around 325 µg/l, 65 times the average concentration of 5 µg/l found in the general population. Because the biochemical and liver function parameters in these workers were not statistically different from those of control collectives, the BAT value for hexachlorobenzene was established at 150 µg/l plasma (Henschler and Lehnert 1984).

Workers who were exposed until 1980 to hexachlorobenzene concentrations of 2.1 to 10.8 mg/m$^3$ and during the study period between 1983 and 1990 to 0.012 to 0.022 mg/m$^3$ still had serum hexachlorobenzene levels of more than 500 µg/l even 9 years after the exposure concentrations had been reduced (1980: 534 µg/L, 1989: 575 µg/L). Significant changes, but within the normal biological range, were found in the levels of immunoglobulin (IgG, IgA, IgM and IgE increased) and transport proteins (including transferrin) and there was evidence of effects on kidney function and of reduced immune responses (Richter *et al.* 1994).

### 4.1.2 General population

In the years between 1955 and 1959, the use of seed corn treated with hexachlorobenzene for flour and bread production resulted in poisoning of more than 3000 persons. The seed had been treated with 0.2 % of a fungicide containing 10 % hexachlorobenzene. The exact composition of the fungicide is, however, unknown. The symptoms included photosensitive lesions on skin areas exposed to light, bald skin areas, arthritic hands and fingers, hypertrichosis, enlargement of the liver and thyroid, weight loss, general weakness, anorexia and, mainly in men, porphyria (*porphyria cutanea tarda*). The daily dose of hexachlorobenzene was estimated to be 50 to 200 mg (about 0.7–4 mg/kg body weight). About 10 % of the affected persons died. Late sequelae such as scarring, hyperpigmentation, arthritis, osteoporosis, myotonia, paresthesia and enlargement of the thyroid gland were still recognizable in some cases 30 years later. Children under 4 years of age, especially breast-fed children of mothers exposed to hexachlorobenzene, were particularly sensitive (BUA 1994).

## 4.2 Reproductive toxicity

Breast-fed babies and children whose mothers had eaten seed corn treated with hexachlorobenzene in the years 1955 to 1959 (Section 4.1) had more severe symptoms of intoxication (including cutaneous erythema ("pink sores"), weakness, tremor and convulsions) than did adults. Mortality was as high as 95 % in some areas (BUA 1994).

## 4.3 Genotoxicity

In 41 workers who were exposed to tetrachloromethane, tetrachloroethylene and hexachlorobenzene, the incidence of micronuclei in the peripheral lymphocytes was increased significantly above that in 28 control persons but was not correlated with the serum hexachlorobenzene concentration (da Silva Augusto *et al.* 1997).

## 4.4 Carcinogenicity

A hepatocellular carcinoma was diagnosed in a worker who was employed from 1955 to 1981 in an aluminium smelter. Until 1973, hexachloroethane was added to degas the molten aluminium (700°C, open container); this resulted in the formation of large amounts of gas with a chlorine-like odour. Ventilation and protective clothing were not available. In another plant it was demonstrated that during this process not only chlorobenzenes, chlorophenols, dioxins and dibenzofurans are formed but also particularly high concentrations of hexachlorobenzene (Seldén *et al.* 1989). However, because it must be assumed that the man was also exposed to the other chlorinated compounds, it is not possible to ascribe the tumour formation to the hexachlorobenzene exposure alone.

In a town with a population of 5003 which was close to an old factory producing chlorinated organic compounds and chlorinated solvents, significantly increased hexachlorobenzene concentrations were found in the air (average 35 µg/m$^3$, reference value 0.3 µg/m$^3$); in earlier years the concentrations must have been much higher, at times over 100 µg/m$^3$. The concentrations of polychlorinated biphenyls (PCB), dichlorobischlorophenylethene (DDE), chloroform (trichloromethane), tetrachloromethane, trichloroethylene and tetrachloroethylene in the air were not increased above the reference concentration. The concentrations of chlorinated compounds were measured in the serum of 21 persons. The concentration of hexachlorobenzene was 26 μg/l (7.5–69 μg/l) and was significantly increased above the values in a reference population (4.8 μg/l (1.5–15 μg/l). The average concentration in the 9 men who were studied was 32 μg/l and was higher than the 22 μg/l found in the 12 women. The concentrations of DDE (12 μg/l) were increased relative to those in the control population (1.6 μg/l) but were not significantly higher than those in other populations (e.g. 1978–1982 in the Netherlands 3.9–22 μg/l). Concentrations of other substances were not increased. In the study period from 1980 to 1989, the incidences of "unknown tumours" (ICD 199), tumours of the thyroid (ICD 193), the brain (ICD 191) and of soft-tissue sarcomas (ICD 171) were increased. The increase was significant or of borderline significance for males (Table 1). All the men with tumours were employed in the factory (Grimalt *et al.* 1994). The authors suggest that there could be an association between the observed tumours and the exposure to chlorinated organic compounds, especially hexachlorobenzene, because the thyroid gland is a target organ of hexachlorobenzene in experimental animals and soft tissue sarcomas have been associated with the exposure to chlorinated compounds.

**Table 1.** Incidences of tumours in the years 1980–1989 in a population exposed to hexachlorobenzene (from Grimalt *et al.* 1994)

| Tumour localization (ICD-9) | Men number | Men SIR (95% CI) | Women number | Women SIR (95% CI) |
|---|---|---|---|---|
| thyroid gland (193) | 2 | 6.7 (1.6–28)* | 1 | 1.0 (0.14–7.4) |
| soft tissue (171) | 3 | 5.5 (1.7–17.5)* | 1 | 2.2 (0.3–16) |
| brain (191) | 4 | 2.7 (0.99–7.2) | 1 | 0.93 (0.13–6.7) |
| unknown (199) | 10 | 2.35 (1.25–4.4)* | 4 | 1.2 (0.45–3.2) |
| urinary bladder (188) | 12 | 1.3 (0.75–2.3) | 2 | 1.1 (0.28–4.6) |
| female breast (174) | 0 | – | 19 | 1.3 (0.84–2.1) |
| colo-rectum (153–154) | 10 | 1.0 (0.54–1.9) | 7 | 0.82 (0.39–1.7) |
| lungs (162) | 11 | 0.76 (0.42–1.4) | 2 | 1.6 (0.39–1.7) |
| lymphomas (196) | 1 | 0.56 (0.08–4.02) | 1 | 0.67 (0.09–4.8) |
| liver (155) | 0 | – | 1 | 0.82 (0.11–5.8) |
| Total (140–208) | 74 | 0.96 (0.76–1.2) | 55 | 0.93 (0.72–1.2) |

* significant; level of significance not specified
ICD: International Classification of Disease
SIR: standardized incidence ratio
CI: confidence interval

There are no studies which have investigated the development of tumours in the population which was poisoned with the hexachlorobenzene-containing seed corn (Section 4.1).

# 5 Animal Experiments and *in vitro* Studies

## 5.1 Acute toxicity

The acute toxicity of hexachlorobenzene is low.

In inhalation studies, the $LC_{50}$ in the rat was found to be 3600 mg/m$^3$, in the mouse 4000 mg/m$^3$, in the rabbit 1800 mg/m$^3$ and in the cat 1600 mg/m$^3$.

The oral $LD_{50}$ values reported for the rat after administration of hexachlorobenzene were 3500 and 10000 mg/kg body weight, for the mouse 4000 mg/kg body weight, for the rabbit 2600 mg/kg body weight and for the cat 1700 mg/kg body weight. The symptoms of intoxication include weakness, convulsions, tremor and ataxia (BUA 1994).

A single oral hexachlorobenzene dose of 50 mg/kg body weight caused a significant reduction in the T4 level in female rats (Foster *et al.* 1993).

## 5.2 Subacute, subchronic and chronic toxicity

There are no inhalation studies available with hexachlorobenzene.

### 5.2.1 Ingestion

A description of all the studies with repeated administration of hexachlorobenzene to experimental animals may be found in the BUA review (BUA 1994). Here only a few selected studies are described.

Oral administration of hexachlorobenzene doses of 485 or 750 mg/kg body weight to male WAG-RIJ rats, three times weekly for 2 weeks, resulted respectively in slight and significant reductions in the blood T4 levels. At 256 mg/kg body weight, no changes were observed. Administration of hexachlorobenzene doses of about 1000 mg/kg body weight for 4 weeks resulted also in increased TSH concentrations (van Raaij *et al.* 1993b) and in significantly reduced T3 concentrations (Kleimann de Pisarev *et al.* 1995).

Porphyrinogenic effects of hexachlorobenzene have been observed in the rat, mouse, rabbit and pig. The NOEL for the porphyrinogenic effects was found to be 0.5 mg/kg body weight and day for the rat (treated for 9 months) and 0.05 mg/kg body weight and day for the pig (treated for 13 weeks) (BUA 1994). In the hamster too, especially in males, after administration of hexachlorobenzene in the diet in a concentration of 100 ppm (about 10 mg/kg body weight and day) for a period of 5 or 10 months, the relative liver weight was increased and the relative spleen weight reduced. In both organs deposition of iron was observed. The porphyrin level in the liver was increased (Ríos de Molina *et al.* 1996).

After administration of hexachlorobenzene to Sprague-Dawley rats and golden Syrian hamsters in concentrations of 0, 200 or 400 ppm in the diet (about 0, 20 and 40 mg/kg body weight and day) and to Swiss mice in concentrations of 0, 100 or 200 ppm (about 0, 15 and 30 mg/kg body weight and day) for up to 13 weeks, the pathological examination revealed mainly effects on the lymphohaematopoietic system (hyperplasia with lymphocytic infiltration in the liver, kidneys and in some other organs, lymphosarcomas), the liver (hepatitis, liver cirrhosis, metaplasia, preneoplastic foci, hepatomas) and the kidneys (severe hyperaemia, degeneration, nephritis, renal cell adenomas) (Ertürk *et al.* 1986). In the publication it is not clear which effects are associated with which concentrations; the study can therefore not be used in the present assessment.

In Sprague-Dawley rats given hexachlorobenzene in the diet in doses of 0, 0.5, 2, 8 and 32 mg/kg body weight and day for up to 15 weeks, the liver proved to be the most sensitive organ. Female rats were more affected than males. In the females of the group given 0.5 mg/kg body weight and day, the liver porphyrin levels were increased during the recovery period in week 19. The NOEL for porphyrin excretion was therefore below 0.5 mg/kg body weight and day. In animals given 2 mg/kg body weight and day or more, the induction of xenobiotic-metabolizing enzymes was observed. From doses of 8 mg/kg body weight and day, relative liver weights were increased and ultrastructural changes were detected in the liver (increased amount of smooth endoplastic reticulum). In addition, the porphyrin concentrations in the liver, kidneys and spleen were increased. In the animals given 32 mg/kg body weight and day, mortality was increased and clinical symptoms such as tremor, irritability, ataxia and hair loss were seen. Relative weights of kidneys, spleen and adrenal glands were increased (Kuiper-Goodman *et al.* 1977, Kuiper-Goodman and Grant 1986).

In a 2-generation study with Sprague-Dawley rats, hexachlorobenzene was administered in the diet in concentrations of 0, 0.32, 1.6, 8.0 and 40 ppm (about 0.03, 0.16, 0.8 and 4 mg/kg body weight and day) to the parent animals ($F_0$) for 3 months before mating and to the progeny ($F_1$) for 130 weeks. In the $F_0$ generation males given doses of 0.8 mg/kg body weight and day or more, statistically significant increases in relative liver weights and in absolute heart and brain weights were recorded. In the $F_1$ generation males given doses from 0.03 mg/kg body weight and day, the incidences of peribiliary lymphocytosis and fibrosis and biliary distention were increased. These effects were, however, not dose-related and can therefore not be evaluated. In animals given doses from 0.8 mg/kg body weight and day, the incidence of centrilobular basophilic chromogenesis in the liver was increased. With doses from 4 mg/kg body weight and day increased incidences of severe chronic nephrosis, parathyroid adenomas, neoplastic nodules in the liver (female animals) and phaeochromocytomas of the adrenal glands (female animals) were observed (Arnold *et al.* 1985, Arnold and Krewski 1988). The publication does not state whether the excretion of porphyrins or the induction of xenobiotic-metabolizing enzymes was determined. The NOAEL (no observed adverse effect level) can be given as 1.6 ppm hexachlorobenzene in the diet, that is about 0.16 mg/kg body weight and day.

Groups of 5 male pigs were given hexachlorobenzene in the diet for 13 weeks in doses of 0, 0.05, 0.5, 5 and 50 mg/kg body weight and day. At doses of 0.5 mg/kg body weight and day or more, the excretion of porphyrins was increased as was the induction of microsomal liver enzymes, hypertrophy and alterations in liver cells. From 5 mg/kg body weight and day, weights of liver, kidneys and thyroid glands were increased and the histopathological examination revealed changes in the liver. At 50 mg/kg body weight and day, anorexia, tremor, shortness of breath, and mortality were recorded and the histopathological examination revealed changes in the kidneys, testes and lymph nodes. The NOEL found in this study was 0.05 mg/kg body weight and day (den Tonkelaar *et al.* 1978).

In several studies by one research group, hexachlorobenzene was administered for 90 days in capsules to groups of 4 cynomolgus monkeys. At all doses from the lowest dose of 0.01 mg/kg body weight and day, the mitochondria in the developing egg cells were condensed and in the follicles deep indentations and an abnormal accumulation of lipid droplets in the cytoplasm were seen. The authors described these changes as reversible. In animals given 0.1 mg/kg body weight and day or more, degeneration of follicles was observed (Bourque *et al.* 1995). In the egg cell membranes, nuclei and cytoplasm, alterations were seen which became more severe with increasing dose (Jarrell *et al.* 1993). There were changes in the surface epithelium of the ovary such as stratification of the epithelial cells, columnar cells and irregular shapes, accumulation of lysosomes and lipid vesicles and signs of degeneration (Babineau *et al.* 1991, Sims *et al.* 1991). Histopathological examination of animals given doses of 1 mg/kg body weight and day or more revealed changes in the liver. At 10 mg/kg body weight and day the weights of the liver and adrenal glands were increased. Increased excretion of porphyrins was not detected. The number of primordial follicles was reduced significantly. There were no adverse effects on menstruation or fertility (Jarrell *et al.* 1993). A NOEL cannot be derived from the results of this study.

## 5.3 Local effects on skin and mucous membranes

Hexachlorobenzene has been described as having little irritating effect on the skin and none in the eyes, but this claim is inadequately documented (BUA 1994).

## 5.4 Allergenic effects

In a study by Landsteiner and Jacobs (1936), hexachlorobenzene (0.05 mg in 0.05 ml olive oil) was administered intracutaneously to guinea pigs (numbers not specified), once weekly for 10 weeks. After a 2-week pause, the provocation treatment was carried out with "1 drop of a 20 % solution of the substance in alcohol". Sensitization was not detected (BUA 1994).

## 5.5 Reproductive and developmental toxicity

### 5.5.1 Fertility

**Fertility of males**

In a dominant lethal test, male rats were treated with hexachlorobenzene doses of 70 or 221 mg/kg body weight and day for 5 days; the only effect was dose-dependent reduction in the number of matings (Simon *et al.* 1979), presumably a result of the fact that the doses were in the toxic range.

**Fertility of females**

Administration of oral doses of hexachlorobenzene of 0, 1, 10 or 100 mg/kg body weight and day for 30 days to adult female Sprague-Dawley rats resulted in reduced concentrations of corticosterone in the blood of ovarectomized animals given 1 mg/kg body weight. There were no effects on the levels of progesterone or aldosterone (Foster *et al.* 1995b). In superovulated animals (21 days exposure, ovulation stimulated with gonadotropin) the ovary weights were increased only in the animals given 1 mg/kg body weight and day. However, the serum progesterone level was increased in all dose groups (Foster *et al.* 1992). Treatment with 10 mg/kg body weight and day led to ultrastructural changes in the granulosa-lutein cells (MacPhee *et al.* 1993).

In cynomolgus monkeys given oral doses of hexachlorobenzene for 13 weeks and then treated with follicle stimulating hormone (FSH) and luteinizing hormone (LH) to stimulate follicle development and gonadotropin to stimulate ovulation, even the lowest hexachlorobenzene dose of 0.01 mg/kg body weight and day produced ultrastructural changes in the developing egg cells and follicles (Bourque *et al.* 1995). In the animals given 0.1 mg/kg body weight and day, ultrastructural changes in the epithelium of the ovary and in the oocytes were seen (Babineau *et al.* 1991, Jarrell *et al.* 1993, Sims *et al.* 1991). In addition, the length of the menstrual cycle increased with the dose. After doses of 10 mg/kg body weight and day, the concentration of oestradiol during ovulation was significantly reduced (Foster *et al.* 1995a) but fertility was not affected (Jarrell *et al.* 1993; see also Section 5.2.2).

## 5.5.2 Developmental toxicity

**Prenatal toxicity**

Hexachlorobenzene readily passes the placental barrier. Accumulation in foetal tissue has been demonstrated in the rat (Nakashima *et al.* 1997), rabbit, pig and monkey (BUA 1994).

Prenatal toxic effects such as an increased incidence of unilateral and bilateral 14th ribs and sternal defects were seen in the progeny of rats given hexachlorobenzene doses of 40 mg/kg body weight and day for at least 10 days but not in those given doses up to 80 mg/kg body weight and day for up to 8 days (in both cases from day 6 of gestation). Therefore these results require confirmation. Toxic effects in the dams in the form of reduced body weights were first seen after administration of 80 mg/kg body weight and day for at least 16 days (Khera 1974). In the mouse, only maternally toxic doses have been studied. Doses of 100 mg/kg body weight administered to pregnant mice on days 7 to 16 of gestation caused increased liver weights in the dams and an increased incidence of kidney anomalies in the progeny, in one litter also an increase in the incidence of cleft palate (Courtney *et al.* 1976). In dams given doses of 125 mg/kg body weight and day for 5 days before mating and also on days 8 to 12 of gestation, mortality was increased. Of

30 litters, 2 were totally resorbed. Only 6 dams gave birth to pups; the litter sizes were significantly reduced (Kavlock *et al.* 1987).

**Perinatal and postnatal toxicity; multigeneration studies**

Because of the mobilization of lipids during lactation, a large amount of the hexachlorobenzene in the dams is transferred via the milk to the suckling pups and can result in toxicity and increased postnatal mortality. The available studies have been reviewed by BUA (1994). Therefore only the most important and more recent studies are described below.

In lactating sheep it was shown that, 37 days after administration of hexachlorobenzene in a single intramuscular injection of 0.2 mg/kg body weight, the concentrations of hexachlorobenzene in the milk (84 ng/g lipid) and blood (76 ng/g lipid) were similar, but the concentration in the blood of the lambs (216 ng/g lipid) was much higher (Vrecl *et al.* 1996).

In a 4-generation study, Sprague-Dawley rats were given hexachlorobenzene in the diet in concentrations up to 640 ppm (about 1, 2, 4, 8, 16, 32, 64 mg/kg body weight and day). In the dams given 16 mg/kg body weight or more, mortality was increased and from 32 mg/kg body weight the fertility was decreased (not clear whether male or female fertility). In the $F_1$ generation, birth weights were reduced from doses of 8 mg/kg body weight and survival during the lactation period from 16 mg/kg body weight; 32 mg/kg body weight resulted in an increased incidence of resorptions. In later generations, these effects were also seen in the groups given 4 mg/kg body weight. No significant effects were seen at 2 mg/kg body weight (BUA 1994).

In a study to examine neurobehavioural functions in rats after prenatal and postnatal exposure, female Wistar rats were given hexachlorobenzene in the diet in concentrations of 0, 4, 8, or 16 mg/kg diet (about 0, 0.4, 0.8, 1.6 mg/kg body weight and day) from 90 days before mating until the end of the lactation period. The progeny was given hexachlorobenzene in the diet until they were 150 days old. The relative liver weights in the dams were increased during the lactation period. In the progeny exposed to hexachlorobenzene, neither open-field activity (postnatal day 21) or active avoidance learning (postnatal day 90) was different from that of the control animals. There were significant changes in trained operant behaviour (from postnatal day 150) for animals given 0.8 mg/kg body weight or more (Lilienthal *et al.* 1996).

In sows given 1 or 20 ppm hexachlorobenzene in the diet (about 0.025 and 0.5 mg/kg body weight and day) from 1 week before mating until the birth (maximum 170 and 190 days), body weight gains were increased and mild haematological effects (increased total leukocyte count, reduced lymphocyte count, increased incidence of neutrophilia in the high dose group), a slight dose-dependent increase in relative liver weight and evidence of stomach irritation were seen. Mortality of the piglets was increased during the first week of lactation in both dose groups; the incidence of stillbirths was increased in the high dose group (BUA 1994).

## 5.6 Genotoxicity

### 5.6.1 *In vitro*

The studies of *in vitro* genotoxicity of hexachlorobenzene are listed in Table 2.

**Table 2.** *In vitro* genotoxicity studies with hexachlorobenzene

| Test system | Test organism | Concentration | Results without S9 | Results with S9 | References |
|---|---|---|---|---|---|
| *Salmonella* mutagenicity test (Ames test) | *S. typhimurium* TA98, TA100, TA1535, TA1537, TA1538 | up to 10000 µg/plate | – | – | BUA 1994 |
| mutation, DNA repair | *Escherichia coli* WP2, WPuvrA | up to 1000 µg/plate | – | | BUA 1994 |
| SOS chromotest | *E. coli* K12 PQ37 | no data | – | – | Raabe *et al.* 1993 |
| HPRT (8-azaguanine) | V79 cells | about 2–8 µg/ml | (+) | | Kuroda 1986 |
| $Na^+K^+$-ATPase (ouabain) | V79 cells | about 2–8 µg/ml | – | | Kuroda 1986 |
| chromosomal aberrations | human lymphocytes | 0.001–0.1 mM | – | | Siekel *et al.* 1991 |
| micronucleus test | rat hepatocytes | 0.10–0.18 mM<br>0.32 mM | –<br>+ | | Canonero *et al.* 1997 |
| micronucleus test | human hepatocytes | 0.10 mM<br>0.18 mM<br>0.32 mM | –/–[1]<br>+/(+)<br>(+)/(+) | | Canonero *et al.* 1997 |
| DNA strand breaks | rat hepatocytes | 0.18–0.56 mM | – | | Canonero *et al.* 1997 |
| DNA strand breaks | human hepatocytes | 0.18-0.56 mM | (+)[1,2] | | Canonero *et al.* 1997 |

+ positive and significant, (+) weakly positive (not significant), – negative
[1] results obtained with hepatocytes from two donors
[2] designated by the authors as biologically significant

In the *Salmonella* mutagenicity test, hexachlorobenzene proved not to be mutagenic (BUA 1994). An SOS chromotest with *Escherichia coli* yielded negative results (Raabe *et al.* 1993). Selection of V79 cells with ouabain revealed no increase in mutation frequency, with 8-azaguanine a minimal increase (Kuroda 1986). Hexachlorobenzene did not induce chromosomal aberrations in human lymphocytes (Siekel *et al.* 1991). An increased incidence of micronuclei in rat hepatocytes was seen at the highest hexachlorobenzene concentration tested. In human hepatocytes the incidence was slightly but not significantly increased. The incidence of DNA single strand breaks was not increased in rat hepatocytes but in human hepatocytes it was slightly (from 24 % to 32 % of eluted DNA) but not significantly increased (Canonero *et al.* 1997).

### 5.6.2 *In vivo*

Two dominant lethal tests with rats given hexachlorobenzene doses of 0, 20, 40 or 60 mg/kg body weight and day for 10 days (Khera 1974) or 0, 70 or 221 mg/kg body weight and day for 5 days (Simon *et al.* 1979) revealed no significant effects on pregnancy, fertility, number of implantations or resorptions (BUA 1994).

In one inadequately documented study, a test for sister chromatid exchange in the bone marrow cells of male mice yielded negative results. In the rat liver, a slight dose-dependent increase in DNA strand breaks was reported (Górski *et al.* 1986).

A test for replicative DNA synthesis in the livers of $B6C3F_1$ mice given single oral hexachlorobenzene doses of 500 or 1000 mg/kg body weight and examined after 24, 39 or 48 hours revealed a slight but not significant increase in the incidence of replicative DNA synthesis in the 500 mg/kg group animals from 0.18 % to 0.36 %, 0.35 % and 0.29 %, respectively. In the 1000 mg/kg group, the incidences were 0.16 %, 0.66 % and 0.01 %; the 0.66 % incidence proved statistically significant (Miyagawa *et al.* 1995).

In male mice given hexachlorobenzene in doses of 170 mg/kg body weight and day for 5 days, the liver weights were significantly increased. The incidence of DNA single-strand breaks and of 8-hydroxydeoxyguanosine formation were not increased (Umegaki *et al.* 1993).

## 5.7 Carcinogenicity

### 5.7.1 Short-term studies

In initiation-promotion studies, initiating potential of hexachlorobenzene in the rat liver could not be detected but a tumour-promoting effect was found (Table 3). This was evident after initiation with a single diethylnitrosamine dose of 10 mg/kg body weight followed by hexachlorobenzene doses of 2 to 5 mg/kg body weight and day administered for 11 weeks (Oesterle and Deml 1994, 1998).

The development of the observed tumours can be explained in terms of known mechanisms (see Sections 2 and 6).

**Table 3.** Initiation-promotion studies with hexachlorobenzene

| Species | Initiation (I) | Promotion (P) | Results | References |
|---|---|---|---|---|
| *Initiation studies* | | | | |
| rat, F344 10 or 4 ♂ | PH, after 12 h oral HCB dose 5000 mg/kg body weight | week 2–12: 0.15 % cholic acid in the diet; week 4: one oral $CCl_4$ dose 1 ml/kg body weight | **number of GST-positive liver foci per cm$^2$:**<br>P: 0.72 ± 0.41<br>I (HCB): 0.79 ± 0.41<br>I (HCB) + P: 0.91 ± 0.65 | Tsuda *et al.* 1993 |
| *Promotion studies* | | | | |
| rat, F344 at least 10 ♂ per group | DEN: 200 mg/kg body weight, i.p., once | 6 weeks 0.6, 3, 15, 75, 150 ppm HCB in the diet (about 0.06, 0.3, 1.5, 7.5, 15 mg/kg body weight and day); after 1 week PH | **number of GST-positive liver foci per cm$^2$:**<br>I: 8.8 ± 2.00<br>I + P (HCB 0.06): 9.95 ± 2.60<br>I + P (HCB 0.3): 10.84 ± 2.19*<br>I + P (HCB 1.5): 11.86 ± 1.97**<br>I + P (HCB 7.5): 16.96 ± 3.40**<br>I + P (HCB 1.5): 21.18 ± 4.53** | Cabral *et al.* 1996 |
| rat, SD at least 5 ♀ per group | 1 × 10 mg DEN oral | 11 weeks (3 × weekly) HCB: 2, 5, 10 mg/kg body weight and day, oral | **number of GGT-positive liver foci per cm$^2$:**<br>0: 0.2–0.3<br>I: 1.0<br>I + P (HCB 2): 1.3<br>I + P (HCB 5): 2.6<br>I + P (HCB 10): 2.5 | Oesterle and Deml 1994, 1998 |
| rat F344, 10 ♂, 10 ♀ | 3 weeks 0.015 % DEN in the drinking water | 30 weeks 0.02 % HCB in the diet (12 mg/kg body weight and day) | **% liver foci which were GGT-positive:**<br>0: < 0.1 (♂), < 0.1 (♀)<br>I: 3.4 (♂), 5.5 (♀)<br>P (HCB 12): < 0.1 (♂), < 0.1 (♀)<br>I + P (HCB 12): 18.8 (♂), 43.4 (♀) | Stewart *et al.* 1989 |
| | DEN: 20 mg/kg body weight, i.p., once | 30 weeks 0.02 % HCB in the diet (12 mg/kg body weight and day) | **% liver foci per cm$^2$ which were GGT-positive:**<br>0: < 0.01 (♂), 0.32 (♀)<br>I: 0.06 (♂), 0.75 (♀)<br>P (HCB 12): 1.41 (♂), 1.71 (♀)<br>I + P (HCB 12): 17.9 (♂), 12.36 (♀) | |

DEN: diethylnitrosamine, F344: Fischer-344, GGT: γ-glutamyl transpeptidase, GST: glutathione-*S*-transferase, HCB: hexachlorobenzene, I: initiation, P: promotion, PH: partial hepatectomy

* $p < 0.05$, ** $p < 0.01$

## 5.7.2 Long-term studies

After long-term administration of hexachlorobenzene in the diet, liver tumours were found in all the animal species studied (rat, mouse, golden Syrian hamster) (Table 4). In rats, especially in female rats, the incidence of neoplastic nodules was significantly increased from hexachlorobenzene concentrations of 40 ppm in the diet, equivalent to about 4 mg/kg body weight and day (Arnold *et al.* 1985, Arnold and Krewski 1988) and that of haemangiomas and hepatocarcinomas at 75 ppm, equivalent to about 7.5 mg/kg body weight and day (Ertürk *et al.* 1986). In the mouse, liver tumours were seen in animals given 100 ppm or more (about 6 mg/kg body weight and day; Cabral *et al.* 1979) and in golden Syrian hamsters from 50 ppm (about 6 mg/kg body weight and day; Cabral *et al.* 1977).

In addition, in the rats, especially in the females, phaeochromocytomas in the adrenal glands and bile duct tumours were found and, especially in the males, adenomas in the parathyroid glands and the kidneys (Arnold *et al.* 1985, Arnold and Krewski 1988, Ertürk *et al.* 1986). In golden Syrian hamsters, especially in the male hamsters, the incidence of thyroid tumours was increased (Cabral *et al.* 1977).

**Table 4.** Carcinogenicity studies with hexachlorobenzene

| | |
|---|---|
| Author: | Arnold *et al.* 1985, Arnold and Krewski 1988 |
| Substance: | hexachlorobenzene (analytical grade) |
| Species: | rat, Sprague-Dawley, at least 40 ♂, 40 ♀ per group |
| Administration route: | (*in utero*, via lactation) in the diet |
| Concentration: | 0, 0.32, 1.6, 8.0, 40 ppm<br>(about 0, 0.032, 0.16, 0.8, 4 mg/kg body weight and day) |
| Duration: | (*in utero*, via lactation) 130 weeks in the diet |
| Toxicity: | at concentrations of 8 ppm or more: slight effects on the liver (centrilobular basophilic chromogenesis)<br>40 ppm: peribiliary lymphocytosis and fibrosis (♂), chronic nephrosis (♂) |

| Tumours: | | Concentration (ppm in the diet) | | | | |
|---|---|---|---|---|---|---|
| | | 0 | 0.32 | 1.6 | 8 | 40 |
| number of animals | ♂ | 48 | 48 | 48 | 49 | 49 |
| | ♀ | 49 | 49 | 50 | 49 | 49 |
| liver, | ♂ | 2 (4 %) | 0 (–) | 0 (–) | 2 (4 %) | 1 (2 %) |
| neoplastic nodules | ♀ | 0 (–) | 0 (–) | 2 (4 %) | 2 (4 %) | 10 (20 %)** |
| adrenal glands, | ♂ | 10 (21 %) | 12 (25 %) | 7 (15 %) | 13 (27 %) | 17 (35 %) |
| phaeochromocytomas | ♀ | 2 (4 %) | 4 (8 %) | 4 (8 %) | 5 (10 %) | 17 (35 %)** |
| parathyroid glands, | ♂ | 2 (4 %) | 4 (8 %) | 2 (4 %) | 1 (2 %) | 12 (24 %)* |
| adenomas | ♀ | 0 (–) | 0 (–) | 0 (–) | 1 (2 %) | 2 (4 %) |

* $p < 0.05$, ** $p < 0.01$

**Table 4.** continued

| | |
|---|---|
| Author: | Ertürk *et al.* 1986 |
| Substance: | hexachlorobenzene (> 99.5 %) |
| Species: | rat, Sprague-Dawley, at least 50 ♂, 50 ♀ per group |
| Administration route: | diet |
| Concentration: | 0, 75, 150 ppm (about 0, 7.5, 15 mg/kg body weight and day) |
| Duration: | 104 weeks |
| Toxicity: | no data |

| Tumours: | | Concentration (ppm in the diet) | | |
|---|---|---|---|---|
| | | 0 | 75 | 150 |
| number of animals | ♂ | 54 | 52 | 56 |
| | ♀ | 52 | 56 | 55 |
| liver, haemangiomas* | ♂ | 0 (–) | 10 (20 %) | 11 (20 %) |
| | ♀ | 0 (–) | 23 (41 %) | 35 (64 %) |
| liver, carcinomas* | ♂ | 0 (–) | 3 (6 %) | 4 (7 %) |
| | ♀ | 0 (–) | 36 (66 %) | 48 (87 %) |
| bile duct tumours* | ♂ | 0 (–) | 2 (4 %) | 2 (4 %) |
| | ♀ | 1 (2 %) | 19 (34 %) | 29 (53 %) |
| kidney, | ♂ | 7 (13 %) | 41 79 %) | 42 (75 %) |
| renal-cell adenomas* | ♀ | 1 (2 %) | 7 (13 %) | 15 (27 %) |

* the authors claimed that the incidence of these tumours was increased; level of significance not stated

| | |
|---|---|
| Author: | Smith *et al.* 1985 |
| Substance: | hexachlorobenzene (analytical grade) |
| Species: | rat, Fischer 344, 15 ♂, 15 ♀ |
| Administration route: | diet |
| Concentration: | 0.02 % (about 20 mg/kg body weight and day) |
| Duration: | 90 weeks |
| Toxicity: | 0.02 %: decreased body weight gains, increased liver and kidney weights, decreased spleen weights in ♀ |

| Tumours: | | Concentration (ppm in the diet) | |
|---|---|---|---|
| | | 0 | 0.02 % |
| number of animals | ♂ | 10 | 12 |
| | ♀ | 10 | 10 |
| liver tumours, neoplastic nodules | ♂ | 0 | 0 |
| | ♀ | 0 | 5 |
| hepatocellular carcinomas | ♂ | 0 | 0 |
| | ♀ | 0 | 5 |

**Table 4.** continued

| | |
|---|---|
| Author: | Smith and Cabral 1980 |
| Substance: | hexachlorobenzene (> 99.5 %) |
| Species: | rat, Agus, 14 ♀, Wistar, 6 ♀ |
| Administration route: | diet |
| Concentration: | 100 ppm: (about 6–8 mg/kg body weight and day) |
| Duration: | 90 weeks (Agus), 75 weeks (Wistar) |
| Toxicity: | 100 ppm: decreased body weight gains, porphyria |

| Tumours: | Concentration (ppm in the diet) | | | |
|---|---|---|---|---|
| | Agus | | Wistar | |
| | 0 | 100 | 0 | 100 |
| number of animals | 12 | 14 | 4 | 6 |
| liver cell tumours | 0 | 14 | 0 | 4 |

| | |
|---|---|
| Author: | Cabral *et al.* 1979 |
| Substance: | hexachlorobenzene (> 99.5 %) |
| Species: | mouse, Swiss,<br>50 ♂ and 50 ♀ (0, 200 ppm) and 30 ♂ and 30 ♀ (50, 100 ppm) |
| Administration route: | diet |
| Concentration: | 0, 50, 100, 200 ppm (0, 6, 12, 24 mg/kg body weight and day) |
| Duration: | 101–120 weeks |
| Toxicity: | at concentrations of 50 ppm or more: decreased body weight gains (♀)<br>at concentrations of 100 ppm or more: decreased body weight gains (♀)<br>200 ppm: increased mortality |

| Tumours: | | Concentration (ppm in the diet) | | | |
|---|---|---|---|---|---|
| | | 0 | 50 | 100 | 200 |
| number of animals | ♂ | 47 | 30 | 29 | 44 |
| | ♀ | 49 | 30 | 30 | 41 |
| survival (90 weeks) | ♂ | 50 % | 30 % | 27 % | 4 % |
| | ♀ | 48 % | 40 % | 30 % | 0 % |
| liver tumours | ♂ | 0 (–) | 0 (–) | 3 (10 %)* | 7 (16 %)* |
| | ♀ | 0 (–) | 0 (–) | 3 (10 %)* | 14 (34 %)* |

* increased incidence, level of significance not specified

| | |
|---|---|
| Author: | Cabral *et al.* 1977 |
| Substance: | hexachlorobenzene (> 99.5 %) |
| Species: | golden Syrian hamster, 40 ♂ and 40 ♀ (0 ppm),<br>30 ♂ and 30 ♀ (50, 100 ppm), 60 ♂ and 60 ♀ (200 ppm) |
| Administration route: | diet |
| Concentration: | 0, 50, 100, 200 ppm (0, 4, 8, 16 mg/kg body weight and day) |
| Duration: | for life (> 70 weeks) |
| Toxicity: | no data |

**Table 4.** continued

| Tumours: | | Concentration (ppm in the diet) | | | |
|---|---|---|---|---|---|
| | | 0 | 50 | 100 | 200 |
| number of animals | ♂ | 40 | 30 | 30 | 57 |
| | ♀ | 39 | 30 | 30 | 60 |
| tumour-bearing animals | ♂ | 3 (8 %) | 18 (60 %) | 27 (90 %) | 56 (98 %) |
| | ♀ | 5 (13 %) | 16 (53 %) | 18 (60 %) | 52 (87 %) |
| liver, hepatomas* | ♂ | 0 (–) | 14 (47 %) | 26 (87 %) | 49 (86 %) |
| | ♀ | 0 (–) | 14 (47 %) | 17 (57 %) | 51 (85 %) |
| haemangioendotheliomas* | ♂ | 0 (–) | 1 (3 %) | 6 (20 %) | 20 (35 %) |
| | ♀ | 0 (–) | 0 (–) | 2 (7 %) | 7 (12 %) |
| thyroid gland, tumours | ♂ | 0 (–) | 0 (–) | 1 (3 %) | 8 (14 %)** |
| | ♀ | 0 (–) | 2 (7 %) | 1 (3 %) | 3 (5 %) |
| spleen, haemangioendotheliomas | ♂ | 0 (–) | 1 (3 %) | 3 (10 %) | 4 (7 %) |
| | ♀ | 1 (3 %) | 0 (–) | 3 (10 %) | 4 (7 %) |
| other tumours | ♂ | 3 (8 %) | 11 (37 %) | 9 (30 %) | 6 (11 %) |
| | ♀ | 4 (10 %) | 5 (17 %) | 9 (30 %) | 8 (13 %) |

* increased incidence, level of significance not specified; ** $p < 0.05$

## 5.8. Other effects

### 5.8.1 Effects on the immune system

In persons who had eaten bread contaminated with hexachlorobenzene for a long period (see Section 4.1) enlarged thyroid glands and arthritis were observed, symptoms suggestive of an autoimmune reaction (Vos and Van Loveren 1995).

In rats, hexachlorobenzene caused amplification of the humoral and cell-mediated immune responses; the developing immune system was shown to be particularly sensitive (see also BUA 1994).

In Wistar rats exposed prenatally and postnatally to hexachlorobenzene, even the lowest concentration tested (4 ppm in the diet; about 0.4 mg/kg body weight and day) caused an increase in the immune response (IgM, IgG) to tetanus toxoid. The weight of the lymph nodes was slightly increased at 4 ppm and significantly increased from 20 ppm. In the lymph nodes the endothelial veins were enlarged. The spleen weights increased with the dose. B-lymphocyte hyperplasia was detected (BUA 1994), a sign of selective activation of B-1 cells (Schielen *et al.* 1995b). Macrophage accumulation was observed in the lungs. From 100 ppm the IgM level was increased (BUA 1994). There is also evidence that hexachlorobenzene causes effects like autoimmune reactions in rats (IgM directed, for example, against the autoantigens single strand DNA and native DNA), the significance of which is still unclear (Schielen *et al.* 1993, Vos and Van Loveren 1995).

The observed dermal lesions resulting from exposure to hexachlorobenzene are attributed at least in part to autoimmune reactions (Schielen *et al.* 1995a).

In mice, on the other hand, a suppression of immune responses was generally observed (BUA 1994).

### 5.8.2 Neurotoxicity

Weakness, tremor, seizures and high mortality were seen in breast-fed babies and small children whose mothers were exposed to hexachlorobenzene. In exposed adults, weakness but no tremor or seizures developed (see Section 4.1). Suckling rat pups whose dams were exposed to hexachlorobenzene died in convulsions. In adult rats seizures were rarely seen. Epileptiform seizures were, however, induced with ultrasound in 3 of 11 rats given oral hexachlorobenzene doses of 100 mg/kg body weight, daily for 13 days (Mylchreest and Charbonneau 1994).

## 6 Manifesto (MAK value/classification)

An epidemiological study of a population exposed to hexachlorobenzene has provided evidence of carcinogenic effects in man; clarification is still required.

Hexachlorobenzene is a substance with typical tumour-promoting properties. The liver, bile duct and kidney tumours observed in experimental animals may be attributed to the cytochrome P450-mediated formation of oxygen radicals resulting in cell damage and subsequent compensatory hyperplasia. Thyroid tumours result from induction of glucuronyl transferases which causes increased glucuronidation and excretion of thyroxine and subsequent enlargement of the thyroid gland. As for the majority of substances with tumour-promoting properties, if hexachlorobenzene has any clastogenic effects they are very weak. Because of its tumour-promoting properties, hexachlorobenzene is classified in Section III, Category 4 of the *List of MAK and BAT Values*.

Derivation of a MAK value for the substance on the basis of the data available for effects in man is not possible. Extrapolation from the results of animal studies to the human situation is associated with considerable uncertainty because of the different half-lives of the substance in man and animals (about 5 months in the rat, more than 2 years in man). Because of the highly cumulative effects of the substance in man, it is not to be expected that reliable data which would permit the establishment of a MAK value will be available in the future. There are, however, data available for the systemic levels of hexachlorobenzene in workers from which the BAT value of 150 µg hexachlorobenzene per litre plasma was established (Henschler and Lehnert 1984). Because no statistically significant changes in biochemical parameters such as markers of porphyrin metabolism or liver function were found in these workers, it may be assumed that observance of the BAT value also provides protection from the tumour-promoting effects of the substance.

Hexachlorobenzene can penetrate the skin and is therefore designated with an "H". The data available from animal studies yield no evidence of sensitizing effects of hexachlorobenzene. The substance has therefore not been designated with an "S". There is no evidence that the substance has potential to cause germ cell mutations.

# 7 References

Almeida MG, Fanini F, Davino SC, Aznar AE, Koch OR, Barros SBM (1997) Pro- and antioxidant parameters in rat liver after short term exposure to hexachlorobenzene. *Hum Exp Toxicol 16*: 257–261

Arnold DL, Krewski D (1988) Long-term toxicity of hexachlorobenzene. *Food Chem Toxicol 26*: 169–174

Arnold DL, Moodie CA, Charbonneau SM, Grice HC, McGuire PF, Bryce FR, Collins BT, Zawidzka ZZ, Krewski DR, Nera EA (1985) Long-term toxicity of hexachlorobenzene in the rat and the effect of dietary vitamin A. *Food Chem Toxicol 23*: 779–793

Babineau KA, Singh A, Jarrell JF, Villeneuve DC (1991) Surface epithelium of the ovary following oral administration of hexachlorobenzene to the monkey. *J Submicrosc Cytol Pathol 23*: 457–464

den Besten C, Bennik MM, van Iersel M, Peters MA, Teunis C, van Bladeren PJ (1994) Comparison of the urinary metabolite profiles of hexachlorobenzene and pentachlorobenzene in the rat. *Chem Biol Interact 90*: 121–137

Billi de Catabbi S, Sterin-Speziale N, Fernandez MC, Minutolo C, Aldonatti C, San Martin de Viale L (1997) Time course of hexachlorobenzene-induced alterations of lipid metabolism and their relation to porphyria. *Int J Biochem Cell Biol 29*: 335–344

Bourque AC, Singh A, Lakhanpal N, McMahon A, Foster WG (1995) Ultrastructural changes in ovarian follicles of monkeys administered hexachlorobenzene. *Am J Vet Res 56*: 1673–1677

BUA (Beratergremium für umweltrelevante Altstoffe der Gesellschaft deutscher Chemiker) (1994) *Hexachlorbenzol, Bericht Nr. 119*, Hirzel Verlag, Stuttgart

Cabral JRP, Shubik P, Mollner T, Raitano F (1977) Carcinogenic activity of hexachlorobenzene in hamsters. *Nature 269*: 510–511

Cabral JRP, Mollner T, Raitano F, Shubik P (1979) Carcinogenesis of hexachlorobenzene in mice. *Int J Cancer 23*: 47–51

Cabral R, Hoshiya T, Hakoi K, Hasegawa R, Ito N (1996) Medium-term bioassay for the hepatocarcinogenicity of hexachlorobenzene. *Cancer Lett 100*: 223–226

Canonero R, Campart GB, Mattioli F, Robbiano L, Martelli A (1997) Testing of *p*-dichlorobenzene and hexachlorobenzene for their ability to induce DNA damage and micronucleus formation in primary cultures of rat and human hepatocytes. *Mutagenesis 12*: 35–39

Carthew P, Smith AG (1994) Pathological mechanisms of hepatic tumour formation in rats exposed chronically to dietary hexachlorobenzene. *J Appl Toxicol 14*: 447–452

Courtney KD, Copeland MF, Robbins A (1976) The effects of pentachloronitrobenzene, hexachlorobenzene, and related compounds on fetal development. *Toxicol Appl Pharmacol 35*: 239–256

Ertürk E, Lambrecht RW, Peters HA, Cripps DJ, Gocmen A, Morris CR, Bryan GT (1986) Oncogenicity of hexachlorobenzene. *IARC Scientific Publ No 77*: 417–423

Foster WG, Pentick JA, McMahon A, Lecavalier PR (1992) Ovarian toxicity of hexachlorobenzene (HCB) in the superovulated female rat. *J Biochem Toxicol 7*: 1–4

Foster WG, Pentick JA, McMahon A, Lecavalier PR (1993) Body distribution and endocrine toxicity of hexachlorobenzene (HCB) in the female rat. *J Appl Toxicol 13*: 79–83

Foster WG, McMahon A, Younglai EV, Jarrell JF, Lecavalier P (1995a) Alterations in circulating ovarian steroids in hexachlorobenzene-exposed monkeys. *Reprod Toxicol 9*: 541–548

Foster WG, Mertineit C, Yagminas A, McMahon A, Lecavalier P (1995b) The effects of hexachlorobenzene on circulating levels of adrenal steroids in the ovariectomized rat. *J Biochem Toxicol 10*: 129–135

Górski T, Górska E, Górecka D, Sikora M, (1986) Hexachlorobenzene is non-genotoxic in short-term tests. *IARC Scientific Publ No 77*: 399–401

Grimalt JO, Sunyer J, Moreno V, Amaral OC, Sala M, Rosell A, Anto JM, Albaiges J (1994) Risk excess of soft-tissue sarcoma and thyroid cancer in a community exposed to airborne organochlorinated compound mixtures with a high hexachlorobenzene content. *Int J Cancer 56*: 200–203

Gulyaeva LF, Grishanova AY, Lyakhovich VV (1994) Induction of cytochromes P-4501A and P-4502B in various organs of rats treated with hexachlorobenzene and Arochlor-1254. *Biochemistry (Moskow) 59*: 383–387

Henschler D, Lehnert G (Eds) (1994) Hexachlorbenzol. in: *Biologische Arbeitsstoff-Toleranz-Werte (BAT-Werte) und Expositionsäquivalente für krebserzeugende Arbeitsstoffe (EKA)*, 7th issue (1994), and in English translation in: *Biological Exposure Values for Occupational Toxicants and Carcinogens Volume 2*, VCH-Verlagsgesellschaft, Weinheim

Jarrell JF, McMahon A, Villeneuve D, Franklin C, Singh A, Valli VE, Bartlett S (1993) Hexachlorobenzene toxicity in the monkey primordial germ cell without induced porphyria. *Reprod Toxicol 7*: 41–47

Kavlock RJ, Short Jr RD, Chernoff N (1987) Further evaluation of an *in vivo* teratology screen. *Teratogen Carcinogen Mutagen 7*: 7–16

Khera KS (1974) Teratogenicity and dominant lethal studies on hexachlorobenzene in rats. *Food Cosmet Toxicol 12*: 471–477

Kleiman de Pisarev DL, Ferramola de Sancovich AM, Sancovich HA (1995) Hepatic indices of thyroid status in rats treated with hexachlorobenzene. *J Endocrinol Invest 18*: 271–276

Koss G, Koransky W (1975) Studies on the toxicology of hexachlorobenzene. I. Pharmacokinetics. *Arch Toxicol 34*: 203–212

Kuiper-Goodman T, Grant DL (1986) Subchronic toxicity of hexachlorobenzene in the rat: clinical, biochemical, morphological and morphometric findings. *IARC Scientific Publ No. 77*: 343–346

Kuiper-Goodman T, Grant DL, Moodie CA, Korsrud GO, Munro IC (1977) Subacute toxicity of hexachlorobenzene in the rat. *Toxicol Appl Pharmacol 40*: 359–375

Kuroda Y (1986) Genetic and chemical factors affecting chemical mutagenesis in cultured mammalian cells. *Basic Life Sci 39*: 359–375

Landsteiner BK, Jacobs J (1936) Studies on the sensitization of animals with simple chemical compounds II. *J Exp Med 64*: 625–639

Lilienthal H, Benthe C, Heinzow B, Winneke G (1996) Impairment of schedule-controlled behavior by pre- and postnatal exposure to hexachlorobenzene in rats. *Arch Toxicol 70*: 174–181

MacPhee IJ, Singh A, Wright GM, Foster WG, LeBlanc NN (1993) Ultrastructure of granulosa lutein cells from rats fed hexachlorobenzene. *Histol Histopathol 8*: 35–40

Michielsen CP, Bloksma N, Ultee A, van Mil F, Vos JG (1997) Hexachlorobenzene-induced immunomodulation and skin and lung lesions: a comparison between brown Norway, Lewis, and Wistar rats. *Toxicol Appl Pharmacol 144*: 12–26

Miyagawa M, Takasawa H, Sugiyama A, Inoue Y, Murata T, Uno Y, Yoshikawa K (1995) The *in vivo–in vitro* replicative DNA synthesis (RDS) test with hepatocytes prepared from male B6C3F$_1$ mice as an early prediction assay for putative nongenotoxic (Ames-negative) mouse hepatocarcinogens. *Mutat Res 343*: 157–183

Mylchreest E, Charbonneau M (1994) Ultrasound-induced epileptiform activity in rats treated with hexachlorobenzene. *Neurotoxicology 15*: 273–278

Mylchreest E, Charbonneau M (1997) Studies on the mechanism of uroporphyrinogen decarboxylase inhibition in hexachlorobenzene-induced porphyria in the female rat. *Toxicol Appl Pharmacol 145*: 23–33

Nakashima Y, Ohsawa S, Umegaki K, Ikegami S (1997) Hexachlorobenzene accumulated by dams during pregnancy is transferred to suckling rats during early lactation. *J Nutr 127*: 648–654.

Oesterle D, Deml E (1994) Different potency of promoting agents in three rat liver foci bioassays. 35th Spring Meeting of the German Society for Experimental and Clinical Pharmacology and Toxicology, Mainz, Germany, March 15–17, *Naunyn-Schmiedeberg's Arch Pharmacol 349, Suppl*: R127

Oesterle D, Deml E (1998) Personal communication to the secretariat of the Commission, 19.05.1998

van Ommen B, Hendriks W, Bessems JGM, Geesink G, Müller F, van Bladeren PJ (1989) The relation between the oxidative biotransformation of hexachlorobenzene and its porphyrinogenic activity. *Toxicol Appl Pharmacol 100*: 517–528

Raabe F, Janz S, Wolff G, Merten H, Landrock A, Birkenfeld T, Herzschuh R (1993) Genotoxicity assessment of waste products of aluminum plasma etching with the SCS chromotest. *Mutat Res 300*: 99–109

van Raaij JAGM, Kaptein E, Visser TJ, van den Berg KJ (1993a) Increased glucuronidation of thyroid hormone in hexachlorobenzene-treated rats. *Biochem Pharmacol 45*: 627–631

van Raaij JAGM, Frijters CM, van den Berg KJ (1993b) Hexachlorobenzene-induced hypothyroidism. Involvement of different mechanisms by parent compound and metabolite. *Biochem Pharmacol 46*: 1385–1391

Richter J, Landa K, Reznicek J (1994) Immune response in persons occupationally exposed to hexachlorobenzene. *Prac Lek 46*: 151–154

Rietjens IM, Steensma A, den Besten C, van Tintelen G, Haas J, van Ommen B, van Bladeren PJ (1995) Comparative biotransformation of hexachlorobenzene and hexafluorobenzene in relation to the induction of porphyria. *Eur J Pharmacol 293*: 293–299

Ríos de Molina MC, Iglesias S, Lauria L, San Martin de Viale LC (1996) Precancerous pathology evoked by hexachlorobenzene treatment. *Acta Physiol Pharmacol Ther Lat Am 46*: 71–81

Schielen P, Schoo W, Tekstra J, Oostermeijer HH, Seinen W, Bloksma N (1993) Autoimmune effects of hexachlorobenzene in the rat. *Toxicol Appl Pharmacol 122*: 233–243

Schielen P, den Besten C, Vos JG, van Bladeren PJ, Seinen W, Bloksma N (1995a) Immune effects of hexachlorobenzene in the rat: role of metabolism in a 13-week feeding study. *Toxicol Appl Pharmacol 131*: 37–43

Schielen P, Van Rodijnen W, Pieters RHH, Seinen W (1995b) Hexachlorobenzene treatment increases the number of splenic B-1-like cells and serum autoantibody levels in the rat. *Immunology 86*: 568–574

Schielen P, van der Pijl A, Bleumink R, Pieters RH, Seinen W (1996) Local popliteal lymph node reactions to hexachlorobenzene and pentachlorobenzene: comparison with systemic effects. *Immunopharmacology 31*: 171–181

Seldén A, Jacobson G, Berg P, Axelson O (1989) Hepatocellular carcinoma and exposure to hexachlorobenzene: a case report. *Br J Ind Med 46*: 138–140

Siekel P, Chalupa I, Beno J, Blasko M, Novotny J, Burian J (1991) A genotoxicological study of hexachlorbenzene and pentachloroanisole. *Teratogen Carcinogen Mutagen 11*: 55–60

Simon GS, Tardiff RG, Borzelleca JF (1979) Failure of hexachlorobenzene to induce dominant lethal mutations in the rat. *Toxicol Appl Pharmacol 47*: 415–419

da Silva Augusto LG, Lieber SR, Ruiz MA, Souza CAD (1997) Micronucleus monitoring to assess human occupational exposure to organochlorides. *Environ Mol Mutagen 29*: 46–52

Sims DE, Singh A, Donald A, Jarrell J, Villeneuve DC (1991) Alteration of primate ovary surface epithelium by exposure to hexachlorobenzene: a quantitative study. *Histol Histopathol 6*: 525–529

Smith AG, Cabral JR (1980) Liver-cell tumors in rats fed hexachlorobenzene. *Cancer Lett 11*: 169–172

Smith AG, Francis JE, Dinsdale D, Manson MM, Cabral JRP (1985) Hepatocarcinogenicity of hexachlorobenzene in rats and the sex difference in hepatic iron status and development of porphyria. *Carcinogenesis 6*: 631–636

Soprena de Kracoff YE, Ferramola de Sancovich AM, Sancovich HA, Kleiman de Pisarev DL (1994) Effect of thyroidectomy and thyroxine on hexachlorobenzene induced porphyria. *J Endocrinol Invest 17*: 301–305

Stewart FP, Manson MM, Cabral JRP, Smith AG (1989) Hexachlorobenzene as a promoter of diethylnitrosamine-initiated hepatocarcinogenesis in rats and comparison with induction of porphyria. *Carcinogenesis 10*: 1225–1230

den Tonkelaar EM, Verschuuren HG, Bankovska J, de Vries T, Kroes R, van Esch GJ (1978) Hexachlorobenzene toxicity in pigs. *Toxicol Appl Pharmacol 43*: 137–145

Tsuda H, Matsumoto K, Ogino H, Ito M, Hirono I, Nagao M, Sato K, Cabral R, Bartsch H (1993) Demonstration of initiation potential of carcinogens by induction of preneoplastic glutathione *S*-transferase P-form-positive liver cell foci: possible *in vivo* assay system for environmental carcinogens. *Jpn J Cancer Res 84*: 230–236

Umegaki K, Ikegami S, Ichikawa T (1993) Hepatic DNA damage in mice given organochlorine chemicals. *J Food Hyg Soc Jpn 34*: 68–73

Vos JG, van Loveren H (1995) Markers for immunotoxic effects in rodents and man. *Toxicol Lett 82–83*: 385–394

Vrecl M, Jan J, Pogacnik A, Bavdek SV (1996) Transfer of planar and non-planar chlorobiphenyls, 4,4′-DDE and hexachlorobenzene from blood to milk and to suckling infants. *Chemosphere 33*: 2341–2346

completed 31.03.1998

# 2-Hydroxyethyl acrylate

| | |
|---|---|
| **MAK value** | – |
| **Peak limitation** | – |
| **Absorption through the skin** | – |
| **Sensitization (1998)** | **Sh** |
| **Carcinogenicity** | – |
| **Prenatal toxicity** | – |
| **Germ cell mutagenicity** | – |
| **BAT value** | – |
| Synonyms | acrylic acid 2-hydroxyethyl ester<br>ethylene glycol monoacrylate |
| Chemical name (CAS) | 2-propenoic acid 2-hydroxyethyl ester |
| CAS number | 818-61-1 |

2-Hydroxyethyl acrylate is used above all in the production of acrylate polymers, e.g. for lacquers.

## 1 Allergenic Effects

### 1.1 Effects in man

That 2-hydroxyethyl acrylate has sensitizing effects on the skin is apparent from several case reports of reactions to the substance in patch tests (see Table 1). It is not always possible, however, to identify 2-hydroxyethyl acrylate as the cause of sensitization, as acrylates and methacrylates are usually used in mixtures and thus a clear demonstration of exposure to one substance is rarely possible. In addition, as a result of sensitization to other acrylates and methacrylates, cross-reactions also seem to occur.

**Table 1.** Positive results with 2-hydroxyethyl acrylate in patch tests on patients with suspected acrylate allergy

| Persons tested | Concentration (vehicle) | Results | Contact / comments | References |
|---|---|---|---|---|
| 1 hearing aid manufacturer with conjunctivitis | 0.1 % (petrolatum) | 2+ | UV-cured lacquers | Estlander *et al.* 1996 |
| 1 patient | 0.5 % (petrolatum) | 2+ (read after 72 and 96 hours) | glue containing 2-hydroxyethyl acrylate for an electrosurgical earthing plate / duration of primary skin changes > 4 weeks | Hausen 1996 |
| 4 patients | 0.1 % (petrolatum) | 4/4 (3+) | glue containing 2-hydroxyethyl acrylate for an electrosurgical earthing plate | Jagtman 1998 |
| 5 patients (1 dentist and 4 dental assistants) | 0.17 % (petrolatum) | 2/5 (2+ and 3+) | patients with suspected sensitization to acrylates in synthetic resins | Kanerva *et al.* 1989 |
| 6 patients (3 dentists and 3 dental assistants) | 0.1 % (petrolatum) | 2/6 (2+ and 3+) | patients with allergy to glues containing 2-hydroxyethyl methacrylate | Kanerva *et al.* 1991 |
| 1 dentist with pharyngitis | 0.1 % (petrolatum) | 3+ | patient without skin symptoms; reactions to 16 acrylates and methacrylates | Kanerva *et al.* 1992b |
| 124 patients | 0.1–0.5 % (petrolatum) | 14/124 (no other details) | anamnesis: exposure to acrylates | Kanerva *et al.* 1995 |
| 8 patients | 0.1 % (petrolatum) | 2/8 (no other details) | anamnesis: exposure to acrylates | Kanerva *et al.* 1997 |
| 23 patients | 0.1 % (petrolatum) | 2/23 (1+ and 3+) | anamnesis: exposure to acrylates, reactions in 2 dentists | Kiec-Swierczynska 1996 |
| 13 patients | 0.5 % (petrolatum) | 8/13 (no other details) | reactions in 7/9 patients with artificial finger nails; also irritative reactions possible | Koppula *et al.* 1995 |
| 1 patient | 0.1 % (petrolatum) | 3+ (read after 96 hours) | glue containing 2-hydroxyethyl acrylate for an electrosurgical earthing plate | Miranda-Romero *et al.* 1998 |
| 2 persons employed in the production of contact lenses | 0.01–0.1 % (petrolatum) | 2/2 (both 2+ with 0.01 %, 1 × also 3+ with 0.1 %) | UV-cured plastic containing 2-hydroxyethyl acrylate / in one case irritative reaction to 1 % even after 144 hours | Peters and Andersen 1986 |

Table 1. continued

| Persons tested | Concentration (vehicle) | Results | Contact / comments | References |
|---|---|---|---|---|
| 1 patient | 1 % (petrolatum) | not specified | irritative reaction not excluded | Taylor 1989 |
| 1 laboratory assistant | 0.5 % and 1 % (ethanol) | 1+/2+ (read after 48 and 72 hours) | embedding resins for electron microscopy / 2 other employees with skin changes were not tested; in 3 control persons no reaction to 0.5 % and 1 % | Tobler *et al.* 1990 |

With a test concentration of not more than 0.1 %, 2-hydroxyethyl acrylate is not likely to cause irritation. Higher test concentrations, however, can lead to false positive reactions. In addition, in several persons sensitization was induced by patch tests with 2-hydroxyethyl acrylate. Testing with 0.5 % 2-hydroxyethyl acrylate in petrolatum caused skin irritation in 8/24 patients. In three of these patients a reaction was observed at the reading after 18–21 days, which indicates sensitization as a result of the patch test. Renewed testing of 2 of the patients confirmed this suspicion (no other details) (Kanerva *et al.* 1988). In another case, a patch test with 0.1 % 2-hydroxyethyl acrylate in petrolatum also resulted in sensitization. When the patch test was repeated, the patient produced a marked crescendo reaction: even after 13 days a strong reaction could be observed and after 21 days a clear reaction was still visible. As a result of this further case of iatrogenic sensitization, if allergy to acrylates is suspected, the authors no longer carry out routine testing with 2-hydroxyethyl acrylate (Kanerva *et al.* 1992a).

There are no data available for allergenic or irritative effects on the respiratory passages.

## 1.2 Animal studies

In the Magnusson and Kligman maximization test, as a result of intradermal induction (injections in duplicate: 50 µl of each of FCA, FCA and the substance, and 0.5% 2-hydroxyethyl acrylate in water) and epicutaneous induction (on day 7, 250 mg 10 % sodium dodecylsulfate in petrolatum applied for 24 hours non-occlusively, on day 8, 400 µl 25 % 2-hydroxyethyl acrylate, applied for 48 hours occlusively) all 12 treated female SSc:AL guinea pigs were sensitized and reacted on provocation on day 21 to both 25 µl 0.3 % 2-hydroxyethyl acrylate and 25 µl 0.3 % hydroxypropyl acrylate. 7/12 and 8/12 animals reacted to 25 % 2-hydroxyethyl methacrylate and 25 % hydroxypropyl methacrylate. In the controls pretreated with FCA (number of animals not specified) no reactions were observed (Clemmensen 1984).

According to the incompletely documented results of another maximization test, 70 % of the animals (no other details) were sensitized. Intradermal induction was carried out by treatment with 0.25 % 2-hydroxyethyl acrylate in isotonic NaCl solution, epicutaneous induction with 5 % and provocation with 1 % 2-hydroxyethyl acrylate in each case in acetone/polyethylene glycol 400 (7:3). None of the 4 control animals reacted on provocation after pretreatment with the vehicle carried out in the same way (Basketter and Scholes 1992, Scholes *et al.* 1992).

In a maximization test, none of 10 guinea pigs treated intradermally for induction with 5 % 2-hydroxypropyl methacrylate in olive oil:acetone 10:1 and epicutaneously with 25 % 2-hydroxypropyl methacrylate reacted on provocation with 0.2 % 2-hydroxyethyl acrylate (Björkner 1984).

In a split adjuvant test, sensitizing effects could not be detected in any of the 8 Hartley guinea pigs used. In this experiment the animals were treated epicutaneously 4 times within 10 days with 2-hydroxyethyl acrylate at the maximum non-irritative concentration (no other details) (Rao *et al.* 1981). As the animals, however, were also subject to intradermal treatment with FCA at the third induction treatment, and the test protocol therefore possibly provoked the development of tolerance, and as neither the induction nor provocation concentrations are documented, this study cannot be included in the present evaluation.

Positive results were obtained also in the local lymph node assay (LLNA) in the mouse with 10 % and 25 % 2-hydroxyethyl acrylate in acetone:olive oil 4:1 (Basketter and Scholes 1992).

## 2 Manifesto (sensitization)

The sensitizing effects of 2-hydroxyethyl acrylate in man have been demonstrated in several cases of contact dermatitis, in some cases after occupational exposure, and in individual cases of sensitization induced by the patch test. These findings are supported by positive results in several experiments with animals both with and without the use of adjuvant. There are no data available for sensitizing effects on the respiratory passages.

2-Hydroxyethyl acrylate is therefore designated with “Sh”.

## 3 References


Basketter DA, Scholes EW (1992) Comparison of the local lymph node assay with the guinea-pig maximization test for the detection of a range of contact allergens. *Food Chem Toxicol 30*: 65–69

Björkner B (1984) Contact allergy to 2-hydroxypropyl methacrylate (2-HPMA) in an ultraviolet curable ink. *Acta Derm Venereol Stockh 64*: 264–267

Clemmensen S (1984) Cross-reaction patterns in guinea pigs sensitized to acrylic monomers. *Drug Chem Toxicol 7*: 527–540

Estlander T, Kanerva L, Kari O, Jolanki R, Mölsä K (1996) Occupational conjunctivitis associated with type IV allergy to methacrylates. *Allergy 51*: 56–59

Hausen BM (1996) Allergische Kontaktdermatitis auf eine Neutral-Elektrode. *Allergologie 19*: 409–410

Jagtman BA (1998) Contact dermatitis from acrylates in an electrosurgical earthing plate. *Contact Dermatitis 38*: 280–281

Kanerva L, Estlander T, Jolanki R (1988) Sensitization to patch test acrylates. *Contact Dermatitis 18*: 10–15

Kanerva L, Estlander T, Jolanki R (1989) Allergic contact dermatitis from dental composite resins due to aromatic epoxy acrylates and aliphatic acrylates. *Contact Dermatitis 20*: 201–211

Kanerva L, Turjanmaa K, Estlander T, Jolanki R (1991) Occupational allergic contact dermatitis caused by 2-hydroxyethyl methacrylate (2-HEMA) in a new dentin adhesive. *Am J Contact Dermatitis 2*: 24–30

Kanerva L, Estlander T, Jolanki R (1992a) Double active sensitization caused by acrylics. *Am J Contact Dermatitis 3*: 23–26

Kanerva L, Estlander T, Jolanki R, Pekkarinen E (1992b) Occupational pharyngitis associated with allergic patch test reactions from acrylics. *Allergy 47*: 571–573

Kanerva L, Estlander T, Jolanki R, Tarvainen K (1995) Statistics on allergic patch test reactions caused by acrylate compounds, including data on ethyl methacrylate. *Am J Contact Dermatitis 6*: 75–77

Kanerva L, Estlander T, Jolanki R (1997) 10 years of patch testing with the (meth)acrylate series. *Contact Dermatitis 37*: 255–258

Kiec-Swierczynska M (1996) Occupational allergic contact dermatitis due to acrylates in Lodz. *Contact Dermatitis 34*: 419–422

Koppula SV, Fellman JH, Storrs FJ (1995) Screening allergens for acrylate dermatitis associated with artificial nails. *Am J Contact Dermatitis 6*: 78–85

Miranda-Romero A, Martínez M, Sanchez-Sambucety P, Aragoneses H, Muñoz CMG (1998) Allergic contact dermatitis from the acrylic adhesive of a surgical earthing plate. *Contact Dermatitis 38*: 279–280

Peters K, Andersen KE (1986) Allergic hand dermatitis from 2-hydroxyethyl-acrylate in contact lenses. *Contact Dermatitis 15*: 188–189

Rao KS, Betso JE, Olson KJ (1981) A collection of guinea pig sensitization test results – grouped by chemical class. *Drug Chem Toxicol 4*: 331–351

Scholes EW, Basketter DA, Sarll AE, Kimber I, Evans CD, Miller K, Robbins MC, Harrison PTC, Waite SJ (1992) The local lymph node assay: results of a final inter-laboratory validation under field conditions. *J Appl Toxicol 12*: 217–222

Taylor JS (1989) Acrylic reactions – ten-years' experience. in: Frosch PJ, Dooms-Goossens A, Lachapelle JM (Eds) *Current topics in contact dermatitis*, Springer, Berlin, 346–351

Tobler M, Wüthrich B, Freiburghaus AU (1990) Contact dermatitis from acrylate and methacrylate compounds in Lowicryl embedding media for electron microscopy. *Contact Dermatitis 23*: 96–102

completed 31.03.1998

# Hydroxypropyl acrylate (all isomers)

| | |
|---|---|
| **MAK value (1998)** | **not yet established; see Section IIb of the *List of MAK and BAT Values*** |
| **Peak limitation** | **–** |
| **Absorption through the skin** | **–** |
| **Sensitization (1998)** | **Sh** |
| **Carcinogenicity** | **–** |
| **Prenatal toxicity** | **–** |
| **Germ cell mutagenicity** | **–** |
| **BAT value** | **–** |

| Isomer | hydroxypropyl acrylate (all isomers) | 2-hydroxy-1-propyl acrylate | 1-hydroxy-2-propyl acrylate | 1-hydroxy-3-propyl acrylate |
|---|---|---|---|---|
| Synonyms | acrylic acid hydroxypropyl ester<br>2-propenoic acid hydroxypropyl ester<br>propylene glycol monoacrylate | 1,2-propanediol-1-acrylate | 1,2-propanediol-2-acrylate | 1,3-propanediol-1-acrylate |
| Chemical name (CAS) | acrylic acid monoester with propanediol | 2-propenoic acid 2-hydroxypropyl ester | 2-propenoic acid 2-hydroxy-1-methylethyl ester | 2-propenoic acid 3-hydroxypropyl ester |
| CAS number | 25584-83-2 | 999-61-1 | 2918-23-2 | 2761-08-2 |
| Structural formula R: $H_2C{=}CHCOO$ | | $R{-}CH_2CH(OH)CH_3$ | $R{-}CH(CH_3)CH_2OH$ | $R{-}(CH_2)_2CH_2OH$ |

| | |
|---|---|
| Molecular formula | $C_6H_{10}O_3$ |
| Molecular weight | 130.14 |
| Melting point* | < –50°C |
| Boiling point* | 83°C |
| Density at 20°C* | 1.054 g/cm$^3$ |
| Vapour pressure at 20°C* | 0.0001 Pa |
| **1 ml/m$^3$ (ppm) ≙ 5.32 mg/m$^3$** | **1 mg/m$^3$ ≙ 188 ml/m$^3$ (ppm)** |

* data for CAS number 25584-83-2; there are no other data available

Hydroxypropyl acrylates are used in the industrial production of acrylate polymers for lacquers and artificial resins, usually as isomer mixtures. Toxicological studies have been carried out with all the hydroxypropyl acrylate isomers, mainly, however, with the commercially available hydroxypropyl acrylate isomer mixtures with 75 % 2-hydroxy-1-propyl acrylate and 25 % 1-hydroxy-2-propyl acrylate or 66 % 2-hydroxy-1-propyl acrylate, 33 % 1-hydroxy-2-propyl acrylate and 1 % free acrylic acid.

# 1 Toxic Effects and Mode of Action

The acute oral toxicity of hydroxypropyl acrylates is low. Hydroxypropyl acrylates are highly irritating to the skin and mucous membranes. In a 30-day inhalation study, irritative effects and their sequelae were observed in the upper and lower respiratory passages of dogs, rabbits and rats exposed to a mixture of hydroxypropyl acrylate isomers in a concentration of 5 $ml/m^3$; the dog was found to be the most sensitive species. Hydroxypropyl acrylates cause contact allergy in man and animals. Tests with various strains of bacteria to detect the induction of point mutations in adenine-thymine base pairs yielded inconsistent results.

There are no data available for the mode of action, toxicokinetics or metabolism of hydroxypropyl acrylates.

# 2 Effects in Man

The only known effects of hydroxypropyl acrylates in man are allergenic effects.

It is clear from several case reports of reactions to the substance in patch tests that hydroxypropyl acrylates have sensitizing effects on the skin. With test concentrations of not more than 0.1 %, hydroxypropyl acrylates are not likely to cause irritation (see Table 1). Higher test concentrations, however, can lead to false positive reactions. In several cases, sensitization was induced by patch tests with 0.5–1 % hydroxypropyl acrylate in petrolatum. Testing with 0.5 % hydroxypropyl acrylate caused slight irritation in 6/24 patients. In three of these patients a reaction was observed at a reading after 18–21 days, which indicates sensitization as a result of the patch test. Renewed testing of 2 of the patients confirmed this suspicion (no further details) (Kanerva *et al.* 1988). In another case, a patch test with 0.1 % hydroxypropyl acrylate in petrolatum also resulted in sensitization. When the patch test was repeated, the patient produced a marked crescendo reaction: even after 13 days a strong reaction could be observed and after 21 days a clear reaction was still visible. As a result of this further case of iatrogenic sensitization, if allergy to acrylates is suspected, the authors no longer carry out routine testing with hydroxypropyl acrylate (Kanerva *et al.* 1992a).

**Table 1.** Positive results with hydroxypropyl acrylates in patch tests on patients with suspected acrylate allergy

| Persons tested | Concentration (vehicle) | Results | Contact / comments | References |
|---|---|---|---|---|
| 1 patient with conjunctivitis | not specified | (2+) | hearing aid manufacturer who used also UV-cured lacquers | Estlander *et al.* 1996 |
| 5 patients | 5 % (olive oil) | 2/5 (both 2+) | sensitization from sticking plaster / test substance produced moderate irritation in "most" control persons | Jordan 1975 |
| 5 patients (4 dental assistants, 1 dentist) | 0.17 % (petrolatum) | 2/5 (2+ and 3+) | patients with suspected sensitization from acrylates in synthetic resins | Kanerva *et al.* 1989 |
| 1 dentist | 0.1 % (petrolatum) | (3+) | patient with pharyngitis, no skin symptoms / reactions to 16 acrylates and methacrylates | Kanerva *et al.* 1992b |
| 124 patients | 0.1–0.5 % (not specified) | 12/124 (no other details) | anamnesis: 'exposure to acrylates' | Kanerva *et al.* 1995 |
| 8 patients | 0.1 % (not specified) | 2/8 (no other details) | anamnesis: 'exposure to acrylates' | Kanerva *et al.* 1997 |
| 23 patients | 0.1 % (petrolatum) | 1/23 (2+) | anamnesis: 'exposure to acrylates' | Kiec-Swierczynska 1996 |
| 13 patients | 0.5 % (petrolatum) | 6/13 (no other details) | reactions in 5/10 patients with artificial finger nails / also irritative reaction possible | Koppula *et al.* 1995 |
| 1 patient from the production of hydroxypropyl acrylate | 1 % (petrolatum) | (2+; read after 48 and 96 hours) | initially marked irritation after contact with hydroxypropyl acrylate / renewed skin reactions on returning to work after 6 months; in 6 control persons no reactions | Lovell *et al.* 1985 |
| 1 patient | 0.1 % (petrolatum) | (2+, read after 48 and 96 hours) | acrylate lacquer of a leg prosthesis / reactions to 14 acrylates and methacrylates | Romaguera *et al.* 1990 |

There are no data available for allergenic or irritative effects on the respiratory passages.

# 3 Animal Experiments and *in vitro* Studies

There are no data available for the toxic effects of hydroxypropyl acrylates on reproduction or its carcinogenicity.

## 3.1 Acute toxicity

### 3.1.1 Inhalation

Inhalation exposure of 6 rats for 8 hours to an atmosphere saturated at room temperature with a hydroxypropyl acrylate isomer mixture of 75 % 2-hydroxy-1-propyl acrylate and 25 % 1-hydroxy-2-propyl acrylate led to transient loss of co-ordination (Union Carbide 1971). Under the same conditions, after exposure of 12 rats to a mixture of 66 % 2-hydroxy-1-propyl acrylate, 33 % 1-hydroxy-2-propyl acrylate and 1 % free acrylic acid, dyspnoea and marked irritation of the mucous membranes were observed (BASF 1974). All of 5 rats survived exposure for 7 hours to an atmosphere saturated at room temperature with 75 % 2-hydroxy-1-propyl acrylate and 25 % 1-hydroxy-2-propyl acrylate (total concentration about 650 ml/m$^3$). Pathological examination of one animal revealed changes in the lungs, liver and kidneys (no other details). An atmosphere saturated at 100°C with the same isomer mixture (about 6450 ml/m$^3$) caused the death of all animals within 4.5 hours (Dow 1983).

### 3.1.2 Ingestion, dermal absorption and intraperitoneal administration

Oral $LD_{50}$ values for hydroxypropyl acrylate isomer mixtures (75 % 2-hydroxy-1-propyl acrylate and 25 % 1-hydroxy-2-propyl acrylate or 66 % 2-hydroxy-1-propyl acrylate, 33 % 1-hydroxy-2-propyl acrylate and 1 % free acrylic acid) of 950–1130 mg/kg body weight were determined for the rat (BASF 1974, Union Carbide 1971). The dermal $LD_{50}$ value in rabbits after occlusive application of a mixture of 75 % 2-hydroxy-1-propyl acrylate and 25 % 1-hydroxy-2-propyl acrylate was found to be 112 µl/kg body weight (118 mg/kg body weight). The severe skin irritation (see Section 3.3) led to ecchymosis and oedema at the application site, after doses of 200 µl/kg body weight (211 mg/kg body weight) also to necrosis (Union Carbide 1971). For 1-hydroxy-3-propyl acrylate a dermal $LD_{50}$ value between 1000 and 1600 mg/kg body weight was found in the rat after occlusive application of the substance for 24 hours; marked primary irritation was observed locally and later developed into necrosis (BASF 1979). After intraperitoneal administration, the $LD_{50}$ value in the mouse was 950 mg/kg body weight (mixture of 66 % 2-hydroxy-1-propyl acrylate, 33 % 1-hydroxy-2-propyl and 1 % free acrylic acid) (BASF 1974). The symptoms of intoxication were given as dyspnoea, trembling, jerking, unsteady gait, latero-abdominal position, slight apathy and salivation.

## 3.2 Subacute, subchronic and chronic toxicity

There is no information available about the toxic effects after ingestion or dermal absorption of hydroxypropyl acrylates.

### 3.2.1 Inhalation

Beagle dogs, New Zealand White rabbits, Sprague-Dawley rats and Swiss-Webster mice were exposed to a hydroxypropyl acrylate isomer mixture (74 % 2-hydroxy-1-propyl acrylate, 25 % 1-hydroxy-2-propyl acrylate, 0.73 % free acrylic acid) in concentrations of 0, 5 or 10 ml/m$^3$ (0, 27 or 53 mg/m$^3$) for 6 hours/day on 5 days/week for 30 days. The effects on body weight, and the clinical, gross pathological and histopathological findings are listed in Table 2. No exposure-related changes were found in organ weights, or in the results of haematological, clinico-chemical and urine analyses. Concentrations of 10 ml/m$^3$ led in dogs and mice to weight loss over the days with exposure, which was partially reversible during the exposure-free weekends. On the basis of the severity and incidence of irritation and its sequelae in the eye, nasal mucosa, trachea and lungs from concentrations of 5 ml/m$^3$, the dog was found to be the most sensitive species, followed by the rabbit and rat, while the low exposure concentration did not yield any pathological findings in mice. The changes in the testes, prostate gland, thymus and lymphatic tissues observed only in dogs were interpreted as secondary stress-induced effects, as this species suffered severe toxic effects on the respiratory passages (Dow 1983). A NOEL (no observed effect level) cannot be derived from this study as severe effects on the mucous membranes occurred in some dogs, rabbits and rats even at the lowest concentration tested of 5 ml/m$^3$.

## 3.3 Local effects on skin and mucous membranes

An undiluted hydroxypropyl acrylate isomer mixture (75 % 2-hydroxy-1-propyl acrylate and 25 % 1-hydroxy-2-propyl acrylate) applied without occlusion to the intact rabbit skin produced moderate irritation (Union Carbide 1971). Occlusive application of a mixture of 66 % 2-hydroxy-1-propyl acrylate, 33 % 1-hydroxy-2-propyl acrylate and 1 % free acrylic acid or of 1-hydroxy-3-propyl acrylate for 15 minutes and 4 hours, respectively, led within 8 days to severe, deep-seated necrosis (BASF 1974, 1980). Also in the rabbit eye these test substances were highly irritative to corrosive (BASF 1974, 1978, Union Carbide 1971).

**Table 2.** Results of the 30-day inhalation study with a hydroxypropyl acrylate isomer mixture (74 % 2-hydroxy-1-propyl acrylate, 25 % 1-hydroxy-2-propyl acrylate and 0.73 % free acrylic acid) (Dow 1983)

| Species, number per group, sex | Concentration [ml/m³] | Clinical findings, effects on body weight | Gross pathological and histopathological findings |
|---|---|---|---|
| beagle dog, 2, ♂ | 10 | in 1/2 bilateral purulent conjunctivitis and keratitis, nasal discharge<br>in 2/2 bronchopneumonia<br>weight loss during exposure, in some cases reversible during the exposure-free weekend | **eye:** in 1/2 bilateral keratitis and conjunctivitis<br>**upper respiratory passages:** in 2/2 continual purulent discharge from nose, marked inflammation, squamous metaplasia and hyperplasia and focal ulceration in the mucous membranes of the nasal concha<br>**trachea:** in 2/2 tracheitis, changes in the mucous membranes as in upper respiratory passages<br>**lung:** in 2/2 purulent bronchopneumonia<br>**testis:** in 1/2 seminiferous tubules with reduced spermatogenic activity, in some cases with the formation of giant cells<br>**prostate gland:** in 1/2 reduced in size<br>**lymphatic and thymus tissue:** in 1/2 cellular elements reduced |
| | 5 | no adverse effects | **eye:** in 1/2 chronic focal conjunctivitis<br>**upper respiratory passages:** in 2/2 continual purulent discharge from nose, marked inflammation, squamous metaplasia and hyperplasia and focal ulceration in the mucous membranes of the nasal concha<br>**lung:** in 1/2 focal interstitial pneumonia<br>**testis:** in 2/2 seminiferous tubules with reduced spermatogenic activity, in some cases with the formation of giant cells<br>**prostate gland:** reduced in size |
| | 0 | no adverse effects | no adverse effects |
| New Zealand White rabbit, 4, ♂ | 10 | in 1/4 bilateral conjunctivitis<br>in 3/4 bilateral nasal discharge | **upper respiratory passages:** in 4/4 mucopurulent rhinitis, squamous metaplasia and ulceration in the mucous membranes of the nasal concha; in 2/4 thickening of the mucous membranes of the nasal concha; in 1/4 subacute laryngitis<br>**trachea:** in 3/4 subacute–chronic tracheitis; in 1/4: focal ulceration<br>**lung:** in 2/4 subacute bronchiolitis, focal pneumonitis |
| | 5 | in 4/4 nasal discharge | **upper respiratory passages:** in 3/4 squamous metaplasia and ulceration in the mucous membranes of the nasal concha, mucopurulent rhinitis; in 1/4 subacute inflammation in the nasal mucosa, focal mononuclear cells; in 1/4 acute laryngitis<br>**trachea:** in 3/4 subacute–chronic focal tracheitis<br>**lung:** in 2/4 subacute focal pneumonitis; in 1/4 subacute bronchitis |

**Table 2.** continued

| Species, number per group, sex | Concentration [ml/m$^3$] | Clinical findings, effects on body weight | Gross pathological and histopathological findings |
|---|---|---|---|
| New Zealand White rabbit, 4, ♂ | 0 | in 1/4 minimal bilateral nasal discharge | **upper respiratory passages:** in 4/4 focal mononuclear cells in nasal mucosa<br>**trachea:** in 1/4 chronic focal tracheitis |
| Sprague-Dawley rat, 10, ♂ | 10 | in 8/10 focal bilateral clouding of the cornea | only 5 animals examined<br>**eye:** in 5/5 focal infiltration of mononuclear cells into lacrimal gland; in 1/5 subacute focal keratitis<br>**upper respiratory passages:** in 1/5 subacute inflammation and focal mononuclear cells in the nasal mucosa<br>**trachea:** in 5/5 chronic focal tracheitis<br>**lung:** subacute focal pneumonitis in 2/5 minimal, in 2/5 moderate |
| | 5 | in 1/10 focal unilateral clouding of the cornea | only 5 animals examined<br>**eye:** in 5/5 focal infiltration of mononuclear cells into lacrimal gland; no subacute focal keratitis<br>**upper respiratory passages:** in 3/5 subacute inflammation of the nasal mucosa; no focal mononuclear cells<br>**trachea:** in 3/5 chronic focal tracheitis<br>**lung:** subacute focal pneumonitis in 1/5 minimal, in 1/5 moderate |
| | 0 | no adverse effects | only 5 animals examined<br>**eye:** in 1/5 focal infiltration of mononuclear cells into lacrimal gland; no subacute focal keratitis<br>**upper respiratory passages:** nasal mucosa: in 3/5 focal mononuclear cells; no subacute inflammation<br>**trachea:** in 5/5 chronic focal tracheitis<br>**lung:** in 1/5 minimal focal subacute pneumonitis |
| Swiss-Webster mouse, 20, ♂ | 10 | in 3/20 eye irritation (no other details)<br>weight loss during exposure, reversible during the exposure-free weekend | **lung:** in 1/5 minimal focal subacute pneumonitis |
| | 5 | no adverse effects | **lung:** in 3/5 minimal subacute focal pneumonitis |
| | 0 | no adverse effects | **lung:** in 2/5 minimal subacute focal pneumonitis |

## 3.4 Allergenic Effects

In the Magnusson and Kligman maximization test all 12 female SSc:AL guinea pigs used were sensitized by intradermal (injections in duplicate, 50 µl of FCA, FCA and the substance, and 0.5 % hydroxypropyl acrylate in water) and epicutaneous induction (on day 7 250 mg 10 % sodium dodecylsulfate in petrolatum, 24 hours non-occlusive, on day 8 400 µl 25 % hydroxypropyl acrylate, 48 hours occlusive), and reacted on provocation on day 21 to both 0.3 % hydroxypropyl acrylate and 0.3 % 2-hydroxyethyl acrylate (25 µl of each substance in petrolatum). In each group only 2/12 animals reacted to 25 % 2-hydroxyethyl methacrylate and 25 % hydroxypropyl methacrylate. In the controls pretreated with FCA (number of animals not specified) no reactions were observed (Clemmensen 1984).

In a split adjuvant test, sensitizing effects of hydroxypropyl acrylate could not be detected in any of the 9 Hartley guinea pigs used. In this incompletely documented experiment the animals were treated epicutaneously 4 times within 10 days with hydroxypropyl acrylate at the maximum non-irritative concentration (no other details) (Rao *et al.* 1981). As the animals, however, were also treated intradermally with 0.2 ml FCA at the third induction treatment and the test protocol therefore possibly provoked the development of tolerance, and as neither the induction nor provocation concentrations are documented, this study cannot be included in the present evaluation.

In a maximization test, none of 10 guinea pigs treated intradermally for induction with 5 % 2-hydroxypropyl methacrylate in olive oil/acetone (10:1) and epicutaneously with 25 % 2-hydroxypropyl methacrylate reacted on provocation with 0.2 % hydroxypropyl acrylate (Björkner 1984).

## 3.5 Genotoxicity

### 3.5.1 *In vitro*

Tests with various strains of bacteria to detect the induction of point mutations in adenine-thymine base pairs yielded inconsistent results. In the mutagenicity test with *Salmonella typhimurium* TA102 and TA2638, after metabolic activation with S9 mix 2-hydroxypropyl acrylate up to cytotoxic concentrations was not found to cause mutagenicity. In *Escherichia coli* WP2/pKM101 and WP2 uvrA/pKM101 after metabolic activation, however, an increase in mutations was found from 1250 and 2500 µg/plate (Watanabe *et al.* 1996).

There is no information available about the effects of hydroxypropyl acrylates *in vivo.*

# 4 Manifesto (MAK value/classification)

The critical effects of dermal and inhalation exposure to hydroxypropyl acrylates are the irritative effects on the skin and mucous membranes. There are, however, no case reports available. A scientifically grounded MAK value cannot be derived from the only well-documented study—a 30-day inhalation study with dogs, rabbits, rats and mice—as even at the lowest concentration tested (5 ml/m$^3$) irritation of the mucous membranes was observed in the upper and in some animals also in the lower respiratory passages. Hydroxypropyl acrylate is therefore listed in Section IIb of the *List of MAK and BAT Values*. To prevent adverse effects on health, exposure concentrations in the air at the workplace should be kept far below 5 ml/m$^3$.

In the studies of acute dermal toxicity, marked irritation and in some cases also necrosis were observed, effects which may increase the ability of the substance to penetrate the skin. As there are no studies with non-irritative concentrations available, absorption of the substance through the skin cannot be evaluated. Hydroxypropyl acrylate is therefore not designated with an "H".

The sensitizing effects of hydroxypropyl acrylate in man have been demonstrated in several cases of contact dermatitis, in some cases after occupational exposure, and in individual cases of sensitization induced by the patch test. These findings are supplemented by positive results in an animal study with the use of adjuvant. There are no data available, however, for sensitizing effects on the respiratory passages. Hydroxypropyl acrylate is therefore designated with "Sh".

The toxic effects on reproduction, and the carcinogenic and germ cell mutagenic potential of the substance cannot be evaluated as there is insufficient data available.

# 5 References

BASF (1974) *Hydroxypropylacrylat. Ergebnis der gewerbetoxikologischen Vorprüfung*, unpublished report from 27.11.74

BASF (1978) *Primary skin and eye irritation tests with hydroxypropylacrylate in albino rabbits.* Centraal Institut voor Voedingsonderzoek, Report No. R 5916

BASF (1979) *Bericht über die Prüfung der akuten dermalen Toxizität von Hydroxypropylacrylat an der Ratte*, unpublished report from 8.1.79

BASF (1980) *Bericht über die Prüfung der Ätzwirkung von "Hydroxypropylacrylat" am Kaninchen im 4-Stunden- und 1-Stunden-Test*, unpublished report from 21.1.80

Björkner B (1984) Contact allergy to 2-hydroxypropyl methacrylate (2-HPMA) in an ultraviolet curable ink. *Acta Derm Venereol Stockh 64*: 264–267

Clemmensen S (1984) Cross-reaction patterns in guinea pigs sensitized to acrylic monomers. *Drug Chem Toxicol 7*: 527–540

Dow (1983) *Initial submission: Repeated inhalation toxicity of* hydroxypropylacrylate (final report) with cover letter dated 042192. NTIS/OTS 0537352, EPA/OTS Doc ID 88-920002542

Estlander T, Kanerva L, Kari O, Jolanki R, Mölsä K (1996) Occupational conjunctivitis associated with type IV allergy to methacrylates. *Allergy 51*: 56–59

Jordan Jr WP (1975) Cross-sensitization patterns in acrylate allergies. *Contact Dermatitis 1*: 13–15

Kanerva L, Estlander T, Jolanki R (1988) Sensitization to patch test acrylates. *Contact Dermatitis 18*: 10–15

Kanerva L, Estlander T, Jolanki R (1989) Allergic contact dermatitis from dental composite resins due to aromatic epoxy acrylates and aliphatic acrylates. *Contact Dermatitis 20*: 201–211

Kanerva L, Estlander T, Jolanki R (1992a) Double active sensitization caused by acrylics. *Am J Contact Dermatitis 3*: 23–26

Kanerva L, Estlander T, Jolanki R, Pekkarinen E (1992b) Occupational pharyngitis associated with allergic patch test reactions from acrylics. *Allergy 47*: 571–573

Kanerva L, Estlander T, Jolanki R, Tarvainen K (1995) Statistics on allergic patch test reactions caused by acrylate compounds, including data on ethyl methacrylate. *Am J Contact Dermatitis 6*: 75–77

Kanerva L, Estlander T, Jolanki R (1997) 10 years of patch testing with the (meth)acrylate series. *Contact Dermatitis 37*: 255–258

Kiec-Swierczynska M (1996) Occupational allergic contact dermatitis due to acrylates in Lodz. *Contact Dermatitis 34*: 419–422

Koppula SV, Fellman JH, Storrs FJ (1995) Screening allergens for acrylate dermatitis associated with artificial nails. *Am J Contact Dermatitis 6*: 78–85

Lovell CR, Rycroft RJ, Williams DM, Hamlin JW (1985) Contact dermatitis from the irritancy (immediate and delayed) and allergenicity of hydroxypropyl acrylate. *Contact Dermatitis 12*: 117–118

Rao KS, Betso JE, Olson KJ (1981) A collection of guinea pig sensitization test results – grouped by chemical class. *Drug Chem Toxicol 4*: 331–351

Romaguera C, Vilaplana J, Grimalt F, Ferrando J (1990) Contact sensitivity to meth(acrylates) in a limb prosthesis. *Am J Contact Dermatitis 1*: 183–185

Union Carbide (1971) *Initial submission: Range finding toxicity studies of hydroxypropyl-acrylate in rats and rabbits with cover letter dated 072992*. NTIS/OTS 0537960, EPA/OTS Doc ID 88-920004579

Watanabe K, Sakamoto K, Sasaki T (1996) Comparisons on chemically-induced mutagenicity among four bacterial strains, *Salmonella typhimurium* TA102 and TA2638, and *Escherichia coli* WP2/pKM101 and WP2 uvrA/pKM101: Collaborative study I. *Mutat Res 361*: 143–155

completed 30.03.1998

# 2-Hydroxypropyl methacrylate

| | |
|---|---|
| **MAK value** | **–** |
| **Peak limitation** | **–** |
| **Absorption through the skin** | **–** |
| **Sensitization (1998)** | **Sh** |
| **Carcinogenicity** | **–** |
| **Prenatal toxicity** | **–** |
| **Germ cell mutagenicity** | **–** |
| **BAT value** | **–** |
| Synonyms | 2-hydroxypropyl 2-methyl-2-propenoate<br>methacrylic acid 2-hydroxypropyl ester<br>2-methyl-2-propenoic acid 2-hydroxy propyl ester |
| Chemical name (CAS) | 2-methyl-2-propenoic acid 2-hydroxy propyl ester |
| CAS number | 923-26-2 |

2-Hydroxypropyl methacrylate is used in the production of acrylate polymers for UV-cured lacquers, in industry sometimes also as an isomer mixture with 1-hydroxypropyl methacrylate (CAS number 27813-02-1).

## 1 Allergenic Effects

### 1.1 Effects in man

That hydroxypropyl methacrylate has sensitizing effects on the skin is apparent from numerous case reports of reactions to the substance in patch tests. Only in a few cases is it possible, however, to identify 2-hydroxypropyl methacrylate as the cause of sensitization, as acrylates and methacrylates are usually used in mixtures and thus a clear demonstration of exposure to a single substance is rarely possible. Often, as a result of sensitization to other acrylates and methacrylates, cross-reactions also seem to occur (Table 1).

**Table 1.** Positive results with hydroxypropyl methacrylate in patch tests on patients with suspected acrylate allergy

| Persons tested | Concentration (vehicle) | Results | Contact / comments | References |
|---|---|---|---|---|
| 1 employee from paint production | 2 % (petrolatum) | positive (no other details) | hydroxypropyl methacrylate in UV-curing paints | Björkner 1984 |
| 1 hearing aid manufacturer and 1 dental technician with conjunctivitis | 2 % (petrolatum) | 2/2 (2+ and 3+) | UV-curing lacquers and plastics | Estlander *et al.* 1996 |
| 45 patients with contact dermatitis from shoes | not specified | 1/45 (no other details) | plastic materials for shoes | Grimalt and Romaguera 1975 |
| 5 patients | 0.6 % (petrolatum) | 5/5 (2 × 1+, 2 × 2+, 1 × 3+) | UV-curing plastics for artificial finger nails / in most cases also (strong) reactions to hydroxyethyl methacrylate, triethylene glycol dimethacrylate and ethylene glycol dimethacrylate | Hemmer *et al.* 1996 |
| 1 patient with contact dermatitis | 2 % (petrolatum) | positive (2+) | UV-curing glue / reactions to 11 acrylates and methacrylates | van den Hove *et al.* 1996 |
| 1 employee from the production of printed circuit boards | 2 % (petrolatum) | positive (3+) | UV-curing screen printing ink / marked reaction to 6 acrylates and methacrylates | Jolanki *et al.* 1994 |
| 46 patients | 1 %, 2 % (petrolatum) | 5/46 (no other details) | anamnesis: exposure to acrylates | Kanerva *et al.* 1988 |
| 5 patients | 2 % (petrolatum) | 3/5 (2 × 2+, 1 × 3+) | patients with suspected sensitization to acrylates | Kanerva *et al.* 1989 |
| 2 dental assistants | 2 % (petrolatum) | 2/2 (both 3+) | primary sensitization probably from glue containing 2-hydroxyethyl methacrylate | Kanerva *et al.* 1991 |
| 1 patient | 2 % (petrolatum) | positive (2+ from day 8) | primary test sensitization from 2-hydroxyethyl acrylate and hydroxypropyl acrylate | Kanerva *et al.* 1992a |
| 1 dentist with pharyngitis | 2 % (petrolatum) | positive (3+) | patient without skin symptoms; reactions to 16 acrylates and methacrylates | Kanerva *et al.* 1992b |
| 1 dental technician | 2 % (petrolatum) | positive (3+) | reactions to 7 acrylates and methacrylates | Kanerva *et al.* 1993 |

**Table 1.** continued

| Persons tested | Concentration (vehicle) | Results | Contact / comments | References |
|---|---|---|---|---|
| 124 patients | 2 % (petrolatum) | 15/124 (no other details) | anamnesis: exposure to acrylates | Kanerva *et al.* 1995a |
| 1 employee in the automobile industry | 2 % (petrolatum) | positive (3+, also after 96 hours) | glue with 24.6 % hydroxyethyl methacrylate, reactions to 10 acrylates and methacrylates | Kanerva *et al.* 1995b |
| 1 dentist with dermatitis of the finger tips | 2 % (petrolatum) | positive (2+, read after 48, 72 and 96 hours) | dental plastics containing acrylates / marked reactions to 5 acrylates and methacrylates | Kanerva *et al.* 1996a |
| 1 beautician | 2 % (petrolatum) | positive (2+) | UV-curing plastics for artificial finger nails / marked reactions to 10 acrylates and methacrylates | Kanerva *et al* 1996b |
| 1 dental assistant | 2 % (petrolatum) | positive (2+, read after 48 and 96 hours) | UV-curing plastics containing 2-hydroxyethyl methacrylate; primary sensitization probably from 2-hydroxyethyl methacrylate | Kanerva *et al* 1997a |
| 118 patients | 2 % (petrolatum) | 14/118 (no other details) | anamnesis: exposure to acrylates | Kanerva *et al* 1997b |
| 1 dental assistant with dermatitis on the finger tips | 2 % (petrolatum) | positive (2+) | plastics containing 2-hydroxyethyl methacrylate / primary sensitization probably from 2-hydroxyethyl methacrylate / reactions to 11 acrylates and methacrylates | Kanerva *et al* 1998 |
| 23 patients | 2 % (petrolatum) | 2/23 (2+ and 3+) | anamnesis: exposure to acrylates / reactions in 2 dentists | Kiec-Swierczynska 1996 |
| 13 patients | 2 % (petrolatum) | 8/13 (no other details) | reactions in 5/9 patients with artificial finger nails | Koppula *et al.* 1995 |
| 1 employee from the production of hydroxypropyl (meth)acrylate | 2 % (petrolatum) | positive (1+; read after 48 and 96 hours) | initially marked irritation after contact with hydroxypropyl acrylate; also reactions to hydroxyethyl methacrylate (1+) and hydroxypropyl acrylate (2+) | Lovell *et al.* 1985 |
| 93 dental technicians | not specified | 16/93 (no other details) | patients with clinical skin diseases tested | Peiler *et al.* 1996 |

**Table 1.** continued

| Persons tested | Concentration (vehicle) | Results | Contact / comments | References |
|---|---|---|---|---|
| 1 patient | 2 % (petrolatum) | positive (2+, read after 48 and 96 hours) | acrylate lacquer of a leg prosthesis / reactions to 13 acrylates and methacrylates | Romaguera *et al.* 1990 |
| 35 dental technicians | 2 % (petrolatum) | 7/35 (no other details) | patients with allergic contact dermatitis on the hands tested | Rustemeyer and Frosch 1996 |
| 6 patients | 5 % (petrolatum) | 3/6 (no other details) | patients with suspected allergy to acrylates tested | Taylor 1989 |

There are no data available for allergenic or irritative effects on the respiratory passages.

## 2 Animal studies

Experiments with animals yielded positive results in the maximization test and a modified FCA test only when a concentration of at least 10 % was used for intradermal induction (Table 2).

In a modified split adjuvant test, sensitizing effects could not be detected in any of the 7 Hartley guinea pigs used. In this experiment the animals were treated epicutaneously 4 times within 10 days with hydroxypropyl methacrylate at the maximum non-irritative concentration (no further details) (Rao *et al.* 1981). As the animals, however, were also treated intradermally with FCA at the third induction treatment and the test protocol therefore possibly provoked the development of tolerance, and as neither the induction nor provocation concentrations are documented, this study cannot be included in the present evaluation.

Negative results were obtained in the local lymph node assay in the mouse with 10 %, 25 % and 50 % hydroxypropyl methacrylate in acetone:olive oil 4:1 (Basketter and Scholes 1992, Scholes *et al.* 1992).

Cross-reactions have been described in various studies. Of 15 animals treated with 2-hydroxyethyl methacrylate (intradermal induction with 25 %, epicutaneous induction with 100 %), 5 animals reacted also to 25 % hydroxypropyl methacrylate. Of 12 animals pretreated with 2-hydroxyethyl acrylate and 12 pretreated with hydroxypropyl acrylate (intradermal induction with 0.5%, epicutaneous induction with 25% in each group), 8 and 2 animals, respectively, reacted also to 25 % hydroxypropyl methacrylate (Clemmensen 1984).

**Table 2.** Sensitizing effects of hydroxypropyl methacrylate (HPMA) in the guinea pig

| Method, strain, sex | **Intradermal induction** injection volume, concentration (vehicle); amount/animal | **Epicutaneous induction** day, application, concentration (vehicle) | **Provocation** day, application, concentration (vehicle) | Number of animals which produced reactions | Comments | References |
|---|---|---|---|---|---|---|
| maximization test, Dunkin-Hartley, not specified | not specified, 1 % (isotonic saline solution); not specified | days 6–8, 48 hours occlusive, 100 % | day 21, 24 hours occlusive, 10 % (acetone/propylene glycol 400, 7:3) | **0/10** | contradictory data for provocation concentration | Besketter and Scholes 1992, Scholes *et al.* 1992 |
| maximization test, Dunkin-Hartley, ♀ | injections in duplicate: 100 µl of HPMA, HPMA in FCA and FCA, 5 % (olive oil:acetone 10:1); 20 mg | day 7, 48 hours occlusive, 25 % (petrolatum); non-occlusive pretreatment with 10 % sodium dodecylsulfate in water | day 21, 24 hours occlusive, 2 % (petrolatum) | **1/10** | | Björkner 1984 |
| maximization test, SSc:Al, ♀ | injections in duplicate: 50 µl of HPMA, HPMA in FCA and FCA, 10 % (water); 20 mg | day 7, 24 hours non-occlusive, 250 mg 10 % sodium dodecylsulfate in petrolatum; day 8, 48 hours occlusive, 400 µl 100 % | day 21, 24 hours occlusive, 25 % (water) | **3/12** | no reactions in the control animals pretreated with FCA, no data for the number of control animals | Clemmensen 1984 |
| modified FCA test, not specified, ♀ and ♂ | 50 µl of each substance in FCA in both flanks and ears and 100 µl in the neck, 1 M (about 14 %) (water); 42 mg | – | day 14, non-occlusive, 50 % (DMSO:ethanol 4:1) | **15/16** | mean reaction score after 24 and 48 hours: 0.9–1.0; reactions reproducible after 3 months; no irritative effects with 50 % HPMA | Rustemeyer *et al.* 1998 |

An animal sensitized to hydroxypropyl methacrylate reacted on provocation also to 2 % 2-hydroxyethyl methacrylate (Björkner 1984). In a modified FCA test, animals sensitized to hydroxypropyl methacrylate reacted on provocation in most cases also to 50 % 2-hydroxyethyl methacrylate (15/15, mean reaction score 0.7) and 50 % ethylene glycol dimethacrylate (11/11, mean reaction score 1.0) in DMSO/ethanol (4:1) and to a much smaller extent to methyl methacrylate (7/11, mean reaction score 0.3). Of 11 animals sensitized by intradermal induction treatment with 12 % 2-hydroxyethyl methacrylate, 8 reacted on provocation also to 50 % hydroxypropyl methacrylate (Rustemeyer *et al.* 1998; see also Table 2).

## 3 Manifesto (sensitization)

The sensitizing effect of hydroxypropyl methacrylate in man has been demonstrated in numerous cases of contact dermatitis, some of which were work-related. These findings are supplemented by borderline and clearly positive results in two animal studies with the use of adjuvant. There are no data available for sensitizing effects on the respiratory passages.

Hydroxypropyl methacrylate is therefore designated with "Sh".

## 4 References

Basketter DA, Scholes EW (1992) Comparison of the local lymph node assay with the guinea-pig maximization test for the detection of a range of contact allergens. *Food Chem Toxicol 30*: 65–69

Björkner B (1984) Contact allergy to 2-hydroxypropyl methacrylate (2-HPMA) in an ultraviolet curable ink. *Acta Derm Venereol 64*: 264–267

Clemmensen S (1984) Cross-reaction patterns in guinea pigs sensitized to acrylic monomers. *Drug Chem Toxicol 7*: 527–540

Estlander T, Kanerva L, Kari O, Jolanki R, Mölsä K (1996) Occupational conjunctivitis associated with type IV allergy to methacrylates. *Allergy 51*: 56–59

Grimalt F, Romaguera C (1975) New resin allergens in shoe contact dermatitis. *Contact Dermatitis 1*: 169–174

Hemmer W, Focke M, Wantke F, Götz M, Jarisch R (1996) Allergic contact dermatitis to artificial fingernails prepared from UV light-cured acrylates. *J Am Acad Dermatol 35*: 377–380

van den Hove J, Jacobs M-C, Tennstedt D, Lachapelle J-M (1996) Allergic contact dermatitis from acrylates in insulin pump infusion sets. *Contact Dermatitis 35*: 108

Jolanki R, Kanerva L, Estlander T, Tarvainen K (1994) Concomitant sensitization to triglycidyl isocyanurate, diaminodiphenylmethane and 2-hydroxyethyl methacrylate from silk-screen printing coatings in the manufacture of circuit boards. *Contact Dermatitis 30*: 12–15

Kanerva L, Estlander T, Jolanki R (1988) Sensitization to patch test acrylates. *Contact Dermatitis 18*: 10–15

Kanerva L, Estlander T, Jolanki R (1989) Allergic contact dermatitis from dental composite resins due to aromatic epoxy acrylates and aliphatic acrylates. *Contact Dermatitis 20*: 201–211

Kanerva L, Turjanmaa K, Estlander T, Jolanki R (1991) Occupational allergic contact dermatitis caused by 2-hydroxyethyl methacrylate (2-HEMA) in a new dentin adhesive. *Am J Contact Dermatitis 2*: 24–30

Kanerva L, Estlander T, Jolanki R (1992a) Double active sensitization caused by acrylics. *Am J Contact Dermatitis 3*: 23–26

Kanerva L, Estlander T, Jolanki R, Pekkarinen E (1992b) Occupational pharyngitis associated with allergic patch test reactions from acrylics. *Allergy 47*: 571–573

Kanerva L, Estlander T, Jolanki R, Tarvainen K (1993) Occupational allergic contact dermatitis caused by exposure to acrylates during work with dental prostheses. *Contact Dermatitis 28*: 268–275

Kanerva L, Estlander T, Jolanki R, Tarvainen K (1995a) Statistics on allergic patch test reactions caused by acrylate compounds, including data on ethyl methacrylate. *Am J Contact Dermatitis 6*: 75–77

Kanerva L, Jolanki R, Leino T, Estlander T (1995b) Occupational allergic contact dermatitis from 2-hydroxyethyl methacrylate and ethylene glycol dimethacrylate in a modified acrylic structural adhesive. *Contact Dermatitis 33*: 84–89

Kanerva L, Estlander T, Jolanki R, (1996a) False negative patch test reaction caused by testing with dental composite acrylic resin. *Int J Dermatol 35*: 189–192

Kanerva L, Lauerma A, Estlander T, Alanko K, Henriks-Eckerman M-L, Jolanki R (1996b) Occupational allergic contact dermatitis caused by photobonded sculptured nails and a review of (meth)acrylates in nail cosmetics. *Am J Contact Dermatitis 7*: 109–115

Kanerva L, Estlander T, Jolanki R (1997a) Occupational allergic contact dermatitis caused by acrylic tri-cure glass ionomer. *Contact Dermatitis 37*: 49–50

Kanerva L, Estlander T, Jolanki R (1997b) 10 years of patch testing with the (meth)acrylate series. *Contact Dermatitis 37*: 255–258

Kanerva L, Mikola H, Henriks-Eckerman M-L, Jolanki R, Estlander T (1998) Fingertip paresthesia and occupational allergic contact dermatitis caused by acrylics in a dental nurse. *Contact Dermatitis 38*: 114–116

Kiec-Swierczynska M (1996) Occupational allergic contact dermatitis due to acrylates in Lodz. *Contact Dermatitis 34*: 419–422

Koppula SV, Fellman JH, Storrs FJ (1995) Screening allergens for acrylate dermatitis associated with artificial nails. *Am J Contact Dermatitis 6*: 78–85

Lovell CR, Rycroft RJG, Williams DMJ, Hamlin JW (1985) Contact dermatitis from the irritancy (immediate and delayed) and allergenicity of hydroxypropyl acrylate. *Contact Dermatitis 12*: 117–118

Peiler D, Rustemeyer T, Frosch PJ (1996) Dermatosen bei Zahntechnikern – Irritantien und Allergene. *Allergologie 19*: 93–94

Rao KS, Betso JE, Olson KJ (1981) A collection of guinea pig sensitization test results – grouped by chemical class. *Drug Chem Toxicol 4*: 331–351

Romaguera C, Vilaplana J, Grimalt F, Ferrando J (1990) Contact sensitivity to met(acrylates) in a limb prosthesis. *Am J Contact Dermatitis 1*: 183–185

Rustemeyer T, Frosch PJ (1996) Occupational skin diseases in dental laboratory technicians (I). Clinical picture and causative factors. *Contact Dermatitis 34*: 125–133

Rustemeyer T, de Groot J, von Blomberg BME, Frosch PJ, Scheper RJ (1998) Cross-reactivity patterns of contact-sensitizing methacrylates. *Toxicol Appl Pharmacol 148* 83–90

Scholes EW, Basketter DA, Sarll AE, Kimber I, Evans CD, Miller K, Robbins MC, Harrison PTC, Waite SJ (1992) The local lymph node assay: results of a final inter-laboratory validation under field conditions. *J Appl Toxicol 12*: 217–222

Taylor JS (1989) Acrylic reactions – ten years' experience. in: Frosch PJ, Dooms-Goossens A, Lachapelle JM (Eds) *Current topics in contact dermatitis*, Springer, Berlin 346–351

completed 31.03.1998

# Lindane

| | |
|---|---|
| **MAK value (1998)** | **0.1 mg/m$^3$ I (inhalable dust)** |
| **Peak limitation (1983)** | **Category III** |
| **Absorption through the skin (1998)** | **H** |
| **Sensitization** | **–** |
| **Carcinogenicity (1998)** | **Category 4** |
| **Prenatal toxicity (1998)** | **Pregnancy risk group C** |
| **Germ cell mutagenicity** | **–** |
| **BAT value (1988)** | **20 µg/l blood or 25 µg/l serum<br>Sampling time: end of exposure or end of shift** |
| Synonyms | $\gamma$-1,2,3,4,5,6-hexachlorocyclohexane<br>$\gamma$-HCH |
| Chemical name (CAS) | (1$\alpha$,2$\alpha$,3$\beta$,4$\alpha$,5$\alpha$,6$\beta$)-1,2,3,4,5,6-hexachlorocyclohexane |
| CAS No. | 58-89-9 |
| Structural formula | 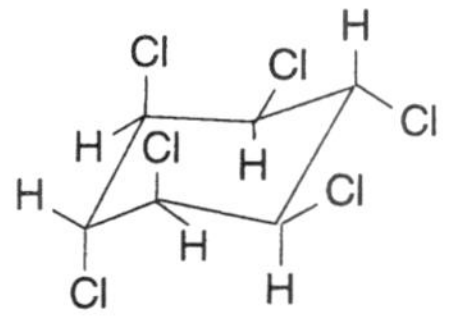 |
| Molecular formula | $C_6H_6Cl_6$ |
| Molecular weight | 290.8 |
| Melting point | 112°C |
| Boiling point | 323.4°C |
| log $P_{ow}$* | 3.7 |
| **1 mg/m$^3$ ≙ 0.083 ml/m$^3$ (ppm)** | **1 ml/m$^3$ (ppm) ≙ 12.1 mg/m$^3$** |

The present document is based on the WHO Environmental Health Criteria Document "Lindane" (WHO 1991).

* *n*-octanol/water distribution coefficient

# 1 Toxic Effects and Mode of Action

Lindane, a colourless crystalline solid, is the only *gamma*-isomer of technical hexachlorocyclohexane (HCH) which has insecticidal activity.

After inhalation, or oral or dermal administration, lindane is readily absorbed. The dermal absorption reported for man varies from 10 % to 90 %, depending on the solvent used. The half-life of lindane in the blood of man and rat was about 20 hours after short-term exposure; after longer exposures it was 3 to 4 days in the rat and 10 days in man.

The 4-hour $LC_{50}$ determined for the rat was 1600 mg/m$^3$. The oral $LD_{50}$ for lindane administered to rats and mice is in the range between 100 and 200 mg/kg body weight. The $LD_{50}$ for dermal application to the rabbit was found to be 75 mg/kg body weight; acute intoxication causes central nervous symptoms; most frequent are torpidity, restlessness, anxiety, tonic-clonic cramps and unconsciousness. Likewise in man, especially after ingestion of large amounts of lindane, central nervous symptoms even including tonic-clonic cramps have been reported.

Lindane does not cause skin irritation or skin sensitization detectable in the maximization test but can have slight irritative effects on mucous membranes.

Long term exposure to lindane results in liver damage. The induction of cytochrome P450-dependent enzymes correlates with the liver damage. In workers exposed to lindane, the induction of liver enzyme activities has been observed when the serum lindane concentrations were above 10 µg/l. In rats exposed for 3 months to a lindane concentration of 4.5 mg/m$^3$, slight clinical symptoms and the induction of cytochrome P450-dependent monooxygenases were observed. In rats given lindane in the diet for a period of 3 months, increased cytochrome P450-dependent monooxygenase activities were seen after doses of 0.8 mg/kg diet (about 0.06 mg/kg body weight and day) or more and histopathological changes in the liver and kidneys after 20 mg/kg diet (about 1.5 mg/kg body weight and day).

In addition, immunomodulating effects of lindane have been demonstrated in mice, rats and rabbits. In a study with mice, immunomodulating effects were seen at lindane doses down to 0.012 mg/kg body weight.

Lindane has oestrogenic activity. In adult male rats given lindane only as pups via mother's milk (the dams were given lindane during the lactation period either as one dose of 6 mg/kg body weight or as 6 doses of 1 mg/kg body weight on 6 days) the testosterone levels were reduced, and the testis weights, sperm count and spermatid count increased. Fertility was, however, not affected. Long-term oral administration of lindane doses of 10 mg/kg body weight to female rats had effects on oestrus; doses of 30 mg/kg body weight resulted in reduced readiness to mate. In a 3-generation study no effects on fertility were observed.

Developmental toxicity studies with several species given oral, subcutaneous or intraperitoneal doses of lindane yielded no evidence of teratogenic effects of the substance. Maternal toxicity was observed after doses of 10 mg/kg body weight and day or more, prenatal toxic effects not until doses of 60 mg/kg body weight.

In standardized mutagenicity tests with bacteria, mammalian cells and *in vivo* test systems, neither mutagenic nor clastogenic effects were detected. However, both *in vitro* and *in vivo* an increase in DNA strand breaks was observed and 2 of 4 dominant lethal tests yielded questionably positive results. Weak binding of lindane to mouse liver DNA was also demonstrated.

Carcinogenicity studies revealed neoplastic foci in the rat liver and an increased incidence of liver tumours in the mouse. Initiation-promotion studies with rats revealed no initiating activity but a tumour-promoting effect of lindane.

## 2 Mechanism of Action

Acute intoxication with lindane causes central nervous symptoms. The alterations detected in the electroencephalogram (EEG) can even include seizures of 'grand mal' type (McNamara and Krop 1948); disorders in the permeability of lipid-containing membranes are probably involved. It has also been shown that lindane inhibits the uptake of chloride ions by the inhibitory synapses of the brain. This seems to be the main cause of seizure-induction by lindane. Because of its structural similarity with picrotoxin, lindane can bind to the picrotoxin binding-site in the exterior part of the chloride canal. This blocks the effects of the neurotransmitter γ-aminobutyric acid (GABA), the mediator of entry of chloride into the canal. Dose-dependent GABA-inhibiting effects have also been demonstrated in the rat *in vivo* in the lindane concentration range from 5 to 40 mg/kg body weight. Lindane also increases the excitability of presynaptic cholinergic neurones in central and peripheral synapses. Applied to nerve-muscle preparations, lindane concentrations of $5 \times 10^{-5}$ or $10^{-4}$ mol/l caused a massive increase in the spontaneous release of the neurotransmitter acetylcholine. Lindane also inhibits $Na^+$,$K^+$-ATPase, $Ca^{2+}$,$Mg^{2+}$-ATPase and $Mg^{2+}$-ATPases and this too can affect calcium homeostasis. Thus, these effects together cause the excitatory activity in the central nervous system (WHO 1991).

In several animal species, repeated administration of lindane causes liver enlargement which is associated with the induction of xenobiotic-metabolizing enzymes. Induction of various cytochrome P450-dependent monooxygenases (mixed type: CYP1A, CYP2B2, CYP3A1) has been demonstrated (Dubois *et al.* 1997). In addition, the induction of epoxide hydrolase, glutathione-*S*-transferase and UDP-glucuronosyl transferase (Oesch *et al.* 1982) has been shown to have a morphological equivalent in the form of an increase in the amount of smooth endoplasmic reticulum. Functional changes detected sometimes even after a single dose of lindane include reductions in the hexobarbital sleep time and in the half-lives of antipyrine and phenylbutazone, and increases in urinary ascorbic acid excretion. Evidence of the induction of liver enzymes in persons exposed to lindane has also been obtained (WHO 1991).

Under experimental conditions, lindane causes the formation of toxic oxygen radicals. Evidence that the substance can cause oxidative stress is available from *in vivo* (Antunes-Madeira *et al.* 1990, Barros *et al.* 1991) and from *in vitro* (Bainy *et al.* 1994) studies. This can perhaps account for the positive results obtained with toxic concentrations of lindane in two studies of chromosomal aberrations *in vitro* and the questionably positive results of the two dominant lethal tests and the comet assay.

# 3 Toxicokinetics and Metabolism

Lindane is absorbed rapidly from the gastrointestinal tract of the rat and distributed in all organs and tissues within a few hours (WHO 1991). The fraction of the inhaled substance which is absorbed is given as 50 % (Sagunski and Perger 1994). The rate of uptake of lindane through the skin depends on the solvent present. High absorption rates are achieved with lipophilic carriers. In man, the level of dermal absorption reported under various conditions ranges from 10 % to 90 %. In excretion studies carried out with $^{14}C$-labelled lindane in acetone, doses of 4 μg/cm$^2$ (1, 2, or 5 μCi) were applied to the skin of the forearms of 6 volunteers; about 10 % of the administered radioactivity appeared in the urine within 5 days (WHO 1991).

The highest concentrations were found in adipose tissue and in the skin. In various studies in which lindane was administered orally to rats, the fat:blood ratio for lindane was shown to be between 150 and 200, the liver:blood ratio between 5.3 and 9.6 and the brain:blood ratio between 4 and 6.5. The ratios were higher in female animals. Similar results were obtained in a study in which rats were exposed by inhalation (WHO 1991).

Lindane is metabolized mainly in the liver. Phase I reactions such as hydroxylation, dehydrogenation, dechlorination and oxidation are involved as well as phase II reactions, that is, conjugation reactions. Particularly cytochrome P450-dependent monooxygenases are responsible for the metabolism. In the rat and mouse, lindane is converted to chlorobenzenes and chlorophenols. The chlorophenols are mostly tetrachlorophenols (e.g. 2,3,4,6-tetrachlorophenol and 2,3,4,5-tetrachlorophenol), trichlorophenols (e.g. 2,4,5-trichlorophenol and 2,4,6-trichlorophenol) and dichlorophenol. In man as well as various monochlorophenols, dichlorophenols and tetrachlorophenols, mainly 2,4,6-chlorophenol, 2,3,5-chlorophenol and 2,4,5-trichlorophenol have been detected (WHO 1991). 2,4,6-Trichlorophenol causes bone marrow hyperplasia and leukaemia in the rat, and liver adenomas and carcinomas in the mouse (NCI 1978). In the rat, alcohols such as tetrachloro-2-cyclohexen-1-ol and pentachloro-2-cyclohexen-1-ol have been detected. Glucuronide and sulfate conjugates of some of the chlorophenols and of the alcohols have been detected in urine. The dichlorophenols are also conjugated with glutathione and excreted as the mercapturic acids (WHO 1991).

Lindane is excreted with the urine and faeces, mostly in the form of metabolites. After administration of a single lindane dose of 6 mg/kg body weight to lactating rats, the half-life of the substance in the organism was found to be 22 hours (Dalsenter *et al.* 1997b). Rats were treated with daily doses of radioactively labelled lindane of 8 mg/kg body weight for 10 days. Within 24 hours of the end of treatment 17 % of the dose had been recovered in the urine and about 14 % in the faeces. After another 48 hours, the fraction excreted with the urine had increased to 24.5 % and with the faeces to about 21 %. The elimination half-time was found to be 3 to 4 days. Accumulation of lindane in adipose tissue was not detected in the rat even after continuous feeding (WHO 1991).

In man, the half-time for the elimination of lindane from the serum was found to be 20 hours after brief exposure (Henschler and Lehnert 1988) and 9 to 12 days after long-term exposure (Angerer *et al.* 1983). Unlike *β*-hexachlorohexane, which has been shown to accumulate in exposed workers, lindane seems not to accumulate (Baumann *et al.* 1980).

# 4 Effects in Man

There are no data available for the local effects of lindane on skin or mucous membranes, nor for the reproductive toxicity of the substance.

## 4.1 Single exposures

Clinical symptoms of intoxication develop after intake of toxic doses of lindane within a period of a few minutes to several hours, depending on the route of uptake and the formulation. Mild intoxications are characterized clinically by a feeling of being unwell, nausea, dizziness, restlessness, frontal headaches and vomiting. Sometimes muscular fasciculation, balance disorders, ataxia and tremor develop as well. Testing of reflexes mostly reveals increased irritability. At the same time the patients often suffer from upper abdominal pain, diarrhoea and uresiesthesis. Severe cases of poisoning are characterized by tonic-clonic seizures which begin after a very short and sometimes—especially in children—absent prodromal stadium, persist mostly for several minutes and are repeated during a period of some hours, more rarely days. Optic, tactile or acoustic stimulation can induce a seizure. In fatal poisonings, death occurs some hours to days after intake of lindane and is often observed to take place after a seizure, or in coma as a result of central paralysis or after acute cardiocirculatory failure (Herbst and Bodenstein 1973, WHO 1991).

There are no useful reports available of the effects of single inhalation exposures to lindane.

### 4.1.1 Ingestion

The reactions described for test persons or patients given oral doses of lindane are not always identical, presumably because the substance was administered in different formulations or because of differences in the physical condition of the persons. On the one hand, in persons given single oral doses of 45 mg lindane (0.64 mg/kg body weight for a 70 kg person) in a highly disperse lipid-containing emulsion for antiparasitic therapy, vomiting and seizures were described, but on the other, repeated administration of doses of 100 mg per day (1.43 mg/kg body weight and day for a 70 kg person) was tolerated without symptoms. Doses from about 2.5 to 3 mg/kg body weight caused dizziness and vomiting. Under certain conditions, lindane doses of 10 to 20 mg/kg body weight could be lethal for man. However, much higher doses taken orally with suicidal intent (up to about 300 mg/kg body weight) were not fatal in spite of marked clinical symptoms such as seizures and severe liver damage. After intensive therapy, children have also survived the intake of large amounts of lindane, 100 to 300 mg/kg body weight, in spite of very severe symptoms of poisoning (Herbst and Bodenstein 1973, WHO 1991).

### 4.1.2 Dermal absorption

In particular because of the use of lindane in the therapy of scabies and louse infections, we have many years of experience of the dermal toxicity of lindane for man. In general, 1 % solutions are used (Kramer *et al.* 1980). Clinical reports indicate that dermal application of doses of about 5 mg/kg body weight and day to adults does not usually result in acute neurological symptoms (WHO 1991). There are, however, occasional reports of symptoms of poisoning resulting from topical use of a 1 % lindane solution (ACGIH 1992).

Lindane concentrations of 3 µg/l were found in the serum of persons not exposed to the substance (Lange *et al.* 1981). Application of 30 g of a 0.3 % lindane emulsion (about 1.3 mg/kg body weight for a 70 kg person) to the whole of the body apart from the head and the hollow of the elbow joint resulted after 5.5 hours in average serum concentrations of 5 µg/l in healthy volunteers (Zesch *et al.* 1982) and in maximum concentrations of 50 µg/l. After 143 hours, the serum concentrations were in the range between 0.2 and 8.9 µg/l (Lange *et al.* 1981). In adults with scabies, the serum lindane concentration after a single application of 50 g of the 0.3 % lindane emulsion was 200 µg/l as little as 4 hours after the application (Zesch *et al.* 1982).

In children with head lice, the maximum concentration of lindane in serum 2 to 4 hours after use of a shampoo containing 1 % lindane was 1.4 µg/l. In children who had used the shampoo repeatedly, the concentrations increased to 3.6 µg/l (WHO 1991). Six hours after dermal application of a lotion containing 1 % lindane to healthy children and children with scabies, the maximum lindane concentrations in blood were 28 µg/l. The half-life of the substance in the organism was 17.9 to 21.4 hours (Ginsburg *et al.* 1977).

## 4.2 Repeated exposure

A study of 60 hexachlorocyclohexane (HCH) production workers with an average age of 42 years and exposure periods of 1 to 30 years (on average 10 ± 9 years) was carried out. Concentrations of lindane in the workplace air were found to be 0.004 to 0.15 mg/m$^3$. In 57 workers, the serum concentrations of γ-HCH (lindane) averaged 36.9 ± 39.9 µg/l (median 23 µg/l), those of α-HCH averaged 69.6 ± 51.5 µg/l (median 59 µg/l) and β-HCH 190.3 ± 153.4 µg/l (median 160 µg/l). In 21 workers from the "pure γ department", the serum lindane concentrations averaged 64.3 ± 53.5 µg/l, median 42 µg/l. In 8 workers, the average lindane concentrations in subdermal adipose tissue were determined to be 3.1 µg/g extractable lipid (Baumann *et al.* 1980). As some of the workers were exposed dermally to large amounts of the crystalline substance (Angerer 1998), it is not possible to establish a direct correlation between the concentrations measured in air and serum. No evidence of adverse effects on the health of these 60 workers was obtained. The medical examination included neurological status, ECG, clinical chemistry, haematology and general clinical examination (Brassow *et al.* 1981) as well as tests of reflexes, sensitivity, electromyogram, nerve conduction velocity, neuromuscular signal transfer and EEG (Baumann *et al.* 1981). The changes observed in clinical parameters were slight and not in the pathological range. The authors draw attention to the fact that effects on these parameters of the simultaneous exposure to benzene cannot be excluded. There was no evidence of lindane-induced liver damage such as increased γ-glutamyl transpeptidase levels (Brassow *et al.* 1981). The blood levels of the hormones, testosterone and follicle stimulating hormone (FSH), in 54 exposed workers and 20 control persons did not differ significantly, whereas the level of luteinizing hormone (LH) was significantly increased, on average from 6.1 ± 2.1 to 9.6 ± 4.2 mIU/ml. The authors suggested that this could be caused by induction of steroid metabolism by cytochrome P450-dependent monooxygenases. Preliminary studies with 5 workers demonstrated that blood cortisol concentrations and 24-hour urinary 6-β-hydroxycortisol levels were unchanged whereas excretion of 17-hydroxycorticosteroids and 17-ketosteroids were reduced and the D-glucaric acid excretion increased, as is typical after enzyme induction (Tomczak *et al.* 1981).

In another study of 64 HCH production workers (average exposure period more than 19 years), the serum lindane concentrations were 16 to 57 µg/l (α-HCH 22–100 µg/l, β-HCH 97–413 µg/l). Most workers suffered from paresthesia of the face and extremities, headaches and giddiness. Some of the workers also suffered from malaise, vomiting, tremor, apprehension, confusion, loss of sleep, impaired memory and loss of libido. The workers were exposed simultaneously to benzene (Nigam *et al.* 1986).

In comparison with 22 control persons, 45 workers were found to suffer from paresthesia of the face and the extremities, headaches, giddiness, vomiting, apprehension, loss of sleep and changes in some liver function parameters (e.g. γ-glutamyl transpeptidase (γ-GT), lactate dehydrogenase, ornithine carbamyl transaminase). The effects were particularly marked in persons who had direct contact with lindane (Chattopadhyay *et al.* 1988).

In a study of 54 spraymen exposed to 4% lindane and other insecticides in the form of aerosols and mists and who also occasionally diluted stock solutions, average plasma lindane concentrations between 6.4 and 9.9 µg/l were revealed; the maximum was 87 µg/l. Their exposure varied from once daily to once weekly, and the length of exposure was 1 to 20 years; they did not always wear protective gloves and respirators. In 26 workers exposed mainly to lindane, the mean antipyrine plasma half-life was significantly shorter than in 33 controls (7.7 ± 2.6 h compared with 13.1 ± 7.5 h for control persons). Enzyme induction occurred in persons with plasma levels above 10 µg/l; workers who were exposed to lindane had shorter antipyridine and phenylbutazone half-lives when their serum levels were above this value. Hypercholesterolaemia and hyperphospholipidaemia were detected in 40°% of the spraymen (Kolmodin-Hedman 1974, WHO 1991).

Clinical symptoms or EEG changes were seen in 19 workers who had been exposed to lindane for up to 2 years and had blood lindane concentrations of more than 14 µg/l (maximum 340 µg/l). Clinical symptoms or EEG changes were detected in only 2 of 18 workers with blood lindane concentrations in the range between 2 and 13 µg/l (Czeglédi-Jankó and Avar 1970).

In forestry workers involved with preparation and planting of conifer seedlings treated with a 1.6 % aqueous lindane suspension in the warmer season of the year, plasma lindane concentrations up to a maximum of 123 nmol/l, that is 36 µg/l, were determined. In some individuals unspecific symptoms such as flu-like illness, sore throat, nausea and tiredness developed but no changes in liver-specific enzymes (alkaline phosphatase, γ-GT, aspartate aminotransferase, alanine aminotransferase). About 5 to 6 weeks after the end of exposure, all plasma lindane concentrations were below the detection limit of 5 nmol/l, that is below about 1.5 μg/l (Drummond *et al.* 1988).

In numerous case reports, effects on the bone marrow such as aplastic anaemia, leukaemia, thrombopenia and panmyelopathy were described in persons exposed to lindane or lindane and other chemicals (IARC 1979, WHO 1991). Studies carried out as a result of these reports for periods of several weeks to several years and assessed by the Environmental Protection Agency (EPA) yielded no evidence of a causal relationship between exposure to lindane and any abnormal or pathologic condition of the blood (Herbst and Bodenstein 1973, WHO 1991).

## 4.3 Allergenic effects

Several authors have reported the occurrence of allergic reactions such as dermatitis, rhinitis, conjunctivitis and eczema mainly in factories producing HCH (WHO 1991). In several test series, patch tests were carried out on both men and women and on persons who had contact with the substance in agriculture. Negative results were obtained in patch tests carried out with 1 % lindane in petrolatum on 93 patients with dermatological disorders (Lisi *et al.* 1986). In a subsequent multicentre study, lindane (1 % in petrolatum) was patch tested on 335 patients, 95 of whom worked or had worked in agriculture; neither irritative nor allergic reactions were observed (Lisi *et al.* 1987).

## 4.4 Genotoxicity

Lindane production workers who had been potentially exposed in a ventilated process room during a period of at least 6 months, 8 hours daily, during which they were supposed to wear face masks, were examined for the presence of chromatid and labile chromosome aberrations in peripheral lymphocytes. The incidence of aberrations was not significantly different from that in control persons (Király *et al.* 1979).

## 4.5 Carcinogenicity

In the years 1970 to 1975, there were reports of an increased incidence of lung tumours among 285 workers who sprayed various pesticides including lindane. It is not possible to assess the carcinogenicity of lindane on the basis of this study (IARC 1979).

# 5 Animal Experiments and *in vitro* Studies

## 5.1 Acute toxicity

The 4-hour $LC_{50}$ determined for lindane in the rat was 1600 mg/m$^3$ (WHO 1991).

$LD_{50}$ values for oral doses of lindane in oil were found to be 88 to 225 mg/kg body weight for the rat, 72 to 299 mg/kg body weight for the mouse, 100 to 127 mg/kg body weight for the guinea pig, 40 to 75 mg/kg body weight for the rabbit, 30 to 200 mg/kg body weight for the dog and 25 mg/kg body weight for the cat. For an aqueous solution, the $LD_{50}$ values of 170 to 300 mg/kg body weight for the rat were of the same order of magnitude (Herbst and Bodenstein 1973).

The main clinical symptoms of lindane poisoning are central nervous effects which are manifested in the mouse, rat, cat and dog mainly as excitation and in the guinea pig and rabbit as depressive phases. Most frequent are torpidity, restlessness, anxiety, tremor, tonic-clonic cramps and unconsciousness. The most common organ changes are finely granular fatty degeneration in the CNS, vacuolar or fatty degeneration and necrosis in the liver, degeneration of the epithelium of the renal tubules, oedema and inflammatory exudation and bronchopneumonia in the lungs after inhalation of the substance, and bleeding in the gastrointestinal tract (Herbst 1983, Herbst and Bodenstein 1973).

The dermal $LD_{50}$ values obtained for lindane dissolved in oil are 50 to 400 mg/kg body weight for the rat, about 180 mg/kg body weight for the mouse and 75 mg/kg body weight for the rabbit. When lindane was applied in a "vanishing cream", the $LD_{50}$ for the rabbit was 50 mg/kg body weight, when applied in acetone 300 mg/kg body weight (Herbst and Bodenstein 1973).

A single, non-occlusive application of a commercially available formulation containing 1 % lindane to the entire skin of shaved 6-week-old New Zealand White rabbits (dose: 60 mg/kg body weight) caused excitation and seizures. For animals whose skin was also depilated or depilated and abraded, the treatment was lethal. In one shaved and depilated animal, the lindane concentration determined in blood was 650 µg/l and in two animals whose skin had been shaved, depilated and abraded, the blood levels were 1670 and 2480 µg/l. Young adult rabbits were less sensitive than the 6-week-old animals. No effects were seen in shaved animals and in those with depilated and abraded skin only excitation but no seizures and no deaths (Hanig *et al.* 1976).

## 5.2 Subacute, subchronic and chronic toxicity

There are a number of studies in which lindane was administered repeatedly to experimental animals (see WHO 1991). Only those relevant for the present assessment are described below.

### 5.2.1 Inhalation

In a 3-month inhalation study with a 6-week recovery period, groups of 12 male and 12 female Wistar rats were exposed for 6 hours daily to lindane concentrations of 0, 0.02, 0.12, 0.6 or 4.5 mg/m$^3$ in inhalation chambers. Only in the group exposed to 4.5 mg/m$^3$ were there transient mild clinical effects (poorer general condition, slight diarrhoea, ruffled fur) and clearly increased hepatic cytochrome P450 values which returned to normal within 4 weeks of the end of exposure. The numerous functional, laboratory diagnostic and morphological tests revealed no effects of the substance. The highest concentrations of lindane in serum, liver, brain and adipose tissue increased linearly with increasing exposure concentration; the highest levels were found in serum: 390 µg/kg serum in males and 270 mg/kg serum in females. The no observed effect level (NOEL) found in this study was 0.6 mg/m$^3$ (CIEL 1983b).

Groups of 45 male and 45 female CD-1 mice were exposed in inhalation chambers to lindane in concentrations of 0, 0.25, 1 and 5 or 10 mg/m$^3$, 6 hours/day, 5 days/week for 14 weeks. Because 2 male and 12 female mice from the 10 mg/m$^3$ group died in the first week of the study, this concentration was reduced to 5 mg/m$^3$ from week 2; another 3 females and 3 males in this group died. Of the mice exposed to 1 mg/m$^3$, one female and one male died; the causes of death were not established. Changes in clinicochemical or haematological parameters were not detected. Nor were there any histopathological changes. Recording of a myelogram also revealed no changes. The serum lindane concentrations in the female animals in week 14 were found to be 41, 85 and 290 µg/l; the concentrations in the females were about twice those in the males. Because the causes of death in the 1 mg/m$^3$ group could not be established, the no observed adverse effect level (NOAEL) was given as 0.3 mg/m$^3$ (CIEL 1988b). Data for P450 activities which would permit the derivation of a NOEL are not available.

### 5.2.2 Ingestion

In a 3-month feeding study, Wistar rats were given lindane in concentrations of 0, 0.2, 0.8, 4, 20 or 100 mg/kg diet. In the females given 0.8 mg/kg diet (about 0.06 mg/kg body weight and day) or more and in the males of the highest dose group, the cytochrome P450 activities in the liver were increased. In animals given 20 mg/kg diet or more, liver and kidney weights were increased and the histopathological examination revealed signs of liver cell hypertrophy and degenerative changes in the epithelium of the renal tubules. At 100 mg/kg diet, body weight gains were reduced. The NOAEL found in this study for adverse histopathological effects was 4 mg/kg diet, that is, 0.3 mg/kg body weight and day (CIEL 1983a). The NOEL for the induction of cytochrome P450 was 0.2 mg/kg diet, that is, about 0.015 mg/kg body weight and day.

In earlier studies, rats were given crystalline lindane in the diet in concentrations of 0, 10, 100 and 800 mg/kg diet or lindane dissolved in oil in concentrations of 0, 5, 10, 50, 100, 400, 800 or 1600 mg/kg diet. Mortality was increased from 400 mg/kg diet (about 30 mg/kg body weight and day). From 100 mg/kg diet (about 7.5 mg/kg body weight and day) liver weights were increased in a dose-dependent manner and histopathological changes were found in the liver (hepatocellular hypertrophy, fatty degeneration, necrosis) and kidneys (nephritis). The NOEL was given as 50 mg/kg diet (about 3.5 mg/kg body weight and day). In another study lasting 2 years, liver cell hypertrophy was also seen at 50 mg/kg diet and fatty degeneration of the liver cells at concentrations of 100 mg/kg diet or more. At 25 mg/kg diet (about 1.8 mg/kg body weight and day) no changes were detected (WHO 1991).

In another 2-year study, beagle dogs were given lindane in concentrations of 25, 50 or 100 mg/kg diet. Whereas the high dose (corresponding to 2.92 mg/kg body weight and day) caused increased serum alkaline phosphatase activity and dark coloured fragile livers, at the two lower lindane concentrations no substance-related changes were detected in the numerous clinical, functional and morphological tests (Rivett *et al.* 1978).

### 5.2.3 Dermal absorption

Groups of 49 male and 49 female Wistar rats were treated dermally with lindane doses of 0, 10, 60 or 400 mg/kg body weight applied as a suspension in carboxymethyl cellulose under an occlusive bandage for 6 hours daily, 5 days per week for 13 weeks. From each dose group, 23 animals of each sex were examined at the end of the treatment period and groups of 13 animals of each sex 6 and 12 weeks later. In animals treated with doses of 60 mg/kg body weight or more, reversible hepatocellular hypertrophy was diagnosed; the kidney changes (increased kidney weights, degeneration and necrosis in the tubules, basophilia in the tubule epithelium, increased incidence of cylinders) seen at these doses were only partially reversible. A dose of 10 mg/kg body weight was considered to be non-toxic (CIEL 1988a).

## 5.3 Local effects on skin and mucous membranes

In a study of skin irritation carried out according to OECD criteria, lindane was applied to the intact skin of New Zealand White rabbits; no evidence of primary skin irritating properties was found (CIEL 1986b).

A similar study in which the substance was administered into the conjunctival sac of New Zealand White rabbits revealed mild irritation of the conjunctival mucosa (CIEL 1986a).

## 5.4 Allergenic effects

Lindane and various formulations containing the substance were tested for skin sensitizing potential in Dunkin-Hartley guinea pigs with the Magnusson-Kligmann maximization test carried out according to OECD guidelines. None of the tests (intracutaneous first injection, 2 challenge injections) revealed contact allergenic properties of lindane (CIEL 1986b, 1987, WHO 1991).

## 5.5 Reproductive and developmental toxicity

In a 3-generation study, groups of 10 male and 10 female Crl:CD rats were given lindane in concentrations of 0, 25, 50 and 100 mg/kg diet, corresponding to doses of about 0, 7, 14, 28 mg/kg body weight and day. Histological examination, carried out only with the 3-week-old pups of the second litter of the third generation, revealed significantly increased liver weights in the females of all dose groups and in males only in the highest dose group. Fertility disorders or effects on litter size, weights of the pups, lactation, malformations or development were not detected (Palmer *et al.* 1978b).

### 5.5.1 Fertility

#### Fertility of males

In male rats given lindane doses of 10 or 20 mg/kg body weight and day from day 21 to days 100 to 105, kidney weights were increased from doses of 10 mg/kg body weight and day and also liver weights at 20 mg/kg body weight. No effects on fertility were observed (Gray *et al.* 1988). This study is inadequately documented.

In a study of the oestrogenic effects of lindane, groups of 9 lactating Wistar rats were given single oral lindane doses of 6 mg/kg body weight on day 9 or day 14 after parturition or daily doses of 1 mg/kg body weight from days 9 to 14. Four hours after the first dose on day 9 (6 mg/kg body weight) the lindane level in plasma was found to be about 300 μg/l, in the milk 8 to 12 hours after the dose it was about 22 mg/l. To study the effect of the treatment on the reproductive physiology of the male pups, these were examined either at age 65 days (puberty) or 140 days ("adult"). The testis weights and the numbers of spermatozoa and spermatids were increased slightly on day 65 and significantly in all groups on day 140. The testosterone levels were significantly decreased on day 65 in all treated animals and on day 140 in those treated once with 6 mg/kg body weight. Histological examination of the testes revealed especially a reduction in the numbers of Leydig cells and in some cases the absence of spermiogenesis in the tubules. Adverse effects were found in tests for several parameters of sexual behaviour but indices of pregnancy and fertility were similar in all the treated groups and not different from the control values. The numbers of litters, implantations per litter, foetuses per litter and resorptions were not different in the treated groups and the control groups included in the study (Dalsenter *et al.* 1997b).

In another study by the same research group, groups of 20 Wistar rats were given single oral lindane doses of 30 mg/kg body weight on day 15 of gestation or only the vehicle, arachis oil. Of the 45 male pups raised in each group, 15 were examined at the age of 100 days and another 15 at 130 days. Effects on testis weights were not found. On day 100, less spermatids were counted in the group treated with lindane but no less spermatozoa. On day 130 the counts were no longer different from the control values. The remaining 15 pups were mated after 7 months in the ratio 1:1 with untreated female rats. Observation of mating behaviour revealed a significant decrease in libido. Quantitative details were not given. The loss of libido was considered to be causally associated with the significantly lower blood testosterone levels in these animals (0.94 ng/ml; 1.65 ng/ml in the control group). The number of pregnancies and thus the fertility of the animals was not affected by the lindane treatment (Dalsenter *et al.* 1997a).

### Fertility of females

There are numerous studies available which demonstrate that longer-term oral administration of lindane or administration of single injections of the substance to rats in doses of 10 to 20 mg/kg body weight at specific times during the oestral cycle results in lengthening of oestrus, delay in vaginal opening and reduced readiness for mating, but not in histopathological changes in the uterus, cervix or vagina. Reduced litter sizes have also been observed (Chadwick *et al.* 1988, Cooper *et al.* 1989, Gray *et al.* 1988, Raizada *et al.* 1980, Uphouse 1987, Uphouse and Williams 1989, WHO 1991). In rats which were not yet sexually mature and in ovariectomized adult rats, however, subcutaneous injection of lindane doses of 40 mg/kg body weight daily for 7 days or oral administration of lindane doses of 10, 20 or 40 mg/kg body weight daily for 5 days had no effect on the oestrogen receptor or on the oestrogen-dependent induction of progesterone receptors (Laws *et al.* 1994).

In pregnant Swiss mice given lindane doses of 0.26 mg per animal (about 10.8 mg/kg body weight and day) on days 1 to 4 of gestation, no implantations were observed. Simultaneous administration of oestrogen and progesterone prevented the adverse effects on implantation but not those on further foetal development. Administration of lindane doses of 0.15 mg per animal (about 6.25 mg/kg body weight) on days 6 to 12 of gestation did not reduce the average number of implantations; however, the incidence of resorptions was 100 %. Doses of 0.09 mg per animal (about 3.75 mg/kg body weight) on days 14 to 19 of gestation did not result in changes in the number of progeny but the body weights of the pups were reduced and all died within the first 4 days after birth. The authors concluded that lindane has not only foetotoxic effects but also causes steroid hormone deficiency which leads to further toxic effects on reproduction and development (Sircar and Lahiri 1989).

Groups of 10 and 15 female mink were given lindane doses of 1 mg/kg body weight and day with the diet for 3 or 6 weeks before mating and until the pups were weaned 8 weeks after birth. The animals were mated twice at intervals of 7 and 2 days, respectively. No effects on body weight gain or food consumption were seen in the dams. Fertility was reduced and there were embryonic deaths after implantation (Beard *et al.* 1997).

## 5.5.2 Developmental toxicity

Groups of 15 to 19 CFY rats were given oral lindane doses of 5, 10 or 20 mg/kg body weight on days 6 to 15 of gestation. Doses of 5 mg/kg body weight and day were not toxic for the dams or embryos. In dams given 10 and 20 mg/kg body weight and day, body weight gains were reduced, and at 20 mg/kg two animals died. The incidence of malformations was not increased in the progeny but there was a dose-dependent increase in the incidence of 14th ribs (control 12.7 %; exposed 21.0 %; 31.7 %; 40.6 %) which was significant in the highest dose group (Palmer *et al.* 1978a). This effect, which could not be reproduced in the other studies, is considered to be unspecific.

Groups of 20 Wistar rats were given oral lindane doses of 6.25, 12.5 or 25 mg/kg body weight on days 6 to 15 of gestation. Doses of 6.25 and 12.5 mg/kg body weight and day were not toxic for the dams or embryos. In dams given 25 mg/kg body weight, maternal body weights were slightly reduced; embryotoxic effects were not seen (Khera *et al.* 1979).

Groups of 20 Sprague-Dawley rats were given lindane doses of 5, 15 or 30 mg/kg body weight on days 6 to 15 of gestation by subcutaneous injection. Doses of 5 mg/kg body weight and day were not toxic for the dams or embryos. Body weight gains of the dams given 15 or 30 mg/kg body weight were reduced; two dams in the high dose group died. Even in the group given 30 mg/kg body weight and day, there were no effects on the numbers of living foetuses or on their development (e.g. on skeletal or visceral anomalies) (CIEL 1976a).

Groups of 25 NMRI mice were given oral lindane doses of 12, 30 or 60 mg/kg body weight on days 11 to 13 or days 6 to 15 of gestation. Doses of 12 mg/kg body weight and day were not toxic for the dams or embryos. In the 30 mg/kg group, one dam died. In the 60 mg/kg groups, 12 and 4 dams, respectively, died and the body weight gains were significantly reduced. The incidence of miscarriages was increased, and the number of foetuses and the foetal weights were decreased. The incidence of malformations was not increased (CIEL 1972).

Administration of single oral lindane doses of 30 or 45 mg/kg body weight on day 12 of gestation to groups of 5 to 7 C57BL/6J or DBA/2J mice resulted in an increase in deaths of the dams, in reduced placenta weights and reduced foetal body weights. In the high dose group, the number of dead or resorbed foetuses was also increased and the number of implantations decreased (Hassoun and Stohs 1996).

Groups of 25 NMRI mice were given lindane doses of 6 mg/kg body weight on days 11 to 13 or 6 to 15 of gestation by subcutaneous injection. The treatment was not toxic for dams or embryos. The incidence of malformations was not increased (WHO 1991).

Groups of 10 to 12 pregnant New Zealand White rabbits were given oral lindane doses of 5, 10 or 20 mg/kg body weight on days 6 to 18 of gestation. Lindane had toxic effects on the dams in all dose groups (slight lethargy, tachypnoea and reduced body weight gains in all rabbits) but was not embryotoxic or teratogenic (Palmer *et al.* 1978a).

Groups of 15 rabbits were given subcutaneous doses of lindane of 5, 15 or 45 mg/kg body weight on days 6 to 18 of gestation. Whereas 5 mg/kg body weight was tolerated, 15 mg/kg body weight and day resulted in marked maternal toxicity (body weight reduction). The highest dose of 45 mg/kg body weight and day was reduced on day 9 to 30 mg/kg body weight and day because of severe toxicity for the dams. Nevertheless, 14 of 15 rabbits died. Embryotoxicity or teratogenicity was not detected (CIEL 1976b).

Groups of 6 pigs were given lindane concentrations in the diet of 50 or 500 mg/kg diet from 30 days before mating until 30 days afterwards. The treatment was not toxic for the dams or embryos nor teratogenic (WHO 1991).

Groups of 13 and 14 beagle dogs were given oral lindane doses of 7.5 or 15 mg/kg body weight from day 1 or 5 until the end of gestation. There was a dose-independent increase in stillbirths. The number of living pups was not affected in comparison with the values for the 5 control animals. Lindane was not teratogenic (WHO 1991).

## 5.6 Genotoxicity

### 5.6.1 *In vitro*

Lindane has been tested in numerous genotoxicity studies in a variety of different tests (Pool-Zobel *et al.* 1993, WHO 1991).

Neither the tests for mutagenicity (Rec assay with *Bacillus subtilis*, test for mutation in *Escherichia coli*, the *Salmonella* mutagenicity test, hypoxanthine guanine phosphoribosyl transferase (HPRT) test with V79 and CHO cells (a cell line from Chinese hamster ovary)) nor tests for clastogenicity (sister chromatid exchange (SCE) with CHO cells) nor other appropriate genotoxicity tests (UDS (unscheduled DNA synthesis) test with human lymphocytes, an SV-40 transformed human fibroblast cell line and rat hepatocytes) yielded evidence of genotoxic properties of lindane *in vitro*. Tests for chromosomal aberrations with CHL cells (a cell line from Chinese hamster lung) and human lymphocytes revealed gaps and chromosome breaks at toxic concentrations. Because of the toxicity, this is considered irrelevant.

In foetal rat hepatocytes and in the cell line HepG2, which is derived from human liver cells, the formation of not further identified DNA adducts was demonstrated (Dubois *et al.* 1997). Because of the low level of DNA adducts and because of inadequacies in the methods used (e.g. no standard adducts) this study cannot be included in the present assessment.

With the comet assay, a single cell microgel electrophoresis, an increased incidence of DNA strand breaks was detected in nasal, colon and gastric epithelial cells from rats after exposure *in vitro* for various times. The results were confirmed in another *in vitro* study with nasal epithelial cells from the rat and man and with rat gastric epithelial cells but not with human gastric epithelial cells (Pool-Zobel *et al.* 1993, 1994).

### 5.6.2 *In vivo*

*In vivo* mutagenicity studies (host-mediated assay in the mouse with *S. typhimurium* G46 and *Serratia merescens*), clastogenicity studies (tests for chromosomal aberrations in the bone marrow cells of the rat, Chinese hamster and Syrian hamster, SCE in bone marrow cells of the mouse, micronuclei in mouse erythrocytes and in the bone marrow of the mouse, rat and hamster), one dominant lethal test in the rat and one in the mouse, and a DNA binding study in the mouse yielded no evidence of genotoxic effects of lindane *in vivo* (Dzwonkowska and Hübner 1989, Pool-Zobel *et al.* 1993, WHO 1991). Questionably positive results were obtained in a dominant lethal test in the rat and one in the mouse with five oral doses of the substance, but the methods used in these studies were inadequate (WHO 1991).

The comet assay carried out on nasal, colon and gastric epithelial cells from rats exposed *in vivo* for various periods of time (oral 60 mg/kg body weight, inhalation of concentrations up to 3 mg/$m^3$) revealed—as *in vitro*—an increase in the incidence of DNA strand breaks (Pool-Zobel *et al.* 1993). In rats given single lindane doses of 30 mg/kg body weight, lipid peroxidation and DNA single strand breaks were detected in the liver (Hassoun *et al.* 1993).

Weak binding of lindane to the liver DNA of NMRI mice was demonstrated 10 hours after oral administration of doses of radioactively labelled lindane of 12 and 13 mg/kg body weight. From an experimentally determined covalent binding index (CBI) of about 0.2 and a "true CBI" of 0.02 to 0.1, calculated because of technical problems (50 % of the radioactivity was not covalently associated with the DNA) the authors concluded that the DNA binding is so weak that a non-genotoxic mechanism must be assumed for the development of the observed liver tumours (Sagelsdorff *et al.* 1983).

## 5.7 Carcinogenicity

### 5.7.1 Initiation and promotion studies

In an initiation/promotion study, female Wistar rats were subjected to partial hepatectomy and initiation by administration of lindane doses of 30 mg/kg body weight, daily for 2 weeks; after the subsequent promotion with phenobarbital doses of 50 mg/kg body weight, daily for 15 weeks, no initiating effects of lindane could be detected in the liver (Schröter *et al.* 1987).

However, lindane has promoting effects in the rat liver. After initiation by gavage of a single dose of *N*–nitrosomorpholine of 250 mg/kg body weight and promotion for up to 20 weeks by administration of lindane in the diet in doses of 0, 0.1, 0.5, 2.5, 10, or 30 mg/kg body weight and day, the number and size of preneoplastic foci in the liver was significantly increased at doses of 2.5 mg/kg body weight or more. In the animals of the 30 mg/kg dose group, the changes were more marked than in the positive control treated with phenobarbital (50 mg/kg body weight and day). In animals given lindane doses of 10 mg/kg body weight and day or more, increased monooxygenase activities (demethylation of aminopyrine and benzphetamine) were found. The NOEL for preneoplastic foci was calculated to be 0.5 mg/kg body weight and day or 10 µg/g adipose tissue and that for the induction of monooxygenases 1 mg/kg body weight and day or 12 µg/g adipose tissue (Schröter *et al.* 1987).

### 5.7.2 Long-term studies

#### Rat

In an early study by the National Cancer Institute (NCI 1977), the administration of lindane in average concentrations of 236 or 472 mg/kg diet (about 18 or 35 mg/kg body weight and day) to male Osborne-Mendel rats and of 135 or 275 mg/kg diet (about 10 or 20 mg/kg body weight and day) to female animals did not result in significant dose-dependent increases in tumour incidence. In both sexes, the incidence of neoplastic nodules in the liver was increased but not significantly. In addition, in male animals the incidences of adenomas and carcinomas of the thyroid gland were slightly but not significantly increased. In the female animals of the low dose group, the incidence of C cell adenomas of the thyroid gland was significantly increased. Slight but not significant increases in tumour incidences were also seen in female rats for adrenal adenomas and for mammary adenomas and carcinomas and in animals of both sexes for subcutical fibromas and chromophobic adenomas of the pituitary gland (Table 1).

In other long-term studies with rats given lindane in concentrations up to 100 mg/kg diet (about 7.5 mg/kg body weight and day) (WHO 1991) or 500 mg/kg diet (about 38 mg/kg body weight and day) (Ito *et al.* 1975) or 1600 mg/kg diet (about 120 mg/kg body weight and day) (Fitzhugh *et al.* 1950) no increased tumour incidence was found (Table 1).

## Mouse

In mice given lindane in concentrations up to 500 mg/kg diet (about 75 mg/kg body weight and day) for periods between 24 and 32 weeks, no liver tumours were observed (Ito *et al.* 1973); in those given 600 mg/kg diet (about 90 mg/kg body weight and day) the liver tumour incidence was increased (Goto *et al.* 1972, Hanada *et al.* 1972; Table 1).

No increase in tumour incidence was found in NMRI mice given lindane in concentrations up to 50 mg/kg diet (about 7.5 mg/kg body weight and day) for at least 80 weeks (Herbst *et al.* 1975, Weisse and Herbst 1977). At lindane concentrations of 80 and 160 mg/kg diet (about 12 and 24 mg/kg body weight and day) liver tumours developed in B6C3F$_1$ mice, but the incidence was significantly increased only in the male animals of the low dose group (NCI 1977). The incidence of hepatocellular adenomas was increased in 2 of 3 phenotypes of (YS×VY)F$_1$ hybrid mice given lindane in a concentration of 160 mg/kg diet (about 24 mg/kg body weight and day) and lung tumours were diagnosed as well (Wolff and Morrissey 1986, Wolff 1993). The administration of lindane in a concentration of 400 mg/kg diet (about 60 mg/kg body weight and day) to Carsworth mice for 110 weeks resulted in malignant liver tumours and metastases in the lungs (Thorpe and Walker 1973; Table 1).

No tumours were observed after application of 0.5 % lindane in acetone to the skin of mice, twice weekly for 15 months (Orr 1948).

Administration of 2,4,6-trichlorophenol, a metabolite of lindane, in concentrations of 5000 or 10000 mg/kg diet to male F344 rats for 106 or 107 weeks resulted in a dose-dependent increase in the incidence of lymphoma and leukaemia. In male and female rats, leukocytosis and monocytosis of the peripheral blood, and bone marrow hyperplasia were observed as well. In B6C3F$_1$ mice, 2,4,6-trichlorophenol caused a dose-dependent increase in the incidence of hepatocellular adenomas and carcinomas (NCI 1978). Comparative determination of the carcinogenic potencies of lindane and 2,4,6-trichlorophenol has demonstrated, however, that the metabolite makes only a very small, not significant contribution to the carcinogenic effects of lindane (WHO 1991).

**Table 1.** Long-term carcinogenicity studies with lindane

| | |
|---|---|
| Author: | NCI (1977) |
| Substance: | lindane (purity 100 %) |
| Species: | rat (Osborne-Mendel); treated groups 50 ♂ and 50 ♀, control group 10 ♂ and 10 ♀ (study control = matched control) and control group 50 ♂ and 50 ♀ (pooled control; control animals also from other studies with other substances) |
| Administration route: | in the diet |
| Dose: | ♂: 236 mg/kg diet: 320 mg/kg (38 weeks) + 160 mg/kg (42 weeks)<br>472 mg/kg diet: 640 mg/kg (38 weeks) + 320 mg/kg (42 weeks)<br>♀: 135 mg/kg diet: 320 mg/kg (2 weeks) + 160 mg/kg (49 weeks) + 29 mg/kg (29 weeks)<br>275 mg/kg diet: 640 mg/kg (2 weeks) + 320 mg/kg (49 weeks) + 160 mg/kg (29 weeks) |
| Duration: | 80 weeks |
| Toxicity: | clinical symptoms in all treated groups |
| Tumours: | no significant dose-dependent increase in tumour incidence |

| | | Matched control | | Pooled control | | Low concentration | | High concentration | |
|---|---|---|---|---|---|---|---|---|---|
| **liver** | | | | | | | | | |
| neoplastic nodules | ♂ | 0/10 | 0 % | 0/49 | 0 % | 3/45 | 7 % | 2/45 | 4 % |
| | ♀ | 0/10 | 0 % | 1/49 | 2 % | 4/48 | 8 % | 2/45 | 4 % |
| **thyroid gland** | | | | | | | | | |
| follicle cell adenomas | ♂ | 1/6 | 17 % | 3/42 | 7 % | 5/37 | 14 % | 0/37 | 0 % |
| | ♀ | 0/8 | 0 % | 0/48 | 0 % | 1/44 | 2 % | 1/42 | 2 % |
| follicle cell carcinomas | ♂ | 0/6 | 0 % | 1/42 | 2 % | 1/37 | 3 % | 4/37 | 11 % |
| | ♀ | 0/8 | 0 % | 0/48 | 0 % | 1/44 | 2 % | 0/42 | 0 % |
| C cell adenomas | ♂ | 1/6 | 17 % | 2/42 | 5 % | 3/37 | 8 % | 1/37 | 3 % |
| | ♀ | 0/8 | 0 % | 0/48 | 0 % | 4/44 | 9 %* | 3/42 | 7 % |
| **adrenal glands** | | | | | | | | | |
| adenomas | ♀ | 0/9 | 0 % | 0/51 | 0 % | 3/42 | 7 % | 2/44 | 5 % |
| **subcutis** | | | | | | | | | |
| fibromas | ♂ | 0/10 | 0 % | 0/49 | 0 % | 1/48 | 2 % | 3/49 | 6 % |
| | ♀ | 0/10 | 0 % | 0/54 | 0 % | 1/50 | 2 % | 0/50 | 0 % |
| **pituitary gland** | | | | | | | | | |
| chromophobic adenomas | ♂ | 0/10 | 0 % | 6/47 | 13 % | 3/32 | 9 % | 1/35 | 3 % |
| | ♀ | 3/7 | 43 % | 6/46 | 13 % | 14/45 | 31 %* | 8/41 | 20 % |
| **mammary glands** | | | | | | | | | |
| adenomas | ♀ | 0/10 | 0 % | 0/47 | 0 % | 3/50 | 6 % | 1/50 | 2 % |
| adenomas + carcinomas | ♀ | 1/10 | 10 % | 1/47 | 2 % | 4/50 | 8 % | 1/50 | 2 % |

* $p < 0.05$

**Table 1.** continued

| | |
|---|---|
| Author: | WHO (1991) |
| Substance: | lindane (purity not specified) |
| Species: | rat (strain not specified); groups of 10 ♂, 10 ♀ |
| Administration route: | in the diet |
| Dose: | 0, 25, 50, 100 mg/kg diet |
| Duration: | "for life" |
| Toxicity: | 25 mg/kg diet: no effects observed<br>from 50 mg/kg diet: liver cell hypertrophy<br>100 mg/kg diet: fatty degeneration in the liver |
| Tumours: | no increased tumour incidence |
| Author: | Ito *et al.* (1975) |
| Substance: | lindane (purity not specified) |
| Species: | rat (Wistar); 6 ♂ treated (24 weeks),<br>8 ♂ treated (48 weeks), 8 ♂ controls |
| Administration route: | in the diet |
| Dose: | 0, 500 mg/kg diet |
| Duration: | 24 or 48 weeks |
| Toxicity: | reduced body weight gains, increased liver weights |
| Tumours: | no hyperplasia or carcinomas in the liver |
| Author: | Fitzhugh *et al.* (1950) |
| Substance: | lindane (purity > 98 %) |
| Species: | rat (Wistar), groups of 10 ♂, 10 ♀ |
| Administration route: | in the diet |
| Dose: | 0, 5, 10, 50, 100, 400, 800 1600 mg/kg diet (technical grade in oil)<br>0, 10, 100, 800 mg/kg diet (crystalline) |
| Duration: | 107 weeks |
| Toxicity: | from 100 mg/kg diet: reduced body weight gains, histopathological changes in the liver and kidneys, increased liver weights<br>from 800 mg/kg diet: increased mortality, nervous symptoms, seizures |
| Tumours: | no tumours observed |
| Author: | Goto *et al.* (1972) |
| Substance: | lindane (purity not specified) |
| Species: | mouse (ICR-JCL); groups of 20 ♂ |
| Administration route: | in the diet |
| Dose: | 0, 300, 600 mg/kg diet |
| Duration: | 6 months |
| Toxicity: | 300 mg/kg diet: body weight gains and organ weights not "unusually" altered, no histopathological damage in the liver<br>600 mg/kg diet: increased liver weights |
| Tumours: | 300 mg/kg diet: no liver tumours<br>600 mg/kg diet: 5/10 animals with liver tumours |

**Table 1.** continued

| | |
|---|---|
| Author: | Ito *et al.* (1973) |
| Substance: | lindane (purity 99 %) |
| Species: | mouse (dd); groups of 20 ♂ |
| Administration route: | in the diet |
| Dose: | 0, 50, 100, 250, 500 mg/kg diet |
| Duration: | 24 weeks |
| Toxicity: | from 250 mg/kg diet: liver cell hypertrophy |
| Tumours: | no hyperplasia or carcinomas in the liver |

| | |
|---|---|
| Author: | Hanada *et al.* (1973) |
| Substance: | lindane (purity not specified) |
| Species: | mouse (dd); groups of 10–11 ♂, 10–11 ♀ |
| Administration route: | in the diet |
| Dose: | 0, 100, 300, 600 mg/kg diet |
| Duration: | 32 weeks |
| Toxicity: | from 300 mg/kg diet: atypical proliferation in the liver<br>600 mg/kg diet: increased mortality |
| Tumours: | |

| Concentration (mg/kg diet) | | 0 | 100 | 300 | | 600 | |
|---|---|---|---|---|---|---|---|
| **liver** | | | | | | | |
| atypical proliferation | ♂ | 0/14 | 0/10 | 5/9 | 56 % | 4/4 | 100 % |
| | ♀ | 0/15 | 0/8 | 1/7 | 14 % | 3/3 | 100 % |
| hepatomas | ♂ | 0/14 | 0/10 | 0/9 | 0 % | 3/4 | 75 % |
| | ♀ | 0/15 | 0/8 | 0/7 | 0 % | 1/3 | 33 % |

significance not specified

| | |
|---|---|
| Author: | Herbst *et al.* (1975); Weisse and Herbst (1977) |
| Substance: | lindane (purity not specified) |
| Species: | mouse (NMRI); treated groups 50 ♂ and 50 ♀,<br>100 ♂ and 100 ♀ in the control group |
| Administration route: | in the diet |
| Dose: | 0, 12.5, 25, 50 mg/kg diet |
| Duration: | 80 weeks |
| Toxicity: | up to 50 mg/kg diet: no hepatocellular alterations visible in the electron microscopic examination |
| Tumours: | no increased tumour incidence (NMRI mice have a low spontaneous incidence of hepatomas) |

**Table 1.** continued

| | |
|---|---|
| Author: | NCI (1977) |
| Substance: | lindane (purity 100 %) |
| Species: | mouse ($B6C3F_1$); treated groups 50 ♂, 50 ♀, control group 10 ♂, 10 ♀ (study control = matched control) and control group 50 ♂, 50 ♀ (pooled control; control animals also from other studies with other substances) |
| Administration route: | in the diet |
| Dose: | 0, 80, 160 mg/kg diet |
| Duration: | 80 weeks |
| Toxicity: | clinical symptoms in all treated groups |
| Tumours: | |

| Concentration (mg/kg diet) | | 0 (Matched control) | | 0 (Pooled control) | | 80 | | 160 | |
|---|---|---|---|---|---|---|---|---|---|
| **liver** | | | | | | | | | |
| carcinomas | ♂ | 2/10 | 20 % | 5/49 | 10 % | 19/49 | 39 %** | 9/46 | 20 % |
| | ♀ | 0/10 | 0 % | 2/47 | 4 % | 2/47 | 4 % | 3/46 | 7 % |

** $p < 0.01$

| | |
|---|---|
| Author: | Wolff and Morrissey (1986), Wolff (1993) |
| Substance: | lindane (purity not specified) |
| Species: | mouse [$(YS\times VY)F_1$ hybrids: 3 phenotypes: yellowAvy/a; pseudoaguti Avy/a; black a/a)]; ♀ |
| Administration route: | in the diet |
| Dose: | 0, 160 mg/kg diet |
| Duration: | 24 months |
| Toxicity: | no data |
| Tumours: | |

| Concentration (mg/kg diet) | 0 | 160 |
|---|---|---|
| **hepatocellular adenomas** | | |
| yellowAvy/a: | 9 % | 35 % |
| pseudoaguti Avy/a: | 5 % | 12 % |
| black a/a: | 6 % | 3 % |
| **lung tumours** | | |
| yellowAvy/a: | 4 % | 19 % |
| pseudoaguti Avy/a: | 6 % | 14 % |
| black a/a: | 2 % | 3 % |

significance not specified

**Table 1.** continued

| | |
|---|---|
| Author: | Thorpe and Walker (1973) |
| Substance: | lindane (purity > 99.5 %) |
| Species: | mouse (Carsworth CF1); treated groups 30 ♂, 30 ♀, control group 45 ♂, 45 ♀ |
| Administration route: | in the diet |
| Dose: | 0, 400 mg/kg diet |
| Duration: | 110 weeks |
| Toxicity: | increased mortality; liver enlargement |
| Tumours: | in animals treated with lindane the liver tumours developed earlier than in the controls and the animals died earlier |

| Concentration (mg/kg diet) | | 0 | | 400 | |
|---|---|---|---|---|---|
| survival | ♂ | 44 % | | 17 % | |
| | ♀ | 32 % | | 3 % | |
| **liver tumours** | | | | | |
| "benign tumour" | ♂ | 9/45 | 20 % | 11/28 | 39 % |
| | ♀ | 10/44 | 23 % | 10/21 | 48 %* |
| "malignant tumour" | ♂ | 2/45 | 4 % | 16/28 | 57 %** |
| | ♀ | 0/44 | 0 % | 20/21 | 95 %** |
| metastases in the lungs | ♂ | 0/45 | 0 % | 3/28 | 11 % |
| | ♀ | 0/44 | 0 % | 1/21 | 5 % |

* $p < 0.05$, ** $p < 0.01$

| | |
|---|---|
| Author: | Orr (1948) |
| Substance: | lindane (purity not specified) |
| Species: | mouse (strain not specified); numbers not specified |
| Administration route: | dermal |
| Dose: | 0.5 % lindane in acetone |
| Duration: | twice weekly for 15 months |
| Toxicity: | no effects at the application site |
| Tumours: | no tumours |

## 5.8 Other effects

### 5.8.1 Immunotoxic effects

Young Wistar rats were given lindane in concentrations of 0, 5, 20 or 30 mg/kg diet (about 0, 0.25, 1 or 1.5 mg/kg body weight and day) for up to 22 weeks. 20 days before the test the animals received a subcutaneous injection of 0.2 ml tetanus toxoid in Freund's complete adjuvant. In rats fed with a lindane concentration of 5 mg/kg diet, the concentration of toxoid-specific antibodies was slightly but not significantly reduced. At 20 mg/kg diet or more, the titre of toxoid-specific antibodies and that of total IgM and IgG were significantly reduced and the migration of leukocytes and peritoneal macrophages inhibited (Saha and Banerjee 1993).

Male and female Charles-Forster rats were given oral doses of 0, 6.25 or 25 mg/kg body weight daily for 35 days and intramuscular injections of *Salmonella typhimurium* and *Salmonella paratyphimurium* A and B antigens on days 7 and 14; the specific antibody titre was reduced by the treatment with lindane (Dewan *et al.* 1980).

Groups of 6 female Swiss mice were given lindane in the diet in doses of 0, 0.012, 0.12 and 1.2 mg/kg body weight and day for a maximum of 24 weeks. Both the cell-mediated immune response (delayed type hypersensitivity reaction: induction and provocation with sheep red blood cells (SRBC); lymphoproliferative response to concanavalin A) and the humoral immune response (IgM plaque formation to T-cell-dependent antigen (SRBC) and T cell-independent antigen (SRBC coated with lipopolysaccharides)) were dose-dependent and biphasic with initial stimulation and subsequent suppression. The observed biphasic course of the immune response is in accordance with the histopathological findings in the lymphoid organs. The initial increase in the lymphoid follicle activity was followed in the course of the study by a reduction in cell numbers which progressed until cell depletion was evident in the thymus, spleen and lymph nodes. Effects on the phagocytic activity of peritoneal macrophages or on the lymphocyte reaction after administration of mitomycin C *in vitro* were not detected (Meera *et al.* 1992).

Male Hissar mice were given lindane in the diet in concentrations of 0, 10, 30 or 50 mg/kg diet (about 0, 1.5, 4.5 or 7.5 mg/kg body weight and day) for up to 12 weeks. The primary humoral antibody response to SRBC was impaired in animals given concentrations of 50 mg/kg diet. At 30 mg/kg diet or more, the level of the secondary immune response was reduced. No effects on the immune response were seen after lindane concentrations of 10 mg/kg diet (about 1.5 mg/kg body weight and day) (Banerjee *et al.* 1996).

Groups of 20 adult BALB/c mice were fed lindane concentrations of 0 or 150 mg/kg diet (about 22.5 mg/kg body weight and day) for one month. After the subsequent immunization with a single intraperitoneal injection of 0.5 ml of a 2 % suspension of SRBC, no significant difference in the IgM production in the two groups could be detected. Treatment with lindane and four intragastric doses of $4 \times 10^9$ SRBC did not cause changes in the titres of IgA, IgG1, IgG2a or IgG3. However, the titre of IgG2b was significantly increased. In animals treated with lindane in a concentration of 150 mg/kg diet and subsequent oral administration of 1000 *Giardia muris* cysts, the *Giardia* infection 28 days later was more persistent and more severe than in the animals which were not treated with lindane. In the small intestine the formation of trophozoites was seen; increased levels of specific antibodies were found (André *et al.* 1983).

Groups of 6 male rabbits were given lindane in oral doses of 1.5, 3, 6 or 12 mg/kg body weight in gelatine capsules on 5 days per week for 6 weeks. The control animals received oil in gelatine capsules. Once weekly, the animals were given an intravenous injection of *Salmonella typhimurium* "$Ty_3$". In the animals treated with lindane, the antibody titres were reduced in a dose-dependent manner (Dési *et al.* 1978).

The significance for man of the immune-modulating effect of lindane which has been demonstrated in animal studies, in mice even at daily doses as low as 0.012 mg/kg body weight, is still not clear. At the present time, this end point has not been used in the establishment of the MAK value. However, further examination of these results is necessary, especially with respect to their relevance for man.

### 5.8.2 Neurotoxic effects

Lindane has neurotoxic effects on the central and peripheral nervous systems. Such studies have been reviewed by WHO (1991). The induction of seizures in rats was seen after administration for 12 days of lindane doses of 12 mg/kg body weight and day but not after 5 mg/kg body weight and day. Changes in the EEG and in behavioural parameters were seen after administration of 5 mg/kg body weight for 40 days but not after 2.5 mg/kg body weight for 22 days (WHO 1991).

## 6 Manifesto (MAK value/classification)

There are no studies of the carcinogenicity of lindane in man.

Liver carcinomas were observed in mice fed lindane in concentrations of 600 mg/kg diet (about 90 mg/kg body weight and day) for 6 months and also in those given concentrations of at least 80 mg/kg diet (about 12 mg/kg body weight and day) for two years. In rats fed lindane in the diet in a long-term study, neoplastic foci were observed in the liver. In an initiation-promotion study with rats given the substance in the diet for 20 weeks, lindane was shown to have tumour-promoting effects on the liver; liver foci developed in animals given lindane doses of at least 2.5 mg/kg body weight and day. In addition, a long-term feeding study with rats provided evidence of a carcinogenic effect of lindane on the thyroid gland. This finding requires further clarification. The numerous reports of genotoxicity studies carried out appropriately with "classical" methods *in vitro* and *in vivo* have yielded no evidence of genotoxic activity of lindane. However, according to some studies, a weak clastogenic effect like that known for other substances with tumour-promoting effects cannot be excluded. In the main, the results suggest an epigenetic mechanism of tumour development. Because of its tumour-promoting effects on the liver of the rat and mouse, lindane is classified in Section III, Category 4 of the *List of MAK and BAT Values*. The tumour promotion in the liver can be explained by receptor-mediated, hormone-like effects and the inhibition of apoptosis. In the case of lindane, the induction of cytochrome P450-dependent monooxygenases correlates with the area of foci with modified enzyme activity. Because the dose-response relationships are non-linear, a NOEL can be determined for both processes. The currently available data indicate that tumour-promoting effects may be avoided by ensuring that no induction of cytochrome P450-dependent monooxygenases takes place. The induction of monooxygenase activities has been observed in workers whose serum lindane concentrations were above 10 µg/l. Reliable data for the concentrations of lindane in air which correlate with these serum concentrations or which would permit the derivation of a NOEL are not available. In workers exposed to lindane concentrations of 0.004 to 0.15 mg/m$^3$, there was evidence of the induction of monooxygenase activities. Average serum concentrations of 36.9 µg/l were determined Adverse effects on the health of these workers were, however, not seen. The description of the exposure conditions indicates that these workers had intensive skin contact with lindane and that uptake of lindane through the skin made a significant contribution to the systemic dose. A direct correlation of the serum concentrations with the concentrations in the air is therefore not possible. In an inhalation study with rats, the NOEL for the induction of monooxygenases was found to be 0.6 mg/m$^3$. From the data available for man and the results of animal experiments it is concluded that, provided skin contact is avoided, exposure to lindane concentrations of 0.1 mg/m$^3$ does not induce monooxygenases in man. In 1998, the MAK value was therefore reduced provisionally to 0.1 mg/m$^3$. However, more studies of the effects in man are required to confirm this value. The classification in Peak limitation category III has been retained.

The available reproductive toxicity studies with the rat, mouse, rabbit, dog, and pig have yielded no evidence for embryotoxic or teratogenic effects of maternally tolerated doses. Lindane is therefore classified in Pregnancy risk group C.

Because the substance is readily absorbed through the skin, lindane is designated with an "H".

The available data indicate that designation with an "S" (for substances which cause sensitization) is not necessary.

There is no evidence of germ cell mutagenicity.

# 7 References

ACGIH (American Conference of Governmental Industrial Hygienists) (1992) Lindane. in: *Documentation of threshold limit values and biological exposure indices*, American Conference of Governmental Industrial Hygienists, Cincinnati, Ohio

André F, Gillon J, André C, Lafont S, Jourdan G (1983) Pesticide-containing diets augment anti-sheep red blood cell nonreaginic antibody responses in mice but may prolong murine infection with *Giardia muris. Environ Res 32*: 145–150

Angerer (1998) Communication to the secretariat of the Commission, 09.07.1998

Angerer J, Behling K, Brassow H-L, Lehnert G (1983) Elimination von Hexachlorcyclohexan aus dem menschlichen Körper. in: Stalder K (Ed.) *Bericht über die 23. Jahrestagung der Deutschen Gesellschaft für Arbeitsmedizin e.V.* unter Mitwirkung des Verbandes Deutscher Betriebs- und Werksärzte e.V. in Verbindung mit dem Hauptverband der Gewerblichen Berufsgenossenschaften e.V., Bonn, Immunologische Aspekte in der Arbeitsmedizin, Medizinische und berufliche Rehabilitation, Unfallverhütungsvorschriften und ihre Bedeutung für den Betriebsarzt, Betriebsärztliche Untersuchunen an Sinnesorganen unter Berücksichtigung ihrer Zuverlässigkeitskriterien, Gentner Verlag, Stuttgart, 445–450

Antunes-Madeira MC, Almeida LM, Madeira VMC (1990) Effects of lindane on membrane fluidity: intramolecular excimerization of a pyrene derivative and polarization of diphenylhexatriene. *Biochim Biophys Acta 1022*: 110–114

Bainy ACD, Silva MAS, Kogake M, Videla LA, Junqueira VBC (1994) Influence of lindane and paraquat on oxidative stress-related parameters of erythrocytes *in vitro*. *Hum Exp Toxicol 13*: 461–465

Banerjee BD, Koner BC, Ray A, Pasha ST (1996) Influence of subchronic exposure to lindane on humoral immunity in mice. *Indian J Exp Biol 34*: 1109–1113

Barros SBM, Simizu K, Junqueira VBC (1991) Liver lipid peroxidation-related parameters after short-term administration of hexachlorocyclohexane isomers to rats. *Toxicol Lett 56*: 137–144

Baumann K, Angerer J, Heinrich R, Lehnert G (1980) Occupational exposure to hexachlorocyclohexane. I. Body burden of HCH-isomers. *Int Arch Occup Environ Health 47*: 119–127

Baumann K, Behling K, Brassow H-L, Stapel K (1981) Occupational exposure to hexachlorocyclohexane. III. Neurophysiological findings and neuromuscular function in chronically exposed workers. *Int Arch Occup Environ Health 48*: 165–172

Beard AP, McRae AC, Rawlings NC (1997) Reproductive efficiency in mink (*Mustela vison*) treated with the pesticides lindane, carbofuran and pentachlorophenol. *J Reprod Fertil 111*: 21–28

Brassow H-L, Baumann K, Lehnert G (1981) Occupational exposure to hexachlorocyclohexane. II. Health conditions of chronically exposed workers. *Int Arch Occup Environ Health 48*: 81–87

Chadwick RW, Cooper RL, Chang J, Rehnberg GL, McElroy WK (1988) Possible antiestrogenic activity of lindane in female rats. *J Biochem Toxicol 3*: 147–158

Chattopadhyay P, Karnik AB, Thakore KN, Lakkad BC, Nigam SK, Kashyap SK (1988) Health effects among workers involved in the manufacture of hexachlorocyclohexane. *J Soc Occup Med 38*: 77–81

CIEL (Centre International d'Etudes du Lindane) (1972) *Lindane - testing for teratogenic effects in mice following oral administration*. E. Merck, Darmstadt, CIEL-No. 451-005, unpublished report

CIEL (Centre International d'Etudes du Lindane) (1976a) *Teratology study in rats. Lindane (gamma benzene hexachloride USP)*. Final Report, Hazleton Laboratories America, Inc, CIEL-No. 451-003, unpublished report

CIEL (Centre International d'Etudes du Lindane) (1976b) *Teratology study in rabbits. Lindane (gamma benzene hexachloride USP)*. Final Report, Hazleton Laboratories America, Inc, CIEL-No. 451-004, unpublished report

CIEL (Centre International d'Etudes du Lindane) (1983a) *3-Months toxicity study in rats with lindane*. RCC (Research & Consulting Company AG), CIEL Doc-No. 433-001, unpublished report

CIEL (Centre International d'Etudes du Lindane) (1983b) *90-Day inhalation study with lindane*. Fraunhofer Institut für Toxikologie und Aerosolforschung, Schmallenberg, CIEL Doc-No. 435-001, unpublished report

CIEL (Centre International d'Etudes du Lindane) (1986a) *Primary eye irritation study with lindane in rabbits*. RCC (Research & Consulting Company AG), CIEL Doc-No. 466-001, unpublished report

CIEL (Centre International d'Etudes du Lindane) (1986b) *Test for delayed contact hypersensitivity in the albino guinea pig with lindane (maximization test)*. RCC (Research & Consulting Company AG), CIEL Doc-No. 467-001, unpublished report

CIEL (Centre International d'Etudes du Lindane) (1987) *Contact hypersensitivity to Nexit flüssig (CME 11129) in albino guinea pigs (maximization test)*. RCC (Research & Consulting Company AG), CIEL Doc-No. 467-002, unpublished report

CIEL (Centre International d'Etudes du Lindane) (1988a) *Lindane: 13 week dermal toxicity study (with interim kill and recovery period) in the rat*. Hazleton UK, unpublished report

CIEL (Centre International d'Etudes du Lindane) (1988b) *Lindane technical - fourteen week dust aerosol inhalation study in mice*. Bushy Run Research Center, unpublished report

Cooper RL, Chadwick RW, Rehnberg GL, Goldman JM, Booth KC, Hein JF, McElroy WK (1989) Effect of lindane on hormonal control of reproductive function in the female rat. *Toxicol Appl Pharmacol 99*: 384–394

Czeglédi-Jankó G, Avar P (1970) Occupational exposure to lindane: clinical and laboratory findings. *Br J Ind Med 27*: 283–286

Dalsenter PR, Faqi AS, Chahoud I (1997a) Serum testosterone and sexual behavior in rats after prenatal exposure to lindane. *Bull Environ Contam Toxicol 59*: 360–366

Dalsenter PR, Faqi AS, Webb J, Merker H-J, Chahoud I (1997b) Reproductive toxicity and toxicokinetics of lindane in the male offspring of rats exposed during lactation. *Hum Exp Toxicol 16*: 146–153

Dési I, Varga L, Farkas I (1978) Studies on the immunosuppressive effect of organochlorine and organophosphoric pesticides in subacute experiments. *J Hyg Epidemiol Microbiol Immunol 22*: 115–122

Dewan A, Gupta SK, Jani JP, Kashyap (1980) Effects of lindane on antibody response to typhoid vaccine in weanling rats. *J Environ Sci Health B 15*: 395–402

Drummond L, Gillanders EM, Wilson HK (1988) Plasma γ-hexachlorocyclohexane concentrations in forestry workers exposed to lindane. *Br J Ind Med 45*: 493–497

Dubois M, Grosse Y, Thome JP, Kremers P, Pfohlleszkowicz A (1997) Metabolic-activation and DNA-adducts detection as biomarkers of chlorinated pesticide exposures. *Biomarkers 2*: 17–24

Dzwonkowska A, Hübner H (1989) Mutagenicity testing of commercial insecticides administered *per os*. *Pol J Occup Med 2*: 286–293

Fitzhugh OG, Nelson AA, Frawley JP (1950) The chronic toxicities of technical benzene hexachloride and its alpha, beta and gamma isomers. *J Pharmacol Exp Ther 100*: 59–66

Ginsburg CM, Lowry W, Reisch JS (1977) Absorption of lindane (gamma benzene hexachloride) in infants and children. *Pediatrics 91*: 998–1000

Goto M, Hattori M, Miyagawa T, Enomoto M (1972) Beiträge zur ökologischen Chemie. II. Hepatoma-Bildung in Mäusen nach Verabreichung von HCH-Isomeren in hohen Dosen. *Chemosphere 6*: 279–282

Gray Jr LE, Ostby J, Sigmon R, Ferrell J, Rehnberg G, Linder R, Cooper R, Goldman J, Laskey J (1988) The development of a protocol to assess reproductive effects of toxicants in the rat. *Reprod Toxicol 2*: 281–287

Hanada M, Yutani C, Miyaji T (1973) Induction of hepatoma in mice by benzene hexachloride. *Gann 64*: 511–513

Hanig JP, Yoder PD, Krop S (1976) Convulsion in weanling rabbits after a single topical application of 1% lindane. *Toxicol Appl Pharmacol 38*: 463–469

Hassoun EA, Stohs SJ (1996) Comparative teratological studies on TCDD, endrin and lindane in C57BL/6J and DBA/2J mice. *Comp Biochem Physiol C 113*: 393–398

Hassoun E, Bagchi M, Bagchi D, Stohs SJ (1993) Comparative studies on lipid peroxidation and DNA-single strand breaks induced by lindane, DDT, chlordane and endrin in rats. *Comp Biochem Physiol C 104*: 427–431

Henschler D, Lehnert G (1988) *Biologische Arbeitsstoff-Toleranz-Werte (BAT-Werte) und Expositionsäquivalente für krebserzeugende Arbeitsstoffe (EKA)*, 5th issue 1990, 7th issue 1994, γ-1,2,3,4,5,6-Hexachlorcyclohexan, VCH-Verlagsgesellschaft, Weinheim and in English translation in Volume 2 of *Biological Exposure Values*.

Herbst M (1983) Toxizität von gamma-HCH (Lindan). in: *Hexachlorcyclohexan als Schadstoff in Lebensmitteln*, DFG, Verlag Chemie, Weinheim, 191–200

Herbst M, Bodenstein G (1973) C. Toxicology of lindane. in: Ulmann E (Ed.) *Lindane, Monograph of an Insecticide*, K. Schillinger, Freiburg, 23–82

Herbst M, Weisse I, Koellmer H (1975) A contribution to the question of the possible hepatocarcinogenic effects of lindane. *Toxicology 4*: 91–96

IARC (International Agency for Research on Cancer) (1979) Hexachlorcyclohexane. in: *IARC monographs on the evaluation of carcinogenic risks of chemicals to humans, Vol 20, Some halogenated hydrocarbons*, IARC, Lyon, 195–239

Ito N, Nagasaki H, Aoe H, Sugihara S, Miyata Y, Arai M, Shirai T (1975) Brief communication: Development of hepatocellular carcinomas in rats treated with benzene hexachloride. *J Nat Cancer Inst 54*: 801–805

Ito N, Nagasaki H, Arai M, Sugihara S, Makiura S (1973) Histologic and ultrastructural studies on the hepatocarcinogenicity of benzene hexachloride in mice. *J Nat Cancer Inst 51*: 817–826

Khera KS, Whalen C, Trivett G, Angers G (1979) Teratogenicity studies on pesticide formulations of dimethoate diuron and lindane in rats. *Bull Environ Contam Toxicol 22*: 522–529

Király J, Szentesi I, Ruzicska M, Czeizel A (1979) Chromosome studies in workers producing organophosphate insecticides. *Arch Environ Contam Toxicol 22*: 522–529

Kolmodin-Hedman (1974) *Exposure to lindane and DDT and its effects on drug metabolism and serum lipoproteins.* Doctoral Thesis from Department of Clinical Pharmacology (at Huddinge Hospital) and Pharmacology, Karolinska Institutet, Toxicology (Swedish Medical Research Council) and Division of Occupational Medicine, National Board of Occupational Safety and Health, Sweden

Kramer MS, Hutchinson TA, Rudnick SA, Leventhal JM, Feinstein AR (1980) Operational criteria for adverse drug reactions in evaluating suspected toxicity of a popular scabicide. *Clin Pharmacol Ther 27*: 149–155

Lange M, Nitzsche K, Zesch A (1981) Percutaneous absorption of lindane in healthy volunteers and scabies patients. *Arch Dermatol Res 271*: 387–399

Laws SC, Carey SA, Hart DW, Cooper RL (1994) Lindane does not alter the estrogen receptor or the estrogen-dependent induction of progesterone receptors in sexually immature or ovariectomized adult rats. *Toxicology 92*: 127–142

Lisi P, Caraffini S, Assalve D (1986) A test series for pesticide dermatitis. *Contact Dermatitis 15*: 266–269

Lisi P, Caraffini S, Assalve D (1987) Irritation and sensitization potential of pesticides. *Contact Dermatitis 17*: 212–218

McNamara BP, Krop S (1948) Observations on the pharmacology of the isomers of hexachlorocyclohexane. *J Pharmacol Exp Ther 92*: 140–146

Meera P, Rao PR, Shanker B, Tripathi O (1992) Immunomodulatory effects of γ-HCH (lindane) in mice. *Immunopharmacol Immunotoxicol 14*: 261–282

NCI (National Cancer Institute) (1977) *Bioassay of lindane for possible carcinogenicity.* Technical Report Series No. 14

NCI (National Cancer Institute) (1978) *Bioassay for 2,4,6-trichlorophenol for possible carcinogenicity.* Report No. NIH 79-1711

Nigam SK, Karnik AB, Majumder SK, Visweseariah K, Suryanarayana Raju G, Maktha Bai K, Lakkad BC, Thahore KN, Chatterjee BB (1986) Serum hexachlorocyclohexane residues in workers engaged at a HCH manufacturing plant. *Int Arch Occup Environ Health 57*: 315–320

Oesch F, Friedberg T, Herbst M, Paul W, Wilhelm N, Bentley P (1982) Effects of lindane treatment on drug metabolizing enzymes and liver weight of CFI mice in which is evoked hepatomas and in non-susceptible rodents. *Chem Biol Interact 40*: 1–14

Orr JW (1948) Absence of carcinogenic activity of benzene hexachloride ("gammexane"). *Nature (London) 162*: 189

Palmer AK, Bottomley AM, Worden AN, Frohberg H, Bauer A (1978a). Effect of lindane on pregnancy in the rabbit and rat. *Toxicology 9*: 239–247

Palmer AK, Cozens DD, Spicer EJF, Worden AN (1978b) Effects of lindane upon reproductive function in a 3-generation study in rats. *Toxicology 10*: 45–54

Pool-Zobel BL, Guigas C, Klein R, Neudecker Ch, Renner HW, Schmezer P (1993) Assessment of genotoxic effects by lindane. *Food Chem Toxicol 31*: 271–283

Pool-Zobel BL, Lotzmann N, Knoll M, Kuchenmeister F, Lambertz R, Leucht U, Schröder HG, Schmezer P (1994) Detection of genotoxic effects in human gastric and nasal mucosa cells isolated from biopsy samples. *Environ Mol Mutagen 24*: 23–45

Raizada RB, Misra P, Saxena I, Datta KK, Dikshith TS (1980) Weak estrogenic activity of lindane in rats. *J Toxicol Environ Health 6*: 483–492

Rivett KF, Chesterman H, Kellett DN, Newman AJ, Worden AN (1978) Effects of feeding lindane to dogs for periods of up to 2 years. *Toxicology 9*: 273–289

Sagelsdorff P, Lutz WK, Schlatter C (1983) The relevance of covalent binding to mouse liver DNA to the carcinogenic action of hexachlorocyclohexane isomers. *Carcinogenesis 4*: 1267–1273

Sagunski H, Perger G (1994) Biozide. Hexachlorcyclohexan (HCH). in: Marquardt H, Schäfer SG (Eds) *Lehrbuch der Toxikologie*, BI Wissenschaftsverlag Mannheim, 447–449

Saha S, Banerjee BD (1993) Effect of sub-chronic lindane exposure on humoral and cell-mediated immune responses in albino rats. *Bull Environ Contam Toxicol 51*: 795–802

Schröter C, Parzevall W, Schröter H, Schulte-Hermann R (1987) Dose-response studies on the effect of α-, β-, and γ-hexachlorocyclohexane on putative preneoplastic foci, monooxygenases, and growth in rat liver. *Cancer Res 47*: 80–88

Sircar S, Lahiri P (1989) Lindane (γ-HCH) causes reproductive failure and fetotoxicity in mice. *Toxicology 59*: 171–177

Thorpe E, Walker AIT (1973) The toxicology of dieldrin (HEOD). II. Comparative long-term oral toxicity studies in mice with dieldrin, DDT, phenobarbitone, β-BHC and γ-BHC. *Food Cosmet Toxicol 11*: 433–442

Tomczak S, Baumann K, Lehnert G (1981) Occupational exposure to hexachlorocyclohexane. IV. Sex hormone alterations in HCH-exposed workers. *Int Arch Occup Environ Health 48*: 283–287

Uphouse L (1987) Decreased rodent sexual receptivity after lindane. *Toxicol Lett 39*: 7–14

Uphouse L, Williams J (1989) Diestrus treatment with lindane disrupts the female reproductive cycle. *Toxicol Lett 48*: 21-28

Weisse I, Herbst M (1977) Carcinogenicity study of lindane in the mouse. *Toxicology 7*: 233–238

WHO (World Health Organisation) (1991) *Lindane. Environmental Health Criteria 124*, International Programme on Chemical Safety - IPCS, World Health Organisation, Genf

Wolff GL (1993) Multiple levels of response in carcinogenicity bioassays: regulational variation among viable yellow (Avy//-) mice. *J Exp Anim Sci 35*: 221–231

Wolff GL, Morrissey RL (1986) Increased responsiveness of lean pseudoagouti Avy/a female mice to lindane enhancement of lung and liver tumorgenesis. *Proc Am Assoc Cancer Res 27*: 138

Zesch A, Nitzsche K, Lange M (1982) Demonstration of the percutaneous resorption of a lipophilic pesticide and its possible storage in the human body. *Arch Dermatol Res 273*: 43–49

completed 29.01.1998

# Methanol

| | |
|---|---|
| **MAK value (1958)** | **200 ml/m³ (ppm) ≙ 270 mg/m³** |
| **Peak limitation (1999)** | **Category II,2** |
| **Absorption through the skin (1969)** | **H** |
| **Sensitization** | **–** |
| **Carcinogenicity** | **–** |
| **Prenatal toxicity (1995)** | **Pregnancy risk group C** |
| **Germ cell mutagenicity** | **–** |
| **BAT value (1983)** | **30 mg/l urine**<br>**Sampling time: end of exposure or end of shift; for long-term exposures after several shifts** |
| Synonyms | carbinol<br>methyl alcohol |
| Chemical name (CAS) | methanol |
| CAS number | 67-56-1 |
| Structural formula | $CH_3OH$ |
| Molecular formula | $CH_4O$ |
| Molecular weight | 32.04 g/mol |
| Melting point | –94°C |
| Boiling point | 64.7°C |
| Density at 20°C | 0.79 g/cm³ |
| Vapour pressure at 20°C | 128 hPa |
| log $P_{ow}$* | –0.77 |
| **1 ml/m³ (ppm) ≙ 1.33 mg/m³** | **1 mg/m³ ≙ 0.752 ml/m³ (ppm)** |

The present document is based on the TLV documentation for methanol (ACGIH 1999), the EHC report (WHO 1997), the documentation of the BAT value (Henschler and

* *n*-octanol/water distribution coefficient

Lehnert 1983) and other reviews of toxicological data for methanol (ECB 1995, Kavet and Nauss 1990, Lington and Bevan 1994).

# 1 Toxic Effects and Mode of Action

Ingested methanol is toxic for man. The lethal dose lies between 30 and 120 ml. Characteristic symptoms of methanol intoxication in man include CNS depression and, after a latency period of several hours to some days, visual impairment or even complete blindness. Persons exposed by long-term inhalation of methanol in concentrations of about 400 ml/m$^3$ report headaches and, from 1000 ml/m$^3$, also eye irritation, blurred vision and nausea. Symptoms of methanol intoxication are also observed in persons exposed dermally.

Depending on the duration of contact, methanol is either not irritating to the skin or, because of its defatting action, moderately irritating. In the rabbit eye, undiluted methanol is a mild to moderate irritant.

Prenatal toxic effects were seen in rats exposed by continuous inhalation of maternally toxic concentrations of 5000 ml/m$^3$ or more and in mice even at concentrations of 2000 ml/m$^3$ or more which were not toxic for the dams.

In the *in vitro* genotoxicity tests and *in vivo* studies which are relevant for the present assessment, methanol was not genotoxic. In a study of limited validity, the carcinogenic effects of methanol were studied in F344 rats and B6C3F$_1$ mice exposed for 24 and 18 months, respectively; there was no evidence of carcinogenic potential.

# 2 Mechanism of Action

The high toxicity of methanol in man is ascribed to the metabolic formation and accumulation of formate, which results in metabolic acidosis. In rodents, whose formate detoxification is more efficient as a result of their higher tetrahydrofolate levels, such effects do not occur; they may be produced in animals with folate-deficiency. The oxidation of formate to $CO_2$ is more rapidly saturated in primates than in rodents because of the lower level of the cosubstrate tetrahydrofolate in the primate liver. For this reason, the rate of elimination of formate in primates is only half that in rats (Medinsky *et al.* 1997; see Section 3.2). The ocular toxicity is put down to the formation of formate in the retina (Garner *et al.* 1995) in the Müller's cells (Martinasevic *et al.* 1996).

Methanol modifies the equilibrium concentrations of monoamine neurotransmitters in the *corpus striatum* and hypothalamus of rats. As the changes in the monoamine levels are well correlated with the methanol levels in blood and brain, the authors suggest that methanol and not its metabolites interact with the monoaminergic neuronal membranes (Jeganathan and Namasivayam 1989a, 1989b).

The prenatal toxic effects seen in the rat and mouse are considered to be mainly effects of methanol itself and less so of its metabolites (Dorman and Welsch 1996, Dorman *et al.* 1995a, 1995b; see Section 5.5). The mechanism of the prenatal toxicity is, however, not yet understood. It is conceivable that a role is played by methanol-induced changes in monoamine levels or changes in membrane structures during development (Dorman *et al.* 1995a).

# 3 Toxicokinetics and Metabolism

## 3.1 Absorption and distribution

Methanol is readily absorbed after inhalation, ingestion and dermal application. The pulmonary retention in man has been given as 57.7 % (53.4–61.3 %) (Sedivec *et al.* 1981), the pulmonary absorption as 60 to 85 % (Medinsky *et al.* 1997). Also from the gastrointestinal tract, a dose of methanol is absorbed rapidly within 30 to 60 minutes (Medinsky *et al.* 1997). The rate of dermal absorption in persons exposed for up to 60 minutes via an exposure chamber mounted on the forearm was found to be 0.192 mg/cm$^2$ and minute (11.52 mg/cm$^2$ and hour): to obtain this result, the difference between the amount of methanol applied and the amount remaining in the chamber at the end of the exposure period was determined (Dutkiewicz *et al.* 1980).

In 12 test persons who were exposed for 2 to 16 minutes to liquid methanol via one hand, the maximum blood methanol concentration at the end of exposure was 11.5 ± 2.3 µg/ml. This value was corrected by subtraction of the methanol level before the exposure (1.7 ± 0.9 µg/ml). The maximum concentrations found in the 7 men were about 24 % higher than those in the 5 women. Maximum blood methanol concentrations were attained 1.9 ± 1.0 hours after the exposure, independent of the exposure duration. Methanol continued to be taken up from the skin for four hours after the end of exposure. The average rate of absorption was 8.1 ± 3.7 mg/cm$^2$ and hour (Batterman and Franzblau 1997). The maximum blood methanol levels achieved by dipping one hand into methanol for 16 minutes and by inhalation of a methanol concentration of 400 ml/m$^3$ for 8 hours were similar (Franzblau *et al.* 1995).

The results of Batterman and Franzblau (1997) reveal that 16200 mg methanol is taken up through the skin in one hour when both hands and forearms (about 2000 cm$^2$) are in contact with the substance. In comparison, during inhalation of a methanol concentration of about 270 mg/m$^3$ for 8 hours, 1620 mg methanol are taken up (10 m$^3$ inhaled air volume, 60 % pulmonary retention). Thus methanol may be assumed to penetrate the skin very readily.

Independent of the administration route, methanol is distributed rapidly and evenly in all organs and tissues of the body in amounts depending on their water content. The highest concentrations are found in muscle, blood, gastrointestinal tract and liver (Medinsky *et al.* 1997).

## 3.2 Metabolism and excretion

Absorbed methanol is metabolized or excreted unchanged with the urine and faeces or exhaled. In rats and monkeys given low doses of methanol (2 mg/kg body weight) about 90 % of the dose is metabolized (Kavet and Nauss 1990).

The enzymes involved in the metabolism of methanol in rodents and primates are not all identical. In rodents, methanol is converted to formaldehyde mainly by the catalase-peroxidase enzyme system, in man and monkey mainly by alcohol dehydrogenase; the rates of metabolism are, however, similar (Kavet and Nauss 1990, Medinsky *et al.* 1997). In rodents and primates, formaldehyde is converted by the $NAD^+$-dependent formaldehyde dehydrogenase in the presence of reduced glutathione first to *S*-hydroxymethyl-glutathione and then to *S*-formylglutathione. The hydrolysis of *S*-formylglutathione to formic acid and reduced glutathione is catalysed by the thiolase associated with $NAD^+$-dependent formaldehyde dehydrogenase (Tephly 1991). Formate released on dissociation of formic acid is oxidized via 10-formyltetrahydrofolate to $CO_2$. For these steps the co-substrate tetrahydrofolate and the enzymes formyltetrahydrofolate synthetase and formyltetrahydrofolate dehydrogenase are necessary. The oxidation of formate to $CO_2$ is more rapidly saturated in primates than in rodents because of the lower level of the co-substrate tetrahydrofolate in primates. The tetrahydrofolate level in man is only 25 % and 50 % of that in the mouse and rat, respectively. The maximum rate of elimination of formate is about 34 mg/kg body weight and hour in primates and about 73 mg/kg body weight and hour in the rat. Thus after intake of large amounts of methanol, formate can accumulate in the blood of man or monkey, but not of rodents. This accumulation of formate in primates leads to metabolic acidosis and ocular toxicity (Johlin *et al.* 1987, Kavet and Nauss 1990).

Rats given intraperitoneal doses of radioactively labelled methanol of 25 mg/kg body weight exhaled 62.6 % of the dose as $CO_2$ and 1.37 % as unchanged methanol within 6 hours; 4.30 % was recovered in the urine, 2.06 % in the faeces. 48 hours after the injection 85.8 % of the radioactivity had been eliminated. Monkeys given intraperitoneal injections of radioactively labelled methanol of 25 mg/kg body weight exhaled 33.0 % as $CO_2$ and 0.74 % as unchanged methanol within 6 hours; 4.28 % was excreted with the urine and 0.18 % with the faeces. After 48 hours, a total of 56.9 % of the dose had been eliminated. With increasing dose, the proportion of the dose exhaled as unchanged methanol increased in both species (NEDO 1987).

The $V_{max}$ for the elimination of methanol in the mouse is twice that in the rat. However, after inhalation of methanol, the blood methanol concentration in the mouse is two to three times that in the rat; this may be accounted for by the fact that the respiration rate of the mouse is more rapid than that of the rat and the absorption of methanol more efficient (Perkins *et al.* 1995a, Pollack and Brouwer 1996).

Toxicokinetic studies of the metabolic elimination of methanol in the monkey have revealed a change from first-order kinetics to zero-order kinetics (saturation kinetics) at a blood methanol concentration of about 10 mmol/l (about 320 μg/ml) (Noker *et al.* 1980).

The urine methanol concentration is proportional to and higher than the blood methanol concentration (Ferry *et al.* 1980). The elimination half-time for inhaled methanol in human urine is about 3.5 hours (Sedivec *et al.* 1981), that for ingested methanol in human blood 2 to 2.25 hours (Ferry *et al.* 1979). For the monkey exposed by inhalation, the half-life of methanol in blood was found to be less than one hour (Medinsky *et al.* 1997). The elimination half-time from the blood of the monkey for intravenously administered formate is also less than one hour (Clay *et al.* 1975, McMartin *et al.* 1977).

## 3.3 Levels of methanol and formate in blood and urine

### 3.3.1 Rat

In rats exposed for 6 hours to a methanol concentration of 200 ml/m$^3$, the blood methanol level (3.1 ± 0.4 µg/ml) (Horton *et al.* 1992) was not markedly different from the endogenous level of about 2.4 µg/ml (NEDO 1987). The blood methanol level in the rat was increased more than proportionately to 76 µg/ml after continuous exposure of the animals to methanol concentrations of 1000 ml/m$^3$ or more (NEDO 1987) and to 80 µg/ml after exposure to 2000 ml/m$^3$ for 6 hours (Horton *et al.* 1992). In rats which had inhaled methanol in concentrations between 50 and 2000 ml/m$^3$ for 6 hours, the blood formate levels were not increased (Horton *et al.* 1992).

### 3.3.2 Monkey

In cynomolgus monkeys which had inhaled methanol in concentrations up to 900 ml/m$^3$ for 2 hours, the blood methanol levels were increased to 3.4 µg/ml and the blood formate to 0.13 µg/ml. In animals given a folate-deficient diet, the blood methanol level increased to 6.75 µg/ml and the formate to 0.44 µg/ml. The authors considered that the increases in blood methanol levels which they had measured were not toxicologically relevant because the endogenous blood methanol levels in not exposed animals were mostly higher. Likewise, the blood formate level resulting from the inhaled methanol was only 1/10 to 1/1000 of the endogenous formate even in folate-deficient animals. In the authors' opinion, the folate reserves of the body are sufficient to detoxify efficiently the amounts of formate formed during brief exposure to the methanol concentrations which occur at the workplace (Dorman *et al.* 1994, Medinsky *et al.* 1997).

In rhesus monkeys exposed for 6 hours to methanol concentrations of 200, 1200 or 2000 ml/m$^3$, blood methanol levels of 3.9 ± 1.0, 38 ± 8.5 and 64.4 ± 10.7 µg/ml were found. At methanol concentrations up to 2000 ml/m$^3$, the blood formate levels were not significantly increased during the 18 hours after the exposure. During exposure to a methanol concentration of 2000 ml/m$^3$, 6 hours daily, 5 days per week for 2 weeks, neither formate nor methanol accumulated in the blood (Horton *et al.* 1992).

### 3.3.3 Man

Endogenous methanol comes mainly from the diet and is also a product of normal metabolism. The published endogenous blood methanol levels are in the range from less than 1 to 2 μg/ml (Chuwers *et al.* 1995, Lee *et al.* 1992). Regular ingestion of alcoholic drinks can result in blood levels of 11 to 27 μg/ml or even more. The increase is put down to the accumulation of endogenously formed methanol as a result of competitive inhibition of methanol oxidation by ethanol (Henschler and Lehnert 1983).

The concentrations of endogenous formate, which also comes from the diet and from normal metabolism, are in the range between 3 and 19 μg/ml (Kavet and Nauss 1990).

**Test persons**

In test persons given oral methanol doses of 71 to 84 mg/kg body weight (about 6.6–7.4 ml), blood methanol concentrations of 47 to 76 μg/ml were found 2 to 3 hours later (Leaf and Zatman 1952).

In persons exposed for 8 hours to a methanol concentration of 400 ml/m$^3$, a blood methanol level of 13.4 ± 4.8 μg/ml was found (Franzblau *et al.* 1995).

Three test persons were exposed to methanol at average concentrations of about 300 mg/m$^3$ (225 ml/m$^3$) for 8 hours on each of 4 consecutive days and the amounts of methanol excreted with the urine were determined. The methanol concentrations in urine, which increased during the course of the day to a maximum of 12 ± 1 μg/ml (read from the graph), had returned to the normal range by the next morning; that is, no accumulation of methanol was found during the 4 days of the study (Sedivec *et al.* 1981).

Twelve male test persons, who also served as their own controls, were exposed for 75 minutes to a methanol concentration of 250 mg/m$^3$ (about 190 ml/m$^3$). The methanol levels in blood and urine, 1.9 ± 0.5 μg/ml and 2.2 ± 0.6 μg/ml, were significantly higher than those before the exposure, 0.6 ± 0.3 μg/ml and 1.0 ± 0.4 μg/ml; the plasma formate level (about 0.08 ± 0.02 mmol/l, i.e., 3.6 μg/ml) remained unchanged (Cook *et al.* 1991).

Fifteen men and eleven women aged 35.7 ± 6.8 years, who served as their own controls, were exposed at rest for 4 hours to a methanol concentration of 200 ml/m$^3$. For the first 8 hours after the start of the exposure, the serum methanol concentration of maximally 6.5 ± 2.7 μg/ml was significantly increased above the starting concentration of 0.9 ± 0.6 μg/ml. The serum formate level at the end of exposure was 14.3 ± 8.9 μg/ml and not significantly increased above the control value of 12.7 ± 6.4 μg/ml (d'Alessandro *et al.* 1994, Chuwers *et al.* 1995, Osterloh *et al.* 1996). The urinary formate excretion was slightly but not significantly increased (d'Alessandro *et al.* 1994).

In one of the test persons, however, the increase in the serum formate concentration from 14.7 to 44.4 μg/ml was unusually large. The urinary excretion of formate was like that of the other test persons. The authors were unable to explain this phenomenon; they suggested an unusual biological response which could be expected in a small proportion of the population (d'Alessandro *et al.* 1994).

In another study with test persons exposed for 6 hours to a methanol concentration of 200 ml/m$^3$, the average blood methanol concentration increased from 1.8 to 7.0 μg/ml. When the test persons were also physically active, the methanol level increased to 8.1 μg/ml. The blood formate level, which was between 5.4 and 10.8 μg/ml before the exposure, remained in both cases unchanged (Lee *et al.* 1992).

Thus in controlled studies in which persons were exposed for up to 6 hours to methanol concentrations of 200 ml/m$^3$, no accumulation of formate in blood was detected.

On the basis of data from the literature and with a physiological toxicokinetic model it has been predicted that in persons exposed for a maximum of 4 hours to a methanol concentration of 500 ml/m$^3$ the blood methanol level would be 19 µg/ml, for 1000 ml/m$^3$ 38.5 µg/ml and for 5000 ml/m$^3$ 224 µg/ml (Perkins *et al.* 1995b).

**Workers**

In 20 workers exposed to average methanol concentrations between 85 and 134 ml/m$^3$, the blood formate concentrations increased significantly from 3.2 ± 2.4 µg/ml before the shift to 7.9 ± 3.2 µg/ml after the end of the shift. In 36 persons who were not exposed to methanol, blood formate concentrations between 0 and 20 µg/ml were found. The urine formate increased in the exposed workers from 13.1 ± 3.9 to 20.2 ± 7 µg/ml whereas in a control group a slight decrease after the end of the shift was observed (Baumann and Angerer 1979).

In 20 persons exposed occupationally to methanol concentrations between 37 and 231 ml/m$^3$, 15 % of the urine formate levels were above the normal range. No correlation with the external exposure levels could be established (Heinrich and Angerer 1982).

A linear relationship between the methanol concentration in the air and the formate concentration in the post-shift urine was established for male workers exposed to methanol in concentrations up to an 8-hour average of 2000 ml/m$^3$ and for female workers up to 4000 ml/m$^3$. The regression equations for the urine methanol concentration ($y$) in men and women, respectively, were given as: $y$ (µg/ml) = 0.089 × $x$ + 25.16 (r = 0.798) and $y$ = 0.049 × $x$ + 28.84 (r = 0.888) where $x$ is the 8-hour time-weighted average methanol concentration in ml/m$^3$. For non-exposed persons, the urine formate level was given as 26.2 ± 12.2 µg/ml (Yasugi *et al.* 1992). On the basis of the regression curves published by these authors, the 8-hour exposure to a methanol concentration of 200 ml/m$^3$ was calculated to result in additional formate in urine of 17 µg/ml for men and 9.8 µg/ml for women.

# 4 Effects in Man

Reports of intoxications after acute oral, dermal or inhalative exposure to methanol or after long-term exposure have been reviewed elsewhere (e.g. ECB 1995, Kavet and Nauss 1990). Mostly the methanol doses were not known.

The symptoms of methanol poisoning include CNS-depression and, after a latency period of between several hours and a few days during which formate accumulates in the blood and leads to metabolic acidosis, visual impairment or even total blindness, headaches, dizziness, nausea, vomiting, abdominal pains, Kussmaul breathing and coma, sometimes ending lethally (Kavet and Nauss 1990). As a sequel to acute methanol poisoning, further CNS disorders and especially parkinsonism-like extrapyramidal symptoms have been described (Henschler and Lehnert 1983). Morphological changes, necrosis in the white substance of the brain and in the putamen, have been demonstrated (Anderson *et al.* 1997).

## 4.1 Single exposures

### 4.1.1 Inhalation

Twelve male test persons, who also served as their own controls, were exposed for 75 minutes to a methanol concentration of 250 mg/m$^3$ (about 190 ml/m$^3$). The results of clinical, psychological and physiological tests were in the normal range. Most neurophysiological and behavioural parameters were unchanged. There were slight but statistically significant changes in the latency period of the P200 component of event-related-potentials and the reaction time in the Sternberg memory test. It was suggested that the changes observed in P200 latency are consistent with the performance changes observed in the Sternberg memory task. When they were exposed to methanol, subjects reported higher levels of fatigue and there was a trend toward poorer concentration and less vigour. These subjective changes were also considered by the authors to be associated with the changes in the ability to perform the Sternberg reaction task. To what extent the test persons knew when they were being exposed to methanol is not absolutely clear. The authors were of the opinion that, because of methodical limitations (small number of test persons, only one concentration, only one exposure time), their study was not sufficiently conclusive to determine a threshold concentration for the induction of neurotoxic effects by methanol (Cook *et al.* 1991).

In the study by Chuwers *et al.* (1995) which was described above (see Section 3), 26 test persons, who served as their own controls, were exposed for 4 hours to a methanol concentration of 200 ml/m$^3$ and then subjected to neurobehavioural, neurophysiological and visual performance tests. In neurobehavioural tests for memory, attention, interference, and tests for contrast sensitivity and colour discrimination to detect effects on visual function, no changes were found. In a study of auditory-evoked potentials in the EEG for the detection of preclinical electrophysiological changes provoked by auditory stimuli, the area under the serum methanol curve from 0 to 4 hours varied inversely with the difference in the change in P300-amplitude. Similarly, with an associative task, the symbol digit substitution test, a learning effect was seen in the comparison of pre-exposure and post-exposure test results when the persons were exposed to water vapour but not when they were exposed to methanol. The observed changes were considered by the authors not to be of significance (Chuwers *et al.* 1995).

### 4.1.2 Ingestion

There are within-population differences in the amount of methanol required to produce symptoms of intoxication. In one case, ingestion of as little as 15 ml of a 40 % solution of methanol was fatal whereas another person survived ingestion of 500 ml of the same solution. There are also large individual differences in the length of the latency period. Symptoms of methanol poisoning can develop within a few hours or not until 72 hours after intake of the substance (WHO 1997). A dose of 95 g methanol (about 120 ml or 1400 mg/kg body weight for a 70 kg person) was lethal in about half of the cases (Kavet and Nauss 1990).

### 4.1.3 Dermal absorption

Signs of intoxication developed in persons who had cleaned their skin with methanol; to what extent it was also taken up by inhalation is not known (Kavet and Nauss 1990).

Eye damage and metabolic acidosis developed in a 31-year-old man who had cleaned a tank with methanol for 2 to 3 hours. The man had worn a positive pressure breathing apparatus but no protective clothing; he went on wearing his methanol-saturated clothing for another hour after finishing the job. The authors ascribed the symptoms to percutaneous uptake of methanol (Downie *et al.* 1992).

A worker who had tipped methanol over his leg and then changed neither his clothing nor his shoes went blind one day after the accident. Seven months later the man was still blind although there had been transient improvement (ECB 1995).

Of 21 children aged between 1.5 months and 4 years who were treated for gastrointestinal complaints by wrapping their bodies for several hours in compresses soaked in methanol, 12 died from cardiac or respiratory failure between 2 and 10 days after admission to hospital. The survivors became well again without any lasting damage. An early sign of intoxication was CNS depression alternating with excitation; in 13 of 21 cases, severe respiratory depression was observed. In 3 cases erythema was recorded and in 2 cases flaking of the skin. Wrapping the body in a compress soaked in alcohol was a household remedy for gastrointestinal complaints which was widely used in Argentina (Giménez *et al.* 1968).

## 4.2 Repeated exposure

### 4.2.1 Inhalation

In 1938, in a study of 19 workers exposed for 9 to 24 months to concentrations of acetone (96–108 mg/m$^3$) and methanol (29–33 mg/m$^3$, about 25 ml/m$^3$) producing a strong odour of solvents, no CNS symptoms or visual disorders were observed (Kavet and Nauss 1990).

Operators of spirit duplicators who were exposed to average methanol concentrations above 260 mg/m$^3$ (about 200 ml/m$^3$) and maximum concentrations of 490 mg/m$^3$ (about 370 ml/m$^3$) complained of headaches (no other details; Kavet and Nauss 1990). Teachers and other staff who operated spirit duplicators using a 99 % methanol fluid in schools, developed symptoms such as blurred vision, headaches, nausea and eye irritation. The average methanol concentrations determined with open doors and windows were given as 1330 mg/m$^3$ (about 1000 ml/m$^3$). It has also been suggested that there was epicutaneous exposure via the still moist methanol-soaked paper (Kavet and Nauss 1990). Methanol concentrations in the breathing zone of persons operating spirit duplicators were in the range from 365 to 3080 ml/m$^3$, averaging about 1040 ± 570 ml/m$^3$. In the group of persons (n = 66) who had operated such spirit duplicators for up to 3 years, complaints of blurred vision, headaches, dizziness and nausea were significantly more frequent than in the control group of teachers (Frederick *et al.* 1984).

Subjective complaints and clinical findings were compared for workers exposed to average concentrations of 459 ml/m$^3$ (n = 22) or 31 ml/m$^3$ (n = 11). The highest 8-hour average concentrations in the breathing zone of the workers were between 3050 and 5500 ml/m$^3$. The workers exposed to higher concentrations reported more frequently blurred vision, irritation of the nasal mucosa and an unusual smell during work as well as headaches, blurred vision and an increased sensitivity of the skin of the extremities after work. None of the exposed persons reported photophobia; examination of the fundus of the eye revealed no abnormalities. Two of three workers who were exposed to 953 to 1626 ml/m$^3$, 1058 to 1585 ml/m$^3$, and 119 to 3577 ml/m$^3$ (all 8-hour averages determined on 2 different days) were shown to have slow pupillary reflexes to light and one mild mydriasis. The affected workers had been employed in the factory for 0.3 to 7.8 years. The authors pointed out that detailed examination of the eyes with visual field measurement was not carried out and that the described clinical findings were probably of limited relevance (Kawai *et al.* 1991). As the workers were also exposed during the process to sodium hydroxide which was heated together with the methanol, it is conceivable that some of the subjective complaints—such as the irritation of the nasal mucosa—were a result of the combined exposure.

### 4.2.2 Ingestion

The literature contains many reports of repeated ingestion of methanol and the resulting typical symptoms of methanol intoxication (CNS symptoms, blindness), but mostly without details of doses (ECB 1995, Kavet and Nauss 1990).

### 4.2.3 Dermal absorption

In 1901 the case was reported of a man who went blind after repeated exposure to varnish dissolved in methanol and use of methanol to clean his face and hands (no other details; Kavet and Nauss 1990).

## 4.3 Local effects on skin and mucous membranes

Repeated or prolonged skin contact with liquid methanol or its vapour results in dry, rough or cracked skin. Liquid methanol or its vapour can cause a burning sensation in the eyes or lacrimation. In one case, liquid methanol in the eyes caused severe conjunctival oedema and surface lesions on the cornea which regressed within a few days after treatment (ECB 1995).

Because of their defatting action, alcohols can cause "subjective irritation" of the skin (Ophaswongse and Maibach 1994).

After repeated exposure of test persons for 3-hour periods to a methanol concentration up to 1000 ml/m$^3$, irritant effects were not reported (Leaf and Zatman 1952).

## 4.4 Allergenic effects

There are only a few reports of positive results obtained in patch tests with methanol. It seems probable that they were the result of cross-reactions because in these very few patients ethanol was very likely the sensitizing agent. In patch tests with such patients, positive results are frequently obtained with various primary and secondary alcohols and with acetone. In some of the cases, skin reactions also developed after the consumption of alcoholic drinks, also erythema, aphthous lesions and a burning sensation in the oral mucosa (Fisher 1983, Fregert *et al.* 1969, Van Ketel and Tan-Lim 1975, Ophaswongse and Maibach 1994). It cannot be concluded from these results that methanol has a significant sensitizing potential.

## 4.5 Reproductive toxicity

### 4.5.1 Fertility

Studies of whether methanol has adverse effects on human fertility—which might be expected from the analogy with ethanol—have not been published.

### 4.5.2 Developmental toxicity

There are no relevant epidemiological studies or case reports which describe an increase in the incidence of malformations in children of mothers exposed to methanol during pregnancy.

In a retrospective study of women who were exposed to solvents during the first trimester of pregnancy, the number of children with malformations of the central nervous system was significantly increased above that in a comparable control group. The mother of a child with anencephaly had been exposed at work to a mixture of petrol, dichloromethane, ether and methanol (Holmberg 1979). An association between the methanol exposure and the malformation of the child cannot be derived from these data.

In a Russian publication which was available only as an abstract, an association between occupational exposure of pregnant women to a mixture of formaldehyde, methanol and ammonia and reduced birth weights of the offspring was described. However, these women were also subject to other adverse conditions such as noise, heat and poor living standards (Pushkina *et al.* 1981). Because of the inadequacy of the available documentation, the study cannot be used for the present evaluation.

A 26-year-old woman who ingested 250 to 500 ml methanol during week 38 of pregnancy was treated and gave birth to a healthy child 6 days later without complications. During an observation period of more than 10 years, no visual impairment was detected in the child. The methanol and formate concentrations in the mother's blood at the time of admission, 5 hours after ingestion of the methanol, were 2300 µg/ml and 336 µg/ml (no other details; Hantson *et al.* 1997).

## 4.6 Genotoxicity

In a study of petrol station attendants, available only as an abstract, the incidence of micronuclei in the cells of the oral mucosa was determined after the introduction of a new fuel containing 33 % methanol, 60 % ethanol and 7 % petrol. Additives were not mentioned. Immediately after introduction of the new fuel, the incidence of micronuclei in the control group was 1.2/2000 and in the exposed group 3.2/2000. Four years later and after reduction of the exposure to methanol, the incidence of micronuclei was in the control range (Gattas *et al.* 1997). Because the persons were exposed to mixtures and because there is no information about additives or confounders, these results cannot be seen as evidence of a genotoxic effect of methanol.

## 4.7 Carcinogenicity

There are no data available. The development of tumours like those observed after long-term ingestion of large amounts of ethanolic drinks is not to be expected with methanol because of its high toxicity which makes long-term ingestion of large amounts of the substance unlikely.

# 5 Animal Experiments and *in vitro* Studies

Under normal laboratory conditions, the toxic effects which are seen in man after exposure to methanol, such as metabolic acidosis and degeneration of the optic nerve, are not observed in experimental animals such as mice and rats because of their different metabolism. However, in animals bred to be folate-deficient or animals given a folate-reduced diet, such effects can be produced. For the prediction of effects in man, monkeys are more suitable experimental animals.

## 5.1 Acute toxicity

### 5.1.1 Inhalation

Symptoms of acute toxicity were not detectable in rats exposed for 8 hours to methanol concentrations up to 4800 ml/m$^3$. At 8800 ml/m$^3$ or more, lethargy was recorded and at 22500 ml/m$^3$ narcosis (Lington and Bevan 1994). The 8-hour $LC_{50}$ for methanol is given as more than 22500 ml/m$^3$ (ACGIH 1999). During 24 hours exposure to 13000 ml/m$^3$, debilitation of the animals was observed. Exposure for 18 to 20 hours to 31600 ml/m$^3$ was lethal for the animals but they survived exposure for one hour to 50000 ml/m$^3$ and for 2.5 hours to 60000 ml/m$^3$ (Lington and Bevan 1994).

The 6-hour $LC_{50}$ for the mouse is above 40000 ml/m$^3$ (ACGIH 1999). $RD_{50}$ values for the mouse have been given as 25300 ml/m$^3$ and 41514 ml/m$^3$ (no other details; Lington and Bevan 1994).

### 5.1.2 Ingestion

After oral administration of methanol, $LD_{50}$ values of 7300 to 10000 mg/kg body weight have been determined for the mouse, 6200 to 12900 mg/kg body weight for the rat and 14400 mg/kg body weight for the rabbit. Doses of 2000 to 7000 mg/kg body weight were lethal for some monkeys (Lington and Bevan 1994, Smyth *et al.* 1941, WHO 1997).

### 5.1.3 Dermal absorption

In rabbits the $LD_{50}$ after dermal application was 20 ml/kg body weight (15800 mg/kg body weight) (Lington and Bevan 1994).

## 5.2 Subacute, subchronic and chronic toxicity

### 5.2.1 Inhalation

The studies of long-term exposure of experimental animals to inhaled methanol which were available as original publications are shown in Table 1. No evidence of systemic toxic effects was obtained with the various species up to the highest concentrations tested (5000–10000 ml/m$^3$). Likewise, in a study from the year 1944 in which dogs were exposed 8 hours daily for 379 days to a methanol concentration of about 500 ml/m$^3$, no adverse effects were seen in the clinicochemical, histological or opthalmologic examinations (Lington and Bevan 1994).

These results are not in agreement with those of a study carried out in Japan (NEDO 1987) which, however, require confirmation because the methods and documentation were inadequate. In rats exposed almost continuously (19.5 hours daily) for 24 months to a methanol concentration of 100 ml/m$^3$, hyperplastic changes in the adrenal glands of the females were found. In the males exposed to 1000 ml/m$^3$, hyperplastic changes were detected in the lungs and testes (NEDO 1987). After exposure of mice for 18 months, no adverse effects were observed (NEDO 1987). In another study from the same laboratory, cynomolgus monkeys were exposed 22 hours daily for 29 months to methanol concentrations of 10, 100 or 1000 ml/m$^3$. Even at the lowest concentration of 10 ml/m$^3$, slight effects on the CNS were reported. Because the documentation of these effects is inadequate, the report does not permit a statement as to whether or not the nervous system was damaged by methanol at the concentrations used. Therefore this study of monkeys cannot be included in the present assessment.

**Table 1.** Effects of long-term inhalation of methanol by experimental animals

| Species | Exposure conditions | Effects | References |
|---|---|---|---|
| **rat**, Sprague-Dawley, 5 ♂, 5 ♀ | **28 days** (6 h/day, 5 days/week), 500, 2000, 5000 ml/m³ | **from 500 ml/m³**: nasal discharge increased in a concentration-dependent manner, results of gross pathological, histological and ophthalmologic examinations normal | Andrews *et al.* 1987 |
| **rat**, Sprague-Dawley, 4 ♂ per group | **6 weeks** (6 h/day, 5 days/week), 200, 2000, 10000 ml/m³ | **up to 10000 ml/m³**: histological and biochemical parameters in the lungs (surfactant, proteins, DNA, enzymes) not changed (other organs not studied) | White *et al.* 1983 |
| **monkey**, cynomolgus, 3 ♂, 3 ♀ | **4 weeks** (6 h/day, 5 days/week), 500, 2000, 5000 ml/m³ | **up to 5000 ml/m³**: no effects on body weights, organ weights, histopathological or ophthalmologic findings | Andrews *et al.* 1987 |

### 5.2.2 Ingestion

Administration of methanol to rats for 6 months as a 1 % solution in the drinking water did not result in any significant changes in physiological parameters. In rats given oral doses of methanol of 10, 100 or 500 mg/kg body weight and day for one month, liver changes such as degeneration of the hepatocyte cytoplasm, changes in the activities of microsomal enzymes, or hepatocyte enlargement were seen (no other details; Lington and Bevan 1994).

## 5.3 Local effects on skin and mucous membranes

Methanol applied to the dorsal skin and ear of the rabbit under occlusive conditions for up to 20 hours (no other details) was not irritating (BASF 1975). When applied to rabbit skin for 24 hours, also under occlusive conditions, a dose of 500 mg methanol caused moderate skin irritation (NIOSH 1983); this irritation was probably a result of the defatting action of methanol on prolonged contact, in analogy to that of other alcohols.

In three studies, one of which was carried out according to OECD guideline 405, methanol proved to be slightly to moderately irritating in the rabbit eye (ECB 1995, Jacobs 1990, Lington and Bevan 1994). In contrast, in another test no irritant effects of 50 mg methanol were recorded (BASF 1975).

## 5.4 Allergenic effects

In a modified Magnusson-Kligman maximization test, methanol was shown not to be sensitizing in guinea pigs (WHO 1997).

## 5.5 Reproductive and developmental toxicity

### 5.5.1 Fertility

Studies on the effects of methanol on the fertility of the rat and mouse are summarized in Table 2. In rats exposed for 7 days, twice daily for 10 minutes to methanol vapour (concentration not specified), the weights of the reproductive organs, testosterone levels, sperm count and the acid phosphatase activity in the prostate were unchanged (Yamada 1993). Exposure to a methanol concentration of 200 ml/m$^3$ for 7 days had no effect on the levels of testosterone, luteinizing hormone or corticosterone (Cameron *et al.* 1985). Exposure to the same concentration for 13 weeks had no effect on testis weights (Lee *et al.* 1991). In animals exposed for 13 weeks to a methanol concentration of 800 ml/m$^3$, an increased incidence of testis degeneration was seen only in 18-month old rats, and not in those which were 10 months old and also folate-deficient (Lee *et al.* 1991). Likewise, in animals which had inhaled methanol at a concentration of 5000 ml/m$^3$ for 6 hours, the levels of luteinizing hormone, testosterone and follicle stimulating hormone were unchanged (Cooper *et al.* 1992).

In mice given oral methanol doses of 1000 mg/kg body weight by gavage, daily for 5 days, no significant increase in the incidence of morphologically altered sperm was detected (Ward *et al.* 1984).

### 5.5.2 Multiple generation studies

In a two-generation study, Sprague-Dawley rats were exposed continuously from 8 weeks of age to methanol concentrations of 10, 100 and 1000 ml/m$^3$. In none of the exposure groups could an effect on the $F_0$ generation be detected. In the $F_1$ and $F_2$ generations, reduced brain weights were found after 8 weeks in the 1000 ml/m$^3$ group in both sexes. In the male animals an early *descensus testis* was described. In the $F_2$ generation, reduced thymus and pituitary weights were also found. No effects on either parents or offspring were detected in the 100 ml/m$^3$ group (NEDO 1987; see Table 3).

### 5.5.3 Developmental toxicity

It has been shown in the mouse and rat that the maximum rate of elimination of methanol is reduced during pregnancy to 65 % to 80 % of the value in non-pregnant animals and that the methanol concentration in the foetus is about 25 % higher than in the dam. In addition, methanol reduces the rate of uptake of $^3H_2O$ by the foetus in a concentration-dependent manner. The authors suggest that methanol reduces the uteroplacental blood flow, which could result in foetal hypoxia (Ward and Pollack 1996a, 1996b). The results of studies on the prenatal toxicity of methanol are shown in Table 3.

**Table 2.** Effects of methanol on the fertility of the male rat and mouse

| Species, strain, number | Administration (duration) concentration or dose, administration route | Effects | References |
|---|---|---|---|
| rat, Sprague-Dawley, 5 per group | 7 days (2 × 10 min/day, or until loss of motor coordination), "methanol vapour", inhalation | **methanol vapour**: no effect on the weights of the reproductive organs, testosterone levels, sperm count, acid phosphatase activity in the prostate | Yamada 1993 |
| rat, Sprague-Dawley, 5 per group | 7 days (6 h/day), 200 ml/m$^3$, inhalation | **200 ml/m$^3$**: only after the first exposure decrease in the serum testosterone level; normalization within 18 hours; no other effects on testosterone, luteinizing hormone or corticosterone | Cameron *et al.* 1985 |
| rat, Long Evans, 9 per group | 6 weeks (8 h/day, 5 days/week), 200 ml/m$^3$, inhalation | **200 ml/m$^3$**: no effects on testosterone levels or testis weights | Lee *et al.* 1991 |
| rat, Long Evans, 9 per group | 13 weeks (20 h/day, 7 days/week), 50, 200, 800 ml/m$^3$, inhalation | **up to 200 ml/m$^3$**: no changes in the testes<br>**800 ml/m$^3$**: no effects on 10-month-old and 18-month-old rats on a normal diet; in rats on a folate-deficient diet increased incidence of testicular degeneration only in the 18-month-old animals | Lee *et al.* 1991 |
| rat, Long Evans, 10 per group | 1, 3, 6 hours, 5000 ml/m$^3$, inhalation | **5000 ml/m$^3$**: no significant effects on luteinizing hormone, testosterone or follicle-stimulating hormone | Cooper *et al.* 1992 |
| mouse, B6C3F$_1$, 10 per group | 5 days, 1000 mg/kg body weight and day, oral, gavage | **1000 mg/kg**: no morphological changes in the sperm | Ward *et al.* 1984 |

**Table 3.** Prenatal and postnatal effects of methanol in experimental animals

| Species | Exposure | Effects | References |
|---|---|---|---|
| **Inhalation** | | | |
| **mouse**, CD-1, 20–61 per group | days 6–15 of gestation (7 h/day) 1000–15000 ml/m$^3$ | **1000 ml/m$^3$**: $F_1$: **NOEC**<br>**from 2000 ml/m$^3$**: $F_1$ increased incidence of cervical ribs, exencephaly<br>**from 5000 ml/m$^3$**: $F_0$: (MeOH 2000 µg/ml serum); $F_1$: increased incidence of cleft palate, complete resorptions<br>**from 7500 ml/m$^3$**: $F_1$: increased prenatal mortality<br>**from 10000 ml/m$^3$**: $F_1$: decreased body weights<br>**15000 ml/m$^3$**: $F_0$: (MeOH 8000 µg/ml serum) no effects | Rogers *et al.* 1991, 1993 |
| **mouse**, CD-1, 5–12 per group | days 6–15 of gestation (6 h/day) 10000 ml/m$^3$ | **10000 ml/m$^3$**: $F_0$: no effects; $F_1$: foetal weights decreased on day 17 of gestation, increased incidence of malformations (neural tube defects, cleft palate, anomalies of the toes) | Bolon *et al.* 1992, 1993 |
| **mouse**, CD-1, 20–27 per group | days 7–9 of gestation (6 h/day) 5000, 10000, 15000 ml/m$^3$ | **from 5000 ml/m$^3$**: $F_0$: no effects; $F_1$: increased incidence of litters with at least one resorption, increased number of animals with dilation of the renal pelvis<br>**from 10000 ml/m$^3$**: $F_0$: no effects; $F_1$: increased incidence of litters with at least one resorption, increased incidence of malformations (dilation of the renal pelvis, cleft palate, eye defects, tail anomalies, neural tube defects (not significant))<br>**15000 ml/m$^3$**: $F_0$: decreased body weights, neurological disorders such as ataxia and decreased motor activity; $F_1$: increased incidence of litters with at least one resorption, increased number of resorptions/litter, increased incidence of malformations (neural tube defects, dilation of the renal pelvis, cleft palate, eye defects, tail anomalies) | Bolon *et al.* 1992, 1993 |
| **mouse**, CD-1, 5–17 per group | days 9–11 of gestation (6 h/day) 10000, 15000 ml/m$^3$ | **10000 ml/m$^3$**: $F_0$: no effects, $F_1$: increased number of animals with cleft palate<br>**15000 ml/m$^3$**: $F_0$: neurological disorders such as ataxia and decreased motor activity, $F_1$: increased incidence of malformations (dilation of the renal pelvis, cleft palate, tail and limb anomalies), no neural tube defects | Bolon *et al.* 1992, 1993 |
| **mouse**, CD-1 | day 7 of gestation (1–7 h) 2000–15000 ml/m$^3$ | **from 15000 ml/m$^3$ × h**: $F_1$: increased incidence of cervical ribs<br>**70000 ml/m$^3$ × h**: $F_1$: increased embryo mortality, increased incidence of cleft palate, multiple skeletal defects | Rogers *et al.* 1995 |

**Table 3.** continued

| Species | Exposure | Effects | References |
|---|---|---|---|
| **Inhalation** | | | |
| **mouse**, CD-1 | day 8 of gestation (6 h) 10000, 15000 ml/m$^3$ | **control**: $F_0$: (formate in plasma: 0.5 ± 0.3 mM, decidua: 1.1 ± 0.2 mM); $F_1$: no exencephaly<br>**10000 ml/m$^3$**: $F_0$: (plasma MeOH: 65 ± 25 mM); $F_1$: exencephaly<br>**15000 ml/m$^3$**: $F_0$: (MeOH in plasma: 223 ± 56 mM; formate in plasma 0.75 ± 0.1 mM, in decidua: 2.2 ± 0.3 mM); $F_1$: exencephaly | Dorman *et al.* 1995a, 1995b |
| | day 8 of gestation 750 mg Na formate/kg body weight (1× oral) | **750 mg Na formate/kg**: $F_0$: (formate in plasma: 1.05 ± 0.2 mM, in decidua: 2.0 ± 0.2 mM); $F_1$: no exencephaly; therefore exencephaly caused by methanol | |
| **mouse**, CD-1, 12–19 per group | 1× during gestation days 5–9 or 2× during days 6–13 (7 h) 10000 ml/m$^3$ | **10000 ml/m$^3$**: $F_0$: (maximum MeOH 4000 µg/ml blood), decreased body weights, complete resorptions (day 7 of gestation); $F_1$: increased foetal mortality, cleft palate (days 7–8 of gestation), exencephaly (days 6–7), skeletal defects (days 5–7) | Rogers and Mole 1997 |
| **rat**, SD, 30 ♂ and 30 ♀ per group | 2 generations (continuous) 10, 100, 1000 ml/m$^3$ | **up to 100 ml/m$^3$**: $F_0$: no effects; $F_1$: no effects<br>**1000 ml/m$^3$**: $F_0$: no effects; $F_1$: decreased brain weights, early *descensus testis*; $F_2$: decreased brain, thymus and pituitary weights, early *descensus testis* | NEDO 1987 |
| **rat**, SD, 36 per group | days 7–17 of gestation (continuous) 200, 1000, 5000 ml/m$^3$ | **up to 1000 ml/m$^3$**: $F_0$: no effects; $F_1$: no effects (NOEC)<br>**5000 ml/m$^3$**: $F_0$: decreased body weights, prolonged gestation period, 1/36 died; $F_1$: increased incidence of skeletal and visceral malformations in the foetuses, decreased birth weights, increased mortality during the first 4 postnatal days, early appearance of the incisors, early eye opening, early *descensus testis*, decreased brain, thyroid and thymus weights in 8-week old animals, no histological changes; when the $F_1$ generation was mated, no effects on fertility and no prenatal toxicity | NEDO 1987 |
| **rat**, SD, 14–15 per group | days 1–19 of gestation (7 h/day) 5000, 10000 ml/m$^3$ or days 7–15 of gestation (7 h/day) 20000 ml/m$^3$ | **5000 ml/m$^3$**: $F_0$: no effects; $F_1$: **NOAEC**<br>**10000 ml/m$^3$**: $F_0$: no effects; $F_1$: decreased body weights<br>**20000 ml/m$^3$**: $F_0$: unsteady gait; $F_1$: increased incidence of cervical ribs, defects of cardiovascular system and urinary tract, encephalocele, exencephaly | Nelson *et al.* 1985, 1990 |

**Table 3.** continued

| Species | Exposure | Effects | References |
|---|---|---|---|
| **Inhalation** | | | |
| **rat**, LE, 10–12 per group | day 6 of gestation till day 21 *post partum* (6 h/day) 4500 ml/m$^3$ | **4500 ml/m$^3$**: $F_0$: (500–800 µg/ml blood); $F_1$: (1260 µg/ml blood postnatally) no neuropathological effects, impaired motor activity (day 18 decreased activity, day 24 increased activity), slightly altered operant conditioning of the pups when adult | Stern *et al.* 1996, 1997, Weiss *et al.* 1996 |
| **rat**, LE, number n.s. | days 7–19 of gestation (7 h/day) 15000 ml/m$^3$ | **15000 ml/m$^3$**: $F_0$: slight decrease in body weights from days 8–10 of gestation, blood methanol concentration 3 mg/ml; $F_1$: slight decrease in body weights on postnatal days 1 and 3, no effects on motor activity, learning or behaviour | Stanton *et al.* 1991 |
| **rat**, LE, 7 | days 7–19 of gestation (7 h/day) 15000 ml/m$^3$ | **15000 ml/m$^3$**: $F_0$: (MeOH 3800 µg/ml blood), transient decrease in body weights on days 8–10 of gestation (4–7 %); $F_1$: body weights decreased on days 1, 21, 35; malformations (unilateral anophthalmia) in 2 pups from one litter, postnatal development and behaviour normal | Stanton *et al.* 1995 |
| **Oral or intraperitoneal administration** | | | |
| **mouse**, CD-1, 4–8 per group | days 6–15 of gestation 2× 2000 mg/kg body weight and day, gavage | **4000 mg/kg**: $F_0$: decreased body weights, blood MeOH 3.7–4.1 mg/ml; $F_1$: 3/8 litters completely resorbed, increased postimplantation losses, decreased foetal weights, increased incidence of malformations (cleft palate in 43%, exencephaly in 29%) | Rogers *et al.* 1993 |
| **mouse**, CD-1, 6–7 per group | day 7 of gestation 4000, 5000 mg/kg body weight (oral) | **from 4000 mg/kg**: $F_1$: changed segment patterning | Connelly and Rogers 1997 |
| **mouse**, CD-1, 21–24 per group | days 6–10 of gestation 5000 mg/kg body weight and day (oral) ± folate | **5000 mg/kg**: $F_0$: increased liver weights after folate-deficient diet; $F_1$: decreased body weights, increased incidence of cleft palate and exencephaly (especially after folate-deficient diet) | Fu *et al.* 1995, 1996 |

**Table 3.** continued

| Species | Exposure | Effects | References |
|---|---|---|---|
| **mouse**, CD-1 | days 6–15 of gestation 5000 mg/kg body weight and day (oral) ± folate | **5000 mg/kg**: $F_0$: (increased serum formate: control: about 0.4 mM; +folate: 3.9 ± 0.9 mM; –folate: 5.13 ± 0.7 mM), decreased body weights; $F_1$: decreased number of viable implants, body weights and crown-rump length decreased most after folate-deficient diet, incidence of cleft palate highest after folate-deficient diet | Hong *et al.* 1997 |
| **mouse**, C57Bl | day 7 of gestation 1700, 2450 mg/kg body weight (2× i.p.) | **2450 mg/kg**: $F_1$: holoprosencephaly, resorptions, decreased body weights | Rogers 1995 |
| **rat**, Holtzmann 8 per group | days 1–8 of gestation, 1600, 2400, 3200 mg/kg body weight and day, gavage | **from 1600 mg/kg**: $F_0$: delayed development of the decidua on day 9 of gestation, $F_1$: no early implantation losses, no effects on embryonal viability, growth, malformations, foetal survival at the examinations on days 11 and 20 of gestation<br>**3200 mg/kg**: $F_0$: decreased body weights; $F_1$: see above | Cummings 1993 |
| **rat**, LE, 10 per group | days 15–17 or 17–19 of gestation, 2% (v/v) methanol in the drinking water (2500 mg/kg body weight and day) | **2500 mg/kg**: $F_0$: no effects; $F_1$: independent of the time of methanol administration, no foetotoxic effects, no effects on postnatal growth, evidence of behavioural disorders (delayed suckling begin on day 1 *post partum*, delay in finding nest material from their home cages on day 10 post partum) | Infurna and Weiss 1986 |
| **rat**, LE, number n.s. | days 8–21 of gestation 0.6 %, 0.9 %, 1.6 % v/v methanol in liquid diet (333, 500, 800 mg/kg body weight and day) | **from 333 mg/kg**: $F_0$: decreased body weights; $F_1$: dose-dependent decrease in litter size from 12 to 5 pups/litter, perinatal mortality increased from 4 % to 25 %, postnatal mortality increased from 0 to 100 %, decreased body weights during the lactation period | Abel and Bilitzke 1992 |
| **rat**, LE, 10–13 per group | day 10 of gestation 1.3, 2.6, 5.2 ml/kg body weight (oral) | **from 1.3 ml/kg**: $F_0$: no effects<br>$F_1$: decreased body weights; increased incidence of anomalies (undescended testes, exophthalmia, anophthalmia) and haemorrhage<br>**5.2 ml/kg**: $F_0$: decreased body weights | Youssef 1991, Youssef *et al.* 1991, 1997 |
| **rat**, LE, 3–6 per group | days 6–15 of gestation 0.5, 1, 2 % in the drinking water, – folate | **0.5 %**: $F_0$, $F_1$: no effects described<br>**from 1 %**: $F_0$: decreased body weights; $F_1$: increased incidence of resorptions, increased mortality, decreased body weights<br>**2 %**: $F_0$: decreased food and water consumption, malformations not studied | Amoco Corporation 1991 |

MeOH: methanol concentration; SD: Sprague-Dawley; LE: Long-Evans

### Inhalation

In rats exposed to methanol for 7 hours daily from days 7 to 15 of gestation, signs of maternal toxicity (unsteady gait) were seen only at 20000 ml/m$^3$, the highest concentration tested. At 10000 ml/m$^3$ (equivalent to a blood level of 2040 µg/ml), body weights of the progeny were reduced without any visible effects on the dams. At 20000 ml/m$^3$ (equivalent to a blood level of 6950 µg/ml) the incidence of malformations (additional cervical ribs, defects of the cardiovascular system and urinary tract, encephalocele, exencephaly) was increased. The NOEC (no observed effect concentration) for developmental toxicity was given as 5000 ml/m$^3$ (Nelson *et al.* 1985, 1990).

Maternal toxicity (reduced body weight) was seen in rats exposed continuously to methanol from days 7 to 17 of gestation at a concentration as low as 5000 ml/m$^3$, which also produced reduced body weights and skeletal and visceral malformations in the foetuses and effects on the development of the new-born progeny. The NOEC for developmental toxicity was about 1000 ml/m$^3$ (NEDO 1987).

In a more recent study with Long-Evans rats exposed for 7 hours daily on days 7 to 19 of gestation, there were no postnatal changes in the behaviour of the progeny of dams exposed to a methanol concentration of 15000 ml/m$^3$ (Stanton *et al.* 1995; Table 3). In contrast, exposure of dams to a methanol concentration of 4500 ml/m$^3$ for 6 hours daily beginning on day 6 of gestation and continuing until postnatal day 21—whereby the progeny were also exposed by inhalation during lactation—resulted in slight changes in motor activity of the neonates and slightly altered operant conditioning when they reached the adult stadium (Stern *et al.* 1997, Weiss *et al.* 1996).

In mice exposed to methanol for 7 hours daily from days 6 to 15 of gestation, no signs of maternal toxicity were seen even at the highest concentration of 15000 ml/m$^3$ but one animal in each of the groups exposed to concentrations above 7500 ml/m$^3$ died. In the animals exposed to 2000 ml/m$^3$ or higher concentrations, additional cervical ribs were seen in the foetuses and from 5000 ml/m$^3$ the incidences of cleft palate, exencephaly, and defects of the sternum and internal organs were increased. The NOEC for developmental toxicity was given as 1000 ml/m$^3$ (Rogers *et al.* 1991, 1993).

The studies with CD-1 mice which have appeared since the studies by Rogers *et al.* confirm these findings (Table 3) and also confirm that methanol, and not its metabolite formate, is responsible for the teratogenic effects (Dorman *et al.* 1995a, 1995b).

In a study of the effects of methanol on the neural tube, mice inhaled methanol for 6 hours daily on gestation days 6 to 15 or 7 to 9, the critical phase for the development and closure of the neural tube. Defects of the neural tube, especially exencephaly, were seen at exposure concentrations of 10000 ml/m$^3$ or more (Bolon *et al.* 1992, 1993). Histopathological examination of foetuses whose mothers had been exposed to a methanol concentration of 15000 ml/m$^3$ on days 7 to 9 of gestation revealed marked reduction of cell density and mitotic index in the embryonal neuroepithelium even after only 8.5, 9.5 and 10.5 days of gestation and displacement of groups of neural crest cells to the neural folds dorsal to the foregut after 8.5 days of gestation (Bolon *et al.* 1994). Effects on the neural tube were not seen after exposure on days 9 to 11 of gestation, a phase of re-opening of the neural tube. Other effects such as dilation of the renal pelvis, cleft palate and tail anomalies developed independently of the exposure time even after exposure to 5000 ml/m$^3$ (Bolon *et al.* 1992, 1993).

## Ingestion

After oral administration of methanol doses of 1600, 2400 or 3200 mg/kg body weight to Holtzmann rats on days 1 to 8 of gestation, a dose-dependent reduction in decidua formation, and reduced uterus weights were detected. The examination of the embryos and foetuses on days 11 and 20 of gestation revealed no treatment-related effects on prenatal development or survival (Cummings 1993). Single oral methanol doses of 1014 mg/kg body weight or more administered to rats on day 10 of gestation produced a significant increase in the incidence of malformations (Youssef 1991, Youssef *et al.* 1991). After administration of methanol doses of 2500 mg/kg body weight and day in the drinking water on days 15 to 17 or 17 to 19 of gestation, no foetotoxic effects and no effects on postnatal growth were found but evidence of behavioural disorders (Infurna and Weiss 1986). Administration of methanol to folate-deficient rats in the drinking water from days 6 to 15 of gestation produced maternal toxic effects (reduced body weight) from doses of about 1250 mg/kg body weight and day and an increase in early deaths of embryo and foetuses (increased incidence of resorptions, increased mortality). Neither maternal nor foetal toxicity was detected in animals given 625 mg/kg body weight and day (Amoco Corporation 1991). Methanol doses of 333 mg/kg body weight and more, administered to rats in the drinking water on days 8 to 21 of gestation, resulted in dose-dependent reductions in litter size, increases in perinatal and postnatal mortality and reductions in body weight gain of surviving animals during lactation (Abel and Bilitzke 1992).

In mice given oral methanol doses of 4000 mg/kg body weight from day 6 to day 15 of gestation, blood methanol levels of 4000 µg/ml were determined. The incidences of resorption and malformation of the foetuses were increased markedly and were like those seen after inhalation of a methanol concentration of 10000 ml/m$^3$, at which comparable blood levels were attained (Rogers *et al.* 1993).

### 5.5.4 *In vitro* studies

In rat embryos explanted on day 9 of gestation and incubated in rat serum containing methanol concentrations of 4 to 16 mg/ml, a dose-dependent decrease in the number of somites, embryonal deaths and dysmorphogenesis (delayed development of the limbs, abnormal brains, open neural tube) was found at concentrations of 8 mg/ml and more (Andrews *et al.* 1991, 1993). This serum methanol concentration corresponded with the blood methanol concentrations in rats exposed to a methanol concentration in air of 20000 ml/m$^3$ (Nelson *et al.* 1985, 1990). In mouse embryos, a reduction in crown-rump length was detected even at the lowest serum methanol concentration tested, 2 mg/ml (Andrews *et al.* 1991, 1993), a concentration which can be attained in mouse blood after exposure to about 5000 ml/m$^3$ (Rogers *et al.* 1991, 1993). With increasing methanol concentration, reductions in the length of the head, and number of somites, increases in embryonic death and dysmorphogenesis were found (Andrews *et al.* 1991, 1993).

The fusion of explanted palates of 12-day old mouse embryos was inhibited in a dose-dependent and time-dependent manner by incubation in a medium containing methanol (0–20 mg/ml). The most sensitive parameter was found to be the reduction in DNA content, from which the authors concluded that cell proliferation is a sensitive target.

Thymidine incorporation in mesenchymal cells of the palate was significantly reduced after 4 days exposure to methanol (Abbott *et al.* 1994, Wilke and Abbott 1993).

## 5.6 Genotoxicity

### 5.6.1 *In vitro*

The studies of *in vitro* genotoxicity of methanol are reviewed in Table 4.

In the published genotoxicity tests with *in vitro* test systems, mainly negative results were obtained with methanol. The occasional positive results are not to be interpreted as evidence of genotoxicity. In the bacterial test systems with *Escherichia coli* WP67 and CM871 and no metabolic activation, weak positive results were obtained with methanol. As it was conceivable that the findings were caused by impurities, the authors classify methanol as an uncertain or questionable mutagen (De Flora *et al.* 1984). In a test for forward mutation in *E. coli* SA500, methanol concentrations above 23 % were cytotoxic, reducing survival of the cells to less than 40 %. A 23 % methanol solution produced no mutagenic effects; cell survival was 53 % (Hayes *et al.* 1990). In the fungus *Aspergillus nidulans*, in a test without metabolic activation, methanol concentrations in the cytotoxic range were shown to induce aneuploidy (Crebelli *et al.* 1989). At the highest methanol concentration tested, which caused 50 % growth inhibition, positive results were obtained in a test for sister chromatid exchange (SCE) in Chinese hamster lung cells without an added metabolic system (NEDO 1987). Negative results were obtained in a test with mouse lymphoma cells with and without S9 mix at standard concentration (2.5 mg/ml). However, a dose-dependent increase in mutants was obtained when a higher concentration of S9 mix (7.5 mg/ml) was used; this concentration of S9 mix on its own had little effect on cytotoxicity or the incidence of mutations (McGregor *et al.* 1988b). Details of the kind of mutant colonies induced—which would indicate whether the effect was mutagenic or clastogenic—were not given. It is conceivable that this positive result was caused by the presence of the metabolite formaldehyde. The relevance of the result is unclear and requires further investigation.

### 5.6.2 *In vivo*

The available data from *in vivo* genotoxicity tests with methanol do not indicate that the substance has genotoxic potential.

No increase in sex-linked lethal mutations in either the wild-type or the Basc strain of *Drosophila melanogaster* was found after feeding methanol (1000 mM) (Gocke *et al.* 1981).

Groups of 2 male and 2 female NMRI mice were given methanol in doses of 1920, 3200 or 4480 mg/kg body weight (60, 100 or 140 mmol/kg body weight) as single intraperitoneal injections. In the polychromatic erythrocytes in bone marrow smears prepared 30 hours later, no effect on the incidence of micronuclei was found (Gocke *et al.* 1981).

**Table 4.** *In vitro* genotoxicity studies with methanol

| Test system<br>Test | Species, strain | Concentration | Results[a]<br>–S9 | <br>+S9 | References |
|---|---|---|---|---|---|
| *Salmonella* mutagenicity test | *S. typhimurium* TA98, TA100, TA1535, TA1537, TA1538 | 2.5 mmol/plate | – | – | De Flora 1981 |
| *Salmonella* mutagenicity test | *S. typhimurium* TA98, TA100, TA1535, TA1537, TA1538 | n.s. | – | – | De Flora *et al.* 1984 |
| *Salmonella* mutagenicity test (closed system) | *S. typhimurium* TA98, TA100, TA1535, TA1537, TA1538 | n.s. | – | – | Simmon *et al.* 1977 |
| *Salmonella* mutagenicity test | *S. typhimurium* TA98, TA100, TA1535, TA1537 | 3 μmol/plate | – | – | Florin *et al.* 1980 |
| *Salmonella* mutagenicity test | *S. typhimurium* TA98, TA100, TA1535, TA1537, TA1538 | 3.6 mg/plate | – | – | Gocke *et al.* 1981 |
| forward mutation | *Escherichia coli* SA500(λ*c*I857Δ431) | 23–31 % | + | n.d. | Hayes *et al.* 1990 |
| DNA repair: | *E. coli* WP2, WP67, CM871 | n.s. | | | De Flora *et al.* 1984 |
| microtitre plate | | | (+) | – | |
| preincubation | | | – | – | |
| "spot" test | | | – | n.d. | |
| prophage induction | *E. coli* $WP2_s(\lambda)$ | 0.15–5 % | – | – | DeMarini *et al.* 1991 |
| chromosomal malsegregation: | *Aspergillus nidulans* P1 | 5.2–7.0 % v/v | | | Crebelli *et al.* 1989 |
| aneuploidy | | | + | n.d. | |
| crossing over | | | – | n.d. | |
| mutation | *Saccharomyces cerevisiae* ATCC 26422 | 2–8 % | – | n.d. | Hamada *et al.* 1988 |
| forward mutation | *Schizosaccharomyces pombe* | 5 % | – | –[b] | Abbondandolo *et al.* 1980 |
| $tk^+/tk^-$ | L5178Y mouse lymphoma cells | 5–50 μl/ml (about 3.95–39.5 mg/ml) | – | –/+[c] | McGregor *et al.* 1988a, 1988b |

**Table 4.** continued

| Test system<br>Test | Species, strain | Concentration | Results[a] –S9 | +S9 | References |
|---|---|---|---|---|---|
| mutation:<br>8-azaguanine, 6-thioguanine, ouabain | V79 cells | 15.8–63.3 mg/ml | – | – | NEDO 1987 |
| chromosomal aberrations | Chinese hamster lung cells | 7.1, 14.3, 28.5 mg/ml | – | – | NEDO 1987 |
| sister chromatid exchange | Chinese hamster lung cells | 7.1, 14.3 mg/ml | – | – | NEDO 1987 |
| | | 28.5 mg/ml | + | – | |
| sister chromatid exchange | CHO cells | 0.1 % v/v<br>8 days | – | n.d. | Obe and Ristow 1977 |

[a] – = negative,
+ = positive,
(+) = weakly positive
[b] mouse S10
[c] after addition of S9 mix in standard concentration (2.5 mg/ml) test result negative, with higher concentrations of S9 mix (from 7.5 mg/ml) test result positive
n.s. not specified
n.d. not determined

Positive results after oral or intraperitoneal administration of methanol to mice have been described in poorly documented studies. Thus, after oral administration of methanol doses of 1000 mg/kg body weight, an increase in the incidence of chromosomal aberrations in bone marrow cells (Pereira *et al.* 1982, Ward *et al.* 1983) and in micronuclei in polychromatic erythrocytes was reported. After oral administration of $^{14}$C-labelled methanol, the authors reported alkylation of haemoglobin because they had detected amino acid adducts (Pereira *et al.* 1982). A dose-dependent increase in structural aberrations, especially centric fusions, was found in the bone marrow cells of mice given methanol on three days by intraperitoneal injection of a total dose of 75 to 300 mg/kg body weight (Chang *et al.* 1983). These studies must be questioned critically, on the one hand because the methods were not described in detail and on the other because in oral studies only one dose was used. In later studies which are better validated, such positive results could not be confirmed.

Groups of 10 male C57Bl/6J mice were exposed to methanol concentrations of 800 or 4000 ml/m$^3$, 6 hours daily for 5 days. In the polychromatic and normochromatic erythrocytes of these mice, the incidence of micronuclei was not increased. In primary cultures of lung cells, the incidences of micronuclei, SCE and structural chromosomal aberrations were also not increased. In the spermatocytes, the incidence of aberrations at the synaptonemal complex was not increased after exposure of the animals to methanol (Campbell *et al.* 1991).

Groups of 10 male Swiss-Webster mice which had been kept on either a normal or a folate-deficient diet for 9 weeks before dosing, were given intraperitoneal methanol doses of 300, 600, 1200 or 2500 mg/kg body weight on 4 consecutive days. 24 hours after the last dose, the incidence of micronuclei in the RNA-positive erythrocytes of the animals was determined. The treatment with methanol did not significantly increase the incidence of micronuclei above the control values. In the high dose group, 4 of the 10 animals on the folate-deficient diet died on days 2 or 3 of the treatment period. In these animals the preterminal clinical symptoms included rough fur, humped backs, laboured breathing and uncoordinated movement (US EPA 1991).

CD-1 mice fed a normal or folate-deficient diet from 5 weeks before mating until the end of gestation (day 18) were given methanol doses of 2500 mg/kg body weight by gavage, twice daily from day 6 to day 10 of gestation. The incidence of micronuclei was not significantly increased in the maternal or foetal reticulocytes harvested 48 hours after the last dose (Fu *et al.* 1996).

In another study, there was also no increase in micronuclei in the polychromatic erythrocytes in the bone marrow smear prepared 24 hours after administration of single oral methanol doses of 1050, 2110, 4210 or 8410 mg/kg body weight to groups of 6 ICR mice (NEDO 1987).

## 5.7 Carcinogenicity

In an inhalation study in which groups of 52 male and 52 female F344 rats and B6C3F$_1$ mice were exposed to methanol concentrations of 10, 100 or 1000 ml/m$^3$ continuously for 24 or 18 months, respectively, a significant increase in tumour incidence was not reported (NEDO 1987).

The study with rats which was carried out by NEDO, like the inhalation study with monkeys (Section 5.2.1), is inadequately documented and so cannot be used for a conclusive evaluation of the carcinogenic potential of methanol.

# 6 Manifesto

There are no epidemiological studies of the carcinogenic effects of methanol. An (inadequate) carcinogenicity study in which methanol was inhaled by rats and mice did not yield evidence of carcinogenic potential. The usable animal studies have yielded no evidence of genotoxic potential.

After exposure of test persons for 4 hours by inhalation of methanol in a concentration of 200 ml/m$^3$, no adverse effects were observed. After long-term inhalation of methanol in concentrations of 365 to 3080 ml/m$^3$ (but average concentration about 1040 ml/m$^3$) at the workplace, effects such as headaches and blurred vision developed; information as to adverse effects of lower concentrations is not available. The effect of peak concentrations on the symptoms is not known. Animal studies with rodents are not suitable for the derivation of a MAK value because of differences in metabolism. The only available study of the chronic toxicity of methanol in the monkey is also not usable for this evaluation, because of inadequate methods.

Thus the existing MAK value of 200 ml/m$^3$ does not have an adequate scientific basis. However, the available experience of effects at the workplace, although limited, and the results of studies with test persons do not suggest that the MAK value must be changed because they indicate that local and systemic toxic effects do not develop at low concentrations. Therefore the MAK value of 200 ml/m$^3$ is retained provisionally. Further studies to determine the NOEC for neurotoxic effects in man are urgently necessary.

After repeated exposure of test persons for 3-hour periods to methanol concentrations up to 1000 ml/m$^3$, irritant effects were not reported (Leaf and Zatman 1952). In a study of workers exposed to methanol in Japan, the 8-hour average exposure concentration and the concentration of formate, measured as the ethyl ester in urine at the end of the shift, were determined (Yasugi *et al.* 1992). A linear relationship between the methanol concentration in the air and the formate concentration in the post-shift urine was established for male workers exposed to methanol in concentrations up to an 8-hour average of 2000 ml/m$^3$ and for female workers up to 4000 ml/m$^3$. There was no evidence of saturation of formate metabolism in the studied concentration range. In this study, the urine formate level for non-exposed persons was given as 26.2 ± 12.2 µg/ml. On the basis of the regression curves published by these authors, the 8-hour exposure to a methanol concentration of 200 ml/m$^3$ was calculated to result in additional formate in urine of 17 µg/ml for men and 9.8 µg/ml for women. Exposure to a methanol peak concentration of 1000 ml/m$^3$, twice for 30 minutes, would result in additional formate in the urine of

11.25 or 6.13 μg/ml, which is within the range of the formate body burden resulting during observance of the MAK value of 200 ml/m$^3$. Therefore, methanol is classified in Peak limitation category II,2.

In 1995, methanol was classified in Pregnancy risk group C because no prenatal toxicity was to be expected provided that the MAK value of 200 ml/m$^3$ was observed. The data published since then support this classification.

Because methanol is readily absorbed through the skin and because of the toxic effects observed in persons exposed dermally to methanol, the designation "H" has been retained. The available data indicate that designation with an "S" (for substances which cause sensitization) is not necessary. There are no studies of germ cell mutagenic effects of methanol. Therefore, a classification of the substance is not possible.

# 7 References

Abbott BD, Logdson TR, Wilke TS (1994) Effects of methanol on embryonic mouse palate in serum-free organ culture. *Teratology 49*: 122–134

Abbondandolo A, Bonatti S, Corsi C, Corsi G, Fiorio R, Leporini C, Mazzaccaro A, Nieri R (1980) The use of organic solvents in mutagenicity testing. *Mutat Res 79*: 141–150

Abel EL, Bilitzke PJ (1992) Effects of prenatal exposure to methanol and *t*-butanol in Long Evans rats. *Am J Obstet Gynecol 166*: 433

ACGIH (American Conference of Governmental Industrial Hygienists) (1999) Methanol. in: *Documentation of TLVs and BEIs*, ACGIH, Cincinnati, OH, USA,

d'Alessandro A, Osterloh JD, Chuwers P, Quinlan PJ, Kelly TJ, Becker CE (1994) Formate in serum and urine after controlled methanol exposure at the threshold limit value. *Environ Health Perspect 102*: 178–181

Amoco Corporation (1991) Initial submission: Letter submitted. *Preliminary information from a pilot developmental toxicity study on methanol with attachment.* NTIS/OTS 0534807, EPA/OTS Doc ID 88-920000356

Anderson CA, Rubinstein D, Filley CM, Stears JC (1997) MR enhancing brain lesions in methanol intoxication. *J Comput Assist Tomogr 21*: 834–836

Andrews JE, Ebron-McCoy M, Rogers JM (1991) Embryotoxic effects of methanol in whole embryo culture. *Teratology 43*: 461

Andrews JE, Ebron-McCoy M, Lodsdon TR, Mole LM, Kavlock RJ, Roger JM (1993) Developmental toxicity of methanol in whole embryo culture: a comparative study with mouse and rat embryos. *Toxicology 81*: 205–215

Andrews LS, Clary JJ, Terrill JB, Bolte HF (1987) Subchronic inhalation toxicity of methanol. *J Toxicol Environ Health 20*: 117–124

BASF (1975) *Gewerbetoxikologische Vorprüfung, Hautreizung und Schleimhautreizung*, Study No. XXIV 436, 03.07.1975, unpublished report

Batterman SA, Franzblau A (1997) Time-resolved cutaneous absorption and permeation rates of methanol in human volunteers. *Int Arch Occup Environ Health 70*: 341–351

Baumann K, Angerer J (1979) Occupational chronic exposure to organic solvents. VI. Formic acid concentration in blood and urine as an indicator of methanol exposure. *Int Arch Occup Environ Health 42*: 241–249

Bolon B, Dorman DC, Morgan KT, Welsch F (1992) Methanol-induced exencephaly in mice is caused by a neural tube closure defect. *Toxicologist 12*: 101

Bolon B, Dorman DC, Janszen D, Morgan KT, Welsch F (1993) Phase-specific developmental toxicity in mice following maternal methanol inhalation. *Fundam Appl Toxicol 21*: 508–516

Bolon B, Welsch F, Morgan KT (1994) Methanol-induced neural tube defects in mice: pathogenesis during neurulation. *Teratology 49*: 497–517

Cameron AM, Zahlsen K, Haug E, Nilson OG, Eik-Nes KB (1985) Circulating steroids in male rats following inhalation of *n*-alcohols. *Arch Toxicol, Suppl 8*: 422–424

Campbell JA, Howard DR, Backer LC, Allen JW (1991) Evidence that methanol inhalation does not induce chromosome damage in mice. *Mutat Res 260*: 257–264

Chang LW, McMillan L, Wynne BR, Pereira MA, Colley RA, Ward JB, Legator MS (1983) The evaluation of six different monitors for the exposure to formaldehyde in laboratory animals. *Environ Mutagen 5*: 381–387

Chuwers P, Osterloh J, Kelly T, d'Alessandro A, Quinlan P, Becker C (1995) Neurobehavioral effects of low-level methanol vapor exposure in healthy human volunteers. *Environ Res 71*: 141–150

Clay KL, Murphy RC, Watkins WD (1975) Experimental methanol toxicity in the primate: analysis of metabolic acidosis. *Toxicol Appl Pharmacol 34*: 49–61

Connelly LE, Rogers JM (1997) Methanol causes posteriorization of cervical vertebrae in mice. *Teratology 55*: 138–144

Cook MR, Bergman FJ, Cohen HD, Gerkovich MM, Graham C, Harris RK, Siemann LG (1991) *Effects of methanol vapor on human neurobehavioral measures*. Research Report No. 42, Health Effects Institute, Cambridge, MA, USA

Cooper RL, Mole ML, Rehnberg GL, Goldman JM, McElroy WK, Hein J, Stoker TE (1992) Effect of inhaled methanol on pituitary and testicular hormones in chamber acclimated and non acclimated rats. *Toxicology 71*: 69–81

Crebelli R, Conti G, Conti L, Carere A (1989) A comparative study on ethanol and acetaldehyde as inducers of chromsome malsegregation in *Aspergillus nidulans*. *Mutat Res 215*: 187–195

Cummings AM (1993) Evaluation of the effects of methanol during early pregnancy in rat. *Toxicology 79*: 205–214

De Flora S (1981) Study of 106 organic and inorganic compounds in the *Salmonella*/microsome test. *Carcinogenesis 2*: 283–298

De Flora S, Zanacchi P, Camoirano A, Bennicelli C, Badolati GS (1984) Genotoxic activity and potency of 135 compounds in the Ames reversion test and in a bacterial DNA-repair test. *Mutat Res 133*: 161–198

DeMarini DM, Lawrence BK, Brooks HG, Houk VS (1991) Compatibility of organic solvents with the Microscreen prophage-induction assay: solvent-mutagen interactions. *Mutat Res 263*: 107–113

Dorman DC, Welsch F (1996) Developmental toxicity of methanol in rodents. *CIIT Act 16*: 1–14

Dorman DC, Moss OR, Farris GM, Janszen D, Bond JA, Medinsky MA (1994) Pharmacokinetics of inhaled [$^{14}$C]methanol and methanol-derived [$^{14}$C]formate in normal and folate-deficient cynomolgus monkeys. *Toxicol Appl Pharmacol 128*: 229–238

Dorman DC, Bolon B, Struve MF, LaPerle KMD, Wong BA, Elswick B, Welsch F (1995a) Role of formate in methanol-induced exencephaly in CD-1 mice. *Teratology 52*: 30–40

Dorman DC, Struve MF, LaPerle KMD, Wong BA, Welsch F (1995b) The role of formate in methanol-induced exencephaly in CD-1 mice. *Teratology 51*: 183–184

Downie A, Khattab TM, Malik MIA, Samara IN (1992) A case report of percutaneous industrial methanol toxicity. *Occup Med 42*: 47–49

Dutkiewicz B, Konczalik J, Karwacki W (1980) Skin absorption and *per os* administration of methanol in men. *Int Arch Occup Environ Health 47*: 81–88

ECB (European Chemicals Bureau) (1995) *IUCLID data sheet, Methanol*, 23.10.1995, ECB, Ispra, Italien

Ferry DG, Temple WA, McQueen EG (1979) *Toxicity of methanol/petrol mixtures. Alcohol fuels technology*, 3rd Int Symp, Asimolar, California, May 28–31, III 65.1–III 65.10

Ferry DG, Temple WA, McQueen EG (1980) Methanol monitoring. Comparison of urinary methanol concentration with formic acid excretion rate as a measure of occupational exposure. *Int Arch Occup Environ Health 47*: 155–163

Fisher AA (1983) Topically applied alcohol as a cause of contact dermatitis. *Cutis 31*: 588–600

Florin I, Rutberg L, Curvall M, Enzell CR (1980) Screening of tobacco smoke constituents for mutagenicity using the Ames' test. *Toxicology 18*: 219–232

Fregert S, Groth O, Hjorth N, Magnusson B, Rorsman H, Övrum P (1969) Alcohol dermatitis. *Acta Derm Venereol 49*: 493–497

Franzblau A, Batterman S, D'Arcy JB, Sargent NE, Gross KB, Schreck RM (1995) Breath monitoring of inhalation and dermal methanol exposure. *Appl Occup Environ Hyg 10*: 833–839

Frederick LJ, Schulte PA, Apol A (1984) Investigation and control of occupational hazards associated with the use of spirit duplicators. *Am Ind Hyg Assoc J 45*: 51–55

Fu SS, Sakanashi TM, Robergs JM, Keen CL (1995) Influence of dietary folic acid on the developmental toxicity of methanol and the frequency of chromosomal breakage in the CD-1 mouse. *Teratology 51*:162–163

Fu SS, Sakanashi TM, Rogers JM, Hong KH, Keen CL (1996) Influence of dietary folic acid on the developmental toxicity of methanol and the frequency of chromosomal breakage in the CD-1 mouse. *Reprod Toxicol 10*: 455–463

Garner CD, Lee EW, Terzo TS, Louis-Ferdinand RT (1995) Role of retinal metabolism in methanol-induced retinal toxicity. *J Toxicol Environ Health 44*: 43–56

Gattas GJF, Cardoso LA, Faria MM, Zaher VL, Doppenschmitt E, Pereira CAB, Saldanha PH (1997) Mutagenic effect of methanol in gas station operators from Sao Paulo – Brazil. *Environ Mol Mutagen 29, Suppl 28*: 60

Giménez ER, Vallejo NE, Roy E, Lis M, Izurieta EM, Rossi S, Capuccio M (1968) Percutaneous alcohol intoxication. *Clin Toxicol 1*: 39–48

Gocke E, King MT, Eckhardt K, Wild D (1981) Mutagenicity of cosmetics ingredients licensed by the European Communities. *Mutat Res 90*: 91–109

Hamada H, Yamashita M, Kashige N, Kojima M, Morita T (1988) Effect of methanol on the frequency of respiration-deficient mutation induced by ethidium bromide in yeast under growing and non-growing conditions. *Chem Pharm Bull (Tokyo) 36*: 4865–4870

Hantson P, Lambermont JY, Mahieu P (1997) Methanol poisoning during late pregnancy. *Clin Toxicol 35*: 187–191

Hayes S, Hayes C, Duncan D, Bennett V, Blushke J (1990) Stimulation of mutations suppressing the loss of replication control by small alcohols. *Mutat Res 231*: 151–163

Heinrich R, Angerer J (1982) Occupational chronic exposure to organic solvents. X. Biological monitoring parameters for methanol exposure. *Int Arch Occup Environ Health 50*: 341–349

Henschler D, Lehnert G (Eds) (1983) Methanol. in: *Biologische Arbeitsstoff-Toleranz-Werte (BAT-Werte) und Expositionsäquivalente für krebserzeugende Arbeitsstoffe (EKA)*, VCH-Verlagsgesellschaft, Weinheim and in English translation in Volume 1 of *Biological Exposure Values*

Holmberg PC (1979) Central-nervous-system defects in children born to mothers exposed to organic solvents during pregnancy. *Lancet 2*: 177–179

Hong KH, Sakanashi TM, Fu SS, Rogers JM, Keen CL (1997) Plasma formate levels in methanol treated dams are not markedly influenced by dietary folate. *Teratology 55*: 57

Horton VL, Higuchi MA, Rickert DE (1992) Physiologically based pharmacokinetic model for methanol in rats, monkeys, and humans. *Toxicol Appl Pharmacol 117*: 26–36

Infurna R, Weiss B (1986) Neonatal behavioral toxicity in rats following prenatal exposure to methanol. *Teratology 33*: 259–265

Jacobs GA (1990) OECD eye irritation test on three alcohols. *J Am Coll Toxicol, Acute toxicity data 1*: 56–57

Jeganathan PS, Namasivayam A (1989a) Methanol-induced monoamine changes in hypothalamus and striatum of albino rats. *Alcohol 6*: 451–454

Jeganathan PS, Namasivayam A (1989b) Brain biogenic amine levels after methanol administration: possible mechanism of action on central monoaminergic neurons in discrete areas of brain in Wistar rat. *Indian J Physiol Pharmacol 33*: 151–156

Johlin FC, Fortman CS, Nghiem DD, Tephly TR (1987) Studies on the role of folic acid and folate-dependent enzymes in human methanol poisoning. *Mol Pharmacol 31*: 557–561

Kavet R, Nauss KM (1990) The toxicity of inhaled methanol vapors. *Crit Rev Toxicol 21*: 21–50

Kawai T, Yasugi T, Mizunuma K, Horiguchi S, Hirase Y, Uchida Y, Ikeda M (1991) Methanol in urine as a biological indicator of occupational exposure to methanol vapor. *Int Arch Occup Environ Health 63*: 311–318

Leaf G, Zatman LJ (1952) A study of the conditions under which methanol may exert a toxic hazard in industry. *Br J Ind Med 9*: 19–31

Lee E, Brady AN, Brabec MJ, Fabel T (1991) Effects of methanol vapors on testosterone production and testis morphology in rats. *Toxicol Ind Health 7*: 261–275

Lee EW, Terzo RS, D'Arcy JB, Gross KB, Schreck RM (1992) Lack of blood formate accumulation in humans following exposure to methanol vapor at the current permissible exposure limit of 200 ppm. *Am Ind Hyg Assoc J 53*: 99–104

Lington AW, Bevan C (1994) Alcohols. in: Clayton GD, Clayton FE (Eds) *Patty's Industrial Hygiene and Toxicology, Vol 2D*. John Wiley & Sons, New York, 2597–2609

Martinasevic MK, Green MD, Baron J, Tephly TR (1996) Folate and 10-formyltetrahydrofolate dehydrogenase in human and rat retina: relation to methanol toxicity. *Toxicol Appl Pharmacol 141*: 373–381

McGregor DB, Brown A, Cattanach P, Edwards I, McBride D, Caspary WJ (1988a) Responses of the L5178Y $tk^{+}/tk^{-}$ mouse lymphoma cell forward mutation assay. II: 18 Coded chemicals. *Environ Mol Mutagen 11*: 91–118

McGregor DB, Edwards I, Riach CG, Cattanach P, Martin R, Mitchell A, Caspary WJ (1988b) Studies of an S9-based metabolic activation system used in the mouse lymphoma L5178Y cell mutation assay. *Mutagenesis 3*: 485–490

McMartin KE, Martin-Amat G, Makar AB, Tephly TR (1977) Methanol poisoning. V. Role of formate metabolism in the monkey. *J Pharmacol Exp Ther 201*: 564–572

Medinsky MA, Dorman DC, Bond JA, Moss OR, Janszen DB, Everitt JI (1997) *Pharmacokinetics of methanol and formate in female cynomolgus monkeys exposed to methanol vapors.* Research Report No. 77, Health Effects Institute, Cambridge, MA, USA

NEDO (New Energy Development Organization) (1987) *Toxicological research of methanol as a fuel for power station. Summary report on tests with monkeys, rats and mice.* Unpublished report

Nelson BK, Brightwell WS, MacKenzie DR, Khan A, Burg JR, Weigel WW, Goad PT (1985) Teratological assessment of methanol and ethanol at high inhalation levels in rat. *Fundam Appl Toxicol 5*: 727–736

Nelson BK, Brightwell WS, Krieg Jr EF (1990) Developmental toxicology of industrial alcohols a summary of 13 alcohols administered by inhalation to rats. *Toxicol Ind Health 6*: 373–387

NIOSH (National Institute of Occupational Safety and Health) (1983) Methanol, *Registry of toxic effects of chemical substances (RTECS)*, NIOSH, Cincinnati, OH, USA

Noker PE, Eells JT, Tephly TR (1980) Methanol toxicity: treatment with folic acid and 5-formyltetrahydrofolic acid. *Alcohol Clin and Exp Res 4*: 378–383

Obe G, Ristow H (1977) Acetaldehyde, but not ethanol, induces sister chromatid exchanges in Chinese hamster cells *in vitro*. *Mutat Res 56*: 211–213

Ophaswongse S, Maibach HI (1994) Alcohol dermatitis: allergic contact dermatitis and contact urticaria syndrome. *Contact Dermatitis 30*: 1–6

Osterloh JD, d'Alessandro A, Chuwers P, Mogadeddi H, Kelly TJ (1996) Serum concentrations of methanol after inhalation at 200 ppm. *J Occup Environ Med 38*: 571–576

Pereira MA, Chang LW, McMillan L, Ward JB, Legator MS (1982) Battery of short-term tests in laboratory animals to corroborate the detection of human population exposures to genotoxic chemicals. *Environ Mutagen 4*: 317

Perkins RA, Ward KW, Pollack GM (1995a) Comparative toxicokinetics of inhaled methanol in the female CD-1 mouse and Sprague-Dawley rat. *Fundam Appl Toxicol 28*: 245–254

Perkins RA, Ward KW, Pollack GM (1995b) A pharmacokinetic model of inhaled methanol in humans and comparison to methanol disposition in mice and rats. *Environ Health Perspect 7–8*: 726–733

Pollack GM, Brouwer KLR (1996) *Maternal-fetal pharmacokinetics of methanol.* Research Report No. 74, Health Effects Institute, Cambridge, MA, USA

Pushkina IK, Bereznyak IV, Tsirkova NL, Garkavenk OS (1981) Effect of working conditions and family life on the health status and reproductive function of women in the production and use of urea resins (Russian). *Vop Gig Tr Nov Tekhnol Protsessami Mashinostoirt Khim Promsti*, 35th Meeting, 87. cited from: *Chem Abstr 110*: 96892v, 1987

Rogers JM (1995) Methanol causes the holoprosencephaly spectrum of malformations at lower dosages than ethanol in C57BL/6J mice. *Teratology 51*: 195

Rogers JM, Mole ML (1997) Critical periods of sensitivity to the developmental toxicity of inhaled methanol in the CD-1 mouse. *Teratology 55*: 364–372

Rogers JM, Chernoff N, Mole ML (1991) Developmental toxicity of inhaled methanol in mice. *Toxicologist 11*: 344

Rogers JM, Mole ML, Chernoff N, Barbee BD, Turner CI, Logsdon TR, Kavlock RJ (1993) The developmental toxicity of inhaled methanol in the CD-1 mouse. With quantitative dose-response modeling for estimation of benchmark doses. *Teratology 47*: 175–188

Rogers JM, Barbee BD, Mole ML (1995) Exposure concentration and time (C×T) relationships in the developmental toxicity of methanol in mice. *Toxicologist 15*:164

Sedivec V, Mráz M, Flek J (1981) Biological monitoring of persons exposed to methanol vapours. *Int Arch Occup Environ Health 48*: 257–271

Simmon VF, Kauhanen K, Tardiff RG (1977) Mutagenic activity of chemicals identified in drinking water. in: Scott D, Bridges BA, Sobels FH (Eds) *Progress in Genetic Toxicology, Vol 2*, Elsevier/North Holland Press, Amsterdam, 249–268

Smyth HF, Seaton J, Fischer L (1941) The single dose toxicity of some glycols and derivatives. *J Ind Hyg Toxicol 23*: 259–268

Stanton ME, Crofton KM, Gray LE, Gordon CM, Bushnell PJ, Mole ML, Peele DB (1991) Assessment of offspring development and behavior following gestational exposure to inhaled methanol in the rat. *Toxicologist 11*: 118

Stanton ME, Crofton KM, Gray LE, Gordon CJ, Boyes WK, Mole ML, Peele DB, Bushnell PJ (1995) Assessment of offspring development and behavior following gestational exposure to inhaled methanol in the rat. *Fundam Appl Toxicol 28*: 100–110

Stern S, Reuhl K, Soderholm S, Cox C, Sharma A, Balys M, Gelein R, Yin C, Weiss B (1996) Perinatal methanol exposure in the rat. I. Blood methanol concentration and neural cell adhesion molecules. *Fundam Appl Toxicol 34*: 36–46

Stern S, Cox C, Preston R, Sharma A, Inglis GB, Balys M, Weiss B (1997) Perinatal methanol exposure in the rat. II. Behavioral effects in neonates and adults. *Fundam Appl Toxicol 36*: 163–176

Tephly TR (1991) The toxicity of methanol. *Life Sci 48*: 1031–1041

US EPA (Environmental Protection Agency) (1991) *Effect of methanol on the frequency of micronuclei in erythrocytes of normal and folate-deficient Swiss mice* (final) with cover letter dated 062191. NTIS/OTS0000812, EPA/OTS Doc# FYI-OTS-0791-0812

Van Ketel WG, Tan-Lim KN (1975) Contact dermatitis from ethanol. *Contact Dermatitis 1*: 7–10

Ward JB, Hokanson JA, Smith ER, Chang LW, Pereira MA, Whorton Jr EB, Legator MS (1984) Sperm count, morphology and fluorescent body frequency in autopsy service workers exposed to formaldehyde. *Mutat Res 130*: 417–424

Ward KW, Pollack GM (1996a) Comparative toxicokinetics of methanol in pregnant and non-pregnant rodents. *Drug Metab Dispos 24*: 1062–1070

Ward KW, Pollack GM (1996b) Use of intrauterine microdialysis to investigate methanol-induced alterations in uteroplacental blood flow. *Toxicol Appl Pharmacol 140*: 203–210

Ward JB, Legator MS, Pereira MA, Chang LW (1983) Evaluation in man and animals of tests for the detection of population exposures to genotoxic chemicals. in: Waters MC, Sanghu SS, Lewtas J, Claxton L, Chernoff N, Nesnow S (Eds) *Short-term bioassays in the analysis of complex environmental mixtures III*, Plenum Press, New York, 461–484

Weiss B, Stern S, Soderholm SC, Cox C, Sharma A, Inglis GB, Preston R, Balys M, Reuhl KR, Gelein R (1996) *Developmental neurotoxicity of methanol exposure by inhalation in rats.* Research Report No. 73, Health Effects Institute, Cambridge, MA, USA

White LR, Marthinsen ABL, Richards RJ, Eik-Nes KB, Nilsen OG (1983) Biochemical and cytological studies of rat lung after inhalation of methanol vapour. *Toxicol Lett 17*: 1–5

WHO (World Health Organization) (1997) *Methanol. IPCS – Environmental health criteria 196*, WHO, Genf

Wilke TS, Abbott BD (1993) Analysis of the effects of methanol on embryonic mouse palates in serum-free organ culture. *Teratology 47*: 438–439

Yamada K (1993) Influence of lacquer thinner and some organic solvents on reproductive and accessory reproductive organs in the male rat. *Biol Pharm Bull 16*: 425–427

Yasugi T, Kawai T, Mizunuma K, Horiguchi S, Iwami O, Iguchi H, Ikeda M (1992) Formic acid excretion in comparison with methanol excretion in urine of workers occupationally exposed to methanol. *Int Arch Occup Environ Health 64*: 329–337

Youssef AF (1991) Teratogenicity of methanol by the oral route. *Toxicologist 11*: 344

Youssef AF, Baggs RB, Weiss B, Miller RK (1991) Methanol teratogenicity in pregnant Long-Evans rats. *Teratology 43*: 467

Youssef AF, Baggs RB, Weiss B, Miller RK (1997) Teratogenicity of methanol following a single oral dose in Long-Evans rats. *Reprod Toxicol 11*: 503–510

completed 02.03.1999

# Methyl acrylate

| | |
|---|---|
| **MAK value (1985)** | **5 ml/m$^3$ (ppm) ≙ 18 mg/m$^3$** |
| **Peak limitation (1985)** | **Category 1 (excursion factor 1)** |
| **Absorption through the skin** | **–** |
| **Sensitization (1985)** | **Sh** |
| **Carcinogenicity** | **–** |
| **Prenatal toxicity (1994)** | **see Section IIc of the *List of MAK and BAT Values*** |
| **Germ cell mutagenicity** | **–** |
| **BAT value** | **–** |
| Synonyms | acrylic acid methyl ester |
| Chemical name (CAS) | 2-propenoic acid methyl ester |
| CAS number | 96-33-3 |

Methyl acrylate is an acrylate monomer apparently used only rarely in the production of polymers for paints and lacquers or for glues. Its main use is as a co-monomer with up to 85 % acrylonitrile in the production of acrylic and modacrylic fibres. Studies of the sensitizing effects on the skin were reviewed in the MAK documentation from 1985 (Volume 6 of the present series), and the substance was designated at that time with an "Sh". This supplement to the previous documentation presents more information on the sensitizing effects of methyl acrylate.

## 1 Allergenic Effects

### 1.1 Effects in man

There are early reports of occasional skin reactions to methyl acrylate, which were probably allergic reactions, in Russian workers (see "Methyl acrylate" in Volume 6 of the present series). According to a Polish publication, in employees working in the production of artificial fibres (copolymer consisting of acrylonitrile and methyl acrylate) skin diseases resulting from contact with methyl acrylate were frequently observed and verified by (weak) reactions in patch tests. Three patients tested with 3 %, 6 % and 25 % methyl acrylate in the patch test produced delayed reactions after 8–14 days. Two of these patients were retested 2 and 11 months later; a reaction was already observed after 24 hours (Sokolowski 1977).

In another early study of the irritative and allergenic effects of acrylates and methacrylates, skin irritation was observed after epicutaneous application of 20 % methyl acrylate in olive oil for 48 hours to 10/30 persons examined by dermatologists for various reasons. The treatment was repeated 4 weeks later on 22 persons with 20 %, 5 % and 1 % preparations. Four of the 10 patients who originally reacted to methyl acrylate produced irritative reactions to 20 % methyl acrylate, 3 of these also to 5 % and one of them even to 1 % methyl acrylate in olive oil. In one case the skin reaction observed during the second testing was classified on the basis of the histological findings as an allergic reaction. Also the reactions observed in one patient 2 weeks after a third test were attributed on the basis of the histological findings to an immunological mechanism. Tests with lower concentrations were, however, not carried out (Cavellier *et al.* 1981).

Within 15 minutes after accidental contact with undiluted methyl acrylate, an employee of a petrochemical company cleaned the contaminated areas of skin on his leg and foot. After 8 days an initially circumscribed reddening occurred on the affected ankle, which developed into a bullous skin lesion, but healed after topical treatment. 17 days after the accident the employee again worked with methyl acrylate for one hour and one day later developed an eczematous skin reaction on the neck and those areas of the face not protected by his face mask. In addition, there was a renewed bullous reaction on the ankle contaminated during the first contact, although the patient was not wearing the originally contaminated boots. The sensitization caused by the accident was confirmed 8 weeks later in 24-hour patch tests with 1 %, 0.33 % and 0.1 % methyl acrylate in petrolatum in which the patient reacted strongly to all concentrations after 48 hours. In addition, after 96 hours strong reactions (3+) were observed in patch tests with 0.167 % butyl acrylate, 2-hydroxyethyl acrylate and 2-hydroxypropyl acrylate, and moderately strong reactions (2+) to 2 % methyl methacrylate and 0.1 % diethylene glycol diacrylate (Kanerva *et al.* 1993).

Another case of contact dermatitis possibly caused by methyl acrylate contained in a nail varnish cannot be included in the evaluation as the documentation was contradictory and the test concentration used (1.5 % in petrolatum) was too high. In the patch test, however, the patient also produced a marked reaction to 2 hardened nail varnishes containing methyl acrylate (9 % and 15 % methyl acrylate) (Kanerva *et al.* 1995).

There are no data available for allergenic effects on the respiratory passages.

## 1.2 Animal studies

With a modified Draize test and with various test protocols using adjuvant (modified maximization test, Polak test, split adjuvant test) sensitizing effects were detected in animals (Parker and Turk 1983, Parker *et al.* 1985) (Table 1).

**Table 1.** Sensitizing effects of methyl acrylate in the guinea pig

| Method, strain, sex | **Intradermal induction:** injection volume, concentration (vehicle); amount/animal **Epicutaneous induction:** treatment days, volume, application, concentration (vehicle) | **Provocation** day, application, concentration (vehicle) | Number of animals with reactions | Comments | References |
|---|---|---|---|---|---|
| **Tests with intradermal induction** | | | | | |
| modified maximization test, Hartley, ♂♀ | injections in duplicate: 100 µl of substance, substance in FCA and FCA, 1 % (ethanol-physiological NaCl); 4 mg | from day 14, one non-occlusive 20 µl dose per week, non-irritative concentration, max. 5 % | **2/6** | incompletely documented; reactions at the 2nd provocation on day 21 | Parker and Turk 1983 |
| Polak test, Hartley, ♂♀ | subcutaneous doses of 100 µl in each paw and in the neck area, probably 0.2 % (physiological NaCl:ethanol 4:1 in FCA); about 1 mg | days 7 and 14, non-occlusive 20 µl, 1 % or 5 % (acetone:olive oil 4:1) | mean reaction score: day 7: 0.8 and 2.0 day 14: 1.1 and 2.2 | n = 5, no. of animals with reactions not specified | Bull *et al.* 1985 |
| Polak test, Hartley, ♂♀ | subcutaneous doses of 100 µl in each paw and in the neck area, 0.2 % (physiological NaCl:ethanol 4:1 in FCA); 1 mg | day 7, non-occlusive 20 µl, 0.5 %, 1 % or 5 % (acetone:olive oil 4:1) | mean reaction score: 0, 0.5 and 1.2 | n = 15, no. of animals with reactions not specified | Parker and Turk 1983, Parker *et al.* 1985 |
| split adjuvant test, Hartley, ♂♀ | on day 0: 5 doses of 50 µl FCA and on day 1: 5 doses of 100 µl of 0.2 % (physiological NaCl:ethanol 100:1) in the dorsal flank; 1.5 mg | from day 14, one non-occlusive 20 µl dose per week, non-irritative concentration, max. 5 % | **4/6** | incompletely documented | Parker and Turk 1983 |
| **Tests with epicutaneous induction** | | | | | |
| patch test, Hartley, ♂♀ | days 0, 2, 4, 7, 9 and 11: 100 µl non-occlusive on the shaved flanks, 2.6 % (ethanol:2-methoxyethanol:Tween 80 9:9:2) | from day 28, one non-occlusive 20 µl dose per week, non-irritative concentration, max. 5 % | **0/6** | incompletely documented | Parker and Turk 1983 |
| modified Draize test, Hartley, ♂♀ | 10 non-occlusive doses: each 100 µl on days 0-4 and 7-11, 10 % (acetone:olive oil 1:1) | from day 21, one non-occlusive 20 µl dose per week, non-irritative concentration, max. 5 % | **4/6** | incompletely documented | Parker and Turk 1983 |

With repeated non-occlusive epicutaneous application there was no evidence of sensitizing effects: with a cotton wool swab, repeated doses of about 0.1 ml 20 % or 50 % methyl acrylate in alcohol and the undiluted substance were applied to the depilated flank skin of guinea pigs until irritation was observed in the treated area. After a 9-day treatment-free interval the skin lesions had healed, and the treated animals and 3 control animals were treated with a single dose of a solution of 5 % methyl acrylate in chloroform. No skin reactions were observed after either 8, 12 or 24 hours (BASF 1958).

Hartley guinea pigs (number of animals not specified) treated once epicutaneously on the ear with 50 µl 1 M methyl acrylate in acetone:olive oil 1:1 produced on provocation after 7 and 14 days marked reactions to 20 µl 1 % and 5 % methyl acrylate in acetone: olive oil 4:1. The weight of the homolateral auricular lymph nodes of 5 animals was not increased on days 4 and 6 after the induction treatment. In the lymph nodes the number of large antibody-producing cells was increased above that in the control animals, and histological examination revealed moderate lymphocyte proliferation (Bull *et al.* 1985).

## 2 Manifesto (sensitization)

Methyl acrylate has been described as the cause of several cases of occupationally-related contact dermatitis; an allergic genesis, however, was only demonstrated in one case. Single exposures to high concentrations of methyl acrylate can lead to sensitization. These findings are supported by positive results in several experiments with animals both with and without the use of adjuvant, although these experiments were not carried out according to present-day standards. Designation with "Sh" has therefore been retained. There are no data available for sensitizing effects on the respiratory passages.

## 3 References

BASF (1958) *Bericht über die Prüfung der Hautwirkung verschiedener Acrylsäureester. Gewerbehygienisch-pharmakologisches Institut*, Report No. VII/308–310, unpublished

Bull JE, Parker D, Turk JL (1985) Predictive value of assessment of lymph node weight and T-lymphocyte proliferation in contact sensitivity in acrylates. *J Invest Dermatol 85*: 403–406

Cavellier C, Jelen G, Hervé-Bazin B, Foussereau J (1981) Irritation et allergie aux acrylates et méthacrylates. Première partie: monoacrylates et monométhacrylates simples. *Ann Dermatol Venereol (Paris) 108*: 549–556

Kanerva L, Jolanki R, Estlander T (1993) Accidental occupational sensitization caused by methyl acrylate. *Eur J Dermatol 3*: 195–198

Kanerva L, Lauerma A, Jolanki R, Estlander T (1995) Methyl acrylate: a new sensitizer in nail lacquer. *Contact Dermatitis 33*: 203–204

Parker D, Turk JL (1983) Contact sensitivity to acrylate compounds in guinea pigs. *Contact Dermatitis 9*: 55–60

Parker D, Long PV, Bull JE, Turk JL (1985) Epicutaneous induction of tolerance with acrylates and related compounds. *Contact Dermatitis 12*: 146–154

Sokolowski F (1977) Allergie auf Methylacrylat nach Hauttests (talk). *Berufs-Dermatosen 25*: 92

completed 02.03.1999

# Methyl methacrylate

| | |
|---|---|
| **MAK value (1988)** | **50 ml/m³ (ppm) ≙ 210 mg/m³** |
| **Peak limitation (1988)** | **Category 1** |
| **Absorption through the skin** | – |
| **Sensitization (1984)** | **Sh** |
| **Carcinogenicity** | – |
| **Prenatal toxicity** | – |
| **Germ cell mutagenicity** | – |
| **BAT value** | – |
| Synonyms | methacrylic acid methyl ester<br>methyl α-methyl acrylate<br>methyl 2-methylpropenoate<br>methyl 2-methyl-2-propenoate<br>MMA |
| Chemical name (CAS) | 2-methyl-2-propenoic acid methyl ester |
| CAS number | 80-62-6 |

The MAK documentation for methyl methacrylate (MMA) from 1984 reveals that the substance has sensitizing effects on the skin. The MAK value valid in 1984 was lowered in 1988 (see Volume 3 of the present series). This supplement to the previous documentation presents more recent data for the sensitizing effects of methyl methacrylate on the skin and respiratory passages.

## 1 Allergenic Effects

### 1.1 Effects in man

#### 1.1.1 Sensitizing effects on the skin

Despite the widespread use of methyl methacrylate, in the literature there are only few case reports of persons with contact dermatitis caused by methyl methacrylate. This could be due to the fact that contact for longer periods, or repeated contact with large amounts or high concentrations of the monomer is rare, and that where short-term non-occlusive skin contact does occur the methyl methacrylate quickly evaporates as a result of its relatively high volatility. In addition, in cases of suspected (meth)acrylate allergy, patch tests are often not carried out because the causality is usually clear.

A group particularly at risk are employees in the dental sector, e.g. dental technicians, as during manual work on dental prostheses contact with methyl methacrylate or other acrylates is particularly intensive. Young dental technicians and those still in training are often affected, and if they cannot tolerate the skin problems they are forced to change their occupation (Rustemeyer and Frosch 1996). Of 1132 dental technicians questioned, 55 persons had suspected work-related skin diseases. Of these, 9 reacted to 2 % methyl methacrylate in the patch test, i.e. 0.8 % of the potentially exposed persons. About twice as many persons reacted to 1 % 2-hydroxyethyl methacrylate (ethylene glycol monomethacrylate) and ethylene glycol dimethacrylate. One of the persons tested reacted to hydroquinone, which is often used in methyl methacrylate preparations as a stabilizer, and two persons to benzoyl peroxide (both 1 % in petrolatum) (Rustemeyer and Frosch 1996). Of 93 dental technicians with manifest dermatosis, 17 reacted to methyl methacrylate in the patch test, i.e. about 18 % of the persons tested (Peiler *et al.* 1996). Among 725 persons with suspected contact dermatitis who were selectively tested in the clinics of the IVDK (Information Network of Departments of Dermatology), 2 produced positive results and 11 weak positive results (1.8 % of the collective) to 2 % methyl methacrylate in petrolatum. There were 42 dental technicians in the collective. In this subgroup there was one positive result and 3 weak positive results (Gebhardt and Geier 1996). One of 92 dentists reacted in the patch test to 1 % methyl methacrylate in petrolatum (no further details; Rudzki *et al.* 1989).

Because methyl methacrylate can cause skin irritation, in the patch test a clear differentiation between an irritative reaction and an allergic reaction is not always possible. Even 5 % solutions can in individual cases lead to false positive, irritative reactions (see documentation from 1984). Table 1 describes probable allergic reactions to methyl methacrylate in cases where the patch test was usually carried out with non-irritative 1 % or 2 % test preparations.

In these and other case reports the data for the composition of the products in question are often incomplete, and simultaneous reactions to other (meth)acrylates were frequently observed. Also there is rarely data available for the stabilizers contained in the products and the composition of the hardeners/accelerators used. The positive test results in other studies are, furthermore, of only limited significance as data for the test concentration were often lacking, test concentrations were too high or the results were of uncertain relevance. The products probably responsible for the observed skin reactions were, above all, glues containing acrylates (Dempsey 1982, Kanerva *et al.* 1995c), glues or materials for making artificial finger nails (Conde-Salazar *et al.* 1986, Daecke *et al.* 1994) and plastics for dental prostheses (Farli *et al.* 1990, Kanerva *et al.* 1988, 1989, 1992, 1993). Less frequently described were reactions to plastics in hearing aids (Meding and Ringdahl 1990, Önder *et al.* 1994) or reactions of the mucous membranes to plastics, e.g. in dental prostheses (Corazza *et al.* 1992, Fowler 1992, Kanerva *et al.* 1995a, Kanzaki *et al.* 1989). Small amounts of residual monomer released from some plastics is apparently responsible for these reactions (see e.g. Austin and Basker 1980, 1982, Marx *et al.* 1983). An urticarial reaction was reported in one case and confirmed in a prick test (Tinkelman and Tinkelman 1979). In a predictive patch test (modified Draize test, induction and provocation with 10 % methyl methacrylate, vehicle not specified), an allergic skin reaction could not be induced in any of the 184 volunteers (Marzulli and Maguire 1982).

**Table 1.** Positive results with methyl methacrylate (MMA) in patch tests

| Persons tested | Concentration (vehicle) | Results | Comments | References |
|---|---|---|---|---|
| 199 patients, in some cases with suspected allergy to dental materials, in other cases tested to exclude allergy | 1 % (70 % ethanol) 10 % (petrolatum) | 1 %: 6 × 1+, 4 × 2+ 10 %: 1 × 1+, 2 × 2+, 2 × 3+ | allergic symptoms only in a few cases | Axéll 1983 |
| 1 dental technician and 1 hearing aid manufacturer with dermatitis of the finger tips and conjunctivitis | 2 % (petrolatum) | 1 × 2+ 1 × 3+ | also reactions to the solutions containing MMA used at work and to other acrylates | Estlander *et al.* 1996 |
| 1 patient with a leg prosthesis and with acute vesicular dermatitis at the site of amputation | 2 % (petrolatum) | 2+ on day 2 | patch tests with scrapings from the prosthesis yielded negative results; with 2 synthetic resins containing MMA (0.5 %) and with triethylene glycol dimethacrylate (2 %) positive results | Foussereau *et al.* 1989 |
| 4 patients with dermatitis on the nail beds, finger tips, hands, face or neck probably caused by artificial finger nails | 2 % (petrolatum) | 2 × strong positive | one of the patients who reacted to MMA also reacted to ethylene glycol dimethacrylate (2 %) | Freeman *et al.* 1995 |
| 82 patients with suspected work-related acrylate allergy | 5 % (petrolatum) | 1 × positive | 11 positive results with the acrylates tested; only a printer reacted to MMA: assumed cause: MMA in photopolymer | Guerra *et al.* 1993 |
| 5 patients with perionychial and onychial dermatitis, caused by UV-light cured artificial finger nails | 2 % (petrolatum) | 1 × 2+ 1 × 3+ | all patients reacted also to other acrylates | Hemmer *et al.* 1996 |
| 4684 patients tested in the clinics of the IVDK | not specified | 10 × 2+ 2 × 3+ | at the reading on day 3 also 47 weak positive results (1+) | IVDK 1997 |
| 4 patients with dentures and burning mouth syndrome | 25 % (petrolatum) | 1 × 4+ | also 2+ reaction to polymethyl methacrylate (25 %); 8 control persons did not react to polymethyl methacrylate (not tested with MMA) | van Joost *et al.* 1988 |

**Table 1.** continued

| Persons tested | Concentration (vehicle) | Results | Comments | References |
|---|---|---|---|---|
| 124 patients who claimed exposure to acrylates in their anamnesis | 2 % (not specified) | 8 × positive | testing with 30 different acrylate compounds; 23 patients reacted to at least one substance | Kanerva *et al.* 1995b |
| 23 patients who claimed exposure to acrylates in their anamnesis | 2 % (petrolatum) | 2 × 2+ 2 × 3+ | the 4 persons sensitized to MMA were 2 dental technicians and 2 dentists; reactions also to other acrylates | Kiec-Swiercynska 1996 |
| 1 patient with dentures | 2 % (petrolatum) | purpuric reaction | dentures containing acrylates and coated with a palladium-silver alloy | Koch *et al.* 1996 |
| 22 patients with known allergy to acrylates | 5 % (petrolatum) | 7 × positive | testing with 33 different acrylates, positive results with several acrylates | Koppula *et al.* 1995 |
| 1 dentist with dermal reaction to artificial finger nails | 1 %, 5 % (petrolatum) | 1 %: 2+ 5 %: 3+ (both read on days 2 and 4) | patch tests with the substances of the European standard series yielded negative results | Macedo *et al.* 1995 |
| 3472 patients of the clinics of the North American Contact Dermatitis Group | 2 % (petrolatum) | 38 × positive | in about 80 % of the patients who reacted to MMA there was known exposure to MMA | Marks *et al.* 1995 |
| 1 patient who wore dentures containing acrylates and suffered from perioral dermatitis | 2 % (petrolatum) | 2+ (on days 3 and 4) | positive results also with polymeric MMA, nickel sulfate, formaldehyde and mercuric chloride; free of symptoms after wearing a porcelain prosthesis | Ölveti 1991 |
| 1 patient with a hip prosthesis who suffered from osteomyelitis | 2 % (petrolatum) | 3+ | positive results also with 5 % polymeric MMA and with a 5 % copolymer made of methacrylate and MMA; during implantation, orthopaedic materials containing MMA were used | Romaguera *et al.* 1985 |
| 1 patient with a prosthesis who developed dermal reactions after the prosthesis was lacquered with a new resinous varnish containing MMA | 2 % (petrolatum) | 2+ (on days 2 and 4) | positive results with various (meth)acrylates | Romaguera *et al.* 1990 |

Other case descriptions, above all earlier ones, have been summarized elsewhere (ECETOC 1995).

### 1.1.2 Sensitization of the respiratory passages

In a 40-year-old dental assistant with long-term occupational exposure to polymethyl methacrylate powder and monomeric methyl methacrylate (level of exposure not specified), tightness of the chest, shortness of breath and coughing developed after every new contact with methyl methacrylate. A challenge test with 20-minute simulation of the work process which probably induced the reaction yielded a reduction in the peak expiratory flow of 20 %. This change was reversible after 2 hours. When the test was repeated a week later, an immediate asthmatic reaction was again observed. A challenge test with pure methyl methacrylate was not carried out (Lozewics *et al.* 1985).

After repeated occupational contact with a glue containing methyl methacrylate (level of exposure not specified), a 52-year-old railway worker (smoker for many years) developed symptoms such as headaches, perspiration and fatigue. Respiratory symptoms that developed later were not unequivocally work-related. Inhalation challenge tests with histamine or with the glue did not cause a reduction in the forced expiratory volume in the first second ($FEV_1$). Prick tests with common inhalation allergens (no further details) yielded negative results (Lozewicz *et al.* 1985).

Of 108 workers exposed for at least 5 years to methyl methacrylate, 20 % suffered from a chronic cough. In the control group of 105 persons only 1 % were affected in this way. In 40 persons from the exposed group and 45 persons from the control group, the usual spirometric parameters were determined before and after the workshift. In the exposed persons only the maximum expiratory flow at 50 % exhalation ($MEF_{50}$) after the workshift was significantly reduced relative to the control value. The authors concluded that methyl methacrylate causes slight respiratory obstruction (Marez *et al.* 1993).

A patient who wore dentures made of acrylates suffered repeated asthmatic attacks. To find the reason for these symptoms dentures made of other materials were tested without the patient knowing which material had been used. With polymethyl methacrylate a severe asthmatic attack with a reduction in the maximum expiratory flow occurred after 4 hours. A prick test was not carried out as it was expected to cause another asthma attack. The authors concluded from the course of the symptoms that the patient had an allergy of immediate type caused by polymethyl methacrylate (Basker *et al.* 1990).

A 56-year-old operating theatre nurse who had worked several times a week for 11 years with orthopaedic bone cement made of polymethyl methacrylate and liquid methyl methacrylate, suffered from continual coughing, rales with each breath and shortness of breath. The symptoms were found to be clearly workplace-related. A bronchial challenge test carried out by simulation of the work process in question produced a delayed asthmatic reaction beginning 6 hours later with a maximum reduction in the $FEV_1$ of 25 % after 13 hours. During the 2-minute production of the cement, the maximum methyl methacrylate concentration of 374 ml/m³ was reached after 45 seconds. In the control experiment, in which water was used instead of methyl methacrylate for mixing the cement, no asthmatic reaction was observed (Pickering *et al.* 1986).

Contact with orthopaedic bone cement containing methyl methacrylate (level of exposure not specified) was also found to be responsible for respiratory changes in another nurse (patient known to be atopic). A bronchial challenge test with the bone cement yielded a reduction in the $FEV_1$ of 25 % within 30 minutes after the beginning of exposure (Reynaud-Gaubert *et al.* 1991).

## 1.2 Animal studies

In various studies, weak to moderate sensitizing effects of methyl methacrylate have been detected in guinea pigs (Table 2).

In preliminary studies carried out during the development of the maximization test, sensitizing effects of methyl methacrylate were also observed. After intradermal and epicutaneous induction, in some cases carried out simultaneously during the first 3 days of the study (intradermal: substances injected in duplicate: 50 µl FCA, MMA and FCA, MMA, 5 % in paraffin oil, in all 10 mg MMA per animal; epicutaneous: 5 %, 48 hours occlusive after previous 24-hour non-occlusive application of 10 % sodium dodecyl sulfate in petrolatum) 21/25 animals produced reactions on provocation on day 16. The exact test protocol, the applied dose, the duration of the provocation treatment and the vehicle used are not, however, apparent from the documentation. When only intradermal induction was carried out, 14/25 animals reacted, with only epicutaneous induction 15/25 (Magnusson and Kligman 1970).

No sensitizing effects were found with methyl methacrylate in a study in which tests both without added FCA (Buehler test, Draize test) and tests with FCA (maximization test, cyclophosphamide-FCA test, split adjuvant test) were carried out (Marzulli and Maguire 1982). The sensitivity of the test systems was possibly decreased compared with that in the studies listed in Table 2 because ethanol was used as the vehicle for topical application.

Also in another study, in which various sensitization tests were carried out in guinea pigs (patch test, maximization test, Polak test, split adjuvant test) with a number of substances including methyl methacrylate, no sensitizing effects were found for the substance (Parker and Turk 1983). However, the amounts of methyl methacrylate used for intradermal induction, 0.25–4 mg per animal, were far below the amounts used in tests which yielded positive results (see Table 2). Also for epicutaneous induction and for provocation comparatively small amounts of the substance were used.

In a mouse ear sensitization test, 50 % methyl methacrylate in ethanol was applied twice to both sides of one ear for induction. With the second application also FCA was injected subcutaneously at a different site. For provocation, one week later the test substance was applied to the other ear non-occlusively in the same concentration. The resulting increase in the thickness of the ear by 14 % was regarded as evidence of a sensitizing effect (Descotes 1988). Another mouse ear-swelling test with comparable induction and provocation concentrations and also using FCA led in 44 % of the animals to results which were considered positive. Induction was carried out by means of topical application on the abdomen, provocation on one ear (Gad *et al.* 1986).

**Table 2.** Sensitizing effects of methyl methacrylate in the guinea pig

| Method, strain, sex | **Intradermal induction** injection volume, concentration (vehicle); amount/animal | **Epicutaneous induction** day, application, concentration (vehicle); amount/animal | **Provocation** day, application, concentration (vehicle); amount/animal | Number of animals which produced reactions | Comments | References |
|---|---|---|---|---|---|---|
| maximization test, Dunkin-Hartley, not specified | injections in duplicate: 100 µl of substance, substance in FCA, and FCA; 0.1, 0.3, 1, 3, 10 % (sesame oil); 0.4–40 mg | day 7, 48 hours occlusive, 1 %, (10 %?), 100 % (sesame oil); 2.5, (25?), 250 mg | day 20, 24 hours occlusive with Finn chamber, 100 %; 15–30 mg | 1 %/1 %: **1/5** 3 %/100 %: **2/5** 10 %/1 % (or 10 %?): **5/5** | 24 hours before epicutaneous induction non-occlusive application of 10 % sodium dodecyl sulfate in petrolatum; contradictory data for 10 % epicutaneous induction concentration | Boman *et al.* 1996 |
| maximization test, SSc:Al, ♀ | injections in duplicate: 50 µl of substance, substance in FCA, and FCA; 5 % (soybean oil); 10 mg | day 8, 48 hours occlusive, 100 %, about 380 mg | day 21, 24 hours occlusive with Finn chamber, 3 % in petrolatum; 0.75 mg | **9/10** | 24 hours before epicutaneous induction non-occlusive application of 250 mg 10 % sodium dodecyl sulfate in petrolatum; control animals included (no further details) | Clemmensen 1984 |
| maximization test, not specified | injections in duplicate: 100 µl of substance, substance in FCA, and FCA, 5 % ($H_2O$); 20 mg | day 7, 48 hours filter paper patch, 100 %; amount not specified, applied on 8 $cm^2$ piece of filter paper | day 21, occlusive, 5 % or 100 %; not specified | 5 %: **4/26** 100 %: **20/26** | no reactions in 25 control animals (not pretreated with FCA) | Nyquist *et al.* 1972 |

**Table 2.** continued

| Method, strain, sex | **Intradermal induction** injection volume, concentration (vehicle); amount/animal | **Epicutaneous induction** day, application, concentration (vehicle); amount/animal | **Provocation** day, application, concentration (vehicle); amount/animal | Number of animals which produced reactions | Comments | References |
|---|---|---|---|---|---|---|
| maximization test, not specified | injections in duplicate: 50 µl of substance, substance in FCA, and FCA; 0.2, 1, 5 % (acetone:olive oil 3:7); 0.4, 2, 10 mg | day 7, 48 hours occlusive, 100 %, about 190 mg | day 21, 24 hours occlusive on 4 $cm^2$, 100 %; about 100 mg | **6/6** | 24 hours before epicutaneous induction non-occlusive application of 10 % sodium dodecyl sulfate in petrolatum; no details of control animals; relevant induction concentration unclear | Katsuno *et al.* 1995 |
| modified maximization test, Hartley, not specified | injections in duplicate: 50 µl of substance, substance in FCA, and FCA; 5, 10, 15, 25 % (propylene glycol); 20–100 mg | day 7, 48 hours occlusive, same concentration as for intradermal induction (petrolatum); amount not specified, applied on 8 $cm^2$ piece of filter paper | day 21, 24 hours occlusive non-irritative concentration (no further details) in petrolatum | 5 %: **0/5**<br>10 %: **2/5**<br>15 %: **2/5**<br>25 %: **4/5** | 10 control animals to determine the maximum non-irritative concentration | Nethercott *et al.* 1983 |
| maximization test, Dunkin-Hartley, ♀ | injections in duplicate: 50 µl of substance, substance in FCA, and FCA, 5 % (arachis oil); 20 mg | day 7, 24 hours occlusive, 10 %, 100 % (80 % ethanol); 95, 950 mg | day 21, 24 hours occlusive on 6 $cm^2$ or day 35, 24 hours non-occlusive, each 30 % in arachis oil:2-butanone 1:2; about 7.5 mg | 10 %: **2/10** (days 21, 35)<br>100 %: **3/10** (days 21, 35) | provocation concentration not specified exactly, may be higher; 6 control animals | van der Walle *et al.* 1982 |
| FCA test, Dunkin-Hartley, ♀ | 100 µl on days 0, 2, 4, 7, 9; 1, 5, 10 % (FCA); 5, 25, 50 mg | | day 21 or 35, 24 hours non-occlusive, each 30 % in arachis oil: 2-butanone 1:2; about 7.5 mg | 1 %: **2/6** (days 21, 35)<br>5 %: **2/8** (days 21, 35)<br>10 %: **4/8** (day 21), **2/6** (day 35) | MMA contained 1.1 % dimethyl-*p*-toluidine; 4–6 control animals | van der Walle *et al.* 1982 |

## 2 Manifesto (sensitization)

There are several case reports of the allergenic effects of methyl methacrylate on the skin. In various studies with guinea pigs, methyl methacrylate was found to have weak to moderate sensitizing effects. Evidence of an allergenic potential was only obtained when FCA was used and with usually large amounts of the substance for induction.

As a result of allergies seen in man and the sensitizing effects on the skin found in experiments with animals, methyl methacrylate has been designated with an "Sh".

Although there are a few case reports of the occurrence of asthma after exposure to methyl methacrylate, the substance has provisionally not been designated a respiratory allergen, above all because an immunological mechanism cannot be clearly deduced from the studies and the number of cases described is very small.

## 3 References


Austin AT, Basker RM (1980) The level of residual monomer in acrylic denture base materials. *Br Dent J 149*: 281–286

Austin AT, Basker RM (1982) Residual monomer levels in denture bases. *Br Dent J 153*: 424–426

Axéll T (1983) Evaluation of a preliminary standard patch test series for the screening of dental materials. *Contact Dermatitis 9*: 86–87

Basker RM, Hunter AM, Highet AS (1990) A severe asthmatic reaction to poly(methyl methacrylate) denture base resin. *Br Dent J 169*: 250–251

Boman A, Drying Jacobsen S, Klastrup S, Svendsen O, Wahlberg JE (1996) Potency evaluation of contact allergens – II. Dose-response studies using the guinea pig maximization test. in: Nordic Council of Ministers (Eds) *TemaNord 1996*: 570, Kopenhagen

Clemmensen S (1984) Cross-reaction patterns in guinea pigs sensitized to acrylic monomers. *Drug Chem Toxicol 7*: 527–540

Conde-Salazar L, Guimaraens D, Romero LV, Gonzalez MA, Alomar A (1986) Occupational allergic contact dermatitis to artificial nails. *Contact Dermatitis 15*: 242

Corazza M, Virgili A, Martina S (1992) Allergic contact stomatitis from methyl methacrylate in a dental prosthesis, with a persistent patch test reaction. *Contact Dermatitis 26*: 210–211

Daecke C, Schaller J, Goos M (1994) Acrylates as potent allergens in occupational and domestic exposures. *Contact Dermatitis 30*: 190–191

Dempsey KJ (1982) Hypersensitivity to Sta-Lok and Loctite anaerobic sealants. *J Am Acad Dermatol 7*: 779–784

Descotes J (1988) Identification of contact allergens: the mouse ear sensitization assay. *J Toxicol – Cut & Ocular Toxicol 7*: 263–272

ECETOC (European Centre for Ecotoxicology and Toxicology of Chemicals) (1995) *Methyl methacrylate. Joint assessment of commodity chemicals No. 30*, ECETOC, Brussels

Estlander T, Kanerva L, Kari O, Jolanki R, Mölsä K (1996) Occupational conjunctivitis associated with type IV allergy to methacrylates. *Allergy 51*: 56–59

Farli M, Gasperini M, Francalanci S, Gola M, Sertoli A (1990) Occupational contact dermatitis in 2 dental technicians. *Contact Dermatitis 22*: 282–287

Fowler JF (1992) Burning mouth caused by dentures. *Am J Contact Dermatitis 3*: 3–4

Fousssereau J, Cavelier C, Protois JP, Deviller J (1989) Contact dermatitis from methyl methacrylate in an above-knee prosthesis. *Contact Dermatitis 20*: 69–70

Freeman S, Lee MS, Gudmundsen K (1995) Adverse contact reactions to sculptured acrylic nails: 4 case reports and a literature review. *Contact Dermatitis 33*: 381–385

Gad SC, Dunn BJ, Dobbs DW, Reilly C, Walsh RD (1986) Development and validation of an alternative dermal sensitization test: the mouse ear swelling test (MEST). *Toxicol Appl Pharmacol 84*: 93–114

Gebhardt M, Geier J (1996) Evaluation of patch test results with denture material series. *Contact Dermatitis 34*: 191–195

Guerra L, Vincenzi C, Peluso AM, Tosti A (1993) Prevalence and sources of occupational contact sensitization to acrylates in Italy. *Contact Dermatitis 28*: 101–103

Hemmer W, Focke M, Wantke F, Götz M, Jarisch R (1996) Allergic contact dermatitis to artificial fingernails prepared from UV light-cured acrylates. *J Am Acad Dermatol 35*: 377–380

IVDK (Informationsverbund Dermatologischer Kliniken) (1997) *Auswertung der zwischen 1992 und 1995 vom IVDK erfaßten Daten zu Methylmethacrylat, 16.01.1997*

van Joost Th, van Ulsen J, van Loon LAJ (1988) Contact allergy to denture materials in the burning mouth syndrome. *Contact Dermatitis 18*: 97–99

Kanerva L, Estlander T, Jolanki R (1988) Sensitization to patch test acrylates. *Contact Dermatitis 18*: 10–15

Kanerva L, Estlander T, Jolanki R (1989) Allergic contact dermatitis from dental composite resins due to aromatic epoxy acrylates and aliphatic acrylates. *Contact Dermatitis 20*: 201–211

Kanerva L, Estlander T, Jolanki R, Pekkarinen E (1992) Occupational pharyngitis associated with allergic patch test reactions from acrylics. *Allergy 47*: 571–573

Kanerva L, Estlander T, Jolanki R, Tarvainen K (1993) Occupational allergic contact dermatitis caused by exposure to acrylates during work with dental prostheses. *Contact Dermatitis 28*: 268–275

Kanerva L, Tarvainen K, Jolanki R, Estlander T (1995a) Successful coating of an allergenic acrylate-based dental prosthesis. *Am J Contact Dermatitis 6*: 24–27

Kanerva L, Estlander T, Jolanki R, Tarvainen K (1995b) Statistics on allergic patch test reactions caused by acrylate compounds, including data on ethyl methacrylate. *Am J Contact Dermatitis 6*: 75–77

Kanerva L, Jolanki R, Leino T, Estlander T (1995c) Occupational allergic contact dermatitis from 2-hydroxyethyl methacrylate and ethylene glycol dimethacrylate in a modified acrylic structural adhesive. *Contact Dermatitis 33*: 84–89

Kanzaki T, Kabasawa Y, Jinno T, Isayama K (1989) Contact stomatitis due to methyl methacrylate monomer. *Contact Dermatitis 20*: 146–148

Katsuno K, Manabe A, Itoh K, Hisamitsu H, Wakumoto S, Nakayama S, Yoshida T (1995) A delayed hypersensitivity reaction to dentine primer in the guinea-pig. *J Dentistry 23*: 295–299

Kiec-Swierczynska M (1996) Occupational allergic contact dermatitis due to acrylates in Lodz. *Contact Dermatitis 34*: 419–422

Koch P, Baum HP, John S (1996) Purpuric patch test reaction and venulitis due to methyl methacrylate in a dental prosthesis. *Contact Dermatitis 34*: 213–231

Koppula SV, Fellman JH, Storrs FJ (1995) Screening allergens for acrylate dermatitis associated with artificial nails. *Am J Contact Dermatitis 6*: 78–85

Lozewics S, Davison AG, Hopkirk A, Burge PS, Boldy D, Riordan JF, McGivern DV, Platts BW, Davies D, Newman Tylor AJ (1985) Occupational asthma due to methylmethacrylate and cyanoacrylates. *Thorax 40*: 836–839

Macedo NA, Carmona C, Pineyro I (1995) Contact dermatitis from acrylic nails. *Contact Dermatitis 32*: 362

Magnusson B, Kligman AM (1970) *Allergic contact dermatitis in the guinea pig*. CA Thomas Publ, Springfield, Illinois, 102–123

Marez T, Edmé TL, Boulenguez C, Shirali P, Haguenoer JM (1993) Bronchial symptoms and respiratory function in workers exposed to methylmethacrylate. *Br J Ind Med 50*: 894–897

Marks jr JG, Belsito DV, DeLeo VA, Fowler jr JF, Fransway AF, Maibach HI, Mathias CGT, Nethercott JR, Rietschel RL, Rosenthal LE, Sherertz EF, Storrs FJ, Taylor JS (1995) North American Contact Dermatitis Group standard tray patch test results (1992 to 1994). *Am J Contact Dermatitis 6*: 160–165

Marx H, Fukui M, Stender E (1983) Zur Frage der Restmonomer-Untersuchung von Prothesenkunststoffen. *Dtsch Zahnaerztl Z 38*: 550–553

Marzulli F, Maguire HC (1982) Usefulness and limitation of various guinea-pig test methods in detecting human skin sensitizers – validation of guinea pig tests for skin hypersensitivity. *Fd Chem Toxicol 20*: 67–74

Meding B, Ringdahl A (1990) Contact dermatitis from the ear mould of hearing aids. *Contact Dermatitis 23*: 253

Nethercott JR, Jakubovic HR, Pilger C, Smith JW (1983) Allergic contact dermatitis due to urethane acrylate in ultraviolet cured inks. *Br J Ind Med 40*: 241–250

Nyquist G, Koch G, Magnusson B (1972): Contact allergy to medicaments and materials used in dentistry (III). Sensitization and sensitivity study in guinea pigs on methyl methacrylate used in a dentifrice. *Odont Rev 23*: 197–204

Ölveti E (1991) Contact dermatitis from an acrylic-metal dental prosthesis. *Contact Dermatitis 24*: 57

Önder M, Önder T, Özönlü A, Makki SS, Gürer MA (1994) An investigation of contact dermatitis in patients with chronic *otitis externa*. *Contact Dermatitis 31*: 116–117

Parker D, Turk JL (1983) Contact sensitivity to acrylate compounds in guinea pigs. *Contact Dermatitis 9*: 55–60

Peiler D, Rustemeyer Th, Frosch PJ (1996) Dermatosen bei Zahntechnikern – Irritantien und Allergene. *Allergologie 19*: 93–94

Pickering CA, Bainbridge D, Birtwistle IH, Griffiths DL (1986) Occupational asthma due to methyl methacrylate in an orthopaedic theatre sister. *Br Med J 292*: 1362–1363

Reynaud-Gaubert M, Philip-Joet F, Arnaud A (1991) Asthme professionnel au méthyl-méthacrylate. *Presse Médicale 20*: 386

Romaguera C, Grimalt F, Vilaplana J (1985) Methyl methacrylate prosthesis dermatitis. *Contact Dermatitis 12*: 172

Romaguera C, Vilaplana J, Grimalt F, Ferrando J (1990) Contact sensitivity to meth(acrylates) in a limb prosthesis. *Am J Contact Dermatitis 1*: 183–185

Rudzki E, Rebandel P, Grzywa Z (1989) Patch tests with occupational contactants in nurses, doctors and dentists. *Contact Dermatitis 20*: 247–250

Rustemeyer T, Frosch PJ (1996) Occupational skin diseases in dental laboratory technicians (I). Clinical picture and causative factors. *Contact Dermatitis 34*: 125–133

Tinkelman DG, Tinkelman CK (1979) An unusual etiology of urticaria. *Pediatrics 63*: 339

van der Walle HB, Klecak G, Geleick H, Bensink T (1982) Sensitizing potential of 14 mono (meth) acrylates in the guinea pig. *Contact Dermatitis 8*: 223–235

completed 05.05.1997

# Pentaerythritol triacrylate

| | |
|---|---|
| **MAK value** | – |
| **Peak limitation** | – |
| **Absorption through the skin** | – |
| **Sensitization (1998)** | **Sh** |
| **Carcinogenicity** | – |
| **Prenatal toxicity** | – |
| **Germ cell mutagenicity** | – |
| **BAT value** | – |
| Synonyms | acrylic acid pentaerythritol triester |
| Chemical name (CAS) | 2-propenoic acid 2-(hydroxymethyl)-2-[[(1-oxo-2-propenyl)oxy]methyl]-1,3-propanediyl ester |
| CAS number | 3524-68-3 |

Pentaerythritol triacrylate is a trifunctional acrylate monomer used, sometimes also as a mixture with pentaerythritol tetraacrylate (CAS No. 4986-89-4), as reactive thinner and cross-linking agent in UV-curing inks and plastics, e.g. for printing plates or plastic coatings, and in the production of polyfunctional aziridine hardeners.

## 1 Allergenic Effects

### 1.1 Effects in man

There are numerous case reports of positive results obtained in patch tests with pentaerythritol triacrylate. The patch tests were mainly carried out with concentrations which were very probably not irritative. In almost all cases UV-curing inks or plastics containing pentaerythritol triacrylate are responsible for the allergic skin symptoms described (see Table 1).

**Table 1.** Positive results with pentaerythritol triacrylate (PETA) from patch tests on patients with suspected acrylate allergy

| Persons tested | Concentration (vehicle) | Results | Contact / comments | References |
|---|---|---|---|---|
| 6 employees in the printing industry with contact dermatitis | 0.1 % (acetone) | 4/6 (1 × 2+, 3 × 1+) | UV-curing printing inks / 4 of the 30 control persons also reacted | Björkner *et al.* 1980 |
| 1 painter | 0.2 % (petrolatum) | positive (read after 48 hours and 7 days; no other details) | paints containing aziridine hardener / no evidence of exposure to PETA; no control persons tested | Cofield *et al.* 1985 |
| 4 employees with dermatitis on the hands and face | 0.1 %<br>0.01 %<br>(acetone) | 2/4<br>1/2<br>(no other details) | lacquers containing aziridine hardener for plastic flooring | Dahlquist *et al.* 1983 |
| 5 lacquer production workers with contact dermatitis | 0.2 % (petrolatum) | 4/5 (1 × 3+, 3 × 2+) | components of UV-curing printing inks / 5 of 26 potentially exposed employees with skin symptoms | Emmett 1977; Emmett and Kominsky 1977 |
| 4 lacquer production workers with contact dermatitis | 0.2 % (petrolatum) | 2/4 (2 × 2+) | components of UV-curing printing inks / patients with phototoxic reactions to components of the inks | Emmett *et al.* 1977 |
| 1 hearing aid manufacturer with conjunctivitis | 0.1 % (petrolatum) | positive (2+) | UV-curing lacquers | Estlander *et al.* 1996 |
| 1 patient with contact dermatitis | 0.1 % (petrolatum) | positive (2+) | UV-curing glue containing acrylates | van den Hove *et al.* 1996 |
| 369 patients with suspected contact dermatitis | 0.1 % (petrolatum) | 5/369 (2+ read after 72 hours) | only reactions of at least 2+ read after 72 hours included; test period: 1992–1995 | IVDK 1997 |
| 124 patients | 0.1 % (petrolatum) | 2/124 (no other details) | anamnesis: exposure to acrylates | Kanerva *et al.* 1995 |
| 23 patients | 0.1 % (petrolatum) | 1/23 (3+) | anamnesis: exposure to acrylates / reaction in 1 textile worker | Kiec-Swiercynska 1996 |
| 16 patients | 5 % (petrolatum) | 4/16 (no other details) | reactions in 2/9 patients with artificial finger nails / provocation concentration questionable, also irritative reaction probable | Koppula *et al.* 1995 |

**Table 1.** continued

| Persons tested | Concentration (vehicle) | Results | Contact / comments | References |
|---|---|---|---|---|
| 18 printing ink production workers, of these 15 with contact dermatitis and 3 with conjunctivitis | 0.1 % (petrolatum) | 7/18 (4 × 3+, 2 × 2+, 1 × 1+) | UV-curing inks / 15 control persons did not react | Nethercott 1978 |
| 7 printing machine operators with contact dermatitis | 0.1 % (petrolatum) | 1/7 (3+ read after 48 and 96 hours) | UV-curing printing inks for plastic containers / 20 control persons did not react | Nethercott *et al.* 1983 |
| 21 employees in the printing trade with contact dermatitis | 0.1 % (petrolatum) | 3/21 (3 × 2+) | UV-curing inks for producing printing plates | Nethercott and Nosal 1986 |
| 35 dental technicians with allergic contact dermatitis of the hand | 0.1 % (petrolatum) | 2/35 (no other details) | | Rustemeyer and Frosch 1996 |
| 2 printing press operators | 0.1 % (2-butanone) | 2/2 (2 × 1+) | UV-curing inks containing pentaerythritol triacrylate / incompletely documented | Smith 1977 |

In a modified Draize test 11 volunteers were treated occlusively on the arm for 24 hours with 10 % pentaerythritol triacrylate in petrolatum. If there was no inflammatory reaction at the application site, the treatment was repeated (no other details). After 4 weeks, 8 of the volunteers underwent occlusive provocation treatment with 0.1 % pentaerythritol triacrylate in petrolatum for 48 hours. The reading was carried out "after pressure effects had resolved"; one strong reaction, four moderate reactions and one weak reaction were observed. In one patient the patch test could not be carried out because of chronic eczematous dermatitis at the test site. Two other volunteers were no longer available for testing (Nethercott 1978). In a repeated insult patch test with ten applications of a 0.1 % solution of pentaerythritol triacrylate in polyethylene glycol, none of 20 volunteers became sensitized (no other details) (Andrews and Clary 1986).

There are no data available for allergenic or irritative effects on the respiratory passages.

## 1.2 Animal studies

Hartley guinea pigs (number of animals not specified) treated once epicutaneously on the ear with 50 µl 1 M pentaerythritol triacrylate in acetone/olive oil (1:1) produced on provocation after 7 and 14 days no marked reactions to 20 µl 0.1 % or 0.25 % pentaerythritol triacrylate in acetone:olive oil (4:1). The weight of the homolateral, auricular lymph nodes of 5 animals was increased on days 4 and 6 after the induction treatment. In the lymph nodes the number of large pyroninophilic cells was around twice that found in the control animals, and histological examination revealed marked lymphocyte proliferation (Bull *et al.* 1985).

Also with a modified Draize test and with various test protocols using adjuvant (maximization test, Polak test) sensitizing effects were detected in the animals (Table 2). An undocumented Bühler test yielded negative results, however (Andrews and Clary 1986).

Cross-reactions were observed in a maximization test. After induction treatment with trimethylolpropane triacrylate (intradermally 1 % in olive oil, epicutaneously 25 % in petrolatum), 12 of 24 guinea pigs reacted on provocation with 0.1 % pentaerythritol triacrylate, but only 6 of 24 animals reacted to the trimethylolpropane triacrylate used for induction. None of the 24 control animals pretreated with adjuvant reacted to the two acrylates after this provocation treatment. On repeated provocation 7 days later, 10 of 24 animals reacted, however, to 0.1% trimethylolpropane triacrylate (Björkner 1980).

# 2 Manifesto (sensitization)

The sensitizing effects of pentaerythritol triacrylate in man have been demonstrated in several cases of contact dermatitis, some of which were work-related, and in an experimental study with volunteers. These findings are supported by positive results in several experiments with animals both with and without the use of adjuvant. There are no data available for sensitizing effects on the respiratory passages.

Pentaerythritol triacrylate is therefore designated with “Sh”.

# 3 References


Andrews LS, Clary JJ (1986) Review of the toxicity of multifunctional acrylates. *J Toxicol Environ Health 19*: 149–164

Björkner B (1980) Allergenicity of trimethylol propane triacrylate in ultraviolet curing inks in the guinea pig. *Acta Derm Venereol 60*: 528–531

Björkner B (1984) The sensitizing capacity of multifunctional acrylates in the guinea pig. *Contact Dermatitis 11*: 236–246

Björkner B, Niklasson B (1984) Influence of the vehicle on elicitation of contact allergic reactions to acrylic compounds in the guinea pig. *Contact Dermatitis 11*: 268–278

Björkner B, Dahlquist I, Fregert S (1980) Allergic contact dermatitis from acrylates in ultraviolet curing inks. *Contact Dermatitis 6*: 405–409

Bull JE, Parker D, Turk JL (1985) Predictive value of assessment of lymph node weight and T lymphocyte proliferation in contact sensitivity in acrylates. *J Invest Dermatol 85*: 403–406

**Table 2.** Sensitizing effects of pentaerythritol triacrylate (PETA) in the guinea pig

| Method, strain, sex | **Intradermal induction** injection volume, concentration (vehicle); amount/animal | **Epicutaneous induction** day, application, concentration (vehicle) | **Provocation** day, application, concentration (vehicle) | Number of animals which produced reactions | Comments | References |
|---|---|---|---|---|---|---|
| maximization test, Dunkin-Hartley, ♀ | injections in duplicate: 100 µl of PETA, PETA in FCA, and FCA, 1 % (olive oil:acetone 9:1); 3 mg | day 7, 48 hours occlusive, 25 % (petrolatum) | day 21, 24 hours occlusive, 0.2 % (15 mg in petrolatum or 20 µl in acetone) | petrolatum: **10/15**<br>acetone: **0/15** | technical grade product, purity and composition not specified | Björkner 1984; Björkner and Niklasson 1984 |
| maximization test, Hartley-Dalkin, not specified | injections in duplicate: 100 µl of PETA, PETA in FCA, and FCA, 0.05 % (propylene glycol); 0.15 mg | day 7, 48 hours occlusive, 25 % (petrolatum) | day 21, 24 hours occlusive, 10 % (petrolatum) | **13/15** | high provocation concentration; concentration > 10 % caused skin irritation in most control animals (no other details) | Nethercott 1978 |
| maximization test, Hartley-Dalkin, not specified | injections in duplicate: 100 µl of PETA, PETA in FCA, and FCA, 0.25 % and 0.5 % (propylene glycol); 0.75 and 1.5 mg | day 7, 48 hours occlusive, 25 % (petrolatum) | day 21, 24 hours occlusive, 10 % (petrolatum) | 0.25 %: **8/15**<br>0.5 %: **10/10** | high provocation concentration, 10 control animals not pretreated with adjuvant did not react | Nethercott 1981 |
| maximization test, Hartley-Dalkin, ♀ | injections in duplicate: 100 µl of PETA, PETA in FCA, and FCA, 0.001 %, 0.01 %, 0.05 % or 0.5 % (propylene glycol); 0.003, 0.03, 0.15 or 1.5 mg | day 7, 48 hours occlusive, 0.001 %, 0.01 %, 0.05 % or 0.5 % (petrolatum) | day 21, 24 hours occlusive, non-irritative concentration, no other details (petrolatum) | 0.001 %: **2/5**<br>0.01 %: **8/15**<br>0.05 %: **4/5**<br>0.5 %: **10/10** | 10 control animals not pretreated with FCA used to determine the maximum non-irritative concentration | Nethercott *et al.* 1983 |

**Table 2.** continued

| Method, strain, sex | **Intradermal induction** injection volume, concentration (vehicle); amount/animal | **Epicutaneous induction** day, application, concentration (vehicle) | **Provocation** day, application, concentration (vehicle) | Number of animals which produced reactions | Comments | References |
|---|---|---|---|---|---|---|
| Polak test, Hartley, ♂ and ♀ | subcutaneous doses of 0.1 ml in each paw and in the neck area, about 0.7 % (physiological NaCl:ethanol 4:1 in FCA); about 3.5 mg | – | days 7 and 14, 20 µl non-occlusive, 0.1 % or 0.25 % (acetone:olive oil 4:1) | mean reaction score on day 7: 0.8 and 1.5; day 14: 1.2 and 2.1 | no. of animals which reacted not specified, 5 animals used | Bull *et al.* 1985 |
| Polak test, Hartley, ♂ and ♀ | subcutaneous doses of 0.1 ml in each paw and in the neck area, 0.2 % (physiological NaCl:ethanol 4:1 in FCA); 1 mg | – | days 7 and 14, 20 µl non-occlusive, 0.1 % or 0.25 % (acetone:olive oil 4:1) | mean reaction score on day 14: 1.1 and 1.4 | no. of animals which reacted not specified, 6 animals used | Parker and Turk 1983, Parker *et al.* 1985 |
| modified Draize test, Hartley, ♂ and ♀ | – | in all 10 non-occlusive doses of 0.1 ml, applied on days 0–4 and 7–11, 0.25 % (acetone:olive oil 1:1) | from day 21 one non-occlusive dose per week of 20 µl, 0.1 % or 0.25 % (acetone:olive oil 4:1) | **5/6** | results incompletely documented, provocation concentration not specified | Parker and Turk 1983 |

Cofield BG, Storrs FJ, Strawn CB (1985) Contact allergy to aziridine paint hardener. *Arch Dermatol 121*: 373–376

Dahlquist I, Fregert S, Trulson L (1983) Contact allergy to trimethylolpropane triacrylate (TMPTA) in an aziridine plastic hardener. *Contact Dermatitis 9*: 122–124

Emmett EA (1977) Contact dermatitis from polyfunctional acrylic monomers. *Contact Dermatitis 3*: 245–248

Emmett EA, Kominsky JR (1977) Allergic contact dermatitis from ultraviolet cured inks. *J Occup Med 19*: 113–115

Emmett EA, Taphorn BR, Kominsky JR (1977) Phototoxicity occurring during the manufacture of ultraviolet-cured ink. *Arch Dermatol 113*: 770–775

Estlander T, Kanerva L, Kari O, Jolanki R, Mölsä K (1996) Occupational conjunctivitis associated with type IV allergy to methacrylates. *Allergy 51*: 56–59

van den Hove J, Jacobs M-C, Tennstedt D, Lachapelle J-M (1996) Allergic contact dermatitis from acrylates in insulin pump infusion sets. *Contact Dermatitis 35*: 108

IVDK (Informationsverbund Dermatologischer Kliniken) (1997) *Auswertung der zwischen 1992 und 1995 vom IVDK erfaßten Daten zu Pentaerythrittriacrylat*, 16.01.1997

Kanerva L, Estlander T, Jolanki R, Tarvainen K (1995) Statistics on allergic patch test reactions caused by acrylate compounds, including data on ethyl methacrylate. *Am J Contact Dermatitis 6*: 75–77

Kiec-Swierczynska M (1996) Occupational allergic contact dermatitis due to acrylates in Lodz. *Contact Dermatitis 34*: 419–422

Koppula SV, Fellman JH, Storrs FJ (1995) Screening allergens for acrylate dermatitis associated with artificial nails. *Amer J Contact Dermatitis 6*: 78–85

Nethercott JR (1978) Skin problems associated with multifunctional acrylic monomers in ultraviolet curing inks. *Br J Dermatol 98*: 541–552

Nethercott JR (1981) Allergic contact dermatitis due to an epoxy acrylate. *Br J Dermatol 104*: 697–703

Nethercott JR, Nosal R (1986) Contact dermatitis in printing tradesmen. *Contact Dermatitis 14*: 280–287

Nethercott JR, Jakubovic HR, Pilger C, Smith JW (1983) Allergic contact dermatitis due to urethane acrylate in ultraviolet cured inks. *Br J Ind Med 40*: 241–250

Parker D, Turk JL (1983) Contact sensitivity to acrylate compounds in guinea pigs. *Contact Dermatitis 9*: 55–60

Parker D, Long PV, Bull JE, Turk JL (1985) Epicutaneous induction of tolerance with acrylates and related compounds. *Contact Dermatitis 12*: 146–154

Rustemeyer T, Frosch PJ (1996) Occupational skin diseases in dental laboratory technicians. (I). Clinical picture and causative factors. *Contact Dermatitis 34*: 125–133

Smith WDL (1977) Allergic dermatitis due to a triacrylate in ultraviolet cured inks. *Contact Dermatitis 3*: 312–314

completed 29.01.1998

# Trimethylolpropane triacrylate

| | |
|---|---|
| **MAK value** | – |
| **Peak limitation** | – |
| **Absorption through the skin** | – |
| **Sensitization (1998)** | **Sh** |
| **Carcinogenicity** | – |
| **Prenatal toxicity** | – |
| **Germ cell mutagenicity** | – |
| **BAT value** | – |
| Synonyms | acrylic acid 1,1,1-(trihydroxymethyl)-propane triester<br>2-ethyl-2-(hydroxymethyl)-1,3-propanediol triacrylate |
| Chemical name (CAS) | 2-propenoic acid 2-ethyl-2-[[(1-oxo-2-propenyl)oxy]methyl]-1,3-propanediyl ester |
| CAS number | 15625-89-5 |

Trimethylolpropane triacrylate is a trifunctional acrylate monomer used as reactive thinner and cross-linking agent in UV-curing inks and plastics, e.g. for printing plates or plastic coatings, and in the production of polyfunctional aziridine hardeners.

## 1 Allergenic Effects

### 1.1 Effects in man

There are numerous case reports of positive results in patch tests with trimethylolpropane triacrylate which were mainly carried out with concentrations which were very probably non-irritative. In almost all cases UV-curing inks or plastics containing trimethylolpropane triacrylate are responsible for the allergic skin symptoms described (Table 1, see also Garabrant 1985).

**Table 1.** Positive results with trimethylolpropane triacrylate from patch tests on patients with suspected acrylate allergy

| Persons tested | Concentration (vehicle) | Results | Contact / comments | References |
|---|---|---|---|---|
| 6 printing industry employees with contact dermatitis | 0.5 % or 0.1 % (acetone) | 6/6 (0.5 %: 2 × 2+, 2 × 3+; 0.1 %: 1 × 1+, 1 × 2+) | UV-curing printing inks / 4 of the 30 control persons also reacted | Björkner *et al.* 1980 |
| 1 painter | 0.5 % (petrolatum) | positive (read after 48 hours and 7 days) | paints containing aziridine hardener; no reaction to 0.1 % / in 5 of 44 control persons irritative reactions after 48 hours, not after 7 days | Cofield *et al.* 1985 |
| 4 employees with dermatitis on the hands and face | 0.1 %<br>0.03 %<br>0.0001 %<br>(acetone) | 4/4<br>3/4<br>1/4<br>(no other details) | lacquers containing aziridine hardener for plastic flooring | Dahlquist *et al.* 1983 |
| 5 employees | 0.2 % (petrolatum) | 4/5 (1 × 1+, 3 × 2+) | components of UV-curing printing inks / 5 of 26 potentially exposed employees with skin symptoms | Emmett 1977; Emmett and Kominsky 1977 |
| 4 employees | 1 % (petrolatum) | 2/4 (1 × 2+, 1 × 3+) | components of UV-curing printing inks / patients with phototoxic reactions to components of the inks; reaction possibly irritative | Emmett *et al.* 1977 |
| 144 employees in the wood surface-coating industry | 0.1 % (petrolatum) | 1/144 (1+) | acrylates in UV-curing lacquers / anamnesis: previous or current skin symptoms | Fisher *et al.* 1994 |
| 220 patients with suspected contact dermatitis | 0.1 % (petrolatum) | 4/220 (3 × 1+, 1 × 3+ read after 72 hours) | test period: 1992–1995 | IVDK 1997 |
| 124 patients | 0.1 % (petrolatum) | 0/124 (no other details) | anamnesis: exposure to acrylates | Kanerva *et al.* 1995 |
| 1 employee in the production of printed circuit boards | 0.1 % (petrolatum) | positive (1+) | UV-curing inks / patient with conjunctivitis (aerogeneous contact dermatitis?); patch test yielded same result three times | Kanerva *et al.* 1998 |
| 23 patients | 0.1 % (petrolatum) | 1/23 (3+) | anamnesis: exposure to acrylates / reaction in 1 textile worker | Kiec-Swiercynska 1996 |

**Table 1.** continued

| Persons tested | Concentration (vehicle) | Results | Contact / comments | References |
|---|---|---|---|---|
| 18 patients | 0.5 % (petrolatum) | 6/18 (no other details) | reactions in 4/10 patients with artificial finger nails / also irritative reaction possible | Koppula *et al.* 1995 |
| 1 printer, 1 mechanic | 0.1 % (petrolatum) | 2/2 (3+ read after 96 hours) | UV-curing printing ink / 20 control persons did not react | Malten and Seuter 1984 |
| 1 employee in the production of optical fibres | 0.1 % (petrolatum) | positive (2+ after 48 hours; 1+ after 96 hours) | UV-curing acrylate plastic / possible irritative reaction | Maurice and Rycroft 1986 |
| 18 employees with contact dermatitis (15) or conjunctivitis (3) | 0.1 % (petrolatum) | 2/18 (2+) | UV-curing inks; in both persons also 3+ reactions to pentaerythritol triacrylate / 15 control persons did not react | Nethercott 1978 |
| 7 employees with contact dermatitis | 0.1 % (petrolatum) | 1/7 (2+ read after 48 and 96 hours) | UV-curing printing inks for plastic containers / 20 control persons did not react | Nethercott *et al.* 1983 |
| 21 employees in the printing trade with contact dermatitis | 0.1 % (petrolatum) | 1/21 (1+) | UV-curing inks for producing printing plates | Nethercott and Nosal 1986 |
| 2 machine operators of a printing press | 0.1 % (2-butanone) | 2/2 (both 1+) | UV-curing inks containing pentaerythritol triacrylate / incompletely documented | Smith 1977 |
| 5 patients | 0.2 % (petrolatum) | 3/5 (no other details) | artificial finger nails, printing inks | Taylor 1989 |

There are no data available for allergenic or irritative effects on the respiratory passages.

## 1.2 Animal studies

Hartley guinea pigs (number of animals not specified) treated once epicutaneously on the ear with 50 µl 1 M trimethylolpropane triacrylate in acetone:olive oil (1:1) produced—on provocation after 7 and 14 days—no marked reactions to 20 µl 0.25 % or 0.5 % trimethylolpropane triacrylate in acetone:olive oil (4:1). The weight of the homolateral auricular lymph nodes of 5 animals was increased on days 4 and 6 after the induction treatment. In the lymph nodes the number of large pyroninophilic cells was around twice that found in the control animals, and histological examination revealed marked lymphocyte proliferation (Bull *et al.* 1985).

With a modified Draize test and with various test protocols using adjuvant (maximization test, Polak test) sensitizing effects were also detected in the animals (see Table 2). An undocumented Bühler test yielded negative results, however (Andrews and Clary 1986).

Evidence of cross-reactions was found in a maximization test with trimethylolpropane triacrylate (Björkner 1980; see also Table 2), in which 12/24 guinea pigs reacted to 0.1 % pentaerythritol triacrylate on provocation, but only 6/24 animals to the trimethylolpropane triacrylate used for induction.

## 2 Manifesto (sensitization)

The sensitizing effect of trimethylolpropane triacrylate in man has been demonstrated in several cases of contact dermatitis, some of which were work-related. These findings are supported by several positive results in experiments with animals both with and without the use of adjuvant. There are no data available for sensitizing effects on the respiratory passages.

Trimethylolpropane triacrylate is therefore designated with "Sh".

## 3 References


Andrews LS, Clary JJ (1986) Review of the toxicity of multifunctional acrylates. *J Toxicol Environ Health 19*: 149–164

Björkner B (1980) Allergenicity of trimethylol propane triacrylate in ultraviolet curing inks in the guinea pig. *Acta Derm Venereol 60*: 528–531

Björkner B, Dahlquist I, Fregert S (1980) Allergic contact dermatitis from acrylates in ultraviolet curing inks. *Contact Dermatitis 6*: 405–409

Bull JE, Parker D, Turk JL (1985) Predictive value of assessment of lymph node weight and T-lymphocyte proliferation in contact sensitivity in acrylates. *J Invest Dermatol 85*: 403–406

Cofield BG, Storrs FJ, Strawn CB (1985) Contact allergy to aziridine paint hardener. *Arch Dermatol 121*: 373–376

Dahlquist I, Fregert S, Trulson L (1983) Contact allergy to trimethylolpropane triacrylate (TMPTA) in an aziridine plastic hardener. *Contact Dermatitis 9*: 122–124

Emmett EA (1977) Contact dermatitis from polyfunctional acrylic monomers. *Contact Dermatitis 3*: 245–248

Emmett EA, Kominsky JR (1977) Allergic contact dermatitis from ultraviolet cured inks. *J Occup Med 19*: 113–115

Emmett EA, Taphorn BR, Kominsky JR (1977) Phototoxicity occurring during the manufacture of ultraviolet-cured ink. *Arch Dermatol 113*: 770–775

**Table 2.** Sensitizing effects of trimethylolpropane triacrylate (TMPTA) in the guinea pig

| Method, strain, sex | **Intradermal induction** injection volume, concentration (vehicle); amount/animal | **Epicutaneous induction** day, application, concentration (vehicle) | **Provocation** day, application, concentration (vehicle) | Number of animals which produced reactions | Comments | References |
|---|---|---|---|---|---|---|
| maximization test, (no other details) | injections in duplicate: 100 µl of TMPTA, TMPTA in FCA, and FCA, 1 % (olive oil); 3 mg | day 7, 48 hours occlusive, 25 % (petrolatum) | day 21, 24 hours occlusive, 0.1 % or 0.5 % (petrolatum), repeated on day 28 with 0.5 % (petrolatum) | day 21: **6/24** and **16/24** day 28: **7/24** | on day 21 none of 24 controls pretreated with FCA showed reactions to the substance, on day 28, however, 10/24 reacted to 0.5 % | Björkner 1980 |
| maximization test, Hartley-Dalkin, ♀ | injections in duplicate: 100 µl of TMPTA, TMPTA in FCA, and FCA, 0.5 % or 10 % (propylene glycol); 1.5 or 30 mg | day 7, 48 hours occlusive, 0.5 % or 10 % (petrolatum) | day 21, 24 hours occlusive, non-irritative concentration (no other details) (petrolatum) | 0.5 %: **4/10** 10 %: **10/20** | 10 control animals not pretreated with FCA used to determine the maximum non-irritative concentration | Nethercott *et al.* 1983 |
| Polak test, Hartley, ♂ and ♀ | subcutaneous doses of 0.1 ml in each paw and in the neck area, about 0.7 % (physiological NaCl:ethanol 4:1 in FCA); about 3.5 mg | – | days 7 and 14, 20 µl non-occlusive, 0.25 % or 0.5 % (acetone:olive oil 4:1) | mean reaction score on day 7: 1.5 and 1.6; day 14: 1.4 and 1.9 | no. of animals which reacted not specified, 5 animals used | Bull *et al.* 1985 |
| Polak test, Hartley, ♂ and ♀ | subcutaneous doses of 0.1 ml in each paw and in the neck area, 0.2 % (physiological NaCl:ethanol 4:1 in FCA); 1 mg | – | day 7 after induction, 20 µl non-occlusive, 0.1 %, 0.25 % or 0.5 % (acetone:olive oil 4:1) | mean reaction score: 0.7, 1.4 and 1.9 | no. of animals which reacted not specified, 11 animals used | Parker and Turk 1983, Parker *et al.* 1985 |
| modified Draize test, Hartley, ♂ and ♀ | – | 10 non-occlusive doses of 0.1 ml, applied on days 0–4 and 7–11, 0.5 % (acetone:olive oil, 1:1) | from day 21 one non-occlusive dose/week of 20 µl, 0.25 % or 0.5 % (acetone:olive oil 4:1) | **6/6** | results incompletely documented, provocation concentration not specified | Parker and Turk 1983 |

Fischer T, Nylander-French L, Rosén G (1994) Dermatologic risk to workers in the UV-radiation wood surface-coating industry. *Am J Contact Dermatitis 5*: 201–206

Garabrant DH (1985) Dermatitis from aziridine hardener in printing ink. *Contact Dermatitis 12*: 209–212

IVDK (Informationsverbund Dermatologischer Kliniken) (1997) *Auswertung der zwischen 1992 und 1995 vom IVDK erfaßten Daten zu Trimethylolpropantriacrylat*, 16.01.1997

Kanerva L, Estlander T, Jolanki R, Tarvainen K (1995) Statistics on allergic patch test reactions caused by acrylate compounds, including data on ethyl methacrylate. *Am J Contact Dermatitis 6*: 75–77

Kanerva L, Tarvainen K, Jolanki R, Henriks-Eckerman M-L, Estlander T (1998) Airborne occupational allergic contact dermatitis due to trimethylolpropane triacrylate (TMPTA) used in the manufacture of printed circuit boards. *Contact Dermatitis 38*: 292–294

Kiec-Swierczynska M (1996) Occupational allergic contact dermatitis due to acrylates in Lodz. *Contact Dermatitis 34*: 419–422

Koppula SV, Fellman JH, Storrs FJ (1995) Screening allergens for acrylate dermatitis associated with artificial nails. *Am J Contact Derm 6*: 78–85

Malten KE, Seutter E (1984) Contact dermatitis from acrylated resins in UV Ebecryl printing inks. *Contact Dermatitis 11*: 56–58

Maurice PDL, Rycroft RJG (1986) Allergic contact dermatitis from UV-curing acrylate in the manufacture of optical fibres. *Contact Dermatitis 15*: 92–93

Nethercott JR (1978) Skin problems associated with multifunctional acrylic monomers in ultraviolet curing inks. *Br J Dermatol 98*: 541–552

Nethercott JR, Jakubovic HR, Pilger C, Smith JW (1983) Allergic contact dermatitis due to urethane acrylate in ultraviolet cured inks. *Br J Ind Med 40*: 241–250

Nethercott JR, Nosal R (1986) Contact dermatitis in printing tradesmen. *Contact Dermatitis 14*: 280–287

Parker D, Turk JL (1983) Contact sensitivity to acrylate compounds in guinea pigs. *Contact Dermatitis 9*: 55–60

Parker D, Long PV, Bull JE, Turk JL (1985) Epicutaneous induction of tolerance with acrylates and related compounds. *Contact Dermatitis 12*: 146–154

Smith WDL (1977) Allergic dermatitis due to a triacrylate in ultraviolet cured inks. *Contact Dermatitis 3*: 312–314

Taylor JS (1989) Acrylic reactions – ten years' experience. in: Frosch PJ, Dooms-Goossens A, Lachapelle JM (Eds) *Current topics in contact dermatitis*, Springer, Berlin, 346–351

completed 29.01.1998

# Metal-working fluid components

# Citric acid* and its alkali metal salts

| | |
|---|---|
| **MAK value** | **not yet established, see Section IIb of the *List of MAK and BAT Values*** |
| **Peak limitation** | **–** |
| **Absorption through the skin** | **–** |
| **Sensitization** | **–** |
| **Carcinogenicity** | **–** |
| **Prenatal toxicity** | **–** |
| **Germ cell mutagenicity** | **–** |
| **BAT value** | **–** |
| Synonyms | α-hydroxypropanetricarboxylic acid |
| Chemical name (CAS) | 2-hydroxy-1,2,3-propanetricarboxylic acid |
| CAS number | 77-92-9 (anhydrous) |
| | 5949-29-1 (monohydrate) |
| Structural formula | $CH_2$–COOH<br>HO–C–COOH<br>$CH_2$–COOH |
| Molecular formula | $C_6H_8O_7$ |
| Molecular weight | 192.13 (anhydrous) |
| Melting point | 151–157°C (monohydrate) |
| Boiling point | no data |
| Decomposition point | > 175°C |
| pK values | $pK_1$: 3.128, $pK_2$: 4.761, $pK_3$: 6.396 (25°C, Bates and Pinching 1949) |
| pH (20°C, 0.1 % in water) | 2.8 (Blair and Staal 1993) |
| Purity, impurities | no data |

* Unless otherwise specified, the data given in the present document were obtained with the anhydrous acid.

Solubility (% w/w) water: 60 (20°C) (Gebr. Overlack 1977), ethanol: 38.3 (25°C), amyl acetate: 4.2 (25°C), diethyl ether: 1.0 (25°C) (Blair and Staal 1993), propanol: 62.8 (20°C) (Windholz 1989); monohydrate of the free acid: very readily soluble in water, ethanol (Astle and Weast 1979) and amyl alcohol (15.43) (Verhoff 1986), soluble in ether (2.17), amyl acetate (5.98) and ethyl acetate (5.28), insoluble in chloroform (< 0.01) (Verhoff 1986), benzene and toluene (Astle and Weast 1979)

Stability readily oxidized to yield oxalic acid, acetone dicarboxylic acid, $CO_2$ and water; thermal decomposition yields aconitic, citraconic, itaconic and acetone dicarboxylic acids, acetone, $CO_2$ and water (Beyer and Walter 1981, Blair and Staal 1993)

Production by fungal fermentation of carbohydrates, e.g. by *Aspergillus niger* (Blair and Staal 1993) or from hydrocarbons by yeasts such as *Candida* spp. (Verhoff 1986, Yokotani *et al.* 1971)

Uses

- in metal-working fluids (no concentration data) (Hüner *et al.* 1994) and in metal polishes (Guest and Katz 1994)
- in the washing powder and paint industries (Blair and Staal 1993, Guest and Katz 1994, Windholz 1989)
- in electroplating (Guest and Katz 1994), as a chelating agent and stabiliser for emulsions (Blair and Staal 1993), in analytical chemistry and treatment of sewage (Windholz 1989)
- in the pharmaceutical industry, e.g. in medicines (Guest and Katz 1994, Windholz 1989).
- in medicine, e.g. as an anticoagulant, for storage of blood, in electrolyte and buffer solutions, as a food additive (FAO 1974, Gebr. Overlack 1977)
- in cosmetics (≤ 1 %) and topical medications (about 5–50 %) (Köhler and Refior 1997, Röding and Artmann 1995)

## Alkali metal salts

| Name | CAS number | Molecular formula | Molecular weight | pH |
|---|---|---|---|---|
| monosodium citrate | 18996-35-5 | $C_6H_7NaO_7$ | 214.12 | not specified |
| disodium citrate | 144-33-2 | $C_6H_6Na_2O_7$ | 236.10 | 4.9–5.2 (3 % in $H_2O$) |
| trisodium citrate | 68-04-2 | $C_6H_5Na_3O_7$ | 258.08 | about 8 |
| tripotassium citrate | 866-84-2 | $C_6H_5K_3O_7$ | 306.41 | about 8.5 |

Citric acid is an intermediate in the citric acid or Krebs cycle in human metabolism (estimated turnover 100–2000 g/day) (Rechenberger and Benndorf 1957), a natural component of foodstuffs and one of the most used food additives. The average daily intake of citric acid or its salts from natural sources and as a food additive can exceed 400 mg/kg body weight and day (FAO 1974, Wright and Hughes 1975).

With multivalent ions, citric acid forms very stable chelate complexes of which the complex with trivalent iron is the most stable. The chelates formed with divalent ions such as calcium and magnesium are readily soluble in water because of the free third carboxyl group.

# 1 Toxic Effects and Mode of Action

Citric acid is absorbed from the gastrointestinal tract. Metabolism of the substance involves the citric acid cycle, the synthesis of fats and amino acids, and gluconeogenesis. Citrate is mostly excreted in the urine after conversion to bicarbonate and only a small percentage is excreted unchanged.

The acute toxicity of citric acid is very low. After long-term feeding of citric acid (together with an adequate calcium supply) to experimental animals in daily doses exceeding those normally ingested by man, no signs of toxicity were observed.

The irritant effects of citric acid on the skin depend on the pH of the applied solution and are comparable with those of other similar acids. In the rabbit eye, undiluted anhydrous citric acid is severely irritating or corrosive.

There is no evidence that citric acid can cause sensitization.

The available *in vitro* mutagenicity tests provide no evidence of genotoxic effects of citric acid or its sodium salts. Administered to rats, high oral doses of sodium citrate (but not of citric acid) act as promoter for bladder tumours induced by nitrosamines.

# 2 Mechanism of Action

Citric acid forms stable chelate complexes with metal ions. Of particular physiological significance are the water-soluble chelates formed with calcium and magnesium ions because of the importance of these ions for numerous body functions such as excitation and signal transmission, electromechanical coupling, membrane maintenance and blood clotting. Therefore, under certain conditions, a reduction in the plasma concentration of free $Ca^{2+}$ and $Mg^{2+}$ can result in so-called "citrate intoxication" with the neuromuscular and cardiovascular symptoms of hypocalcaemia. Because citric acid is metabolized rapidly and distributed into both the intracellular and extracellular fluid space, citrate intoxication is observed only after rapid intravenous application of very large amounts of the substance, for example after massive transfusion of blood containing citrate buffer or plasmapheresis when at the same time the release of $Ca^{2+}$ from the body's own depots or the metabolism and distribution of citrate was restricted (see Section 4.2.2). Factors other than the reduced level of $Ca^{2+}$ in plasma are probably responsible for abnormalities of blood clotting described after massive transfusions of blood containing citric acid or during plasmapheresis.

Secondary effects which can be caused by the effect of citric acid on the toxicokinetics of other metal ions such as aluminium or lead are not discussed in the present document.

The action of sodium citrate as a promoter of bladder tumours in the rat appears to be a result of the stimulation of DNA synthesis in the bladder epithelial cells as a result of the increased sodium concentration or pH of the rat urine.

# 3 Toxicokinetics and Metabolism

The numerous, mainly early animal studies (reviewed by Dzik and Kirkley 1988, Nordmann and Nordmann 1961) are mostly not reviewed here because there are sufficient data available for man. The concentration data given below from the different studies are often not directly comparable because of differences in the analytical methods used (Dzik and Kirkley 1988).

## 3.1 Absorption, distribution, elimination

The physiological plasma levels of citric acid in adults of middle age range from 9 to 25 mg/l (about 47–130 µM) (citrate-lyase method) (Dzik and Kirkley 1988). During plasmapheresis or massive blood transfusions, the plasma concentration can increase to about 100 times this value (Dzik and Kirkley 1988).

In the body, citric acid is taken up by the cells, distributed extracellularly and stored in the bones (Bunker *et al.* 1955, Denlinger *et al.* 1976). About 70 % of the citrate in the body is bound to calcium in the bones (about one to a few percent of the bone weight) (Dickens 1941, Howland *et al.* 1957, Lussier 1957, Saffran and Denstedt 1951).

In the cat, rabbit and rat given the substance by intravenous injection, elimination obeys first order kinetics (Mårtensson 1940, Saffran and Denstedt 1951). The plasma half-life is given as 15 to 30 minutes in the rat and cat (Mårtensson 1940), and about 30 to 60 minutes in the rabbit (Saffran and Denstedt 1951). In man, one hour after the beginning of a 20-minute intravenous infusion of a citric acid dose of 8.6 or 1 mg/kg body weight and minute, the plasma concentrations in the persons had returned to their initial values (Howland *et al.* 1957, Ludbrook and Wynn 1958). After longer term infusion of larger doses of citric acid, the complete elimination of the administered substance can take several hours (Abbott 1983, Driscoll *et al.* 1987) as re-equilibration out of the large distribution volume, and to a lesser extent from the bones, takes time (Denlinger *et al.* 1976).

Citric acid is mainly excreted via the liver and kidneys after complete metabolic conversion to bicarbonate (Dzik and Kirkley 1988, Mårtensson 1940, Simpson 1983). Only low levels of unchanged citric acid are excreted in human urine, about 200 to 1500 mg per day or 10 % to 35 % of the glomerular filtrate (Simpson 1983). Other authors report that after infusion of sodium citrate into one test person, urinary excretion of 18 % to 20 % was found (no other details) (Ludbrook and Wynn 1958). The excretion in the urine is affected by numerous factors. For example, metabolic alkalosis, or high concentrations of citrate or bicarbonate in the plasma result in drastic increases in citrate excretion in the urine. Acidosis reduces urinary excretion (Dzik and Kirkley 1988, Simpson 1983).

There are no studies available of the cutaneous absorption of citric acid.

## 3.2 Metabolism

Citric acid is rapidly oxidized to $CO_2$ and water in the citric acid (Krebs) cycle in the cells of all the organs of the body, but especially in those with high concentrations of mitochondria and in well-vascularized organs. Therefore, the liver, renal cortex and skeletal muscles are largely involved in the metabolism and elimination of citric acid. In addition, citric acid is used in the synthesis of fatty acids and amino acids and in gluconeogenesis (Baruch *et al.* 1975, Dzik and Kirkley 1988).

# 4 Effects in Man

## 4.1 Single exposures

There are no data available for the effects of single doses of citric acid after ingestion or absorption through the skin.

### 4.1.1 Inhalation

The only available data are from pharmacological studies of the effects of antitussives on coughing induced with citric acid and mechanistic studies of the coughing reflex.

In a study of the effects of antitussives, the inhalation of nebulized citric acid solutions (25 % aqueous citric acid solution, flow rate 5 l per minute, corresponding to a citric acid concentration of 0.045 mg per 200 ml air as a mist with a mean particle size of 2.5 µm in radius) induced a reproducible cough in 15 healthy test persons (9 men and 6 women). The tests involved a series of three inhalations, 2 to 3 minutes apart, repeated 8 times at weekly intervals. In the control groups treated with water or physiological saline, no effects were observed (Bickerman *et al.* 1956). 11 adult non-smokers were exposed to aerosols of citric acid at different inspiratory flow rates. The lowest concentration of citric acid which produced coughing was 21 mg/l water at an inspiratory flow rate of 50 l per minute, 31 mg/l at 100 l per minute and 43 mg/l at 150 l per minute (corresponding to inhaled doses of 5.2, 6.6 and 11.6 mg/person). There was considerable interindividual variation in this cough threshold. Forced expiratory volume in the first second ($FEV_1$), forced vital capacity (FVC) and peak expiratory flow rate (PEFR) were not affected (Barros *et al.* 1990). The effects of respiratory disease and atopy on the cough threshold were studied in healthy school children (n = 17) and in children with idiopathic coughing which was not caused by virus infections in the upper respiratory tract (n = 16) or asthma (n = 13). Children aged five years and, in a later study, seven years (n = 23, 13 and 10) were exposed for 30 seconds to aerosols containing citric acid in concentrations of 5.2 to 675 mg/ml 0.9 % saline (pH 2.5 to 0.25; flow rate 10 l/minute, mass median aerodynamic diameter 2.65 μm). The threshold concentration for induction of coughing in healthy five-year-olds was about 50 mg/ml and slightly lower in seven-year-olds. The lowest threshold was found in children with coughs, the highest in healthy children. However, the differences were not statistically significant. The individual thresholds differed by a factor of up to 80 and there was no correlation with respiratory symptoms, bronchial reactivity, smoking habits of the parents or atopy (Riordan *et al.* 1994).

### 4.1.2 Intravenous injection

Massive citrate intoxication with neuromuscular and cardiovascular symptoms and some deaths has been described in numerous studies and case reports of patients with massive transfusions or plasmapheresis (Collins 1974, Dzik and Kirkley 1988, Howland *et al.* 1957, Köhler 1993, McLellan *et al.* 1984, Perdrix *et al.* 1988). The reason for the intoxications seen during massive transfusions in babies, hypothermic patients, patients with circulation disorders or with functional impairment of liver, kidneys or adrenal glands seemed to be too high infusion rates, as in these patients the release of $Ca^{2+}$ and $Mg^{2+}$ from the depots in the body was impaired and citrate metabolism and distribution were restricted.

## 4.2 Repeated exposures

There are no data for the effects of repeated inhalation of citric acid nor are there any reports of the reproductive toxicity, genotoxicity or carcinogenicity of the substance in man.

### 4.2.1 Ingestion

Citric acid is a natural component of many foodstuffs and is one of the most used food additives (Wright and Hughes 1975). A value for the ADI (acceptable daily intake) of citric acid or its sodium, potassium and calcium salts has not been established because the toxicity is considered to be very low (FAO 1974).

### 4.2.2 Dermal absorption

Citric acid and its salts are used in cosmetics and topical medications. There are no reports of systemic-toxic effects of citric acid used in this way.

## 4.3 Local effects on skin and mucous membranes

### 4.3.1 Skin

In a mechanistic study of the effects of α-hydroxyacids on skin chronically damaged by light, 7 patients applied a lotion containing 25 % citric acid (pH 3.5) to one forearm and a placebo lotion to the other forearm twice daily for 6 months. Inflammatory reactions were not visible nor detected in the histological examination (Ditre *et al.* 1996).

In early publications about chronic toxic eczema in the citrus fruit processing industry in the U.S.A., irritating effects of citrus fruit juices during juice production are described. Lemon juice with its relatively high level of citric acid (about 7 % compared with about 1 % in orange juice and about 2 % in grapefruit juice) was said to be most irritating. However, it should be remembered when assessing these reports that the skin of the workers' hands was chronically damaged as a result, for example, of working under constantly wet conditions, frequent cuts and fungal infections (Birmingham *et al.* 1951, Schwartz 1938).

Mucosal irritation in the upper respiratory tract was reported in persons who had inhaled aerosols of undiluted citric acid in a study of salivary reflexes (no other details) (Lee and Linden 1991).

### 4.3.2 Eyes

Eye irritation must be expected to be correlated with the pH of the solution.

## 4.4 Allergenic effects

Consideration of the chemical structure and the physiological properties of citric acid indicates that contact sensitization by the acid or its alkali metal salts is not to be expected.

Patch tests with 2 % aqueous solutions of sodium citrate carried out on 4 patients with contact allergy to propacetamol hydrochloride solutions containing sodium citrate and on 102 control persons (no other details) yielded negative results (Szczurko *et al.* 1996).

In a Danish study, the tolerance of food additives by atopic and non-atopic school children was investigated. For 14 days before the beginning of the study, all the children were given an additive-free diet. Of the 98 children without a history of atopy, none produced a reaction to the administered lemonade which contained not only citric acid (40 or 60 mg/100 ml) but also a mixture of preservatives, synthetic and natural colours and flavours. One tenth of the dose had been administered three hours earlier. The doses were equivalent to the average daily adult intake in mg/kg body weight extrapolated to 30 kg body weight or to one tenth of this amount. Of the 173 children with a history of atopy, 17 produced positive results in the lemonade test. In the subsequent double blind study with citric acid in opaque gelatine capsules coloured with iron oxide, 1/12 atopic children reacted with eczema (5 children had refused to take part in the second test). The children were checked for symptoms such as asthma, dermatitis, gastrointestinal symptoms, urticaria or rhinitis at one hour intervals for 3 hours by a doctor and after that by the parents (Fuglsang *et al.* 1993).

An early study (Birmingham *et al.* 1951) of numerous patients with dermatitis who were employed in the citrus fruit processing industry in the U.S.A. revealed no evidence of sensitization to grapefruit or orange juice containing up to 2 % citric acid (Wright and Hughes 1975).

Whether the rather rare allergic reactions to citrus fruit juices which have been reported in persons handling or eating citrus fruits and which have been confirmed by sometimes extensive tests to exclude false positive results from reactions to the limonene in the skin of the fruits are really reactions to citric acid has not been demonstrated because the tests were carried out only with juice and not with citric acid solutions (Beerman *et al.* 1938, Janson 1953, Schwartz 1938, Wolfram 1934).

# 5 Animal Experiments and *in vitro* Studies

## 5.1 Acute toxicity

There are no data available for the effects of citric acid after cutaneous absorption.

### 5.1.1 Inhalation

There are no studies of the toxicity of inhaled citric acid.

In pharmacological studies of the effects of antitussives or of the mechanisms of cough induction, guinea pigs were exposed to citric acid aerosols in exposure chambers. The concentrations of the citric acid solutions used to prepare the aerosols were between 2.5 % and 15 %. The concentrations in the air were not determined. Generally the animals were exposed for a few minutes. In addition to rapidly induced coughing, other effects frequently observed included an increase in respiratory volume, bronchoconstriction, reduced respiration rate and increased heart rate (Bolser *et al.* 1991, Callaway and King 1992, Clay and Thompson 1985, Forsberg and Karlsson 1986, Forsberg *et al.* 1988, Lalloo *et al.* 1995, Satoh *et al.* 1993). Whereas the frequency of coughing varied widely within the group of exposed animals ($15.51 \pm 8.02$, $n = 77$), the individual reaction remained constant when the test was repeated at longer intervals (Callaway and King 1992, Clay and Thompson 1985). Pretreatment of the animals with capsazepine, theophylline, papaverine, isoproterenol or ruthenium red inhibited the induction of coughing in a dose-dependent manner (Clay and Thompson 1985, Forsberg and Karlsson 1986, Lou and Lundberg 1992). In another study of groups of adult female guinea pigs (n = 6 or 8) exposed for 30 minutes to citric acid aerosols in concentrations of 31.1 or 81.0 mg/m$^3$ (mass median aerodynamic diameter 4.2 ± 2.8 μm) in an exposure chamber, only in the high dose group was an increase in the frequency of coughing observed, with a maximum 5 to 10 minutes *post applicationem*. Citric acid was about one tenth as effective as sodium lauryl sulfate when the two substances were applied as aerosols with similarly sized particles (Zelenak *et al.* 1982). Exposure to 0.5 M sodium citrate solutions, unlike exposure to the same concentration of citric acid, (mass median aerodynamic diameter in both cases 0.9 μm) did not induce coughing (Lalloo *et al.* 1995). Studies of the mechanism of the persistent hyperreactivity induced by citric acid in the airways of the dog (Hirshman *et al.* 1986, Parker *et al.* 1991) and rat (Hamel and Ford-Hutchinson 1985) are not included in the present assessment because the observed effects appear to be specific for these dog and rat strains.

### 5.1.2 Ingestion

Ingested citric acid is of low toxicity for rodents. The $LD_{50}$ is given as 11700 mg/kg body weight for the rat and as 5790 mg/kg body weight for the mouse. In rats and mice which had ingested lethal doses of purified citric acid, symptoms of intoxication included an increase in spontaneous movement a few minutes *post applicationem* and from 50 minutes *post applicationem*, ataxia, lateral position, mydriasis, reduced respiration rate and heart rate. The cause of death was respiratory failure 20 to 80 minutes *post applicationem*, in a few cases not until after 20 hours. Survivors recovered within 24 hours without permanent damage. Autopsy revealed haemorrhage of the stomach mucosa but no changes in other organs (no other details) (Yokotani *et al.* 1971).

A study of toxicity for the rabbit (Weiss *et al.* 1923) is not included here because the number of animals used was too small.

### 5.1.3 Intraperitoneal injection

After intraperitoneal injection, citric acid and its sodium salts are moderately toxic for rodents, the $LD_{50}$ ranging from 375 to 1770 mg/kg body weight (Table 1).

### 5.1.4 Intravenous injection

After intravenous injection, citric acid and its sodium salts are toxic for the mouse, and after infusion moderately toxic for the rabbit (Table 2). Intravenous injection or infusion of citric acid into mice, rats, rabbits (Saffran and Denstedt 1951) and dogs (Baue *et al.* 1961, Bunker *et al.* 1962, Corbascio and Smith 1967, Drop and Scheidegger 1979, Nakasone *et al.* 1954) resulted in the symptoms of hypocalcaemia. Surviving animals recovered within minutes. Gross pathological examination revealed no effects on the organs. Whereas after slow intravenous injection into rabbits, practically no differences were seen in the toxicity of free citric acid and that of its monosodium, disodium and trisodium salts ($LD_{100}$ 2.08, 2.01, 2.21 and 2.24 mmol/kg body weight, respectively), after rapid injection into the rat, free citric acid, probably because it caused an acute decrease in blood pH, was more toxic than the trisodium salt. Infusion of free citric acid into dogs caused a progressive reduction in blood pressure and death. When sodium citrate was administered in a similar way, the blood pressure remained unchanged until shortly before the animals died (Gruber and Halbeisen 1948).

**Table 1.** Acute intraperitoneal toxicity of citric acid and its sodium salts

| Species | Substance | $LD_{50}$ (mg/kg body weight) | References |
|---|---|---|---|
| rat | citric acid, monohydrate | 375 | NIOSH 1994 |
| | citric acid, monohydrate | 975 | Gebr. Overlack 1977 |
| | citric acid, monohydrate | 725 | Yokotani *et al.* 1971 |
| | citric acid, monohydrate | 884 (4.6 mmol/kg body weight) | Gruber and Halbeisen 1948 |
| | monosodium salt | 1349 (6.3 mmol/kg body weight) | |
| | disodium salt | 1723 (7.3 mmol/kg body weight) | |
| | trisodium salt | 1548 (6.0 mmol/kg body weight) | |
| mouse | citric acid, anhydrous | 787 (4.1 mmol/kg body weight) | Llobet *et al.* 1987 |
| | citric acid, monohydrate | 940[a] and 960[b] | Yokotani *et al.* 1971 |
| | citric acid, monohydrate | 961 (5.0 mmol/kg body weight) | Gruber and Halbeisen 1948 |
| | monosodium salt | 1627 (7.6 mmol/kg body weight) | |
| | disodium salt | 1770 (7.5 mmol/kg body weight) | |
| | trisodium salt | 1419 (5.5 mmol/kg body weight) | |

[a] purity 99.8 %
[b] commercially available substance (no other information)

**Table 2.** Acute intravenous toxicity of citric acid and its sodium salts

| Species | Substance | $LD_{50}$ (mg/kg body weight) | References |
|---|---|---|---|
| mouse[a] | citric acid | 42 (0.22 mmol/kg body weight) | Gruber and Halbeisen 1948 |
| | monosodium salt | 44 (0.23 mmol/kg body weight) | |
| | disodium salt | 58 (0.30 mmol/kg body weight) | |
| | trisodium salt | 127 (0.66 mmol/kg body weight) | |
| rabbit[b] | citric acid | 330 (1.72 mmol/kg body weight) | Gruber and Halbeisen 1948 |
| | monosodium salt | 377 (1.76 mmol/kg body weight) | |
| | disodium salt | 418 (1.77 mmol/kg body weight) | |
| | trisodium salt | 449 (1.74 mmol/kg body weight) | |

[a] rapid injection; [b] injection rate 0.358 mmol/minute

### 5.1.5 Local effects

In experimental studies of the dissolution of kidney stones in rabbits, the urinary bladder was irrigated for 24 hours with 0.1 M citric acid/citrate buffer in isotonic saline (pH 4, catheter in the urethra, flow rate equivalent to maximum flow of urine) or, in the controls, with isotonic saline; histological examination of the bladder mucosa revealed marked damage in the treated animals. Alterations in the mucosa of the urethra were also seen in the control group (Reckler *et al.* 1986).

### 5.1.6 Subcutaneous injection

Subcutaneous citric acid is of low toxicity for rodents. The $LD_{50}$ is given as 5500 mg/kg body weight for the rat and 2700 mg/kg body weight for the mouse (Yokotani *et al.* 1971). The symptoms were the same as seen after oral and intraperitoneal administration; death from respiratory failure or emaciation occurred within 15 minutes to 6 days after treatment. In the high dose groups, the skin of the injection site was necrotic and inflamed, at lower doses correspondingly milder inflammation was observed (Yokotani *et al.* 1971).

## 5.2 Subacute, subchronic and chronic toxicity

There are no data available for the toxicity of citric acid administered by medium to long term inhalation or dermal application nor for sensitization by the substance.

### 5.2.1 Ingestion

Rats were given a diet containing up to 5 % citric acid for up to 200 days (about 200–5000 mg/kg body weight and day) and guinea pigs a diet containing 1 % to 5 % citric acid (about 670–3340 mg/kg body weight and day). In animals with an adequate calcium supply, the only adverse effects were a reduction in the haematocrit and the haemoglobin level in the serum (Wright and Hughes 1976b). Neither phosphorus metabolism, the organs examined (Bonting and Jansen 1956) nor the mineral content of the bones (Wright and Hughes 1976b) were affected (Table 3).

In a 6-week feeding study with male SD-JCL rats given citric acid doses less than or equal to 4670 mg/kg body weight and day, a slight tendency to reduction in food consumption and body weight gains was observed. Whereas the absolute weights of brain and thymus were reduced in the high dose group and the weight of the prostate was increased in the lowest dose group, the relative weights of brain, heart, lungs, liver, kidneys, spleen, testes, prostate, adrenal glands, thyroid gland, hypophysis and thymus in all three dose groups were comparable with the control values. The effects on blood counts and urine parameters were slight; the slight decrease in the blood calcium level was not dose-dependent. In the high dose group, microhaematuria was observed. Autopsy revealed no dose-dependent organ changes which could be attributed to the citric acid treatment (Yokotani *et al.* 1971).

### 5.2.2 Subcutaneous injection

Subcutaneous injections of 7 to 60 ml/kg body weight of a 4 % aqueous sodium citrate solution (corresponds to citric acid doses of 280–2400 mg/kg body weight) daily for a period of up to 5 days increased renal calcium excretion and caused bone alterations like those seen after high doses of parathyroid hormone in young dogs and rats (Gomori and Gulyas 1944).

## 5.3 Local effects on skin and mucous membranes

### 5.3.1 Skin

Applied to intact rabbit skin in the Draize test, 50 % aqueous citric acid was not irritating. The reaction was read 4 hours after occlusive application of the substance to the shaved dorsal skin and also 48 hours *post applicationem* (Miles 1979). When the substance was applied to intact rabbit skin in the form of anhydrous citric acid crystals (0.5 g under damp cotton wool) under conditions otherwise identical to the above, weak irritation was recorded (Draize test) (Miles 1990) (Table 4).

### 5.3.2 Eyes

Applied to the rabbit eye without subsequent rinsing, anhydrous citric acid crystals were severely irritating or corrosive (NIOSH 1994, Miles 1991). A 5 % solution of the monohydrate (pH 2.1) was a moderate irritant whether or not the rabbit eye was rinsed (observation period 7 days; Murphy *et al.* 1982) (Table 5).

**Table 3.** Subacute, subchronic and chronic toxicity of ingested citric acid and its sodium salt

| Species, strain number per dose, sex | Dose or concentration in the diet, application form; duration | **Dose** (mg citric acid per kg body weight and day): symptoms | References |
|---|---|---|---|
| **rat** | | | |
| F344<br>20 ♂, 6 weeks old | 5 % sodium citrate (about 3.9 % citric acid, about 1950 mg/kg body weight and day[c]); 32 weeks | **about 1950**: not lethal,<br>urine: increased pH, $MgNH_4PO_4$ crystals and $Na^+$<br>bladder: no substance-related findings | Fukushima *et al.* 1986 |
| F344<br>5 ♂ | 5 % sodium citrate (about 3.9 % citric acid; about 3900 mg/kg body weight and day[b,d]); 8 weeks | **about 3900**[d]: not lethal, increased water consumption;<br>urine: increased pH, volume, $MgNH_4PO_4$ crystals, $Na^+$, decreased osmolality, $K^+$; bladder: increased DNA synthesis, no other changes; forestomach: no substance-related findings | Shibata *et al.* 1992 |
| albino<br>$F_0$: 6 ♀<br>$F_1$: 16 ♂, 15 ♀ | 5% citric acid (about 2500–5000 mg/kg body weight and day[c,d]); calcium level in the diet 0.15 %; exposure *in utero* and 200 days after birth | **about 2500–5000**: only $F_1$ ♂: decreases in body weight gains, wet weights of fibula and tibia, haematocrit and haemoglobin levels | Wright and Hughes 1976b |
| albino<br>12 ♂ | 4 % citric acid (about 2000–4000 mg/kg body weight[c,d]); calcium level in the diet 0.2 %;<br>60 days | **about 2000–4000**: decreased weight gains and haematocrit, unchanged relative weights of liver, kidneys, spleen, adrenal glands, fibula and tibia, Ca levels in fibula and tibia, blood parameters (cholesterol, alkaline phosphatase and triglycerides) and citrate levels in liver and kidneys | Wright and Hughes 1976b |
| albino<br>10 ♂ | 4 % citric acid (about 2000 mg/kg body weight[c]); calcium level in the diet 0.7 %; 60 days | **about 2000**: no effects | Wright and Hughes 1976b |
| albino<br>$F_0$: 12 ♀, 15 months old<br>$F_1$: 7 ♂, 6 months old | 1.2 % citric acid (about 600 mg/kg body weight and day[c]); $F_0$ animals mated in week 32 (after 29 weeks diet) and again in week 43 | **about 600**: $F_0$ and $F_1$: no changes in blood counts, urinalysis or faeces;<br>liver: decreased $Na^+$, muscle: decreased P, increased $Na^+$, other tissues: no changes in P, $Na^+$ or $K^+$, autopsy: no abnormal findings in the liver, spleen, adrenal glands, kidneys, testes, femur or skeletal muscle in the gross pathological or histological examinations, no decalcification of the bones, no effects on the teeth | Bonting and Jansen 1956 |

**Table 3.** continued

| Species, strain number per dose, sex | Dose or concentration in the diet, application form; duration | **Dose** (mg citric acid per kg body weight and day): symptoms | References |
|---|---|---|---|
| **rat** | | | |
| SD-JCL, 5 ♂, 29–35 days old | 1.2, 2.4, 4.8 % citric acid (monohydrate)[a] (1150, 2260 and 4670 mg/ kg body weight and day); 6 weeks | **1150**: not lethal, food consumption and weight gains decreased by 0.7 % relative to control group;<br>blood: $Ca^{2+}$ level unchanged<br>**2260**: decreased plasma protein, weight gains decreased by 7.4 %, food consumption by 2.6 %<br>**from 4670**: decreased plasma albumin and cholesterol, increased serum GOT, kidneys: increased excretion of protein with increased count of lymphocytoid cells in the urine in 1/5 animals, microhaematuria in 4/5, weight gains decreased by 6 %, food consumption by 4 %<br><br>autopsy (gross pathological and histological examinations: brain, heart, lungs, liver, kidneys, spleen, testes, prostate, adrenal glands, thyroid gland, hypophysis, thymus): yellowish-white miliary tubercles on the inner surface of the right hepatic lobe (dose group not specified), changes in the size of the testes in one animal from each of the 1150 mg/kg and the 2260 mg/kg groups;<br>only 1150 mg/kg group: increase in the absolute weight of the prostate;<br>only 4670 mg/kg group: decreased thymus weight, slight atrophy of the thymus cortex and the lymph follicles in the spleen, no other histopathological changes | Yokotani *et al.* 1971 |
| **mouse** | | | |
| albino (not specified) ♂, young and adult animals | 5 % citric acid (about 7500 mg/kg body weight[e]); 14 months | **about 7500**: decreased body weights and survival, more pronounced in animals which were adult at the start of the study, no other data | Wright and Hughes 1976a |
| **rabbit** | | | |
| New Zealand White 15 ♂ | 7.7 % sodium citrate (about 5% citric acid, equivalent to about 2000–6000 mg/kg body weight); average food consumption 119 g/day; 150 days | **about 2000–6000**: urine and blood parameters (day 60) unaffected,<br>autopsy (gross pathological and histological examinations): no changes in organ weights (adrenal glands, bladder, brain, heart, kidneys, liver, lungs, prostate, spleen, stomach, testes, thyroid gland) | Packman *et al.* 1963 |

**Table 3.** continued

| Species, strain number per dose, sex | Dose or concentration in the diet, application form; duration | **Dose** (mg citric acid per kg body weight and day): symptoms | References |
|---|---|---|---|
| **guinea pig** | | | |
| young animals 10 (sex not specified) | 3 % citric acid (about 1200 mg/kg body weight), also ascorbic acid 10 mg/kg body weight and day; 24 days | **about 1200**: decreased haematocrit | Wright and Hughes 1976b |
| adult 10 (sex not specified) | 1 or 5 % citric acid (about 670 or 3354 mg/kg body weight and day[f]); also ascorbic acid (10 mg/kg body weight and day); 60 days | **about 670 or 3340**: decreased haematocrit | Wright and Hughes 1976b |
| **dog** | | | |
| 3 (sex not specified) | citric acid, in gelatine capsules, daily dose 1380 mg/kg body weight in 2 portions 6 h apart, 112–120 days | **1380**: not lethal, no effects on blood or urine parameters, no symptoms of intoxication, no impairment of renal function; autopsy (1/3): liver, kidneys and lungs normal in gross pathological and histological examinations | Krop *et al.* 1945 |

[a] purity 99.8 %
[b] dose probably lower, because food often scattered
[c] calculated with 0.4 kg body weight und 20 g diet/animal and day
[d] calculated with 0.1 kg body weight and 10 g diet/animal and day
[e] calculated with 0.02 kg body weight and 3 g diet/animal and day
[f] calculated with 0.45 kg body weight and 30 g diet/animal and day
[c-f] calculated in each case on the basis of the body weight at the beginning of the study

**Table 4.** Studies of the irritant effects of citric acid on rabbit skin

| Number of animals | Methods | Effects | References |
|---|---|---|---|
| no data | 500 mg, 24 h, Draize test | slight irritation | NIOSH 1994 |
| 6 | 50 % in water, 0.5 ml, occlusive, 4 h | no irritation (read 4 and 48 h *post applicationem*) | Miles 1979 |
| 6 | 500 mg, moistened pad, semi-occlusive, 4 h, then rinsed with water for 30 sec; GLP study | mild irritation (read 1, 24, 48 and 72 h *post applicationem*, erythema and very mild oedema in 1/6 (1, 24 and 48 h), primary irritation index 0.33 (72 h) | Miles 1990 |

**Table 5.** Studies of the irritant effects of citric acid in the rabbit eye

| Number of animals | Method of application; test; concentration or amount applied, duration | Effects | References |
|---|---|---|---|
| 6 per group | Draize test modified as given in the Federal Hazardous Substances Act Regulations, monohydrate, 5 % solution, pH 2.1, not rinsed or rinsed for 2 min 30 sec *post applicationem* | conjunctivitis persisting for up to 7 days, no corneal clouding (no other details), moderate irritation | Murphy *et al.* 1982 |
| no data | 750 μg, 24 h, standard Draize test | severe irritation (no other details) | NIOSH 1994 |
| 6 | 100 mg, conjunctival sac, not rinsed, study discontinued 24 h *post applicationem* GLP study | severe irritation or corrosion (1 and 24 h *post applicationem*: cornea: dry, detachment of the epithelium, clouding, ulceration; conjunctiva: erythema, oedema, greenish-brown discoloration, mucus, discharge, red tears) | Miles 1991 |

**Table 6.** Reproductive toxicity of citric acid

| **Species** number per dose and sex | Application form, exposure duration | Effects | References |
|---|---|---|---|
| **rat** | | | |
| albino (number not specified) ♀, 3 weeks old at the beginning of the study | 1.2 % in the diet (about 1200 mg/kg body weight and day), from 3 weeks of age, mating in week 32 and, on the same diet, again 11 weeks later | no effects on body weights of $F_0$ ♀ or reproduction (numbers of pups per litter born living and dead, average weight of living pups, number of pups weaned) | Bonting and Jansen 1956 |
| 6 ♀ | 5 % (about 600 mg/kg body weight and day) in the diet before, during and after mating, no other details | no effects on the number of pregnant ♀, numbers of pups per litter born living and dead, number of pups weaned; no details of food consumption or body weight development of dams | Wright and Hughes 1976a |
| **mouse** | | | |
| 6 ♀ | 5 % (corresponds to about 600 mg/kg body weight and day) in the diet before, during and after mating, no other details | no effects on the number of pregnant ♀, numbers of pups per litter born living and dead, number of pups weaned; no details of food consumption or body weight development of dams | Wright and Hughes 1976a |

## 5.4 Reproductive and developmental toxicity

In the early reproductive toxicity studies, administration of citric acid in concentrations up to 5 % in the diet had no effect on the fertility of either mice or rats (Table 6).

There is no information available about the developmental toxicity of the substance.

## 5.5 Genotoxicity

Citric acid, in concentrations up to 2000 µg/plate, was not mutagenic in the *Salmonella* mutagenicity test in the strains TA97, TA98, TA100 and TA104 either in the presence or absence of a metabolic activation system (liver S9 from rats treated with phenobarbital) (Al-Ani and Al-Lami 1988). Likewise, in another test with strains TA92, TA94, TA98, TA100, TA1535 and TA1537, citric acid (anhydrous) and monosodium citrate (anhydrous) in concentrations up to 5 mg/plate (citric acid: 26 µmol/plate, monosodium citrate: 23 µmol/plate) were not mutagenic in the presence or absence of a metabolic activation system (liver S9 from rats treated with polychlorinated biphenyls) (Ishidate *et al.* 1984).

In a test for chromosomal aberrations in V79 cells, citric acid in concentrations up to 1 mg/ml (5.2 µmol/ml) and monosodium citrate up to 3 mg/ml (13.8 µmol/ml) were not clastogenic (Ishidate *et al.* 1984).

In concentrations up to 100 µg/ml (0.4 µmol/ml), the disodium salt of citric acid did not cause mitotic spindle inhibition in a C3H mouse embryo cell line (De Brabander *et al.* 1976).

## 5.6 Carcinogenicity

In a cocarcinogenicity study, bladder tumours were induced with *N*-butyl-*N*-(4-hydroxybutyl)nitrosamine (BBN) given to male F344 rats for 20 weeks in the drinking water (0.025 % BBN) simultaneously with either a normal diet or the normal diet mixed with sodium citrate (6.50 %) or citric acid (4.85 %, about 4850 mg/kg body weight and day). The amount of BBN taken up by the animals during the 20 weeks was increased by 66 % in the rats given sodium citrate because of an increase in drinking water consumption. The exposure to citric acid did not affect the uptake of BBN. The histopathological examination at the end of the 20 weeks demonstrated that the BBN treatment had induced bladder carcinomas in 3 of 15 animals. In the animals treated with BBN plus sodium citrate or citric acid, carcinomas were found in 12 and 3 of 15 animals, respectively. Determination of the amount of *N*-butyl-*N*-(3-carboxypropyl)nitrosamine, the main carcinogenic metabolite of BBN, excreted with the 24-hour urine during week 20 demonstrated that the simultaneous exposure to sodium citrate increased the amount excreted by a factor of 7 to 8 whereas citric acid had no effect. The pH of the urine after 14 weeks administration of BBN was 7.3. The simultaneous exposure to sodium citrate increased the pH to 8.2 whereas the exposure to citric acid reduced the pH to 6.3. When the BBN was administered together with sodium ascorbate (5 % in the diet, equivalent to about 5000 mg/kg body weight and day), the pH increased to 8.0 but the tumour incidence was not increased above that found with BBN alone. Therefore the reason for the cocarcinogenic effect of sodium citrate was considered not to be the increase in urinary pH but the increased uptake of BBN and the resulting increased excretion of the carcinogenic main metabolite in the urine (Inoue *et al.* 1988).

As it is known that ingested sodium salts of various compounds have tumour-promoting effects in the urinary bladder of the rat, it was investigated whether also sodium citrate promotes the development of bladder tumours. For this purpose, 20 male F344 rats were first exposed for 4 weeks to 0.05 % BBN in the drinking water, and then fed a diet containing 5 % sodium citrate (corresponds to about 5000 mg/kg body weight and day) for another 32 weeks. Two control groups were exposed only to BBN or only to sodium citrate. In the histopathological examination at the end of the exposure, neither hyperplasia nor papillomas nor carcinomas were found in the bladders of the animals exposed only to sodium citrate. In the group exposed to BBN alone, hyperplasia was found in 44 % of the animals, papillomas in 20 % and carcinomas in 8 %, in the group treated with BBN and sodium citrate the corresponding incidences were 100%, 95% and 100%. Comparison of the urine from the animals treated with sodium citrate with that of untreated animals between week 4 and week 16 of the study revealed at all times pH values which were significantly increased by 1.3 to 2.0 units, sodium concentrations which were increased by a factor of 4 to 6 and an increase in formation of $MgNH_4PO_4$ crystals. The authors concluded from these results that the changes seen in these urine parameters could be the cause of the observed tumour-promoting effect of sodium citrate, possibly because of a stimulation of DNA synthesis in the bladder epithelial cells (Fukushima *et al.* 1986).

Oral administration of high sodium citrate doses does in fact cause a stimulation of DNA synthesis in the rat bladder: after feeding sodium citrate (5 % in the diet, corresponding to about 5000 mg/kg body weight and day) to 5 male F344 rats, the average incidence of DNA-synthesizing cells increased from 0.08 % in the untreated control animals to 0.82 %. The pH of the urine was increased on average by about 1.4 pH units and the sodium concentration to 390 %; the potassium concentration was reduced to about 40 % of the control value. In one of the five treated rats and none of the untreated animals, hyperplasia of the bladder epithelium was observed. When sodium citrate was administered in the diet together with 5 % ascorbic acid, which is known to be a "co-tumour promoter" in the rat bladder and on its own causes neither hyperplasia nor stimulation of DNA synthesis, the incidence of DNA-synthesizing cells was increased to 2.04 %, and 4 of 5 treated animals developed hyperplasia (Shibata *et al.* 1992).

In all, the available studies of the effects of sodium citrate and citric acid on the bladder of the rat show that sodium citrate, ingested in large quantities, has tumour-promoting effects which appear to result from a stimulation of DNA synthesis in the cells of the bladder epithelium caused by the increase in the urinary sodium concentration or pH.

# 6 Manifesto (MAK value/classification)

The critical effect of citric acid is the local irritation resulting from the acidity of the substance. A MAK value cannot be derived from the threshold concentrations for the induction of coughing because the interindividual differences and the variation with the inspiratory flow rate are too great.

Citric acid is not mutagenic. The tumour-promoting effect seen in the bladder of rats which had ingested sodium citrate, but not in those which had ingested citric acid, appears to be the result of an increase in the urinary sodium concentration or pH. Similar effects in man would be expected, if at all, only after ingestion of very large sodium citrate doses.

The available data are not sufficient for the establishment of a MAK value. Citric acid and its alkali metal salts are therefore listed in Section IIb of the *List of MAK and BAT Values*.

There is no evidence of sensitization by the acid or its salts. There are no studies of absorption through the skin or of germ cell mutagenicity.

# 7 References

Abbott TR (1983) Changes in serum calcium fractions and citrate concentrations during massive blood transfusions and cardiopulmonary bypass. *Br J Anaesth 55*: 753–759

Al-Ani FY, Al-Lami SK (1988) Absence of mutagenic activity of acidity regulators in the Ames *Salmonella*/microsome test. *Mutat Res 206*: 467–470

Astle MJ, Weast RC (Eds) (1979) *CRC Handbook of Chemistry and Physics*, Boca Raton, USA, C-251

Barros MJ, Zammattio SJ, Rees PJ (1990) Importance of inspiratory flow rate in the cough response to citric acid inhalation in normal subjects. *Clin Sci 78*: 521–525

Baruch SB, Burich RL, Eun CK, King VF (1975) Renal metabolism of citrate. *Med Clin North Am 59*: 569–582

Bates RG, Pinching GD (1949) Resolution of the dissociation constants of citric acid at 0 to 50°, and determination of certain related thermodynamic functions. *J Am Chem Soc 71*: 1274–1283

Baue AE, Hermann G, Shaw RS (1961) A study of bank blood toxicity. *Surg Gynaecol Obstet 113*: 40–46

Beerman H, Fondé GH, Callaway JL (1938) Citrus fruit dermatoses. *Arch Dermatol Syphilol (Chicago) 38*: 225–234

Beyer H, Walter W (1981) *Lehrbuch der organischen Chemie*, S. Hirzel Verlag, Stuttgart, 326–327

Bickerman HA, Cohen BM, German E (1956) The cough response of normal human subjects stimulated experimentally by citric acid aerosol: alterations produced by antitussive agents. *Am J Med Sci 232*: 57–66

Birmingham DJ, Campbell PC, Doyle HN, McDonald JM (1951) Investigation of occupational dermatoses in the citrus fruit canning industry. *Arch Ind Hyg Occup Med 3*: 57–63

Blair G, Staal PH (1993) Citric acid. in: Kirk-Othmer (Ed.) *Encyclopedia of Chemical Technology, Vol 6*, Wiley Interscience, New York, 354–380

Bolser DC, Aziz SM, Chapman RW (1991) Ruthenium red decreases capsaicin and citric acid-induced cough in guinea pigs. *Neurosci Lett 126*: 131–133

Bonting SL, Jansen BC (1956) The effect of a prolonged intake of phosphoric acid and citric acid in rats. *Voeding 17*: 137–148

De Brabander M, Van de Veire R, Aerts F, Geuens S, Hoebeke J (1976) A new culture model facilitating rapid quantitative testing of mitotic spindle inhibition in mammalian cells. *J Nat Cancer Inst 56*: 357–363

Bunker JP, Bendixen HH, Murphy AJ (1962) Hemodynamic effects of intravenously administered sodium citrate. *N Engl J Med 266*: 372–377

Bunker JP, Steson JB, Coe RC, Grillo HC, Murphy AJ (1955) Citric acid intoxication. *J Am Med Assoc 157*: 1361–1367

Callaway JK, King RG (1992) Effects of inhaled $\alpha_2$-adrenoceptor and $GABA_B$ receptor agonists on citric acid-induced cough and tidal volume changes in guinea pigs. *Eur J Pharmacol 220*: 187–195

Clay TP, Thompson MA (1985) Irritant induced cough as a model of intrapulmonary airway reactivity. *Lung 163*: 183–191

Collins JA (1974) Problems associated with the massive transfusion of stored blood. *Surgery 75*: 274–295

Corbascio AN, Smith NT (1967) Hemodynamic effect of experimental hypercitremia. *Anesthesiology 28*: 510–516

Denlinger JK, Nahrwold ML, Gibbs PS, Lecky JH (1976) Hypocalcaemia during rapid blood transfusion in anaesthetized man. *Br J Anaesth 48*: 995–999

Dickens F (1941) The citric acid content of animal tissues, with reference to its occurence in bone and tumour. *Biochem J 35*: 1011–1023

Ditre CM, Griffin TD, Murphy GF, Sueki H, Telegan B, Johnson WC, Yu RJ, Van Scott EJ (1996) Effects of alpha-hydroxy acids on photoaged skin: a pilot clinical, histologic, and ultrastructural study. *J Am Acad Dermatol 34*: 187–195

Driscoll DF, Bistrian BR, Jenkins RL, Randall S, Dzik WH, Gerson B, Blackburn GL (1987) Development of metabolic alkalosis after massive transfusion during orthotopic liver transplantation. *Crit Care Med 15*: 905–908

Drop LJ, Scheidegger D (1979) Haemodynamic consequences of citrate infusion in the anaesthetized dog: comparision between two citrate solutions and the influence of beta blockade. *Br J Anaesth 51*: 513–521

Dzik WH, Kirkley SA (1988) Citrate toxicity during massive blood transfusion. *Transfus Med Rev 2*: 76–94

FAO (Food and Agriculture Organization of the United Nations) (1974) *FAO/WHO Nutrition meetings report series No. 53 A. Toxicological evaluation of some food additives including anticaking agents, antimicrobials, antioxidants, emulsifiers and thickening agents.* FAO, Rome, 170–172

Forsberg K, Karlsson J-A (1986) Cough induced by stimulation of capsaicin-sensitive sensory neurons in conscious guinea pigs. *Acta Physiol Scand 128*: 319–320

Forsberg K, Karlsson J-A, Theodorsson E, Lundberg JM, Perrson CGA (1988) Cough and bronchoconstriction mediated by capsaicin-sensitive sensory neurons in the guinea-pig. *Pulmonol Pharmacol 1*: 33–39

Fuglsang G, Madsen C, Saval P, Østerballe O (1993) Prevalence of intolerance to food additives among Danish school children. *Pediatr Allergy Immunol 4*: 123–129

Fukushima S, Thamavit W, Kurata Y, Ito N (1986) Sodium citrate: a promotor of bladder carcinogenesis. *Jpn J Cancer Res 77*: 1–4

Gebr. Overlack (1977) *Zitronensäure monohyd.*, Sicherheitsdatenblatt (safety data sheet) Gebr. Overlack Chemische Fabrik GmbH, Mönchengladbach, 28.10.1977

Gomori G, Gulyas E (1944) Effect of parenterally administered citrate on the renal excretion of calcium. *Proc Soc Exp Biol Med 56*: 226–228

Gruber CM, Halbeisen WA (1948) A study on the comparative effects of citric acid and its sodium salts. *J Pharmacol Exp Ther 94*: 65–67

Guest D, Katz GV (1994) Citric acid. in: Clayton GD, Clayton FE (Eds) *Patty's Industrial Hygiene and Toxicology, Vol II C*, John Wiley & Sons, New York, 3574, 3587–3589

Hamel R, Ford-Hutchinson AW (1985) Pulmonary and cardiovascular changes in hyperreactive rats from citric acid aerosols. *J Appl Physiol 59*: 354–359

Hirshman CA, Austin DR, Klein W, Hanifin JM, Hulbert W (1986) Increased metachromatic cells and lymphocytes in bronchoalveolar lavage fluid of dogs with airway hyperreactivity. *Am Rev Respir Dis 133*: 482–487

Howland WS, Bellville JW, Zucker MB, Boyan P, Clifton EE (1957) Failure to observe citrate intoxication. *Surg Gynaecol Obstet 105*: 529–540

Hüner A, Fartasch M, Hornstein OP, Diepgen TL (1994) The irritant effect of different metal-working fluids. *Contact Dermatitis 31*: 220–225

Inoue T, Imaida K, Suzuki E, Okada M, Fukushima S (1988) Combined effects of L-ascorbic acid, citric acid or their sodium salts on tumor induction by *N*-butyl-*N*-(4-hydroxybutyl)nitrosamine or *N*-ethyl-*N*-(4-hydroxybutyl)nitrosamine in the rat urinary bladder. *Cancer Lett 40*: 265–273

Ishidate M, Sofuni T, Yoshikawa K, Hayashi M, Nohmi T, Sawada M, Matsuoka A (1984) Primary mutagenicity screening of food additives currently used in Japan. *Food Chem Toxicol 22*: 623–636

Janson PH (1953) Citrusfrüchte und Hauterkrankungen. *Z Haut Geschlechtskr 14*: 144–147

Köhler LD, Refior M (1997) Alpha-Hydroxysäuren – eine aktuelle Übersicht. *Kosmet Med 18*: 10–13

Köhler M (1993) Kalziumsubstitution bei Hämotherapie. *Dtsch Med Wochenschr 118*: 438–440

Krop S, Gold H, Paterno CA (1945) On the toxicity of hydroxyacetic acid after prolonged administration: comparison with its sodium salt and citric and tartaric acids. *J Am Pharm Assoc 34*: 86–89

Lalloo UG, Fox AJ, Belvisi MG, Chung KF, Barnes PJ (1995) Capsazepine inhibits cough induced by capsaicin and citric acid but not by hypertonic saline in guinea pigs. *J Appl Physiol 79*: 1082–1087

Lee VM, Linden RW (1991) An olfactory-parotid salivary reflex in humans? *Exp Physiol 76*: 347–355

Llobet JM, Domingo JL, Gomez M, Tomas JM, Corbella J (1987) Acute toxicity studies of aluminium compounds: antidotal efficacy of several chelating agents. *Pharmacol Toxicol 60*: 280–283

Lou YP, Lundberg JM (1992) Inhibition of low pH evoked activation of airway sensory nerves by capsazepine, a novel capsaicin-receptor antagonist. *Biochem Biophys Res Commun 189*: 537–544

Ludbrook J, Wynn V (1958) Citrate intoxication. A clinical and experimental study. *Br Med J*: 523–528

Lussier JP (1957) Incorporation de l'acide citrique radioactif dans la squelette du rat blanc. *Rev Can Biol 16*: 434–444

Mårtensson J (1940) On the citric acid metabolism in mammals. *Acta Physiol Scand 1, Suppl 2*: 34–96

McLellan BA, Reid SR, Lane PL (1984) Massive blood transfusion causing hypomagnesemia. *Crit Care Med 12*: 146–147

Miles (1979) *Skin corrosion potential of Miles liquid citric acid 50%W.* Miles Laboratories Inc, Report No. 5, Elkhart, IN, USA, unpublished report

Miles (1990) *Dermal irritation of citric acid anhydrous in the rabbit.* Miles Laboratories Inc, Report No. MTD0192, Elkhart, IN, USA, unpublished report

Miles (1991) *Ocular irritation of citric acid anhydrous in the rabbit.* Miles Laboratories Inc, Report No. MTD0209, Elkhart, IN, USA, unpublished report

Murphy JC, Osterberg RE, Seabaugh VM, Bierbower GW (1982) Ocular irritancy responses to various pHs of acids and bases with and without irrigation. *Toxicology 23*: 281–291

Nakasone N, Watkins E, Janeway CA, Gross RE (1954) Experimental studies of circulatory derangement following the massive transfusion of citrated blood. *J Lab Clin Med 43*: 184–195

NIOSH (National Institute of Occupational Safety and Health) (1994) Citric acid. *Registry of toxic effects of chemical substances (RTECS)*, extract from the database, NIOSH, Cincinnati, OH, USA

Nordmann J, Nordmann R (1961) Organic acids in blood and urine. *Adv Clin Chem 4*: 53–120

Packman EW, Abbott DD, Harrisson JW (1963) Comparative subacute toxicity for rabbits of citric, fumaric and tartaric acids. *Toxicol Appl Pharmacol 5*: 163–167

Parker SD, Brown RH, Hirshman CA (1991) Differential effect of glucocorticoids on pulmonary responses and eosinophils. *Respir Physiol 83*: 323–331

Perdrix JP, Lepape A, Gronlier T, Grozel JM, Banssillon V (1988) Regional citrate anticoagulation for haemodialysis in the patient at high risk for bleeding. *Ann Fr Anaesth Reanim 7*: 31–35

Rechenberger J, Benndorf S (1957) Über den Citronensäurespiegel des Blutes. I. Methodik, Normalwerte, sowie das Verhalten beim *Diabetes mellitus. Z Klin Med 154*: 648–659

Reckler J, Rodman JS, Jacobs D, Rotterdam H, Marion D, Vaughan Jr ED (1986) Urothelial injury to the rabbit bladder from various alkaline and acidic solutions used to dissolve kidney stones. *J Urol 136*: 181–183

Riordan MF, Beardsmore CS, Brooke AM, Simpson H (1994) Relationship between respiratory symptoms and cough receptor sensitivity. *Arch Dis Child 70*: 299–304

Röding J, Artmann C (1995) The salts of hydroxycarboxylic acids – non-irritant potent active substances. *Seifen, Oele, Fette, Wachse 121*: 1018–1024

Saffran M, Denstedt OF (1951) The effect of intravenously injected citrate on the serum ionized calcium in the rabbit. *Can J Med Sci 29*: 245–254

Satoh H, Lou YP, Lundberg JM (1993) Inhibitory effect of capsazepine and SR 48968 on citric acid-induced bronchoconstriction in guinea-pigs. *Eur J Pharmacol 236*: 367–72

Schwartz L (1938) Cutaneous hazards in the citrus fruit industry. Brief history of citrus fruit in Florida and in California. *Arch Dermatol Syphilol (Chicago) 37*: 631–649

Shibata MA, Fukushima S, Asakawa E, Hirose M, Ito N (1992) The modifying effects of indomethacin or ascorbic acid on cell proliferation induced by different types of bladder tumor promoters in rat urinary bladder and forestomach mucosal epithelium. *Jpn J Cancer Res 83*: 31–39

Simpson DP (1983) Citrate excretion: a window on renal metabolism of citrate. *Am J Physiol 244*: 223–234

Szczurko C, Dompmartin A, Michel M, Castel B, Leroy D (1996) Occupational contact dermatitis from propacetamol. *Contact Dermatitis 35*: 299–301

Verhoff FH (1986) Citric acid. in: Gerhartz W, Yamamoto YS (Eds) *Ullmann's Encyclopedia of Industrial Chemistry, Vol A*, VCH Verlagsgesellschaft, Weinheim, 103–108

Weiss JM, Downs CR, Corson HP (1923) Inactive malic acid as a food acidulent. *Ind Eng Chem 15*: 628–630

Windholz M (Ed.) (1989) *The Merck Index, An Encyclopedia of Chemicals and Drugs*. Merck & Co Inc, Rahway, NJ, USA, 363

Wolfram S (1934) Citronenüberempfindlichkeit. *Z Haut Geschlechtskr 49*: 416

Wright E, Hughes RE (1975) Dietary citric acid. *Nutrition 29*: 367–372

Wright E, Hughes RE (1976a) The influence of a dietary citric acid supplement on the reproduction and survival time of mice and rats. *Nutr Rep Int 13*: 563–566

Wright E, Hughes RE (1976b) Some effects of dietary citric acid in small animals. *Food Cosmet Toxicol 14*: 561–564

Yokotani H, Usui T, Nakaguchi T, Kanabayashi T, Tanda M, Aramaki Y (1971) Acute and subacute toxicological studies of TAKEDA-citric acid in mice and rats. *J Takeda Res Lab 30*: 25–31

Zelenak JP, Alarie Y, Weyel DA (1982) Assessment of the cough reflex caused by the inhalation of sodium lauryl sulfate and citric acid aerosols. *Fundam Appl Toxicol 2*: 177–180

completed 29.01.1998

# 5-Ethyl-3,7-dioxa-1-azabicyclo[3.3.0]-octane (EDAO)

| | |
|---|---|
| **MAK value** | **not yet established, see Section IIb of the *List of MAK and BAT Values*** |
| **Absorption through the skin** | – |
| **Sensitization** | – |
| **Carcinogenicity** | – |
| **Prenatal toxicity** | – |
| **Germ cell mutagenicity** | – |
| **BAT value** | – |
| Chemical name (CAS) | 7a-ethyldihydro-1*H*,3*H*,5*H*-oxazolo-[3,4-*c*]oxazole |
| CAS number | 7747-35-5 |

Since the completion of the previous MAK documentation for EDAO in 1992 (see Volume 9 of the present series) more studies of the allergenic effects of the substance have been published and make a reassessment necessary. For purposes of comparison, the sensitization studies reviewed in the 1992 documentation are also included here.

## 1 Allergenic Effects

### 1.1 Effects in man

There are only few studies available of the allergenic effects of EDAO in exposed persons. There are reviews in which EDAO is described as generally tending to cause sensitization (Senff *et al.* 1991, Uter *et al.* 1992).

EDAO was tested on 100 volunteers as a 0.3 % aqueous solution applied for 24 hours (4 times weekly for 3 weeks) in an occlusive patch test; in one person progressive irritation was observed at the application site. The provocation treatment two weeks later induced no reactions (IMC 1980a). There is also one publication with casuistics of two sensitized patients who produced reactions to a 1 % solution. One of these persons was also allergic to formaldehyde (Dahlquist 1984).

In a more recent publication (Bruze and Gruvberger 1994) it is reported that among 1480 patients with dermatitis, 13 (0.9 %) produced a reaction to EDAO; 8 of these persons were also allergic to formaldehyde. In the other 5 patients who had reacted to EDAO, the test with EDAO was repeated; this time positive results were obtained in only 2 cases. When the test was repeated yet again with decreasing concentrations of the substance, one of these two patients still produced positive results at a test concentration of 0.1 %. In spite of the fact that a maximization test (see below) carried out simultaneously yielded negative results and in spite of the very small number of cases, the authors classified EDAO as sensitizing for man.

### 1.2 Animal studies

In a sensitization study with guinea pigs, the initial intradermal induction was carried out with 0.05 ml of a 0.5 % aqueous EDAO solution. Starting 48 hours later, 0.1 ml aliquots of the solution were injected 3 times weekly until the animals had received a total of 10 injections. 15 days after the last injection, provocation was carried out with 0.1 ml of the solution. In this study too, sensitization reactions were not detected (IMC 1980b).

A modified maximization test with guinea pigs yielded negative results. In a second test, reactions were seen in 4 of 24 animals but the result was considered to be statistically not significant (Bruze and Gruvberger 1994).

## 2 Manifesto (sensitization)

The statement made in the 1992 documentation, that EDAO is sensitizing in individual cases, is correct. However, because the number of cases is so small, it is currently not considered necessary to designate the substance with an “S”.

## 3 References

Bruze M, Gruvberger B (1994) Contact allergy to Bioban CS 1246 in humans and guinea pigs. *Am J Contact Dermatitis 5*: 88–89

Dahlquist I (1984) Contact allergy to the cutting oil preservatives BIOBAN CS-1246 and P-1487. *Contact Dermatitis 10*: 46

IMC (International Minerals and Chemical Corporation) (1980a) *BIOBAN CS-1246: Skin irritation and sensitization potential in human volunteers.* Report No PLR-91, unpublished report

IMC (1980b) *Acute toxicity profile of P-1601 (BIOBAN CS-1246).* Report No PLR-124, unpublished report

Senff H, Köllner A, Tholen S, Frosch PJ (1991) Kontaktallergien gegen neuere Konservierungsmittel. *Hautarzt 42*: 215–219

Uter WJC, Fuchs T, Bahmer F, Bäurle G, Rothe A, Ippen H (1992) Prevalence of occupational type IV contact allergy to industrial preservatives in German metal workers. *Skin Res 34, Suppl 12*: 19–24

completed 24.11.1995

# Hexylene glycol

| | |
|---|---|
| **MAK value (1997)** | **10 ml/m$^3$ (ppm) ≙ 49 mg/m$^3$** |
| **Peak limitation (1997)** | **Category I (excursion factor 2)** |
| **Absorption through the skin** | – |
| **Sensitization** | – |
| **Carcinogenicity** | – |
| **Prenatal toxicity (1997)** | **see Section IIc of the** ***List of MAK and BAT Values*** |
| **Germ cell mutagenicity** | – |
| **BAT value** | – |
| Synonyms | 2-butyne-2,4-diol<br>2,4-dihydroxy-2-methylpentane<br>Diolane<br>Isol |
| Chemical name (CAS) | 2-methyl-2,4-pentanediol |
| CAS number | 107-41-5 |
| Structural formula | $CH_3$–CH(OH)–$CH_2$–C(OH)$(CH_3)_2$ |
| Molecular formula | $C_6H_{14}O_2$ |
| Molecular weight | 118.18 |
| Melting point | –40°C |
| Boiling point | 195–200°C |
| Density at 20°C | 0.920–0.923 g/cm$^3$ |
| Vapour pressure at 20°C | 0.06 hPa |
| log $P_{ow}$* | –0.14 (calculated) |
| **1 ml/m$^3$ (ppm) ≙ 4.904 mg/m$^3$** | **1 mg/m$^3$ ≙ 0.204 ml/m$^3$ (ppm)** |
| Concentration of an atmosphere saturated at 20°C | 60 ml/m$^3$ (Shell 1958a) |

* *n*-octanol/water distribution coefficient

| | |
|---|---|
| Solubility | very readily soluble in water, readily soluble in alcohols, diethyl ether, aliphatic and aromatic hydrocarbons and fatty acids (Anonymous 1985) |
| Stability | does not decompose at temperatures of 260–280°C; stable to organic acids at temperatures up to about 100°C; in the presence of concentrated mineral acids yields methyl pentenol and methyl pentadiene (INRS 1987); reacts with oxidizing agents and isocyanates (Shell 1995a), not with metals or alloys (INRS 1987); in used metal-working fluids, oxidation to diacetone alcohol (4-hydroxy-4-methylpentan-2-one) must be expected (Famulok 1995) |
| Production | hydrogenation of diacetone alcohol in the presence of hydrogenation catalysts (Livengood 1952) |
| Purity, impurities | traces of acetic acid (Anonymous 1985; Shell 1995a) |
| Uses | in metal-working fluids as emulsifier and solvent; in brake fluids and hydraulic fluids; solvent for ink, printing ink, lacquers; in the textile industry; as an additive in cement (Livengood 1952); antifoaming agent and plasticizer (Anonymous 1985) |
| | widely used in cosmetics, e.g. as moisturizer, microbicide (antibacterial and antimycotic) and solvent (Kinnunen and Koskela 1991) in concentrations between 0.1 % and 25 % (Anonymous 1985); in skin medications in concentrations of 12 %, for example (Faergemann and Fredriksson 1980, Kinnunen and Hannuksela 1989) |

# 1 Toxic Effects and Mode of Action

Hexylene glycol is moderately toxic after oral administration and slightly toxic after application to the intact skin. The target tissues are the central nervous system (CNS), liver, kidney, lung and gastrointestinal tract. Hexylene glycol is absorbed through the mucosa of the airways and the gastrointestinal tract. Man excretes about 35 % of a dose of hexylene glycol slowly with the urine, about half as the unchanged substance and half as the glucuronide; rodents excrete 50 % to 60 % of the dose in the urine, mainly as the glucuronide.

On the skin, undiluted hexylene glycol causes at most mild irritation. Exposure of the eye to undiluted hexylene glycol or to the vapour causes marked irritation. The substance has only little potential to cause sensitization in man. Hexylene glycol induces neither gene mutations in bacteria nor structural chromosomal aberrations or polyploidy in hamster cells *in vitro*. No effects on reproduction of exposed male rodents were found.

There are no data available for the mechanism of action of hexylene glycol.

# 2 Toxicokinetics and Metabolism

Five test persons were given oral doses of a 10 % aqueous solution of hexylene glycol, either as single 5 g doses or two 2.5 g doses daily on 5 consecutive days, or 1 g daily for 8 to 11 days followed by 2 g daily for 13 to 14 days; 18 % to 35 % of the dose was excreted in the urine. According to the authors, half of the excreted substance was in the form of the glucuronide of hexylene glycol. Hexylene glycol was detectable in the urine for 5 to 11 days after the last dose. A half-life was not given. After administration of single doses ≤ 2 g or daily doses ≤ 600 mg for 90 days, with the unspecific photometric method which was used (detection limit about 100 mg in the 24-hour urine; Jacobsen 1958a) neither free hexylene glycol nor its glucuronide could be detected in the urine (Jacobsen 1958b).

After ingestion of hexylene glycol doses of 200 mg/kg body weight and day for 60 to 131 days, rats excreted 32 % to 47 % of the administered substance in the urine, 96 % as the glucuronide and 4 % as unchanged hexylene glycol. When the daily dose was 100 mg/kg body weight, 37 % to 60 % was eliminated (Larsen 1958). Rabbits given oral doses of 1 mmol/kg body weight (118 mg/kg body weight) excreted on average 67 % of the dose as the glucuronide (Gessner *et al.* 1960).

The fate of the rest of the administered hexylene glycol was not investigated in these studies.

Studies of the cutaneous absorption of hexylene glycol are not available.

# 3 Effects in Man

## 3.1 Single exposures

### Inhalation

Exposure of 12 male and 12 female test persons to hexylene glycol vapour in a concentration of 50 ml/m$^3$ (approximates the saturation concentration at 25°C) for 15 minutes caused eye irritation in most of the persons but no irritation of the nose or throat (no other details; Silverman *et al.* 1946).

## 3.2 Repeated exposures

There are no data available for the effects of repeated inhalation of hexylene glycol.

### 3.2.1 Ingestion

Five test persons were given either 1 g hexylene glycol daily for 8 to 11 days and then 2 g daily for 13 to 14 days or 5 g daily on 5 consecutive days in the form of a 10 % aqueous solution (< 100 mg/kg body weight and day); the sense of well-being of the persons and the results of urinalysis were normal (Jacobsen 1958b).

### 3.2.2 Dermal application

Among patients with extensive burns who were treated for long periods with antiseptic bandages impregnated with 80 % hexylene glycol, 5 % beeswax, 10 % hydrogenated *Ricinus* oil and 5 % isopropyl myristate, the incidence of coma and fatal kidney failure was increased above that among patients who were not treated with these bandages. When the bandage was removed after the persons became comatose, their state improved. Re-bandaging led to renewed CNS depression. Changes in other aspects of the treatment did not affect the incidence of coma. The authors considered that hexylene glycol, perhaps via the presumed metabolic oxidation product diacetone alcohol, had depressive effects on the CNS and produced kidney damage (Procter 1968). This assumption has not yet been confirmed in valid metabolic studies. It is not clear whether hexylene glycol can be absorbed through intact skin, as data from such studies are not available. However, because the substance is highly soluble in both polar and apolar solvents, it may be assumed that hexylene glycol penetrates the skin readily.

## 3.3 Local effects on skin and mucous membranes

Hexylene glycol applied once to the skin does not cause irritation (Table 1).

**Table 1.** Studies of the irritant effects of hexylene glycol on human skin

| Collective | Test, test conditions | Concentration, solvent | Results | References |
|---|---|---|---|---|
| 37 test persons | patch test<br>24 h, occlusive | 100 % | minimum irritation,<br>PII = 0.02 | Anonymous 1985 |
| 39 test persons | patch test, inside forearm<br>24 h, semi-occlusive | 100 % | minimum irritation,<br>PII = 0.11 | Anonymous 1985 |
| 33 test persons | patch test, inside forearm<br>24 h, occlusive | 20 %<br>in white petrolatum | not irritating<br>(0/33, no other details) | Opdyke 1978 |

**Table 1.** continued

| Collective | Test, test conditions | Concentration, solvent | Results | References |
|---|---|---|---|---|
| 230 workers with metal-working fluid exposure and occupational dermatitis | ICDRG patch test occlusive reading after 48 h | 10 % in water | not irritating (0/230) | Alomar *et al.* 1985 |
| 423 and 400 eczema patients | patch test, occlusive, 48 h, reading 1–4 h after removal of the patch and 48 h later | 50 % or 30 % in water | irritating (in a total of 5.7 % only erythema, in a total of 2.8 % oedema and erythema) | Kinnunen and Hannuksela 1989 |
| 7 eczema patients with reactions to hexylene glycol | twice daily on up to 7 consecutive days without occlusion, daily reading | 5 % or 30 % in water | not irritating in 1/1 highly irritating in 1/6 | Kinnunen and Hannuksela 1989 |
| 11 test persons with healthy skin | twice daily on 8 consecutive days without occlusion | 50 % in water | transepidermal water loss not increased | Kinnunen and Hannuksela 1989 |
| 8 atopic patients with dermatitis | twice daily on 2.5 consecutive days without occlusion | 50 % in water | transepidermal water loss not increased | Kinnunen and Hannuksela 1989 |
| 10 test persons with healthy skin | 24 h, occlusive | 50 % in water | transepidermal water loss slightly increased, significance not specified | Kinnunen and Hannuksela 1989 |

PII = primary irritation index, scale from 0–4

Adverse effects of local application of hexylene glycol on the barrier function of the skin, estimated by determining the increase in the non-immunological reaction to contact with benzoic acid, were not found either in persons with healthy skin or in atopic patients (Kinnunen and Hannuksela 1989).

Eye irritation in volunteers was described in Section 3.1.

## 3.4 Allergenic effects

A maximization test with 20 % hexylene glycol applied to healthy test persons yielded negative results as did a patch test with hexylene glycol concentrations up to 1 % applied to patients with eczema (Table 2). Two other allergy tests carried out on patients with eczema were not included in the present evaluation because the results are contradictory (Alomar *et al.* 1985, Kinnunen and Hannuksela 1989).

**Table 2.** Studies of the sensitizing effects of hexylene glycol in man

| Collective, test, test conditions | Concentration, solvent | Positive reactions/ persons tested | References |
|---|---|---|---|
| volunteers<br>maximization test of Kligman | 20 %<br>in white petrolatum | 0/33 | Opdyke 1978 |
| workers with exposure to metal-working fluids and occupational dermatitis<br>ICDRG patch test, occlusive | ≤ 1 %<br>in white petrolatum | 0/52 | Angelini and Meneghini 1977 |

No data are available for the genotoxicity, reproductive toxicity or carcinogenicity of hexylene glycol in man.

# 4 Animal Experiments and *in vitro* Studies

No data are available for allergenic, developmental or carcinogenic effects of hexylene glycol in experimental animals.

## 4.1 Acute toxicity

### 4.1.1 Inhalation

Inhaled hexylene glycol is of low toxicity for the rat. Six rats which were exposed for one hour to an atmosphere saturated with hexylene glycol at room temperature (about 310 mg/m$^3$) survived without signs of intoxication. The gross pathological examination revealed no organ changes (IBTL 1970). Likewise, when groups of 6 rats were exposed for 8 hours either to air saturated with hexylene glycol at room temperature (about 60 ml/m$^3$) (Carbide and Carbon Chemicals Corporation 1949, Smyth and Carpenter 1948) or to hexylene glycol mist produced by bubbling air through hot (170°C) hexylene glycol and then cooling the mixture (Carbide and Carbon Chemicals Corporation 1949), all animals survived without symptoms (no other details).

### 4.1.2 Ingestion

The acute oral toxicity of undiluted hexylene glycol for rodents, rabbits and guinea pigs is moderate (LD$_{50}$ 2600–4800 mg/kg body weight) (Table 3). The symptoms of intoxication seen in all species after administration of doses in the range of the LD$_{50}$ included muscle weakness and narcosis which persisted for between one hour and several days

(IBTL 1970, Woodard *et al.* 1945); in the rat, conjunctival erythema (IBTL 1970) and blood in the urine were also observed (Carbide and Carbon Chemicals Corporation 1949). Autopsy of rats revealed inflammation in the lungs and gastrointestinal tract (IBTL 1970) and congestion in the kidneys, intestines and lungs with bleeding from the capillaries (Carbide and Carbon Chemicals Corporation 1949). When narcosis persisted for longer than 24 hours in mice, the animals died with respiratory failure. Inflammation was found in the lungs and large intestine but there were no changes in the kidneys, heart or brain (Shell 1958b).

**Table 3.** Acute oral toxicity of hexylene glycol

| Species | $LD_{50}$ (mg/kg body weight) | References |
|---|---|---|
| rat | 4700 | Smyth and Carpenter 1948 |
| rat | 4470 | IBTL 1970 |
| rat | 3700 (4.0 ml/kg body weight) | Woodard *et al.* 1945 |
| mouse | 4200 (4.5 ml/kg body weight) | Woodard *et al.* 1945 |
| mouse | 3500 (3.8 ml/kg body weight) | Shell 1958b |
| mouse | 3100 (26.2 mmol/kg body weight) | Wenzel and Koff 1956 |
| rabbit | 2900 (3.2 ml/kg body weight) | Woodard *et al.* 1945 |
| guinea pig | 2600 (2.8 ml/kg body weight) | Woodard *et al.* 1945 |

### 4.1.3 Intraperitoneal injection

The acute intraperitoneal $LD_{50}$ for the rat and mouse is given as 1300–1500 mg/kg body weight (Sanderson 1959, NIOSH 1994, Woodard *et al.* 1945) or as 4200 mg/kg body weight (5.0 ml/kg body weight) (Shell 1958b). The symptoms include sleepiness and narcosis (Sanderson 1959, Shell 1958b), urinary incontinence and adhesion of the peritoneum to the organs of the abdominal cavity (Sanderson 1959).

### 4.1.4 Dermal absorption

The acute dermal toxicity of hexylene glycol for the rabbit is low ($LD_{50} > 5000$ mg/kg body weight) (Table 4). In one study, marked erythema of the application site, organ damage such as haemorrhage in the lungs, mottled or haemorrhagic livers, kidney damage and congestion in the stomach and intestine were found (purity not specified; Carbide and Carbon Chemicals Corporation 1949). In another study (IBTL 1970), only slight erythema at the application site was observed and no symptoms of intoxication. Gross pathological examination revealed no abnormalities in either the surviving animals or in the one animal which died.

Application of hexylene glycol doses of 1800 mg/kg body weight for 24 hours (no other details) to two rabbits produced merely slight erythema in one animal and slight oedema in the other. The effects regressed within 24 hours and both animals survived (no other details; Shell 1958b).

**Table 4.** Acute toxicity of undiluted hexylene glycol applied to the skin of rabbits

| Application | $LD_{50}$ (mg/kg body weight) | References |
|---|---|---|
| not specified | > 5000 | Opdyke 1978 |
| occlusive, shaved flank, 24 h | 7875 (8.56 ml/kg body weight) | Carbide and Carbon Chemicals Corporation 1949 |
| occlusive, shaved dorsal skin, 24 h | > 10000 | IBTL 1970 |
| occlusive, 24 h | 12200 (13.3 ml/kg body weight) | Smyth and Carpenter 1948 |

## 4.2 Subacute, subchronic and chronic toxicity

### 4.2.1 Inhalation

Groups of 5 male and 5 female Harlan-Wistar rats and one male rabbit were exposed to a hexylene glycol aerosol (pure substance, 700 mg/m$^3$, mean particle diameter 1 μm) for 7 hours daily on five consecutive days and then, after a 2-day break, on another 4 consecutive days. In 2 rats congestion and in one rat a very slight submucosal haemorrhage was found in the trachea. Congestion in the trachea and slight hyperplasia of the tracheal epithelium were seen in the rabbit. Neither macroscopic nor microscopic differences from the control animals were found in the lungs, heart, liver, spleen, kidneys, adrenal glands, thyroid glands, parathyroid glands, oesophagus, bronchi, thymus or cervical lymph nodes. Body and organ weights were also unaffected (Union Carbide Corporation 1976).

### 4.2.2 Ingestion

The results of studies with the rat and mouse are summarized in Table 5. The target tissues were liver, kidney and adrenal gland. Doses up to 40 mg/kg body weight did not induce lesions in any of the studies. A valid no observed effect level (NOEL) cannot, however, be deduced from the available data because the numbers of animals and design of the studies do not meet present day requirements.

### 4.2.3 Dermal absorption

Application of 90 hexylene glycol doses of 0.5 or 1.0 ml/kg body weight and day (460 or 920 mg/kg body weight) on 5 days a week for one hour without occlusion to rabbits did not result in macroscopic or microscopic changes in the livers or kidneys of the surviving animals (7 and 4, respectively). Analogous application of doses of 2.0 ml/kg body weight and day (1850 mg/kg body weight) resulted in slight cloudy swelling of the livers of 5 of the 9 surviving animals. Deaths were not substance-related but caused by pulmonary infections. Weight gains in all treated groups were larger than in the control groups which were treated in an analogous way with glycerol doses of 2.0 and 1.0 ml/kg body weight. Substance-related skin changes were not seen in any of the treated groups (Carbide and Carbon Chemicals Division 1950).

**Table 5.** Subacute, subchronic and chronic toxicity of hexylene glycol ingested by rodents

| Species (strain) number/dose, sex | Dose, interval, application form duration | **Dose (mg/kg body weight and day)**: effects | References |
|---|---|---|---|
| rat<br>not specified | 1840 mg/kg body weight (2 ml/kg body weight);<br>gavage, once daily<br>9 days | **1840**: slight hyperplasia of the liver cells with increased numbers of basophilic granules and slight hyaline alterations in the centrilobular liver cells, no other details | Woodard *et al.* 1945 |
| rat<br>10 ♂ | 45.3, 423.5 and 1740 mg/kg body weight (0.5, 5 and 20 g/kg diet)<br>28 days | **from 45.3**: body weights not different from the controls, not lethal<br>**1740**: increase in relative weights of liver, kidney and adrenal glands, in 2/10 animals congestion in the liver or swollen liver (no histological examination), no other details | IBTL 1970 |
| rat<br>5 ♂/5 ♀ | 43, 78 and 310 mg/kg body weight (0.06 %, 0.12 % and 0.5 % in the diet)<br>90 days | **43, 78 and 310**: no effects on food consumption and body weight gains, not lethal, liver and kidney weights unchanged, kidneys, adrenal glands, lungs, spleen, testes normal in gross pathological and histological examinations; cloudy swelling of the livers of 1 and 2 animals in the two lower dose groups, presumably not substance-related because not observed in the 310 mg/kg group | Carbide and Carbon Chemicals Corporation 1949, Smyth *et al.* 1951 |
| rat (albino)<br>10 ♂<br>age 4 weeks | 315–612 or 500–1000 mg/kg body weight[a] (average 98 or 150 mg per animal and day), dissolved in milk<br>130 days | **315–612**: no abnormal findings but unclear whether histological examination carried out, weight gains higher than in control group (significance not determined)<br>**500–1000**: in 3/10 animals slight haemorrhage, hyperaemia and calcification of the renal medulla, no substance-related findings in liver or testes | Larsen 1958 |
| rat<br>5 ♂/5 ♀ | ♂: 592 mg/kg body weight,<br>♀: 585 mg/kg body weight<br>(0.5 % in the drinking water)<br>8 months | **592/585**: not lethal; weight gains, food and water consumption, erythrocyte and leukocyte counts unchanged, gross pathological and histological examinations revealed no effects on the eyes, brain, heart, kidneys, spleen, gonads, adrenal glands or liver | Brown *et al.* 1955 |

**Table 5.** continued

| Species (strain) number/dose, sex | Dose, interval, application form duration | **Dose (mg/kg body weight and day)**: effects | References |
|---|---|---|---|
| rat (CFE) 10 ♂/10 ♀ | 8, 40, 200, 1000 mg/kg body weight (0.01 %, 0.05 %, 0.25 %, 1.25 % in the diet) 89 days | kidneys, liver, heart, spleen and testes of 4 ♂ per dose examined<br>**in all groups**: food consumption unchanged, gross pathology and histology of heart and testes normal<br>**8**: no effects (no histological examination)<br>**40**: no effects, histological examination revealed no abnormalities<br>**200**: dilated renal tubules in 3/4 ♂ | Union Carbide Chemicals Company 1961 |
| rat (CFE) 10 ♂/10 ♀ | 8, 40, 200, 1000 mg/kg body weight (0.01 %, 0.05 %, 0.25 %, 1.25 % in the diet) 89 days | **200 and 1000**: focal cloudy swelling of the proximal renal tubules with intracellular and free hyaline droplets, glomerular lipoid droplets; mild diffuse cloudy swelling of the central hepatic cords<br>**only 1000**: ♀: weight gains decreased by 15 %, ♂: focal dilation of the renal tubules with flattened epithelial cells, lipoid droplets in proximal renal tubules, ♂ and ♀: relative liver and kidney weights increased | Union Carbide Chemicals Company 1961 |
| mouse 6 ♂ | about 153–175, 344–417 or 571–634 mg/kg body weight[a] (5, 10, 20 mg/animal and day), dissolved in milk, 77–81 days | **153–175 or 344–417**: no effects, no histological examination<br>**571–634**: histological examination: findings like those in rats (see above; Larsen 1958), no other details | Larsen 1958 |

[a] doses calculated from initial and final weights of the animals

## 4.3 Local effects on skin and mucous membranes

### 4.3.1 Skin

After a single 24-hour occlusive application, hexylene glycol was not irritating for intact skin and not or only moderately irritating for abraded or scarified skin. Repeated application to intact skin without occlusion was well tolerated (Table 6). On the rabbit ear, the substance caused acneiform eruptions (Fulton *et al.* 1976).

### 4.3.2 Eyes

Undiluted hexylene glycol produced severe eye irritation which persisted in some cases for 7 days or longer. Aqueous solutions (up to 25 %) were not irritating (Table 6).

**Table 6.** Effects of hexylene glycol on skin and eyes of the rabbit

| Test, application | Concentration, duration | Effects | References |
|---|---|---|---|
| **Skin** | | | |
| GLP study, semi-occlusive, 0.5 ml, 99.66 % | undiluted 4 h | not irritating | Shell 1995b |
| Draize test according to CPSC Standard, occlusive | undiluted 24 h | moderate irritation | Opdyke 1978 |
| Draize test, occlusive, abraded and intact skin, 0.5 ml | undiluted 24 h | moderate irritation | Draize *et al.* 1944 |
| occlusive, scarified and intact skin, 0.5 ml | undiluted 24 h | not irritating | Guillot *et al.* 1982 |
| modified Draize test, occlusive, abraded and intact skin, 0.5 ml | undiluted 24 h | not irritating | IBTL 1970 |
| Draize test, without occlusion | undiluted, once (duration not specified) | slight irritation | Cavender and Sowinski 1994 |
| 1840 mg/kg body weight, no other details | undiluted 24 h | slight irritation | Shell 1958b |
| without occlusion, 2 ml/animal | 10–100 %, once daily, 5 days/week, 6 weeks | "well tolerated" | Guillot *et al.* 1982 |
| **Eyes** | | | |
| Draize test as in the U.S. Federal Register, 0.2 ml | undiluted | highly irritating | Shell 1978 |
| modified Draize test, 0.1 ml, eyes not rinsed | undiluted | highly irritating | IBTL 1970 |
| 0.1 ml | undiluted | highly irritating | Guillot *et al.* 1982 |
| Draize test according to CPSC Standards, 0.1 ml | undiluted | highly irritating | Carbide and Carbon Chemicals Corporation 1949 |
| Draize test according to CPSC Standards, 0.02 ml | undiluted | moderately irritating | Carbide and Carbon Chemicals Corporation 1949 |
| 0.02 ml on the centre of the cornea | undiluted, reading after 24 h | irritating | Smyth and Carpenter 1948 |
| no details | 25 % in water | not irritating | Anonymous 1985 |
| 0.1 ml | 10 % in water | not irritating | Guillot *et al.* 1982 |

## 4.4 Reproductive and developmental toxicity

Daily oral doses of 141 to 196 mg hexylene glycol in milk were administered to 7 male rats for 87 days and then during the mating period which followed for another 47 days;

the treatment had no effect on male fertility, estimated from the number of conceptions and the number of living foetuses. In all males, the testes appeared normal in the histological examination (see Table 5; Larsen 1958).

## 4.5 Genotoxicity

Hexylene glycol in concentrations of 31–4000 µg/plate (0.27–33.85 µmol/plate) was not mutagenic in the *Salmonella* mutagenicity test in the strains TA1535, TA1537, TA1538, TA98 or TA100 or in *Escherichia coli* WP2 uvrA pKM 101 in either the presence or absence of a metabolic activation system (S9 fraction from the livers of rats treated with Aroclor) (Brooks *et al.* 1988, Shell 1985).

In a test for mitotic gene conversion in *Saccharomyces cerevisiae* JD 1 in which the number of prototrophic colonies was recorded, hexylene glycol in concentrations of 10 to 5000 µg/ml (0.09–42.31 µmol/ml) was not genotoxic in either the presence or absence of a metabolic activation system (S9 fraction from the livers of rats treated with Aroclor) (Brooks *et al.* 1988, Shell 1985).

In CHO cells (a cell line from Chinese hamster ovary), hexylene glycol in concentrations of 1250, 2500 and 5000 µg/ml (10.58, 26.45 and 52.89 µmol/ml) did not induce the formation of structural chromosomal aberrations or polyploidy in either the presence or absence of a metabolic activation system (S9 fraction from the livers of rats treated with Aroclor) (Brooks *et al.* 1988, Shell 1985).

In concentrations up to 100 µg/ml (0.847 µmol/ml), hexylene glycol did not cause disorders of spindle function in a C3H mouse embryo cell line (De Brabander *et al.* 1976).

# 5 Manifesto (MAK value/classification)

Carcinogenicity studies with hexylene glycol are not available. However, the genotoxicity studies provide no evidence of a carcinogenic potential.

In animal studies of limited validity, 40 mg/kg body weight and day was found to be the NOEL for systemic toxicity after oral administration of the substance. The critical effect of hexylene glycol, however, is the eye irritation which was observed in animals treated with the undiluted substance. Hexylene glycol does not seem to be a very powerful eye irritant as 25 % solutions were not irritating. In an early study, eye irritation developed in most of the 12 male and female test persons exposed for 15 minutes to hexylene glycol vapour in a concentration of 50 ml/m$^3$ (approximately equivalent to the saturation concentration at 25°C). No NOEL was determined in this study. The eye irritation seems not to have been very severe, a fact which is also suggested by the steepness of the dose-response curve seen in the Draize test data. The MAK value was established provisionally at 10 ml/m$^3$. Because the eye irritation was not severe, for the restriction of exposure peaks hexylene glycol was classified in Peak limitation category I with an excursion factor of 2. Then, even during the permissible peak concentrations up to 20 ml/m$^3$, no marked irritation is expected. Because of the lack of adequate data for the irritant potential of hexylene glycol, the MAK value must be considered provisional and requires confirmation in further studies. Assuming that 100 % of the inhaled substance is absorbed, exposure to concentrations in the range of the MAK value could result in a daily dose of 7.5 mg/kg body weight which is still below the NOEL determined in animal studies with oral administration, assuming 100 % absorption here too.

As there are no reproductive toxicity studies, hexylene glycol is listed in Section IIc of the *List of MAK and BAT Values*.

The sensitizing potential is low and so the substance is not designated with an "Sh".

The solubility of the substance suggests that it would be readily absorbed through the skin, but as the toxicity after repeated dermal application to experimental animals is low, the substance is not designated with an "H".

# 6 References


Alomar A, Conde-Salazar L, Romaguera C (1985) Occupational dermatoses from cutting oils. *Contact Dermatitis 12*: 129–138

Angelini G, Meneghini CL (1977) Dermatitis in engineers due to synthetic coolants. *Contact Dermatitis 3*: 219–220

Anonymous (1985) Final report on the safety assessment of butylene glycol, hexylene glycol, ethoxydiglycol, and dipropylene glycol. *J Am Coll Toxicol 4*: 223–248

Brooks TM, Meyer AL, Hutson DH (1988) The genetic toxicology of some hydrocarbon and oxygenated solvents. *Mutagenesis 3*: 227–232

Brown B, Schaffarzick RW, Dreisbach RH (1955) Anticonvulsant properties of certain secondary and tertiary alcohols. *J Pharmacol Exp Ther 115*: 230–239

Carbide and Carbon Chemicals Corporation (1949) *The acute toxicity of methyl pentane diol.* Mellon Institute of Industrial Research, Report 12-54, unpublished report

Carbide and Carbon Chemicals Division (1950) *Repeated inunction of rabbits with methyl pentane diol.* Mellon Institute of Industrial Research, Report 13-71, unpublished report

Cavender FL, Sowinski EJ (1994) Glycols. in: Clayton GD, Clayton FE (Eds) *Patty's Industrial Hygiene and Toxicology, Volume II F*, John Wiley & Sons, New York, 4645–4719

De Brabander M, Van de Veire R, Aerts F, Geuens S, Hoebeke J (1976) A new culture model facilitating rapid quantitative testing of mitotic spindle inhibition in mammalian cells. *J Nat Cancer Inst 56*: 357–363

Draize JH, Woodard G, Calvery HO (1944) Methods for the study of irritation and toxicity of substances applied topically to the skin and mucous membranes. *J Pharmacol Exp Ther 82*: 377–390

Faergemann J, Fredriksson T (1980) The antimycotic activity *in vitro* of five diols. *Sabouraudia 18*: 287–293

Famulok T (1995) *Entwicklung analytischer Meßverfahren für die Bestimmung von Kühlschmierstoff-Einzelkomponenten und -Summenparametern am Arbeitsplatz.* Doctoral Thesis, Technische Universität München

Fulton JE, Bradley S, Aqundez A, Black T (1976) Non-comedogenic cosmetics. *Cutis 17*: 344–351

Gessner PK, Parke DV, Williams RT (1960) Studies in detoxication. 80. The metabolism of glycols. *Biochem J 74*: 1–5

Guillot JP, Martini MC, Giauffret JY, Gonnet JF, Guyot JY (1982) Safety evaluation of some humectants and moisturizers used in cosmetic formulations. *Int J Cosmet Sci 4*: 67–80

IBTL (Industrial Bio-Test Laboratories) (1970) *Data Sheets 12-4/1970*. BIO-FAX, Industrial Bio-Test Laboratories Inc, Northbrook, IL, unpublished report

INRS (Institut Nationale de Récherche et de Sécurité) (1987) *Cahiers de notes documentaires, sécurité et hygiène du travail, fiches toxicologiques n⁰ 167. 2-Méthyl-2,4-pentanediol*. L'Institut National de Récherche et de Sécurité, 30 rue Olivier-Noyer, 75690 Paris Cedex 14, France

Jacobsen E (1958a) Determination of 2-methyl-2,4-pentane diol (hexylene glycol) in the urine of man and rats. *Acta Pharmacol Toxicol 14*: 195–206

Jacobsen E (1958b) The excretion of hexylene glycol (2-methyl- 2,4-pentane diol) in man. *Acta Pharmacol Toxicol 14*: 207–213

Kinnunen T, Hannuksela M (1989) Skin reactions to hexylene glycol. *Contact Dermatitis 21*: 154–158

Kinnunen T, Koskela M (1991) Antibacterial and antifungal properties of propylene glycol, hexylene glycol and 1,3-butylene glycol *in vitro*. *Acta Derm Venereol (Stockh) 71*: 148–150

Larsen V (1958) The toxicity of 2-methyl-pentan-2,4-diol (hexylene glycol) by chronic oral administration to rats and mice. *Acta Pharmacol Toxicol 14*: 341–349

Livengood SM (1952) Higher glycols. in: Curmo jr GO, Johnstone F (Eds) *Glycols*, American Chemical Society Monograph Series 114, Reinhold, New York, 285–297

NIOSH (National Institute for Occupational Safety and Health) (1994) *Registry of toxic effects of chemical substances (RTECS)*, CD-ROM, NIOSH, Cincinnati, OH

Opdyke DLJ (1978) Fragrance raw materials monographs. *Food Cosmet Toxicol 16, Suppl 1*: 777–778

Procter DSC (1968) Coma in burns – the cause traced to dressings. *S Afr Med J 40*: 1116–1120

Sanderson DM (1959) A note on glycerol formal as a solvent in toxicity testing. *J Pharm Pharmacol 11*: 150–156

Shell (1958a) *Industrial hygiene bulletin, safety data sheet hexylene glycol*, SC: 57-101. Shell Chemical Corporation, April 1958

Shell (1958b) *Industrial hygiene bulletin, toxicity data sheet hexylene glycol*, SC: 57-102. Shell Chemical Corporation, April 1958

Shell (1978) *Toxicity of solvents: eye irritation study on hexylene glycol*. Shell Toxicology Laboratory, GRR-TLGR.0035.78, unpublished report

Shell (1985) *Genotoxicity studies with hexylene glycol*. Shell Research Ltd, London, Group Research Report SBGR.85.253, unpublished report

Shell (1995a) *Sicherheitsdatenblatt, Hexylenglykol*. (Safety Datasheet) Deutsche Shell Chemie GmbH, Abteilung Sicherheit und Umwelt. 01. Januar 1995

Shell (1995b) *Hexylene glycol: skin irritation in the rabbit*. Huntingdon Research Centre, Report HSE 95.1085, unpublished report

Silverman L, Schulte HF, First MW (1946) Further studies on sensory response to certain industrial solvent vapors. *J Ind Hyg Toxicol 28*: 262–266

Smyth HF, Carpenter CP (1948) Further experience with the range finding test in the industrial toxicology laboratory. *J Ind Hyg Toxicol 30*: 63–68

Smyth HF, Carpenter CP, Weil CS (1951) Range-finding toxicity data: list IV. *Arch Ind Hyg 4*: 119–122

Union Carbide Chemicals Company (1961) *Results of three months of inclusion of hexylene glycol in the diet of rats*. Mellon Institute of Industrial Research, Report 24-31, unpublished report

Union Carbide Corporation (1976) *Hexylene glycol: 9-day repeated aerosol inhalation by rats*. Carnegie-Mellon Institute of Research, Report 39-42, unpublished report

Wenzel DG, Koff GY (1956) Anticonvulsant properties of some alkyldiols, alkyldiones and related compounds. *J Am Pharm Assoc 45*: 669–672

Woodard G, Johnson VD, Nelson AA (1945) Acute toxicity of 2-methyl-2,4-pentanediol. *Fed Proc 4*: 142–143

completed 15.11.1996

# 3-Iodo-2-propynyl butylcarbamate

| | |
|---|---|
| **MAK value** | **not yet established, see Section IIb of the *List of MAK and BAT Values*** |
| **Peak limitation** | – |
| **Absorption through the skin** | – |
| **Sensitization** | – |
| **Carcinogenicity** | – |
| **Prenatal toxicity** | – |
| **Germ cell mutagenicity** | – |
| **BAT value** | – |
| Synonyms | butyl-3-iodo-2-propynylcarbamate<br>3-iodo-2-propynyl *N*-butylcarbamate<br>3-iodo-2-propynylcarbamic acid butyl ester |
| Trade names | Troysan® KK-108A<br>Troysan® Polyphase® P100<br>Troysan® Polyphase® Anti-Mildew |
| Chemical name (CAS) | butyl carbamic acid 3-iodo-2-propynyl ester |
| CAS number | 55406-53-6 |
| Structural formula | $I–C≡C–CH_2–O–CO–NH–C_4H_9$ |
| Molecular formula | $C_8H_{12}INO_2$ |
| Molecular weight | 281.09 |
| Melting point | 65–68°C |
| Boiling point | not specified |
| Vapour pressure at 20°C | $2.4 \times 10^{-6}$ hPa |
| log $P_{ow}$* | 2.81 |
| Solubility at 20°C | 156 mg/l water |
| Stability | can release iodine vapour at high temperatures (Troy Chemie GmbH 1993) |

---

* *n*-octanol/water distribution coefficient

Production

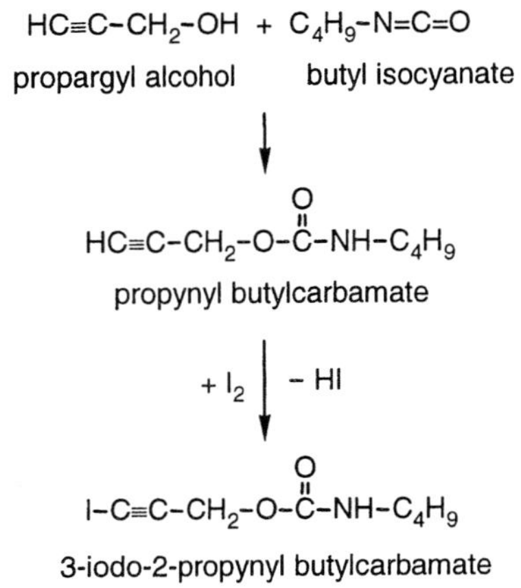


| | |
|---|---|
| Purity | at least 97 % (Troy Chemie GmbH 1993) |
| Impurities | no data |
| Uses | as a fungicide in metal-working fluids and cosmetics: concentration recommended for use in metal-working fluid emulsions 30–60 mg/kg (Troy Chemie GmbH 1994). |

# 1 Toxic Effects and Mode of Action

3-Iodo-2-propynyl butylcarbamate is rapidly absorbed from the gastrointestinal tract and distributed in the organism. The metabolism of the substance is not known. It is exhaled in the form of $CO_2$ and excreted as urinary metabolites which have not been identified; the substance is also stored in adipose tissue both unchanged and as metabolites. 3-Iodo-2-propynyl butylcarbamate is of low toxicity after single inhalation exposures, ingestion or dermal exposure. After repeated ingestion of the substance, the target organ is the liver. In addition, it causes local irritation in the stomach. 3-Iodo-2-propynyl butylcarbamate irritates the respiratory passages. It is a mild skin irritant but is corrosive in the eye. It has no sensitizing effects in the maximization test with guinea pigs. Only after maternally toxic doses is embryotoxicity observed in the rabbit. 3-Iodo-2-propynyl butylcarbamate is not genotoxic in the *Salmonella* mutagenicity test, in the test for unscheduled DNA repair synthesis (UDS) in rat hepatocytes or in the micronucleus test. In a 2-year feeding study with rats, it was not shown to have carcinogenic potential.

# 2 Mechanism of Action

There are no data available. In particular it is not known whether cleavage of iodine under physiological conditions could be responsible for any of the effects of the substance.

# 3 Toxicokinetics and Metabolism

## 3.1 Absorption, distribution, elimination

Groups of 14 male and 14 female CD rats were given 3-iodo-2-propynyl-[1-$^{14}$C]-butyl-[$^{14}$C]-carbamate either as single oral doses (20 or 125 mg/kg body weight) or intravenous injections (0.5 mg/kg body weight), or repeated oral doses (7 × 20 mg/kg body weight). Five minutes after intravenous injection, only 1 % of the dose was detectable in the plasma, evidence that the substance is distributed rapidly. After single oral doses of 20 or 125 mg/kg body weight, the highest levels of radioactivity in the plasma were detected 2 hours after administration of the substance. The highest levels of radioactivity were found in the liver and kidneys. 96 hours after the intravenous injection, 57 % of the dose had been exhaled as $CO_2$ and 32 % excreted in the urine as unidentified polar metabolites. After oral administration, on the other hand, 38 % was eliminated as $CO_2$ and 51 % with the urine. Ten days after a single oral dose, the level of radioactivity in the liver was 2 % of the value determined 3 hours after dosing. When the animals were given repeated doses, the radioactivity in the liver decreased during 10 days to 6 % and in the adipose tissue only to 37 %. In the skin and skeletal muscles the radioactivity also decreased only slowly (Troy Chemical Corporation 1987a).

There is no information available about the metabolism of the substance.

# 4 Effects in Man

There are no data available for the effects of 3-iodo-2-propynyl butylcarbamate in man.

# 5 Animal Experiments and *in vitro* Studies

## 5.1 Acute toxicity

### 5.1.1 Inhalation

In an inhalation limit test, 5 male and 5 female rats were exposed for 4 hours to a 3-iodo-2-propynyl butylcarbamate concentration of 6890 mg/m$^3$. After 1 hour, laboured breathing, salivation and nose bleeding were seen in all animals and during the last hour of exposure lacrimation as well. The animals lost weight for 2 to 4 days after the exposure. Clinical findings during the 14-day recovery period included encrustation of blood around the nose, eyes, muzzle, breast and ears, piloerection, breathing noises and urine-stained fur. In most animals these symptoms had regressed within 11 to 13 days. Enlargement of the renal pelvis was detected in one male and three female animals and three female control animals. In the three males the renal pelvis was also filled with fluid and in the kidneys there were dark red pin-head sized spots. One of the females died on day 5 after exposure, presumably as a result of the treatment (Troy Chemical Corporation 1985).

### 5.1.2 Ingestion

The oral $LD_{50}$ for male Sprague-Dawley rats was 1795 mg/kg body weight and for females 1056 mg/kg body weight. Clinical symptoms included soft faeces, piloerection, red material around the eyes and nose, urine-stained fur, reduced activity, ataxia, exhaustion and laboured breathing. Pathological examination of the surviving animals revealed no substance-related effects. The 16 animals which died during the study had dilated, fluid-filled gastrointestinal tracts and pale livers and kidneys (Troy Chemical Corporation 1984a).

### 5.1.3 Dermal absorption

3-Iodo-2-propynyl butylcarbamate doses of 2000 mg/kg body weight were moistened with physiological saline and applied occlusively for 24 hours to the intact or abraded skin of groups of 5 male rabbits. Immediately after the end of treatment, very slight erythema and severe oedema were seen in all animals. Signs of systemic intoxication were not seen during the 14-day recovery period (Troy Chemical Corporation 1982).

In a repeat study with groups of 5 male and 5 female rabbits, slight to very slight erythema was seen in all animals at the end of exposure but had regressed one day later. Oedema or signs of systemic intoxication were not seen during the 14-day recovery period (Troy Chemical Corporation 1991a).

## 5.2 Subacute, subchronic and chronic toxicity

There are no data available for the effects of repeatedly inhaled 3-iodo-2-propynyl butylcarbamate.

### Ingestion

Groups of 10 male and 10 female rats per dose group were given 3-iodo-2-propynyl butylcarbamate doses of 20, 50 or 125 mg/kg body weight, 5 times weekly for 13 weeks by gavage. No treatment-related changes were detected in food consumption, survival, haematology, biochemistry, urinalysis or autopsy findings. In the animals which had received the highest dose, the liver weights were increased and the hepatocytes enlarged. The males also gained weight less rapidly. These changes had all regressed within 28 days after the end of treatment (data from summary, Troy Chemical Corporation 1984b). In addition, forestomach irritation was found in the highest dose group (Troy Corporation 1994). This study yielded a no observed effect level (NOEL) of 50 mg/kg body weight and day for the rat.

## 5.3 Local effects on skin and mucous membranes

### 5.3.1 Skin

Slight erythema and severe oedema developed on the intact and abraded skin of 4 rabbits, 4 hours after occlusive contact with 0.5 ml 3-iodo-2-propynyl butylcarbamate (concentration not specified) for 4 hours. 48 hours after the treatment, the symptoms had completely disappeared. The substance was classified as not irritating with a primary irritation index of 0.5 (data from summary, Troy Chemical Corporation 1981).

3-Iodo-2-propynyl butylcarbamate doses of 500 mg were mixed to a paste with water and applied semi-occlusively for 4 hours to the skin of rabbits; 24 hours later mild to moderate erythema and mild oedema were seen in 2/6 animals. In one animal slight erythema was still present after 72 hours. The substance was classified as slightly irritating (Troy Chemical Corporation 1990a).

### 5.3.2 Eyes

3-Iodo-2-propynyl butylcarbamate (55 mg) was applied to one eye of each of nine male rabbits. The treated eyes of three of the animals were rinsed with water after 30 seconds. Hyperaemia, chemosis, discharge, corneal clouding, congestion, and a lack of reaction of the iris to light were recorded for the eyes which were not rinsed. Corneal clouding and vascularization, and conjunctivitis were still evident 21 days later. In the eyes which were rinsed, iritis and conjunctival irritation developed and persisted for up to 7 days. The authors classified the substance as highly irritating for the eye (Troy Chemical Corporation 1990b).

3-Iodo-2-propynyl butylcarbamate (100 mg) was applied to one eye of each of 3 male and 3 female rabbits. 48 to 96 hours later, severe corneal clouding, iritis, chemosis, discharge and conjunctival erythema were seen in all animals. The substance was given the highest score of 110 on the Draize scale and classified as corrosive (Troy Chemical Corporation 1991b).

## 5.4 Allergenic effects

3-Iodo-2-propynyl butylcarbamate was tested for sensitizing effects in a maximization test with 20 Dunkin-Hartley guinea pigs. The intradermal induction treatment was two injections of 0.05 ml Freund's complete adjuvant, a 10 % solution of the test substance in propylene glycol, or 20 % test substance together with Freund's complete adjuvant (1:1). Seven days later, 200 mg of the test substance (50 % in petrolatum) was applied topically. Provocation was carried out three weeks after the intradermal induction by occlusive topical application of 0.01 mg test substance (0.01 % in petrolatum). During the induction treatment, erythema and oedema were observed. There was no evidence of sensitization (Troy Chemical Corporation 1989a).

## 5.5 Reproductive and developmental toxicity

In a two-generation study, groups of 25 male and 25 female Sprague-Dawley rats and their progeny were fed 3-iodo-2-propynyl butylcarbamate in the diet in concentrations of 120 or 300 mg/kg (about 10 and 25 mg/kg body weight and day); no effects on mating, fertility or development were seen. In the $F_0$ males given 300 mg/kg diet, body weight gains were delayed before mating. In the group given 750 mg/kg diet (about 60 mg/kg body weight and day) the number of live births was slightly reduced and the food consumption and body weight development of $F_0$ and $F_1$ males before mating were reduced (data from summary; Troy Chemical Corporation 1987b).

Groups of 28 to 33 Sprague-Dawley rats were given daily 3-iodo-2-propynyl butylcarbamate doses of 20, 50 or 125 mg/kg body weight in corn oil by gavage during days 6 to 15 of gestation. Neither embryotoxic nor teratogenic effects were detected in the progeny. Maternal toxicity was seen in the 125 mg/kg group in the form of reduced body weight gains during days 6 to 10 of gestation (data from summary; Troy Chemical Corporation 1986).

Groups of 35 NMRI mice were given daily 3-iodo-2-propynyl butylcarbamate doses of 20, 50 or 125 mg/kg body weight in corn oil by gavage during days 6 to 15 of gestation. No maternal toxicity was seen in the dams nor developmental toxic effects in the progeny (Troy Chemical Corporation 1988b).

On days 6 to 18 of gestation, groups of 20 rabbits were given 3-iodo-2-propynyl butylcarbamate doses of 2, 20 or 50 mg/kg body weight and day in corn oil by gavage. Only in the 50 mg/kg group was there a slight increase in post-implantation losses and the associated reduction in the numbers of living foetuses. In this dose group, severe maternal toxicity was observed: two animals died and one had to be killed in a moribund state. The average body weight of the exposed animals on day 19 of gestation was 90 % of the control value ($p < 0.05$). Also the body weight gains during gestation days 6 to 9, 9 to 12, and 15 to 19 were significantly lower than in the control group. In the other dose groups, no substance-related effects were detected (Troy Chemical Corporation 1992).

This study yielded a NOEL of 20 mg/kg body weight and day for developmental toxicity.

## 5.6 Genotoxicity

### 5.6.1 *In vitro*

3-Iodo-2-propynyl butylcarbamate was not mutagenic in the *Salmonella* mutagenicity test in the strains TA98, TA100, TA1535, TA1537 and TA1538, either with or without a metabolic activation system. The concentrations used were 6.2 to 500 µg/plate; cytotoxicity was seen from 166.7 µg/plate (data from summary; Troy Chemical Corporation 1984c).

In another test with the same *Salmonella typhimurium* strains, mutagenicity was also not detected in the concentration range from 1 to 1000 µg/plate. Cytotoxicity was seen from 333 µg/plate (data from summary; Troy Chemical Corporation 1989b).

3-Iodo-2-propynyl butylcarbamate in concentrations from 3 to 13.5 µg/ml did not induce DNA repair synthesis in cultured primary rat hepatocytes. Cytotoxicity increased with increasing concentration (Troy Chemical Corporation 1988c).

### 5.6.2 *In vivo*

In a micronucleus test, groups of 5 male and 5 female CD-1 mice were given 3-iodo-2-propynyl butylcarbamate doses of 200, 660 or 2000 mg/kg body weight in corn oil by gavage. After 30, 48 and 72 hours, the incidence of polychromatic erythrocytes containing micronuclei was not increased in the bone marrow. Toxic effects were seen at the two higher doses (data from summary; Troy Chemical Corporation 1984d).

## 5.7 Carcinogenicity

In a feeding study, groups of 50 male and 50 female Sprague-Dawley rats were given 3-iodo-2-propynyl butylcarbamate doses of 20, 40 or 80 mg/kg body weight and day for 104 weeks. Body weight gains were significantly reduced in a dose-dependent manner in all the male rats and in the females of the two higher dose groups. In the males of the 20 mg/kg group, there were significant reductions relative to the control values in body weights (reduced by 11 %) and non-significant reductions in food consumption (by 5 %). Food consumption was significantly reduced only in the 80 mg/kg group males. The histopathological examination revealed in both sexes in the medium and high dose groups dose-dependent lobular degeneration of the salivary glands, and in the forestomach submucosal oedema, inflammation, acanthosis, inflammation in the keratin layer, hyperkeratosis, ulceration and basal cell proliferation. Among the males of the 20 mg/kg group, most of these effects were also seen, but their incidence was not significantly higher than in the control group. Tumour incidences were not higher than in the controls. The authors concluded that the NOEL for "significant structural effects" of 3-iodo-2-propynyl butylcarbamate is 20 mg/kg body weight and day (Troy Chemical Corporation 1989c). However, this study provides evidence of local irritative effects and delayed body weight gains in male rats even at the lowest tested dose of 20 mg/kg body weight and day.

## 5.8 Other effects

With rats given intravenous injections of 3-iodo-2-propynyl butylcarbamate in doses of 2, 4, 10 or 16 mg/kg body weight, unlike with rats treated with other carbamates, no evidence of inhibition of erythrocyte cholinesterase was found during the period from 15 minutes to 5 hours after dosing (data from summary; Troy Chemical Corporation 1988a).

# 6 Manifesto (MAK value/classification)

Ingested 3-iodo-2-propynyl butylcarbamate is rapidly absorbed, metabolized and excreted. The acute oral, inhalative and dermal toxicity is low. After repeated oral doses, the target organ is the liver and local irritation is observed in the stomach.

3-Iodo-2-propynyl butylcarbamate is irritating for the skin and respiratory passages and has corrosive effects in the eye. In animal studies (maximization test) sensitization was not demonstrated.

Prenatal toxicity was seen only after maternally toxic doses; rabbits were more sensitive than rats or mice.

Negative results were obtained with the substance in the *Salmonella* mutagenicity test, in a test for unscheduled DNA repair synthesis and in the micronucleus test. No evidence of carcinogenic potential was obtained in a carcinogenicity study.

From a two year feeding study with rats no NOEL can be derived because even at the lowest dose tested (20 mg/kg body weight and day) body weight gains were slightly reduced and local irritation was observed in male rats. The 20 mg/kg body weight dose is, however, very close to the NOEL.

There are no data for effects in man. Studies of repeated inhalation exposure are not available. As mucosal irritation is to be expected during inhalation of 3-iodo-2-propynyl butylcarbamate, a MAK value cannot be derived from the studies available to date in which the substance was administered orally. 3-Iodo-2-propynyl butylcarbamate is therefore listed in Section IIb of the *List of MAK and BAT Values*. There is no evidence of skin penetration or sensitizing properties.

# 7 References

Troy Chemical Corporation (1981) *Primary skin irritation and corrosion test performed on Polyphase anti-mildew # 697-8012.* Bioassay Systems Corporation, 15th April 1981, unpublished report

Troy Chemical Corporation (1982) *Acute dermal toxicity study performed on 100% Polyphase. Project No. 11100,* Bioassay Systems Corporation, unpublished report

Troy Chemical Corporation (1984a) *Acute oral toxicity study in rats. 3-Iodo-2-propynyl butyl carbamate. Project No. 2277-100,* Hazleton Laboratories America Inc, unpublished report

Troy Chemical Corporation (1984b) *90-day subchronic oral toxicity test in rats, test material: Troysan Polyphase.* Bioassay Systems Corporation, 11th May 1984, unpublished report

Troy Chemical Corporation (1984c) *Troysan Polyphase (IPBC): Salmonella typhymurium/mammalian microsome plate incorporation assay.* Hazleton Biotechnologies Corporation, 24th May 1984, unpublished report

Troy Chemical Corporation (1984d) *In vivo micronucleus assay in mice. 3-Iodo-2-propynyl-butyl carbamate (IPBC).* Hazleton Biotechnologies Corporation, 3rd October 1984, unpublished report

Troy Chemical Corporation (1985) *Acute inhalation limit test in rats. 3-Iodo-2-propynyl butyl carbamate. Project No. 2277-101,* Hazleton Laboratories America Inc, unpublished report

Troy Chemical Corporation (1986) *Troysan Polyphase (IPBC): oral (gavage) teratogenicity study in rats.* Hazleton Laboratories Deutschland, 18th December 1986, unpublished report

Troy Chemical Corporation (1987a) *Troysan Polyphase (IPBC): the residue kinetics of 3-Iodo-2-propynyl-N-[1-$^{14}$C]-butyl[$^{14}$C]-carbamate. IRI Project No. 131952,* Inveresk Research Int, unpublished report

Troy Chemical Corporation (1987b) *Troysan Polyphase. Two generation oral (dietary administration) reproduction toxicity study in the rat (one litter per generation). Report No. 548-511/3,* Hazleton Laboratories Deutschland, unpublished report

Troy Chemical Corporation (1988a) *Polyphase cholinesterase inhibition study in rats.* Inveresk Research Int, 22nd April 1988, unpublished report

Troy Chemical Corporation (1988b) *Troysan Polyphase (IPBC): oral (gavage) teratogenicity study in the mouse.* Hazleton Laboratories Deutschland, 27th October 1988, unpublished report

Troy Chemical Corporation (1988c) *IPBC assessment of genotoxicity in an unscheduled DNA synthesis assay using adult rat hepatocyte primary cultures. IRI Project No. 737447,* Inveresk Research Int, unpublished report

Troy Chemical Corporation (1989a) *Test for delayed contact hypersensitivity with Troysan Polyphase P100 in the albino guinea pig. Project No. 11582,* Scantox Biological Laboratory, unpublished report

Troy Chemical Corporation (1989b) *Troysan Polyphase P100 IPBC-97% Batch No. P-2848-8603-P100: testing for mutagenic activity with Salmonella typhymurium TA1535, TA1537, TA1538, TA98 and TA100. IRI Project No. 740140,* Inveresk Research Int, unpublished report

Troy Chemical Corporation (1989c) *3-Iodo-2-propynyl butyl carbamate (IPBC) 104 week dietary carcinogenicity study in rats. Project No. 435580,* Inveresk Research Int, unpublished report

Troy Chemical Corporation (1990a) *Troysan Polyphase P100: primary skin irritation test in rabbits. Report No. 7076,* Inveresk Research Int, unpublished report

Troy Chemical Corporation (1990b) *Primary eye irritation study in rabbits with Troysan Polyphase P100. SLS No. 3228.1,* Springborn Laboratories, unpublished report

Troy Chemical Corporation (1991a) *Acute dermal study. Troysan Polyphase P-100. Projekt No. 91G-0988,* Toxikon Corporation, unpublished report

Troy Chemical Corporation (1991b) *Primary eye irritation study. Troysan Polyphase P-100. Project No. 91G-0305,* Toxikon Corporation, unpublished report

Troy Chemical Corporation (1992) *Teratology study in rabbits with Troysan Polyphase P-100. Study No. 3228.16,* Springborn Laboratories, unpublished report

Troy Chemie GmbH (1993) *EG-Sicherheitsdatenblatt gemäß 91/155 EWG, Troysan Polyphase P100*, (Safety datasheet) 29th November 1993

Troy Chemie GmbH (1994) written communication to the Deutsche Wissenschaftliche Gesellschaft für Erdöl, Erdgas und Kohle e.V., 05.01.1994

Troy Corporation (1994) written communication to the Deutsche Wissenschaftliche Gesellschaft für Erdöl, Erdgas und Kohle e.V., 30.03.1994

completed 24.11.1995

# Isodecyl oleate

| | |
|---|---|
| **MAK value** | **not yet established, see Section IIb of the *List of MAK and BAT Values*** |
| **Peak limitation** | – |
| **Absorption through the skin** | – |
| **Sensitization** | – |
| **Carcinogenicity** | – |
| **Prenatal toxicity** | – |
| **Germ cell mutagenicity** | – |
| **BAT value** | – |
| Synonyms | (*Z*)-(2-methylnonyl)-9-octadecenoate<br>oleic acid isodecyl ester |
| Trade name | ESTOL IDCO 3667 |
| Chemical name (CAS) | (*Z*)-9-octadecenoic acid isodecyl ester |
| CAS number | 59231-34-4 |
| Structural formula | $CH_2-(CH_2)_5-COO-(CH_2)_7-CH(CH_3)_2$<br>\|<br>$CH_2-CH=CH-(CH_2)_7-CH_3$ |
| Molecular formula | $C_{28}H_{54}O_2$ |
| Molecular weight | 422.74 |
| Melting point | see below |
| Boiling point | see below |
| Density at 25°C | 0.87 g/cm$^3$ (Unichema 1991) |
| Vapour pressure | see below |
| log $P_{OW}$* | see below |
| **1 ml/m$^3$ (ppm) ≙ 17.541 mg/m$^3$** | **1 mg/m$^3$ ≙ 0.057 ml/m$^3$ (ppm)** |

* *n*-octanol/water distribution coefficient

| | |
|---|---|
| Solubility | soluble in alcohol and ether, insoluble in water at 20°C |
| Stability | no data |
| Production | esterification of oleic acid with isodecyl alcohol |
| Purity | no data |
| Impurities | maximum 2.5 % oleic acid |
| Uses | – in the cosmetics industry in concentrations of 0.1 % to 25 %<br>– in metal-working fluids (lubricity improver) |

The physical properties of isodecyl oleate are thought to be similar to those of decyl oleate (see *Occupational Toxicants* Volume 9).

Data for the boiling point or vapour pressure of isodecyl oleate were not found in the literature. From the analogy with short-chain aliphatic oleic acid esters, it was estimated that the boiling point of decyl oleate is above 200°C; if it is assumed to be 240°C, from the temperature dependence of the vapour pressure (Clausius-Clapeyron equation) it may be calculated that the vapour pressure at room temperature must be about 0.1 hPa (Henkel 1994a). Decyl oleate and isodecyl oleate are thus not volatile. The log $P_{OW}$ for decyl oleate was calculated to be above 6, the melting point of decyl oleate is 18.2°C (Henkel 1994b).

The metabolism of isodecyl oleate probably yields oleic acid, a substance which has marked tumour-promoting properties on mouse skin. An ester of oleic acid, methyl oleate, has also been shown to be a tumour promoter on mouse skin.

# 1 Toxic Effects and Mode of Action

Isodecyl oleate, with an $LD_{50}$ of more than 34800 mg/kg body weight, is practically non-toxic for the rat. Single doses of the substance applied to the skin or eyes of rabbits did not cause irritation. Repeated dermal application caused slight irritation on rabbit skin but no irritation in man. In animal studies, application of the substance in concentrations up to 15 % did not cause sensitization.

# 2 Mechanism of Action

The irritation observed after repeated application of the substance to the skin of experimental animals could be a result of the defatting action of isodecyl oleate.

# 3 Toxicokinetics and Metabolism

There are no data available for the toxicokinetics or metabolism of isodecyl oleate. The structure of the substance suggests that metabolism would involve hydrolysis to yield isodecanol and oleic acid.

# 4 Effects in Man

There are no data available for the effects on man of single or multiple exposures to isodecyl oleate. Nor are data available for sensitization, genotoxicity, reproductive toxicity or carcinogenicity of isodecyl oleate in man

## 4.1 Local effects on skin and mucous membranes

In patch tests in which undiluted isodecyl oleate was applied occlusively to the skin of 19 test persons for 24 hours, no erythema was observed (Anonymous 1982).

Undiluted isodecyl oleate was applied occlusively to the skin of 9 persons daily for 21 days in the cumulative irritancy assay of Maibach. An overall irritation score of 1 was found (on a scale with a maximum of 756, no other details; Anonymous 1982).

# 5 Animal Experiments and *in vitro* Studies

There are no data available for the reproductive or developmental toxicity, genotoxicity or carcinogenicity of isodecyl oleate (but see the introductory paragraph above).

## 5.1 Acute toxicity

There are no data available for the effects of exposure to isodecyl oleate by inhalation or by application of the substance to the skin.

Two female and three male Wistar rats per dose group were given isodecyl oleate doses of 2.5, 5.0, 10.0, 20.0 or 40.0 ml/kg body weight by gavage. Only one rat in the highest dose group died. Thus, the $LD_{50}$ was above 40 ml/kg body weight (34800 mg/kg body weight). Symptoms of intoxication were not described (Anonymous 1982).

## 5.2 Subacute, subchronic and chronic toxicity

In the 60-day study of Guillot *et al.* (1977; see Section 5.3.1) in which the substance was applied to the skin of rabbits, no systemic effects were reported. For the structurally analogous compound, decyl oleate, administered orally to rats in a 28-day study, a no observed adverse effect level (NOAEL) of 1000 mg/kg body weight was reported (see the 1995 documentation for decyl oleate in Volume 9 of the present series).

## 5.3 Local effects on skin and mucous membranes

### 5.3.1 Skin

Doses of 0.5 ml isodecyl oleate were applied occlusively for 24 hours to the intact and scarified skin of 3 rabbits. The primary irritation score (Draize) was 1.0 (no other details; Anonymous 1982). In another study with rabbits, a primary irritation score of 0.28 was reported for undiluted isodecyl oleate (Anonymous 1982).

In two independent studies with rabbits in which a modified Draize scale was used, the primary irritation index for undiluted isodecyl oleate was given as 0.13 and 0 and for a 15 % aqueous emulsion 0 in both studies (Guillot *et al.* 1977).

Thus, after single applications, isodecyl oleate is not considered to be irritating for the skin.

In a 60-day study, isodecyl oleate was applied daily to the skin of three rabbits, both as the undiluted substance and as a 15 % aqueous emulsion. At the end of the study period, two skin samples from each animal were subjected to histological examination. In one animal, blister formation and congestion with pericapillary inflammatory infiltrates were seen after application of undiluted isodecyl oleate. With the emulsion, only few nodules and blisters developed and no significant histological reaction was detected. Undiluted isodecyl oleate was classified as moderately irritating (Guillot *et al.* 1977).

### 5.3.2 Eyes

Aliquots of 0.1 ml undiluted isodecyl oleate or 0.1 ml of a 15 % solution of the substance in corn oil were applied to rabbits into the conjunctival sac of one eye; the other eye served as the untreated control. For the 15 % solution an average irritation score of 0.3 was obtained; the undiluted substance was tolerated without symptoms (Anonymous 1982).

In a modified Draize test, the irritant effects of undiluted isodecyl oleate in the rabbit eye were recorded, 1 hour and 1, 2, 3, 4 and 7 days after instillation of the substance. The average irritation score of 4 recorded after one hour was reduced to 0 after as little as 24 hours. The irritant effects of the substance were classified as very slight (Guillot *et al.* 1977).

## 5.4 Allergenic effects

In the sensitization test of Landsteiner and Jacobs, a 15 % solution of isodecyl oleate (solvent not specified) was found to have no allergenic effects in the guinea pig (no other details; Anonymous 1982).

In the study by Guillot *et al.* (1977) described in Section 5.3.1, the animals were left untreated for 7 days at the end of the 60-day application period and then a provocation treatment (concentration not specified) was carried out. No evidence of allergic reactions was obtained (no other details).

# 6 Manifesto (MAK value/classification)

The available data are insufficient for the establishment of a MAK value. Isodecyl oleate is therefore listed in Section IIb of the *List of MAK and BAT Values*. Under workplace conditions with the concentrations of about 5 % which are used in metal-working fluids, irritant effects of isodecyl oleate are not to be expected. There are no data available which support the hypothesis that isodecyl oleate, like its structural analogue oleic acid, could have tumour-promoting activity.

In animal studies, no evidence of sensitizing effects of the substance have been found. Reports of effects in users are not available. Isodecyl oleate has therefore not been designated with an "S".

There are no data for the absorption of the substance through the skin.

# 7 References

Anonymous (1982) Final report on the safety assessment of decyl and isodecyl oleates. *J Am Coll Toxicol 1*: 85–95

Guillot JP, Martini MC, Giauffret JY (1977) Safety evaluation of cosmetic raw materials. *J Soc Cosmet Chem 28*: 377–393

Henkel (1994a) written communication to the Deutsche Wissenschaftliche Gesellschaft für Erdöl, Erdgas und Kohle e.V., 20.01.1994

Henkel (1994b) *Grunddatensatz Decyloleat* (Datasheet Decyl oleate)

Unichema (1991) *Product data sheet: Isodecyl oleate*, Unichema International, Gouda, Netherlands

completed 31.03.1998

# 2-Octyl-4-isothiazolin-3-one

| | |
|---|---|
| **MAK value (1997)** | **0.05 mg/m$^3$ I (inhalable aerosol fraction)** |
| **Peak limitation (1997)** | **Category I (excursion factor 2)** |
| **Absorption through the skin (1997)** | **H** |
| **Sensitization (1997)** | **Sh** |
| **Carcinogenicity** | **–** |
| **Prenatal toxicity (1997)** | **Pregnancy risk group C** |
| **Germ cell mutagenicity** | **–** |
| **BAT value** | **–** |
| Synonyms | Octhilinone<br>2-octyl-4-isothiazolin-3-one<br>2-octyl-3(2*H*)-isothiazolone<br>*N*-*n*-octyl-3-isothiazolone<br>2-*n*-octyl-3-isothiazolone |
| Trade names | Acticide® 45<br>Fungicide E®<br>(about 25 % in propylene glycol)<br>Kathon® 893 MW, Skane® M-8,<br>Skane® M-8 HQ<br>(about 45 % in propylene glycol)*<br>Parmetol® F85<br>RH 893 T (88.7 %, technical grade)<br>RH 893 HQ Technical (99.1 %) |
| Chemical name (CAS) | 2-octyl-3(2*H*)-isothiazolone |
| CAS number | 26530-20-1 |
| Structural formula | 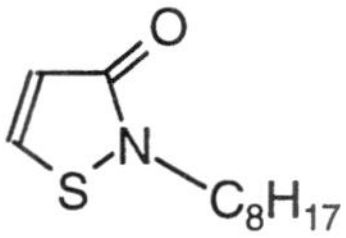 |



**1 ml/m$^3$ (ppm) ≙ 8.85 mg/m$^3$** **1 mg/m$^3$ ≙ 0.11 ml/m$^3$ (ppm)**

---

* The studies described in the present paper were mostly carried out with RH 893, Kathon® 893 and Skane® M-8.

| | |
|---|---|
| Molecular formula | $C_{11}H_{19}NOS$ |
| Molecular weight | 213.34 |
| Melting point | 22°C |
| Boiling point | decomposes at 270–290°C |
| Density at 25°C | 1.024 g/cm$^3$ |
| Vapour pressure | 0.00005 hPa (at 25°C) |
| log $P_{ow}$* | 2.45 |
| Solubility at 25°C | 480 mg/kg water, very highly soluble in methanol, ethanol, propylene glycol, acetone, ethyl ether, ethyl acetate, chloroform, butyl cellosolve, corn oil and mineral oils, insoluble in heptane (Rohm and Haas Company 1985) |
| pH | 2.4 (10 % solution of Kathon® 893 MW) |
| Purity, impurities | 2-Octyl-4-isothiazolin-3-one is commercially available as a solution (Kathon® 893 MW) containing (w/w): 2-octyl-4-isothiazolin-3-one (45–48 %), propylene glycol (52–55 %), monochlorobenzene (max. 0.2 %), related reaction products (not declared, max. 0.5 %), water (max. 0.5 %) (Rohm and Haas Company 1994) |
| Stability | stable in the pH range 2–10 in water and in metal-working fluids, not stable in metal-working fluid concentrates, inactivated by primary and secondary amines, for example (Paulus 1993) |
| Production | reaction of methyl-3-mercaptopropionate with *n*-octylamine, followed by reaction with monochlorobenzene and chlorine (Rohm and Haas Deutschland GmbH 1995b) |
| Uses | in metal-working fluids at a recommended 2-octyl-4-isothiazolin-3-one concentration of 25 mg/kg for routine preventive use, 75 mg/kg to disinfect highly contaminated systems (Rohm and Haas Deutschland GmbH 1995a), in paints as a film fungicide in concentrations up to 1000 mg/kg (Rohm and Haas Company 1991b), in the wood industry as protection from fungal staining, in glues and sealants to prevent microbial decomposition, in tanning of leather and paper production as a fungicide (Paulus 1993) |

* *n*-octanol/water distribution coefficient

# 1 Toxic Effects and Mode of Action

2-Octyl-4-isothiazolin-3-one is readily absorbed through the skin both *in vivo* and *in vitro*. Even 50 % solutions of the substance are corrosive for the skin and eye. Repeated inhalation of 2-octyl-4-isothiazolin-3-one aerosols results in mucosal irritation and macroscopic changes in the lungs, metaplasia of the squamous respiratory epithelium in the nasal cavity, hyperplasia of mucosal secretory cells and the formation of intraepithelial droplets. Both inhalation and ingestion of the substance result in gastrointestinal disorders, irritation of the gastrointestinal mucosa, discoloration of the liver and kidneys and CNS depression. Effects on the CNS are also seen after dermal application. There are a few reports of sensitizations following occupational contact with 2-octyl-4-isothiazolin-3-one. The substance was shown to be sensitizing for the guinea pig in a maximization test and a modified Buehler test.

2-Octyl-4-isothiazolin-3-one is not mutagenic in the *Salmonella* mutagenicity test nor in the hypoxanthine guanine phosphoribosyl transferase (HPRT) test in CHO cells (a cell line from Chinese hamster ovary) either with or without metabolic activation. In a test for unscheduled DNA synthesis (UDS) in cultured hepatocytes, 2-octyl-4-isothiazolin-3-one did not induce repair synthesis. The substance is clastogenic *in vitro* but not *in vivo*. Administered to mice in the diet, 2-octyl-4-isothiazolin-3-one is not carcinogenic.

# 2 Mechanism of Action

The caustic effects of the substance are a result of its low pH and its reactions with nucleophiles. The antimicrobial action of 2-octyl-4-isothiazolin-3-one involves opening of the isothiazolinone ring in reactions with nucleophilic cell components (Paulus 1993).

# 3 Toxicokinetics

Aliquots of 100 µl of a 0.05 % solution of $^{14}C$-labelled 2-octyl-4-isothiazolin-3-one in 80 % aqueous ethanol or in acetone were applied without occlusion to the skin of male Hartley guinea pigs; after three hours 33 % and 29 %, respectively, of the applied radioactivity was found in the skin. After occlusive application of 400 µl aliquots of $^{14}C$-labelled 2-octyl-4-isothiazolin-3-one solutions in concentrations of 0.01 % to 0.2 % in 80 % aqueous ethanol or in acetone for six hours, 35 % to 47 % and 30 % to 36% of the applied radioactivity was retained in the skin. Only $^1/_5$ to $^1/_9$ of these amounts were retained in the skin when 2-octyl-4-isothiazolin-3-one was applied in concentrations up to 0.5 % in paints (Rohm and Haas Company 1990).

The ability of radioactively labelled 2-octyl-4-isothiazolin-3-one to penetrate the skin was investigated in a test with freshly prepared skin from adult Crl:CD®BR rats in Franz diffusion cells. After 3 hours, 87 % of the radioactivity from a 0.03 % aqueous solution had entered the skin. After 24 hours all of the 2-octyl-4-isothiazolin-3-one was either bound to the skin or in the receptor fluid. After 48 hours, 15 % of the applied radioactivity was found in the receptor fluid but intact 2-octyl-4-isothiazolin-3-one could no longer be detected. When propylene glycol was used as the solvent, passage of the substance through the skin was slower at the same concentration: after 24 hours 45 % had entered the skin, after 48 hours 75 %. Most of the radioactivity was bound in or on the skin. In the receptor fluid 10 % was found. The rate of penetration of 2-octyl-4-isothiazolin-3-one from a latex paint into the skin was $^1/_{10}$ to $^1/_{20}$ of that from an aqueous solution. In the receptor fluid, practically no radioactivity could be detected. The authors suggested that the different bioavailability of the substance in the different solutions and emulsions is of significance for the evaluation of the sensitizing potential of 2-octyl-4-isothiazolin-3-one (Rohm and Haas Company 1991b).

There are no data available for distribution, metabolism or excretion of 2-octyl-4-isothiazolin-3-one.

# 4 Effects in Man

There are no data available for the effects of single or repeated exposure of man to 2-octyl-4-isothiazolin-3-one.

## 4.1 Local effects on skin and mucous membranes

See Section 4.2.

## 4.2 Allergenic effects

There are two case reports in which extensive allergic reactions were shown to have resulted from the addition of 2-octyl-4-isothiazolin-3-one to paint. The patch test with 1 % 2-octyl-4-isothiazolin-3-one in olive oil yielded clearly positive results in these persons, but negative results in 20 control persons (Thormann 1982). The allergenic effects of the substance were confirmed in another case where a similar kind of exposure resulted in eczema on the forearm and a bullous reaction in the patch test with 0.045 % Skane® M-8 (Mathias *et al.* 1983).

Undiluted RH 893 Technical (about 89 % 2-octyl-4-isothiazolin-3-one in propylene glycol) was applied to the skin of 50 test persons. Because of the severe irritation which developed during the 24 hours after the application, a 1 % solution in water, which proved not to cause skin irritation, was used for the remaining 5 to 7 applications. In the course of these applications, the substance proved to be sensitizing in all test persons taking part in this section of the study (number not specified). The allergenic effect was confirmed by provocation treatment which resulted in reactions typical of sensitization also on untreated skin sites far from the initially treated area (no other details). Skin sensitization was also reported in 4/15 test persons after exposure for 15 minutes daily, 5 days per week for 3 weeks to vapour produced from RH 893 Technical at 32–37°C (no other details; Rohm and Haas Company 1973a).

The Danish contact dermatitis group tested 671 ambulant eczema patients in occlusive patch tests with 2-octyl-4-isothiazolin-3-one (0.05 % Kathon 893 in petrolatum). No positive results were obtained (Andersen and Veien 1985).

In a cumulative patch test carried out to determine the appropriate concentration for patch tests, 8/20 persons with healthy skin produced allergic reactions to 0.1 % and 1/20 to 0.05 % during the 21 daily applications. Provocation by occlusive application of a 0.025 % preparation for 48 hours produced allergic reactions in 6/6 of these sensitized persons; 0.01 % produced reactions in 1/6. In these studies concentrations up to 0.05 % were not irritating (Emmet *et al.* 1989).

A patient with eczema on the hands produced a positive result in a patch test with a 10 % preparation in olive oil of a commercial product which he used at work in shoe production; the product contained 7.65 % 2-octyl-4-isothiazolin-3-one and 0.14 % of a mixture of chloromethylisothiazolinone and methylisothiazolinone (CMI/MI) as preservative. In tests with the individual substances, reactions were obtained with benzisothiazolinone (1 % in petrolatum, 2+) CMI/MI (0.1 % in petrolatum, 3+) and 2-octyl-4-isothiazolin-3-one (0.1 % in petrolatum, 3+; 1 % in petrolatum, 2+). 2-Octyl-4-isothiazolin-3-one induced “toxic reactions” in 1/20 control persons at 0.1 % concentration and 10/20 at 1 % (Foussereau *et al.* 1984).

The IVDK clinics (“Informationsverbund Dermatologischer Kliniken”: Information Network of German Departments of Dermatology) reported for the years 1990 to 1994 the following results of tests on 928 patients with CMI/MI (0.01 % in water), benzisothiazolinone (0.1 % in petrolatum) and 2-octyl-4-isothiazolin-3-one (0.025 % in petrolatum): allergic reactions to CMI/MI were seen in 36 patients (3.9 %), to benzisothiazolinone in 12 (1.3 %) and to 2-octyl-4-isothiazolin-3-one in 6 (0.6 %). No patient reacted to both CMI/MI and 2-octyl-4-isothiazolin-3-one (Geier and Schnuch 1996).

Of 202 painters (of buildings), 32 reported current eczema and 16 previously suffered eczema; in 16 and 9 cases, respectively, the eczema was on the hands. In allergy tests with the standard series (TRUE test) and painter series, one person reacted to 2-octyl-4-isothiazolin-3-one (and simultaneously to benzisothiazolinone), nickel, cobalt and rosin, 5 to CMI/MI and 7 to benzisothiazolinone. For three of the persons who reacted to CMI/MI, 2 of those reacting to benzisothiazolinone and for those reacting to 2-octyl-4-isothiazolin-3-one, an association between hand eczema and occupational exposure could be constructed (Fischer *et al.* 1995).

In another study, one case of chronic eczema on the foot and one case of acute eczema on the hand, arm and face were reported. In one case the eczema was induced by a shoe treated with a 2-octyl-4-isothiazolin-3-one fungicide, the other by 2-octyl-4-isothiazolin-3-one in a latex paint. Both patients yielded positive results in the patch test with 2-octyl-4-isothiazolin-3-one and also with diphenylthiourea and dibutylthiourea but negative results with the standard and painter test series (Oleaga *et al.* 1992).

# 5 Animal Experiments and *in vitro* Studies

## 5.1 Acute toxicity

### 5.1.1 Inhalation

For Skane® M-8 HQ (purity 95 %), the $RD_{50}$ in the mouse was found to be 19.9 mg/m$^3$. The highest evaluated concentration was 9.4 mg/m$^3$. The substance was used in the form of a 0.1 % to 0.7 % solution in propylene glycol. For the positive control, formaldehyde, an $RD_{50}$ of 5.4 ml/m$^3$ was obtained (Rohm and Haas Company 1991c).

Male and female rats were exposed via the nose for four hours to an aerosol (45 % solution in propylene glycol) in 2-octyl-4-isothiazolin-3-one concentrations of 58, 94, 229 and 671 mg/m$^3$; 61 % to 77 % of the aerosol was inhalable. In the lowest dose group, dyspnoea, bradypnoea and breathing noises developed after the exposure. During the 2-week recovery period, the animals gasped for air, ate little and were in a bad state. In the dose groups exposed to 94 mg/m$^3$ or more, the animals showed signs of CNS depression such as ataxia, prostration and apathy, as well as polyuria, diarrhoea, and swollen, red-tinged noses. The gross pathological examination revealed changes in the lungs and renal medulla in the lowest dose group animals, at 94 mg/m$^3$ also in the liver and at 229 mg/m$^3$ enlargement of the adrenal glands and oedema of the tongue. The only findings which were significantly different from the historical control values were the lung changes in male animals exposed to 229 mg/m$^3$ and in females exposed to 671 mg/m$^3$. Body weight gain was significantly reduced below the historical control values for the first five days after the exposure in all males and until day 7 in the females. For males and females together, the $LC_{50}$ was 586 mg/m$^3$, for the males alone 572 mg/m$^3$. The no observed effect level (NOEL) was lower than the lowest tested concentration of 58 mg/m$^3$ (Rohm and Haas Company 1986a).

### 5.1.2 Ingestion

The results of the acute oral toxicity studies with rats are summarized in Table 1.

In an early study, the $LD_{50}$ for 2-octyl-4-isothiazolin-3-one administered to male rats was given as 708 mg/kg body weight and for females as 562 mg/kg body weight (calculated by the author; purity not specified). In animals given doses of 31.6 mg/kg body weight or more, gastrointestinal and central nervous symptoms developed, and erythema of the stomach, intestine, liver and kidneys from 1000 mg/kg body weight; in the females erythema of the adrenal glands was seen. This dose was lethal for 9/10 animals within 24 hours (Rohm and Haas Company 1967).

**Table 1.** Acute oral toxicity of 2-octyl-4-isothiazolin-3-one

| Species (strain) number, sex | Product, administration form | **Dose of 2-octyl-4-isothiazolin-3-one (mg/kg body weight):** effects | References |
|---|---|---|---|
| rat (Sprague-Dawley), 5 ♂, 5 ♀ per group | RH 893, assumed content of 2-octyl-4-iso-thiazolin-3-one: 100 % 1–10 % in corn oil or undiluted | **31.6 or more:** reduced activity, laboured breathing **100 or more:** thickening of the *pars cardiaca* of the stomach **316 or more:** ataxia, ptosis **1000:** cyanosis, erythema of intestine, liver and kidney; **♂:** 4/5 died, adverse effects on righting reflex, salivation; **♀:** 5/5 died, erythema of the adrenal glands **3160/kg:** cyanosis, erythema of intestine, liver, kid-ney; **♂:** 4/5 died, adverse effects on righting reflex, salivation; **♀:** 5/5 died, erythema of the adrenal glands **$LD_{50}$ ♂:** 708 (447–1129) mg/kg body weight; **$LD_{50}$ ♀:** 562 mg/kg body weight | Rohm and Haas Company 1967 |
| rat (Sprague-Dawley), 5 ♂, 5 ♀ per group | RH 893 Technical, assumed content of 2-octyl-4-iso-thiazolin-3-one: 100 % 10 % in corn oil | **100 and 215 and more:** diarrhoea **464 or more:** CNS depression, laboured breathing **1000:** animals which died: adrenal glands dark red, kidneys discoloured in three zones: black medulla, cortex dark red and white areas with a sharp border, liver blotchy with translucent edges; **♂:** 4/5 died, 1/5 before day 5, CNS depression, posture hunched, discoloured urine, autopsy findings normal; **♀:** 5/5 died **2150: ♂ and ♀:** 10/10 died, stomach distended, pale, *pars pylorica* dark red, adrenal glands black, kidneys as in 1000 mg/kg body weight **$LD_{50}$ ♂:** 794 (584–1080) mg/kg body weight; **$LD_{50}$ ♀:** 681 mg/kg body weight | Rohm and Haas Company 1970c |
| rat (Sprague-Dawley), 5 ♂, 5 ♀ per group | RH 893-50%, assumed content of 2-octyl-4-iso-thiazolin-3-one: 50% 20 % in corn oil | **50–232:** no effects **500 and more:** diarrhoea, CNS depression, posture hunched, breathing difficulties, no mortality **1075 and 2320:** oily fur, red encrustations around eyes and nose, piloerection, tremor, prostration, 100 % mortality **$LD_{50}$:** 735 mg/kg body weight | Rohm and Haas Company 1970d |
| rat, 10 ♂, 10 ♀ per group | Skane M-8 HQ, in propylene glycol, 10 ml/kg body weight | **140: ♀:** passivity, ataxia, abdominal breathing, diar-rhoea, constipation, salivation, distended abdomen, breathing noises **236:** as for 140 mg/kg body weight, 1/10 ♂ and 1/10 ♀ died **374:** kidneys red or pale, stomach mucosa red; **♂:** 7/10 died, **♀:** 4/10 died **560: ♂:** 8/10; **♀:** 9/10 died **934: ♂ and ♀:** 10/10 died **$LD_{50}$ ♂:** 355 (281–433) mg/kg body weight; **$LD_{50}$ ♀:** 358 (281–461) mg/kg body weight | Rohm and Haas Company 1987a |

In another study with RH 893 Technical (content of 2-octyl-4-isothiazolin-3-one assumed to be 100 %), only diarrhoea developed in rats given doses of 100 and 215 mg/kg body weight. From doses of 1000 mg/kg body weight, the adrenal glands and the kidneys were discoloured and the liver blotchy with translucent edges. On day 14, autopsy of the animal which survived revealed no pathological findings. The $LD_{50}$ for the male animals was 794 mg/kg body weight and for the females 681 mg/kg body weight (Rohm and Haas Company 1970c).

Another study with RH 893-50 % (concentration of 2-octyl-4-isothiazolin-3-one not specified, presumably 50 %) revealed for male and female rats an $LD_{50}$ of 1470 mg/kg body weight (735 mg 2-octyl-4-isothiazolin-3-one per kg body weight). The significant symptoms after 2-octyl-4-isothiazolin-3-one doses of 500 mg/kg body weight or more were diarrhoea, CNS depression, laboured breathing and prostration. In animals which died, erythema of the stomach, intestine, kidneys and liver was observed (Rohm and Haas Company 1970d).

The $LD_{50}$ values for 2-octyl-4-isothiazolin-3-one administered as Skane® M-8 HQ to male and female rats were 355 and 358 mg/kg body weight. The test substance was administered in propylene glycol doses of 10 ml/kg body weight. Clinical symptoms were seen in females given 2-octyl-4-isothiazolin-3-one doses of 140 mg/kg body weight and more and in males from 234 mg/kg body weight. In all groups in the animals which died, the mucosa of the stomach and intestine was reddened. Animals subjected to autopsy after 14 days appeared clinically and pathological normal (Rohm and Haas Company 1987a).

In a range-finding study with female rabbits, the $LD_{50}$ for RH 893 Technical (30 % in corn oil) was found to be 280 mg/kg body weight. The animal which survived the 280 mg/kg body weight treatment showed no clinical symptoms. The deaths of the two animals which died were so sudden that no other observations were recorded. The gross pathological examination revealed no abnormalities (Rohm and Haas Company 1971b).

### 5.1.3 Dermal absorption

Groups of 5 male albino rabbits were treated occlusively on the shaved skin for 24 hours with a 45 % solution of 2-octyl-4-isothiazolin-3-one in propylene glycol; the doses applied were 149, 291, 582, 1164 and 2327 mg/kg body weight. Whereas in the lowest dose group no effects were seen, in the animals of all other groups apathy, ataxia, prostration and partial paresis of the back legs developed. Irritation manifested itself as erythema, oedema and scab formation. In the 291 mg/kg body weight group, 2/5 animals died, and in the higher dose groups all animals died on day 1 or day 2 of the study. The $LD_{50}$ was given as 311 mg/kg body weight (Rohm and Haas Company 1977).

The $LD_{50}$ for undiluted RH 893 Technical (assumed to be 100 % 2-octyl-4-isothiazolin-3-one) after occlusive application to rabbit skin for 24 hours was 1.78 ml/kg body weight (about 1800 mg 2-octyl-4-isothiazolin-3-one per kg body weight). The symptoms of intoxication ranged from skin irritation to necrosis and ataxia. A 0.005 % aqueous solution of the test substance applied in doses up to 10 ml/kg body weight proved to have no lethal effects and to be only slightly irritating (Rohm and Haas Company 1970c).

## 5.2 Subacute, subchronic and chronic toxicity

### 5.2.1 Inhalation

The results of studies of the toxicity of the substance after repeated inhalation by rats are summarized in Table 2.

**Table 2.** Subacute and subchronic toxicity of inhaled 2-octyl-4-isothiazolin-3-one

| Species (strain) number per group and sex | Product, exposure duration, conc. 2-octyl-4-isothi-azolin-3-one in mg/m$^3$ (conc. of propylene glycol in mg/m$^3$) | **Conc. 2-octyl-4-isothiazolin-3-one in mg/m$^3$ (conc. of propylene glycol in mg/m$^3$):** effects | References |
|---|---|---|---|
| rat (Sprague-Dawley) 10 ♂, 10 ♀ | RH-893-50% in propylene glycol, 10 days, 1 h/day; 0 (2050), 10 (2070), 100 (1900) | **10 (2070):** ♂ weight loss, tracheitis, reduced leukocyte count<br>**100 (900):** dyspnoea, bradypnoea, pathological breathing noises, nasal secretion, lacrimation, coughing, lungs with red spots, livers blotchy, bronchiolitis, pulmonary congestion, tracheitis with exudate, karyopyknosis of the renal tubule cells; ♂: 9/10 died; ♀: 7/10 died | Rohm and Haas Company 1971a |
| rat (Crl:CD®BR) 22 ♂, 22 ♀ | Skane M-8 in propylene glycol 6 h/day, 5 days/week, 13 weeks 0 (air), 0 (123); 0.05 (141); 0.64 (100); 6.39 (154); 11 animals from each group observed for 14 weeks | **0 (123):** no exposure-related effects<br>**0.05 (141):** no exposure-related effects<br>**0.64 (100):** intraepithelial droplets in the nasal cavity<br>**6.39 (154):** dyspnoea, bradypnoea, pathological breathing noises, nasal secretion, hyperplasia of the secretory cells of the septal mucosa, neutrophilic infiltration of the mucosa, squamous metaplasia of the respiratory epithelium, intraepithelial droplets in the nasal cavity; ♂: decreased body weight gains during whole exposure period, decreased absolute weights of liver, kidneys, spleen and adrenal glands, triglyceride and glucose levels decreased; ♀: body weight gains decreased in weeks 1, 7 and 8 | Rohm and Haas Company 1989 |

Intraepithelial droplets in the planes of section of the nasal cavity

| | after 13 weeks exposure | | | | | | after 14 weeks recovery period | | | | | |
|---|---|---|---|---|---|---|---|---|---|---|---|---|
| | plane II | | plane III | | plane IV | | plane II | | plane III | | plane IV | |
| | ♂ | ♀ | ♂ | ♀ | ♂ | ♀ | ♂ | ♀ | ♂ | ♀ | ♂ | ♀ |
| 0 (air) | 0 | 0 | 4 | 1 | 7 | 1 | 0 | 0 | 7 | 3 | 9 | 10 |
| 0 (123) | 0 | 0 | 5 | 2 | 6 | 1 | 0 | 0 | 4 | 3 | 8 | 8 |
| 0.05 (141) | 0 | 0 | 4 | 2 | 5 | 4 | 0 | 0 | 8 | 8 | 10 | 9 |
| 0.64 (100) | 1 | 2 | 4 | 4 | 5 | 4 | 0 | 2 | 8 | 10 | 10 | 10 |
| 6.39 (154) | 2 | 3 | 7 | 10 | 10 | 8 | 1 | 2 | 7 | 10 | 9 | 10 |

Groups of 10 male and 10 female Sprague-Dawley rats inhaled, one hour daily for ten days, aerosols of 2-octyl-4-isothiazolin-3-one in concentrations of 10 or 100 mg/m$^3$ in propylene glycol or of propylene glycol alone in a concentration of 2080 mg/m$^3$. Neither the purity of the substance nor the average particle diameters in the aerosols were determined. Half of the animals were autopsied at the end of exposure, the rest 14 days later. In the high concentration group the animals lost weight during the exposure; 16/20 died during the exposure period. In the gross pathological examination, alterations were seen in the lungs and livers of some of the animals. The histological examination revealed bronchiolitis, pulmonary congestion, tracheitis with exudate and karyopyknosis in the cells of the renal tubules. Most of the animals with karyopyknosis died. In the males, tracheitis developed even at the lowest concentration. In this group the most severe lung changes were seen: 7/10 animals with tracheitis and 8/10 with pneumonia, in some cases interstitial pneumonia. The authors considered that the lung findings in 4 of these animals could be put down to murine pneumonia, but it seems reasonable to assume that it was an additional effect of the treatment because of the marked neutrophilic response (Rohm and Haas Company 1971a).

Five groups of 22 Crl:CD®BR rats were exposed 6 hours daily, 5 days per week for 90 days via the nose to 2-octyl-4-isothiazolin-3-one aerosols in analytically determined concentrations of 0.05, 0.64 and 6.39 mg/m$^3$ in propylene glycol (about 100 mg/m$^3$) (Table 2). The average particle diameter was 1.4 µm; 68 % of the aerosol was inhalable. The controls were exposed to filtered air or to propylene glycol, 123 mg/m$^3$. Half of the animals were subjected to histological examination at the end of exposure, the other half after a recovery period of 14 weeks. The highest dose resulted in breathing noises and transient dyspnoea, bradypnoea and nasal secretion as well as in reduced body weight gain. At the end of exposure the male animals were 39 g (8 %) lighter than the controls, the females 11 g (4 %) lighter. This reduction in body weight gain was more than compensated during the recovery period. No significant changes in haematological parameters and no changes in clinicochemical parameters were detected. Reduced liver, kidney, spleen and adrenal gland weights were recorded in the males of the highest dose group. As they can be accounted for in terms of the reduced body weight gains and because no histological changes were found in these organs, they are not of toxicological relevance. Histopathological changes were detected only in the nasal cavity. They included minimal to mild squamous metaplasia of the respiratory epithelium of the anterior section of the nose without keratinization after exposure to 6.39 mg/m$^3$, hyperplasia of the secretory cells of the septal mucosa, neutrophilic infiltration of the mucosa and—relative to the negative controls—an increase in the incidence of intraepithelial eosinophilic droplets in the respiratory and olfactory epithelia of the posterior section of the nose. These changes were also detected in male animals at the end of the recovery period when the squamous metaplasia was minimal to moderately severe. In the female rats at this time, the incidence of intraepithelial eosinophilic droplets was also increased in the 0.64 mg/m$^3$ group in which it was not recorded as increased at the end of the exposure period. The authors emphasize, however, that the incidence of intraepithelial eosinophilic droplets was increased at the end of the recovery period in all the groups including the controls. They suggest that this is the result of normal ageing processes on the one hand and an adaptive reaction to a mild nasal irritant on the other. Therefore they consider that in this study the NOEL was shown to be 0.64 mg/m$^3$ (Rohm and Haas Company 1989). However, in one of the anterior planes of section (plane II) of the nasal cavity, eosinophilic droplets were found in 1 to 3 males and 1 to 3 females only at concentrations of 0.64 mg/m$^3$ or more (Table 2) and persisted at the end of the recovery period. Such changes were not found in this plane of section of the nasal cavity in any other groups.

### 5.2.2 Ingestion

CD albino rats were given 2-octyl-4-isothiazolin-3-one in the diet in doses of 6.6, 35.5 and 115 mg/kg body weight and day for 90 days. There were no effects on body weight, blood counts, urinalysis, biochemical parameters, organ weights or mortality. The gross pathological and histological examinations did not reveal changes. The stability of 2-octyl-4-isothiazolin-3-one in the diet was not checked (Rohm and Haas Company 1970a).

Similar results were obtained in a study in which 2-octyl-4-isothiazolin-3-one was fed to beagle dogs for 90 days (concentrations 100, 500 and 2000 mg/kg diet). The dose ingested by the animals of the highest dose group was 50 to 70 mg/kg body weight and day. Reduced food consumption was not detected in any of the dose groups at any time. The stability of 2-octyl-4-isothiazolin-3-one in the diet was not checked (Rohm and Haas Company 1970b).

### 5.2.3 Dermal absorption

In a 90-day study, groups of 10 male and 10 female Crl:CD®BR rats were treated on 5 days per week without occlusion on the intact, shaved skin with 1 ml/kg body weight of about 0.3 %, 0.6 % or 1.5 % solutions of RH 893 HQ in propylene glycol, corresponding to 2-octyl-4-isothiazolin-3-one doses of 2.97, 5.95 and 14.87 mg/kg body weight. In the highest dose group, the body weights, the specific gravity of the urine and the serum triglyceride levels in the male animals were reduced whereas the alkaline phosphatase activity was increased; in the females the erythrocyte count, the haematocrit and the haemoglobin levels were reduced as were serum total protein and albumin and blood glucose levels. These changes were statistically significant. In the highest dose group males, the absolute liver weights were reduced. Gross pathological and histological examination of more than 40 organs including the eyes revealed no treatment-related changes. On the skin of animals treated with 2-octyl-4-isothiazolin-3-one, however, hyperkeratosis, acanthosis, focal necrosis, exudation and scab formation were observed but no oedema. The size and number of sebaceous glands increased. Chronic skin inflammation manifested itself in the form of mononuclear infiltration and occasional fibrotic foci. Similar findings but less severe were also recorded in a few males of the propylene glycol control group. The NOEL for systemic toxicity was given as 5.95 mg/kg body weight (Rohm and Haas Company 1991a).

## 5.3 Local effects on skin and mucous membranes

### 5.3.1 Skin

Applied to shaved rabbit skin for 4 hours as a 45 % to 50 % solution in propylene glycol, 0.5 ml Kathon 893 induced severe oedema and erythema as well as scab formation and tissue damage. The substance was classified as corrosive, with the maximum irritation score of 8.0 (Rohm and Haas Company 1984).

In a test carried out according to the OECD Test Guideline 404, 0.5 ml of a 2.25 %, 4.95 %, 11.25 % or 20.25% solution of 2-octyl-4-isothiazolin-3-one in propylene glycol was applied occlusively for 4 hours to the intact shaved skin of three rabbits per concentration. Treatment with the 2.25 % and 4.95 % solutions caused severe oedema and slight erythema which were rapidly reversible. In the group treated with the 11.25 % solution, slight oedema was still detectable 12 days later. In one animal, the erythema disappeared after 11 days, in another it persisted for 14 days. In animals treated with 20.25 % 2-octyl-4-isothiazolin-3-one, the severe oedema regressed in two days but it had still not disappeared completely after 12 days. 2-Octyl-4-isothiazolin-3-one in concentrations of 2.25 % and 4.95 % was classified as slightly irritating and in concentrations of 11.25 % and 20.25 % as irritating (Rohm and Haas Company 1987b).

### 5.3.2 Eyes

In a Draize test, 0.1 ml of a 45 % solution of 2-octyl-4-isothiazolin-3-one in propylene glycol was introduced into the conjunctival sac of one eye of each of 6 rabbits; the second eye served as control. Severe irritation of conjunctiva, cornea and iris persisted for longer than 7 days (Rohm and Haas Company 1973b). The product was classified as corrosive (Rohm and Haas Company 1985).

In a similar Draize test with a 50 % solution of 2-octyl-4-isothiazolin-3-one, in some of the rabbits the treated eye was rinsed after 2 or 4 seconds. This reduced slightly the severity of the otherwise similar effects. Even after 14 days, corneal damage was detected with fluorescein in all treated animals (Rohm and Haas Company 1970d).

0.1 ml of a 0.5 % aqueous solution of 2-octyl-4-isothiazolin-3-one was tolerated by the rabbit eye without irritation. A 5 % solution caused slight erythema of the conjunctiva when the eyes were rinsed after 2 or 4 seconds; without rinsing the treatment caused corneal clouding and chemosis. One week later no effects could be detected even with fluorescein (Rohm and Haas Company 1971c).

## 5.4 Allergenic effects

In a Magnusson and Kligman maximization test with guinea pigs, the intradermal induction was carried out with a 1 % w/v solution of Kathon 893 (92 % 2-octyl-4-isothiazolin-3-one) in propylene glycol and the topical induction with a 5 % w/w preparation in petrolatum. Provocation was carried out with 0.2 %, 0.1 %, 0.05 % and 0.01 % w/w Kathon 893 in petrolatum on days 21 and 35. On day 21, positive results were obtained in 5/20 animals treated with the 0.2 % solution and in 7/20 treated with 0.1 %. On day 35, positive results were obtained in 8/20 animals treated with 0.1 %, 2/20 with 0.05 % and 0/20 with 0.01 %. The authors classified the substance as moderately sensitizing (Mathias *et al.* 1983).

2-Octyl-4-isothiazolin-3-one was tested for sensitizing effects in a modified Buehler test with Hartley guinea pigs. The topical induction (3 times 6 hours weekly, 10 times in all) was carried out with 0.0025 % to 0.24 % solutions of 2-octyl-4-isothiazolin-3-one in 80 % aqueous ethanol, the provocation 2 weeks later by topical application of 0.01 %, 0.075 % and 0.12 %. The highest concentration tolerated without skin irritation was 0.12 %. The induction concentration ($EC_{50}$) required for positive results after provocation with 0.12 % was found to be 0.013 %, and the $EC_{50}$ for provocation with 0.05 % was 0.04 % (Rohm and Haas Company 1983).

| | Induction (%) | | | | | | | |
|---|---|---|---|---|---|---|---|---|
| Provocation (%) | 0 | 0.0025 | 0.005 | 0.01 | 0.05 | 0.075 | 0.12 | 0.24 |
| 0.01 | 0/10 | 0/10 | 0/10 | 1/10 | 2/10 | 4/10 | 4/10 | 3/10 |
| 0.075 | 0/10 | 0/10 | 1/10 | 4/10 | 6/10 | 10/10 | 10/10 | 9/10 |
| 0.12 | 0/10 | 1/10 | 2/10 | 6/10 | 8/10 | 9/10 | 10/10 | 8/10 |

## 5.5 Reproductive and developmental toxicity

Groups of 25 Sprague-Dawley rats were given 2-octyl-4-isothiazolin-3-one doses of 1, 5 or 30 mg/kg body weight by intubation of a solution of Skane® M-8 HQ in 5 % aqueous methyl cellulose (10 ml/kg body weight) daily from day 6 to day 15 of gestation. The high dose treatment resulted in reduced feed consumption and body weight loss, especially during the first days. This statistically significant effect became less marked in the course of the treatment, and during the treatment-free days (days 16 to 20 of gestation) the treated dams gained more weight than did the controls. There was also a tendency to adverse effects on feed consumption and body weight gain in the middle dose group; therefore the NOEL for maternal toxicity was given as 1 mg/kg body weight. A slight increase in early resorptions without statistically significant postimplantation losses was considered to be an embryotoxic effect of the high dose. There was no evidence of foetotoxicity, and the malformations seen in a few foetuses from the high dose group were accounted for in terms of the historical data of the test laboratory or considered to be a result of the transient significant weight loss in the dams. The NOEL for embryotoxicity under these conditions was 5 mg/kg body weight (Rohm and Haas Company 1987c).

In another developmental toxicity study, groups of 19 rabbits were given daily doses of 2-octyl-4-isothiazolin-3-one of 5, 20 or 80 mg/kg body weight by intubation of a solution of Skane® M-8 or propylene glycol in 5 % aqueous methyl cellulose (5 ml/kg body weight) from day 7 to day 19 of gestation. The rabbits of the high dose group lost 175 g

(5 %) body weight during the treatment. Then they gained weight slightly in the period until they were killed on day 29. Anorexia was reported for this group but data for feed consumption were not shown. The foetuses in this group were significantly lighter (average body weight 42.1 g) than those in the two control groups (48.3 and 45.2 g). Apart from this sign of slight foetotoxicity, there was no evidence of prenatal toxic effects either in this group or in the two lower dose groups. Thus the NOEL for maternal toxicity and embryotoxicity was 20 mg/kg body weight (Rohm and Haas Company 1987d).

## 5.6 Genotoxicity

### 5.6.1 *In vitro*

Because of the high cytotoxicity of 2-octyl-4-isothiazolin-3-one, the substance could be tested for *in vitro* genotoxicity only in very low concentrations and this reduces the meaningfulness of the results obtained in the tests.

In the *Salmonella* mutagenicity test, Skane® M-8 (45.1 % 2-octyl-4-isothiazolin-3-one) was not mutagenic in the strains TA98, TA100, TA1535 and TA1537 either with or without an exogenous metabolic system (S9 from the livers of rats treated with Aroclor). The substance was tested in concentrations up to the toxic range in all strains (Rohm and Haas Company 1988).

Skane® M-8 (46.7 % 2-octyl-4-isothiazolin-3-one) in concentrations of about 0.023 to 3.74 µg/ml (about 0.11–17.5 µM) did not induce DNA repair synthesis in the UDS test with cultured hepatocytes from male Sprague-Dawley rats. Higher concentrations (about 4.20–46.7 µg/ml, that is, about 20–220 µM) were also tested, but because of the high cytotoxicity in these cultures, they were not evaluated (Rohm and Haas Company 1986b).

In an HPRT gene mutation test in CHO cells, Skane® M-8 (46.7 % 2-octyl-4-isothiazolin-3-one) did not induce mutations either in the absence of an exogenous metabolic system (tested concentration range 0.23–1.4 µg/ml, i.e. about 1.1–6.6 µM) or in the presence of S9 from the livers of rats treated with Aroclor (tested concentration range 0.47–2.34 µg/ml, i.e. about 2.2–11 µM). Higher concentrations could not be tested because of their highly cytotoxic effects (Rohm and Haas Company 1986c).

In CHO cells both with and without an added exogenous metabolic system (S9 from the livers of rats treated with Aroclor), Skane® M-8 (43 % 2-octyl-4-isothiazolin-3-one) induced a significant increase in the incidence of structural chromosomal aberrations. The effective concentration range in the absence of S9 was 1.72 to 3.44 µg/ml (about 8.1–16 µM) and in the presence of S9 4.3 to 8.6 µg/ml (about 20–40 µM). At these concentrations cytotoxicity was already apparent. Higher concentrations were all highly toxic (Rohm and Haas Company 1987e).

### 5.6.2 *In vivo*

Groups of 15 male and 15 female Crl:CD®BR-VAF/plus rats were given single 2-octyl-4-isothiazolin-3-one doses of 215 mg/kg body weight as Skane® M-8 (43 %) in propylene glycol by gavage. This dose was equivalent to the $LD_{10}$ for this strain of rats. 6, 27 and 51 hours later, bone marrow preparations were obtained from 5 males and 5 females from each group. At no time was the incidence of structural chromosomal aberrations increased above that in the controls treated only with propylene glycol. The mitotic index was not determined (Rohm and Haas Company 1987f).

## 5.7 Carcinogenicity

For a period of 18 months, groups of 100 male and 100 female mice (C57BL/6 × C3H/Anf) were given a diet containing RH 893 in concentrations of 500 or 1000 mg/kg diet (443 and 887 mg 2-octyl-4-isothiazolin-3-one per kg diet). This was equivalent to 2-octyl-4-isothiazolin-3-one doses of about 75 and 150 mg/kg body weight and day (Rohm and Haas Company 1985). Another group of 100 males and 100 females was given acetylaminofluorene in a concentration of 600 mg/kg diet as positive control and 100 males and 100 females remained untreated. No effects on body weight gains were observed. Survival was 96 % to 98 % in the control and RH 893 groups, in the acetylaminofluorene group 1 % in the females and 37 % in the males. Among the 96 male and 96 female animals of the 1000 mg/kg group which were examined, 3 lymphosarcomas, 3 liver cell carcinomas, 2 fibroadenomas and 1 pulmonary adenoma were found. The incidence of neoplasms was not different from the incidence of spontaneous neoplasms in the control animals. The animals given 500 mg/kg were not examined for lesions or neoplasms. The stability of 2-octyl-4-isothiazolin-3-one in the diet was not checked (Rohm and Haas Company 1975).

# 6 Manifesto (MAK value/classification)

Results of studies of exposed persons which would be suitable for the derivation of a MAK value are not available. Therefore the 90-day inhalation study with the rat was used for this purpose. Clear effects such as squamous metaplasia in the nasal cavity were seen in animals exposed to 6.39 mg/m$^3$. Intraepithelial eosinophilic droplets in the nasal cavity of animals exposed to 0.64 mg/m$^3$ were more extensive 14 weeks after the end of exposure than immediately after. And these changes developed in sections of the nose in which no such changes were detected in the control animals. It must also be remembered that this concentration of the substance in the aerosol was produced with a 0.6 % solution in propylene glycol and that after inhalation of aerosols of more highly concentrated solutions or of the undiluted substance more severe local effects must be expected. At

0.05 mg/m$^3$ no effects were detected, and so the MAK value is provisionally established at this concentration. 2-Octyl-4-isothiazolin-3-one is classified in Peak limitation category I because of its local irritant effects; the excursion factor was set at 2 because the irritation produced during medium-term exposure to 0.64 mg/m$^3$ was marginal.

In mice fed 2-octyl-4-isothiazolin-3-one for 18 months, the substance was not carcinogenic. However, its stability in the diet was not checked. Because of the mainly local effects in the upper respiratory tract, an inhalation study is required for clarification of the carcinogenic potential of 2-octyl-4-isothiazolin-3-one.

2-Octyl-4-isothiazolin-3-one has been shown in two species of experimental animal not to produce prenatal toxicity when administered in doses up to the maternally toxic range. Because the doses which can be taken up during observance of the MAK value are so much smaller than these maternally toxic doses, 2-octyl-4-isothiazolin-3-one is classified in Pregnancy risk group C.

Because of the contact sensitization observed in controlled human and animal studies and the sensitization seen in field studies, 2-octyl-4-isothiazolin-3-one is designated with "Sh".

2-Octyl-4-isothiazolin-3-one is designated with an "H" because, on the one hand, the $LD_{50}$ values obtained in the dermal test with rabbits and the oral test with rats are similar and, on the other hand, the lowest observed effect level (LOEL) for systemic effects in the 90-day dermal test was very low, about 15 mg/kg body weight and day. In addition, it has been demonstrated *in vivo* and *in vitro* that the substance readily penetrates the skin.

# 7 References


Andersen KE, Veien NK (1985) Biocide patch tests. *Contact Dermatitis 12*: 99–103

Emmet EA, No SK, Levy MA, Moss JN, Morici IJ (1989) The irritancy and allergenicity of 2-*n*-octyl-4-isothiazolinon-3-one (Skane M-8), with recommendations for patch test concentration. *Contact Dermatitis 20*: 21–26

Fischer T, Bohlin S, Edling C, Rystedt I, Wieslander G (1995) Skin disease and contact sensitivity in house painters using water-based paints, glues and putties. *Contact Dermatitis 32*: 39–45

Foussereau J, Brändle I, Boujnah-Khouadja A (1984) Allergisches Kontaktekzem durch Isothiazolinon-3-on-Derivate. *Dermatosen Beruf Umwelt 32*: 208–211

Geier J, Schnuch A (1996) No cross-sensitization between MCI/MI, benzisothiazolinone and octylisothiazolinone. *Contact Dermatitis 34*: 148–149

Mathias CGT, Andersen KE, Hamann K (1983) Allergic contact dermatitis from 2-*n*-octyl-4-isothiazolin-3-one, a paint mildewicide. *Contact Dermatitis 9*: 507–509

Oleaga JM, Aguirre A, Landa N, González, Diaz-Pérez JL (1992) Allergic contact dermatitis from Kathon 893. *Contact Dermatitis 27*: 345–346

Paulus W (1993) *Microbicides for the protection of materials*, Chapman & Hall, London

Rohm and Haas Company (1967) *Acute oral LD 50 determination – rats*. Hazleton Laboratories, Project No. 417–210, unpublished report

Rohm and Haas Company (1970a) *RH 893 T, Short term toxicity – 3 month dietary*. Hazleton Laboratories, Report No. 70RC 1053, unpublished summary report

Rohm and Haas Company (1970b) *RH 893 T, Three-month dietary administration – dogs*. Hazleton Laboratories, Report No. 70RC 1052, Project No. 417-334, unpublished report

Rohm and Haas Company (1970c) *RH 893 (Technical) Acute oral – rats, acute dermal – rabbits, acute eye irritation – rabbits, acute inhalation exposure – rats.* Hazleton Laboratories, Project No. 417-306, No. 417-307, No. 417-308, No. 417-310, unpublished report

Rohm and Haas Company (1970d) *RH 893-50%, Acute oral administration – rats, Draize eye irritation – rabbits, primary skin irritation – rabbits, skin sensitization – guinea pigs, acute inhalation exposure – rats.* Hazleton Laboratories, Projects No. 417-323, No. 417-324, No. 417-325, No. 417-326, No. 417-327, unpublished report

Rohm and Haas Company (1971a) *RH 893-50%, Subacute inhalation study – rats.* Hazleton Laboratories, Project No. 417-345, unpublished report

Rohm and Haas Company (1971b) *RH 893 (Technical) Oral dose range – rabbits.* Hazleton Laboratories, Project No. 417-347, unpublished report

Rohm and Haas Company (1971c) *RH 893-50% Draize eye irritation – rabbits.* Hazleton Laboratories, Project No. 417-348, unpublished report

Rohm and Haas Company (1973a) *Evaluation of sensitizing properties of RH 893 technical.* Food And Drug Research Laboratories, 20.07.1973, unpublished report

Rohm and Haas Company (1973b) *Eye irritation – albino rabbits.* Food and Drug Research Laboratories, Ref # 1-3784-850, unpublished report

Rohm and Haas Company (1975) *Eighteen month study on the carcinogenic potential of RH 893 in mice.* Medical College of Virginia, unpublished report

Rohm and Haas Company (1977) *RH 893, Single dermal dose – albino rabbit.* Report No. 77-13, Zusammenfassung, unpublished report

Rohm and Haas Company (1983) *Kathon 893: A study of concentration – dependent delayed contact hypersensitivity in guinea pigs.* Report No. 83R 0143, unpublished summary report

Rohm and Haas Company (1984) *Kathon 893 microbicide: acute skin irritation, definitive, rabbits.* Report No. 84R-0189, unpublished report

Rohm and Haas Company (1985) *Datenblatt, Kathon industrial microbicides, N-octyl-isothiazolones* (Data sheet)

Rohm and Haas Company (1986a) *Skane M-8: acute inhalation toxicity study in rats.* Report No. 85R 0203, unpublished report

Rohm and Haas Company (1986b) *Skane M-8 HQ microbiocide – in vitro unscheduled DNA synthesis assay.* Report No. 86R 0018, unpublished report

Rohm and Haas Company (1986c) *Skane M-8 HQ microbicide – CHO/HGPRT gene mutation assay.* Report No. 86 R 055, unpublished report

Rohm and Haas Company (1987a) *Skane M-8 HQ microbicide – acute oral toxicity study in male and female rats.* Report No. 86 R 178 A und B, unpublished report

Rohm and Haas Company (1987b) *Kathon 893: Skin irritation – rabbits.* Laboratories BIO-TOX, Report No. 87RC 1005, unpublished summary report

Rohm and Haas Company (1987c) *A teratology study in rats with Skane M-8 HQ.* WIL Research Laboratories, Project No. WIL-91003, unpublished report

Rohm and Haas Company (1987d) *Skane M-8 HQ industrial mildewcide: oral (gavage) developmental toxicity study in rabbits.* Report No. 87 R-019, unpublished report

Rohm and Haas Company (1987e) *Clastogenic evaluation of Skane M-8 microbiocide HQ: In an in vitro cytogenetic assay measuring chromosomal aberration frequencies in Chinese hamster ovary (CHO) cells.* Hazleton Biotechnologies Co, Report No. 86 RC 048, unpublished report

Rohm and Haas Company (1987f) *Skane M-8 HQ – in vivo cytogenetic study in rats.* Report No. 86R-218, unpublished report

Rohm and Haas Company (1988) *Skane M-8 HQ: Salmonella typhimurium gene mutation assay.* Report No. 88R-203, unpublished summary report

Rohm and Haas Company (1989) *Skane M-8 HQ microbicide: thirteen-week inhalation toxicity study in rats.* Report No. 87R 013, unpublished report

Rohm and Haas Company (1990) *$^{14}$C-RH 893: effect of vehicle, formulation or occlusion on dermal bioavailability in male guinea pigs.* Report No. 88R-206, unpublished report

Rohm and Haas Company (1991a) *RH 893 HQ Technical: three-month dermal toxicity study in rats.* Report No. 90R-031, unpublished report

Rohm and Haas Company (1991b) *$^{14}$C-RH 893: in vitro dermal absorption study in male rat skin.* Report No. 87R-076, unpublished report

Rohm and Haas Company (1991c) *Skane M-8 technical, HQ: respiratory irritation ($RD_{50}$) study.* Report No. 90RC-180, unpublished report

Rohm and Haas Company (1994) Safety data sheet, *Kathon 893 MW microbicide*

Rohm and Haas Deutschland GmbH (1995a) letter to the DGMK of 16.01.1995

Rohm and Haas Deutschland GmbH (1995b) written communication to the Commission of 21.12.1995

Thormann J (1982) Contact dermatitis to a new fungicide, 2-*n*-octyl-4-isothiazolin-3-one. *Contact Dermatitis 8*: 204

completed 07.04.1997

# Oleoyl sarcosine*

| | |
|---|---|
| **MAK value** | **not yet established, see Section IIb of the *List of MAK and BAT Values*** |
| **Peak limitation** | **–** |
| **Absorption through the skin** | **–** |
| **Sensitization** | **–** |
| **Carcinogenicity** | **–** |
| **Prenatal toxicity** | **–** |
| **Germ cell mutagenicity** | **–** |
| **BAT value** | **–** |
| Synonyms | *N*-methyl-*N*-oleoyl-aminoacetic acid<br>oleic sarcosine<br>oleoyl *N*-methylaminoacetic acid<br>oleoyl *N*-methylglycine<br>*N*-oleoyl sarcosine |
| Trade names | Sarkosyl® O<br>Sarkosyl® OT |
| Chemical name (CAS) | *N*-methyl-*N*-(1-oxo-9-octadecenyl)glycine |
| CAS number | 110-25-8 |
| Structural formula | $CH_3{-}(CH_2)_7{-}CH{=}CH{-}(CH_2)_7{-}C{=}O$<br>$H_3C{-}N{-}CH_2{-}COOH$ |
| Molecular formula | $C_{21}H_{39}NO_3$ |
| Molecular weight | 353.55 |
| Melting point | 0°C (solidification point) |
| Boiling point | 397°C |
| Density at 25°C | 0.95 g/cm$^3$ |
| Vapour pressure at 20°C | $4 \times 10^{-7}$ hPa |
| log $P_{OW}$** | > 6 |
| **1 ml/m$^3$ (ppm) ≙ 14.670 mg/m$^3$** | **1 mg/m$^3$ ≙ 0.682 ml/m$^3$ (ppm)** |

* The physical data given here are those for the commercially available product Sarcosyl®O; the toxicological studies were carried out with FAT 61 004/A which is identical with Sarcosyl®O.

** *n*-octanol/water distribution coefficient

Solubility soluble in organic solvents including paraffin oil and mineral oil, poorly soluble in water (< 0.1 g/l at 23°C) (Ciba-Geigy 1994)

pH 4 (1 % aqueous suspension) (Ciba-Geigy 1994)

Stability thermal decomposition at temperatures above 350°C (Ciba-Geigy 1994)

Production by reaction of oleoyl chloride with sarcosine

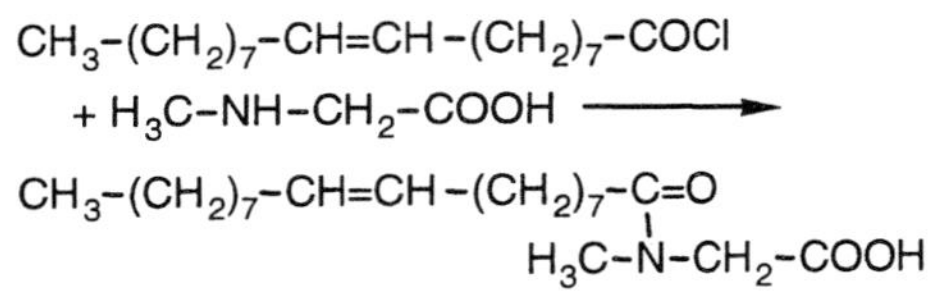

Purity 97 %; for the usual commercially available product a molecular weight range of 340 to 350 is given (Ciba-Geigy, undated)

Impurities 2 % free fatty acids (Ciba-Geigy, undated)

Uses
- corrosion inhibitor in metal-working fluids at recommended concentrations up to 1 % (Ciba-Geigy, undated)
- additive in washing powder
- antistatic treatment of fibres

Oleoyl sarcosine is an amide of oleic acid and sarcosine (*N*-methylglycine), which is a normal metabolite in man.

# 1 Toxic Effects and Mode of Action

Single oral doses of oleoyl sarcosine produce little toxicity in the rat; the $LD_{50}$ is more than 5000 mg/kg body weight. The substance causes severe irritation of the skin and eyes of experimental animals. Oleoyl sarcosine is not phototoxic; it causes contact and photocontact sensitization of guinea pigs. In the *Salmonella* mutagenicity test, oleoyl sarcosine is not mutagenic.

# 2 Metabolism

There are no data available for the mechanism of action, toxicokinetics or metabolism of oleoyl sarcosine. The structure of the substance suggests that metabolism would involve hydrolysis to yield sarcosine and oleic acid.

# 3 Effects in Man

There are no data available for effects of oleoyl sarcosine in man.

# 4 Animal Experiments and *in vitro* Studies

## 4.1 Acute toxicity

There are no data available for the acute effects of oleoyl sarcosine applied to the skin nor for subchronic or chronic toxicity, reproductive or developmental toxicity or for the carcinogenicity of the substance in experimental animals.

### 4.1.1 Inhalation

The 4-hour $LC_{50}$ for rats exposed only via the head and nose was 1370 mg/m$^3$. Inhalation for 8 hours of an atmosphere saturated or enriched with the substance at 20°C was not lethal for rats (BASF 1996).

### 4.1.2 Ingestion

After administration of single oral oleoyl sarcosine doses of 5000 mg/kg body weight to rats, no deaths were observed. The $LD_{50}$ is thus above 5000 mg/kg body weight. The symptoms of intoxication included dyspnoea, exophthalmos, ruffled fur, diarrhoea and hunched posture. All symptoms were reversible and had regressed after 6 days (Ciba-Geigy 1981a). In another study, an $LD_{50}$ of about 9200 mg/kg body weight was reported (BASF 1996).

## 4.2 Local effects on skin and mucous membranes

### 4.2.1 Skin

In a study carried out according to the guidelines of the EPA (Environmental Protection Agency), undiluted oleoyl sarcosine applied to the intact and abraded skin of rabbits proved to be severely irritating with a primary irritation score of 5.6. Erythema and oedema still persisted 7 days after the treatment (Ciba-Geigy 1981b).

In other tests in which substances were applied to rabbit skin occlusively for up to 4 hours, oleoyl sarcosine and its sodium salt were also irritants (BASF 1996).

In a test for phototoxicity, 0.05 ml of a 0.3 %, 1 % or 3 % solution of oleoyl sarcosine in acetone plus ethanol (1:1) was applied to both flanks of each of 6 mice. Half an hour after the application, one flank of each of the animals was irradiated for 100 seconds with UV light of wavelength 320 to 410 nm at an intensity of 25 mW/cm$^2$. Unlike in the positive control group, no phototoxicity was observed (Ciba-Geigy 1981c).

### 4.2.2 Eyes

In a study carried out according to the guidelines of the EPA, oleoyl sarcosine was tested for irritant effects in the rabbit eye. In half of the animals, 30 seconds after application of the substance, the treated eye was rinsed with 10 ml physiological saline. An eye irritation score of 47.0 on the Draize scale was obtained for the unrinsed eyes and 40.0 for the rinsed eyes. The rinsing of the eyes was thus practically without effect. Seven days after application of the substance, the effects in the eyes persisted unchanged. Oleoyl sarcosine was classified as moderately irritating (Ciba-Geigy 1981d).

Detachment and clouding of the cornea was seen in rabbits treated with either oleoyl sarcosine or its sodium salt. After the treatment with the sodium salt, the effects on the cornea had worsened after one week and had not regressed after 15 days (BASF 1996).

## 4.3 Allergenic effects

Oleoyl sarcosine was tested for sensitizing effects in 20 Pirbright white guinea pigs with the Magnusson and Kligman maximization test. Intradermal induction was carried out by injection of two 0.1 ml doses of a 5 % preparation of the test substance mixed with Freund's complete adjuvant (1:1). Seven days later a 30 % preparation of oleoyl sarcosine in petrolatum was applied occlusively to the skin for 48 hours. Provocation was carried out 2 weeks after the epidermal induction by occlusive topical application of 3 % oleoyl sarcosine in petrolatum for 24 hours. Erythema of the skin was seen on 5/20 animals. The substance was classified as a weak contact allergen (Ciba-Geigy 1981e).

Oleoyl sarcosine was tested for photosensitizing effects by means of the method of Maurer *et al.* (1980) with 20 Pirbright white guinea pigs. For epidermal induction, doses of 0.1 ml of a 0.1 % suspension of the test substance in 80 % DAE (40% dimethylacetamide, 30% acetone, 30% ethanol) and 20 % physiological saline were applied topically to the necks of the animals 4 times weekly for 3 weeks. One hour after each application the animals were irradiated for 10 minutes with UVA and visible light, and in the second and third weeks with UVA, UVB and visible light. 24 hours after each of the first four irradiations, the skin reactions were recorded as Draize scores. In the second and third weeks on the Monday and Wednesday, the animals were given 4 injections in the 4 corners of the application site (each 0.1 ml of a suspension of Freund's complete adjuvant in physiological saline); 16 days after the last induction irradiation, 0.1 ml of the test substance suspension was applied non-occlusively to the dorsal skin of the animals on three consecutive days. One hour after each application, the animals were irradiated for 3 minutes with UVA, UVB and visible light for provocation. After a recovery period of 14 days, provocation was repeated by three applications of 0.1 ml of the test substance suspension to the dorsal skin of the animals, followed each time by irradiation with UVA and visible light for 10 minutes. The intensity of the light was chosen to avoid causing erythema. 24 hours after the two provocation treatments, the skin reactions were recorded as Draize scores. Any animal in which the irritation score after provocation was one point or more above that after the first induction week was considered to be photosensitized. Whereas only 3/17 control animals produced slight erythema on provocation, 17/18 treated animals yielded positive results after the first provocation and 15/18 after the second. Oleoyl sarcosine was thus photosensitizing after irradiation with UVB and UVA light and also after UVA light alone (Ciba-Geigy 1981f).

### 4.4 Genotoxicity

Oleoyl sarcosine was studied in the *Salmonella* mutagenicity test in the strains TA98, TA100, TA1535 and TA1537 with and without an added metabolic system; the tested concentrations were 25, 75, 225, 675 and 2025 µg/plate. Bacteriotoxic effects were seen in strain TA1537 at concentrations of 225 µg/plate and above, and from 675 µg/plate precipitates formed and made evaluation of the results impossible. Mutagenicity was not detected (Ciba-Geigy 1981g).

Another test with the same strains of bacteria and oleoyl sarcosine concentrations between 0.8 and 5000 µg/plate also yielded negative results both with and without an added metabolic system. Bacteriotoxic effects were seen at concentrations above 50 to 500 µg/plate (BASF 1996).

## 5 Manifesto (MAK value/classification)

The available data are insufficient for the establishment of a MAK value. Because of the marked local irritant effects, a long-term inhalation study is required. Oleoyl sarcosine is listed in Section IIb of the *List of MAK and BAT Values*.

The structure of oleoyl sarcosine suggests that the photosensitizing effects seen in animal studies are not plausible. To date there are also no reports of such effects in man. Nor could the weak contact sensitization reported in animal studies be confirmed by case reports. Therefore oleoyl sarcosine is not designated with “SP” or “Sh”.

Because data for absorption through the skin are not available, the substance is not designated with an “H”.

# 6 References

BASF (1996) *IUCLID data sheet, (Z)-N-methyl-N-(1-oxo-9-octadecenyl)glycine*, 22.04.1996, unpublished

Ciba-Geigy (undated) *®Sarkosyl-Rostinhibitoren*, Ciba-Geigy Ltd, Basel, Schweiz

Ciba-Geigy (1981a) *Report on the acute oral $LD_{50}$ in the rat of FAT 61 004/A*. Ciba-Geigy, Project No. 801824, unpublished report

Ciba-Geigy (1981b) *Report on skin irritation in the rabbit after single application of FAT 61 004/A*. Ciba-Geigy, Project No. 801826, unpublished report

Ciba-Geigy (1981c) *Skin phototoxicity study in the mouse of FAT 61 004/A*. Ciba-Geigy, Test No. 801827, unpublished report

Ciba-Geigy (1981d) *Report on eye irritation in the rabbit after single application of FAT 61 004/A*. Ciba-Geigy, Project No. 801825, unpublished report

Ciba-Geigy (1981e) *Report on skin sensitizing (contact allergenic) effect in guinea-pigs of FAT 61 004/A: maximization test*. Ciba-Geigy, Test No. 801828, unpublished report

Ciba-Geigy (1981f) *Report on skin photosensitizing (photocontact allergenic) effect in guinea pigs of FAT 61 004/A*. Ciba-Geigy, Test No. 801828, NTIS/OTS 0539883, EPA/OTS Doc #88-920002923

Ciba-Geigy (1981g) *Salmonella/mammalian-microsome mutagenicity test with FAT 61 004/A*. Ciba-Geigy, No. of experiment 801830, unpublished report

Ciba-Geigy (1994) *Material safety data sheet directive 91/155/EEC, Sarkosyl O*, 20.04.1994, Ciba-Geigy Ltd, Basle, Switzerland

Maurer TH, Weirich EG, Hess R (1980) Predictive animal testing for photocontact allergy. *Brit J Dermatol 103*: 593–605

completed 18.06.1997

# Dusts

# Aerosols – Ultrafine aerosol particles, their aggregates and agglomerates

The results of various experimental and epidemiological studies have demonstrated that the adverse effects of a given mass of inhaled poorly soluble particles on the respiratory tract can be more severe if the particles are very small or are aggregates of very small particles than if they are larger, solid particles.

## 1 Definition

For such very small aerosol particles, the term "ultrafine particles" is generally used in the literature. Particles whose mobility-equivalent diameter is less than 0.1 µm should be designated as ultrafine particles.

The mobility-equivalent diameter of a particle of any shape is defined as the diameter of a sphere which has the same mobility (B) as the particle in a given dispersion medium (e.g. air). The mobility-equivalent diameter is identical with the diffusion-equivalent diameter (see the 1997 MAK documentation "Aerosols – Dusts, Fumes and Mists" in Volume 12 of the present series).

## 2 Formation of the particles

Ultrafine particles are mainly produced during burning and in gas phase reactions. Examples include carbon black, welding fumes, laser fumes, exhaust gases, metal fumes, polymer fumes, amorphous silica. Ultrafine particles can also be formed by precipitation (e.g. titanium dioxide, iron oxides).

The primary particles which are formed in these processes have diameters of a few nanometres. Primary particles sometimes coalesce to form agglomerates[1], e.g. in sintering processes, and associate to form aggregates[2] even during the process in which they are generated. Also after particle formation is complete, both primary particles and agglomerates can coalesce to form aggregates. Aggregates consist of loose assemblages

---

[1] agglomerate: assemblage of primary particles rigidly joined together (e.g. by sintering)

[2] aggregate: loosely coherent assemblage of primary particles or agglomerates (held together by adhesion)

of particles held together by adhesion. The process of aggregation depends on the concentration and the size of the particles. With increasing concentration, the rate of aggregation increases.

For characterization of the potential risk associated with ultrafine particles, their agglomerates and aggregates, two parameters are of importance: the deposition mechanisms, which depend on the Brownian motion of the particles, and the effects on the respiratory tract, which depend on the specific surface area[3] and the number concentration of the particles.

# 3 Mechanisms of deposition in the respiratory tract

The movement of ultrafine particles dispersed in a gas is determined entirely by their Brownian motion. Since the motion of such particles in a gas is independent of gravity and, on average, also independent of the density of the particles, their mobility is determined only by their geometric diameter and form. Brownian motion increases with decreasing particle diameter. The mobility of the particles (B) may be described in terms of the diffusion coefficient (D) which has the units $cm^2 \cdot s^{-1}$:

$$B = \frac{D}{kT} = \frac{f(d)}{3\pi\eta d}$$

where:
B is the mobility of the particle,
D is the diffusion coefficient of the particle,
k is the Boltzmann constant,
T is the absolute temperature,
f(d) is the slip correction,
$\eta$ is the viscosity of the dispersion medium (here: air) and
d is the diameter of the particle

As an approximation, the mobility is inversely proportional to the square of the particle diameter (the slip correction f(d) increases with decreasing particle diameter d). Therefore the proportion of particles deposited by diffusion in the respiratory tract increases markedly with decreasing particle diameter. In addition, the proportion of ultrafine particles deposited by diffusion increases with the retention time in the respiratory tract. For particles with diameters of about 0.5 μm, that is, at the borderline between deposition by sedimentation and impaction (as a result of the inertia of the particles) and deposition by diffusion, the probability of deposition is at a minimum. From this point, the deposition of particles in the alveoli increases with decreasing particle diameter to a maximum of about 50 % to 60 % at about 20–50 nm. Below 50 nm the mobility of the

[3] specific surface area: total surface area of the particles divided by their total mass, usually expressed in the units $m^2/g$. (The total surface area can be determined, for example, by adsorption of nitrogen.)

particles becomes so high that even a short period of time in the airways is sufficient for deposition. Thus, particles of less than 1 nm diameter are deposited almost entirely in the extrathoracic airways. Particles with diameters of more than about 20 nm are deposited mainly in the alveolar region. Agglomerates and aggregates of such particles, however, can exceed the threshold value of 0.5 μm so that they are deposited mainly by sedimentation and impaction.

In summary, particles with diameters less than 1 nm are deposited almost entirely in the extrathoracic airways. With increasing diameter, they are increasingly deposited in the lungs, whereby up to 60 % of particles with diameters between 20 and 50 nm are deposited in the alveolar region. At larger particle diameters, the probability of deposition in the lung decreases and reaches a minimum at 500 nm.

# 4 Effects on the respiratory tract

The results of several animal studies suggest that the cytotoxic, inflammatory, proliferative and carcinogenic effects of poorly soluble particles in the rat and mouse increase not so much with the mass concentration of particles but rather with the surface area concentration[4] or number concentration. These studies were carried out mainly with rats and included short-term, medium-term and long-term studies. The dusts used frequently were titanium dioxide ($TiO_2$) and carbon black (Driscoll 1996, Heinrich 1996, Oberdörster 1996, Oberdörster *et al.* 1998). These ultrafine dusts were administered as agglomerates (carbon black) or aggregates ($TiO_2$, carbon black) of ultrafine primary particles (diameters 15–20 nm) by intratracheal instillation and by inhalation. The loose aggregates of ultrafine $TiO_2$, for example, which, unlike agglomerates, can dissociate into ultrafine primary particles after deposition in the lungs, have more severe toxic effects than would be expected from the specific surface area of the aggregates (Heinrich 1996, Oberdörster *et al.* 1998). The high number concentration of particles produced at the site of action by dissociation of the aggregates and the increased specific surface area which results could be important in producing such effects. Comparative studies of ultrafine particles (diameter 20 nm) and fine particles (diameter 250 nm) of $TiO_2$ revealed that the ultrafine particles produced significantly more severe toxic effects such as inflammation, reduction of alveolar clearance, increased translocation of particles into the lung parenchyma and interstitial inflammation than did the fine particles when the comparison was made on the basis of the mass of particles retained in the lungs (Oberdörster *et al.* 1994). When the dose of particles was expressed as particle surface area per gram lung, the inflammatory effects correlated with the particle dose. That is, for the same retained mass of particles, the ultrafine $TiO_2$ particles had more severe toxic effects – probably because of the larger specific surface area and the larger number of particles present – than did the coarser $TiO_2$ particles (Oberdörster *et al.* 1998).

---

[4] surface area concentration: total surface area of the particles in a given volume divided by this volume. The units are $m^2/m^3$.

# 5 Measuring ultrafine particle concentrations

The decisive parameter of effect is considered not to be the concentration expressed in terms of mass of particles but rather the concentration expressed in terms of the number of ultrafine particles and their agglomerates and aggregates.

Until binding conventions have been established, assay procedures which characterize particles in the size range between 5–10 nm and 200 nm in as much detail as possible should be used. In particular, the number concentration should be determined up to levels of about $10^8$ particles/cm$^3$ (concentrations > $10^8$ particles/cm$^3$ do not occur at the workplace because of the high level of aggregate formation) and, in specific cases, also the mass concentration. In addition, it should be possible to carry out a particle size analysis in the given size range.

# 6 References

Driscoll K E (1996) Role of inflammation in the development of rat lung tumors in response to chronic particle exposure. in: Mauderly JL, McCunney RJ (Eds): *Particle overload in the rat lung and lung cancer. Implications for human risk assessment.* Proceedings of a conference held at the Massachusetts Institute of Technology on March 29th and 30th, 1995, 139–153

Heinrich U (1996) Comparative response to long-term particle exposure among rats, mice and hamsters. in: Mauderly JL, McCunney RJ (Eds): *Particle overload in the rat lung and lung cancer. Implications for human risk assessment.* Proceedings of a conference held at the Massachusetts Institute of Technology on March 29th and 30th, 1995, 51–71

Oberdörster G, Ferin J, Lehnert BE (1994) Correlation between particle size, *in vivo* particle persistence, and lung injury. *Environ Health Perspect 102*, 173–179

Oberdörster G (1996) Significance of particle parameters in the evaluation of exposure-dose-response relationships of inhaled particles. in: Mauderly JL, McCunney RJ (Eds): *Particle overload in the rat lung and lung cancer. Implications for human risk assessment.* Proceedings of a conference held at the Massachusetts Institute of Technology on March 29th and 30th, 1995, 73–89

Oberdörster G, Gelein R, Johnston C, Mercer P, Corson N, Finkelstein JN (1998) Ambient ultrafine particles: Inducers of acute lung injury? in: Mohr U, Dungworth DL, Brain JD, Driscoll KE, Grafström RC, Harris CC (Eds): *Relationships between respiratory disease and exposure to air*, ILSI Press, Washington, DC, USA, 216–229

completed 24.06.1998

# Tantalum

| | |
|---|---|
| **MAK value (1999)** | **4 mg/m$^3$ I (inhalable dust fraction) 1.5 mg/m$^3$ R (respirable dust fraction) see Section Vg of the *List of MAK and BAT Values*** |
| **Peak limitation** | – |
| **Absorption through the skin** | – |
| **Sensitization** | – |
| **Carcinogenicity** | – |
| **Prenatal toxicity** | – |
| **Germ cell mutagenicity** | – |
| **BAT value** | – |
| Chemical name (CAS) | tantalum |
| CAS number | 7440-25-7 |
| Structural formula | Ta |
| Density | 16.6 g/cm$^3$ |
| Atomic weight | 180.95 |
| Melting point | about 3020°C |
| Boiling point | about 5500°C |

The present documentation is based on the review of the toxicological data for tantalum by the American Conference of Governmental Industrial Hygienists (ACGIH 1992).

## 1 Chemical properties, occurrence and use

Tantalum is a base metal which, like aluminium, becomes coated with a protective oxide layer when exposed to air. Thus it is extremely resistant to attack by chemical substances, reacting only with hydrofluoric acid (HF), hot fuming sulfuric acid, hot fluorine, chlorine, sulfur and molten alkali metal hydroxides. For this reason, most of the data available are for tantalum oxides.

Tantalum is a rare element. It is found in the crust of the earth together with niobium in a wide variety of deposits all over the world. The main producing countries are Canada, Australia and Brazil. Important reserves of tantalum also occur in tantalum-containing tin slag from Thailand and Malaysia.

The tantalum ore is processed either by leaching with hydrofluoric acid, by molten salt extraction or by chlorination. Tantalum and niobium are separated nowadays mainly by liquid-liquid extraction with methyl isobutyl ketone or by distillation of the penta-chlorides.

Tantalum is used in chemical plant construction as a corrosion-resistant material for container linings, heat exchangers and pump components. Other uses of tantalum include the production of spinning jets, laboratory equipment and cathodes in x-ray and electron tubes. In medicine, tantalum is used for pinning bones and making bone prostheses, joint implants, clips and other instruments. Tantalum powder is used as a contrast medium in human laryngography to ensure better representation of the glottis and the subglottal region (Tegtmeyer *et al.* 1975). Tantalum pentoxide ($Ta_2O_5$) is one of the constituents of optical glass, glass ceramics and electronic components (Falbe and Regitz 1992, Holleman and Wiberg 1985).

## 2 Toxic Effects and Mode of Action

Tantalum and tantalum oxides are practically non-toxic. In experimental animals, the acute toxicity of tantalum oxide is higher than that of the metal. In guinea pigs exposed long-term to tantalum dioxide, transient bronchitis, interstitial pneumonitis and the associated hyperaemia were observed.

## 3 Toxicokinetics and Metabolism

Studies of the clearance of inhaled radioactive tantalum oxide in three workers exposed during a reactor accident demonstrated that, after 7 days, the radioactivity in the lungs had decreased to one percent of the initially deposited amount. The thoracic clearance of this residual contamination took place with a half-time of 1400 days (Newton 1975).

## 4 Effects in Man

Tantalum gauze as used for the repair of hernias seems—even after long periods—to have no adverse effects in comparison with other gauze materials (Dales and Kyle 1958).

Studies of the clearance of inhaled radioactive tantalum oxide in three workers exposed during a reactor accident demonstrated that, after 7 days, the radioactivity in the lungs had decreased to one percent of the initially deposited amount. The thoracic clearance half-time for this residual contamination was 1400 days (Newton 1975).

Dust of tantalum pentoxide ($Ta_2O_5$), which is used in the tungsten carbide industry in the production of hard metals, is not causally associated with the pneumoconiosis which develops in workers in this industry (Miller *et al.* 1953, abstract).

## 5 Animal Studies

There are only a few studies of the toxicity of tantalum and its compounds. No toxic effects were observed in rats given a single oral tantalum dose of 6500 mg/kg body weight in the form of tantalum pentoxide (Cochran *et al.* 1950). Tantalum pentachloride is of higher acute toxicity. The $LD_{50}$ for the rat is 958 mg/kg body weight, expressed as tantalum. After intraperitoneal injection of tantalum pentachloride, the $LD_{50}$ was found to be 38 mg/kg body weight, expressed as tantalum. Potassium tantalum fluoride, however, was less toxic, the intraperitoneal $LD_{50}$ for the rat being 173 mg/kg, expressed as tantalum (no other details) (Cochran *et al.* 1950).

Six guinea pigs were given tantalum dioxide doses of 150 mg ($3 \times 50$ mg at weekly intervals) by intratracheal instillation of a 10% suspension in isotonic solution. The animals were killed 1, 4, 8 and 12 months after treatment and the lungs subjected to histopathological examination. No fibrosis was found although transient bronchitis, interstitial pneumonitis and the associated hyperaemia were observed (Delahant 1995, Schepers 1995).

The passivity of tantalum metal towards biological tissues has been demonstrated by its use in man and animals for surgical implants which are left in the body for long periods. Tantalum plates and screws implanted in animals caused neither inflammation nor oedema. In dogs and rabbits in which tantalum plates had been implanted in the bones, alterations were not found either in the bone or in other tissue (Burke 1940).

# 6 Manifesto (MAK value, classification)

The data available for dusts of tantalum metal or tantalum oxides do not reveal any substance-specific toxicity which would prevent the application of the General threshold limit value for dust.

Therefore, the General threshold limit value for dust of 4 mg/m$^3$ for the inhalable fraction and 1.5 mg/m$^3$ for the respirable fraction applies for tantalum dust and tantalum oxide dust.

# 7 References


ACGIH (American Conference of Governmental Industrial Hygienists) (1992) Tantalum. in: *Documentation of the threshold limit values and biological exposure indices.* ACGIH, Cincinnati, OH, USA

Burke GL (1940) The corrosion of metals in tissues; and an introduction to tantalum. *Can Med Assoc J 43*: 125–128

Cochran KW, Doull J, Mazur M, DuBois KP (1950) Acute toxicity of zirconium, columbium, strontium, lanthanum, cesium, tantalum, and yttrium. *Arch Ind Hyg Occup Med 1*: 637–650

Dales HC, Kyle J (1958) Late results of using tantalum gauze in the repair of large hernias. *Surgery 43*: 294–297

Delahant AB (1995) An experimental study of the effects of rare metals on animal lungs. *Arch Ind Health 12*: 116–120

Falbe J, Regitz M (1992) *Römpp Chemie Lexikon. CD Version 1.* Georg Thieme Verlag, Stuttgart, New York

Holleman AG, Wiberg E (1985) *Lehrbuch der anorganischen Chemie.* Walter de Gruyter Berlin, New York, 1075–1097

Miller CW, Davis MW, Goldman A, Wyatt JP (1953) Pneumoconiosis in the tungsten carbide tool industry. *Arch Ind Hyg Occup Med 8*: 453–463

Newton D (1975) Clearance of radioactive tantalum from the human lung after accidental inhalation. *Am J Roentgenol 129*: 327–328

Schepers GWH (1995) The biological action of tantalum oxide studies on experimental pulmonary histopathology. *Arch Ind Health 12*: 121–123

Tegtmeyer CJ, Smith NJ, El-Mahdi AM, Fitz-Hugh GS, Constable WC (1975) The value of tantalum powder as a contrast medium in laryngography. *Can J Otolaryngol 4*: 81–85


completed 02.03.1999

The highest concentration of fibres (L > 5 μm, D < 3 μm, L:D > 3:1) was found in the packing room, 36 × $10^6$ fibres/$m^3$ (determined with the scanning electron microscope). At the other workplaces the fibre concentrations were in the range of 10 × $10^6$ fibres/$m^3$. The typical range of diameters was given as 0.2–0.3 μm.

Seen as a whole, the exposure situation is unclear because the workers were exposed to mixtures of substances and because no data for the exposure levels in previous years are available. It was stated that shortly before the study, measures for reducing dust exposure had been introduced (Huuskonen *et al.* 1983a).

Apart from that described above, there are no analytical data for wollastonite exposure concentrations, for example from workplaces using the substance.

# 3 Toxic Effects and Mode of Action

In animal studies, wollastonite fibres have been shown to have a relatively short half-life in the lungs, 7 to 15 days, whereby the concentration of thin fibres in lung tissue decreases more rapidly than that of thick fibres. It may be assumed that, unlike the situation with asbestos fibres, significant accumulation of fibres in the lungs of persons exposed for long periods to wollastonite does not occur.

A long-term inhalation study with the rat produced no evidence of fibrogenic effects or progressive lung changes. The known effects in man and the results of animal studies do not suggest that wollastonite has carcinogenic effects.

# 4 Toxicokinetics

Durability in the organism is a prerequisite for carcinogenic activity of fibres (DFG 1996). The results of durability studies with wollastonite fibres are shown in Table 1.

From rats which had inhaled an aerosol containing wollastonite fibres in a concentration of 835 fibres/$cm^3$ (114 mg/$m^3$), 6 hours daily for a period of 5 days, were recovered 1.3 × $10^5$ fibres of critical dimensions (L > 5 μm, L:D > 3:1) per gram dry weight of lung tissue (63.5 μg/mg dry weight). The half-life of the material in the lungs was less than one week when expressed in terms of either numbers of fibres or the mass of particles. The average length of the fibres found in the lung tissue decreased in the course of one month from 11.2 to 6.0 μm and the average diameter increased from 0.53 to 0.95 μm (Wahrheit *et al.* 1994a, 1994b). Thus the concentration of thin fibres decreases rapidly as the fibres are dissolved while thicker fibres may be detected a little longer in the lung tissue.

**Table 1.** Studies of the stability of wollastonite in the organism

| Fibre sample fibre fraction | Fibre dimensions | | | Adminis-tration route | Dose or fibre concentration | Half-life [days] in terms of | | | References |
|---|---|---|---|---|---|---|---|---|---|
| | $L_m$ [µm] | $D_m$ [µm] | MMAD [µm] | | | Total number of fibres | Number of critical fibres (L > 5 µm) | Particle weight | |
| wollastonite NYAD-G [a] | n.s. | D: 0.2–3.0 µm | 2.6 | inhalation 5 days, 6 h/day | 114 mg/m³ 835 fibres/cm³ | < 7 | n.s. | < 7 | Wahrheit *et al.* 1994a, 1994b |
| wollastonite, India, milled [b] | 8.4 | 1.14 | n.s. | intra-tracheal | 2 mg/animal $12.2 \times 10^6$ critical fibres [c] | n.s. | 11 | n.s. | Muhle *et al.* 1991 |
| wollastonite, milled, fractionated [d] | 5.6 | 0.71 | n.s. | intra-tracheal | 2 mg/animal $3.8 \times 10^6$ critical fibres [c] | n.s. | 10 | n.s. | Muhle *et al.* 1991 |
| wollastonite NYAD-G [a], milled | 5.4 | 0.72 | n.s. | intra-tracheal | 2 mg/animal $2.4 \times 10^6$ critical fibres [c] | n.s. | 12 | n.s. | Muhle *et al.* 1991 |
| wollastonite NYAD-G [a] thoracic fraction | 4.3 | 0.85 | 7.95 | intra-tracheal | 2 mg/animal | 17 | 18 | 19 | Bellmann and Muhle 1994a, 1994b |
| wollastonite NYAD-G [a] alveolar fraction | 3.2 | 0.63 | 5.74 | intra-tracheal | 2 mg/animal | 17 | 18 | 18 | Bellmann and Muhle 1994a, 1994b |
| wollastonite NYAD-1250 [a] thoracic fraction | 1.8 | 0.44 | 6.77 | intra-tracheal | 2 mg/animal | 16 | 19 | 16 | Bellmann and Muhle 1994a, 1994b |
| wollastonite NYAD-1250 [a] alveolar fraction | 1.7 | 0.40 | 2.43 | intra-tracheal | 2 mg/animal | 15 | 18 | 17 | Bellmann and Muhle 1994a, 1994b |

**Table 1.** continued

| Fibre sample fibre fraction | Fibre dimensions | | | Adminis-tration route | Dose or fibre concentration | Half-life [days] in terms of | | | References |
|---|---|---|---|---|---|---|---|---|---|
| | $L_m$ [µm] | $D_m$ [µm] | MMAD [µm] | | | Total number of fibres | Number of critical fibres (L > 5 µm) | Particle weight | |
| NYAD-G Wollastocoat[e] 1001 thoracic fraction | 3.7 | 0.77 | 7.98 | intra-tracheal | 2 mg/animal | 16 | 16 | 16 | Bellmann and Muhle 1994a, 1994b |
| NYAD-G Wollastocoat[e] 1001 alveolar fraction | 2.8 | 0.56 | 3.61 | intra-tracheal | 2 mg/animal | 18 | 19 | 17 | Bellmann and Muhle 1994a, 1994b |
| NYAD-G Wollastocoat[e] 2075 thoracic fraction | 4.3 | 0.83 | 9.17 | intra-tracheal | 2 mg/animal | 21 | 22 | 24 | Bellmann and Muhle 1994a, 1994b |
| NYAD-G Wollastocoat[e] 2075 alveolar fraction | 3.5 | 0.70 | 5.26 | intra-tracheal | 2 mg/animal | 19 | 19 | 20 | Bellmann and Muhle 1994a, 1994b |
| crocidolite UICC | 1.2 | 0.2 | 1.6 | intra-tracheal | 2 mg/animal | 236 | ∞ | ∞ | Bellmann and Muhle 1994a, 1994b |

Abbreviations: $D_m$, median fibre diameter; $L_m$, median fibre length; MMAD, mass median aerodynamic diameter

[a] producer: Nyco Inc., Willsboro, USA

[b] administered in the carcinogenicity study by i.p. injection: see Table 4, Pott *et al.* 1989; sample from Osthoff-Petrosch KG, Hamburg, Germany

[c] critical fibres: L > 5 µm, D < 3 µm, L:D > 3:1

[d] administered in the carcinogenicity study by i.p. injection: see Table 4, Muhle *et al.* 1991, Rittinghausen *et al.* 1991, 1992; sample from Eternit, Kapelle, Germany

[e] Wollastocoat: wollastonite NYAD-G, the suface of which is coated with organic substances; producer: Nyco Inc., Willsboro, USA

A study in which fibre concentrations in the rat lung were determined 2 and 30 days after intratracheal administration of three samples of wollastonite fibres from different sources revealed half-lives (expressed in terms of fibre number in lung tissue) of 10 to 12 days. The median fibre dimensions were in the range of 0.71 to 1.1 μm for diameter and 5.4 to 8.4 μm for length (Muhle *et al.* 1991). As the alveolar clearance time for insoluble particles in the rat is about 60 days, these data indicate that wollastonite is dissolved in the lungs.

In another study with size-screened inhalable wollastonite particles from Willsboro, USA, the authors investigated whether the rapid reduction in the numbers of wollastonite fibres in the lungs is also apparent when the process is followed over longer periods (Bellmann and Muhle 1994a, 1994b). The study made use of commercially available wollastonite samples NYAD-G and NYAD-1250 with different particle size distributions and two samples of surface-modified wollastonite fibres NYAD-G Wollastocoat 1001 and 2075 (see Table 1). In order to obtain data relevant for human risk assessment, the thoracic and alveolar fractions, which may be deposited in the whole lung and in the alveolar region, respectively, were separated by sifting from the commercially available fibre samples. The fibre dimensions are shown in Table 1. To determine the stability of these fibres in the organism, 2 mg doses of each fraction were administered intratracheally to rats. Two days after the intratracheal instillation, the distribution of fibres in lung sections was determined with the scanning electron microscope. The distribution proved to be relatively even, with fibres found in the bronchi, bronchioli and alveoli. No clumps or aggregates of fibres were found in the lungs. The elimination of fibres from the lungs was followed quantitatively by counting fibres in serial lung sections under the scanning electron microscope during a period of 6 months. The weight of fibres per lung in the form of fibre fragments with L:D > 3:1 was estimated. For all the various samples of wollastonite, the half time for the decrease in the total number of fibres, the number of all critical fibres (L > 5 μm) and the fibre weight in the lung was found to be 15 to 21 days. For crocidolite fibres (UICC crocidolite), the half time for the decrease in the total number of fibres was found to be 240 days but the number of critical fibres (L > 5 μm) was unchanged after 6 months. Because of the relatively short persistence of wollastonite fibres in the lung, it may be assumed that no significant accumulation of fibres in the lungs takes place in persons handling wollastonite, unlike in those working with crocidolite.

# 5 Effects in Man

The persons working in the wollastonite mine and mill in Willsboro (USA) were subjected to medical examination in 1976 and 1982 (Shasby *et al.* 1979, Hanke *et al.* 1984). The analysis of exposure concentrations (see Section 2) in extraction and processing revealed wollastonite concentrations up to 16.4 mg/m$^3$ (as total dust[a]) and 5.0 mg/m$^3$

[a] nomenclature from Hanke *et al.* 1984, Zumwalde 1977

determined as the inhalable aerosol fraction[a] and fibre concentrations ($L > 5$ µm) up to $30 \times 10^6$ fibres/m$^3$. In the year 1982, 114 persons were still employed in the mine and mill. Of these, 102 (89 %) took part in the study. Of the 24 retired workers, only 6 (25 %) agreed to take part in the medical examination. The average duration of employment of the workers who were studied was 10 years. The control group comprised 86 workers from a neighbouring electronics firm (93 % of the persons employed there).

The medical examination included anamnesis, roentgenograms and spirometry. Three of the 108 exposed workers (3 %) suffered from category 1 pneumoconiosis which had also been diagnosed during the 1976 examination. The lung function parameters $FEV_1$ (forced expiratory volume in the first second), FVC (largest forced vital capacity), $FEV_1$/FVC and PFR (peak flow rate) were subjected to multiple regression analysis. The results for the total collective of 108 exposed workers demonstrated that the lung function parameters $FEV_1$, $FEV_1$/FVC and PFR decreased as a function of the cumulative dust concentration ($\Sigma$ k $\cdot$ t) and that the decrease was statistically significant. For the 32 non-smokers in this collective, a significant association with the cumulative dust concentration was found only for the decrease in $FEV_1$/FVC.

The comparison of workers exposed to a cumulative dust dose (total dust[a]) of at least 30 dust years ("dust years" is the dust concentration in mg/m$^3$ × years) (52 of the 108 exposed workers) with the 86 workers in the control group showed significant differences in $FEV_1$/FVC and PFR between persons of similar ages, heights and smoking habits (see Table 2).

Analysis of the data for the 57 exposed workers who had taken part in the study programmes in both 1976 and 1982 revealed that the PFR decreased between the two study dates as a function of the cumulative dust dose and that the decrease was statistically significant. The authors concluded therefore that long-term exposure to wollastonite can cause airway obstruction which is reflected in changed lung function parameters (Hanke *et al.* 1984). The published exposure data (see Section 2) is not sufficient for the derivation of a dose-response relationship for the described effects which could be used in the establishment of a MAK value. In addition, it may be assumed that at least some of the workers studied were at times also exposed to quartz dust (see Section 2).

A similar study was carried out with 46 Finnish workers who had been employed for at least 10 years at the limestone-wollastonite mine in Lappeenranta (Huuskonen *et al.* 1983a). Of these workers, 15 had worked in the mine for 10 to 19 years, 24 for 20 to 29 years and 7 for more than 30 years. Thus, these workers had been exposed to wollastonite for very much longer than those in the collective studied by Hanke *et al.* (1984). The exposure concentrations determined at the time of the study were up to 27 mg/m$^3$ total dust[b] and $36 \times 10^6$ fibres/m$^3$ ($L > 5$ µm, $D < 3$ µm, L:D > 3:1) (see Section 2). This collective was compared with three different control collectives (lumberjacks, peat cutters and workers from a cigarette-producing firm) with similar age structure. The proportion of smokers including ex-smokers was similar in all groups (about 70 %–80 %).

[a] nomenclature from Hanke *et al.* 1984, Zumwalde 1977

[b] nomenclature from Huuskonen *et al.* 1983a

**Table 2.** Lung function parameters in wollastonite miners and control persons (Hanke *et al.* 1984)

| | Wollastonite miners [i] | Control persons |
|---|---|---|
| Number of workers studied | 52 | 86 |
| $FEV_1/FVC$ | 0.73 ± 0.09** | 0.78 ± 0.08 |
| PFR | 8.59 ± 1.76* | 9.30 ± 1.89 |

* $p < 0.05$
** $p < 0.01$
[i] workers from the wollastonite mine and mill in Willsboro (USA) with a cumulative dose of total dust (nomenclature from Hanke *et al.* 1984; Zumwalde 1977) of 30 dust years ("dust years" is the dust concentration in $mg/m^3$ × years of exposure)
Abbreviations:
PFR peak flow rate
$FEV_1$ forced expiratory volume in the first second
FVC largest forced vital capacity

**Table 3.** X-ray examination of wollastonite miners and control persons (Huuskonen *et al.* 1983a)

| Results | | Miners [i] n | Miners % | Control persons n | Control persons % |
|---|---|---|---|---|---|
| number of workers studied | | 46 | 100 | 138 | 100 |
| workers with only lung fibrosis (≥ 1.0 [ii]) | (= a) | 7 * [iii] | 15 | 5 | 3.6 |
| workers with only pleural thickening (≥ a, 3 [ii]) | (= b) | 6 * | 13 | 6 | 4.3 |
| workers with both changes | (= c) | 7 * | 15 | 1 | 0.7 |
| total workers with lung fibrosis (≥ 1.0 [ii]) | (= a + c) | 14 * [iii,iv] | 30 | 6 | 4.3 |
| total workers with pleural thickening (≥ a, 3 [ii]) | (= b + c) | 13 * [v] | 28 | 7 | 5.1 |

* $p < 0.05$
[i] workers from the wollastonite mine in Lappeenranta, Finland
[ii] according to ILO, 1980
[iii] of these workers 2 had been previously exposed to quartz dust
[iv] duration of employment in the mine: 2 workers < 20 years, 9 workers 20–29 years, 3 workers ≥ 30 years
[v] duration of employment in the mine: 4 workers < 20 years, 6 workers 20–29 years, 3 workers ≥ 30 years

The x-ray examination (see Table 3) revealed lung fibrosis in 14 of the 46 (30 %) mine workers (8 cases with a score of 1/1 and 6 cases 1/0, according to ILO 1980); two of these workers had previously been exposed to quartz dust. In the three control collectives together, 6 of the 138 (4 %) workers had fibrosis ($p < 0.01$) According to the authors, the increased incidence of lung fibrosis cannot be explained entirely by the exposure to quartz in the limestone-wollastonite mine. However, it is still not known how the simultaneous presence of wollastonite and quartz dust affects the development of lung fibrosis.

The incidence of pleural thickening (see Table 3) was also significantly increased above the control values. According to currently available data, pleural changes are induced by fibrous dusts and not by exposure to quartz dust (Woitowitz 1985).

Lung function was studied only in the mine workers. The results showed that $FEV_1$ was reduced in one worker to less than 80 % of the expected value, the maximum expiratory flow (MMEF) was reduced to less than 70 % in four of the exposed persons and, in addition, the carbon monoxide diffusing capacity (DCO) was reduced in two persons. These results of the lung function tests are suggestive rather of pulmonary obstruction than of parenchymal fibrosis. The group of persons studied was, however, too small to make it possible to draw clear conclusions.

The group of 46 exposed persons included 15 non-smokers, three of whom suffered from chronic bronchitis. In the opinion of the authors, the exposure to dust is the only conceivable cause of the bronchitis (Huuskonen *et al.* 1983a). Because the persons were exposed to more than one kind of dust (see Section 2), however, it is not possible to distinguish any effects of wollastonite from those of the other dusts.

Because of the unclear exposure situation and the fact that the workers were exposed to more than one kind of dust, it is not possible to derive a concentration-effect relationship from the results available for the workers exposed to wollastonite in Lappeenranta.

In addition to the above investigations, a cohort study was carried out to investigate the mortality of all workers who had ever been employed for at least one year in the Finnish limestone-wollastonite mine (Huuskonen *et al.* 1983b). Cancer risk was not increased. This study has been described in the MAK documentation for fibrous dust (see Volume 8 of the present series). Because the number of cases was small and the exposure data inadequate, the carcinogenic potential of wollastonite for man could not be evaluated on the basis of this study.

# 6 Animal Experiments and *in vitro* Studies

## 6.1 Toxicity

The results of animal studies are summarized in Table 4.

**Table 4.** *In vivo* studies with fibrous wollastonite

| Fibre sample | Fibre characterization | Species, number of animals | Exposure data | Concentration or dose | Results | Comments | References |
|---|---|---|---|---|---|---|---|
| wollastonite NYAD-G[a] Willsboro, USA | MMAD 5.8 µm | Crl:CDBR rat, number not specified | **inhalation** 6 h observation: 0, 24, 48 h, 1 week, 1 month | 105 mg/m$^3$ 476 × 10$^6$ fibres/m$^3$ D < 3 µm | lung lavage: no substance-related effects | | Warheit *et al.* 1991 |
| | MMAD 5.8 µm | | **inhalation** 3 days, 6 h/day observation: 0, 24, 48 h, 1 week, 1 month | 118 mg/m$^3$ 227 × 10$^6$ fibres/m$^3$ D < 3 µm | 3.9 × 10$^6$ fibres/g lung tissue, lung lavage: no substance-related effects | | |
| | MMAD 4.3 µm | | **inhalation** 5 days, 6 h/day observation: 0, 24, 48 h, 1 week, 1 month | 59 mg/m$^3$ 123 × 10$^6$ fibres/m$^3$ D < 3 µm | lung lavage: 24 h *post expositionem* LDH increased | | |
| | MMAD 2.6 µm | | **inhalation** 5 days, 6 h/day observation: 0, 24, 48 h, 1 week, 1 month | 114 mg/m$^3$ 835 × 10$^6$ fibres/m$^3$ D < 3 µm | 8.9 × 10$^6$ fibres/g lung tissue lung lavage: 1 week *post expositionem* cell count, granulocyte count, protein and LDH increased, macrophage count decreased | | |

**Table 4.** continued

| Fibre sample | Fibre characterization | Species, number of animals | Exposure data | Concentration or dose | Results | Comments | References |
|---|---|---|---|---|---|---|---|
| wollastonite NYAD-G[a] Willsboro, USA | MMAD 2.6 µm | Crl:CDBR rat, 3–4 per observation group | **inhalation** 5 days, 6 h/day observation: 0, 24 h, 1 week, 1, 3 months | 114 mg/m$^3$ 835 × 10$^6$ fibres/m$^3$ D < 3 µm | 0 h *post expositionem*: 1.3 × 10$^5$ fibres/g lung dry weight, 63.5 µg/g lung tissue<br>lung lavage: 1 week *post expositionem* increased granulocyte counts, protein and LDH,<br>BrdU: 0 h and 1 month *post expositionem* no substance-related effects | | Wahrheit *et al.* 1994a, 1994b |
| wollastonite NYAD-G[a] Willsboro, USA | L: 95 % < 8 µm D: 0.1–1.0 µm about 15 % with L > 5 µm | F344 rat, 6, 6 | **inhalation** 3 months | 10 mg/m$^3$ 360 × 10$^6$ fibres/m$^3$ D < 1 µm | fibrosis score: 1.3, 1.3 chrysotile: 2.0, control 1.0 | no fibrogenic effects | McConnell *et al.* 1991 |
| | | 6, 6 | **inhalation** 12 months | | fibrosis score: 2.0, 2.0 chrysotile: 4.0, control 1.0 | | |
| | | 6, 6 | **inhalation** 12 months and observation 12 months or inhalation 24 months | | fibrosis score: 1.3, 1.6 chrysotile 4.0, control 1.0 | | |
| wollastonite NYAD-G[a] Willsboro, USA | L: 95 % < 8 µm D: 0.1–1.0 µm about 15 % with L > 5 µm | F344 rat, 57 | **inhalation** 12 months, observation for life | 10 mg/m$^3$ 360 × 10$^6$ fibres/m$^3$ D < 1 µm | Lu b/m: 0/57 (chrysotile 20/52, control 1/56), fibrosis 0/57 (chrysotile 50/52, control 0/56) | in other organs no substance-related effects | McConnell *et al.* 1991 |

**Table 4.** continued

| Fibre sample | Fibre characterization | Species, number of animals | Exposure data | Concentration or dose | Results | Comments | References |
|---|---|---|---|---|---|---|---|
| | | 60 | **inhalation** 24 months observation for life | | Lu b/m: 1/60 (chrysotile 20/52, control 1/56), fibrosis 1/60 (chrysotile 50/52, control 0/56) | in other organs no substance-related effects | |
| wollastonite "1" Canada | no data | Osborne-Mendel rat, 20 | **intrapleural** | 40 mg $0.14 \times 10^9$ fibres L > 4 µm, D < 2.5 µm[b] | PS 5/20 (25 %) control 2.8 % | no data for mortality, number of critical fibres[c] small, fibre sample inadequately characterized | Stanton *et al.* 1981 |
| wollastonite "2" Canada | no data | 25 | **intrapleural** | 40 mg $0.1 \times 10^9$ fibres L > 4 µm, D < 2.5 µm[b] | PS 2/25 (8 %) control 2.8 % | no data for mortality, number of critical fibres[c] small, fibre sample inadequately characterized | |
| wollastonite "3" Canada | no data | 21 | **intrapleural** | 40 mg $0.2 \times 10^9$ fibres L > 4 µm, D < 2.5 µm[b] | PS 3/21 (14 %) control 2.8 % | no data for mortality, number of critical fibres[c] small, fibre sample inadequately characterized | |
| wollastonite "4" Canada | no data | 24 | **intrapleural** | 40 mg $0.01 \times 10^9$ fibres L > 4 µm, D < 2.5 µm[b] | PS 0/24 (0 %) control 2.8 % | no data for mortality, number of critical fibres[c] small, fibre sample inadequately characterized | |

**Table 4.** continued

| Fibre sample | Fibre characterization | Species, number of animals | Exposure data | Concentration or dose | Results | Comments | References |
|---|---|---|---|---|---|---|---|
| wollastonite, India[d] | $L_m$ 8.1 µm<br>$D_m$ 1.1 µm<br>25 % L > 5 µm,<br>D < 3 µm | Wistar rat<br>54 | **intraperitoneal** | 100 mg (5 × 20)<br>$0.43 \times 10^9$ fibres<br>L > 5 µm,<br>D < 3 µm | MS 0/54 (0 %)<br>control 6 %<br>adhesions grade 1–2 | no substance-related effects on mortality, low durability of the fibres in water | Pott *et al.* 1989 |
| wollastonite[e] | $L_m$ 5.6 µm<br>$D_m$ 0.71 µm | Wistar rat<br>50 | **intraperitoneal** | 30 mg<br>$0.093 \times 10^9$ fibres<br>3.5 %<br>L > 5 µm,<br>D < 2 µm | ME 0/50 (0 %)<br>control 6 % | no data for mortality, low durability of the fibres after intratracheal administration | Muhle *et al.* 1991, Rittinghausen *et al.* 1991, 1992 |
| wollastonite, China | L 12.9 µm<br>D 1.57 µm | Wistar rat<br>18 | **intratracheal** | 25 mg | after 3 months: lung hydroxyproline content increased, wet weight increased | effects more than with wollastonite from USA, fibrogenic effect questionable | Cambelová and Juck 1994 |
| wollastonite NYAD[a], USA | L 10.08 µm<br>D 1.39 µm | Wistar rat<br>18 | **intratracheal** | 25 mg | after 3 months: lung hydroxyproline content increased, wet weight increased | effects less than with wollastonite from China, fibrogenic effect questionable | |

[a] producer: Nyco Inc., Willsboro, USA
[b] Stanton fibres (L > 8, D < 0.25) not present
[c] critical fibres of the dimensions given
[d] for details of the stability of this fibre sample in the organism: see Section 4 and Table 1 (Muhle *et al.* 1991)
[e] sample from Eternit, Kapelle, Germany
abbreviations: LDH lactate dehydrogenase, MMAD mass median aerodynamic diameter, L length, D diameter, $L_m$ median fibre length, $D_m$ median fibre diameter
tumour types: Lu lung tumours, b/m benign/malignant, PS pleural sarcomas, considered identical with mesotheliomas (Stanton *et al.* 1981), MS mesotheliomas or sarcomas of the intestinal cavity excluding animals with only uterus tumours, ME mesotheliomas
fibrosis score: 1–3 cellular changes, > 4 fibrosis

### 6.1.1 Inhalation

Groups of male rats were exposed to commercial wollastonite from Willsboro (NYAD-G) for periods of 6 hours daily for 1, 3 or 5 days. The lungs were examined for alterations 0, 24 and 48 hours and 1 week and 1 month after the end of exposure. The wollastonite concentrations were about 50 and 100 mg/m$^3$ but because the fibre dimensions in the aerosols differed, the different groups were exposed to different fibre concentrations (see Table 4). It was demonstrated that wollastonite caused reversible inflammatory changes in the lungs only when the mass median aerodynamic diameter (MMAD) was not more than about 2.6 μm and the fibre concentration was at least $500 \times 10^6$ fibres/m$^3$ (D < 3 μm), corresponding to about 100 mg/m$^3$ (Warheit *et al.* 1991).

In the lung lavage fluid obtained from rats which had been exposed to an aerosol containing 835 wollastonite fibres per cm$^3$ (wollastonite concentration 114 mg/m$^3$) 6 hours daily for a period of 5 days, increased granulocyte counts, protein levels and enzyme activities were observed for up to one week after the end of exposure. The inflammatory alterations regressed within one month of the end of exposure. At this time and immediately after the end of exposure, determination of the incorporation of bromodeoxyuridine into lung parenchyma cells *in vivo* did not yield evidence of increased cell turnover (Warheit *et al.* 1994a, 1994b). The determinations of the stability of the substance in the organism, which were carried out simultaneously, are described in Section 4.

An inhalation study (McConnell *et al.* 1991) with commercial wollastonite from the deposits in Willsboro (NYAD-G) was carried out with groups of 78 male Fischer 344 rats. The dimensions of 95 % of the wollastonite particles in the aerosol were in the inhalable size range[c]. For the wollastonite fibres, diameters of 0.1 to 1.0 μm were determined. Although 95 % of the fibres were between 1 and 8 μm long, only 15 % were in the critical size range with a length of at least 5 μm.

Two groups of rats were exposed 6 hours daily, 5 days per week for 12 and 24 months to a wollastonite concentration of 10 mg/m$^3$ ($360 \times 10^6$ fibres/m$^3$) and then observed for the rest of their lives. Other groups of animals were kept in clean air as negative control, or exposed for 12 months to chrysotile as positive control at a concentration of 10 mg/m$^3$ (about $1000 \times 10^6$ fibres/m$^3$). In a satellite study, groups of 6 rats were killed after 3, 12 and 24 months. The lungs of these animals were subjected to histopathological examination.

In the lungs of both groups of animals exposed to wollastonite, mild pathological alterations were seen in the form of slight effects on the alveolar macrophages. Although these changes were slightly more severe in the animals exposed for 12 months than in those exposed for 3 months, after 24 months exposure the effects had regressed and were only slightly more severe than in the group exposed for 12 months and observed for another 12 months without exposure. In contrast, in the control group treated with chrysotile, fibrotic changes were seen after as little as 12 months exposure (see Table 4). These results demonstrate that long-term exposure of rats to fibrous wollastonite in a concentration of 10 mg/m$^3$ (95 % of the dust particles in the inhalable size range[c] does not result in fibrotic processes or in progressive lung changes.

---

[c] nomenclature from McConnell *et al.* 1991

In animals exposed to wollastonite for their whole lives, there was no increase in the incidence of lung changes (interstitial fibrosis: 12 months wollastonite 0/57, 24 months wollastonite 1/60, animals exposed to chrysotile 50/52) or of lung tumours (see Table 4). The concentration at which no fibrogenic effects are observed (NOAEC: no observed adverse effects concentration) determined in this study was 10 mg/m$^3$ (95 % of the dust particles in the inhalable size range[c]), which is equivalent to a concentration of critical fibres (L > 5 µm, D < 3 µm, L:D > 3:1) of about 50 × 10$^6$ fibres/m$^3$, and is in the range of the highest fibre concentrations measured during wollastonite extraction and processing (see Section 2). The study can, however, be used for the present evaluation only with reservations as the animals were exposed to only one concentration of wollastonite fibres and the fibre aerosol contained only a small proportion of wollastonite fibrous dust.

### 6.1.2 Intratracheal instillation

Groups of 18 Wistar rats were given 25 mg doses of wollastonite by intratracheal instillation and then exposed for a period of 3 months (no other details; Cambelová and Juck 1994). Wollastonite samples from China and from Willsboro, USA, with similarly thick fibres were used (fibre dimensions given in Table 4). Quartz and crocidolite were used for comparison. In animals treated with wollastonite for a period of 3 months, the authors reported fibrotic reactions reflected in increased hydroxyproline levels in the lungs and increased lung wet weights. Quartz caused very much more pronounced changes in these parameters than did wollastonite. The differences between the effects of wollastonite and crocidolite were much smaller. No histopathological examination was carried out.

These results cannot be used in the present evaluation because the documentation of this study does not make it clear how often the animals were treated during the three month period or how high the individual doses were. The wollastonite samples were not adequately characterized. In particular, there are no details of purity, of concentrations of fibres and other particles, or of the particle size distribution of the samples, so that the relevance of the observed effects for inhalation exposure cannot be determined.

Since fibrotic reactions may be induced in the rat lung by many insoluble dusts at similar concentrations in the lung tissue (Morrow *et al.* 1991), the study by Cambelová and Juck cannot be considered to demonstrate a specific fibrogenic effect of fibrous wollastonite. McConnell (1995) suggested that the changes in the studied parameters could have been a result of granulomatous bronchiolitis in the exposed animals and were not necessarily evidence of fibrogenesis.

## 6.2 Genotoxicity

The genotoxic potential of wollastonite cannot be evaluated because the database is inadequate (see MAK documentation for fibrous dust).

## 6.3 Carcinogenicity

### 6.3.1 Long-term studies

The carcinogenicity studies which have been carried out with wollastonite fibres administered by inhalation, or by intraperitoneal or intrapleural injection are listed in Table 3. A carcinogenic potential of wollastonite cannot be deduced from the available results, which were discussed in detail in the MAK documentation for fibrous dust.

### 6.3.2 Cell transformation studies

Wollastonite fibres (L: 27.5 % > 5 µm, D: 94 % < 1.5 µm) in a concentration of 20 µg/ml induced an increase in cell transformation in SHE (Syrian hamster embryo) cells. A concentration of 50 µg/ml was clearly cytotoxic. Wollastonite in a concentration of 5 µg/ml had transforming effects only in the presence of TPA (12-*O*-tetradecanoylphorbol-13-acetate) or after pretreatment of the cells with MNNG (*N*-methyl-*N'*-nitro-*N*-nitroso-guanidine) (Liu 1993, Liu *et al.* 1993).

## 6.4 Other effects

Cultures of explanted tracheal tissue from Sprague-Dawley rats were exposed for 1 h to wollastonite fibres in a concentration of 3 mg/ml (fibre sample from Warheit; $L_m$ 5.2 µm, $D_m$ 0.52 µm) and the uptake of fibres into the cells was studied. It was demonstrated that pretreatment of the cultures for 10 minutes with cigarette smoke caused a slight but not significant decrease in the uptake of wollastonite fibres. In contrast, cigarette smoke produced a significant increase in the uptake of amosite fibres and of non-fibrous particles of talc and titanium dioxide (4–5 mg/ml) into the cells (Keeling *et al.* 1993). The cause of these different effects was not discussed.

In hepatocytes exposed *in vitro* to three wollastonite samples from an Indian mine (Kemolit ASB-3, Kemolit-N, Kemolit A-60; producer: Wolkem Pvt. Ltd., Udaipur, Rajasthan; no characterization of the fibres), lipid peroxidation and reduced glutathione levels were observed. A sample of chrysotile from India had more pronounced effects. In the presence of vitamin E or methionine the effects were reduced; therefore the involvement of free radicals was suggested (Aslam *et al.* 1992). In another study, these wollastonite samples in concentrations of 1 mg/ml or more caused haemolysis and lipid peroxidation in human erythrocytes *in vitro*. The effects of the wollastonite samples were less marked and developed later than the effects of the Indian chrysotile sample. Less pronounced effects were seen in the presence of various macromolecules such as bovine serum albumin and in the presence of vitamin E or glutathione. Therefore, not only the involvement of free radicals but also an effect of the surface properties of the particles was suggested (Aslam *et al.* 1995).

In a study of 5 different calcium silicate samples, the inhalable fraction[d] of a sample of fibrous American wollastonite and one of Finnish wollastonite with a low proportion of fibres, both in a concentration of 0.5 mg/ml, were shown to have similarly slight haemolytic effects on human erythrocytes *in vitro*. On the other hand, the non-fibrous synthetic wollastonite which was studied in parallel had an activity which was about three times that of the above samples (Skaug and Gylseth 1983).

Five different samples of calcium silicate were also studied by Skaug *et al.* (1984) for their effects on the release of lysosomal enzymes and their cytotoxic effects on macrophages. The Finnish wollastonite had a higher activity than the American. The quartz content of the Finnish product was given as 2 %; in the American wollastonite quartz was not found.

A samples of wollastonite from China and one from America (no other characterization) in concentrations between 100 and 300 µg/ml were less cytotoxic for guinea pig alveolar macrophages than were crocidolite fibres (Hurbánková and Tilkes 1993).

# 7 Manifesto (MAK value/classification)

The carcinogenic potential of fibrous wollastonite was reviewed in the MAK documentation for fibrous dust (Volume 8 of the present series). The fact that the incidence of tumours is not increased in persons exposed to wollastonite and not clearly increased in exposed experimental animals is in accordance with the relatively short persistence of wollastonite fibres in the lung which has been observed in the rat. This was confirmed in more recent studies in which the half-life of wollastonite administered by intratracheal instillation or by inhalation (Bellmann and Muhle 1994a, 1994b, Wahrheit *et al.* 1994a, 1994b) was shown to be less than 7 days or about 15 days, respectively. Therefore it is likely that wollastonite fibres have no carcinogenic potential.

In a long-term inhalation study with rats (McConnell *et al.* 1991), the data from which can be accepted only with reservations, no evidence of fibrotic processes or progressive lung changes was seen in animals exposed to fibrous wollastonite at the only concentration tested, 10 mg/m$^3$ (95 % of particles in the inhalable size range[c]).

In workers from a Finnish limestone-wollastonite mine who had been exposed for long periods to wollastonite, calcite and quartz, the incidences of pulmonary fibrosis and pleural thickening were significantly increased (Huuskonen *et al.* 1983a). In workers from an American wollastonite mine and mill, significant alterations in lung function parameters were reported (Hanke *et al.* 1984). Because the exposure situation is unclear in both reports, it is not possible to evaluate the observed effects or to derive a dose-response relationship for the establishment of a MAK value for wollastonite from either of these studies.

---

[d] nomenclature from Skaug and Gylseth 1983

[c] nomenclature from McConnell *et al.* 1991

Thus, neither from the known effects in man nor from the results of animal studies are sufficient data available for the establishment of a MAK value for wollastonite. Therefore, wollastonite as fibrous dust and as particles of dimensions which do not meet the definition of fibrous dust (see MAK documentation for fibrous dust) is listed in Section IIb of the *List of MAK and BAT Values*.

# 8 References

Anonymous (1975) Wollastonite: USA dominates both production and consumption. *Ind. Miner. 94*: 15–29

Aslam M, Ashquin M, Rahman Q (1992) *In vitro* cytotoxic effects of wollastonites on rat hepatocytes: II. Lipid peroxidation and glutathione depletion. *Bull Environm Contam Toxicol 49*: 547–554

Aslam M, Arif MJ, Rahman Q (1995) Red blood cell damage by wollastonite: *In vitro* study. *J Appl Toxicol 15*: 27–31

Bellmann B, Muhle H (1994a) Investigation of the biodurability of wollastonite and xonotlite. *Environ Health Persp 102, Suppl. 5*: 191–195

Bellmann B, Muhle H (1994b) Untersuchungen zur Biobeständigkeit von Wollastonitfasern. *Zbl Arbeitsmed 44*: 118–123

Cambelova M, Juck A (1994) Fibrogenic effect of wollastonite compared with asbestos dusts and dusts containing quarz. *Occup Environ Med 51*: 343–346

DFG (Deutsche Forschungsgemeinschaft) (1996) *MAK- und BAT-Werte-Liste 1996*. Senatskommission zur Prüfung gesundheitsschädlicher Arbeitsstoffe, Report No. 32 (also available in English translation), VCH Verlagsgesellschaft mbH, Weinheim, Germany

Hanke W, Sepulveda MJ, Watson A, Jankovic J (1984) Respiratory morbidity in wollastonite workers. *Br J Ind Med 41*: 474–479

Hurbánková M, Tilkes F (1993) Use of the test for estimation of LDH activity as one parameter for demonstrating cytotoxicity of fibrous dust. *Pracov Lék 45*: 149–153

Huuskonen MS, Tossavainen A, Koskinen H, Zitting A, Korhonen O, Nickels J, Korhonen K, Vaaranen V (1983a) Wollastonite exposure and lung fibrosis. *Environ Res 30*: 291–304

Huuskonen MS, Järvisalo J, Koskinen H, Nickels J, Räsänen J, Asp S (1983b) Preliminary results from a cohort of workers exposed to wollastonite in a Finnish limestone quarry. *Scand J. Work Environ Health 9*: 169–175

Keeling B, Hobson J, Churg A (1993) Effects of cigarette smoke on epithelial uptake of non-asbestos mineral particles in tracheal organ culture. *Am J Respir Cell Mol Biol 9*: 335–340

Liu J (1993) Studies on the transforming effects of wollastonite on SHE cells (Chinese). *Chung Kuo I Hsueh Yuan Hsueh Pao (Acta Academiae Medicinae Sinicae) 15*: 132–137

Liu J, Li S, Xu D, Shen B, Liang S (1993) Morphological transformation of Syrian hamster embryo cells induced by wollastonite. *Proc Am Assoc Cancer Res 34*: 186

McConnell EE, L Hall, Adkins B (1991) Studies on the chronic toxicity (inhalation) of wollastonite in Fischer 344 rats. *Inhal Toxicol 3*: 323–337

McConnell EE (1995) Fibrogenic effect of wollastonite compared with asbestos dust and dusts containing quartz. *Occup Environ Med 52*: 621

Morrow PE, Muhle H, Mermelstein R (1991) Chronic inhalation study findings as a basis for proposing a new occupational dust exposure limit. *J Am Coll Toxicol 10*: 279–290

Muhle H, Bellmann B, Pott F (1991) Durability of various mineral fibers in rat lungs. in: Brown RC, Hoskins JA, Johnson NF (Eds) *Mechanisms in fibre carcinogenesis*. Proceedings of a NATO advanced research workshop on mechanisms in fibre carcinogenesis, held October 22–25, 1990, Albuquerque, New Mexico. NATO ASI Series A, Life Sciences 223: 181–187

Pott F, Roller M, Ziem FJ, Reiffer B, Bellmann M, Rosenbruch M, Huth F (1989) Carcinogenicity studies on natural and man-made fibres with the intraperitoneal test in rats. in: Bignon J, Peto J, Saracci R (Eds) *Non-Occupational Exposure to Mineral Fibres. IARC Sci Publ No 90*: 173–179

Power T (1986) Wollastonite, performance filler potential. *Ind Miner 220*: 19-34

Rittinghausen S, Ernst H, Muhle H, Fuhst RH, Mohr U (1991) Histopathological analysis of tumour types after intraperitoneal injection of mineral fibres in rats. in: Brown RC, Hoskins JA, Johnson NF (Eds) *Mechanisms in fibre carcinogenesis*. Proceedings of a NATO advanced research workshop on mechanisms in fibre carcinogenesis, held October 22–25, 1990, Albuquerque, New Mexico. NATO ASI Series A, *Life Sciences 223*: 81–89

Rittinghausen S, Ernst H, Muhle H, Mohr U (1992) Atypical malignant mesotheliomas with osseous cartilaginous differentitation after intraperitoneal injection of various types of mineral fibres in rats. *Exp Toxicol Pathol 44*: 55–58

Shasby DM, Peterson M, Hodous T, Boehlecke B, Merchant J (1979) Respiratory morbidity of workers exposed to wollastonite through mining and milling. in: Lemen R, Dement JM (Eds) *Dusts and Disease*, Pathotox Publishers, Park Forest South, IL, 251–256

Skaug V, Gylseth B (1983) Hemolytic activity of five different calcium silicates. *Environ Health Perspect 51*: 195–203

Skaug V, Davies R, Gylseth B (1984) *In vitro* macrophage cytotoxicity of five calcium silicates. *Br J Ind Med 41*: 116–121

Stanton MF, Layard M, Tegeris A, Miller E, May M, Morgan E, Smith A (1981) Relation of particle dimension to carcinogenicity in amphibole asbestoses and other fibrous minerals. *J Nat Cancer Inst 67*: 965–975

Warheit DB, Kimberly AM, Carakostas MC, Hartsky MA (1991) Acute pulmonary effects of inhaled wollastonite fibers are dependent on fiber dimensions and aerosol concentrations. in: Brown RC (Ed.) *Mechanisms in fibre carcinogenesis*. Plenum Press, New York, 143–156

Warheit DB, Hartsky MA, McHugh TA, Kellar KA (1994a) Biopersistence of inhaled organic and inorganic fibers in the lungs of rats. *Environ Health Perspect 102, Suppl 5*: 151–157

Warheit DB, McHugh TA, Kellar KA, Hartsky MA (1994b) Biopersistence of inhaled *p*-aramid or wollastonite fibers following short-term exposures. in: Mohr U, Dungworth DL, Mauderly JL, Oberdörster G (Eds) *Toxic and carcinogenic effects of solid particles in the respiratory tract.* ILSI Monograph, ILSI Press, Washington, DC, 469–472

Woitowitz HJ (1985) Erkrankungen der Atemwege und der Lungen, des Rippenfells und des Bauchfells. in: Valentin H, Lehnert G, Petry H, Weber G, Wittgens H, Woitowitz HJ (Eds) *Arbeitsmedizin*, Vol 2, Thieme Verlag Stuttgart, New York: 172–261

Zumwalde R (1977) *Industrial Hygiene Study of the Interpace Corporation*, Willsboro, NY. National Institute for Occupational Safety and Health, Cincinnati, OH. NTIS PB84-100098

completed 7.02.1996

# Authors of Documents in Volume 16


Dr. R. Bartsch, Senatskommission der DFG zur Prüfung gesundheitsschädlicher Arbeitsstoffe, Technische Universität München, D-85350 Freising-Weihenstephan

Professor Dr. Dr. H.-P. Gelbke, BASF Aktiengesellschaft, Abteilung Toxikologie, 67063 Ludwigshafen

Dr. R. Grebenstein, Robert-Bosch-GmbH, 87542 Immenstadt

Professor Dr. U. Heinrich, Fraunhofer-Institut für Toxikologie und Aerosolforschung, Nikolai-Fuchs-Straße 1, 30625 Hannover

Dr. M. Herbst, SCISERV AG, Froburgstraße 42, CH-4410 Liestal/Bl

Professor Dr. H. Herrmann, Jürgensallee 44m, 22609 Hamburg

Professor Dr. med. R. Hess, Fluhweg 11, CH-4143 Dornach

Dr.med.vet. A. Hofmann, Weimarerstraße 6, 64807 Dieburg

Professor Dipl. Chem. Dr. med. H. Ippen (†), formerly Universitätshautklinik und Poliklinik, von-Siebold-Straße 3, 37077 Göttingen

Frau Dr. P. Kricke-Helling, Welfenweg 1, 37120 Bovenden

Dr. H. Lessmann, Informationsverbund Dermatologischer Kliniken an der Universitätshautklinik, von-Siebold-Straße 3, 37075 Göttingen

Professor Dr. H. Merk, Universitätshautklinik der RWTH Aachen, Pauwelsstraße 30, 52074 Aachen

Frau Dr. C. Pohlenz-Michel, Senatskommission der DFG zur Prüfung gesundheitsschädlicher Arbeitsstoffe, Technische Universität München, D-85350 Freising-Weihenstephan

Professor Dr. H. Muhle, Fraunhofer-Institut für Toxikologie und Aerosolforschung, Nikolai-Fuchs-Straße 1, 30625 Hannover

Frau Dr. U. Reuter, Senatskommission der DFG zur Prüfung gesundheitsschädlicher Arbeitsstoffe, Technische Universität München, D-85350 Freising-Weihenstephan

Dr. G. Scheuch, GSF-Institut für Inhalationsbiologie, Robert-Koch-Allee 6, 82131 Gauting

Professor Dr. Dr. H.-W. Thielmann, Deutsches Krebsforschungszentrum, Im Neuenheimer Feld 280, 69120 Heidelberg

Dr. D. Walter, Institut für Anorganische Chemie der TU-Berlin, Straße des 17. Juni 135, 10623 Berlin

# Index for Volumes 1–16

6-12® **V** 345
abachi **XIII** 288
absolute ethanol **XII** 129
Abstensil **V** 165
Abstinyl® **V** 165
*Acacia melanoxylon* **XIII** 285, 291–293
acetaldehyde **III** 1–9
acetene **X** 91
acetic acid 2-butoxyethyl ester **VI** 53
acetic acid ethenyl ester **V** 229
acetic acid 2-ethoxyethyl ester **VI** 213
acetic acid ethyl ester **XII** 167
acetic acid 2-methoxy-1-methylethyl ester **V** 217
acetic acid pentyl ester **XI** 211
acetic acid 2-propoxyethyl ester **XII** 187
acetic acid vinyl ester **V** 229
acetic anhydride **XIII** 43–46
acetic ether **XII** 167
acetic oxide **XIII** 43
acetic peroxide **VII** 229
acetone **VII** 1–8
1-acetoxyethylene **V** 229
2-acetoxypentane **XI** 211
acetyl hydroperoxide **VII** 229
acetyl oxide **XIII** 43
aclarubicin **I** 5
acquinite **VI** 105
acraldehyde **XVI** 1
acrolein **I** 41–43, 60, **XVI** 1–33
acrylaldehyde **XVI** 1
acrylamide **III** 11–21
acrylic acid *n*-butyl ester **V** 5, **XII** 57, **XVI** 35
acrylic acid ethyl ester **VI** 217, **XVI** 41
acrylic acid 2-ethylhexyl ester **XVI** 47
acrylic acid 2-hydroxyethyl ester **XVI** 89
acrylic acid hydroxypropyl ester **XVI** 95
acrylic acid methyl ester **VI** 253, **XVI** 177
acrylic acid monoester with propanediol **XVI** 95
acrylic acid pentaerythritol triester **XVI** 193
acrylic acid polymer, neutralized, cross-linked **XV** 1–29
acrylic acid 1,1,1-(trihydroxymethyl)propane triester **XVI** 201
acrylic aldehyde **XVI** 1
acrylic amide **III** 11
Acticide® 45 **XVI** 263
actinolite **II** 96, 184, 185
actinomycin D **I** 5
adriamycin **I** 5
AEPD® **IX** 223
aeropur® **I** 125
aerosols **XII** 271–292, **XVI** 289
afara **XIII** 288
African afzelia **XIII** 285
African black walnut **XIV** 299
African blackwood **XIII** 286, 305
African 'cherry' **XIII** 288
African ebony **XIII** 286, **XIV** 287
African mahogany **XIII** 285, 286, 287
African maple **XIII** 288
African satinwood **XIII** 286, **XIV** 291
African whitewood **XIII** 288
*Afrormosia elata* **XIII** 287
*Afzelia* spp. **XIII** 285
AGE **VII** 9
alabaster **II** 117
alcohol **XII** 129
allyl alcohol **XV** 31–40
allyl 2,3-epoxypropyl ether **VII** 9
allyl glycidyl ether **VII** 9–16
1-allyloxy-2,3-epoxypropane **VII** 9
allyl trichloride **IX** 171
altretamine **I** 2
aluminium **II** 69–93
aluminium hydroxide **II** 69–93
aluminium oxide **II** 69–93, **VIII** 141–338
American arborvitae **XIII** 288
American black walnut **XIII** 286
American mahogany **XIII** 287
American walnut **XIII** 286
*Amerimnum ebenus* **XIII** 285, 299
Amine 220® **V** 373
Amine CS-1246 **IX** 275
2-aminoaniline **XIII** 215
3-aminoaniline **VI** 287
4-aminoaniline **VI** 311
*m*-aminoaniline **VI** 287
*o*-aminoaniline **XIII** 215
*p*-aminoaniline **VI** 311
2-aminoanisole **X** 1
*o*-aminoanisole **X** 1
aminobenzene **VI** 17
4-(4-aminobenzyl)aniline **VII** 37
4-aminobiphenyl **I** 257–259
6-aminocaproic acid lactam **IV** 65
1-amino-2-chlorobenzene **III** 31

1-amino-3-chlorobenzene **III** 37
1-amino-4-chlorobenzene **III** 45
*m*-aminochlorobenzene **III** 37
1-amino-3-chloro-6-methylbenzene **VI** 143
2-amino-4-chlorotoluene **VI** 143
2-amino-5-chlorotoluene **VI** 127
3-amino-*p*-cresol methyl ether **IV** 135
*m*-amino-*p*-cresol methyl ether **IV** 135
aminodiglycol **IX** 215
2-aminodimethylethanol **IX** 229
2-aminoethanol **XII** 15–35
2-(2-aminoethoxy)ethanol **IX** 215–222
2-aminoethoxyethanol **IX** 215
2-amino-6-ethoxynaphthalene **VII** 17
6-amino-2-ethoxynaphthalene **VII** 17–19
*β*-aminoethyl alcohol **XII** 15
3-amino-9-ethylcarbazole **V** 1–3
3-amino-*N*-ethylcarbazole **V** 1
2-amino-2-ethyl-1,3-propanediol **IX** 223–227
Aminoform **V** 355
aminoglutethimide **I** 2
6-aminohexanoic acid cyclic lactam **IV** 65
2-aminoisobutanol **IX** 229
*beta*-aminoisobutanol **IX** 229
*alpha*-aminoisopropylalcohol **IX** 237
aminomethane **VII** 145
1-amino-2-methoxybenzene **X** 1
1-amino-2-methoxy-5-methylbenzene **IV** 135
3-amino-4-methoxytoluene **IV** 135
2-amino-4-methylanisole **IV** 135
1-amino-2-methylbenzene **III** 307
1-amino-4-methylbenzene **III** 323
2-amino-1-methylbenzene **III** 307
4-amino-1-methylbenzene **III** 323
1-amino-2-methyl-5-nitrobenzene **VI** 271
2-amino-2-methyl-1-propanol **IX** 229–236
6-aminonaphthol ether **VII** 17
4-amino-2-nitroaniline **IV** 295, **XIII** 207
4-amino-2-nitrophenol **IV** 289
2-amino-4-nitrotoluene **I** 167, **VI** 271
*p*-aminophenyl ether **VI** 277
4-aminophenyl ether **VI** 277
1-amino-2-propanol **IX** 237–244
2-aminotoluene **III** 307
4-aminotoluene **III** 323
*o*-aminotoluene **III** 307
*p*-aminotoluene **III** 323
3-amino-*p*-toluidine **VI** 339
5-amino-*o*-toluidine **VI** 339
2-amino-1,3,4-triazole **IV** 1
3-amino-1,2,4-triazole **IV** 1
3-amino-1*H*-1,2,4-triazole **IV** 1
1-amino-2,4,5-trimethylbenzene **IV** 335
1,2-aminozophenylene **II** 231
amitrol **IV** 1–10
Ammoform **V** 355
ammonia **I** 40, **VI** 1–16, **XIII** 47–48
amosite **II** 95–116, 188
AMP® **IX** 229
AMS **XV** 137
1-amyl acetate **XI** 211
*n*-amyl acetate **XI** 211
*sec*-amyl acetate **XI** 211
*tert*-amyl acetate **XI** 211
*α*-amylase **XI** 1–3
amylcarbinol **IX** 283
anaesthetic ether **XIII** 149
anatase **II** 199
anhydrite **II** 117
anhydrous hydrobromic acid **XIII** 187
aniline **I** 40, **III** 32, 42, 50, 151, 152, 159, 160, **IV** 132, **VI** 17–36
anilinomethane **VI** 263
*Aningeria* spp. **XIII** 285
aningré **XIII** 285
2-anisidine **X** 1
*o*-anisidine **X** 1–13
anone **X** 35
anprolene **V** 181
Antabuse® **V** 165
Antadix **V** 165
anthophyllite **II** 96, 182, 183, 187
apyonine auramine base **IV** 12
aqua fortis **III** 233
aquinite **VI** 105
arborvitae **XIII** 288
arsenic compounds **II** 48, 136
arsenic trihydride **XV** 41
arsine **XV** 41–50
Arubren CP® **III** 81
asbestos **I** 28, **II** 48, 95–116, 136, 181–198, **VIII** 141–338
ASP 47 **XIII** 237
*Aspalatus ebenus* **XIII** 285, 299
Asulgan® K **V** 289
asymmetrical dichloroethylene **VIII** 109
atmospheric pressure **XIV** 11–16
attapulgite **VIII** 141–338
*Aucoumea klaineana* **XIII** 285
auramine **I** 250, **IV** 11–25
auramine base **IV** 11–25
Australian blackwood **XIII** 285, 291
Australian silk(y) oak **XIII** 286, **XIV** 295
avodiré **XIII** 288
ayan **XIII** 286, **XIV** 291
1-aza-2-cycloheptanone **IV** 65

5-azacytidine **I** 4
1-aza-3,7-dioxa-5-ethylbicyclo[3.3.0]octane **IX** 275
azathioprine **I** 4
azimethylene **XIII** 141
azimidobenzene **II** 231
aziminobenzene **II** 231
azotic acid **III** 233
barium chromate **III** 101–122, **VII** 33
basic yellow 2 **IV** 11
basic zinc chromate **XV** 289
battery acid **XV** 165
Bayer E393 **XIII** 237
Baywood **XIII** 287
*α*-BCH **V** 193
*β*-BCH **V** 193
BCNU **I** 2
beech wood dust **IV** 363
benzal chloride **VI** 79
benzenamine **VI** 17
benzene azimide **II** 231
benzenecarbonyl chloride **VI** 80
benzene chloride **XII** 103
1,2-benzenediamine **XIII** 215
1,3-benzenediamine **VI** 287
1,4-benzenediamine **VI** 311, **XIV** 137
*m*-benzenediamine **VI** 287
*o*-benzenediamine **XIII** 215
*p*-benzenediamine **VI** 311
benzene-1,2-dicarboxylic acid **VII** 241
benzene-1,3-dicarboxylic acid **VII** 241
benzene-1,4-dicarboxylic acid **VII** 241
1,2-benzenedicarboxylic acid anhydride **VII** 247
1,2-benzenedicarboxylic acid di-2-propenyl ester **IX** 11
1,4-benzenediol **X** 113
*p*-benzenediol **X** 113
*α*-benzene hexachloride **V** 193
*β*-benzene hexachloride **V** 193
benzenyl chloride **VI** 80
benzenyl trichloride **VI** 80
benzidine **I** 258–259
1,2-benzisothiazol-3(2*H*)-one **II** 221–230
benzohydroquinone **X** 113
benzoic acid chloride **VI** 80
benzoic trichloride **VI** 80
benzoisotriazole **II** 231
benzo[*a*]pyrene **I** 41–43, 58–60, 112, 180
1,2,3-benzotriazole **II** 231
1*H*-benzotriazole **II** 231–238
1*H*-benzotriazole, methyl **II** 295
benzotrichloride **VI** 80
2*H*-3,1-benzoxazine-2,4-[1*H*]-dione **XIII** 97
benzoyl chloride **VI** 80
(benzoyloxy)tributylstannane **I** 315
benzoyl peroxide **III** 249–256
benztriazole **II** 231
benzyl alcohol mono(poly)hemiformal **II** 239–247
benzyl disulfide **II** 283
*α*-(benzyldithio)toluene **II** 283
benzylene chloride **VI** 79
benzyl hemiformal **II** 239
benzyl hydroxymethyl ether **II** 239
benzylidene chloride **VI** 79
benzylidyne chloride **VI** 80
benzyl oxy methanol **II** 239
benzyl trichloride **VI** 80
beryllium **III** 23–29
beryllium compounds **III** 23–29
bété **XIII** 287, **XIV** 299
*N,N'*-bianiline **IX** 83
bianisidine **V** 153
bicarburetted hydrogen **X** 91
1,2-bichloroethane **III** 137
bicyclopentadiene **V** 125
biethylene **XV** 51
big-leaf mahogany **XIII** 287
binitrobenzene **I** 147
Bioban® CS-1246 **IX** 275
Bioban® P-1487 **IX** 295
[1,1'-biphenyl]-2-ol **II** 299
2-biphenylol **II** 299
*o*-biphenylol **II** 299
[1,1'-biphenyl]-2-ol sodium salt **II** 299
2-biphenylol sodium salt **II** 299
(2-biphenyloxy)sodium **II** 299
[1,1'-biphenyl]-3,3',4,4'-tetramine **III** 131
3,3',4,4'-biphenyltetramine **III** 131
bis(4-amino-3-chlorophenyl)methane **VII** 193
bis(4-aminophenyl)ether **VI** 277
bis(4-aminophenyl)methane **VII** 37
bis(*p*-aminophenyl)methane **VII** 37
bis(4-aminophenyl)sulfide **IV** 331
bis(*p*-aminophenyl)sulfide **IV** 331
2,6-bis(*tert*-butyl)phenol **V** 341
bis[(3-carboxyacryloyl)oxy]dioctylstannane diisooctyl ester **VII** 93
bis(*β*-chloroethyl)methylamine **I** 211
*N,N*-bis(2-chloroethyl)-*N*-methylamine **I** 211
bischloroethyl nitrosourea (BCNU) **I** 2
bis(2-chloroethyl)sulfide **IV** 27
bis(*β*-chloroethyl)sulfide **IV** 27–43
biscyclopentadiene **V** 125
bis(3,5-dichloro-2-hydroxyphenyl) sulfide **IX** 245

bis((diethylamino)thioxomethyl)disulfide **V** 165
bis(diethylthiocarbamoyl) disulfide **V** 165
bis(diethylthiocarbamyl) disulfide **V** 165
bis(*N*,*N*-diethylthiocarbamyl) disulfide **V** 165
4,4′-bis(dimethylamino)diphenylmethane **I** 249
*p*,*p*′-bis(dimethylamino)diphenylmethane **I** 249
bis(*p*-(dimethylamino)phenyl)methane **I** 249
bis(*p*-dimethylaminophenyl)methyleneimine **IV** 11
2,6-bis(1,1-dimethylethyl)phenol **V** 341
bis(dimethylthiocarbamoyl) disulfide **XV** 163
bis(dimethylthiocarbamyl) disulfide **XV** 163
1,3-bis(2,3-epoxypropoxy)benzene **VII** 69
*m*-bis(2,3-epoxypropoxy)benzene **VII** 69
*m*-bis(glycidyloxy)benzene **VII** 69
bis(2-hydroxy-3,5-dichlorophenyl) sulfide **IX** 245
bis(2-hydroxyethyl) ether **X** 73
bis(2-hydroxyethyl)methylamine **IX** 291
bis(hydroxymethyl)acetylene **XV** 71
1,3-bis(hydroxymethyl)urea **V** 289–293
*N*,*N*′-bis(hydroxymethyl)urea **V** 289
2,2-bis(4-hydroxyphenyl)propane **XIII** 49
bis(1,4-isocyanatophenyl)methane **VIII** 65
bis(*p*-isocyanatophenyl)methane **VIII** 65
bis(isooctyloxymaleoyloxy)dioctylstannane **VII** 93
bis(2-methoxyethyl)ether **IX** 41
bisphenol A **XIII** 49–87
bis(phenylmethyl)disulfide **II** 283
bis(tri-*n*-butyltin)oxide **I** 315
bithionol **IX** 245–262
bithionol sulfide **IX** 245
4,4′-bi-*o*-toluidine **V** 153
bivinyl **XV** 51
black ebony **XIII** 286
black sucupira **XIII** 285, 295
black walnut **XIII** 286
black wattle **XIII** 285, 291
Bladafum® **XIII** 237
bleomycin **I** 5
BNPD **II** 249
body burden and hyperbaric pressure **XIV** 11–16
Bombay blackwood **XIII** 285, 301
boracic acid **V** 295
boric acid **V** 295–321
boric anhydride **XIII** 89
boric oxide **XIII** 89
boroethane **XIV** 83
Borofax® **V** 295
boron fluoride **XIII** 93
boron oxide **XIII** 89–91
boron sesquioxide **XIII** 89
boron trifluoride **XIII** 93–95
boron trioxide **XIII** 89
*Bowdichia nitida* **XIII** 285, 295–297
Brasilian rosewood **XIII** 286, 309
brilliant oil yellow **IV** 11
broadleaf mahogany **XIII** 287
bromic ether **VII** 115
2-bromo-2-(bromomethyl)glutaronitrile **XIV** 87
2-bromo-2-(bromomethyl)pentanedinitrile **XIV** 87
bromoethane **VII** 115
bromofluoroform **VI** 37
bromoform **VII** 319
bromomethane **VII** 155
2-bromo-2-nitro-1,3-propanediol **II** 249–270
*β*-bromo-*β*-nitrotrimethylene glycol **II** 249
bromotrifluoromethane **VI** 37–45
Bronopol® **II** 249
brookite **II** 199
brown ebony **XIII** 285, 299
brucite **VIII** 141–338
*Brya ebenus* **XIII** 285, 299–300
busulfan **I** 2
*α*,*γ*-butadiene **XV** 51
1,3-butadiene **XV** 51–70
butadiene diamer **XIV** 185
butanesulfone **IV** 45
*δ*-butane sultone **IV** 45
1,4-butane sultone **IV** 45–49
2,4-butane sultone **IV** 45–49
2-butanone **XII** 37–56
2-butanone peroxide **III** 250, 252, 254, 255
*cis*-butenedioic anhydride **IV** 275
1-butene oxide **V** 13
1,2-butene oxide **V** 13
3-buten-2-one **IX** 91
butoxydiethylene glycol **VII** 59
butoxydiglycol **VII** 59
1-*n*-butoxy-2,3-epoxypropane **IV** 51
1-*tert*-butoxy-2,3-epoxypropane **IV** 51
2-butoxyethanol **V** 210, **VI** 47-52
*n*-butoxyethanol **VI** 47
2-butoxyethanol acetate **VI** 53
2-(2-butoxyethoxy)ethanol **VII** 59
2-[2-(2-butoxyethoxy)ethoxy]ethanol **IX** 315
2-butoxyethyl acetate **VI** 53–55
butoxymethyl oxirane **IV** 51
*tert*-butoxymethyl oxirane **IV** 51
butoxytriethylene glycol **IX** 315
butoxytriglycol **IX** 315

buttercup yellow **XV** 289
*n*-butyl acrylate **V** 5–12, **XII** 57–62, **XVI** 35–40
butyl carbamic acid 3-iodo-2-propynyl ester **XVI** 247
butyl carbitol **VII** 59
Butyl Cellosolve® **VI** 47
Butyl Cellosolve® acetate **VI** 53
*O*-butyl diethylene glycol **VII** 59
butyl diglycol **VII** 59
butyl dioxitol **VII** 59
1,2-butylene oxide **V** 13–18
*α*-butylene oxide **V** 13
1,4-butylene sulfone **IV** 45
*O*-butyl ethylene glycol **VI** 47
*n*-butyl glycidyl ether **IV** 51–64
*tert*-butyl glycidyl ether **IV** 51–64
butyl glycol **VI** 47
butyl glycol acetate **VI** 53
*tert*-butyl hydroperoxide **III** 250, 253, 254
butyl-3-iodo-2-propynylcarbamate **XVI** 247
*tert*-butyl peracetate **III** 250
*tert*-butyl peroxyacetate **III** 250
4-*tert*-butylphenol **XI** 5
*p*-*tert*-butyl phenol **XI** 5–25
butyl 2-propenoate **V** 5, **XII** 57
*n*-butyl triethylene glycol **IX** 315
*n*-butyl triglycol **IX** 315
butynediol **XV** 71–74
2-butyne-1,4-diol **XV** 71
2-butyne-2,4-diol **XVI** 233
cadmium **I** 40, **V** 19–50
cadmium compounds **V** 19–50, **VII** 21
cadmium sulfide **VII** 21–25
calcium carbimide **V** 51
calcium cyanamide **V** 51–64
calcium sodium metaphosphate **VIII** 141–338
calcium sulfate **II** 117–128, **VIII** 141–338
Californian redwoood **XIII** 287
*Calocedrus decurrens* **XIII** 285, **XIV** 275–278
CAM **IX** 263
Cameroon ebony **XIV** 287
6-caprolactam **IV** 65
*ε*-caprolactam **IV** 65–78
caproyl alcohol **IX** 283
captan **I** 292, 295
Carbamol® **V** 289
carbinamine **VII** 145
carbinol **XVI** 143
carbon bichloride **III** 271
carbon bisulfide **XII** 63
carbon dichloride **III** 271
carbon disulfide **XII** 63–79
carbon monoxide **I** 40, 42–43, 109, 111, 119, **IV** 79-95, 174, 188
carbon oxide **IV** 79
carbonic oxide **IV** 79
4,4′-carbonimidoylbis(*N*,*N*-dimethylbenzenamine) **IV** 11, 12
4,4′-carbonimidoylbis[*N*,*N*-dimethylbenzenamine] monochloride **IV** 11
carbon sulfide **XII** 63
Carbowax® **X** 247
*N*-carboxyanthranilic anhydride **XIII** 97–98
Caribbean mahogany **XIII** 287
carmustine **I** 2
caustic soda **XII** 195
caviuna **XIII** 287, 321
CCNU **I** 2
CDM **VI** 57
cedar of Lebanon **XIII** 285
*Cedrus* spp. **XIII** 285
Cellosolve **VI** 205
cellosolve acetate **VI** 213
ceramic fibres **VIII** 141–338
cereal flour dusts **XIII** 99–100
Cereclor® **III** 81
Ceylonese ebony **XIII** 286, **XIV** 287
CFC 12 **V** 109
CFC 13 **I** 69
CFC 21 **V** 119
CFC 31 **V** 75
CFC 112 **I** 297, **III** 366
CFC 142b **I** 65
'cherry mahogany' **XIII** 288
chlorambucil **I** 2
chlordimeform **VI** 57–78
chlorfenamidine **VI** 57
chlorinated hydrocarbon waxes **III** 81
chlorinated naphthalenes **XIII** 101–110
chlorinated paraffins **III** 81, **VII** 27–32
chlorinated paraffin waxes **III** 81
*α*-chlorinated toluenes **VI** 79–103
chlormethine **I** 211
chloroacetaldehyde **XII** 81–102
2-chloroacetaldehyde **XII** 81
chloroacetaldehyde monomer **XII** 81
chloroacetamide-*N*-methylol **IX** 263–267
chloroacetic acid methyl ester **IX** 1–8
2-chloroacrylonitrile **X** 15–19
4-chloro-2-aminotoluene **VI** 143
5-chloro-2-aminotoluene **VI** 127
2-chloroaniline **III** 31
3-chloroaniline **III** 37
4-chloroaniline **III** 45
*m*-chloroaniline **III** 37–43

*o*-chloroaniline **III** 31–35, 42
*p*-chloroaniline **III** 42, 45–61, **IV** 132
chlorobenzal **VI** 79
α-chlorobenzaldehyde **VI** 80
2-chlorobenzenamine **III** 31
3-chlorobenzenamine **III** 37
4-chlorobenzenamine **III** 45
chlorobenzene **XII** 103–127
chlorobenzol **XII** 103
*p*-chlorobenzotrichloride **X** 21–30
2-chloro-*N*-(2-chloroethyl)-*N*-methyl-ethanamine **I** 211
1-chloro-2-(*β*-chloroethylthio)ethane **IV** 27
*p*-chlorocresol **II** 271
*p*-chloro-*m*-cresol **II** 271–282
chlorodeoxyglycerol **V** 83
1-chloro-2,2-dichloroethylene **X** 201
chlorodifluoroethane **I** 65
1-chloro-1,1-difluoroethane **I** 65–67
chlorodifluoromethane **I** 69, **III** 63–71
2-chloro-2-(difluoromethoxy)-1,1,1-trifluoro-ethane **VII** 127
2-chloro-1-(difluoromethoxy)-1,1,2-trifluoro-ethane **IX** 51
1-chloro-2,3-dihydroxypropane **V** 83
3-chloro-1,2-dihydroxypropane **V** 83
1-chloro-2,4-dinitrobenzene **XIII** 111–115
2-chloro-1-ethanal **XII** 81
chloroethane **III** 73–79
2-chloroethanol **V** 65–73
*β*-chloroethanol **V** 65
chloroethene **V** 66, 241
chloroethene homopolymer **II** 149
2-chloroethyl alcohol **V** 65
1-(2-chloroethyl)-3-cyclohexyl-1-nitrosourea **I** 2
chloroethylene **V** 241
chloroethylidene fluoride **I** 65
chlorofluoromethane **III** 63, 69, 70, **V** 75–77
chloroform **IV** 173, **XIV** 19–58
*N*-chloroformylmorpholine **V** 79–81
chlorohydric acid **VI** 231
*α*-chlorohydrin **V** 83–98
2-chloro-*N*-(hydroxymethyl)acetamide **IX** 263
6-chloro-3-hydroxytoluene **II** 271
*γ*-chloroisobutylene **IV** 97
chloromethane **I** 69, **VII** 173
3-chloro-6-methylaniline **VI** 143
4-chloro-2-methylaniline **VI** 127
5-chloro-2-methylaniline **VI** 143
4-chloro-2-methylbenzenamine **VI** 127
5-chloro-2-methylbenzenamine **VI** 143
(chloromethyl)benzene **VI** 79
4-chloro-2-methylbenzeneamine **VI** 127
5-chloro-2-methyl-2,3-dihydroisothiazol-3-one **V** 323–339
5-chloro-2-methyl-4-isothiazolin-3(2*H*)-one **V** 323
5-chloro-2-methyl-3-isothiazolone **V** 323
4-chloro-3-methylphenol **II** 271
*N′*-(4-chloro-2-methylphenyl)-*N*,*N*-dimethyl-methanimidamide **VI** 57
3-chloro-2-methylpropene **IV** 97–106
3-chloro-2-methyl-1-propene **IV** 97
1-chloro-2-nitrobenzene **IV** 107
1-chloro-3-nitrobenzene **IV** 115
1-chloro-4-nitrobenzene **IV** 121
2-chloro-1-nitrobenzene **IV** 107
3-chloro-1-nitrobenzene **IV** 115
4-chloro-1-nitrobenzene **IV** 121
*m*-chloronitrobenzene **IV** 107, 115–120, 121
*o*-chloronitrobenzene **IV** 107–114, 115, 121
*p*-chloronitrobenzene **IV** 107, 115, 121–133
1-chloro-1-nitropropane **XI** 27–29
chloroparaffins **III** 81–100, **VII** 27
chlorophenamidine **VI** 57
2-chlorophenylamine **III** 31
3-chlorophenylamine **III** 37
4-chlorophenylamine **III** 45
*m*-chlorophenylamine **III** 37
*o*-chlorophenylamine **III** 31
*p*-chlorophenylamine **III** 45
*p*-chlorophenyl chloride **IV** 141
chlorophenylmethane **VI** 79
*Chlorophora excelsa* **XIV** 279–285
*Chlorophora* spp. **XIII** 285
chloropicrin **VI** 105–111
chloroprene **III** 205
1-chloro-2,3-propanediol **V** 83
3-chloro-1,2-propanediol **V** 83
2-chloro-2-propenenitrile **X** 15
3-chloropropylene glycol **V** 83
chlorothalonil **VI** 113–126
*α*-chlorotoluene **VI** 79
*ω*-chlorotoluene **VI** 79
4-chloro-2-toluidine **VI** 127
4-chloro-*o*-toluidine **VI** 127–141, 144
5-chloro-*o*-toluidine **VI** 143–144
*N′*-(4-chloro-*o*-tolyl)-*N*,*N*-dimethyl-formamidine **VI** 57
1-chloro-4-(trichloromethyl)benzene **X** 21
1-chloro-2,2,2-trifluoroethyldifluoromethyl ether **VII** 127
2-chloro-1,1,2-trifluoroethyldifluoromethyl ether **IX** 51
chlorotrifluoromethane **I** 69–70
Chlorowax® **III** 81

chlorphenamidine **VI** 57
chlorpromazine **IX** 9–10
chlorvinphos **IV** 201
chromates **III** 101–122
chromic acid zinc salt **XV** 289
chromium compounds **III** 101–122
chromium(VI) compounds **II** 47, 48, **III** 101–122, **VII** 33–35
chromium potassium zinc oxide **XV** 289
chrysotile **I** 28, **II** 95–116, 182, 184, **VIII** 141–338
C.I. 23060 **V** 99
C.I. 24110 **V** 139
C.I. 37105 **VI** 271
C.I. 37230 **V** 153
C.I. 41000 **IV** 11
C.I. 76000 **VI** 17
C.I. 76025 **VI** 287
C.I. 76035 **VI** 339
C.I. 76050 **VI** 145
C.I. 76060 **VI** 311
C.I. 76070 **IV** 295, **XIII** 207
C.I. 76555 **IV** 289
C.I. azoic red 83 **IV** 135
cigarette smoke **I** 39–62, **XIII** 3–39
C.I. oxidation base 22 **IV** 295, **XIII** 207
C.I. pigment yellow 36 **XV** 289
cisplatin **I** 6
citric acid **XVI** 209–230
citron yellow **XV** 289
classification of carcinogenic chemicals **XII** 3–12
Cloparin® **III** 81
coastal redwood **XIII** 287
cobalt **III** 123–130, **X** 31–34
cobalt compounds **III** 123–130, **X** 31–34
Cobratec® 99 **II** 231
cocobolo **XIII** 286, 313
cocuswood **XIII** 285, 299
colloidal mercury **XV** 81
colophony **XI** 239
Compound 469 **VII** 127
Contralin **V** 165
coromandel **XIII** 286, **XIV** 287
*p*-cresidine **IV** 135–139
*o*-cresol **XIV** 59
*m*-cresol **XIV** 59
*p*-cresol **XIV** 59
cresol (all isomers) **XIV** 59–82
*o*-cresylic acid **XIV** 59
*m*-cresylic acid **XIV** 59
*p*-cresylic acid **XIV** 59
cristobalite **II** 182, **XIV** 205
crocidolite **II** 95–116, **VIII** 141–338
Cronetal **V** 165
cross-linked polyacrylates **XV** 1
crystalline silicon dioxide **XIV** 205–271
Cuban mahogany **XIII** 287
cumene **XIII** 117–128
cumene hydroperoxide **III** 250, 253–255
cumol **XIII** 117
cyanamide, calcium salt (1:1) **V** 51
cyanoacrylates **XIII** 201
2-cyanoacrylic acid ethyl ester **I** 233, **XIII** 201
2-cyanoacrylic acid methyl ester **I** 233, **XIII** 201
2-cyano-2-propenoic acid ethyl ester **I** 233
2-cyano-2-propenoic acid methyl ester **I** 233
cyclohexane **XIII** 129–140
cyclohexanone **X** 35–51
cyclohexanone iso-oxime **IV** 65
cyclohexenylethylene **XIV** 185
cyclohexyl ketone **X** 35
1,3-cyclopentadiene dimer **V** 125
cyclophosphamide **I** 2
Cystamin **V** 355
Cystogen **V** 355
cytarabine **I** 4
2,4-D **IV** 191, **XI** 61
2,4-DAA **VI** 145
dacarbazine **I** 6
DACPM **VII** 193
dactinomycin **I** 5
*Dalbergia latifolia* **XIII** 285, 301–303
*Dalbergia melanoxylon* **XIII** 286, 305–307
*Dalbergia nigra* **XIII** 286, 309–311
*Dalbergia retusa* **XIII** 286, 313–316
*Dalbergia stevensonii* **XIII** 286, 317–319
DAPM **VII** 37
daunomycin **I** 5
daunorubicin hydrochloride **I** 5
dawsonite **VIII** 141–338
DBDCB **XIV** 87
1,2-DCE **III** 137
DDVP **IV** 201
decyl 9-octadecenoate **IX** 269
decyl oleate **IX** 269–274
deodar cedar **XIII** 285
DGA® **IX** 215
diallyl phthalate **IX** 11–19
diamide **I** 171
diamide monohydrate **XIII** 181
diamine **I** 171
2,4-diamineanisole **VI** 145
2,4-diaminoanisole **VI** 145-156
*m*-diaminoanisole **VI** 145
2,4-diaminoanisole base **VI** 145

2,4-diaminoanisole sulfate **VI** 145–156
1,2-diaminobenzene **XIII** 215
1,3-diaminobenzene **VI** 287
1,4-diaminobenzene **VI** 311
*m*-diaminobenzene **VI** 287
*o*-diaminobenzene **XIII** 215
*p*-diaminobenzene **VI** 311
3,3′-diaminobenzidine **III** 131–135
3,3′-diaminobenzidine tetrahydrochloride **III** 131–135
4,4′-diamino-3,3′-bichlorobiphenyl **V** 99
di(4-amino-3-chlorophenyl)methane **VII** 193
4,4′-diamino-3,3′-dichlorodiphenyl **V** 99
4,4′-diamino-3,3′-dichlorodiphenylmethane **VII** 193
4,4′-diamino-3,3′-dimethoxybiphenyl **V** 139
4,4′-diamino-3,3′-dimethylbiphenyl **V** 153
4,4′-diaminodiphenyl ether **VI** 277
4,4′-diaminodiphenylmethane **VII** 37–57
*p,p*′-diaminodiphenylmethane **VII** 37
4,4′-diaminodiphenyl oxide **VI** 277
4,4′-diaminodiphenylsulfide **IV** 331
*p,p*′-diaminodiphenylsulfide **IV** 331
diaminoditolyl **V** 153
1,3-diamino-4-methoxybenzene **VI** 145
2,4-diamino-1-methoxybenzene **VI** 145
1,3-diamino-4-methylbenzene **VI** 339
2,4-diamino-1-methylbenzene **VI** 339
1,4-diamino-2-nitrobenzene **IV** 295, **XIII** 207
di(4-aminophenyl)methane **VII** 37
di(*p*-aminophenyl)sulfide **IV** 331
2,4-diaminotoluene **VI** 339
dianilinomethane **VII** 37
dianisidine **V** 139
diatomaceous earth **II** 157
1,4-diazacyclohexane **IX** 303, **XII** 177
diazinon **XI** 31–59
diazomethane **XIII** 141–148
dibenzoyl peroxide **III** 249–256
dibenzyl disulfide **II** 283–286
diborane **XIV** 83–85
diborane(6) **XIV** 83
diboron hexahydride **XIV** 83
[2,7-dibromo-9-(*o*-carboxyphenyl)-6-hydroxy-3-oxo-3*H*-xanthen-4-yl] hydroxymercury disodium salt **XV** 75
1,2-dibromo-2,4-dicyanobutane **XIV** 87–90
(2′,7′-dibromo-3′,6′-dihydroxy-3-oxo-spiro-[isobenzofuran-1(3H),9′-[9H]-xanthen]-4′-yl)hydroxymercury disodium salt **XV** 75
1,2-dibromoethane **III** 144
dibromohydroxymercurifluorescein disodium salt XV 75
di-*tert*-butyl hydroperoxide **III** 254
2,6-di-*tert*-butylphenol **V** 341-344
dicarboxybenzene **VII** 241
dichloroacetylene **VI** 157–163
1,2-dichlorobenzene **I** 71–80
1,3-dichlorobenzene **I** 83–89
1,4-dichlorobenzene **I** 71, 73, 83, 196, 347, **IV** 141–171
*m*-dichlorobenzene **I** 83
*o*-dichlorobenzene **I** 71
*p*-dichlorobenzene **I** 71, 73, 83, 196, 347, **IV** 141
3,3′-dichlorobenzidine **V** 99–107
*o,o*′-dichlorobenzidine **V** 99
3,3′-dichloro(1,1′-biphenyl)-4,4′-diamine **V** 99
3,3′-dichlorobiphenyl-4,4′-diamine **V** 99
3,3′-dichloro-4,4′-biphenylenediamine **V** 99
3,3′-dichloro-4,4′-diaminobiphenyl **V** 99
3,3′-dichloro-4,4′-diaminodiphenylmethane **VII** 193
2,2′-dichlorodiethylether **V** 66, 71, 72
2,2′-dichlorodiethyl sulfide **IV** 27
*β,β*′-dichlorodiethyl sulfide **IV** 27
dichlorodifluoromethane **I** 336, 341, **V** 109–117
dichlorodioctylstannane **VII** 91
1,2-dichloroethane **III** 137–147
*α,β*-dichloroethane **III** 137
1,1-dichloroethene **V** 66, 72, **VIII** 109
2,2-dichloroethenyl dimethyl phosphate **IV** 201
2,2-dichloroethenyl phosphoric acid dimethyl ester **IV** 201
1,1-dichloroethylene **VIII** 109
di(2-chloroethyl)methylamine **I** 211
1,2-dichloroethylmethyl ether **I** 91
*α,β*-dichloroethylmethyl ether **I** 91
di-2-chloroethyl sulfide **IV** 27
dichloroethyne **VI** 157
dichlorofluoromethane **I** 69, **V** 119–124
*α*-dichlorohydrin **I** 95
*sym*-dichloroisopropyl alcohol **I** 95
dichloromethane **IV** 173–190
1,2-dichloromethoxyethane **I** 91–93
(dichloromethyl)benzene **VI** 79
2,2′-dichloro-*N*-methyldiethylamine **I** 211
dichloromonofluoromethane **V** 119
dichloronaphthalenes **XIII** 101
2,4-dichlorophenoxyacetic acid **IV** 191–200, **XI** 61–104

2,4-dichlorophenoxyacetic acid salts and esters **XI** 61
dichlorophos **IV** 201
1,2-dichloropropane **IX** 21–39
1,3-dichloro-2-propanol **I** 95–98
*α,α*-dichlorotoluene **VI** 79
2,2-dichloro-1,1,1-trifluoroethane **X** 53–71
2,2-dichlorovinyl dimethyl phosphate **IV** 201
2,2-dichlorovinyl phosphoric acid dimethyl ester **IV** 201
dichlorovos **IV** 201
dichlorvos **IV** 201–216
dicumyl hydroperoxide **III** 252
dicumyl peroxide **III** 251
1,3-dicyanotetrachlorobenzene **VI** 113
dicyclopentadiene **V** 125–133
diesel engine emissions **I** 101–120, **II** 136
diethanolmethylamine **IX** 291
diethylamine **I** 25–28, 32–34
(diethylamino)ethane **XIII** 267
2-diethylaminoethanol **XIV** 91–100
*β*-diethylamino ethyl alcohol **XIV** 91
diethylcarbinol acetate **XI** 211
1,4-diethylenediamine **IX** 303, **XII** 177
diethylene glycol **X** 73–90
diethylene glycol *n*-butyl ether **VII** 59
diethylene glycol dimethyl ether **IX** 41–50
diethylene glycol monobutyl ether **VII** 59–67
*N,N*-diethylethanamine **XIII** 267
diethyl ether **I** 125, **XIII** 149–160
diethylmethylmethane **IV** 269
diethylnitrosamine **I** 23–35, 41, 60, 261-263, 268, 286
diethyl oxide **XIII** 149
Diflamoll® TP **II** 321
difluorochloromethane **III** 63
difluorodichloromethane **V** 109
1,1-difluoroethene **V** 135
1,1-difluoroethylene **V** 135–137
difluoromonochloroethane **I** 65
difluoromonochloromethane **III** 63
1,2-difluoro-1,1,2,2-tetrachloroethane **I** 297
1,3-diformyl propane **VIII** 45, **XVI** 59
diglycidyl ether **IV** 59
1,3-diglycidyloxybenzene **VII** 69
diglycidyl resorcinol ether **VII** 69–74
diglycolamine **IX** 215
Diglycolamine® Agent **IX** 215
diglycol monobutyl ether **VII** 59
diglyme **IX** 41
1,3-dihydro-1,3-dioxo-isobenzofuran **VII** 247
dihydrooxirene **V** 181
1,4-dihydroxybenzene **X** 113
4,4′-dihydroxydiphenylpropane **XIII** 49
*p,p′*-dihydroxydiphenylpropane **XIII** 49
1,2-dihydroxyethane **IV** 225
2,2′-dihydroxyethyl ether **X** 73
*β,β′*-dihydroxyisopropyl chloride **V** 83
2,4-dihydroxy-2-methylpentane **XVI** 233
*N,N′*-dihydroxymethylurea **V** 289
2,3-dihydroxypropyl chloride **V** 83
2,2′-dihydroxy-3,3′,5,5′-tetrachloro-diphenylsulfide **IX** 245
4,4′-diisocyanatodiphenylmethane **VIII** 65
*p,p′*-diisocyanatodiphenylmethane **VIII** 65
diisooctyl[(dioctylstannylene)dithio]diacetate **VII** 92
diisopropyl peroxydicarbonate **III** 250
dilauroyl peroxide **III** 250–255
dimethanol urea **V** 289
3,3′-dimethoxybenzidine **V** 139–152
3,3′-dimethoxy-[1,1′-biphenyl]-4,4′-diamine **V** 139
3,3′-dimethoxy-4,4′-diaminobiphenyl **V** 139
dimethoxyphosphine oxide **I** 133
dimethylamine **I** 25–28, 32–34, **VII** 75–89
*N,N*-dimethylaminobenzene **III** 149
4,4′-dimethylaminobenzophenonimide **IV** 11
dimethylaminosulfonyl chloride **I** 143
*N,N*-dimethylaniline **III** 149–161
*N,N*-dimethylbenzenamine **III** 149
dimethylbenzene **V** 263, **XV** 257
3,3′-dimethylbenzidine **V** 153–164
*α,α*-dimethylbenzyl hydroperoxide **III** 250, 253–255
3,3′-dimethyl-[1,1′-biphenyl]-4,4′-diamine **V** 153
3,3′-dimethylbiphenyl-4 4′-diamine **V** 153
2,2-dimethylbutane **IV** 269
2,3-dimethylbutane **IV** 269
*N,N*-dimethylcarbamoyl chloride **I** 143, 145
dimethyl 2,2-dichlorovinyl phosphate **IV** 201
*O,O*-dimethyl-*O*-(2,2-dichlorovinyl)-phosphate **IV** 201
3,3′-dimethyldiphenyl-4 4′-diamine **V** 153
dimethylene oxide **V** 181
dimethyl ether **I** 125–131
[(1,1-dimethylethoxy)methyl]oxirane **IV** 51
4-(1,1-dimethylethyl)phenol **XI** 5
dimethylformaldehyde **VII** 1
dimethylformamide **VIII** 1–44
*N,N*-dimethylformamide **VIII** 1
1,2-dimethylhydrazine **I** 176
dimethylketal **VII** 1
dimethyl ketone **VII** 1
*N,N*-dimethylmethanamide **VIII** 1
dimethyl monosulfate **IV** 217

dimethylnitromethane **III** 241
dimethylnitrosamine **I** 23–35, 41–44, 59, 261–264, 286
1,3-dimethylolurea **V** 289
*N,N'*-dimethylolurea **V** 289
*N,N*-dimethylphenylamine **III** 149
dimethyl phosphite **I** 133
dimethyl phosphonate **I** 133
1,1-dimethylpropyl acetate **XI** 211
dimethylpropylmethane **IV** 269
*N,N*-dimethylsulfamoyl chloride **I** 143–145
*N,N*-dimethylsulfamyl chloride **I** 143
dimethyl sulfate **IV** 48, 217–223
dimethyl sulfoxide **III** 163–171
dinitrobenzene (all isomers) **I** 147–167
2,4-dinitro-1-chlorobenzene **XIII** 111
dinitrogen monoxide **IX** 115
dinitrophenylmethane **VI** 165
dinitrotoluenes **I** 167, 361, **VI** 165–198
*ar,ar*-dinitrotoluene **VI** 165
2,3-dinitrotoluene **VI** 165
2,4-dinitrotoluene **VI** 165
2,5-dinitrotoluene **VI** 165
2,6-dinitrotoluene **VI** 165
3,4-dinitrotoluene **VI** 165
3,5-dinitrotoluene **VI** 165
dioctyldichlorostannane **VII** 91
2,2-dioctyl-1,3,2-dioxastannepin-4,7-dione **VII** 93
1,3-dioctyl-1,3-dioxodistannoxane **VII** 93
dioctyloxostannane **VII** 91
dioctylstannium dichloride **VII** 91
4,4'-[(dioctylstannylene)bis(oxy)]bis(4-oxo-2-butenoic acid) diisooctyl ester **VII** 93
2,2'-[(dioctylstannylene)-bis(thio)]bis(acetic acid)bis(2-ethylhexyl) ester **VII** 92
2,2'-[(dioctylstannylene)bis(thio)]bis(acetic acid) diisooctyl ester **VII** 92
dioctylstannylene maleate **VII** 93
di-*n*-octyltin bis(2-ethylhexyl mercapto-acetate) **VII** 92
di-*n*-octyltin bis(2-ethylhexyl thioglycolate) **VII** 92
di-*n*-octyltin bis(isooctyl maleate) **VII** 93
dioctyltin-*S,S'*-bis(isooctyl mercaptoacetate) **VII** 92
di-*n*-octyltin bis(isooctyl thioglycolate) **VII** 92
di-*n*-octyltin compounds **VII** 91-114
di-*n*-octyltin dichloride **VII** 91
dioctyltin di(isooctyl thioglycolate) **VII** 92
di-*n*-octyltin dithioglycolic acid 2-ethylhexyl ester **VII** 92
di-*n*-octyltin-2-ethylhexyl-dimercapto-ethanoate **VII** 92
di-*n*-octyltin maleate **VII** 93
di-*n*-octyltin oxide **VII** 91
Diolane **XVI** 233
*Diospyros* spp. **XIII** 286, **XIV** 287–290
2,5-dioxo-dihydrofuran **IV** 275
1,3-dioxophthalan **VII** 247
dipentene **I** 185
1,4-diphenyl-2,3-dithiabutane **II** 283
1,2-diphenylhydrazine **IX** 83
*N,N'*-diphenylhydrazine **IX** 83
4,4'-diphenylmethanediamine **VII** 37
diphenylmethane-4,4'-diisocyanate **VIII** 65
diphenylmethane-*p,p'*-diisocyanate **VIII** 65
di(phenylmethyl)disulfide **II** 283
*o*-diphenylol **II** 299
diphenylolpropane **XIII** 49
dipropylamine **I** 25
dipropylene glycol monomethyl ether **VI** 199–204
dipropyl methane **XI** 165
disodium 2',7'-dibromo-4'-(hydroxy-mercury)fluorescein **XV** 75
*Distemonanthus benthamianus* **XIII** 286, **XIV** 291–293
disulfiram **V** 165–180
disulfuram **V** 165
2,6-ditertiary butyl phenol **V** 341
dithio **XIII** 237
1,1-dithiobis(*N,N*-diethylthioformamide) **V** 165
dithiocarbamates **XV** 147–149
dithiocarbonic anhydride **XII** 63
dithione **XIII** 237
dithiophos **XIII** 237
4,4'-di-*o*-toluidine **V** 153
divinyl **XV** 51
DMF **VIII** 1
DMFA **VIII** 1
DMSO **III** 163
DNCB **XIII** 111
DNT **VI** 165
dolomite **II** 182
DOTC **VII** 91
DOTO **VII** 91
DOTTG **VII** 92
doussié **XIII** 285
Dowanol® TBH **IX** 315
Dowicide® **II** 299
doxorubicin **I** 5
dry-zone mahogany **XIII** 287
*Dumoria heckelii* **XIII** 288

dusts **II** 3–46, 47–57, **XI** 281–301, **XII** 239, 271, **XVI** 287–315
Dymel A® **I** 125
eastern white cedar **XIII** 288
East Indian ebony **XIV** 287
East Indian rosewood **XIII** 285, 301
EDAO **IX** 275, **XVI** 231
elayl **X** 91
enflurane **IX** 51–68
environmental tobacco smoke **XIII** 3
epirubicin **I** 5
epirubicin hydrochloride **I** 5
1,2-epoxy-3-allyoxypropane **VII** 9
1,2-epoxybutane **V** 13
1,2-epoxy-4-(epoxyethyl)cyclohexane **I** 391
1,2-epoxyethane **V** 181
1-(epoxyethyl)-3,4-epoxycyclohexane **I** 391
3-(epoxyethyl)-7-oxabicyclo(4.1.0)heptane **I** 391
4-(epoxyethyl)-7-oxabicyclo(4.1.0)heptane **I** 391
3-(1,2-epoxyethyl)-7-oxabicyclo(4.1.0)-heptane **I** 391
4-(1,2-epoxyethyl)-7-oxabicyclo(4.1.0)-heptane **I** 391
1,2-epoxy-3-isopropoxypropane **VII** 141
1,2-epoxy-3-phenoxypropane **IV** 305
1,2-epoxypropane **V** 221
2,3-epoxypropane **V** 221
2,3-epoxypropyl phenyl ether **IV** 305
(2,3-epoxypropyl)trimethylammonium chloride **IV** 247
erionite **VIII** 141–338
erythrene **XV** 51
Esperal® **V** 165
ESTOL IDCO 3667 **XVI** 257
estramustine **I** 2
Etabus **V** 165
ethanal **III** 1
ethane dichloride **III** 137
1,2-ethanediol **IV** 225
ethaneperoxoic acid **VII** 229
ethanoic acid ethenyl ester **V** 229
ethanol **XII** 129–165
ethanolamine **XII** 15
Ethanox 701® **V** 341
ethene **X** 91
ethene oxide **V** 181
ethenyl acetate **V** 229
9-ethenyl-9*H*-carbazole **XIV** 181
4-ethenyl-1-cyclohexene **XIV** 185
ethenyl ethanoate **V** 229
1-ethenyl-2-pyrrolidinone **V** 249
ether **XIII** 149
ethinyl trichloride **X** 201
ethohexadiol **V** 345
2-ethoxy-6-aminonaphthalene **VII** 17
ethoxyethane **XIII** 149
2-ethoxyethanol **I** 205, 209, **V** 210, **VI** 205–211, **XI** 105–119
2-ethoxyethanol acetate **VI** 213, **XI** 105
2-ethoxyethyl acetate **VI** 213–216, **XI** 105–119
ethyl acetate **XII** 167–176
ethyl acrylate **VI** 217–229, **XVI** 41–46
ethyl alcohol **XII** 129
ethyl aldehyde **III** 1
ethylamine **I** 28, 34
ethyl bromide **VII** 115–120
ethylcellosolve **VI** 205
ethyl chloride **III** 73
ethyl 2-cyanoacrylate **I** 233–245, **XIII** 201–206
ethyl *α*-cyanoacrylate **I** 233, **XIII** 201
7*a*-ethyldihydro-1*H*,3*H*,5*H*-oxazolo-[3,4-*c*]oxazole **IX** 275, **XVI** 231
10-ethyl-4,4-dioctyl-7-oxo-8-oxa-3,5-dithia-4-stannatetradecanoic acid 2-ethylhexyl ester **VII** 92
5-ethyl-3,7-dioxa-1-azabicyclo[3.3.0]octane **IX** 275–281, **XVI** 231–232
ethyldithiurame **V** 165
ethylene **X** 91–107
ethylene carboxamide **III** 11
ethylene chloride **III** 137
ethylene chlorohydrin **V** 65
ethylene dichloride **III** 137
ethylene glycol **IV** 225–245
ethylene glycol *n*-butyl ether **VI** 47
ethylene glycol chlorohydrin **V** 65
ethylene glycol dimethacrylate **XIII** 161–163
ethylene glycol ethyl ether **VI** 205
ethylene glycol ethyl ether acetate **VI** 213
ethylene glycol isopropyl ether **V** 207
ethylene glycol methacrylate **XIII** 193
ethylene glycol methyl ether **VI** 239
ethylene glycol methyl ether acetate **VI** 249
ethylene glycol monoacrylate **XVI** 89
ethylene glycol monobutyl ether **VI** 47
ethylene glycol monobutyl ether acetate **VI** 53
ethylene glycol monoethyl ether **VI** 205
ethylene glycol monoethyl ether acetate **VI** 213
ethylene glycol monoisopropyl ether **V** 207
ethylene glycol monomethacrylate **XIII** 193

ethylene glycol monomethyl ether **VI** 239
ethylene glycol monomethyl ether acetate **VI** 249
ethylene glycol mono-*n*-propyl ether **XII** 179
ethylene glycol monopropyl ether acetate **XII** 187
ethylene monochloride **V** 241
ethylene oxide **V** 181–192, **XIII** 165–168
1-ethyleneoxy-3,4-epoxycyclohexane **I** 391
ethylene tetrachloride **III** 271
ethylene thiourea **XI** 121–163
ethylene trichloride **X** 201
ethyl ether **XIII** 149
ethyl-ethylene oxide **V** 13
ethyl glycol acetate **VI** 213
2-ethylhexane-1,3-diol **V** 345
2-ethyl-1,3-hexanediol **V** 345–353
2-ethylhexyl acrylate **XVI** 47–50
ethyl hexylene glycol **V** 345
2-ethylhexyl propenoate **XVI** 47
2-ethyl-2-(hydroxymethyl)-1,3-propanediol triacrylate **XVI** 201
ethyl(2-mercaptobenzoato-*S*)mercury sodium salt **XV** 249
ethyl methacrylate **XVI** 51–58
ethyl α-methacrylate **XVI** 51
ethyl methyl ketone **XII** 37
ethyl 2-methyl-2-propenoate **XVI** 51
4,4′-(2-ethyl-2-nitro-1,3-propanediyl)-bismorpholine **IX** 295–301
4,4′-(2-ethyl-2-nitro-trimethylene)-dimorpholine **IX** 295
ethyl orthosilicate **III** 299
ethyl oxide **XIII** 149
ethyl oxirane **V** 13
ethyl propenate **VI** 217
ethyl-2-propenoate **VI** 217
2-ethyl-3-propyl-1,3-propanediol **V** 345
ethyl silicate **III** 299
ethyl thiram **V** 165
Ethyl Thiurad **V** 165
ethyltrimethylmethane **IV** 269
etoposide **I** 6
ETS **XIII** 3
European walnut **XIII** 286
excursion limits **I** 15–22
exhaust fumes **I** 101–120
Exhoran® **V** 165
exposure peaks **I** 15–22
fast dark blue base R **V** 153
F-13B1 **VI** 37
F 134a **XIII** 251
FC 11 **I** 335, **III** 366
FC 12 **I** 336, 341, **III** 366, **V** 109
FC 13 **I** 69
FC 21 **V** 119
FC 22 **III** 63
FC 31 **V** 75
FC 112 **I** 297
FC 113 **III** 365, 366
FC 123 **X** 53
FC 142b **I** 65
ferric oxide **II** 135
ferrous oxide **II** 135
fibrous dust **VIII** 141–338
fine dust **II** 53
flint **II** 157
Fluorocarbon 134a **XIII** 251
fluorochloroform **I** 335
fluorodichloromethane **V** 119
fluorotrichloromethane **I** 335
5-fluorouracil **I** 4
flutamide **I** 6
FLX-0012® **III** 81
formaldehyde **I** 41-42, 44, 60, **II** 240, 250, 287, 331, **III** 1-8, 173–189, **IV** 136
formalin **III** 173
formic acid dimethylamide **VIII** 1
Formin **V** 355
*N*-formyldimethylamine **VIII** 1
2-formylfuran **IX** 69
fraké **XIII** 288
Freon® 13B1 **VI** 37
Freon® FE 1301 **VI** 37
fumes **II** 59–67, **XII** 271
Fundal® **VI** 57
Fungicide E® **XVI** 263
2-furanaldehyde **IX** 69
2-furancarbinol **VII** 121
2-furancarbonal **IX** 69
2-furancarboxaldehyde **IX** 69
2,5-furandione **IV** 275
2-furanmethanol **VII** 121
furfural **IX** 69–79
furfural alcohol **VII** 121
furfuraldehyde **IX** 69
furfuryl alcohol **VII** 121–125
furyl alcohol **VII** 121
2-furylaldehyde **IX** 69
2-furylcarbinol **VII** 121
2-furylmethanal **IX** 69
2-furylmethanol **VII** 121
fused silica **II** 157
Gaboon ebony **XIV** 287
Gaboon mahogany **XIII** 285
Galecron® **VI** 57

gasoline engine emissions **I** 109–120
general threshold limit value for dust **II** 3–46, **XII** 239–270
germ cell mutagens **I** 9–13
giant arborvitae **XIII** 288
giant cedar **XIII** 288
glass fibres **VIII** 141–338
glass wool **VIII** 141–338
glauramine **IV** 12
glutaral **VIII** 45, **XVI** 59
glutaraldehyde **VIII** 45–64, **XVI** 59–64
glutardialdehyde **VIII** 45, **XVI** 59
glutaric dialdehyde **VIII** 45, **XVI** 59
glycerin $\alpha$-monochlorohydrin **V** 83
glycerol chlorohydrin **V** 83
*sym*-glycerol dichlorohydrin **I** 95
glycerol-$\alpha,\gamma$-dichlorohydrin **I** 95
glycerol trichlorohydrin **IX** 171
glyceryl monothioglycolate **IX** 81–82
glyceryl trichlorohydrin **IX** 171
glycidyl allyl ether **VII** 9
glycidyl butyl ether **IV** 51
glycidyl isopropyl ether **VII** 141
1-*n*-glycidyloxybutane **IV** 51
glycidyl phenyl ether **IV** 305
glycidyl trimethylammonium chloride **IV** 247–255
glycol **IV** 225
glycol butyl ether **VI** 47
glycol chlorohydrin **V** 65
glycol dichloride **III** 137
glycol ether DB **VII** 59
glycol ethyl ether **VI** 205
glycol methacrylate **XIII** 193
glycol methyl ether **VI** 239
glycol methyl ether acetate **VI** 249
glycol monobutyl ether acetate **VI** 53
glycolmonochlorohydrin **V** 65
glycol monoethyl ether **VI** 205
glycol monoethyl ether acetate **VI** 213
glycol monomethyl ether **VI** 239
glycol monomethyl ether acetate **VI** 249
G-MAC® **IV** 247
gold 'teak' **XIII** 287
*Gonystylus bancanus* **XIII** 286
Grand Bassam mahogany **XIII** 287
graphite **II** 129–134
green ebony **XIII** 285, 299
*Grevillea robusta* **XIII** 286, **XIV** 295–297
Grotan® BK **II** 331
Grotan® HD **IX** 263
gypsum **II** 117, **VIII** 141–338
haematite **II** 136
halloysite **VIII** 141–338
Halon® 1301 **VI** 37
hard metal **III** 123, 124
HCB **XVI** 65
$\alpha$-HCH **V** 193
$\beta$-HCH **V** 193
$\gamma$-HCH **XVI** 113
2-(8-heptadecenyl)-4,5-dihydro-1-*H*-imidazole-1-ethanol **V** 373
2-(8-heptadecenyl)-2-imidazoline-1-ethanol **V** 373
2-heptadecenyl-2-imidazoline-1-ethanol **V** 373
*n*-heptane **XI** 165–176
*Heritiera utilis* **XIII** 288
hexabutyldistannoxane **I** 315
hexachlorobenzene **XVI** 65–88
$\alpha$-1,2,3,4,5,6-hexachlorocyclohexane **V** 193
$\beta$-1,2,3,4,5,6-hexachlorocyclohexane **V** 193
1$\alpha$,2$\alpha$,3$\beta$,4$\alpha$,5$\alpha$,6$\beta$-hexachlorocyclohexane **XVI** 113
1$\alpha$,2$\alpha$,3$\beta$,4$\alpha$,5$\beta$,6$\beta$-hexachlorocyclohexane **V** 193
1$\alpha$,2$\beta$,3$\alpha$,4$\beta$,5$\alpha$,6$\beta$-hexachlorocyclohexane **V** 193
$\alpha$-hexachlorocyclohexane **V** 193–206
$\beta$-hexachlorocyclohexane **V** 193–206
$\gamma$-1,2,3,4,5,6-hexachlorocyclohexane **XVI** 113
hexachloronaphthalenes **XIII** 101
hexahydro-2*H*-azepin-2-one **IV** 65
hexahydro-2*H*-azepin-7-one **IV** 65
hexahydrobenzene **XIII** 129
hexahydro-1,4-diazine **IX** 303, **XII** 177
hexahydrophthalic anhydride **X** 109–111
hexahydropyrazine **IX** 303, **XII** 177
hexahydro-1,3,5-tris(2-hydroxyethyl)-s-triazine **II** 331
hexamethylene **XIII** 129
hexamethyleneamine **V** 355
hexamethylenetetraamine **V** 355
hexamethylenetetramine **V** 355–372
hexamethylmelamine **I** 2
hexamine **V** 355
hexanaphthene **XIII** 129
*n*-hexane **IV** 257–268, **XIV** 101–126
hexane (all isomers except *n*-hexane) **IV** 269–273
6-hexanelactam **IV** 65
1-hexanol **IX** 283–290
*n*-hexanol **IX** 283
hexone **XIII** 169–180
*n*-hexyl alcohol **IX** 283
hexylene glycol **XVI** 233–246
*Heyderia decurrens* **XIII** 285, **XIV** 275

HFC-134a **XIII** 251
HMT **V** 355
HMTA **V** 355
HN2 **I** 211
Honduras mahogany **XIII** 287
Honduras rosewood **XIII** 286, 317
Hordaflex® **III** 81
hydrazine **I** 40, 171–182, **XIII** 181–186
hydrazine anhydrous **XIII** 181
hydrazine hydrate **XIII** 181–186
hydrazine monohydrate **XIII** 181
hydrazine salts **XIII** 181–186
hydrazinobenzene **XI** 225
hydrazobenzene **IX** 83–89
hydrobromic acid **XIII** 187
hydrobromic ether **VII** 115
hydrochloric acid **VI** 231
hydrochloric ether **III** 73
hydrocyanic acid **I** 42
hydrogen arsenide **XV** 41
hydrogen bromide **XIII** 187–191
hydrogen chloride **VI** 231–238
hydrogen chloride gas **VI** 231
hydrogen nitrate **III** 233
hydrogen peroxide **III** 249, 252, 253
hydrogen sulfate **XV** 165
$\alpha$-hydro-$\omega$-hydroxypoly(oxy-1,2-ethanediyl) **X** 247
$\alpha$-hydro-$\omega$-hydroxypoly[oxy(methyl-1,2-ethandiyl)] **X** 271
hydroquinone **III** 195, **X** 113–145
2-hydroxybiphenyl **II** 299
*o*-hydroxybiphenyl **II** 299
4-hydroxy-2-butane sulfonic acid **IV** 45
1-hydroxy-4-*tert*-butylbenzene **XI** 5
hydroxycarbamide **I** 7
2-hydroxyethyl acrylate **XVI** 89–93
2-hydroxyethyl methacrylate **XIII** 193–199
2-hydroxy-1,3-di-*tert*-butylbenzene **V** 341
2-hydroxy-3,5-dichlorophenyl sulfide **IX** 245
2-hydroxydiphenyl **II** 299
hydroxy ether **VI** 205
2-(2-hydroxyethoxy)ethylamine **IX** 215
2,2′-hydroxyethoxyethylamine **IX** 215
2-hydroxyethylamine **XII** 15
1-hydroxyethyl-2-heptadecenylglyoxalidine **V** 373
1-(2-hydroxyethyl)-2-heptadecenylglyoxalidine **V** 373
1-(2-hydroxyethyl)-2-(8-heptadecenyl)-2-imidazoline **V** 373
1-(2-hydroxyethyl)-2-*n*-heptadecenyl-2-imidazoline **V** 373
1-hydroxyethyl-2-heptadecenyl-imidazoline **V** 373–375
$\beta$-hydroxyethyl isopropyl ether **V** 207
2-(*N*-2-hydroxyethyl-*N*-methylamino)ethanol **IX** 291
*N*-(2-hydroxyethyl)-3-methyl-2-quinoxalinecarboxamide 1,4-dioxide **IX** 151
1-hydroxy hexane **IX** 283
1-hydroxyisopropyl-(2-methoxypropyl)ether **VI** 200
1-hydroxy-1′-methoxydiisopropylether **VI** 200
2-hydroxy-2′-methoxydi-*n*-propylether **VI** 200
1-hydroxy-2-methylbenzene **XIV** 59
1-hydroxy-3-methylbenzene **XIV** 59
1-hydroxy-4-methylbenzene **XIV** 59
hydroxymethylene benzyl ether **II** 239
hydroxymethylene mono(poly)oxymethylene benzyl ether **II** 239
2-hydroxymethylfuran **VII** 121
2-hydroxymethyl-*n*-heptan-4-ol **V** 345
2-(hydroxymethyl)-2-nitro-1,3-propanediol **II** 287–293
4-hydroxy-3-nitroaniline **IV** 289
hydroxy octyl oxostannane **VII** 93
2-hydroxypropanamine **IX** 237
3-hydroxy-1-propanesulfonic acid sulfone **IV** 313
3-hydroxy-1-propanesulfonic acid sultone **IV** 313
$\alpha$-hydroxypropanetricarboxylic acid **XVI** 209
2-hydroxy-1,2,3-propanetricarboxylic acid **XVI** 209
hydroxypropyl acrylate (all isomers) **XVI** 95–104
1-hydroxy-2-propyl acrylate **XVI** 95
1-hydroxy-3-propyl acrylate **XVI** 95
2-hydroxy-1-propyl acrylate **XVI** 95
2-hydroxypropylamine **IX** 237
2-hydroxypropyl methacrylate **XVI** 105–111
2-hydroxypropyl-(1-methoxyisopropyl)ether **VI** 200
2-hydroxypropyl 2-methyl-2-propenoate **XVI** 105
1-hydroxy-2-(1*H*)-pyridinethione sodium salt **X** 287
2-hydroxytoluene **XIV** 59
3-hydroxytoluene **XIV** 59
4-hydroxytoluene **XIV** 59
1-hydroxy-2,4,5-trichlorobenzene **XII** 209
2-hydroxytriethylamine **XIV** 91
hydroxyurea **I** 7

hyperbaric pressure and body burden **XIV** 11–16
hyponitrous acid anhydride **IX** 115
hytrol O **X** 35
ifosfamide **I** 2
IGE **VII** 141
2-imidazolidinethione **XI** 121
4,4′-imidocarbonyl-bis(*N,N*-dimethylaniline) **IV** 11
4,4′-imidocarbonyl-bis(*N,N*-dimethylbenzenamine) **IV** 11
inactive limonene **I** 185
incense cedar **XIII** 285, **XIV** 275
Indian ebony **XIII** 286, **XIV** 287
Indian rosewood **XIII** 285, 301
inhalable dust **XII** 239
inhalation kinetics **XIV** 3–10
inorganic fibres **VIII** 141–338
inorganic tin compounds **XIV** 149–165
iodomethane **VII** 219
3-iodo-2-propynylbutylcarbamate **XVI** 247–256
3-iodo-2-propynyl-*N*-butylcarbamate **XVI** 247
3-iodo-2-propynylcarbamic acid butyl ester **XVI** 247
Ionox 99® **V** 341
ipe **XIII** 288, **XIV** 311
ipé peroba **XIII** 287, **XIV** 307
iroko **XIII** 285, **XIV** 279
iron oxides **II** 135–144
isatoic acid anhydride **XIII** 97
isatoic anhydride **XIII** 97
isoamyl acetate **XI** 211
1,3-isobenzofurandione **VII** 247
isobutanol-2-amine **IX** 229
isobutenyl chloride **IV** 97
isobutyltrimethylmethane **I** 347
isocyanic acid methylene di-*p*-phenylene ester **VIII** 65
isocyanic acid polymethylenepolyphenylene ester **VIII** 65
isodecyl oleate **XVI** 257–261
isoflurane **VII** 127–140
isohexane **IV** 269
Isol **XVI** 233
isonitropropane **III** 241
isooctane **I** 347
isoparaffins **IV** 269–273
isopentyl acetate **XI** 211
isophorone **I** 196
isophosphamid **I** 2
isophthalic acid **VII** 241
isopropanolamine **IX** 237
isopropenylbenzene **XV** 137
4-isopropenyl-1-methyl-1-cyclohexene **I** 185
2-isopropoxyethanol **V** 207-211
(isopropoxymethyl)oxirane **VII** 141
isopropylacetone **XIII** 169
isopropylamine **I** 34
isopropylbenzene **XIII** 117
isopropyl cellosolve **V** 207
isopropyl ethylene glycol ether **V** 207
isopropyl glycidyl ether **VII** 141–144
isopropyl glycol **V** 207
4,4′-isopropylidenediphenol **XIII** 49
3-isopropyloxypropylene oxide **VII** 141
isothiourea **I** 301
'jacaranda' **XIII** 287, 321
jeweler's rouge **II** 135
*Juglans* spp. **XIII** 286
kambala **XIII** 285, **XIV** 279
Kathon® **V** 323
Kathon® 893 MW **XVI** 263
Kathon® biocide **V** 323
Kaurit® S **V** 289
ketohexamethylene **X** 35
ketone propane **VII** 1
*β*-ketopropane **VII** 1
*Khaya anthotheca* **XIII** 286
*Khaya grandifoliola* **XIII** 287
*Khaya ivorensis* **XIII** 287
*Khaya senegalensis* **XIII** 287
kieselguhr **II** 157
lapacho **XIII** 288, **XIV** 311
Larvacide 100 **VI** 105
laughing gas **IX** 115
lead chromate **III** 101–122, **VII** 33
lead tetraethyl **XV** 223
lead tetramethyl **XV** 237
*Libocedrus decurrens* **XIII** 285, **XIV** 275
limba **XIII** 288
limitation of exposure peaks **I** 15–22
limonene **I** 185–200, 347
lindane **XVI** 113–141
*α*-lindane **V** 193
*β*-lindane **V** 193
lithium hydride **III** 191–193
loadstone **II** 135
lodestone **II** 135
longleaf pine **XIII** 287
long-term threshold values **XI** 281
Lorol C 6 **IX** 283
low cristobalite **XIV** 205
low quartz **XIV** 205
low tridymite **XIV** 205
Lubrimet® P600, P900 **X** 271
lusamba **XIII** 288

lye **XII** 195
Macassar ebony **XIII** 286, **XIV** 287
*Machaerium scleroxylon* **XIII** 287, 321–323
Mafu® **IV** 201
magnesia **II** 145
magnesium oxide **II** 145–148
magnesium oxide sulfate **VIII** 141–338
magnetite **II** 136
mahogany **XIII** 287
mainstream smoke **XIII** 3
makoré **XIII** 288
maleic acid anhydride **IV** 275, **XI** 177
maleic anhydride **IV** 275–287, **XI** 177–178
manganese **XII** 293–328
manganese compounds **XII** 293
man-made mineral fibres **VIII** 141–338
*Mansonia altissima* **XIII** 287, **XIV** 299–305
maritime pine **XIII** 287
MBOCA **VII** 193
MDEA **IX** 291
MDI **VIII** 65
mechlorethamine **I** 211
mecrylate **I** 233
MEK **XII** 37
melphalan **I** 3
*p*-mentha-1,8-diene **I** 185
1,8(9)-*p*-menthadiene **I** 185
merbromin **XV** 75–79
2-mercaptoimidazoline **XI** 121
6-mercaptopurine **I** 4
mercurialin **VII** 145
mercurochrome **XV** 75
Mercurothiolate **XV** 249
mercury and inorganic mercury compounds **XV** 81–122
mercury, organic compounds **XV** 123–136
mesitylene **IV** 341
metallic mercury **XV** 81
metal-working fluids **II** 207–220, **IX** 195–213, **XVI** 207–286
metaphenylenediamine **VI** 287
methacrylic acid ethyl ester **XVI** 51
methacrylic acid 2-hydroxyethyl ester **XIII** 193
methacrylic acid 2-hydroxypropyl ester **XVI** 105
methacrylic acid methyl ester **XVI** 181
2-methallyl chloride **IV** 97
*β*-methallyl chloride **IV** 97
methanal **III** 173
methanamine **VII** 145
methane base **I** 249
methane dichloride **IV** 173
methanol **XVI** 143–175
methenamine **V** 355
methotrexate **I** 3
4-methoxy-2-aminoanisole **IV** 135
2-methoxyaniline **X** 1
*o*-methoxyaniline **X** 1
2-methoxybenzenamine **X** 1
4-methoxy-1,3-benzenediamine **VI** 145
2-methoxyethanol **V** 210, **VI** 239–248
2-methoxyethanol acetate **VI** 249
2-methoxyethyl acetate **VI** 249–251
methoxyhydroxyethane **VI** 239
2-methoxy-5-methylaniline **IV** 135
2-methoxy-5-methylbenzenamine **IV** 135
(2-methoxymethylethoxy)propanol **VI** 199
2-(2-methoxy-1-methylethoxy)-1-propanol **VI** 200
1-(2-methoxy-1-methylethoxy)-2-propanol **VI** 200
2-(2-methoxy-2-methylethoxy)-1-propanol **VI** 200
1-(2-methoxy-2-methylethoxy)-2-propanol **VI** 200
1-methoxy-2-nitrobenzene **IX** 103
*o*-methoxyphenylamine **X** 1
*p*-methoxy-*m*-phenylenediamine **VI** 145
4-methoxy-*m*-phenylenediamine **VI** 145
1-methoxy-2-propanol **V** 213–216, **XIV** 127–129
1-methoxypropan-2-ol **I** 204, **V** 213
2-methoxy-1-propanol **I** 203, 207
1-methoxy-2-propanol acetate **V** 217
2-methoxy-1-propanol acetate **I** 203, 207
1-(1-methoxy-2-propoxy)-2-propanol **VI** 200
1-(2-methoxy-1-propoxy)-2-propanol **VI** 200
2-(1-methoxy-2-propoxy)-1-propanol **VI** 200
2-(2-methoxy-1-propoxy)-1-propanol **VI** 200
1-methoxypropyl-2-acetate **V** 217–220
2-methoxypropyl-1-acetate **I** 207–209
4-methoxy-*m*-toluidine **IV** 135
methural **V** 289
methyl acrylate **VI** 253–262, **XVI** 177–180
2-methylacrylic acid methyl ester **III** 195
methyl alcohol **XVI** 143
methyl aldehyde **III** 173
methyl allyl chloride **IV** 97
methylamine **I** 28, 34, **VII** 145—153
1-methyl-2-aminobenzene **III** 307
2-methyl-1-aminobenzene **III** 307
(methylamino)benzene **VI** 263
*N*-methylaminobenzene **VI** 263
*N*-methylaminodiglycol **IX** 291
1-methyl-2-aminoethanol **IX** 237
*N*-methylaniline **I** 25, **VI** 263–270
2-methylaniline **III** 307

4-methylaniline **III** 323
*p*-methylaniline **III** 323
5-methyl-*o*-anisidine **IV** 135
*p*-methylanisidine **IV** 135
1-methylazacyclopentan-2-one **X** 147
2-methylbenzenamine **III** 307
4-methylbenzenamine **III** 323
4-methyl-1,3-benzenamine **VI** 339
*N*-methylbenzenamine **VI** 263
methylbenzene **VII** 257
*p*-methylbenzeneamine **III** 323
methylbenzotriazole **II** 295
methyl-1*H*-benzotriazole **II** 295–298
*N*-methyl-bis(2-chloroethyl)amine **I** 211–230
methylbis(2-hydroxyethyl)amine **IX** 291
methyl bromide **VII** 155–172
2-methyl-1-butanol acetate **XI** 211
2-methyl-2-butanol acetate **XI** 211
3-methyl-1-butanol acetate **XI** 211
1-methylbutyl acetate **XI** 211
2-methylbutyl acetate **XI** 211
3-methylbutyl acetate **XI** 211
*β*-methylbutyl acetate **XI** 211
methyl carbinol **XII** 129
Methyl Cellosolve® **VI** 239
Methyl Cellosolve® acetate **VI** 249
methyl chloride **VII** 173–191
methyl chloroacetate **IX** 1
2-methyl-4-chloroaniline **VI** 127
3-methyl-4-chlorophenol **II** 271
methyl 2-cyanoacrylate **I** 233–245, **XIII** 201–206
methyl *α*-cyanoacrylate **I** 233, **XIII** 201
methyldibromoglutaronitrile **XIV** 87
*N*-methyl-2,2′-dichlorodiethylamine **I** 211
methyldiethanolamine **IX** 291–294
*N*-methyldiethanolamine **IX** 291
*N*-methyldiethanolimine **IX** 291
2-methyl-2,3-dihydroisothiazol-3-one **V** 323–339
methyldinitrobenzene **VI** 165
1-methyl-2,3-dinitrobenzene **VI** 165
1-methyl-2,4-dinitrobenzene **VI** 165
1-methyl-3,4-dinitrobenzene **VI** 165
1-methyl-3,5-dinitrobenzene **VI** 165
2-methyl-1,3-dinitrobenzene **VI** 165
2-methyl-1,4-dinitrobenzene **VI** 165
4-methyl-1,2-dinitrobenzene **VI** 165
methylene acetone **IX** 91
methylene base **I** 249
methylene bichloride **IV** 173
4,4′-methylenebis(aniline) **VII** 37
4,4′-methylenebis(benzenamine) **VII** 37
4,4′-methylene(bis)chloroaniline **VII** 193
4,4′-methylenebis(2-chloroaniline) **VII** 193–218
*p,p′*-methylenebis(*alpha*-chloroaniline) **VII** 193
4,4′-methylenebis(2-chlorobenzenamine) **VII** 193
4,4′-methylene-bis(*N,N*-dimethylaniline) **I** 249-256, **IV** 15
4,4′-methylene-bis(*N,N*-dimethyl)benzenamine **I** 249
1,1′-methylenebis(4-isocyanatobenzene) **VIII** 65
methylenebis(4-phenylene isocyanate) **VIII** 65
methylenebis(*p*-phenylene isocyanate) **VIII** 65
methylene bisphenyl isocyanate **VIII** 65
methylene chloride **IV** 173
4,4′-methylenedianiline **VII** 37
*p,p′*-methylenedianiline **VII** 37
methylene dichloride **IV** 173
methylenedi-4-phenylene diisocyanate **VIII** 65
methylenedi-*p*-phenylene diisocyanate **VIII** 65
methylene di(phenylene isocyanate) **VIII** 65
4,4′-methylenediphenylene isocyanate **VIII** 65
4,4′-methylenediphenyl diisocyanate **VIII** 65
4,4′-methylene diphenyl isocyanate **VIII** 65, **XIV** 131–133
methylene glycol **III** 173
methylene oxide **III** 173
(1-methylethenyl)-benzene **XV** 137
methyl ether **I** 125
methyl ethoxol **VI** 239
2-(1-methylethoxy)ethanol **V** 207
[(1-methylethoxy)methyl]oxirane **VII** 141
(1-methylethyl)benzene **XIII** 117
methylethylene oxide **V** 221
4,4′-(1-methylethylidene)bisphenol **XIII** 49
methyl ethyl ketone **XII** 37
methyl ethyl ketone peroxide **III** 250, 252, 254, 255
methyl glycol **VI** 239
methyl glycol acetate **VI** 249
methyl glycol monoacetate **VI** 249
2,2′-(methylimino)bisethanol **IX** 291
2,2′-(methylimino)diethanol **IX** 291
*N*-methyliminodiethanol **IX** 291
*N*-methyl-2,2′-iminodiethanol **IX** 291
methyl iodide **VII** 219–228
methyl isobutyl ketone **XIII** 169
1-methyl-4-isopropenyl-1-cyclohexene **I** 185

2-methyl-4-isothiazolin-3(2*H*)-one **V** 323
2-methyl-3(2*H*)-isothiazolone **V** 323
methyl ketone **VII** 1
*N*-methyl-2-ketopyrrolidine **X** 147
*N*-methyl lost **I** 211
methyl methacrylate **III** 195–231, **XVI** 181–191
methyl α-methacrylate **III** 195
*N*-methylmethanamine **VII** 75
methyl methylacrylate **III** 195, **XVI** 181
1-methyl-4-(1-methylethenyl)cyclohexene **I** 185
methyl 2-methyl-2-propenoate **III** 195, **XVI** 181
methyl monochloroacetate **IX** 1
2-methyl-5-nitroaniline **VI** 271
6-methyl-3-nitroaniline **VI** 271
2-methyl-5-nitrobenzenamine **VI** 271
1-methyl-2-nitrobenzene **VIII** 97
2-methyl-1-nitrobenzene **VIII** 97
*o*-methylnitrobenzene **VIII** 97
2-methyl-5-nitrobenzeneamine **VI** 271
(*Z*)-(2-methylnonyl)-9-octadecenoate **XVI** 257
*N*-methyl-*N*-oleoyl-aminoacetic acid **XVI** 281
4-methyl-3-oxapentan-1-ol **V** 207
3-methyl-1,2-oxathiolane 2,2-dioxide **IV** 45
methyloxirane **V** 221
*N*-methyl-*N*-(1-oxo-9-octadecenyl)glycine **XVI** 281
*N*-methyl-2-oxypyrrolidine **X** 147
2-methylpentane **IV** 269
3-methylpentane **IV** 269
2-methyl-2,4-pentanediol **XVI** 233
4-methyl-2-pentanone **XIII** 169
2-methylphenol **XIV** 59
3-methylphenol **XIV** 59
4-methylphenol **XIV** 59
*N*-methylphenylamine **VI** 263
4-methyl-*m*-phenylenediamine **VI** 339
1-methyl-1-phenylethylene **XV** 137
1-methyl-1,3-propane sultone **IV** 45
methyl propenate **VI** 253
methyl-2-propenoate **VI** 253
2-methyl-2-propenoic acid ethyl ester **XVI** 51
2-methyl-2-propenoic acid 2-hydroxyethyl ester **XIII** 193
2-methyl-2-propenoic acid 2-hydroxypropyl ester **XVI** 105
2-methyl-2-propenoic acid methyl ester **III** 195, **XVI** 181
*N*-methyl-2-pyrrolidinone **X** 147
1-methyl-2-pyrrolidinone **X** 147
1-methyl-5-pyrrolidinone **X** 147
*N*-methylpyrrolidone **X** 147
1-methyl-2-pyrrolidone **X** 147
*N*-methyl-2-pyrrolidone **X** 147–170
α-methyl styrene **XV** 137–140
methylsulfinylmethane **III** 163
methyl sulfoxide **III** 163
methyltetrahydrophthalic anhydride **X** 109–111
*N*-methyl-*N*,2,4,6-tetranitroaniline **XI** 179–185
*N*-methyl-*N*,2,4,6-tetranitrobenzenamine **XI** 179
methyl toluene **V** 263
methyl tribromide **VII** 319
1-methyl-2,4,6-trinitrobenzene **I** 359
2-methyl-1,3,5-trinitrobenzene **I** 359
methyl vinyl ketone **IX** 91–101
Michler's base **I** 249
Michler's hydride **I** 249
Michler's ketone **IV** 12, 13, 20, 21
Michler's methane **I** 249
*Microberlinia* spp **XIII** 287
*Mimusops heckelii* **XIII** 288
mineral fibres **II** 185
MIPA **IX** 237
mists **XII** 271
mithramycin **I** 5
mitobronitol **I** 7
mitomycin **I** 5
mitoxantrone **I** 5
MMA **III** 195, **XVI** 181
MME **III** 195
4-MMPD **VI** 145
MOCA **VII** 193
monobromoethane **VII** 115
monobromomethane **VII** 155
monobutyl glycol ether **VI** 47
monochloroacetaldehyde **XII** 81
monochloroacetic acid methyl ester **IX** 1
monochlorobenzene **XII** 103
monochlorodifluoromethane **III** 63
monochloroethane **III** 73
2-monochloroethanol **V** 65
monochloroethene **V** 241
monochloroethylene **V** 241
monochlorofluoromethane **V** 75
α-monochlorohydrin **V** 83
monochloromethane **VII** 173
monochloromonofluoroethane **V** 75
monochloronaphthalenes **XIII** 101
monochloropropanediol **V** 83
monochlorotrifluoromethane **I** 69
monoethanolamine **XII** 15

monoethyleneglycol **IV** 225
mono-isopropanolamine **IX** 237
monomethylamine **VII** 145
monomethylaniline **VI** 263
monomethyl glycol acetate **VI** 249
mono-*n*-octyltin compounds **VII** 91–114
monooctyltin hydroxide **VII** 93
mono-*n*-octyltin oxide **VII** 93
monooctyltin thioglycolate **VII** 94
mono-*n*-octyltin trichloride **VII** 93
mono-*n*-octyltin tris(2-ethylhexyl thioglycolate) **VII** 94
mono-*n*-octyltin tris(2-ethylhexylmercaptoacetate) **VII** 94
mono-*n*-octyltin tris(isooctylthioglycolate) **VII** 94
mopidamol **I** 4
morpholine **I** 25, 28, 33–34
4-morpholinecarbonyl chloride **V** 79
morpholinylcarbamoyl chloride **V** 79
morpholinylcarbonyl chloride **V** 79
MOTC **VII** 93
MOTO **VII** 93
MOTTG **VII** 94
*Morus excelsa* **XIV** 279
muriatic acid **VI** 231
mustard gas **IV** 27
mustargen **I** 211
myleran **I** 2–3
Mystox WFA® **II** 299
nadone **X** 35
naphthalene **XI** 187–210
2-naphthylamine **I** 258–259
Natriphene® **II** 299
Natrium-Pyrion® **X** 287
natural rubber latex (*Hevea* species) **XV** 141–143
NCI-C56417 **V** 295
2-NDB **XIII** 207
nemalite **VIII** 141–338
neohexane **IV** 269
niangon **XIII** 288
nickel compounds **I** 40, **II** 47, 48, 136
nicotine **I** 28, 40, 42–44, 59
Nigerian ebony **XIV** 287
Nigerian satinwood **XIII** 286, **XIV** 291
Nigerian walnut **XIV** 299
nimustine **I** 3
nitramine **XI** 179
nitric acid **III** 233–240
nitrite **I** 27–34, 59
2-nitro-4-aminophenol **I** 167, **IV** 289–293
*o*-nitro-*p*-aminophenol **IV** 289
4-nitro-2-aminotoluene **VI** 271
nitroaniline **I** 149–150, 166
2-nitroanisole **IX** 103–114
*o*-nitroanisole **IX** 103
2-nitro-1,4-benzenediamine **IV** 295, **XIII** 207
4-nitrobiphenyl **I** 257–259
4-nitro-1,1′-biphenyl **I** 257
*p*-nitrobiphenyl **I** 257
4-(2-nitrobutyl)morpholine **IX** 295–301
*N*-(2-nitrobutyl)morpholine **IX** 295
*o*-nitrochlorobenzene **IV** 107
*m*-nitrochlorobenzene **IV** 115
*p*-nitrochlorobenzene **IV** 121
nitrochloroform **VI** 105
2-nitro-1,4-diaminobenzene **IV** 295, **XIII** 207
4-nitrodiphenyl **I** 257
*p*-nitrodiphenyl **I** 257
nitrogen mustard **I** 211
nitrogen oxide **IX** 115
nitrogen oxides **I** 28, 40, 42–44, 58
nitro-isobutylglycerol **II** 287
nitroisopropane **III** 241
3-nitro-6-methylaniline **VI** 271
1-nitro-2-methylbenzene **VIII** 97
2-nitro-1,4-phenylenediamine **IV** 295, **XIII** 207
2-nitro-*p*-phenylenediamine **IV** 295–303, **XIII** 207–210
*o*-nitro-*p*-phenylenediamine **IV** 295, **XIII** 207
*o*-nitrophenyl methyl ether **IX** 103
1-nitropropane **XIII** 211–213
2-nitropropane **III** 241–247
nitropyrenes **I** 112, 118, 258
*N*-nitrosamines **I** 23–35, 42, 58–59, 261–286, **II** 250
nitrosation of amines **I** 23–35
*N*-nitrosodi-*n*-butylamine **I** 24, 29, 261–262, 274–276, 286
*N*-nitrosodiethanolamine **I** 27, 29, 261–262, 277–278, 286
*N*-nitrosodiethylamine **I** 23–35, 41, 60, 261–262, 268–271, 286
*N*-nitrosodiisopropylamine **I** 261–262, 274, 286
*N*-nitrosodimethylamine **I** 3–35, 41–44, 59, 261–267, 286
*N*-nitrosodiphenylamine **I** 27
*N*-nitrosodi-*n*-propylamine **I** 261–262, 271–273, 286
*N*-nitrosoethylphenylamine **I** 261–262, 280, 286
*N*-nitrosomethylethylamine **I** 261–262, 267, 286

*N*-nitrosomethylphenylamine **I** 29, 261–262, 279, 286
*N*-nitrosomorpholine **I** 29, 34, 261–262, 280–282, 286
*p*-nitrosophenol **XIV** 135
4-nitrosophenol **XIV** 135–136
*N*-nitrosopiperidine **I** 261–262, 282–284, 286
*N*-nitrosopyrrolidine **I** 40, 261–262, 284–285, 286
2-nitrotoluene **VIII** 97–108
*o*-nitrotoluene **VIII** 97
5-nitro-*o*-toluidine **VI** 271–275
nitrotrichloromethane **VI** 105
nitrous oxide **IX** 115–150
$NO_x$ **I** 28
northern white cedar **XIII** 288
Noxal **V** 165
1-NP **XIII** 211
2-NP **XIII** 207
2-NPPD **XIII** 207
Nuran® **IV** 201
oak wood dust **IV** 363, **XIII** 287
obeche **XIII** 288
octachloronaphthalene **XIII** 101
9-octadecenoic acid (*Z*) decyl ester **IX** 269
(*Z*)-9-octadecenoic acid isodecyl ester **XVI** 257
Octhilinone **XVI** 263
octylene glycol **V** 345
2-octyl-4-isothiazolin-3-one **XVI** 263–280
*N*-*n*-octyl-3-isothiazolone **XVI** 263
2-*n*-octyl-3-isothiazolone **XVI** 263
2-octyl-3(2*H*)-isothiazolone **XVI** 263
2,2′,2″-[(octylstannylidyne)tris(thio)]tris-(acetic acid)tri-isooctyl ester **VII** 95
2,2′,2″-[(octylstannylidyne)tris(thio)]tris-(acetic acid)tris(2-ethylhexyl) ester **VII** 95
[(octylstannylidyne)trithio]tri(acetic acid)-tris(2-ethylhexyl ester) **VII** 94
octyltin trichloride **VII** 93
octyltin tris(isooctyl thioglycolate) **VII** 94
octyltrichlorostannane **VII** 93
octyltris(2-ethylhexyloxycarbonylmethylthio)-stannane **VII** 94
oil of turpentine **XIV** 173
oil of vitriol **XV** 165
OKO® **IV** 201
okoumé **XIII** 285
olaquinox **IX** 151–169
olefiant gas **X** 91
oleic acid decyl ester **IX** 269
oleic acid isodecyl ester **XVI** 257
oleic sarcosine **XVI** 281
oleoyl *N*-methylaminoacetic acid **XVI** 281
oleoyl *N*-methylglycine **XVI** 281
oleoyl sarcosine **XVI** 281–286
*N*-oleoyl sarcosine **XVI** 281
Onyxide® 500 **II** 249
opal **II** 157
organic fibres **VIII** 205
orthoboric acid **V** 295
oxacyclopropane **V** 181
3-oxa-1-hepatanol **VI** 47
oxane **V** 181
3-oxapentane-1,5-diol **X** 73
1,2-oxathiane 2,2-dioxide **IV** 45
1,2-oxathiolane 2,2-dioxide **IV** 313
Oxazolidine E **IX** 275
oxidation base 25 **IV** 289
*α,β*-oxidoethane **V** 181
oxiran **V** 181
oxirane **V** 181, **XIII** 165
oxirane methane ammonium-*N*,*N*,*N*-trimethyl chloride **IV** 247
3-oxiranyl-7-oxabicyclo[4.1.0]heptane **I** 391
Oxitol **VI** 205
3-oxo-1,2-benzisothiazoline **II** 221
2-oxobutane **XII** 37
oxodioctylstannane **VII** 91
2-oxo-hexamethylenimine **IV** 65
oxomethane **III** 173
oxybis(4-aminobenzene) **VI** 277
4,4′-oxybisaniline **VI** 277
4,4′-oxybisbenzenamine **VI** 277
1,1′-oxybisethane **XIII** 149
oxybismethane **I** 125
1,1′-oxybis(2-methoxyethane) **IX** 41
4,4′-oxydianiline **VI** 277–286
2,2-oxydiethanol **X** 73
4,4′-oxydiphenylamine **VI** 277
oxyfume **V** 181
oxymethurea **V** 289
oxy methylene **III** 173
ozone **X** 171–199
Pacific red cedar **XIII** 288
PAH **I** 42, 58–59, 111, 112
palygorskite **VIII** 141–338
pâo ferro **XIII** 287, 321
paraphenylenediamine **VI** 311
*Paratecoma peroba* **XIII** 287, **XIV** 307–309
Parmetol® F85 **XVI** 263
Paroil® **III** 81
passive smoking at work **I** 39–62, **XIII** 3–39
PCM **I** 291
PCMC **II** 271
pencil cedar **XIII** 285, **XIV** 275
pentachloronaphthalenes **XIII** 101

pentachlorophenol **III** 261–270
pentaerythritol triacrylate **XVI** 193–199
1,5-pentanedial **VIII** 45, **XVI** 59
1,5-pentanedione **VIII** 45
3-pentanol acetate **XI** 211
2-pentanol acetate **XI** 211
pentyl acetate **XI** 211–223
1-pentyl acetate **XI** 211
3-pentyl acetate **XI** 211
pentylcarbinol **IX** 283
per **III** 271
peracetic acid **VII** 229–239
perchlorobenzene **XVI** 65
perchloroethylene **III** 271
perchloromercaptan **I** 291
perchloromethylmercaptan **I** 291–295
periclase **II** 145
*Pericopsis elata* **XIII** 287
peroba do(s) campo(s) **XIII** 287, **XIV** 307
peroxides **III** 249–260
peroxoacetic acid **VII** 229
peroxyacetic acid **III** 250, 253, 254, **VII** 229
petrol engine emissions **I** 109–120
3-phenoxy-1,2-epoxypropane **IV** 305
(phenoxymethyl)oxirane **IV** 305
phenoxypropene oxide **IV** 305
phenoxypropylene oxide **IV** 305
phenylamine **VI** 17
phenyl chloroform **VI** 80
2,2′-[1,3-phenylenebis(oxymethylene)]-bisoxirane **VII** 69
1,2-phenylenediamine **XIII** 215
1,3-phenylenediamine **VI** 287
1,4-phenylenediamine **VI** 311
*m*-phenylenediamine **VI** 287–299
*o*-phenylenediamine **XIII** 215–235
*p*-phenylenediamine **VI** 311–337, **XIV** 137–141
*p*-phenylenediamine compounds **XV** 151–153
phenyl 2,3-epoxypropyl ether **IV** 305
phenyl glycidyl ether **IV** 59, 305–311
phenylhydrazine **XI** 225–234
phenylmethane **VII** 257
(phenylmethoxy)methanol **II** 239
[(phenylmethoxy)methoxy]methanol **II** 239
[[(phenylmethoxy)methoxy]methoxy]-methanol **II** 239
*N*-phenylmethylamine **VI** 263
1-phenyl-1-methylethylene **XV** 137
phenyl mono(poly)methoxy methanol **II** 239
4-phenylnitrobenzene **I** 257
*p*-phenylnitrobenzene **I** 257
2-phenylphenol **II** 299
*o*-phenylphenol **II** 299–316
*o*-phenylphenol sodium salt **II** 299
2-phenylpropane **XIII** 117
2-phenylpropene **XV** 137
phenyl trichloromethane **VI** 80
phosphonic acid dimethyl ester **I** 133
phosphoric acid 2,2-dichloroethenyl dimethyl ester **IV** 201
phosphoric acid 2,2-dichlorovinyl dimethyl ester **IV** 201
phosphoric acid triphenyl ester **II** 321
phosphorothioic acid *O,O*-diethyl *O*-[6-methyl-2-(1-methylethyl)-4-pyrimidinyl] ester **XI** 31
phosphorous acid dimethyl ester **I** 133
1,3-phthalandione **VII** 247
phthalic acid **VII** 241–246
phthalic acid anhydride **VII** 247, **XI** 235
phthalic acid diallyl ester **IX** 11
phthalic anhydride **VII** 247–255, **XI** 235–237
physical activity and inhalation kinetics **XIV** 3–10
Pic-clor **VI** 105
Picfume® **VI** 105
Picride **VI** 105
picrylmethylnitramine **XI** 179
picrylnitromethylamine **XI** 179
pimelic ketone **X** 35
*Pinus* spp. **XIII** 287
piperazidine **IX** 303, **XII** 177
piperazine **IX** 303–313, **XII** 177–178
piperidine **I** 25
pitch pine **XIII** 287
plaster **II** 117
plaster of Paris **II** 117
plicamycin **I** 5
Pluracol® P410, P1010, P2010, P4010 **X** 271
Pluriol® P600, P2000 **X** 271
PMDI **VIII** 65–95, **XIV** 131
PNB **I** 257
polyethylene glycol **X** 247–270
polyethylene oxide **X** 247
polyglycol **X** 247
Polyglycol P425, P1200, P2000, P4000 **X** 271
"polymeric MDI" **VIII** 65–95, **XIV** 131
polyoxyethylene **X** 247
poly(propane-1,2-diol) **X** 271
polypropylene glycol **X** 271–285
polypropylene oxide **X** 271
Poly-Solv® TB **IX** 315
polyvinyl chloride **II** 149–156
portland cement **XI** 303–333
potassium citrate **XVI** 209

potassium titanates **VIII** 141–338
potassium zinc chromate **XV** 289
PPG **X** 271
Preventol® CMK **II** 271
Preventol® D 2 **II** 239
Preventol® O Extra **II** 299
Preventol® ON Extra **II** 299
procarbazine **I** 7
1,2-propanediol-1-acrylate **XVI** 95
1,3-propanediol-1-acrylate **XVI** 95
1,2-propanediol-2-acrylate **XVI** 95
1,3-propane sultone **IV** 46, 313–321
2-propanone **VII** 1
2-propenal **XVI** 1
2-propenamide **III** 11
propene oxide **V** 221
2-propenoic acid butyl ester **V** 5, **XII** 57, **XVI** 35
2-propenoic acid ethyl ester **VI** 217, **XVI** 41
2-propenoic acid 2-ethylhexyl ester **XVI** 47
2-propenoic acid 2-ethyl-2-[[(1-oxo-2-propenyl)oxy]methyl]-1,3-propane-diyl ester **XVI** 201
2-propenoic acid homopolymer, sodium salt **XV** 1
2-propenoic acid 2-hydroxyethyl ester **XVI** 89
2-propenoic acid 2-hydroxy-1-methylethyl ester **XVI** 95
2-propenoic acid 2-(hydroxymethyl)-2-[[(1-oxo-2-propenyl)oxy]methyl]-1,3-propanediyl ester **XVI** 193
2-propenoic acid 2-hydroxypropyl ester **XVI** 95
2-propenoic acid 3-hydroxypropyl ester **XVI** 95
2-propenoic acid hydroxypropyl ester **XVI** 95
2-propenoic acid methyl ester **VI** 253, **XVI** 177
1-propenol-3 **XV** 31
2-propen-1-ol **XV** 31
[(2-propenyloxy)methyl]oxirane **VII** 9
2-propoxyethanol **XII** 179–186
2-propoxyethanol acetate **XII** 187
2-propoxyethyl acetate **XII** 187–193
propyl cellosolve **XII** 179
propylene chloride **IX** 21
propylene dichloride **IX** 21
propylene epoxide **V** 221
propylene glycol 1-methyl ether **V** 213, **XIV** 127
propylene glycol 2-methyl ether **I** 203
propylene glycol 1-methyl ether 2-acetate **V** 217
propylene glycol 2-methyl ether 1-acetate **I** 207
propylene glycol monoacrylate **XVI** 95
propylene glycol monomethyl ether **V** 213
1,2-propylene oxide **V** 221–228
*n*-propyl glycol **XII** 179
Protectol® DMU **V** 289
Proxel® **II** 221
pruno **XIII** 287, **XIV** 299
pseudothiourea **I** 301
PVC **II** 149–156
pyrazine hexahydride **IX** 303, **XII** 177
2-pyridinethiol 1-oxide sodium salt **X** 287
pyroacetic acid **VII** 1
pyroacetic ether **VII** 1
pyrrolylene **XV** 51
QUAB 151® **IV** 247
quartz **I** 113, **II** 129–134, 182–195, **XIV** 205–271
quartz glass **II** 157
*Quercus* spp **XIII** 287
quinone monoxime **XIV** 135
quinone oxime **XIV** 135
R 11 **I** 335, **III** 366
R 12 **V** 109
R 13 **I** 69
R 21 **V** 119
R 22 **III** 63
R 31 **V** 75
R 112 **I** 297
R 113 **III** 365, 366
R 142b **I** 65
radon **II** 136
ramin **XIII** 286
RDGE **VII** 69
red cedar **XIII** 288
red pine **XIII** 287
reduced Michler's ketone **I** 249
Remol TRF® **II** 299
Reomet® SBT 75 **II** 317
resorcinol bis(2,3-epoxypropyl) ether **VII** 69
resorcinol diglycidyl ether **VII** 69
resorcinyl diglycidyl ether **VII** 69
respirable dust **XII** 239
Rewopon® IM OA **V** 373
RH 893 HQ Technical **XVI** 263
RH 893 T **XVI** 263
Ribeclor® **III** 81
rock wool **VIII** 141–338
rosewood **XIII** 286, 313
rosin **XI** 239–242
Ro-sulfiram® **V** 165
rubber components **XV** 145–164
Rutgers 612® **V** 345

rutile **II** 199
rye flour dust **XIII** 99–100
*Samba scleroxylon* **XIII** 288
Santos rosewood **XIII** 287, 321
sapupira (da mata) **XIII** 285, 295
Sarkosyl® O **XVI** 281
Sarkosyl® OT **XVI** 281
Scots pine **XIII** 287
selenite **II** 117
sepiolite **VIII** 141–338
*Sequoia sempervirens* **XIII** 287
serpentine **II** 182, 183
shinglewood **XIII** 288
short-term exposures **I** 15–22
sidestream smoke **XIII** 3
silica, amorphous **II** 157–179
silica, crystalline **XIV** 205–271
silica fume **II** 160, 165
silica gel **II** 157
silica glass **II** 157
silicic acid tetraethyl ester **III** 299
silicon carbide **VIII** 141–338
silicon dioxide, crystalline **XIV** 205–271
silk oak **XIII** 286
silky oak **XIII** 286
silver oak **XIII** 286, **XIV** 295
Skane® M-8 **XVI** 263
Skane® M-8 HQ **XVI** 263
slag wool **VIII** 141–338
soapstone **II** 181
soda lye **XII** 195
sodium (1,1′-biphenyl)-2-olate **II** 299
sodium 2-biphenylolate **II** 299
sodium citrate **XVI** 209
sodium ethylmercurithiosalicylate **XV** 249
sodium hydrate **XII** 195
sodium hydroxide **XII** 195–207
sodium 2-hydroxybiphenyl **II** 299
sodium 1-hydroxy-2-(1*H*)-pyridinethione **X** 287
Sodium Omadine® **X** 287
sodium *o*-phenylphenate **II** 299
sodium *o*-phenylphenol **II** 299–316
sodium *o*-phenylphenolate **II** 299
sodium *o*-phenylphenoxide **II** 299
sodium pyridinethione **X** 287
sodium pyrithione **X** 287–311
solvent yellow 34 **IV** 11
South American mahogany **XIII** 287
Southern silk(y) oak **XIII** 286, **XIV** 295
spanon **VI** 57
spinel **III** 128
spirit of hartshorn **VI** 1
spirit of turpentine **XIV** 173
spirits of salt **VI** 231
spirits of wine **XII** 129
steatite **II** 181
Stopetyl **V** 165
Stopmold B® **II** 299
succinic acid peroxide **III** 254
sucupira **XIII** 285, 295
sulfate of lime **II** 117
sulfate turpentine **XIV** 173
sulfinylbismethane **III** 163
sulfotep(p) **XIII** 237
sulfuric acid **XV** 165–222
sulfuric acid calcium salt **II** 117
sulfuric acid, dimethyl ester **IV** 217
sulfuric ether **XIII** 149
sulfur mustard **IV** 27
sulourea **I** 301
$\gamma$-sultone **IV** 45
*Swietenia* spp. **XIII** 287
2,4,5-T **XI** 243
*Tabebuia* spp. **XIII** 288, **XIV** 311–315
talc **II** 181–198
tantalum **XVI** 293–296
*Tarrietia utilis* **XIII** 288
Tasmanian blackwood **XIII** 285, 291
TBP **IX** 245
teak **XIII** 288, **XIV** 317
*Tecoma peroba* **XIII** 287
*Tectona grandis* **XIII** 288, **XIV** 317–323
TEDP **XIII** 237–249
tegafur **I** 4
TEL **XV** 223
teniposide **I** 6
terephthalic acid **VII** 241
*Terminalia superba* **XIII** 288
3,3′,4,4′-tetraaminobiphenyl **III** 131
1,3,5,7-tetraazaadamantane **V** 355
1,3,5,7-tetraazatricyclo[3.3.1.1$^{3,7}$]decane **V** 355
2,4,5,6-tetrachloro-1,3-benzene-dicarbonitrile **VI** 113
2,4,5,6-tetrachloro-3-cyanobenzonitrile **VI** 113
1,1,2,2-tetrachloro-1,2-difluoroethane **I** 297–299
tetrachloroethene **III** 271
tetrachloroethylene **III** 271–297
2,4,5,6-tetrachloroisophthalonitrile **VI** 113
tetrachloronaphthalenes **XIII** 101
*m*-tetrachlorophthalodinitrile **VI** 113
*m*-tetrachlorophthalonitrile **VI** 113
$\alpha,\alpha,\alpha$,4-tetrachlorotoluene **X** 21
Tetradine **V** 165
tetraethoxysilane **III** 299

tetraethyl dithiopyrophosphate **XIII** 237
tetraethyllead **XV** 223–235
tetraethyl orthosilicate **III** 299
tetraethylplumbane **XV** 223
tetraethyl silicate **III** 299–305
tetraethylthioperoxydicarbonic diamide **V** 165
tetraethylthiram disulfide **V** 165
*N,N,N',N'*-tetraethylthiuram disulfide **V** 165
Tetraetil **V** 165
1,1,1,2-tetrafluoroethane **XIII** 251–266
4,5,6,7-tetrahydro-1*H*-1,2,3-benzotriazole **II** 317
4,5,6,7-tetrahydro-1*H*-benzotriazole **II** 317–319
3*a*,4,7,7*a*-tetrahydro-4,7-methano-1*H*-indene **V** 125
1,2,3,4-tetrahydrostyrene **XIV** 185
tetralit **XI** 179
tetralite **XI** 179
tetramethyldiaminodiphenylacetimine **IV** 11
*N,N,N',N'*-tetramethyl-*p,p'*-diaminodiphenylmethane **I** 249
tetramethyl-*p*-diamino-imido-benzophenone **IV** 11
tetramethyllead **XV** 237–248
tetramethylplumbane **XV** 237
tetramethylthioperoxydicarbonic diamide **XV** 163
tetramethylthiuram disulfide **XV** 163–164
tetranitromethane **IV** 323–330
Tetrosin OE® **II** 299
tetryl **XI** 179
teturamin **V** 165
thiazols **XV** 155–157
thimerosal **XV** 249–255
thioaniline **IV** 331
4,4'-thiobis(aniline) **IV** 331
4,4'-thiobisbenzenamine **IV** 331
1,1'-thiobis(2-chloroethane) **IV** 27
2,2'-thiobis(4,6-dichlorophenol) **IX** 245
thiocarbamide **I** 301, **XIV** 143
thiocarbonyl tetrachloride **I** 291
4,4'-thiodianiline **IV** 331–334
*p,p'*-thiodianiline **IV** 331
thiodi-*p*-phenylenediamine **IV** 331
thiodiphosphoric acid tetraethyl ester **XIII** 237
thioguanine **I** 4
Thiomersal **XV** 249
Thiomersalate **XV** 249
thiopyrophosphoric acid tetraethyl ester **XIII** 237
thiotepa **I** 3
thiotepp **XIII** 237
thiourea **I** 301–312, **XIV** 143–148
thiram **XV** 163
thiurams **XV** 159–164
Thiuranide **V** 165
THT **IX** 323
*Thuja* spp. **XIII** 288
*Tieghemella heckelii* **XIII** 288
tin **XIV** 149–165
tin compounds, inorganic **XIV** 149–165
tin compounds, organic **I** 317
titanium dioxide **II** 199–204
titanium oxide **II** 199
TMTD **XV** 163
TNT **I** 359
tobacco smoke **I** 28, 39–62, **XIII** 3
2-tolidine **V** 153
3,3'-tolidine **V** 153
*o*-tolidine **V** 153
tolite **I** 359
toluene **VII** 257–318
2,4-toluenediamine **VI** 339
toluene-2,4-diamine **VI** 339–352
*m*-toluenediamine **VI** 339
toluene trichloride **VI** 80
2-toluidine **III** 307
4-toluidine **III** 323
*o*-toluidine **III** 307–322, 329, **VI** 144
*p*-toluidine **III** 323–330
toluol **VII** 257
2,4-toluylenediamine **VI** 339
*m*-toluylenediamine **VI** 339
*o*-tolylamine **III** 307
*p*-tolylamine **III** 323
tolyl chloride **VI** 79
tolyltriazole **II** 295
Topane® **II** 299
total dust **II** 52
tremolite **II** 96, 182–187
treosulfan **I** 3
tretamine **I** 3
tri **X** 201
triatomic oxygen **X** 171
1,2,3-triazaindene **II** 231
1,3,5-triazin-1,3,5(2*H*,4*H*,6*H*)-triethanol **II** 331
1*H*-1,2,4-triazol-3-amine **IV** 1
tribromomethane **VII** 319–332
tributylchlorostannane **I** 315
tributylfluorostannane **I** 315
tributyl-(2-methyl-1-oxo-2-propenyl)oxystannane **I** 316
tributylmono(naphthenoyl-oxy)stannane derivatives **I** 316

tributyl-(1-oxo-9,12-octadecadienyl)oxy-(Z,Z)-stannane **I** 316
tri-*n*-butyltin benzoate **I** 315
tri-*n*-butyltin chloride **I** 315
tri-*n*-butyltin compounds **I** 315–331
tri-*n*-butyltin fluoride **I** 315
tri-*n*-butyltin linoleate **I** 316
tri-*n*-butyltin methacrylate **I** 316
tri-*n*-butyltin naphthenate **I** 316
tributyltin oxide **I** 315
1,2,3-trichlorobenzene **III** 331–363
1,2,4-trichlorobenzene **I** 72, **III** 331–363, **XIV** 167–172
1,3,5-trichlorobenzene **III** 331–363
1,1,2-trichloroethane **V** 66, 72
trichloroethene **X** 201
trichloroethylene **III** 273, 369, **X** 201–244
trichlorofluoromethane **I** 69, 335–344
trichlorohydrin **IX** 171
trichloromethane **XIV** 19
trichloromethanesulfenyl chloride **I** 291
(trichloromethyl)benzene **VI** 80
trichloromethylsulfenyl chloride **I** 291
trichloromonofluoromethane **I** 335
trichloronaphthalenes **XIII** 101
trichloronitromethane **VI** 105
trichlorooctylstannane **VII** 93
2,4,5-trichlorophenol **XII** 209–221
2,4,5-trichlorophenoxyacetic acid **XI** 243–278
2,4,5-trichlorophenoxyacetic acid salts and esters **XI** 243
trichlorophenylmethane **VI** 80
1,2,3-trichloropropane **IX** 171–192
$\alpha,\alpha,\alpha$-trichlorotoluene **VI** 80
$\omega,\omega,\omega$-trichlorotoluene **VI** 80
1,1,2-trichloro-1,2,2-trifluoroethane **III** 365–370
tridymite **XIV** 205
triethanolamine **I** 27
triethoxybutanol **IX** 315
triethylamine **I** 26, 33, **XIII** 267–280
triethylene glycol *n*-butyl ether **IX** 315–321
triethylene glycol monobutyl ether **IX** 315
1,3,5-triethylol-hexahydro-s-triazine **II** 331
trifluoroborane **XIII** 93
trifluorobromomethane **VI** 37
trifluorochloromethane **I** 69
1,1,1-trifluoro-2,2-dichloroethane **X** 53
trifluoromethyl chloride **I** 69
trifluoromonochlorocarbon **I** 69
trifluorotrichloroethane **III** 365
triglycol monobutyl ether **IX** 315
trihydroxymethylnitromethane **II** 287
trimethylamine **I** 26, 28
1,2,4-trimethyl-5-aminobenzene **IV** 335
2,4,5-trimethylaniline **IV** 335–340
2,4,5-trimethylbenzenamine **IV** 335
*sym*-trimethylbenzene **IV** 341
1,3,5-trimethylbenzene **IV** 341–348
2,4,5-trimethylbenzeneamine **IV** 335
trimethylolmelamine **I** 7
trimethylolnitromethane **II** 287
trimethylolpropane triacrylate **XVI** 201–206
*N,N,N*-trimethyl-oxiranemethanaminium chloride **IV** 247
2,2,4-trimethylpentane **I** 196, 347–356
2,4,6-trinitrophenylmethylnitramine **XI** 179
2,4,6-trinitrophenyl-*N*-methylnitramine **XI** 179
trinitrotoluene (all isomers) **I** 148, 359–360
3,6,9-trioxa-1-tridecanol **IX** 315
triphenyl phosphate **II** 321–330
*Triplochiton scleroxylon* **XIII** 288
tris(2-hydroxyethyl)-hexahydro-1,3,5-triazine **II** 331
*N,N',N''*-tris($\beta$-hydroxyethyl)-hexahydro-1,3,5-triazine **II** 331–339, **IX** 323–324
tris($\omega$-hydroxyethyl)-hexahydro-1,3,5-triazine **II** 331
tris(hydroxymethyl)nitromethane **II** 287
Tris Nitro® **II** 287
tris(oxymethyl)nitromethane **II** 287
tritol **I** 359
triton **I** 359
trofosfamide **I** 3
trotyl **I** 359
Troysan® KK-108A **XVI** 247
Troysan® Polyphase® Anti-Mildew **XVI** 247
Troysan® Polyphase® P100 **XVI** 247
true teak **XIII** 288
TTD **V** 165
Tumescal OPE® **II** 299
turpentine **XIV** 173–179
turpentine oil **XIV** 173
*Turraeanthus africanus* **XIII** 288
ultrafine aerosol particles **XVI** 289–292
Uniclor® **III** 81
Uritone® **V** 355
Urotovet® **V** 355
Urotropin® **V** 355
Ursol D® **VI** 311
$\delta$-valerosultone **IV** 45
vanadic anhydride **IV** 349
vanadium compounds **IV** 349–361
vanadium(V) oxide **IV** 349
vanadium pentoxide **IV** 349–361
Vapona® **IV** 201

Varine® O **V** 373
VDC **VIII** 109
vinblastine sulfate **I** 6
vincristine sulfate **I** 6
vindesine **I** 6
vinegar naphtha **XII** 167
vinyl acetate **V** 229–239
vinyl alcohol 2,2-dichloro dimethyl phosphate **IV** 201
vinylbutyrolactam **V** 249
vinylcarbazole **XIV** 181–183
*N*-vinylcarbazole **XIV** 181
9-vinylcarbazole **XIV** 181
vinyl carbinol **XV** 31
vinyl chloride **I** 40, **II** 149–156, **V** 66, 71, 241–248
1-vinylcyclohex-3-ene **XIV** 185
4-vinylcyclohexene **XIV** 185–202
vinyl cyclohexene diepoxide **I** 391
4-vinyl-1,2-cyclohexene diepoxide **I** 391
4-vinyl-1-cyclohexene dioxide **I** 391–395
1-vinyl-3-cyclohexene dioxide **I** 391
vinyl ethanoate **V** 229
vinylethylene **XV** 51
vinylidene chloride **V** 66, 72, **VIII** 109–139
vinylidene dichloride **VIII** 109
vinylidene fluoride **V** 135
1-vinyl-2-pyrrolidinone **V** 249
*N*-vinylpyrrolidinone **V** 249
1-vinyl-2-pyrrolidone **V** 249
*N*-vinyl-2-pyrrolidone **V** 249–261
vitreous silica **II** 157
walnut **XIII** 286
wawa **XIII** 288
western red cedar **XIII** 288
wheat flour dust **XIII** 99–100
white acajou **XIII** 287
white afara **XIII** 288
white cedar **XIII** 288
white peroba **XIII** 287, **XIV** 307
wishmore **XIII** 288
Witaclor® **III** 81
wollastonite **VIII** 141–338, **XVI** 297–315
wood dust **IV** 363–373, **XIII** 283–323, **XIV** 273–323
wood ether **I** 125
woods **IV** 363–373, **XIII** 283–323, **XIV** 273–323
wood turpentine **XIV** 173
xylene **V** 263–285, **XV** 257–288
xylol **V** 263
yellow pine **XIII** 287
zebrano **XIII** 287
zebrawood **XIII** 287
zeolites **VIII** 141–338
zinc chromate **III** 101–122, **XV** 289–294
zinc chromate hydroxide **XV** 289
zinc potassium chromate **XV** 289
zinc yellow **XV** 289
zirconium **XII** 223–236
zirconium alloys and compounds **XII** 223
Zoldine® ZE **IX** 275